Advances in Turbulence

Advances in Turbulence

Proceedings of the
First European Turbulence Conference
Lyon, France, 1–4 July 1986

Editors: G. Comte-Bellot and J. Mathieu

With 437 Figures

Springer-Verlag Berlin Heidelberg New York
London Paris Tokyo

Professor Dr. Geneviève Comte-Bellot
Professor Dr. Jean Mathieu
Ecole Centrale de Lyon, 36, Avenue Guy de Collongue,
F-69131 Ecully Cedex, France

Scientific Committee

Prof. G. Comte-Bellot, France
Prof. H. H. Fernholz, Germany
Dr. U. Frisch, France
Dr. J. C. R. Hunt, UK *(Chairman)*
Prof. E. Krause, Fed. Rep. of Germany
Prof. M. Landahl, Sweden
Prof. J. Mathieu, France
Dr. H. Tennekes, Netherlands
Prof. A. M. Yaglom, USSR

Local Organizing Committee

Dr. E. Alcaraz
Dr. J. P. Bertoglio
Prof. G. Comte-Bellot *(Co-Chairman)*
Prof. J. N. Gence
Dr. D. Juvé
Prof. J. Mathieu *(Co-Chairman)*
Dr. R. Morel
Mme E. Roche

ISBN-13: 978-3-642-83047-1 e-ISBN-13: 978-3-642-83045-7
DOI: 10.1007/978-3-642-83045-7

Softcover reprint of the hardcover 1st edition 1987

2153/3150-543210

Foreword

Since 1964 the main function of the European Mechanics Committee has been to arrange Euromech Colloquia. These are three- or four-day meetings for the discussion of current research on a specified and relatively narrow topic in mechanics, by about 50 specialists chosen for their active involvement in research in that topic. The organization of each Euromech Colloquium is entrusted by the Committee to one or two selected scientists of repute in the field, and these organizers are enjoined to achieve a friendly and informal forum for discussion, with a minimum of paper work and expenditure. Over 220 Euromech Colloquia have been held since 1964 (about 40 each in France, West Germany and Britain and the remainder in 18 countries in both western and eastern Europe) on a wide range of topics drawn from the mechanics of solid materials, hydrodynamics, gas dynamics and mechanical systems. The Committee believes that collectively, Euromech Colloquia have made a significant contribution to the exchange of ideas on topics in mechanics within Europe and have thereby helped to overcome the barriers to easy scientific communication in that sorely divided continent.

A few years ago the European Mechanics Committee turned its attention to the possible need for European conferences on a larger scale than Euromech Colloquia. It was put to the Committee that European scientists working on certain broad topics are not well provided with opportunities for surveying developments in the field as a whole, and that in particular turbulence specialists would welcome the opportunity to discuss all aspects of their subject – laboratory observations, mathematical theory, numerical simulations, geophysical flow systems, and engineering applications – and in more depth than is possible at large-scale conferences. The Committee was impressed by the evidence, and set up a Turbulence Sub-Committee to advise it on the desirability of holding a European Turbulence Conference. This Sub-Committee recommended that such a conference be held, and they also reported that Professor Mathieu and Professor Comte-Bellot were willing to organize it in Lyon. The European Mechanics Committee consequently decided to hold this medium-size conference on turbulence as the first of what for the Committee is a new type of scientific meeting in Europe, characterized by the same general principles as Euromech Colloquia but on a broader topic and with larger participation.

In the event, the expectation that a European conference on turbulence would be found enjoyable, informative and stimulating has been confirmed, as the pages of this volume bear witness. For this success, we are indebted to Professors Mathieu and Comte-Bellot and their colleagues on the local organizing committee for their efficient handling of all the local arrangements, and to Dr. Hunt and his colleagues on the Turbulence Sub-Committee for their wise scientific planning of the conference. It is the general wish that this Turbulence Conference be followed at regular intervals by others of the same type, and I am glad to be able to announce that the Second European Turbulence Conference will be held in West Berlin in September 1988.

G.K. Batchelor
Chairman, European Mechanics Committee

Preface

This European Turbulence Conference was held at the Ecole Centrale de Lyon, France, 1–4 July 1986, under the auspices of the European Mechanics Committee. The scientific committee was constituted by the Euromech Turbulence Sub-Committee whose present Chairman is Dr. J.C.R. Hunt.

The Conference was devoted to the fundamental aspects of turbulence of interest to engineers, physicists and mathematicians. The research topics discussed concern the chaotic behaviour of non-linear systems, instability and transition, coherent structures, large eddy simulation, new techniques in measurements, flow visualisation and image analysis. Engineering applications of fundamental research in turbulence were also included.

The Conference was opened by Professor G.K. Batchelor as Chairman of the European Mechanics Committee. There were 6 general invited lectures and 57 other contributions, with 152 participants from 19 countries. Sufficient time was provided for discussions, particularly for the poster sessions. There was no implied distinction in merit between oral and poster presentations; the posters were on subjects that were excluded from oral contributions. There were several social gatherings, including one in the City Hall of Lyon where a welcome address was given by the Mayor.

The organizing Committee of the Symposium wish to express their thanks to la Direction des Recherches, Etudes et Techniques du Ministère des Armées (DRET), le Centre National de la Recherche Scientifique (CNRS), le Ministère de la Technologie (MRT), l'Electricité de France (EDF), the United States Army (European Research Office), l'Association Universitaire de Mécanique (AUM) and le Conseil Général du Rhône for financial support.

J. Mathieu
G. Comte-Bellot

STANLEY CORRSIN

April 20, 1920–June 2, 1986

Professor Stanley Corrsin died on June 2, 1986, one month before this conference. Though expected, his death was nonetheless a severe blow to the turbulence community around the world.

His many honors include election to the National Academy of Engineering, the Fluid Dynamics Prize of the American Physical Society, the von Karman Medal of the American Society of Civil Engineers, and the title of Docteur Honoris Causa of the University of Lyon.

Stan Corrsin's legacy will remain with us through his numerous Ph. D. and post doctoral students. He will be remembered for his multifaceted personality: his untiring pursuit of knowledge, his strength of convictions, his love of academic life and his sense of humor.

Contents

Part I Instability and Transition. Theory and Experiments

Part II Chaotic Behaviour of Non-Linear Systems and Turbulent Fields

Titles printed in *italics* refer to invited papers

Part III Direct and Large Eddy Simulation of Turbulence

Part IV Coupling of Fourier Modes. Spectral Analysis of Turbulence and Related Problems

Part V Two-Dimensional Velocity Fields. Geophysical and Astrophysical Turbulence

Part VI **Coherent Structures in Turbulent Flows. Conditional Averaging. Pattern Recognition**

Part VII Experimental Techniques: Hot-Wire Anemometry, Vorticity Meters, Electrochemical Methods, Image Analysis

Part VIII Engineering Applications of Turbulence and the Effect of External Disturbances

Part I

Instability and Transition. Theory and Experiments

Instability and Transition

J.T. Stuart

Department of Mathematics, Imperial College of Science and Technology, Queen's Gate, London SW7 2BZ, UK

1. Introduction

Fluid flows can be in an ordered or laminar state, or they can be in a relatively-disordered, chaotic or turbulent state. The theory of hydrodynamic stability, which has a history of some 100 years, is concerned with the evolution between these two separate basic states of laminar flow and turbulent flow. Early theoretical work in the subject, associated with such names as Reynolds, Kelvin, Rayleigh, Orr, Sommerfeld, Prandtl, Taylor, Heisenberg, Tollmien, Schlichting, Squire, Jeffreys, Lin and Meksyn, was concerned mainly, though not entirely, with the development of infinitesimal disturbances and with the associated concept of exponential stability or instability. Crucial experiments were made by a number of scientists and engineers, including Reynolds, Bénard, Taylor and Schubauer and Skramstad, and good agreement was established between the observations and the results of theory. This stage may be said to cover the period 1880 to 1950 approximately.

During the last three or four decades, theoretical attention has been more closely focussed on non-linear aspects of the problem of flow instability, starting perhaps with Landau's paper of 1944 although some quite central ideas of a weakly non-linear formulation were given by Heisenberg as early as 1925. The adjective "weakly" is used to indicate a small magnitude of flow disturbances, albeit not infinitesimal, usually in association with a small parameter. Developments of this kind have been made during the past 35 years up to the present day. Moreover, during the latter period especially, occasional papers have attempted to deal with disturbances of larger magnitude by use of special mathematical techniques. Crucial experiments have been made by Coles, Donnelly, Swinney, Benjamin, Mullin, Klebanoff, Kovasznay, Gaster, Nishioka and Saric and by Kachanov and Levchenko.

Other theoretical distinctions are necessary and desirable:

(I) In some cases (a) there is a clearly defined body force, centrifugal, gravitational or Coriolis, or some combination of them, which can generate an instability of a laminar flow or of a rest state, examples occurring in Taylor-Couette flow between rotating cylinders, Bénard-Rayleigh convection between parallel planes and Eady-Charney baroclinic instability of an atmospheric flow. On the other hand, (b) in flows of aerodynamic or technological significance there is often no such body force, but independently Prandtl (and his school) and Heisenberg exposed an instability mechanism driven by viscosity through its effect on the phase difference between velocity components and therefore on the Reynolds stress.

(II) In some cases (a) the flow is confined within a fixed geometrical configuration, such as in the annulus between two cylinders of finite length. The problem of instability of laminar flow is then concerned with a temporal evolution between different possible states. Thus, at a given point, we may refer to an instability being absolute. In an aerodynamic boundary layer (b), on the other hand, a disturbance propagates as it develops; the problem of instability is one of spatial evolution of a wave packet or wave system, and any instability may be said to be of a convective nature.

Some of the important results already obtained, and some of the crucial problems of the present day, can be exposed by reference to two key examples:

(A) Taylor-Couette flow, which is of type I(a) according to the distinctions drawn above;

(B) Plane Poiseuille flow, which is of type II(b), and in many respects exhibits characteristics typical of boundary-layer flows.

We now give some attention to each of these prototype flows. More detailed discussions of these matters are to be found in references 1, 2 and 3.

2. Taylor-Couette Flow

In the configuration adopted by G.I. Taylor, the two concentric cylinders were very long, of order several hundred times the radial gap between them. If the inner cylinder speed is high enough, with the outer one at rest, the familiar toroidal (or Taylor) vortices, which are periodic along the axis, occur. For infinitesimal perturbations, Taylor's own work gave good agreement between theory and experiment for the neutral states (of speed and axial wavelength). Weakly non-linear theory was used first by Davey and later by Kirchgassner and Sorger to predict the velocity amplitudes of the circulations in the vortices, while Donnelly and his co-workers have shown good agreement with experiment.

Later experimental work of Coles, who used cylinders of length some thirty gap widths, showed that doubly-periodic flows are also possible, the vortices having also the character of a wave propagating in the azimuthal direction. Moreover, Coles' work made clear that a multiplicity of states of flow were possible at a given speed, with different combinations of wave numbers along the axial and azimuthal directions.

Coles' pioneering experimental, and associated conceptual, contributions led to theoretical activity which has continued to the present day. Davey, DiPrima and Stuart were led to discuss the instability of a Taylor-vortex flow against doubly periodic, travelling-wave perturbations also periodic in the azimuthal direction. This gave predictions qualitatively in agreement with Coles' observations, and with reasonable quantitative agreement also. There is, however, a difficulty in the analysis of Davey et al., which is important at least from the point of view of mathematical rigour. This arises from the fact of the axisymmetric and non-axisymmetric modes having different critical speeds or bifurcation points. Davey et al. needed to assume that the weakly non-linear development of the Taylor-vortex flow extended to the bifurcation point for the non-axisymmetric mode.

As DiPrima noticed, however, it is possible to choose a case of counter-rotating cylinders so as to set the two critical speeds, or bifurcation points, at the same value. In this situation the mathematical basis for the work is on a sounder footing. DiPrima, Sijbrand and Eagles have pursued this line of work, as have Golubitsky and Iooss.

Further important developments have taken place on the experimental side with long cylinders, through work originated by Swinney and Gollub and later pursued further by themselves, by Donnelly and by others. Here we refer to the development of a broad band spectrum and of an additional frequency peak (additional to that of the Coles' wavy vortex) as the inner cylinder speed is raised to a level some tens of times the critical value. Chaos appears to reign here, but with <u>some</u> order, but theoretical understanding has so far eluded us.

A separate, but perhaps equally important, attack on Taylor-Couette flow comes from experiments with very short cylinders, as initiated by Benjamin and continued by himself and Mullin. A multiplicity of steady states is possible, and bifurcation boundaries between such states have been tracked experimentally and related to general bifurcation theory. Substantial agreement has been obtained between the observations and the analytical and computational work of Schaeffer, Hall and Cliffe. It may be said that the Taylor-Couette problem forms a fine prototype example of a dynamical system, and one which will repay much further study.

An account of this subject has been published in reference 2.

3. Plane Poiseuille Flow

Plane Poiseuille flow is typical of boundary-layer flows, in that the ranges of unstable Reynolds numbers and wave numbers are qualitatively very similar. A mathematical advantage of plane Poiseuille flow is that there is no flow divergence, although this is a disadvantage in terms of aerodynamic reality. Many of the crucial phenomena can, however, best be explained in terms of the simpler parallel flow.

The independent work of Heisenberg and of Prandtl (followed by Tollmien and Schlichting especially) exposed the instability mechanism associated with viscosity and concomitant phase shifts of velocity components, more mathematically in

Heisenberg's case and more physically (and perhaps more clearly at the time) in Prandtl's. Tollmien's theoretical predictions for the growth of infinitesimal perturbations in Blasius boundary-layer flow were brilliantly confirmed by Schubauer and Skramstad in Washington in the early 1940's.

For plane Poiseuille flow the work of Lin, Meksyn and others on infinitesimal perturbations led to the excellent experiments of Nishioka and his colleagues in Osaka in the 1970's, and to fine agreement between theory and experiment. On the theoretical side, the recognition by F.T. Smith for instability problems of the rôle of the triple-deck method has been of immense importance, and has been followed by significant applications. Moreover, Gaster's theoretical and experimental work on the evolution of wave packets (spots of oscillations) in boundary layers has been a seminal contribution, with good agreement between the two approaches.

From a weakly non-linear point of view, a major theoretical contribution has been the appreciation that a threshold (or subcritical) instability phenomenon occurs in plane Poiseuille flow, at speeds below the critical value for infinitesimal perturbations. Nishioka has brilliantly confirmed this prediction both qualitatively and quantitatively. Moreover, it has been known from the work of many authors since 1951 that, in plane Poiseuille flow, two-dimensional equilibrium oscillatory fields are possible for a range of speeds down to about half that for the existence of neutral or growing infinitesimal perturbations.

Orszag has shown by numerical calculations, moreover, that such states are unstable against three-dimensional velocity perturbations. This is not at all surprising in the light of known experimental evidence in boundary layers, a point to which we return shortly. In the present situation, however, the work of Herbert seems to show that this three-dimensional instability is an example of a resonance mechanism between interacting triads of waves of appropriate wave numbers and frequencies. Such resonances have been known in detail for some 25 years, partly through work on water waves; but they were discovered in the instability field by Raetz in California in 1959. Related theoretical work followed from Kelly, independently, and from Craik, the latter especially developing Raetz's work for boundary layers. In the boundary-layer case, patterns arising from such resonances have been detected in boundary layers by Saric and his co-workers and by Kachanov and Levchenko. More work is needed and anticipated on this possible route for the transition process.

In many ways, however, the most central experiments in boundary-layer and shear flows may be those conducted by Klebanoff and his co-workers about 1960, in essentially the same laboratories as those of Schubauer and Skramstad some 20 years earlier. A two-dimensional instability wave is found readily and naturally to develop a three-dimensional pattern periodic across the span. At certain locations in the spanwise and downstream coordinates, local shear layers develop and shed spots of higher frequency fluctuations, which grow, agglomerate and develop further downstream into turbulence. This work has developed much theoretical activity, partly of a weakly non-linear kind and partly of a more strongly non-linear type. Further developments, both theoretical and experimental, are anticipated and are awaited with interest. Further details of the work on plane Poiseuille flow and on boundary-layer transition are to be found in references 1 and 3.

4. Literature

1. J.T. Stuart: Stewartson Memorial Lecture: Hydrodynamic stability and turbulent transition. In: Numerical and Physical Aspects of Aerodynamic Flows, Proc. 3rd Symposium, Long Beach, California 1985, ed. by T. Cebeci. To be published by Springer, New York (1986).
2. J.T. Stuart: Taylor-vortex flow: a dynamical system. SIAM Review 28, No. 3, 315-342 (1986). See also Proc. Conf. Math. Appl. Fluid Mech. Stability (In memory of Richard C. DiPrima), 9-11 September 1985, ed. by D.A. Drew and J. Flaherty, SIAM publication (1986).
3. J.T. Stuart: Instability of flows and their transition to turbulence. 29th Ludwig Prandtl Memorial Lecture, 1 April 1986, Dortmund. Z. Flugwiss. to appear (1986).

An Experimental Investigation of Transition to Turbulence in Taylor-Couette Flow Using Digital Image Processing

J. Brindley[1] *and F.R. Mobbs*[2]

[1]Department of Applied Mathematical Studies, University of Leeds, Leeds LS2 9JT, UK
[2]Department of Mechanical Engineering, University of Leeds, Leeds LS2 9JT, UK

1. Introduction

The rapidly proliferating use during the last ten years of LDV techniques in analysing complex laminar or turbulent flows has focussed attention on the time variation of a particular fluid property at a fixed point in the fluid. It has largely superseded the earlier hot film probes or thermistors, which were used to obtain similar information, with the recognised risk that the very presence of the probe might itself modify the flow.

The output of either technique is typically a time series, whose character is analysed spectrally in order to detect changes in the underlying flow pattern. This form of analysis has been remarkably successful in advancing our understanding of the process of transition to turbulence, especially in the somewhat special configurations of Rayleigh-Bénard convection and Taylor-Couette flow [1,2]. A major shortcoming of this essentially Eulerian analysis at a single point is, however, its inability to reveal adequate information on the form and evolution of the spatial structures whose temporal behaviour is so well resolved.

In this paper we describe an attempt to address this shortcoming by the use of systematic image processing techniques in flow visualisation experiments. An immediate and obvious difficulty with almost any form of flow visualisation is of course that the characteristics of the flow are deduced, at second hand as it were, from the observed behaviour of tracer particles suspended in the fluid. It is indisputable that such particles will have motion relative to the fluid, and, often more seriously, any method of observation using the reflection of light from the tracer particles will be affected by their orientation. Any attempts to deduce from such observations the actual detailed motion of the fluid would be fraught with difficulties and dangers of misinterpretation. We are certainly on much safer ground, however, in associating steady or periodic particle behaviour with steady or periodic behaviour of the fluid, and qualitative changes in particle behaviour with qualitative changes in the underlying fluid motion. Note that we do not, nor would it be valid to,imply from aperiodic behaviour of the particle suspensions any corresponding aperiodic behaviour of the fluid; it has been shown [3] that the motion of a suspended particle can exhibit chaotic characteristics even in a steady flow. Moreover we interpret significant local differences in particle behaviour as reflecting local differences in the characteristics of the fluid motion.

The particular flow configuration we discuss is that usually known as Taylor-Couette flow, in which the fluid contained between two concentric circular cylinders is driven in relative motion by differential rotation of the cylinders. We consider the best known case, in which the inner cylinder rotates with angular velocity Ω and the outer cylinder is at rest. The approach to turbulence in any particular apparatus is characterised by a somewhat complex sequence of flow transitions as Ω is increased; this sequence is sensitively dependent on the detailed geometry of the apparatus as well as on Ω. It is convenient to introduce four non-dimensional parameters, T, η, λ, and Γ, in terms of which to "map" the flow characteristics. We define these parameters in Fig. 1.

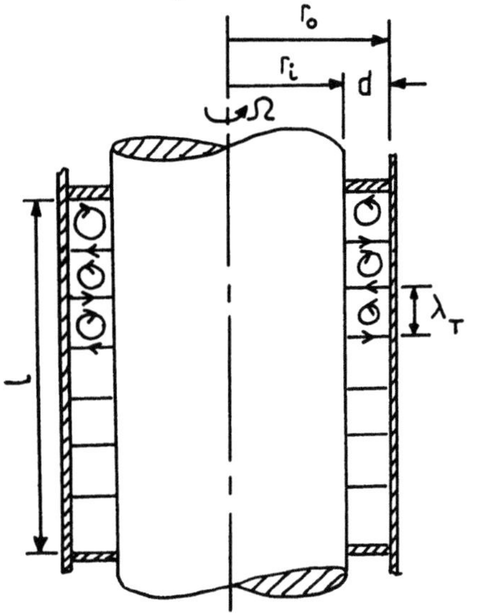

Fig. 1. Schematic diagram of apparatus and definition of parameters.

Taylor number $T = \dfrac{2\Omega^2 r_i^2 d^3}{\nu^2 (r_i + r_o)}$,

radius ratio $\eta = \dfrac{r_i}{r_o}$,

cell aspect ratio $\lambda = \dfrac{\lambda_T}{d}$,

annulus aspect ratio $\Gamma = \dfrac{\ell}{d}$.

The value of Γ, specially when it is small, is very influential in the process of state selection, i.e. the number of Taylor cells and the wavenumber of azimuthally propagating wavy disturbances to those cells. However, at large values of Γ, when the number of Taylor vortices is large, (typically greater than 40), the direct sensitivity to Γ is reduced and it is more useful to relate changes in flow structure to changes in the cell aspect ratio λ (λ is of course an internal parameter, and it is not easy to control its value directly; a typical experiment, in which Ω is changed while η and Γ remain fixed, follows a particular and repeatable path through "λ-space", a path which is however very sensitive to initial conditions and the rate of change of Ω). This approach is suggested by the observation that time varying behaviour in otherwise similar conditions appears to depend on λ in a well-defined and repeatable way.

We describe the experimental and observational details in sections 2 and 3, in section 4 we interpret some of the results in terms of a simple model of modal superposition and synchronisation, and suggest that evidence from many other experiments indicates the ubiquity of this phenomenon, and its possible importance in the laminar turbulent transition process.

2. Apparatus and Experimental Techniques

The experiments were carried out on four rigs, each of which had a fixed, transparent, outer cylinder and a rotating, concentric inner cylinder. By use of inner cylinders of differing radii, eight values of η were obtained. The actual dimensions used are listed in table 1, together with the end conditions.

Table 1. Cylinder dimensions (all in mm)

R_2	R_1	$\eta = \frac{R_1}{R_2}$	Maximum value of annulus aspect ratio Γ	Stationary adjustable upper end plate
89.4	72.0	0.804	34.8	Yes
89.4	53.15	0.595	16.9	Yes
57.15	48.51	0.85	115	No
152.5	138.5	0.908	65	Yes
152.5	106.75	0.7	26.6	Yes
152.5	82.35	0.54	17.4	Yes
57.16	28.41	0.497	20.5	No
57.16	40.76	0.713	35.9	No

Mineral oils of different viscosities were used, as appropriate, to obtain the desired ranges of Taylor number. Annulus aspect ratio was varied either by adjusting the position of a stationary end plate at the top end of the annulus or by adjusting the free surface level. At any aspect ratio a range of axial vortex cell lengths was accessed by accelerating the inner cylinder to its final angular velocity Ω at different rates, sometimes in combination with end plate adjustment.

The frequencies associated with wavy Taylor vortex flow were determined by spectral analysis of hot film probe signals, or by digital image processing. Two radially traversible probes were mounted with a known azimuthal separation, at the same distance from the ends of the annulus. The two probe signals were digitized and processed on a VAX 11/780 computer. Systematic variation of the sampling rate and the number of samples suggested that a sampling interval of 10 ms and a total of 2048 samples was satisfactory in most cases. Frequency spectra were determined using an FFT, and wave numbers obtained from the phase difference between the two signals at each spectral frequency.

A large number of measurements, covering a wide range of Taylor number, were made. Each experiment was carried out at fixed Ω and λ, accessed in various ways in order to attain a range of λ values. A cut off factor (COF), equal to some fraction of the power associated with the dominant frequency in each spectrum, was applied, as illustrated in Fig. 2. The frequencies which exceeded the cut off factor were then plotted against the ratio of Taylor number to critical Taylor number, as in Fig. 3.

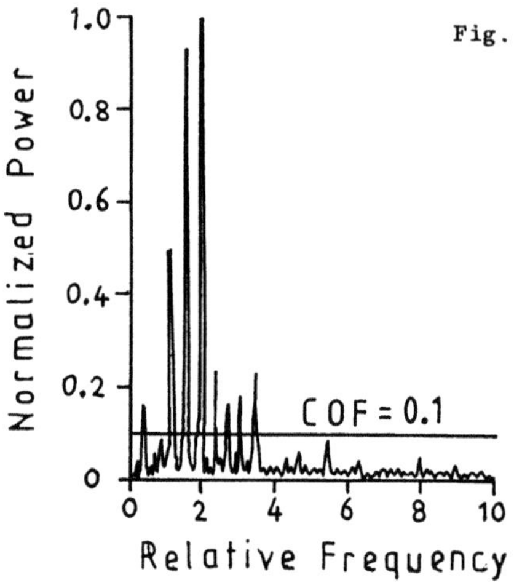

Fig. 2. Frequency spectrum, normalized power against relative frequency (frequency/ cylinder speed), illustrating the use of a cut off factor.

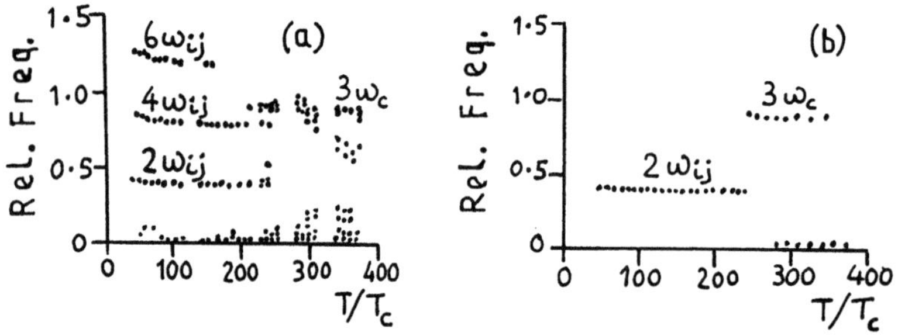

Fig. 3. Relative frequency against T/T_c for radius ratio 0.713

(a) Hot film probe measurements. Cut off factor = 0.001.

(b) Digital image processing. Cut off factor = 0.5

ω_c and ω_{ij} are defined in Fig. 7.

In applying digital image processing, flow visualisation was effected by the addition of small quantities of aluminium paint pigment or hollow glass spheres to the fluid. The output of a standard monochrome TV camera was connected to a frame grabber, which in turn was plugged into the Q-bus of an LSI 11/23 microcomputer. The frame grabber dignitizes the light intensity of 256 points on each of 256 lines. Each of these 256 x 256 pixels is digitized to an 8 bit resolution which gives 256 levels of grey.

The resulting data is stored in a frame store which reconstitutes the video signal from the contents of its memory and this is displayed on a video monitor. The frame store is addressable by the LSI 11/23 enabling picture data to be accessed, processed and re-displayed if required. The microcomputer also controls the frame grabbing function and the clearing and scrolling of the frame store. A direct memory access link between the LSI 11/23 and the VAX 11/780 allows for high speed data transfer between the two computers. The serial line is used to load the operating software into the microcomputer. Data received by the VAX is transferred to disc, from where it can be accessed by programs running on the VAX.

In the experiments the camera was turned through 90° so that picture lines were parallel to the cylinder axes. Every 40 ms, two selected lines of data were extracted and stored in one of two buffers, each of 16 K bytes, in the LSI 11/23. This enabled one buffer to be filled while the contents of the other were being transferred to the VAX 11/780. Time series were constructed from the light intensity variations of chosen pixels and analysed in the same way as the digitized hot film probe signals.

Objections may be raised to both these procedures; it is possible that the introduction of hot film probes may interfere with the flow, while the exact relationship between light intensity fluctuations and the underlying behaviour of solid particles in the flow is not well understood. However, Gorman et al [4] have shown that the fundamental frequency components obtained from scattered light intensity measurements are identical to those obtained from laser Doppler measurements. In Fig. 3, the results obtained by digital image processing, simultaneously with those from hot film probe signals show a similar correspondence

3. Frequency and Wave-number Observations

The distinctive feature of the results, which we shall report more fully elsewhere, is the contrast between the complex behaviour of the measured "wave number" defined as rate of change of phase with azimuthal angle, and the relatively simple behaviour of the associated frequencies. This is illustrated by the typical set of results shown in Figs. 4a and 4b. The spectral frequencies (Fig. 4a) lie in distinct bands which at first descend with increase in Taylor number, but eventually become independent of Taylor number. Similar results have been obtained by many previous workers. The corresponding wave numbers, however, have some unexpected features. Firstly they are, in general, non-integer, and vary with Taylor number, even when

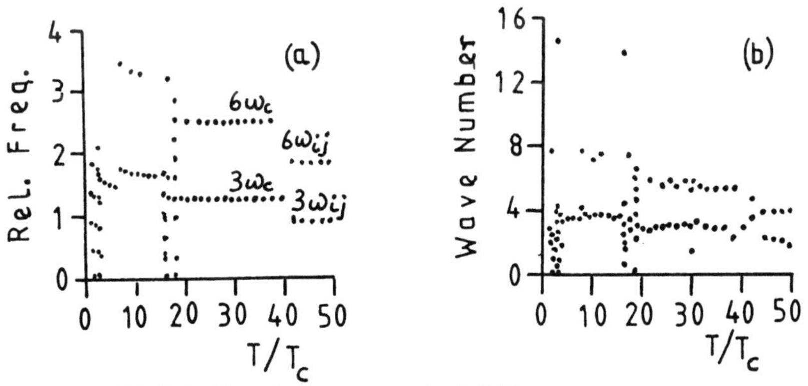

Fig. 4. (a) Relative frequency against T/T_c.
(b) Wave number against T/T_c. Radius ratio = 0.85, cut off factor = 0.05.

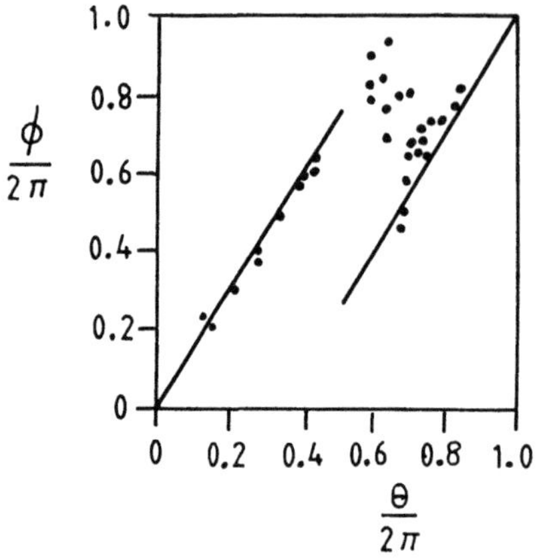

Fig. 5. Variation of phase difference ϕ with azimuthal angle θ assuming $m_1 = 1$, $m_2 = 2$ and $a_1 = a_2$, —— theoretical model, • experimental observations of Ahlers et al [5].

the frequency is constant. Moreover the wave number corresponding to the first harmonic of a frequency is not, as expected, double the wave number of the fundamental frequency. A non-integer or fractional wave number has been reported by Ahlers et al (1982), who use scattered light intensity measurements to determine the phase difference between oscillations at one fixed point and a second, variable, point at an angular displacement θ, which was varied over the range 0° to 360° (Fig. 5). They suggested the presence of a mode with wave number 3/2, continued around the annulus by means of a phase jump π stationary in the laboratory frame.

A second unexpected feature, spatial dependence of the wave number, is revealed by using one of the main advantages of digital image processing, the possibility of measuring the wave number simultaneously in different parts of the flow. Fig. 6a shows the variation of wave number, rounded to the nearest integer, with Taylor number for four different locations across a vortex cell, while Fig. 6b shows the variation across a pair of cells at a fixed Taylor number.

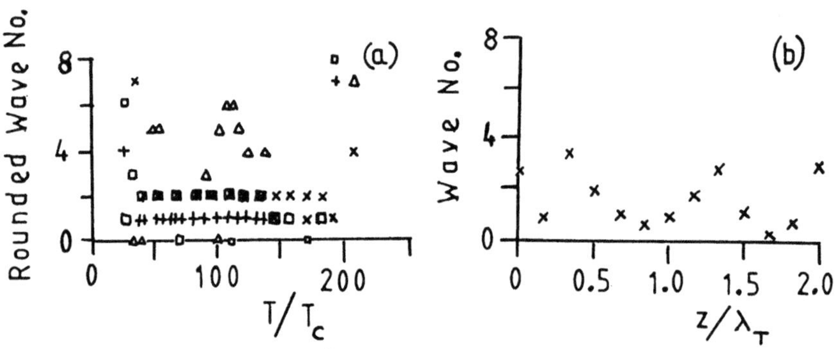

Fig. 6. (a) Variation of azimuthal wave number (rounded to the nearest integer) with T/T_c at four positions across a vortex cell. Radius ratio = 0.713
X at the inward jet, 0.2 λ_T from inward jet,
Δ 0.68 λ_T from inward jet, + at outward jet.
(b) Variation of wave number across a pair of cells starting from an outward jet. Radius ratio = 0.54, T/T_c = 89.55.

A final, important, feature of the spectral frequencies is their strong dependence on the Taylor vortex aspect ratio λ. This was first remarked on by Lorenzen et al (1983) using an apparatus with η = 0.497 and Γ = 17. We have obtained similar results for η = 0.497 over a range of Γ. The initial decrease in frequency with increase in Taylor number T is due to an increase in λ until a steady or plateau value if reached. We also obtained analogous results for a range of values of η; these will be fully reported elsewhere, but a summary is presented in Fig. 7, in which we have indicated that the various plateau frequencies may be associated with "bands" of wave propagation speed. One such band is clearly related to the vortex core speed; the other has no such easy interpretation but is always strongly associated with the inward jet. We should stress that the points plotted in Fig. 7 represent wave speeds calculated by dividing the observed frequency by an integer, m, say. This "wave number" m does not necessarily coincide with the apparent wave number reported by the authors; the simple model described in the next section provides an explanation of this anomaly.

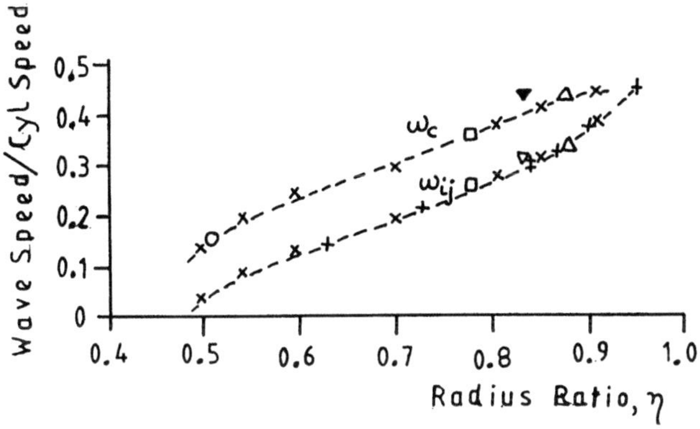

Fig. 7. Variation of vortex core related wave speed ω_c and inward jet related wave speed with radius ratio.
Observed plateau frequencies are integer multiples of these wave speeds, but these integer values do not necessarily conform to the observed azimuthal wave numbers.

x	Present results	□	Park & Jeong [9]
+	King et al [6]	▼	Numerical Prediction- Core Velocity
Δ	Shaw et al [7]		
○	Lorenzen et al [8]	▽	Numerical Prediction - Velocity at mid-point of inward jet

4. A Simple Model

We conclude this paper with a simple conceptual model for the frequency and phase behaviour we have reported. We propose that any fluid variable, e.g. pressure, may be represented by a superposition of two synchronised modes, i.e. waves with different wave numbers m_1 and m_2 but with the same frequency.

Thus we write

$$y = a_1 \sin m_1(\theta + m_2\omega t) + a_2 \sin m_2(\theta + m_1\omega t)$$

$$= p \cos m_1 m_2 \omega t + q \sin m_1 m_2 \omega t$$

$$= r \sin(m_1 m_2 \omega t + \phi)$$

where $p = a_1 \sin m_1\theta + a_2 \sin m_2\theta$

$$q = a_1 \cos m_1\theta + a_2 \cos m_2\theta$$

$$r = \sqrt{p^2 + q^2}$$

$$\phi = \sin^{-1} \frac{p}{r} = \cos^{-1} \frac{q}{r}$$

and the amplitudes a_1, a_2 vary in space and with the parameters T, η and λ. The special case of $m_1 = 1$, $m_2 = 2$, $a_1 = a_2$ is shown in Fig. 5, superimposed on the results of Ahlers et al. The apparent wave number is 3/2 and there is a phase jump π similar to that observed experimentally.

The observed variations, with T, and in space, of the wave number may clearly be reproduced in this way (see e.g. Fig. 8), and the anomalous wave numbers associated with harmonic frequencies can be explained in terms of different amplitude ratios for the harmonics of the combining waves.

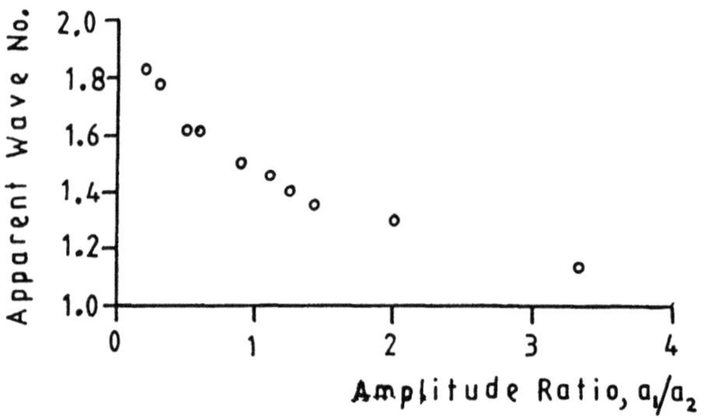

Fig. 8. Effect of wave amplitude ratio on the apparent wave number, based on the phase difference which would be observed if measurements were to be made at $\theta = 0°$ and $\theta = 20°$.

The compatibility of many observations with such a simple linear superposition model should motivate the development of nonlinear models having equilibrated states in which two such modes co-exist at finite amplitudes. The synchronisation (a frequency locking) and subsequent desynchronisation of these modes as parameters are varied will be clear indicators of the progress of the flow towards turbulence. Moreover, our ability to identify the interacting modes with spatial features and hence physical mechanisms, may have important implications for flow transition in less special geometries.

References

1. Busse, F. H., 1981. Transition to Turbulence in Rayleigh-Benard Convection in Hydrodynamic Instabilities and the Transition of Turbulence, Topics in Applied Physics, 45, Springer-Verlag.
2. Di Prima, R. C. and H. L. Swinney, 1981. Instabilities and Transition in Flow between Concentric Rotating Cylinders in Hydrodynamic Instabilities and the Transition of Turbulence, Topics in Applied Physics, 45, Springer-Verlag.
3. Broomhead, D. and S. Ryrie. Private communication.
4. Gorman, M., L. A. Reith and H. L. Swinney, 1980. Modulation Patternsm Multiple Frequencies and Other Phenomena in Circular Couette Flow. Ann. N.Y. Acad. Sci. 357, 10.
5. Ahlers, G., D. S. Cannell and M. A. Dominguez-Lerma, 1982. Fractional Mode Numbers in Wavy Taylor Vortex Flow. Phys. Rev. Lett. 49, 368.
6. King, G., Y. Li, W. Lee, H. Swinney and P. Marcus, 1984. Wave Speeds in Wavy Taylor Vortex Flow. J. Fluid Mech. 141, 365-390.
7. Shaw, R. S., C. D. Andereck, L. A. Reith and H. L. Swinney, 1982. Superposition of Travelling Waves in the Circular Couette System. Phys. Rev. Lett. Vol. 48, No. 17, 1172-1175.
8. Lorenzen, A., G. Pfister and T. Mullin, 1983. End Effects on the Transition to Time-Dependent Motion in the Taylor Experiment. Phys. Fluids 26(1), 10-13.
9. Park, K. and K. Jeong, 1984. Stability Bounding of the Taylor Vortex Flow. Phys. Fluids, 27(9), 2201-2203.

Dynamical Instabilities and Transition to Turbulence in Spherical Gap Flows

K. Bühler and J. Zierep

Institut für Strömungslehre und Strömungsmaschinen, Universität (TH), Kaiserstraße 12, D-7500 Karlsruhe 1, Fed. Rep. of Germany

The transition from laminar to turbulent spherical gap flow is investigated by analysing the dynamical behaviour of the occuring instabilities. The structure of the secondary instabilities is observed by flow visualization. Time-dependent wall shear stress signals are measured at different positions on the outer sphere and analysed to obtain the power spectra. Starting from the different bifurcation solutions different routes of the laminar-turbulent transition are considered.

Introduction

Turbulent flows are important for many reasons in nature and technology. The physical behaviour is investigated by studying problems with full developed turbulence and during the laminar-turbulent transition. The last way becomes important through new results from investigations in the dynamics of nonlinear systems and new experimental methods to obtain the power spectra of the time-dependent flow field. A survey is given by Eppler and Fasel [1], Swinney and Gollub [2], Barenblatt, Ioss and Joseph [3], Tatsumi [4] and Kozlov [5]. It has been shown by Fenstermacher, Swinney and Gollub [6], that in circular Couette flows the transition takes place by time-dependent modulations of the Taylor vortices. In this work the dynamical behaviour of the spherical gap flow with and without Taylor vortices is investigated for small gap size. Increasing the Reynolds number quasistatically, the three-dimensional boundary layer flow becomes unstable with respect to time-dependent shear waves near the outer sphere. The superposition with Stuart vortices near the poles leads than to a stochastic motion, that occupays for increasing Reynolds number finally the whole flow field. A direct transition to turbulence is realized by a sudden start of the inner sphere to a high supercritical Reynolds number. This investigation is a continuation of the work presented at the "IUTAM-Symposium on Laminar-Turbulent Transition" [7].

The three-dimensional basic flow

We investigate the flow in the gap between two concentric spheres. The inner rotating sphere is sourronded by a transparent outer one being at rest. Silicon oil with a kinematic viscosity of $\nu = 5 \cdot 10^{-6}$ m²/s is used as test fluid. Small aluminium particles were served into the fluid for flow visualization. The basic flow in spherical gaps is always three-dimensional. With increasing Reynolds number the flow becomes non-unique to different spatial steady states with and without Taylor vortices. The influence of initial and time-dependent boundary conditions on this non-uniqueness was shown experimentally by Sawatzki and Zierep [8] and Wimmer [9]. Numerical results to this bifurcation phenomenon are known from Bartels [10], Tuckerman [11] and Bühler [12]. Figure 1 shows the spatial structure of the steady supercritical flow states Mode I, III and IV with lines of constant streamfunction in the meridional plane.

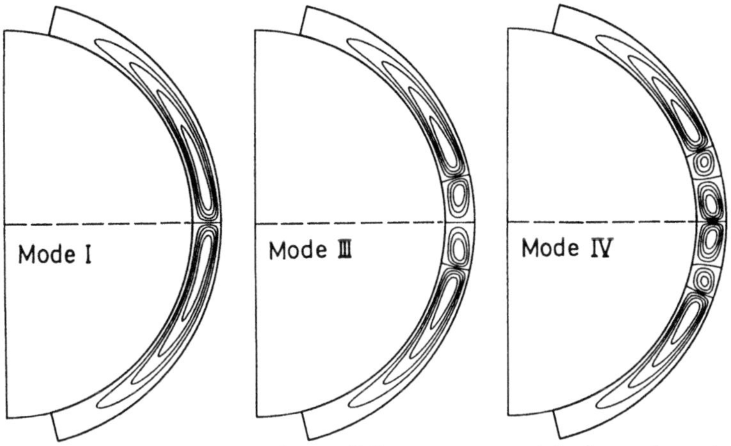

Fig. 1: Structure of the different supercritical steady solutions at the same Reynolds number Re = 2600, σ = 0.154, shown with lines of constant streamfunction

Starting with these different flow states, time-dependent instabilities occur at higher Reynolds numbers. The existence regions of the steady and time-dependent flow regions are shown in Figure 2. The solid lines represent laminar flows while the dashed lines correspond to turbulent flow. The instabilities are strongly connected to the different modes of flow. Figure 3 shows the circumferential velocity distribution at $\vartheta = 60°$ for the three modes taken from numerical calculations of Bühler [12]. These velocity distributions demonstrate the boundary layer character of the flow. In the case of separated boundary layers we find in the middle of the gap a nearly rigid body rotation with $\omega = k \cdot \omega_1$ as angular velocity. The boundary layer thickness near the rotation inner and fixed outer sphere can be deduced from a fractional force analysis:

$$\delta_i = \frac{a}{\sqrt{\cos\vartheta}} \cdot \sqrt{\frac{\nu(1-k)}{\omega_1}}\,, \qquad \delta_a = \frac{a}{\sqrt{\cos\vartheta}} \cdot \sqrt{\frac{\nu}{k \cdot \omega_1}}$$

It follows with $k \approx 0.5$, that the thickness of the outer boundary layer is twice that of the inner one. The condition for separated boundary layers $\delta_a + \delta_i < s$ is fullfilled first near the poles, if the Reynolds number exceeds a characteristic

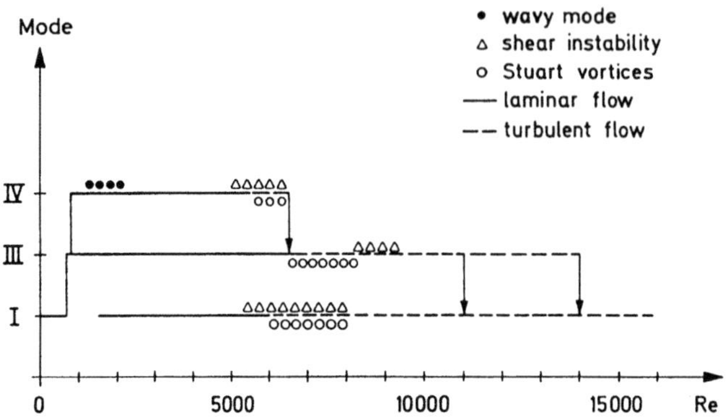

Fig. 2: Existence regions and kind of instabilities as function of Reynolds number, relativ gap width $\sigma = s/R_1 = 0.1779$

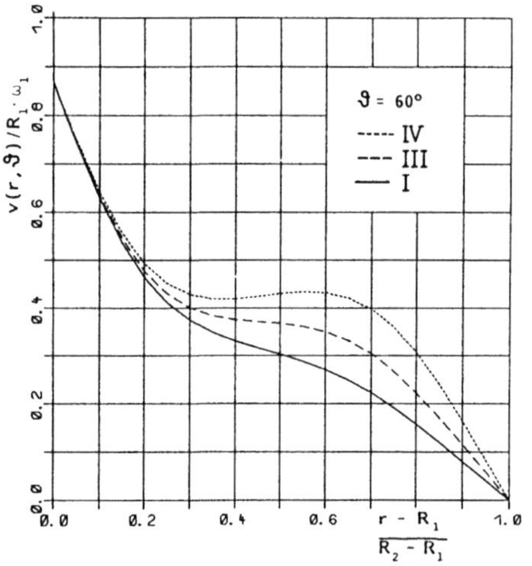

Fig. 3: Circumferential velocity distribution for the three modes at $\vartheta = 60°$, $\sigma = s/R_1 = 0.154$, $Re = (R_1^2\ \omega_1)/\nu = 2600$

value:

$$Re > \frac{4a^2}{\sigma^2}\ ,\quad Re = \frac{R_1^{\ 2}\ \omega_1}{\nu}\ ,\quad \sigma = \frac{s}{R_1}$$

The constant a depends on the flow configuration. For rotating systems the mean values are $1 < a < 3$. The meridional dependence of the boundary layers is the same as obtained by Moore [13] for the Ekman boundary layer at curved boundaries.

Time-dependent Instabilities

Mode I: This supercritical flow state without vortices is realized by rapidly accelerating the inner sphere from rest. Three-dimensional shear waves appear as spirals in both hemispheres for Reynolds numbers $Re > 5400$. Their interaction at the equator give rise to a wavy interface which moves circumferentially as shown on

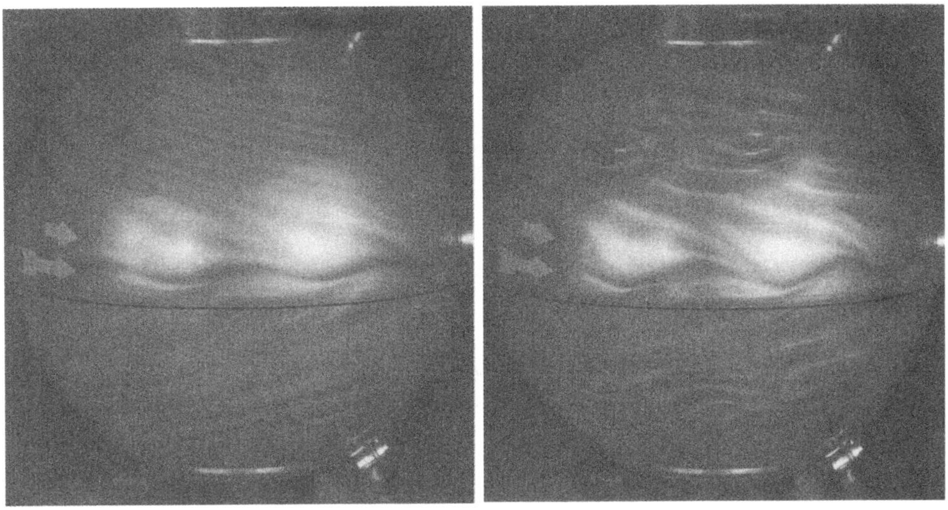

Fig. 4: Mode I with shear instabilities Re = 5400 (left side) and superimposed Stuart vortices at Re = 6050 (right side)

the left in Figure 4. The structure of these instabilities in the meridional plane is visualized by a light cut method. The streaklines in Figure 5, right side, demonstrate the structure of these boundary layer instability near the outer sphere. A complete principle sketch of the spherical gap flow with the shear instability is given in Figure 5, left side. The circumferential velocity v_w of the shear waves depends for a given gap width only on the Reynolds number.

$$\frac{v_w}{R_1\,\omega_1} = k(Re) \;\wedge\; \frac{f_1}{f_r} = n\,\frac{R_1}{R_2}\,k(Re) = n \cdot k^*(Re)$$

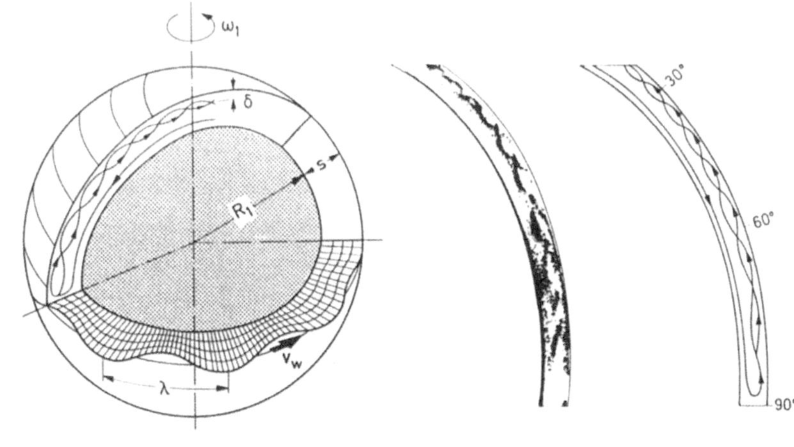

Fig. 5: Principle sketch of the shear waves (left side) and visualization of the structure in the meridional plane by streaklines (right side)

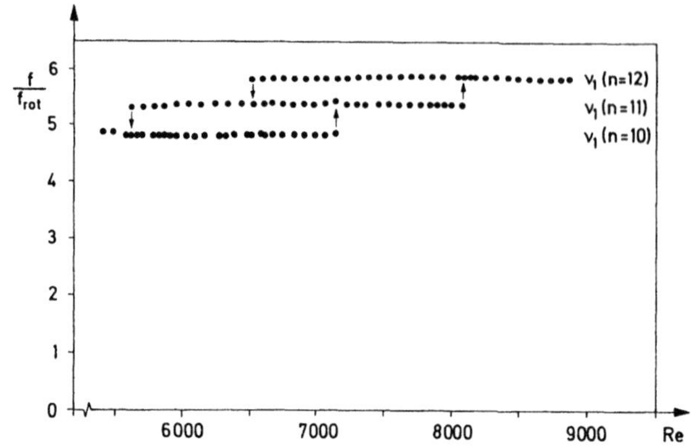

Fig. 6: Frequencies of the shear waves as function of the Reynolds number for Mode I

The dimensionless frequency $\nu_1 = f_1/f_r$ is therefore directly proportional to the discrete wave number n. The measured shear wave frequency is plotted over the Reynolds number in Figure 6. For a given wave number n, the frequency ν_1 is nearly constant. With increasing Reynolds number the wave number increases stepwise, corresponding to a decreasing wave length λ. These transitions are accompanied by strong hysteresis effects. The frequency jumps $\Delta\nu$ correspond to the angular velocity of the shear waves. Characteristic time-dependent signals and corresponding power spectras of the local wall shear stress are given in Figure 7. At the equator we find a harmonic variation with the frequency f_1 and near the pole at $\vartheta = 30°$ are the frequency f_1 and the rotation frequency f_r dominant. $f_0 = f_1/n$ corresponds to the basic frequency of the shear waves. Increasing the Reynolds number, the formation of Stuart vortices takes place near the poles. These spiral vortices are oriented in the direction of the secondary motion near the outer sphere. The superposition of the Stuart vortices and the shear waves leads to a stochastic motion with a coherent structure as shown in Figure 4, right side. For Reynolds numbers Re > 8000, the whole flow becomes turbulent. The waviness of the equatorial plane is conserved for even higher Reynolds numbers, as indicated by the frequency f_1 in the lowest spectra of Figure 7.

Mode III: This mode of flow with two Taylor vortices near the equator is show in Figure 8. This flow becomes unstable with respect to Stuart vortices for Re > 6500 near the poles as shown in Figure 8, left side. A secondary instability within the Taylor vortices occurs for Re > 7800. Increasing the Reynolds number further, the development of shear waves is obtained. The superposition between the shear waves and the Stuart vortices leads then to a turbulent motion as shown in Figure 8, right side. The Taylor vortices are conserved up to Reynolds numbers 11000 < Re < 14000, at which a transition into Mode I takes place. Characteristic time-dependent signals and corresponding power spectras are given in Figure 9.

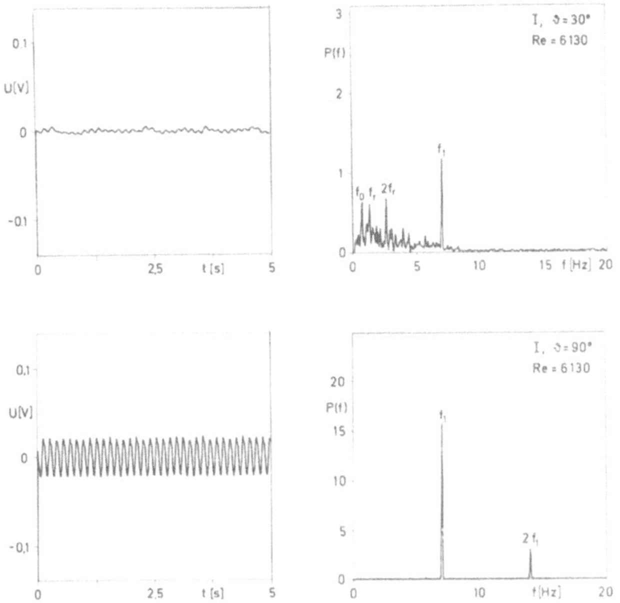

<u>Fig. 7:</u>

Time dependence of the local wall shear stress and corresponding power spectras of Mode I

$U \sim \sqrt{\tau_w'}$ [V]

$P(f) \sim \sqrt{\overline{\tau_w'^2}}$ [mV]

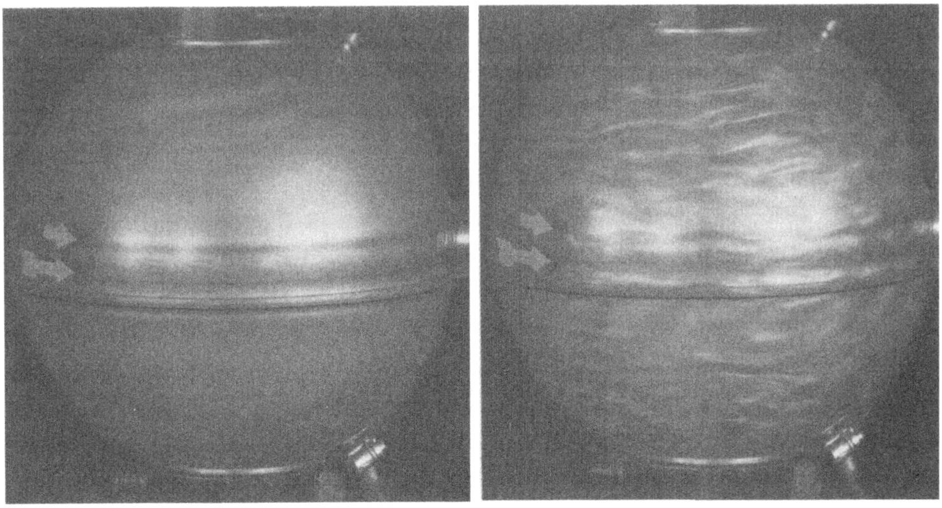

<u>Fig. 8:</u> Mode III with Taylor and Stuart vortices at Re = 7800 (left side), turbulent flow with Taylor vortices at Re = 10 700 (right side)

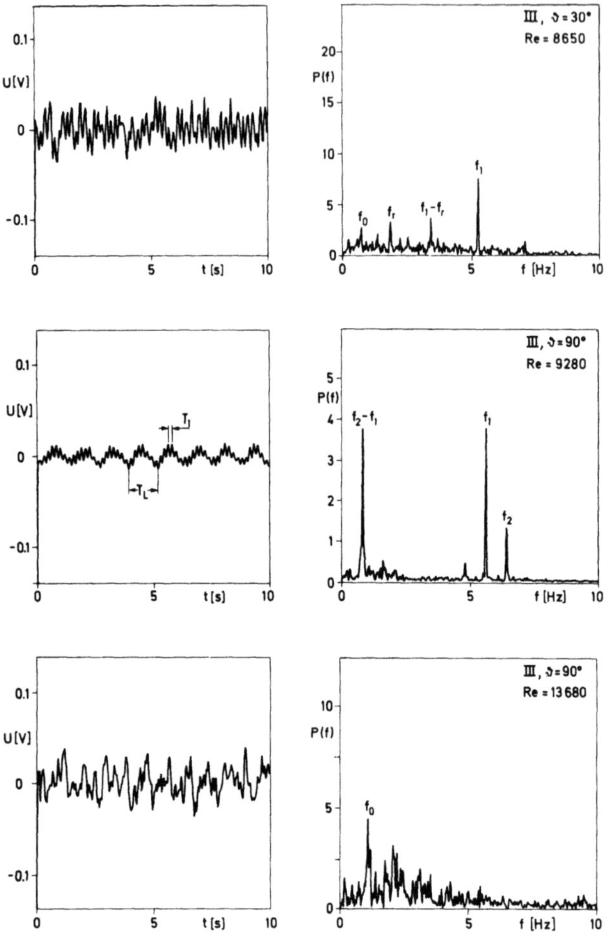

Fig. 9:

Time dependence of the local wall shear stress and corresponding power spectras of Mode III

$U \sim \sqrt{\tau_w'}$ [V]

$P(f) \sim \sqrt{\tau_w'^2}$ [mV]

Near the pole at $\vartheta = 30°$ at shear wave frequency f_1, the basic frequency $f_0 = f_1/n$ and the rotation frequency f_r are important. A broadband level of oscillations with small amplitude is also indicated. At the equator we obtained in addition to the frequency f_1 the frequency f_2 of the secondary instability within the Taylor vortices. In dimensionless form, these frequency $\nu_1 = f_1/f_r$ and $\nu_2 = f_2/f_r$ are nearly independent of the Reynolds number. With increasing Reynolds number the amplitude level of the broadband spectra increases, so that at Re = 13680 only the basic frequency f_0 can be detected.

Mode IV: This flow state with four Taylor vortices becomes unstable in the low Reynolds number range 1200 < Re < 2200 with respect to the time-dependent wavy mode. After that range the flow is steady up to Reynolds numbers Re = 4900. Then, near the poles the shear waves are formed as shown in Figure 10, left side. As can be deduced from the structure in the meridional plane in Figure 10, right side, this is also a instability of the three-dimensional boundary layer near the

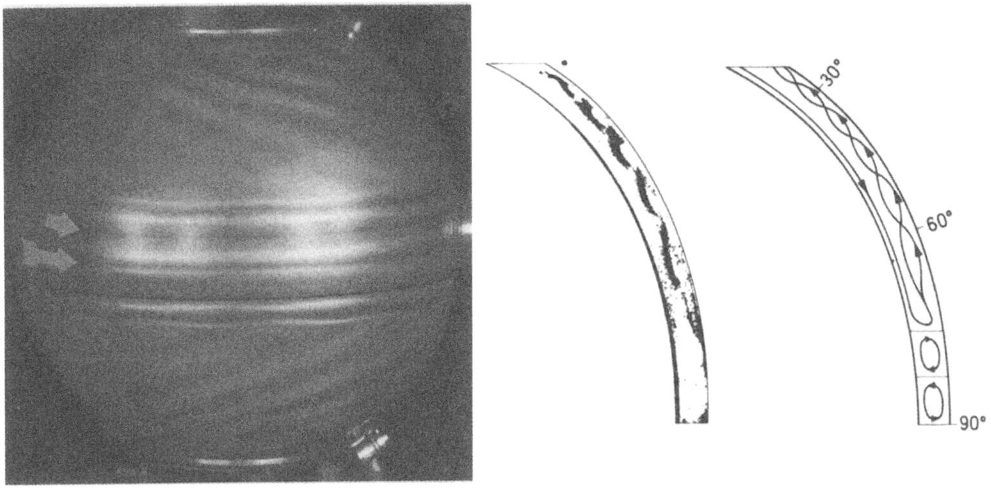

Fig. 10: Mode IV with Taylor vortices and shear waves at Re = 5550 (left side) and streaklines of the structure in the meridional plane (right side)

outer sphere. The interaction of these shear waves with the Stuart vortices near the poles at higher Reynolds numbers leads then to a stochastic motion. A transition into the laminar Mode III takes place, even when the stochastic motion reaches the vortex boundaries. The time-dependent signals and the corresponding power spectras in Figure 11 demonstrate that for Re = 5400 near the poles at ϑ = 30° the shear wave frequency f_1 and the rotation frequency f_r are important, while at the equator only the shear wave frequency f_1 occurs. The additional modulation frequencies at Re = 6750 indicate the transition into Mode III by the disappearence of two vortices. The dimensionless shear wave frequency $\nu_1 = f_1/f_r$ increases slightly with the Reynolds number.

Direct Transition to Turbulence

A direct transition from laminar to turbulent motion can be realized by a sudden start of the inner sphere to a high supercritical Reynolds number Re > 8000. The turbulence appears first at the equator and than for different times the transition fronts are spreading toward the poles as shown in Figure 12. A simple description can be given by applying the analogy to the Rayleigh-Stokes plate problem [7]. By the high acceleration of the inner sphere a diffusion process in radial direction takes place. After a characteristic time t_1 a sudden transition from laminar to turbulent motion occurs. For this characteristic time we get the following dependence for the final Reynolds number $t_1 \sim 1/\omega_1{}^n$ with n = 4/3 by fractional force analysing and n = 1 by energy considerations. The transition to turbulence is performed in the meridional plane with time relaxation because the circumferential velocity decreases with ϑ^*:

$$v(R_1, \vartheta^*) = R_1\,\omega_1\,\cos\vartheta^*$$

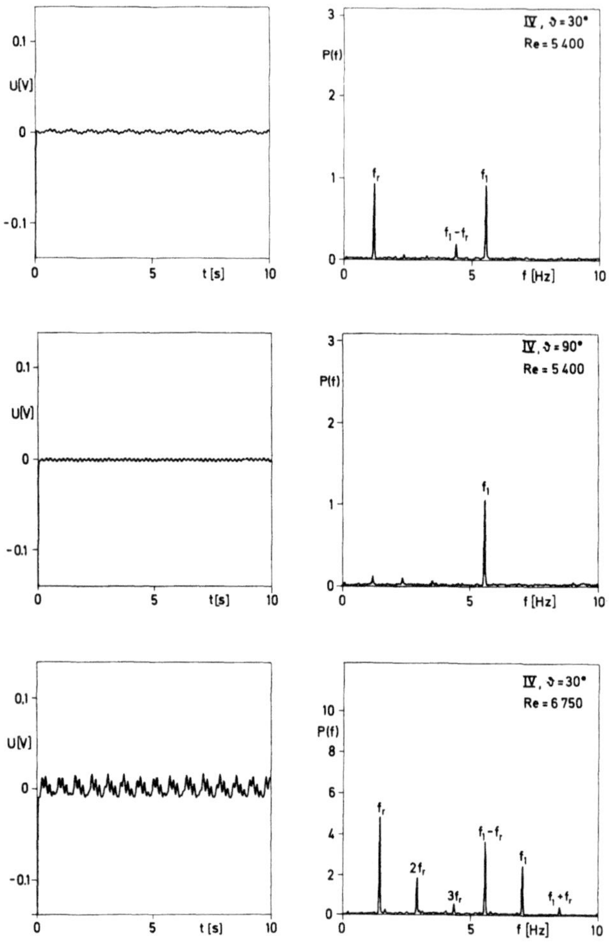

Fig. 11:

Time dependence of the local wall shear stress and corresponding power spectras of Mode IV

$U \sim \sqrt{\tau_w'}$ [V]

$P(f) \sim \sqrt{\tau_w'^2}$ [mV]

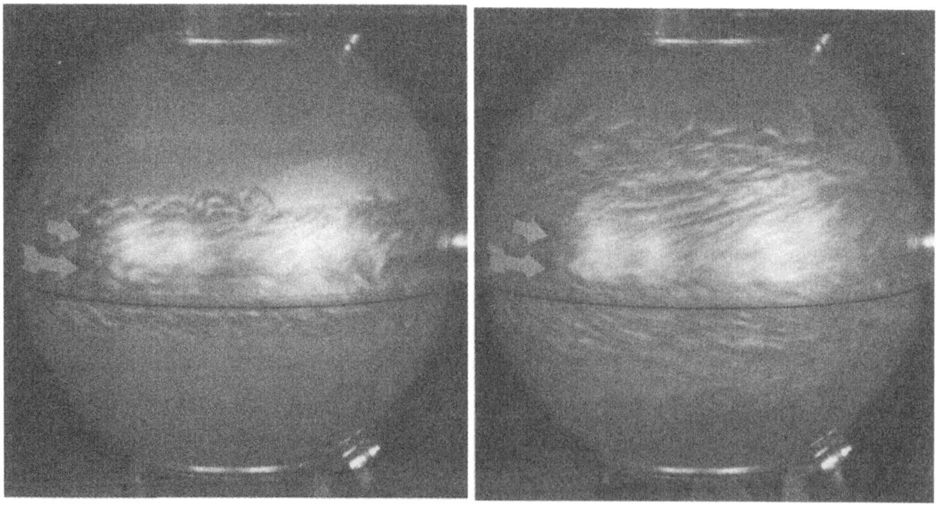

Fig. 12: Spreading turbulent flow region after a sudden acceleration of the inner sphere, $Re_E = 3.9 \cdot 10^4$

We assume now a high constant acceleration b of the inner sphere. The development of the boundary layer in radial direction is then in analogy to the result from Howarth [14] for the plate given approximately by

$$v(r^*, \vartheta^*, t) = b \cdot t \left[\left(1 + \frac{r^*}{4\sqrt{\nu t}}\right) \cdot \operatorname{erfc} \frac{r^*}{2\sqrt{\nu t}} - \frac{r^*}{\sqrt{\pi \nu t}} \cdot e^{-\frac{r^{*2}}{4\nu t}} \right] \cdot \cos \vartheta^* .$$

r^* is the coordinate in the gap and erfc the complementary error function. For short times $t \to t_1$ we find with this expression the same asymptotic solution as that for the sudden-start of the inner sphere:

$$\cos \vartheta^* \cdot \sqrt{\tau} = 1 \quad , \quad \tau = \frac{t}{t_1}$$

It is important, that the acceleration b has no influence in this asymptotic solution.

The meridional coordinate ϑ^* is measured from the equator and indicates the place of the spreading turbulence front. A comparison between this analytical result and the experiments for final Reynolds numbers $8 \cdot 10^3 < Re < 4 \cdot 10^4$ is given in Figure 13.

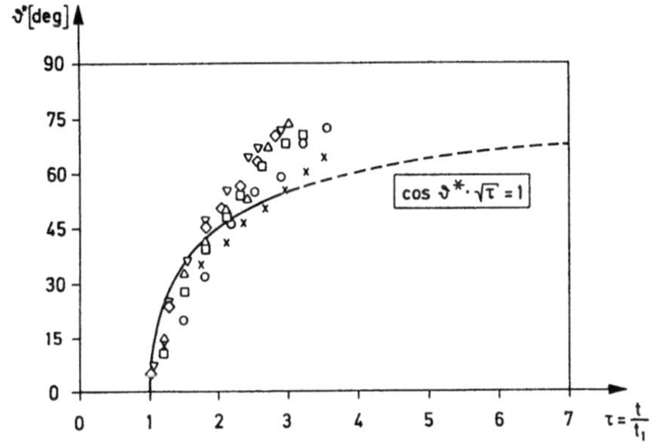

Fig. 13: Comparison between theoretical and experimental results

Conclusion

The investigation of the different routes from laminar to turbulent spherical gap flows gives a deeper insight into the process of turbulence development. The time-dependent shear waves have an important influence in this transition process. Analytical considerations are used to explain the physical nature of these shear waves and the spreading process of the turbulence front. The disappearence of the Taylor vortices at high Reynolds numbers indicates the characteristic difference between cylindrical and spherical Couette flow in the physical phenomena during the laminar-turbulent transition and the final turbulent flow. The relative gap width is also of important influence, as follows from a comparison with results of Belayev et al. [15] for wide spherical gaps.

References

1. R. Eppler, H. Fasel: Laminar-Turbulent Transition. Berlin 1980
2. H.L. Swinney, J.P. Gollub: Hydrodynamic Instabilities and the Transition to Turbulence. Top. in Appl. Phys. 45, Berlin 1981
3. G.I. Barenblatt, G. Ioss, D.D. Joseph: Nonlinear Dynamics and Turbulence. Boston 1983
4. T. Tatsumi: Turbulence and Chaotic Phenomena in Fluids. Amsterdam 1984
5. V.V. Kozlov: Laminar-Turbulent Transition. Berlin 1985
6. P.R. Fenstermacher, H.L. Swinney, J.P. Gollub: Dynamical instabilities and the transition to chaotic Taylor vortex flow. J.F.M. 94, 103-128 (1979)
7. K. Bühler, J. Zierep: New Secondary Instabilities for High Re-Number Flow Between Two Rotating Spheres. Laminar-Turbulent Transition, ed. Kozlov, Berlin, 677-685 (1985)
8. O. Sawatzki, J. Zierep: Das Stromfeld im Spalt zwischen zwei konzentrischen Kugelflächen, von denen die innere rotiert. Acta Mechanica 9, 13-35 (1970)
9. M. Wimmer: Experiments on a viscous fluid flow between concentric rotating spheres. J.F.M. 78, 317-335 (1976)
10. F. Bartels: Taylor vortices between two concentric rotating spheres. J.F.M. 119, 1-25 (1982)
11. L. Tuckerman: Formation of Taylor vortices in spherical Couette flow. Ph.D.Thesis, MIT Boston 1983
12. K. Bühler: Strömungsmechanische Instabilitäten zäher Medien im Kugelspalt. VDI-Fortschritt Berichte, Reihe 7, Nr. 96, Düsseldorf 1985
13. D.W. Moore: Viscous effects. Rotating Fluids in Geophysics, ed. P.H.Roberts, A.M. Soward, London, 29-66, 1978
14. L. Howarth: Laminar Boundary Layers. Handbuch der Physik Bd.VIII/1, ed. S. Flügge, p.290, 1959
15. Yu.N. Belayev, I.M. Yavorskaya: Transition to Stochasticity of Viscous Flow between Rotating Spheres. Nonl. Dyn. and Turb., ed. Barenblatt et al., Boston, 61-70 (1983)

On Transition to Turbulence in Boundary Layers

F.T. Smith

Mathematics Department, University College, Gower St.,
London, WC1E 6BT, UK

1. INTRODUCTION

This article is concerned with transition to turbulence in boundary layers and, by implication, in similar kinds of wall-bounded shear flows. There appear to be many types of boundary-layer transition observable in practice, depending on the disturbance environment, e.g. on how rough the solid surface supporting the boundary layer is upstream or on how unsteady the oncoming free stream is. Two possible extremes of transition are the following. First there is the "fast", savage, transition that is epitomised by the wake-passing effect, from an upstream row of rotor blades, on the boundary layers on stator blades in turbines. Broadly, a characteristic measure of the stator boundary-layer properties, such as the heat transfer at a fixed station on the surface, shows a very rapid change taking place there, apparently from a fully laminar to a fully turbulent state almost immediately as the highly turbulent wake of a typical rotor hits the stator blade. When the wake moves on there is a slightly less rapid adjustment, back to laminar conditions, until the violent wake of the next rotor blade hits the stator and the rapidly changing temporal cycle repeats itself. The other possible extreme of boundary-layer transition is "slow" (civilised) transition. This is exemplified by extensions of the classical experiments by Schubauer & Skramstad in the 1940's, on tiny unsteady controlled disturbances introduced upstream into a basic planar laminar boundary layer, the extensions being carried out in the recent experiments of Saric, Gaster, Kachanov and their co-workers, and others, who examine the effects of gradually increasing the size of the small input disturbance upstream. Typically, say for an input sizeable disturbance due to a ribbon, vibrating at a fixed frequency (approximately) and situated ahead of the lower-branch linear neutral position corresponding to that frequency, the Tollmien-Schlichting travelling 2D disturbance observed in the boundary-layer flow decays spatially in x up to the neutral position, then it exhibits relatively slow growth in amplitude just beyond the neutral position, and (unless the disturbance is infinitesimal) that is followed by the gradual development of a nonlinear travelling-wave state, still predominantly two-dimensional in nature. For sufficiently large, although still small, amplitudes of disturbance the occurrence of the nonlinear planar state just mentioned is itself then followed by very significant three-dimensional action appearing in the unsteady motion downstream, with the three-dimensionality often developing a preferred cross-stream wavenumber dependence which corresponds to oblique and quite fast-growing waves at angles of about 55°-65° from the free-stream direction. Associated with this, the amplitude spectrum exhibits a strong growth in its sub-harmonic components, particularly the half harmonic, then multiples thereof, and thereafter the spectrum starts to fill up relatively rapidly. Turbulent flow ensues some way downstream and, although it may be difficult to determine exactly how far downstream, the nominally full turbulent state eventually reached may be regarded as virtually the same as that reached via the fast type of by-pass transition described earlier.

The aim in the present article is to report on some theoretical research in progress which is directed towards theoretically understanding the transition process somewhat more, be it fast or slow transition, and ultimately of course towards a rational account of turbulence-modelling if possible, granted that the latter step is without doubt a formidable one. We concentrate here mostly on the slow types of transitions although the fast by-pass type is referred to later on.

As a guide for the theory, we should remark here that the Schubauer-Skramstad experimental results on infinitesimal disturbances agree well on the whole with classical linear Orr-Sommerfeld theory; that linear triple-deck theory based on assuming the Reynolds number to be large also agrees fairly well with the above experiments; that the typical spatial regions of interest in the earlier noted experiments on slow transition straddle the linear neutral curve but they are especially concerned with the nonlinear features emerging as the unsteady flow approaches the linear upper branch downstream; and that the characteristic Reynolds number (local R_{δ} or global Re) is relatively large in those transition experiments. The above aspects of the experiments motivate the present rational theory, which addresses nonlinear effects as the unsteady boundary-layer flow progresses downstream. The only main assumption in the theory is that the typical Reynolds number is large.

Section 2 below considers the theoretical two-dimensional flow properties, as a starting point and to gain some overall ideas on the downstream progress of nonlinear travelling disturbances. Then section 3 moves on to the more realistic case of three-dimensional flow theory. Certain aspects of the planar case cross over directly to the three-dimensional case, but the latter also brings in some very significant responses of its own. In particular, nonlinear interactions can occur sooner, i.e. at lower amplitudes, in the three- compared with the two-dimensional version. It is found that these theoretical nonlinear three-dimensional interactions agree well quantitatively with the experimental findings mentioned previously. Further, the theory seems to be quite promising so far, overall, although there is still a very long way to go in the required three-dimensional context.

2. TWO-DIMENSIONAL FLOW PROPERTIES

To start the nonlinear unsteady-flow theory we consider two-dimensional features first, as a guideline. Given that the Reynolds number can be taken to be a large parameter we address the Navier-Stokes equations but scale the nondimensional velocities $\bar{u}$, $\bar{v}$ in the x , y directions, and the nondimensional pressure $\bar{p}$, in the Tollmien-Schlichting (TS) fashion, which is equivalent to the triple deck scalings

$$(\bar{u},\bar{v}) = (Re^{-1/8}\lambda^{1/4}u,\ Re^{-3/8}\lambda^{3/4}v) + \ldots \tag{2.1a}$$

$$\bar{p} = Re^{-1/4}\lambda^{1/2}p(X,T) + \ldots \tag{2.1b}$$

$$(x - x_o,\ y,\ t) = (x_o + Re^{-3/8}\lambda^{-5/4}X,\ Re^{-5/8}\lambda^{-3/4}Y,\ Re^{-1/4}\lambda^{-3/2}T)\ . \tag{2.1c}$$

Here the scalings (2.1a-c) apply in the so-called lower deck, a viscous sublayer close to the surface, while the factor $\lambda = \lambda(x_o)$ is the reduced skin friction of the oncoming undisturbed boundary layer, e.g. in Blasius flow $\lambda(x_o) \propto x_o^{-\frac{1}{2}}$, at the typical $O(1)$ station $x = x_o$. With (2.1) acting, the flow problem comes down to the unsteady viscous-inviscid interactive one of solving the nonlinear boundary-layer equations

$$\frac{\partial u}{\partial X} + \frac{\partial v}{\partial Y} = 0\ , \tag{2.2a}$$

$$\frac{\partial u}{\partial T} + u\frac{\partial u}{\partial X} + v\frac{\partial u}{\partial Y} = -\frac{\partial p}{\partial X} + \frac{\partial^2 u}{\partial Y^2}\ , \tag{2.2b}$$

subject to the conditions

$$u = v = 0 \quad \text{at} \quad Y = 0\ , \tag{2.2c}$$

$$u \sim Y + A(X,T) \quad \text{as} \quad Y \to \infty\ , \tag{2.2d}$$

$$p(X,T) = \frac{1}{\pi}\int_{-\infty}^{\infty} \frac{\partial A}{\partial \xi}(\xi,T)\,\frac{d\xi}{(X - \xi)}\ , \tag{2.2e}$$

representing in turn the no-slip constraint at the surface, the displacement effect $(\propto -A)$ on the motion, and the interaction between the unknown pressure p and the unknown displacement $(-A)$ caused by the outer potential flow. Other rather obvious initial or spatial boundary conditions are assumed depending on the particular flow configuration.

The two main alternatives to the nonlinear triple-deck version (2.2) of the unsteady flow at high Reynolds numbers are the Navier-Stokes equations and interactive boundary-layer versions, both requiring numerical treatments. The former tends to be hindered in general by grid-resolution difficulties, among others, whereas interactive boundary-layer methods, which are preferable and involve sensible interpretation of asymptotic properties such as (2.2) at finite Reynolds numbers, have still to be developed seriously for unsteady flows. With the present rational approach beginning with the nonlinear TS problem (2.2) and its extensions later on the hope is that the small- and large-scale structures of transitional boundary layers can be understood in a more definite way, including the interaction between those scales and the possibility of coherent structures emerging in line with experimental findings in transitional and turbulent boundary layers.

Some properties of the viscous-inviscid unsteady nonlinear system (2.2) are quite well known (see, e.g. in SMITH & BURGGRAF /1/). The linearized version, linearized about the original steady flow $u = Y$, $v = p = A = 0$, compares well with Orr-Sommerfeld results as noted before. It gives the neutral value $\Omega = \Omega_N \doteqdot 2.30$, with Ω denoting the disturbance frequency i.e. $\partial/\partial T \to -i\Omega$ or, as an approximate generalization for subsequent working,

$$\Omega \equiv |\partial/\partial T| , \tag{2.3}$$

schematically. At $\Omega = \Omega_N$ there is a supercritical bifurcation, which yields stable nonlinear travelling states for Ω just exceeding Ω_N (equivalent to $x_o > x_{oN}$ or $R_\delta > R_{\delta N}$, in view of the λ factor, "N" standing for neutral), as opposed to the basic steady flow which is unstable there. Computations of the nonlinear travelling states for a wide range of values of Ω greater than Ω_N are currently in progress, by O.R. Burggraf, A.T. Conlisk and the author, while DUCK /2/ and SMITH /3/ have recently presented numerical time-marching solutions of (2.2) starting from initial conditions at a scaled time T of zero.

Of probably most concern next is what happens downstream as the station x_o increases. This corresponds to increasing Ω since the skin-friction factor $\lambda(x_o)$ in (2.1) accentuates the time-dependence. The same effect of increasing our parameter Ω results, at a fixed station x_o, from an increase in the input frequency upstream, again from the scalings in (2.1). In either case, our interest is steered towards investigating the properties of the nonlinear TS system (2.2) when the reduced frequency parameter defined in (2.3) is large,

$$\Omega \gg 1 . \tag{2.4}$$

In that regime it is found that an initially small disturbance passes through at least two main stages I, II as it progresses downstream. These successive stages are conveniently defined by their representative pressure levels,

$$p = O(\Omega^{\frac{1}{2}})\,(\text{stage I}) \text{ and } p = O(\Omega)\,(\text{stage II}). \tag{2.5a,b}$$

During stage I, the scaled form of the unsteady pressure response turns out to be governed by a generalized cubic Schrödinger equation,

$$\frac{\partial P}{\partial \hat{T}} - i\frac{\partial^2 P}{\partial \hat{X}^2} = \frac{P}{\sqrt{2}} - iP|P|^2 \tag{2.6}$$

from analysis of (2.2) (SMITH /3/), in slow variables $\hat{X}$, $\hat{T}$ and in a suitable frame moving fast downstream with the group velocity. The principal features of the solution to (2.6) are that the disturbance broadens spatially in an exponentially fast form, as $\hat{T}$ increases, and the maximum amplitude also grows exponentially rapidly, with the amplitude solution then taking on an elliptical shape.

In stage II, further downstream say, where the pressure p rises to $O(\Omega)$, the system (2.2) becomes fully nonlinear but viscous forces appear at first sight to become negligible. As a result the unsteady motion there is governed by the Benjamin-Ono equation

$$\frac{\partial A}{\partial T} + A\frac{\partial A}{\partial X} = -\frac{1}{\pi}\int_{-\infty}^{\infty} \frac{\partial^2 A}{\partial \xi^2} \frac{(\xi,T)d\xi}{(X-\xi)} , \tag{2.7}$$

for $A(X,T)$. This matches with stage I earlier on and can admit nonlinear travelling waves and solitary waves as solutions. The flow structure here is subject to the existence of a viscous unsteady sublayer, however, close to the surface, and that sublayer can itself break down in a singular form to provide eruptions of vorticity into the outer inviscid zone where (2.7) applies.

The other major features so far in the development of the sizeable disturbance are: (a) the amplitude of the nonlinear disturbance is much greater than that of the basic steady flow (recall that this is for the flow nearest to the surface); (b) the mean shear stress at the surface is increased by an order of magnitude above its original steady laminar value; (c) there are large oscillations of the unsteady shear stress about that mean; (d) secondary instability of the inflexion-point kind may well also be present. These features (a) - (d) continue to hold in the subsequent development of the disturbance even further downstream.

The next stage is the Euler stage, which finds the disturbance stronger still, further downstream. Formally the larger disturbance amplitude and increased frequency parameter Ω ($\to Re^{\frac{1}{4}}$) make the dominant spatial scales of the triple-deck structure compress, so that now typically x, y, t are all fast, of order $Re^{-\frac{1}{2}}$, while the velocities $\bar{u}$, $\bar{v}$ and the pressure $\bar{p}$ rise to $O(1)$. Hence the current Euler structure is controlled predominantly by the inviscid nonlinear system

$$\frac{\partial \bar{u}}{\partial x} + \frac{\partial \bar{v}}{\partial y} = 0 \tag{2.8a}$$

$$\frac{\partial \bar{u}}{\partial t} + \bar{u}\frac{\partial \bar{u}}{\partial x} + \bar{v}\frac{\partial \bar{u}}{\partial y} = -\frac{\partial \bar{p}}{\partial x} \tag{2.8b}$$

$$\frac{\partial \bar{v}}{\partial t} + \bar{u}\frac{\partial \bar{v}}{\partial x} + \bar{v}\frac{\partial \bar{v}}{\partial y} = -\frac{\partial \bar{p}}{\partial y} \tag{2.8c}$$

and appropriate initial and boundary conditions. The unsteady Euler equations (2.8) allow the necessary merging with the earlier stage II - the equation (2.7) for instance is a reduced-amplitude long-wave case of a solution to (2.8), in fact. Like the earlier stage, though, the validity of (2.8) is subject to the behaviour of the nonlinear viscous unsteady sublayer, of $O(Re^{-\frac{3}{4}})$ thickness, adjoining the surface and that sublayer can act as a source of vorticity bursts, occurring possibly in a random fashion, which can regenerate the disturbance as it continues to move downstream. This combined nonlinear structure, featuring both larger-scale properties governed mainly by the Euler equations and smaller-scale properties due to viscous action and bursting, has some promising connections with the full computations and/or modelling and/or experiments on transitional or turbulent boundary layers by WALKER & ABBOTT /4/, FASEL /5/, ORSZAG & PATERA /6/ and Blackwelder and co-workers, among others, especially if the extension to three-spatial dimensions is allowed for (see section 3 below). Again, we may also expect to be able to describe a "savage" transition (see introduction) as one containing sufficiently high amplitudes $(O(1))$ and frequency range that it by-passes the earlier stages I, II and enters the fully nonlinear Euler stage (2.8) directly.

Much more needs to be known about this stage, possibly from further computational studies, to increase our structural understanding, and the same goes for related interactive boundary-layer and composite methods aimed at incorporating the important viscous effects present also. Some encouragement can be drawn nevertheless from the two-dimensional theory so far, because of the connections with other approaches as noted previously and because the various stages yield both spectral focussing, from (2.6), and spectral broadening, from inflexional instabilities and/or sublayer bursting, aspects that are observed in practice. The major extra element to be acknowledged in the theory is three-dimensionality, which is considered next.

3. THREE-DIMENSIONAL FEATURES

We aim now to extend the above Re-large theory to allow for three-dimensional effects, as experimental findings tell us we must. Some of the above theory and conclusions can go straight through, in the various stages, but much more volatile events can take place even at relatively low amplitudes, in three dimensions.

Once again we start with the triple-deck version since it allows for linear and nonlinear TS disturbances, but now in the three-dimensional form

$$\frac{\partial u}{\partial X} + \frac{\partial v}{\partial Y} + \frac{\partial w}{\partial Z} = 0 \tag{3.1a}$$

$$\frac{\partial u}{\partial T} + u\frac{\partial u}{\partial X} + v\frac{\partial u}{\partial Y} + w\frac{\partial u}{\partial Z} = - \frac{\partial p}{\partial X} + \frac{\partial^2 u}{\partial Y^2} \tag{3.1b}$$

$$\frac{\partial w}{\partial T} + u\frac{\partial w}{\partial X} + v\frac{\partial w}{\partial Z} + w\frac{\partial w}{\partial Z} = - \frac{\partial p}{\partial Z} + \frac{\partial^2 w}{\partial Y^2} \tag{3.1c}$$

$$\text{with}\quad p(X,Z,T) = -\frac{1}{2\pi}\int_{-\infty}^{\infty}\int_{-\infty}^{\infty}\frac{\partial^2 A}{\partial\xi^2}(\xi,\eta,T)\,\frac{d\xi\, d\eta}{[(X-\xi)^2+(Z-\eta)^2]^{\frac{1}{2}}} \tag{3.1d}$$

$$u \sim Y + A(X,Z,T),\; w \to 0 \quad \text{as} \quad Y \to \infty \tag{3.1e}$$

$$\text{no-slip at} \quad Y = 0\,, \tag{3.1f}$$

as generalized from (2.1), (2.2).

The lowest-amplitude three-dimensional response, i.e. that which can arise first, occurs through the "resonant triad" mechanism [CRAIK /7/], again as the disturbance moves downstream, in line with experiments (see introduction). When (2.4) holds the triad can come into play in the form of the pressure expansion

$$\begin{aligned} p = {} & P_1 \exp\,[i\alpha X - i\Omega T] \\ & + P_2 \exp\left[\frac{i\alpha X}{2} + i\beta Z - \frac{i\Omega T}{2}\right] \\ & + P_3 \exp\left[\frac{i\alpha X}{2} - i\beta Z - \frac{i\Omega T}{2}\right] + O(\Omega^{-\frac{1}{2}}) \end{aligned} \tag{3.2}$$

and similarly for the other flow variables, all expanded in inverse powers of Ω. Here the P_1-wave is the incoming two-dimensional disturbance, whereas the P_2- and P_3-waves are three-dimensional parasitic ones which can start up provided that $\beta = \sqrt{3}\,\alpha$, i.e. the three-dimensional waves are inclined at angles

$$\theta_{res} = 60^{\circ} \tag{3.3}$$

to the free-stream direction. This angle compares favourably with the observed values mentioned in the introduction. At second order the nonlinear interaction between the three waves takes effect and yields from (3.1) the governing equations (SMITH & STEWART /8/)

$$\frac{\partial P_1}{\partial T} + 2\frac{\partial P_1}{\partial X} = (1-i)\frac{P_1}{\sqrt{2}} - 2iP_2P_3 \tag{3.4a}$$

$$\frac{\partial P_2}{\partial T} + \frac{5}{4}\frac{\partial P_2}{\partial X} + \frac{\sqrt{3}}{4}\frac{\partial P_2}{\partial Z} = (1-i)\frac{P_2}{2} - \frac{i}{4}P_1P_3{}^* \tag{3.4b}$$

$$\frac{\partial P_3}{\partial T} + \frac{5}{4}\frac{\partial P_3}{\partial X} - \frac{\sqrt{3}}{4}\frac{\partial P_3}{\partial Z} = (1-i)\frac{P_3}{2} - \frac{i}{4}P_1P_2{}^*\,, \tag{3.4c}$$

the nonlinearity coming from the coupling of P_2, P_3 which reproduces P_1, for example. In (3.4) the asterisk denotes the complex conjugate and X, Z, T are to be regarded as relatively slow coordinates, compared with the faster dependence in the waves of (3.2). The terms involving $(1 - i)$ in (3.4) are effects due to viscosity and are responsible for all the significant growth in the system.

Various computations of (3.4) are in progress but the central one we report here concerns the special case of purely temporal growth, where $\partial/\partial X$ and $\partial/\partial Z$ are identically zero, and for definiteness we take $P_3 \equiv P_2$, so that (3.4) reduces to the coupled nonlinear ordinary differential equations

$$P_{1T} = (1 - i) \frac{P_1}{\sqrt{2}} - 2i P_2^2 \tag{3.5a}$$

$$P_{2T} = (1 - i) \frac{P_2}{2} - \frac{i}{4} P_1 P_2^* . \tag{3.5b}$$

Pure spatial dependence, where $\partial/\partial T$ and $\partial/\partial Z$ are identically zero instead, results in essentially the same features in fact. Sample computations of (3.5) are presented in /8/ which are typical of the nonlinear growth, enhanced spikiness and apparently chaotic response produced as the flow solution develops from almost any initial condition. Of most interest is what happens as $T \to \infty$ or, in the spatial context, as $X \to \infty$ downstream. A close-up view of the numerical solution helps to confirm the subsequent analysis of the large-time asymptotic behaviour /8/, which consists on the larger scale of a predominantly inviscid balance (a "coherent structure" in some sense) but interrupted increasingly by a quick viscous "rejuvenation" process that adds energy either to the two- or the three-dimensional part of the disturbance. Along with this the effective phase function, which plays a significant role, becomes nearly but not quite phase-locked before rapidly swinging, through an amount π, to become nearly phase-locked again, and so on.

The nonlinear large-time/far-downstream structure of the solution to (3.5), inferred first from the computations, has the schematic form

$$|P_n| \sim \exp(\sigma T) F_n(\exp(\sigma T)) \tag{3.6a}$$

where the F_n are approximately periodic functions, and the nonlinear growth rate is

$$\sigma = \frac{1}{\sqrt{2}} \tag{3.6b}$$

and stems from the viscous forces. The limiting structure is a cycle of four states which goes on repeating itself indefinitely in the present regime. Further details here are given by SMITH & STEWART /8/.

Comparisons given in the last-named reference between the above three-dimensional theory and the experiments of KACHANOV & LEVCHENKO /9/ are rather encouraging, yielding both small- and large-scale agreement in quantitative terms.

It is worth observing here that the first possible three-dimensional nonlinear interaction, in (3.2), occurs at scaled pressure levels p of $O(1)$, much less than in the corresponding two-dimensional version summarized in (2.5). Moreover, the interaction is occurring theoretically in the same range, viz. Ω increasing, station x_o increasing, as is observed experimentally.

To maintain the above relevance to experimental findings and go further into transition requires theoretical investigations of pressure levels slightly higher than in (3.2). Some such investigations for three-dimensional effects have been started by the author and co-workers and the current state of the investigations is summarized briefly below. First, if the incoming planar disturbance has a pressure level p raised to the order $\Omega^{\frac{1}{2}}$, then three-dimensional interaction can come in nonlinearly through a combination of the amplitude-cubed influence in (2.6) from the planar wave and all the amplitude - squared influences present in (3.4) due to three-dimensionality. This is felt to be analogous to the secondary instabilities described by HERBERT /10/ but now, in contrast, allowing for nonlinear effects; and indeed parasitic three-dimensional growth of the fast form exp (exp (time))(schematically) is possible. Second, if both two- and three-dimensional pressure levels of order $\Omega^{\frac{1}{2}}$ are provoked, then amplitude-squared and -cubed nonlinear interactions come into operation, in a very complicated way. In particular, even just a single three-dimensional disturbance with wavenumbers α,β exhibits some surprising features then, since in scaled oblique coordinates ξ,η instead of X, Z, and with an assumption made about the mean-flow correction, the amplitude-cubed response takes the form

$$\frac{\partial P}{\partial T} - i\left\{\frac{\partial^2 P}{\partial \xi^2} \mp \frac{\partial^2 P}{\partial \eta^2}\right\} = \frac{P}{\sqrt{2}} - i\, P|P|^2 \tag{3.7}$$

for $\beta/\alpha \gtrless \sqrt{2}$. There is therefore a crossover from effectively planar behaviour, for wave angles $\theta < \theta\text{-crit}$ (c.f. (2.6)), to a new form involving sideband instability for $\theta > \theta\text{-crit}$, where

$$\theta\text{-crit} = \tan^{-1}(\sqrt{2}) = 54.7^\circ . \tag{3.8}$$

Again we note the comparison with the experiments referred to in the introduction. Computations of (3.7) for $\theta > \theta\text{-crit}$ show a pronounced growth and focussing or spikiness phenomenon appearing, together with apparent chaos, as time T increases (SMITH /11/). The growth of the spikes exceeds the growth in (3.6) for instance. Third, and subsequent to the stages above, the three-dimensional disturbances tend to the three-dimensional analogue of (2.7), which then requires increased computational effort. Fourth, and even further downstream, or at higher input amplitudes, the unsteady flow enters an Euler stage which is the three-dimensional analogue of (2.8), and that can lead to extra secondary instability aspects as well as the sublayer bursting process described in section 2. Further research is necessary and has been started for these later, more nonlinear, stages.

At present, or so it appears, the two-dimensional flow theory (section2) suggests that nonlinear disturbances of various amplitudes and frequencies can be followed all the way through their development downstream, ending up at the Euler stage which then requires Euler and viscous combined computations; and the three-dimensional flow theory shows good and encouraging agreement so far with the experiments and should be pursued further into the realms of increased nonlinearity occurring further downstream.

1. F.T. Smith, O.R. Burggraf: Proc. Roy. Soc. A399, 25 (1985)
2. P.W. Duck: J. Fluid Mech. 160, 465 (1985)
3. F.T. Smith: United Tech. Res. Cent., E. Hartford, Ct.,U.S.A., Rept. UTRC85-36 (1985) (see also A.I.A.A. paper no. 84-1582 (1984) and J. Fluid Mech., in press (1986))
4. J.D.A. Walker, D.E. Abbott: in Turb. in Internal Flows (ed. S.N.B. Murthy), Hemisph. Pub. Corp., Wash. (1977)
5. H. Fasel: Proc. Symp. on Turb. & Chaotic Phen. in Fluids (ed. T.Tatsumi), Elsevier (1984)
6. S.A. Orszag, A.T. Patera: J. Fluid Mech. 128, 347 (1983)
7. A.D.D. Craik: Proc. IUTAM Symp. on Lam.-Turb. transition (ed. V.V. Kozlov), Springer-Verlag (1985)
8. F.T. Smith, P.A. Stewart: United Tech. Res. Cent. E. Hartford, Ct., U.S.A., Rept. 86-26 (1986) (also to appear, J. Fluid Mech. (1987))
9. Y.S. Kachanov, V.Y. Levchenko: J. Fluid Mech. 138, 209 (1984)
10. T. Herbert: A.I.A.A. paper no. 84-0009 (1984)
11. F.T. Smith: United Tech. Res. Cent., E. Hartford, Ct., U.S.A., Rept. 86-10 (1986)

Instability and Transition of Flow in a Two-Dimensional Channel

P.G. Drazin[1] *and I.J. Sobey*[2]

[1]School of Mathematics, University Walk, Bristol BS8 1TW, UK
[2]Schlumberger Cambridge Research, P.O. Box 153, Cambridge CB3 0HG, UK

1. Introduction

Consider the flow of a uniform incompressible fluid of kinematic viscosity ν at a constant flux Q through an indentation, i.e. a 'throat', in a two-dimensional channel, as illustrated in Fig. 1. For rigid channels of a given shape, the conditions of flow may be specified by

(i) the Reynolds number $R = Q/2\nu$,

(ii) the expansion ratio D, i.e. the ratio of the maximum height of the channel to the minimum height, and

(iii) the channel length L,

where, by convention, we define the unit of length so that the height of the channel at its minimum is two units and we suppose that the flow is always from left to right.

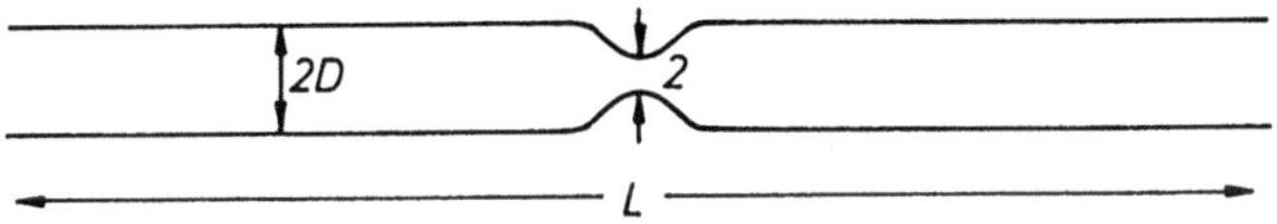

Fig. 1. Configuration of the channels

Although the conditions may be specified by the values of R, D and L and by the shape of the channel, we shall show that many flows are possible with the same conditions. We have studied steady and time-periodic flows in the early stages of transition to turbulence, finding, in particular, many bifurcations and multiple equilibria of steady flows.

We did numerical experiments for channels of the form illustrated in Fig. 1. Note that the channel is symmetric up and down, by reflection in the centre line, $z = 0$ say. These experiments were for periodic boundary conditions such that the inlet is the outlet. Thus the channel is equivalent to one of the same shape and size with the inlet at the middle of the indentation, where the height of the channel is a minimum. We also did laboratory experiments in a similar channel with its inlet at the middle of the indentation, so that the flow enters the expanding channel.

This class of flows in a channel is

(i) of practical importance, and

(ii) a good examplar of the physical and mathematical properties of transition to turbulence.

(In the latter respect, it may be noted that the set of channel flows is not invariant under a continuous group of transformations, so that transition of channel flows is more 'typical' than of Couette flow between rotating coaxial cylinders, whose conditions are invariant under rotation, or of plane Poiseuille flow, whose conditions are invariant under translation.)

To study channel flows we have used not only numerical and laboratory experiments, but also whatever else seemed helpful:

(i) physical arguments,

(ii) qualitative theory of bifurcation and transition to turbulence, and

(iii) asymptotic analysis.

This last has led to an asymptotic theory of two-dimensional flow in a channel with walls of small curvature, by use of a Jeffery-Hamel flow as the local approximation at each station downstream. SOBEY and DRAZIN [1] reported these results in detail. BANKS et al. [2] have developed the asymptotic theory for walls of small curvature. Here we review the numerical results and make some new observations on their relevance to transition to turbulence.

2. Numerical Results

We have integrated the time-dependent vorticity equation of two-dimensional motion of an inviscid incompressible fluid with the appropriate boundary conditions to find the streamfunction for a variety of initial conditions as well as a variety of values of R, D and L. (The numerical methods can be found elsewhere [1].) This gave various steady and time-periodic solutions at moderate values of R. The results are thought to be inaccurate when R is more than 200, so they can tell us little about the later stages of transition.

The essential pattern of the flows as R increases is as follows. A unique steady solution prevails for small values of R; it is stable and symmetric about the centre line, $z = 0$, of the channel. As R increases above a critical value, R_c say, the steady solution becomes unstable at a pitchfork bifurcation (this bifurcation is imperfect in our numerical as well as our laboratory experiments). Thus the stable symmetric flow is replaced by two stable asymmetric flows and one unstable symmetric flow, each asymmetric flow being the reflection of the other in the centre line of the channel. The solution settles down at one or other of these two asymmetric steady solutions according to the initial conditions. The value of R_c depends upon the shape of the channel as well as D and L, but may be quite low. We found, in particular, that $R_c \approx 5$ when $D = 3$ and $R_c \approx 20$ when $D = 2$ for the channel of Fig. 1. The asymptotic theory for walls of small curvature gives R_c as a specific function of the greatest angle between the tangents to the two walls at the same station. Some of the symmetric and asymmetric steady flows are illustrated in Fig. 2.

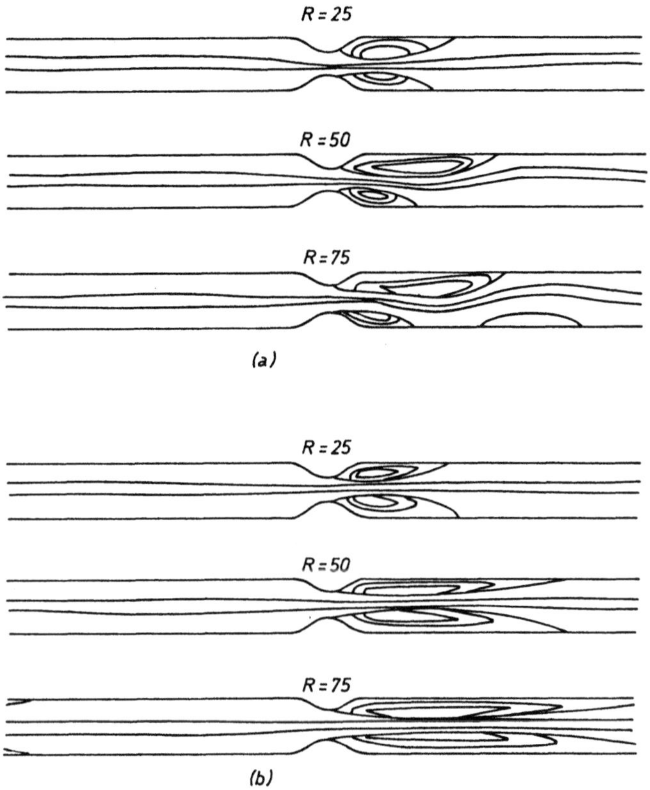

Fig.2. Some streamlines of steady flows calculated for $D = 3$, $L = 80$ and various values of R. Note that flow is from left to right. (a) Asymmetric (stable) flows. Note that, by symmetry, for each asymmetric flow there exists also its reflection in the centre line. (b) Symmetric (unstable) flows. These were calculated over half the channel by use of symmetry conditions on the centre line

As R increases above R_c a succession of bifurcations occur to give multiple equilibria of steady solutions. These bifurcations seem to be turning points. They give as many as eight stable steady solutions co-existing (for $D = 1.35$ and $R = 80$) and, by inference, seven unstable steady solutions. There also may be Hopf bifurcations at $R = R_c$ or at larger values of R. Indeed, the theory of the nonlinear instability of plane Poiseuille flow gives a subcritical Hopf bifurcation with $R_c = 3848$, and the channel flow becomes plane Poiseuille flow when $D = 1$, so we must have a Hopf bifurcation at $R = R_c$ when D is close to one.

Various state variables may be used to describe these bifurcations. In particular, define ζ such that

$$\zeta^2 = \iint_S \{\psi(x,z) + \psi(x, -z)\}^2 dxdz \Big/ \iint_S dxdz,$$

$\zeta > 0$ if the main stream is deflected towards the upper wall, and $\zeta < 0$ if it is directed towards the lower; where S is the domain of the flow in the xz-plane, ψ is the stream function, and $z = 0$ is the centre line of symmetry of the channel.

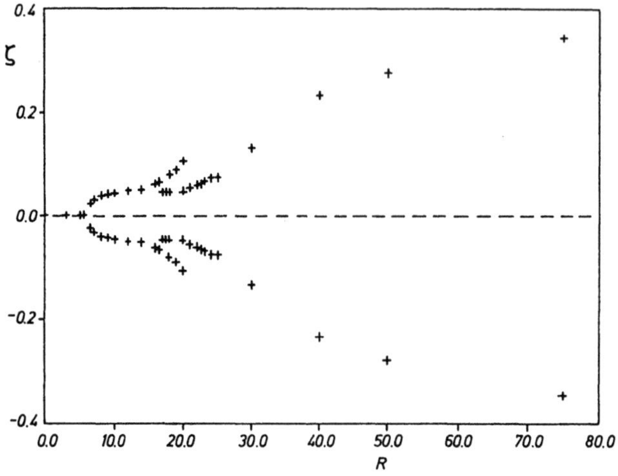

Fig. 3. Calculated bifurcation diagram for the channel with $D = 3$ and $L = 80$

Note that $\zeta = 0$ if and only if the flow is symmetric. Then the results may be encapsulated in a bifurcation diagram of the $R\zeta$-plane, as illustrated in Fig.3 for $D = 3$ and $L = 80$. It can be seen that there are four asymmetric stable steady solutions when $R = 20$, for example. In fact, the two co-existing asymmetric flows with positive (or negative) ζ at the same values of R, D and L are very similar.

SOBEY and DRAZIN [1] have also plotted the stations x where the direction of flow near each wall of the channel changes sign. This gives the extent of the separation 'bubbles', i.e. the closed regions of reversed flow, as a function of time. It reveals clearly how the flow approaches a steady or time-periodic one. In the latter case, the periodic shedding of a part of a separation bubble as an eddy is seen as an early stage of the transition to turbulence.

3. Conclusions

Perhaps the most striking result is the multiplicity, and the richness of the structure, of the steady flows in a given channel at given values of R, D and L. The multiplicity might, following the discovery of the multiplicity of the steady and periodic flows between co-axial rotating rigid cylinders, be regarded as normal in the transition to turbulence. The existence of a large number of steady solutions is also recognized as being associated with the onset of chaos in difference equations and in ordinary differential equations; the unstable steady solutions suggest the existence of homoclinic orbits which break up into chaos. The richness of the structure of the steady flows in a channel is seen better as the value of D as well as R is varied [1].

Another important property of channel flows, one not discussed above, was shown by our laboratory experiments. These were, unlike our numerical experiments but like all laboratory experiments, three-dimensional. They were made in channels with rectangular sections of sizes 15mm × 3mm and 15mm × 1mm in the y- and z-directions respectively. Symmetry breaking in the spanwise y-direction occurred at values of R depending on D. These values are quite low; for example $R \approx 80$ when $D = 3$. This suggests that three-dimensional effects first arise in practice not from obscure secondary or tertiary instabilities at large values of R but from symmetry breaking as a secondary instability at moderate values of R. We observed three-dimensional flow always first as a steady flow following symmetry breaking in the spanwise direction; and this symmetry breaking in the y-direction always occurred at a larger value of R than the symmetry breaking in the z-direction, as discussed in §2.

The asymptotic theory for walls of small curvature which we have mentioned is better assessed elsewhere [1,2] because it is more relevant to weakly nonlinear instability and to separation than to turbulence.

References

1. I.J. Sobey, P.G. Drazin: J.Fluid Mech.: in the press.
2. W.H.H. Banks, P.G. Drazin, M.B. Zaturska: to be published.

The Entrainment of Particles by a Turbulent Spot in a Laminar Boundary Layer

F.G.J. Absil and G. Ooms***

Laboratory for Aero- and Hydrodynamics, Delft University of Technology, Rotterdamseweg 145, Delft, The Netherlands

The entrainment of solid particles from a flat surface by an artificially generated turbulent spot in a laminar boundary layer in a wind tunnel was studied. High-speed films of the entrainment process revealed two different entrainment forms: direct entrainment for larger particles and entrainment after piling of the leading edge of the particle bed for small, strongly cohesive particles. A theoretical model for the entrainment process was developed based on the presence of turbulent bursts in the spot. Good agreement was found between model predictions and the results of particle sampling experiments.

1. Introduction

The entrainment of particles from a surface by a fluid flow is an important phenomenon. In nature it can be observed in many situations, e.g. sand- and dust storms. In industry the entrainment and dispersion of particles, when handling or stockpiling ores, is an economical loss and may lead to pollution problems. In the past a criterion for the critical wall shear stress τ_{cr} at which entrainment starts, has been developed by Phillips/1/. In this criterion three particle size ranges are discerned. In SI-units it is given by

- small particles ($d_p < 30\mu m$): $\tau_{cr} = 5.10^{-8}\, d_p^{-4/3}$ (1)
- intermediate sized particles ($30\mu m < d_p < 200\mu m$): $\tau_{cr} = 0.049$ (2)
- large particles ($d_p < 200\ \mu m$): $\tau_{cr} = 0.014\,(\rho_p - \rho_g)\, g\, d_p$, (3)

in which d_p is the particle diameter, ρ_p the particle density, ρ_g the fluid density and g the acceleration due to gravity. In Fig. 1 the dotted lines represent (1), (2) and (3) for particles with ρ_p = 2500 kg/m3 in an air flow. As can be seen there is a minimum value of the wall shear stress, below which entrainment does not take place.

We start in this paper with the results of a detailed experimental investigation of the entrainment process. Due to the random nature of turbulence in space and time it is difficult to study its effect on the entrainment of a particle or

* Present address: TNO Institute of Applied Physics (TPD), P.O. Box 155, 2600 AD Delft, The Netherlands

** Present address: Koninklijke/Shell-Laboratorium Amsterdam, P.O. Box 3003, 1003 AA Amsterdam, The Netherlands

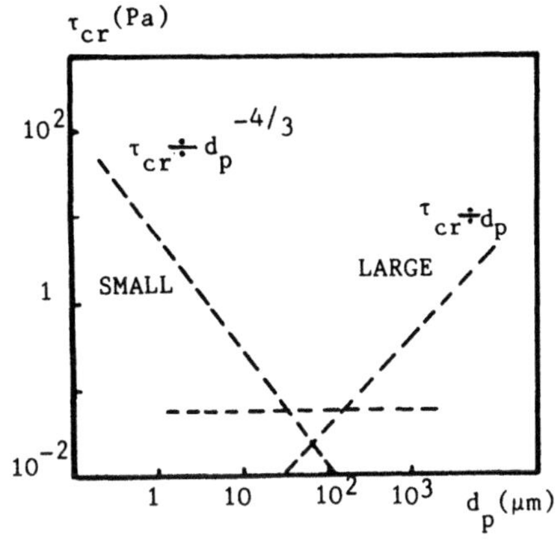

Fig. 1 - The critical wall shear stress versus particle diameter for particles in an air flow.

cluster of particles in a fully developed turbulent boundary layer. To avoid this problem we used a turbulent spot in a laminar boundary layer as a kind of building unit of a fully developed turbulent boundary layer. Visualization studies and hot-wire measurements have shown the analogy between the turbulence structure of such spots and of the fully developed turbulent boundary layer. We generated a spot at an appropriate position and time by disturbing a laminar boundary layer. In our detailed experimental investigation of the entrainment process both small and large particles were used.

Experiments have revealed the presence of turbulent bursts in a boundary layer, giving rise to fluid motions towards and away from the wall. In this paper we give some results of a theoretical model for the entrainment process by a turbulent spot based on the occurrence of bursts within the spot.

2. Experimental Investigation of the Entrainment Process

In a low speed open circuit wind tunnel (cross section 0.3 x 0.4m^2) a flat plate with a sharp leading edge was placed horizontally at 0.1m above the test section floor; see Fig. 2. At 0.3m downstream of its leading edge a trip wire was

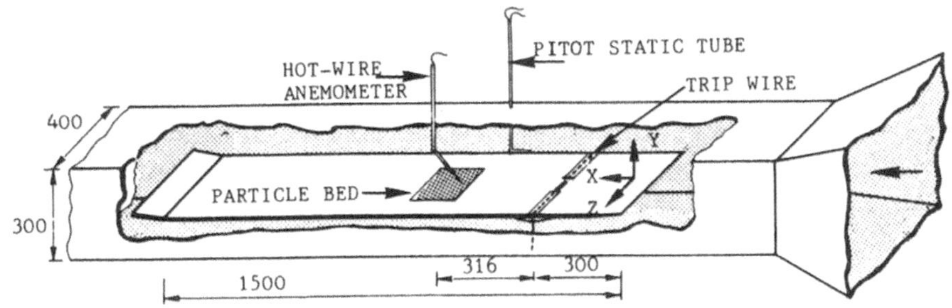

Fig. 2 - Particle wind tunnel test section (dimensions in mm)

mounted perpendicular to the flow. Only the centre 15mm of the wire was exposed to the flow. Underneath the plate was a row of magnets and as a current through the wire was switched on, the wire was pulled to the wall, thereby generating a point-like flow disturbance that developed downstream into a turbulent spot. At a distance of 0.6m, a sheet of emery paper (simulating the roughness of a particle bed) was glued to the plate. From a 15 x 2 mm^2 strip (at several lateral positions) the emery grains were scratched. Here a bed of particles was allowed to settle after being injected through a hole in the top of a cylinder, placed vertically over the plate. This method produced reasonably homogeneous particle beds, being nearly flush with its surroundings (average bed thickness 0.2mm). Particle size distributions used in the experiment are shown in Fig. 3. In this figure Durcal stands for $CaCO_3$, a filler material used in the dye industry with ρ_p = 2750 kg/m^3.

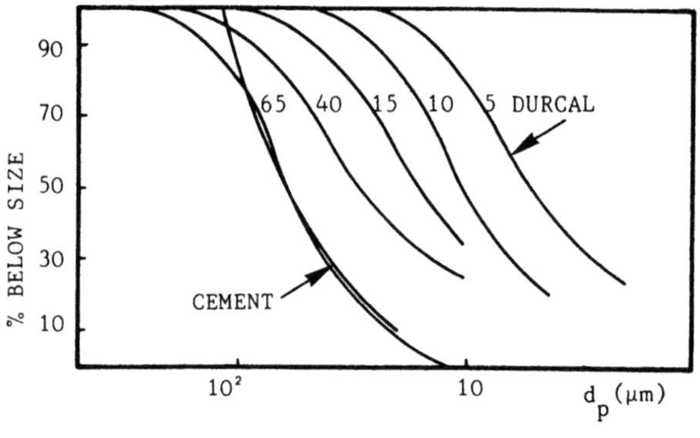

Fig. 3 - Cumulative particle size distribution.

After preparation of the particle bed the laminar flow velocity was set at 7.3 m/s (just below the critical value) and a turbulent spot was generated, which in its turn initiated particle entrainment when passing over the bed. A high speed camera filmed the entrainment process at a magnification factor of 3 (field of view 8 x $6mm^2$). A shadow light technique was used with flash frequency of 2000 Hz and flash duration of 10^{-8}s. An example of the filmed entrainment process is given in Fig.4.

Two different types of entrainment were observed, which are sketched in Fig. 5. In the case of the largest particles used, individual particles or clumps of particles were lifted from the bed (Fig. 5a). But in the case of the smallest particle fractions a shifting and piling of the bed leading edge region occurred (Fig. 5b). Due to the particle cohesion this pile nearly retains its shape until a height of about 0.25mm is reached. At the start of entrainment, particles are lifted mainly from the top of the hill, leading to entrainment further downstream. This second form leads to a much higher particle flux than the individual transport form. Using a motion analyser about 400 particle motions were measured and the average trajectories of the two types of entrainment are given in Fig. 6.

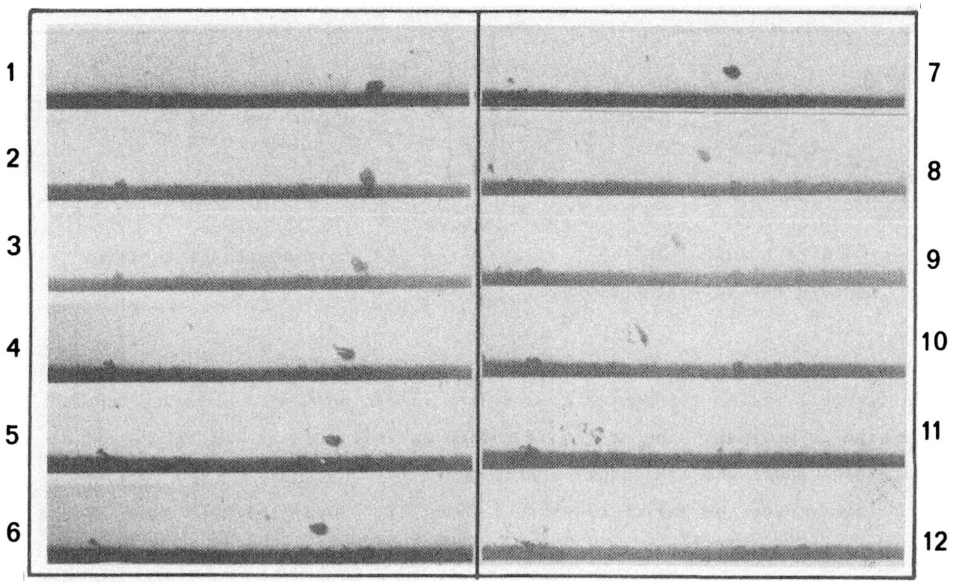

Fig. 4 - Entrainment of particles by a turbulent spot.

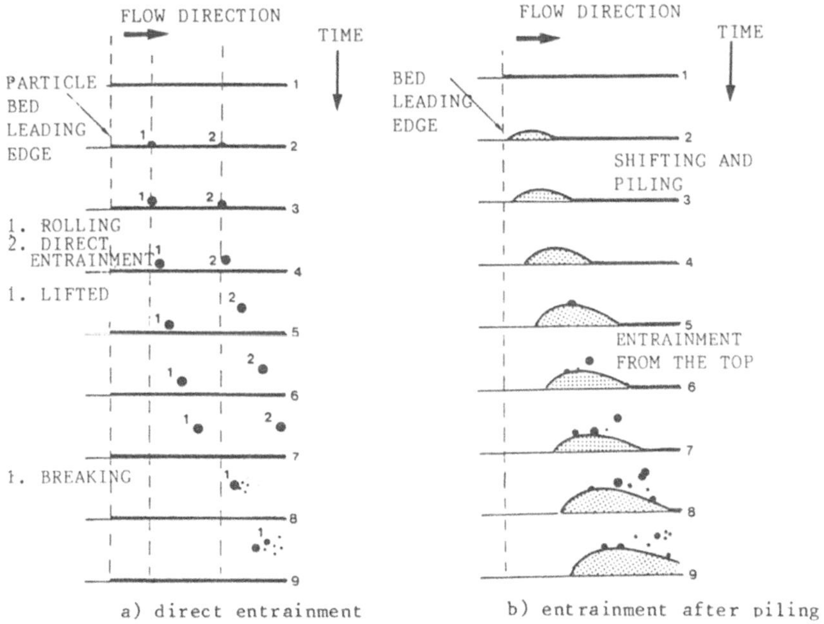

Fig. 5 - Types of entrainment.

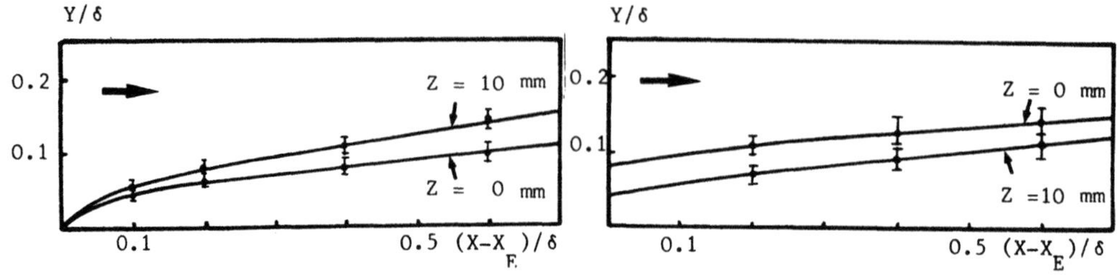

a) direct entrainment b) entrainment after piling

Fig. 6 - Average particle trajectory.

In this figure X is the streamwise co-ordinate, Y the normal co-ordinate and Z the spanwise co-ordinate; the origin is chosen as indicated in Fig. 2. X_E is the value of X where the trajectory starts and δ the thickness of the laminar boundary layer. Over the region observed ($-30mm < Z < 30mm$) the wall shear stress within the spot is sufficient to initiate particle entrainment.

3. Theoretical Model for the Entrainment Process

A theoretical model was developed to calculate the entrainment flux by a turbulent spot when passing a particle bed of finite dimensions. In the model a self-similar shape for the spot is assumed; its growth rate being chosen in accordance with experimental data. From the size of the spot and an assumed velocity profile the wall shear stress underneath the spot is calculated. Next, using the criterion given by equation (1) to (3), it is determined whether entrainment takes place. If entrainment occurs the flux of particles entrained by the spot can be calculated from the number and size of bursts present inside the spot, and from the relation for the particle flux per burst. The number and size of the bursts follows directly from the wall shear stress. The deposition of particles from the boundary layer back onto the surface is calculated from the settling velocity of the particles and from the particle concentration profile.

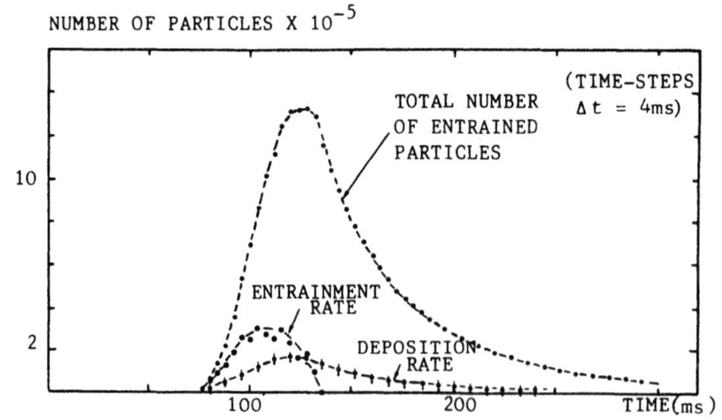

Fig. 7 - The entrainment process as a function of time.

The model is described in detail by Absil /2/. It contains two empirical parameters: an entrainment parameter present in the relation for the particle flux per burst, and a parameter present in the particle concentration profile.

In Fig.7 some theoretical results are given. It shows the entrainment rate and deposition rate as a function of time. The entrainment starts when the spot arrives above the particle bed and ends when it has passed it. Deposition continues until all the particles are left on the surface behind the spot. The entrainment rate and deposition rate determine the total number of entrained particles as a function of time, also given in Fig. 7.

4. Particle Sampling Experiments

Sampling experiments were carried out of particles entrained by one turbulent spot when passing over a 15 x 15 cm^2 bed with its leading edge at X = 0.6 m. Particles were sampled using a frame with 190 electrostatic precipitators. The frame was placed at X = 0.85, 0.95 and 1.05 m. Ensemble-averaged sample weight distributions were measured and compared with predictions of the theoretical model. In Fig. 8 the lateral weight distributions as measured and calculated at X = 0.85 and 1.05 m are given. As can be seen the agreement is reasonably good.

After each experiment photograhs were taken to study the planwise entrainment forms. In Fig. 9 some examples are given.

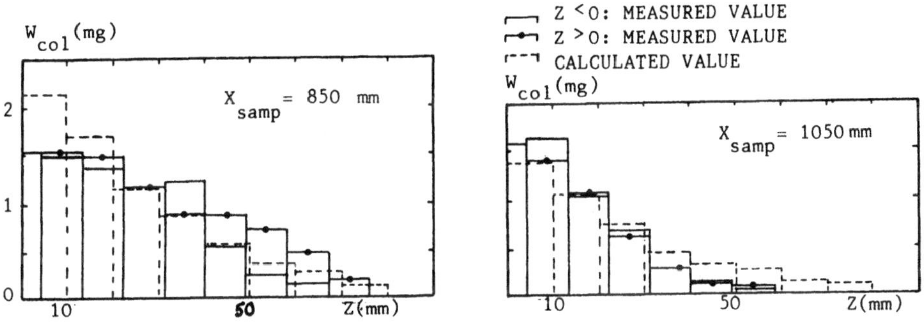

Fig. 8 - The lateral weight distribution.

5. Conclusions

From our study the two following conclusions can be drawn:

1) Two different entrainment forms can occur. One consists of individual particle motions (for the large particles); the other is a kind of massive particle transport after piling of the bed leading edge (for the small particles).
2) The overall characteristics of the entrainment process can reasonably well be described by a simple theoretical model based on the idea of turbulent bursts being present inside a turbulent spot.

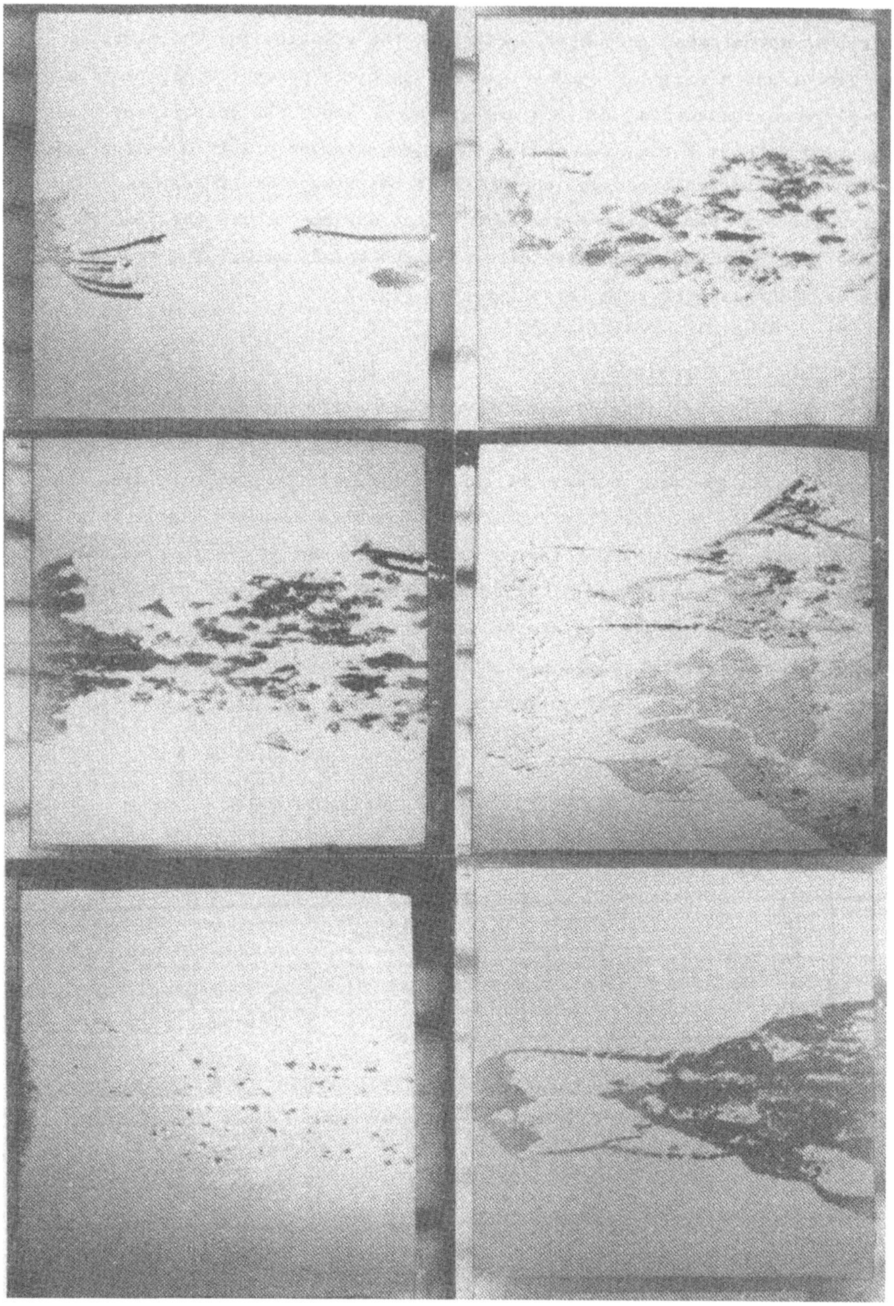

Fig. 9 - Entrainment patterns on a 15 x 15 cm^2 particle bed.

6. References

1. M. Phillips: Appl. Phys. 13, 221 (1980)
2. F.G.J. Absil: The Entrainment of Particles by a Turbulent Spot in a Laminar Boundary Layer, Thesis, Delft University of Technology, (1986).

Modal Approach to Low Prandtl Number Thermal Convection

J.M. Massaguer[1], *I. Mercader*[2], *and S. Blázquez*[1]
[1]E.T.S.E.C.C.P, Universitat Politècnica de Catalunya, E-08034 Barcelona, Spain
[2]F.I.B., Universitat Politècnica de Catalunya, E-08034 Barcelona, Spain

1. Introduction

Thermal convection in low Prandtl number fluids can develop large Reynolds number regimes near the threshold of the convective instability. The reason is that the velocity for the primary flow near the onset of instability scales with the thermal time, with the implication that the associated Reynolds number increases with decreasing Prandtl number. Therefore, low Prandtl number thermal convection provides an excellent setting to explore shear type instabilities in well organized, confined flows.

One example of shear instability in low Prandtl number convection has been considered by the authors in Massaguer & Mercader [1,2]. That instability is associated with the bending of a roll axis and it looks very similar to the production of axial flow in a bent filament. The agreement between the numerical results for convection in mercury, with Prandtl number $\sigma = 0.025$, and the corresponding experimental measurements stresses our confidence on the modal approach used, at least for modeling the first transition to time dependence. Unfortunately, the agreement concerns only fluxes and frequencies but not the geometry of the flow: the modal truncation used there was too crude, though it might contain the modes dominant in that range of parameters.

The aim of the present paper is to reexamine such an instability, called *swirl instability* in [2], in the light of an old experiment done by Mitchell & Quinn [3] that suits rather nicely the toroidal geometry of our modal approach. These authors explored thermal convection in air ($\sigma = 0.7$) and other gases of similar Prandtl number value, confined in a cylinder of aspect ratio Γ close to one, with Γ being defined as the ratio of radius to height. Because of the small aspect ratio, the geometry for the steady flows was not roll-like but four different patterns were observed, all them displaying a pattern dominated by one of the four lowest eigenvalues of the laplacian operator. All these patterns bifurcated towards an oscillatory regime with the new perturbations spiralling around the axis of the primary flow.

2. About the Model

As a crude aproximation for the velocity we have assumed in [2] a divergence free field defined as.

$$v = \nabla \times \nabla \times \Phi k + \nabla \times \Psi k \qquad (1.1)$$

where k is a unit vertical vector and the potentials Φ and Ψ have been taken to be

$$\Phi(x,y,z,t)=a^{-2}f(x,y)\,W(z,t)$$

$$\Psi(x,y,z,t)=a^{-2}f(x,y)\,\xi(z,t) \qquad (1.2)$$

where

$$(\partial^2_{xx}+\partial^2_{yy})f=-a^2f$$

and the function f is a periodic planform of wavenumber a with zero mean value and normalized to unity over a cell. Then, the vertical components of velocity and vorticity can be written as.

$$v_z=f(x,y)\,W(z,t)$$

$$\omega_z=f(x,y)\,\xi(z,t)$$

with the horizontal average of both, the vertical mass flux and the vertical component of the angular momentum being zero.

An axisymmetric flow is the most simple and ilustrative case, for the function $f(x,y)$ is now the Bessel function $J_0(ar)$. The trajectories for that flow can be obtained by integration of the following system.

$$\frac{dr}{dt}=-a^{-1}J_1(ar)\frac{dW(z,t)}{dz}$$

$$\frac{d\theta}{dt}=a^{-1}J_1(ar)\,\xi(z,t)/r \qquad (2.1)$$

$$\frac{dz}{dt}=J_0(ar)\,W(z,t)$$

where r,θ,and z are the cylindrical coordinates. As expected, the trajectories for a steady flow project in the in the r-z plane as closed orbits, and they project on the horizontal plane r-θ as displayed in fig. 2, where the condition of zero mean vertical vorticity can be realized. Therefore, for a steady solution with $\xi\neq 0$ or even for a time dependent one but near the bifurcating point, the trajectories spiral on a torus, as in fig. 1, showing a geometry fully consistent with the pattern displayed by the so-called mode 3 in [3].

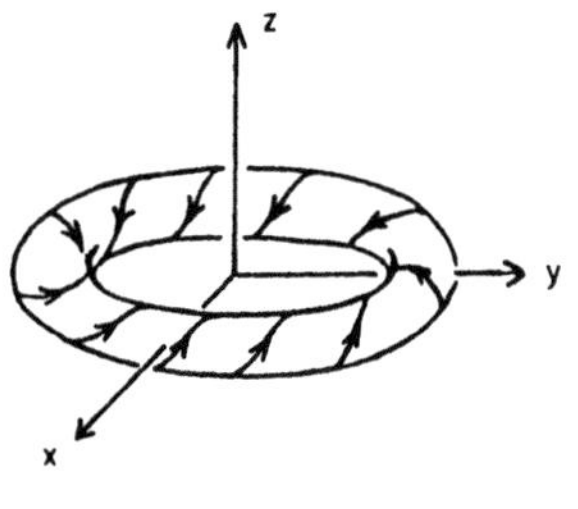

Figure 1. Sketch of the toroidal flow described by (2.1). A trajectory has been drawn on the torus for the case $\xi\neq 0$

Figure 2. Projection of a trajectory on the horizontal plane. The projection can be viewed as the superposition of a net flow in the azimuthal direction plus a local rotation, back and forth, giving zero mean vertical vorticity.

3. On the Instability of Swirl

If the velocity field defined in (1.1) is substituted in the full system of equations for thermal convection and then projected acording to the Galerkin procedure, we obtain a system of partial diferential equations to be integrated. The system can be found in [1] and the numerical procedure has been discused at large in [2]. For the case of air and fluids of Prandtl number values close to one we have found there that increasing the Rayleigh number, with fixed wavenumber a, the system goes first through a steady bifurcation producing swirl and, next, through a Hopf bifurcation. Therefore, in the case of fig.1 the swirl grows steadily from an axisymmetric toroidal flow, beginning to oscillate for larger Rayleigh number values.

From the experimental results [3] we can expect the transition to time-dependence to be highly sensitive to the aspect ratio, as it will be shown. In fact, the major contribution from the esperimental work in [3] concerns that problem. As a rough comparison between theory and experiments we have computed the Rayleigh number and the frequencies for the Hopf bifurcation as a function of the wavenumber a and the self-interaction coefficient C, defined as $C = 1/2\overline{f^3}$ with an overbar meaning horizontal average. Defined in that way, the parameter C is the only hint about the shape of the planform that can be found in the equations. Information about the aspect ratio is contained in the wavenumber a. As an example, we can mention that in the axisymmetric case $f = J_0(ar)$, $C = 0.6$, $\Gamma = j_1/a$, where j_1 is the first zero of the Bessel function J_1.

In fig.3 we have plotted the Rayleigh number against the frequency ω for the Hopf bifurcation. We have plotted the results for different wavenumbers a corresponding to a planform with $C = 0.6$. Such a value has been choosen for the main features on that plot are independent on C and this is the value that suits the best the experimental results in [3].The numerical results show a reasonable agreement with experiments on Rayleigh number values and frequencies - if dimensions are properly taken - but the most remarkable point is the presence of two branches, one for large one for small aspect ratio, as in the experiments, suggesting that our modal approach, in spite of being very crude, retains enough structure to deal properly with the instability of swirl.

4. Physical Mechanism

As mentioned before, the modal approach used in (1.2) gives quite reasonable results even in situations where the geometry of the flow is not toroidal. This is, for instance, the

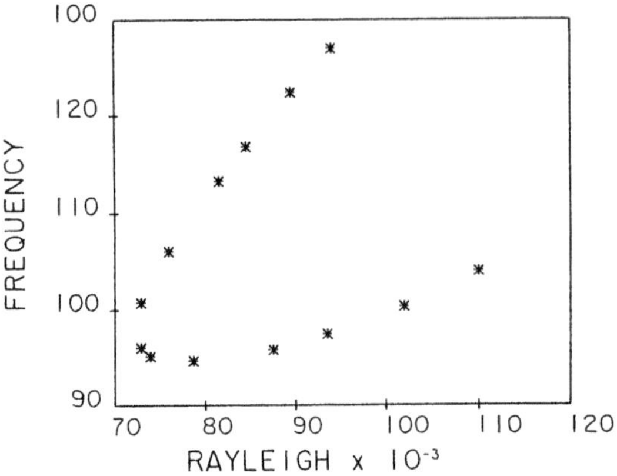

Figure 3. Frequency ($\omega = 2\pi d^2/\kappa T$) versus Rayleigh number for the Hopf bifurcation with C = 0.6. Each $*$ is for a different wavenumber a. Upper and lower branches correspond to $a>3.117$ and $a<3.117$ respectively.

case for convection in a large aspect ratio rectangular container filled with mercury. If numerical results are compared with the experiments done by Krishnamurti [4] and Rossby [5] the agreement between frequencies and fluxes near the bifurcation is quite satisfactory [2] in spite of the flow patterns being roll-like. In our opinion, the reason is that the relevant Reynolds number for that instability depends on the bending of roll axis. We can easily explain our statement by writing the θ-component of the Navier-Stokes equation for an axisymmetric flow -i.e.: $\partial_\theta = 0$

$$\partial_t v_\theta + v.\nabla v_\theta + \frac{v_r v_\theta}{r} = \sigma \pounds v_\theta \tag{4.1}$$

where £ includes all the viscous terms. In the absence of pressure and buoyancy the driving in (4.1) is to be provided by the term $v_r v_\theta / r$, a term that can be thought as the Coriolis force acting on a particle in its local frame of reference. To be precise, in the absence of vertical vorticity we have from (2.1) that a particle describes circles in the r-z plane. If its velocity is slightly tilted on the torus the particle will be acted upon by a Coriolis force in the azimuthal direction, which sign will depend on v_r being centrifugal or centripetal. Therefore, the restoring force will be stabilizing in a fraction of the trajectory and destabilizing in the remainder, and the flow will be unstable if the average work done by such a Coriolis force overcomes the viscous disipation. Obviously, the total work can be zero if the flow is symmetric against a mid horizontal plane.

From the expression of the driving term in (4.1) we can conclude that large velocities or small radius of curvature are needed to trigger the instability of swirl. This result seems to be consistent with the quoted experiments. In the case of air the bifurcation producing swirl takes place for values of the Rayleigh number much larger than for mercury, to compensate for the Prandtl number values being also larger. In addition, the larger the curvature the smaller the Rayleigh number value, as suggested by fig.3 were the upper branch corresponds to the larger aspect ratio.

5. Conclusions

We have examined a shear-type instability associated with the bending of the primary flow - roll or cell type structure. Results have been found to be consistent with experiments on convection in small aspect ratio containers and they look very similar to the production of swirl in toroidal vortices or even to the presence of axial flow in filaments, as the model has been found elsewhere to be consistent with situations where no toroidal structure can be expected [2]. So, we conjecture that our modal approach contains most of the dominant modes for describing such a shear-type instability. In that respect it can be found in Leonard [6] a very nice phenomenological description about how the instability of axial flow in bent filaments can be related to the instability of swirl in a toroidal flow. Finally, we have shown that the instability mechanism requires a flow not showing any symmetry against an horizontal plane at mid height. That might explain why no trace of such an instability has been found before in the literature on thermal convection and is the subject of some work in progress [7].

Acknowlkedgements :Thanks are due to Dr. R. Krishnamurti for calling our attention on the experiments done by W.T.Mitchell and J.A.Quinn. This work has been supported by a grant from Comision Asesora Científica y Técnica, Spain.

References

1. J.M.Massaguer, I.Mercader: In Cellular Structures in Instabilities. ed. J.E.Wesfreid and S.Zaleski. Lecture Notes in Physics vol. 210 (Springer, Berlin Heidelberg 1984)

2. J.M.Massaguer, I.Mercader: Instability of Swirl in Low Prandtl Number Thermal Convection. Preprint

3. W.T.Mitchell, J.A.Quinn: A.I.Ch.E.Journal, 12,1116 (1966)

4. R.Krishnamurti: J.Fluid Mech. , 60, 285 (1973)

5. H.T.Rossby: J.Fluid Mech. 36, 309 (1969)

6. A.Leonard:Ann.Rev.Fluid Mech. 17, 523 (1985)

7. S.Blázquez: Tesis de licenciatura.Univ. Barcelona (1986)

Part II

Chaotic Behaviour of Non-Linear Systems and Turbulent Fields

Chaotic Behaviour in a Non-Linear System: Turbulence in Rayleigh Benard Convection

P. Bergé

Service de Physique du Solide et de Résonance Magnétique, CEN-Saclay, F-91191 Gif-sur-Yvette Cedex, France

In this paper we will show that turbulent motions in confined convecting fluid are well understood in the frame of the theory of non-linear dynamical systems with a small number of degrees of freedom.

DISSIPATIVE DYNAMICAL SYSTEM

An example of dynamical system, is for example, a forced (and sustained) pendulum. The *dissipative* character is due to the presence of unavoidable friction and the existence of permanent oscillations is allowed by a convenient *sustaining*. In a general manner, the time evolution of a dynamical system can be described through a set of ordinary differential equations. In all that follows, we will consider that these equations are *non linear*, these non linearities giving all the richness in the behaviour.

For example we will consider a system of 3 equations:

$$\begin{cases} dX/dt = f(X,Y,Z) \\ dY/dt = g(X,Y,Z) \\ dZ/dt = h(X,Y,Z) \end{cases} \tag{1}$$

where X, Y, Z are the *variables* of the system or *degrees of freedom*. The corresponding dynamics is represented by the solutions of (1):

$$X = F(t) \quad Y = G(t) \quad Z = H(t)$$

which give the time evolution of the various variables allowing to determine the state of our system.

In an alternative description of the dynamics, one considers the trajectory of the point whose coordinates are X, Y, Z; doing so, we define the so called *phase space*, the coordinates of which are the variables of the system. The prime importance of this particular space will appear all along this paper.

As an illustration, let us look at a simple example: that of the sustained pendulum as modeled by Van der Pol whose second order non-linear differential equation is:

$$d^2X/dt^2 - (\varepsilon - X^2)\, dX/dt + X = 0 \tag{2}$$

Equivalently, we can describe the system by a system of 2 first order-equations:

$$\begin{cases} dX/dt = Y \\ dY/dt = (\varepsilon - X^2)Y - X \end{cases}$$

in which the two degrees of freedom, i.e. the position X and velocity Y of the pendulum, clearly appear. The motion of the point of coordinate X, Y in the phase space tends to the asymptotic trajectory which is a closed loop, named the *"limit cycle"*. Note that this limit cycle is only reached after a long time during which the trajectory is progressively *attracted* towards the limit cycle, and named for this reason *attractor* (see figure 1).

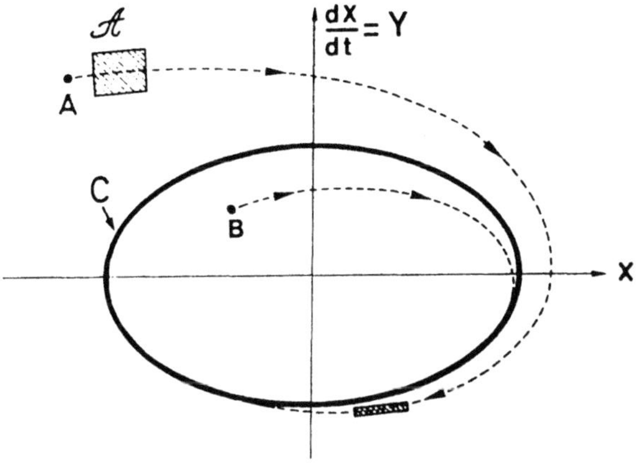

Fig.1 : Phase space of a periodic regime.
The trajectories originating from all initial points A,B, etc... are all folded back to the attractor C, the limit cycle. Note the correlative area contraction of A for example.

The existence of an attractor arises from the dissipation of the system represented here by the second term of equation (2). In the same manner, dissipation produces area (or volume) contraction in phase space and, as such, the dimension of the attractor is lower than that of the phase space. In the example quoted above the attractor has a dimension 1 in a two dimensional phase space.

THE SIMPLEST ATTRACTORS AND THEIR POINCARE SECTION

The simplest attractors are the *fixed point* corresponding to the state of rest, the *limit cycle* to the periodic regime, and the *tori* which correspond to the multiperiodic regimes. Furthermore, instead of considering the full trajectories, one can consider its intersection with a plane (or hyperplane). The representation obtained this way, the *Poincaré section*, has a dimension lowered by 1 in comparison with the original representation. In particular, this procedure allows one to draw on a sheet of paper the Poincaré section of an attractor immersed in a three-dimensional phase space (see figure 2).

PERIODIC REGIME VERSUS CHAOTIC REGIME

A chaotic regime has a disordered and erratic *time evolution* which contrasts with the ordered evolution of the periodic regime. Another striking difference is the presence of a broad part in the *spectrum* of a chaotic signal, when, in the case of a periodic or multiperiodic signal, only sharp lines are present. But the most interesting property consists in the unpredictability associated with a chaotic

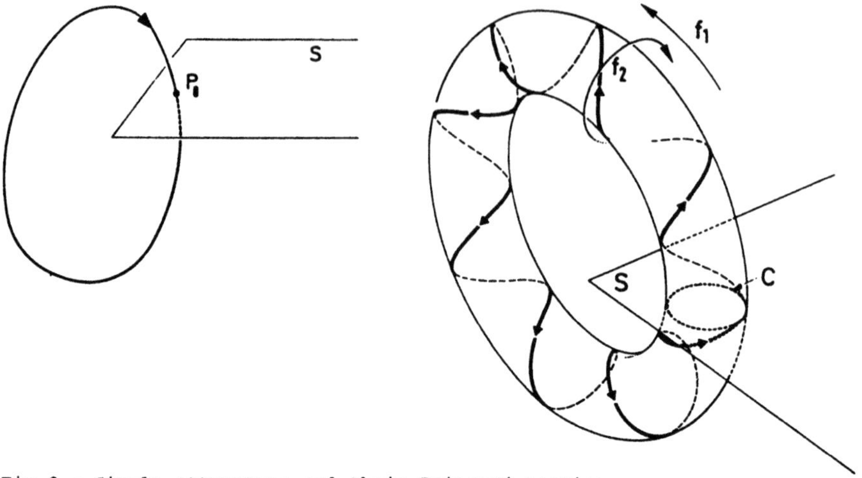

Fig.2 : Simple attractors and their Poincaré section.
On the left the limit cycle whose Poincaré section by the plane S is a point P. On the right, the torus, whose Poincaré is the closed loop C.

behaviour: this unpredictability is due to the loss of similarity of the signal with itself (i.e. loss of self-similitude) when the time evolves (more precisely, the autocorrelation function decays to zero with time). As a matter of fact, prediction is based on the possibility to extrapolate the past towards the future. This works very well in the case of periodic signals because of the repetition of a pattern (permanence of the self-similarity). The loss of self-similarity does not allow this extrapolation to be done, except for very short time intervals. In the phase space, the question of permanence (periodic) or loss (chaotic) of the self-similitude has a particular and important consequence. In the case of a periodic (or multi periodic) signal, two neighbouring trajectories will remain so when the time runs (conservation of similitude). On the contrary, they will diverge in the case of a chaotic signal (loss of similarity) see figure 3. This very important property of the trajectories of a chaotic regime is called Sensitivity on Initial Condictions or S.I.C. Once more, the consequence of S.I.C. is the following: even a very small (infinitesimal) differençe on the knowledge of the exact condition of the system will be (exponentially) amplified and, as such, has dramatic consequence in the subsequent behaviour, producing impredictability. Then, it suffices that a system presents the property of S.I.C. to be chaotic, whatever the number of its degrees of freedom may be.

GEOMETRY OF A CHAOTIC ATTRACTOR

Returning to figure 1, we have yet seen that all the points corresponding to different initial conditions are folded back to the limit cycle, the attractor of the periodic regime. Let us see, by comparison, how it is possible to construct the attractor of a chaotic regime. We take the simple case of a three-dimensional phase space; see figure 4. let us start with initial conditions (t=0) lying the rectangle ABCD. When the time runs, we have two opposite effects; in one direction, let us say X X' we have contraction towards the atractor, but in the other Y Y' we have divergence (S.I.C.). So, the rectangle is, at the same time, contracted (area contraction due to dissipation) and stretched (S.I.C.). Then, the

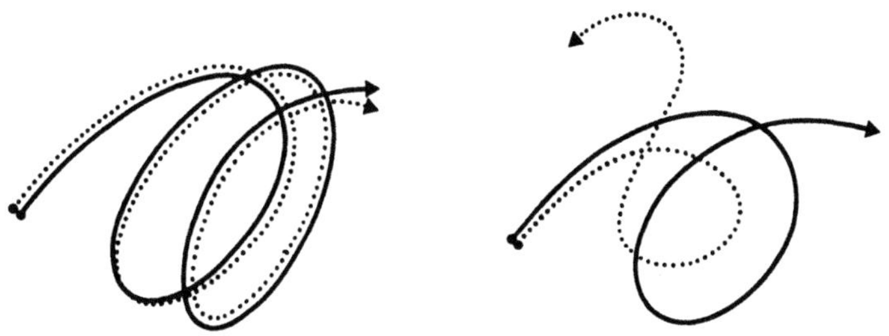

Fig.3 : Behaviour of neighbouring trajectories.
On the left, the two trajectories remain neighbour and similar when time runs (non chaotic regime) but they diverge rapidly becoming completely dissimilar in a chaotic regime.

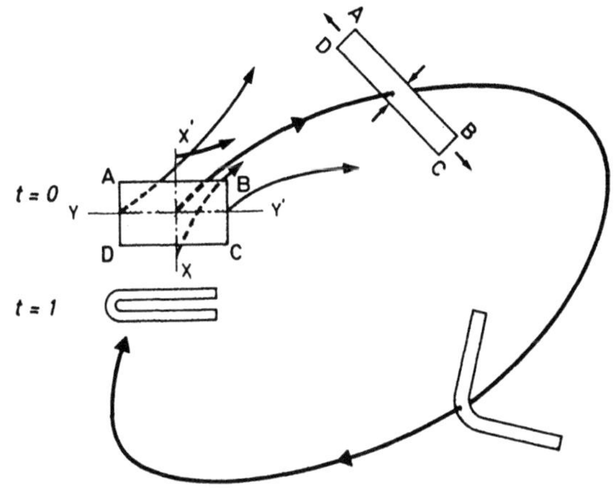

Fig.4 : The first step of the construction of a chaotic (or strange) attractor.
The ensemble of the initial conditions lie in the rectangle ABCD which is contracted along xx' (attraction) and stretched along yy'(S.I.C.), the "mean" trajectory being drawn as a double line.

rectangle becomes narrow and elongated; but in order to remain in a bounded region of the phase space it has to fold. So, after one turn (t=1), we are back to the starting plane which we can choose as the Poincaré plane (in the following, we will not consider the full attractor but only its Poincaré section). Due to the combined effects of contraction, stretching and folding, the initial rectangle has been transformed into some kinds of horseshoe. With time evolution, contraction stretching and folding continue to act on this horsehoe, giving some kind of double hairpin after two turns (t=2) see figure 5. When many turns will be made, one obtain, in the Poincaré plane, a complicated structure with an infinite number of sheets. One can see on figure 5 that the transverse structure of the Poincaré section is formed according to the same manner than a Cantor fractal ensemble. So, a chaotic attractor is a fractal object with a non integer dimension D (here $2 < D < 3$). Because of the very particular and complex topology of these chaotic attractors, they were named "strange attractors".

Among the conclusions of this geometrical reasoning, one can point out the fact that a strange attractor (i.e. an attractor with S.I.C.) can exist in a low dimensional phase space (minimum 3 dimensions). That means that chaos may appear in a system with only 3 variables (or, in this context, 3 degrees of freedom). In

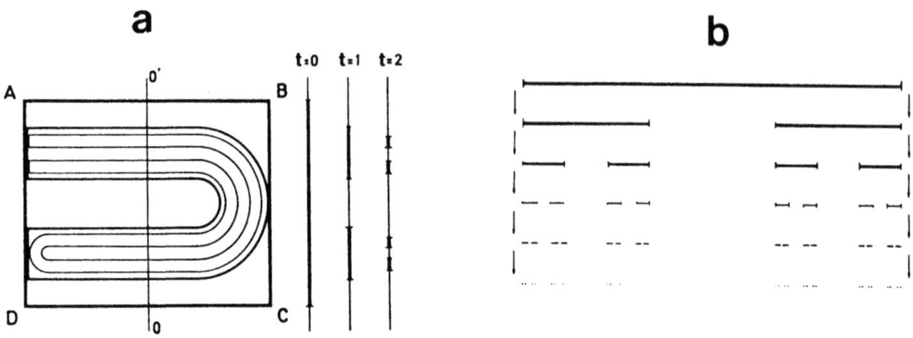

Fig.5 : Successive Poincaré sections corresponding to the construction of the strange attractor schematically drawn in fig.4.
Note the similar between of the successive steps of the transverse structure of the strange attractor (a) and those of the Cantor fractal set (b).

other words, the system of 3 differential equations (1) may have chaotic solutions in spite of its deterministic character. This is the reason why this kind of chaos has been named "deterministic chaos" (or deterministic turbulence if we deal with an hydrodynamical system); see general references (1) (2) (3).

FROM THE THEORY.......TO EXPERIMENTS

An hydrodynamical system, almost ideal to study turbulent chaos and to check the ideas on strange attractors, is the Rayleigh-Benard thermal convection (R.B.). We refer the reader to specialized articles about R.B. convection (4) (5). Let us just recall that when a fluid layer is confined between two horizontal conductive plates, the bottom one being at a higher temperature than the top one, convective motions appear when the temperature difference ΔT exceeds a certain critical value ΔT_c (more precisely, one use the Rayleigh number Ra which measures a-dimensionally the temperature difference and whose critical value is Ra_c). The fluid motions spatially organize themselves under the form of convective rolls arranged in an almost regular convective structure. But, in order to prevent spatial problems due to the presence of defects in the structure, we work on confined geometry i.e. in

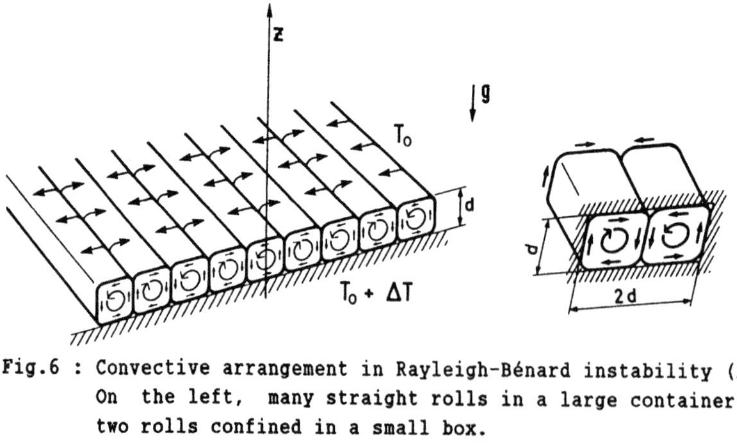

Fig.6 : Convective arrangement in Rayleigh-Bénard instability (schematic).
On the left, many straight rolls in a large container, and on the right, two rolls confined in a small box.

small containers of horizontal extent twice the depth, so only 2 (or at least a very few) rolls are present (see figure 6). Under that conditions, the first instability encoutered (after the onset of steady convection at Ra) is that of the thermal boundary layers. In these boundary layers, thermal oscillators develop which can have different forms. The important fact is that, by increasing Ra, one obtains, first, only oscillator whose pulsations are easily visualized through simple optical means (see figure 7). The corresponding regime is periodic (i.e. the temperature and velocity of the convective fluid vary periodically with time), with a frequency perfectly well determined and stable. Indeed the spectrum consists of only one frequency together with its harmonics. The corresponding attractor is a limit cycle: the hydrodynamical system behaves exactly like an oscillator with 2 degrees of freedom.

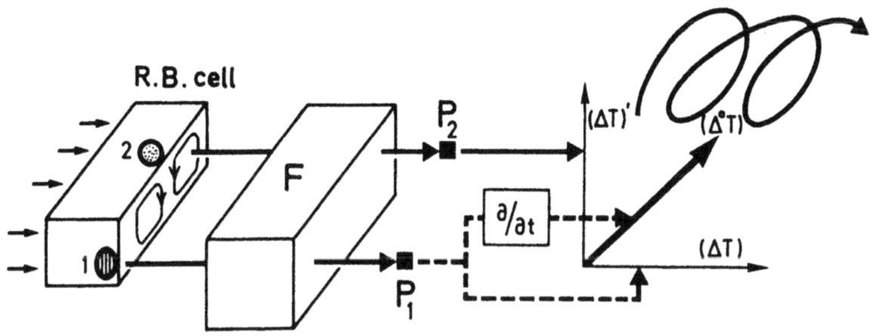

Fig.7 : The experimental arrangement (schematic).
A parallel beam is sent through the cell where 1 and 2 schematize the two thermal oscillators. Through a Foucault method (the corresponding optical arrangement is schematized as F), a contrasted image of the thermal gradients is formed on a plane in which the photodiodes P_1 and P_2 detect the time dependence of these gradients. From these signals one can directly construct the attractor in a three dimensional phase space, as well as its Poincaré section.

By increasing the control parameter Ra, the regime becomes more complicated according to many different manners (6) (7). Among these routes to chaos, one passes through the intermediate of the quasiperiodicty, more precisely the multiperiodicity with two frequencies (or biperiodicity). Practically, that means that a second thermal oscillator appears easily visualisable like the first one (8) (9). Now, both temperature and velocity of the convective fluid oscillate with two characteristic frequencies f_1 and f_2 and the spectrum, though complex, contains only sharp lines all indexable as mf_1+nf_2 (see figure 8). The fluid behaves now like 2 coupled (non linear) oscillators and, as such, has 4 degrees of freedom. Optical images formed through the method of Foucault (sometimes called Schlieren method) are well adapted to the visualisation of both oscillators. Photodiodes suitably located on the image allow to detect modulations originating from either one or the other oscillator. It is then possible, through simple electronic arangement, to detect directly Poincaré section of the attractor (10). One can see on figure 9 the Poincaré section of such a biperiodic regime. By a further increase of the Rayleigh number, the regime becomes chaotic: a broad component appears in the spectrum and the time evolution becomes erratic. That means that, now, the temperature and velocity of the fluid vary erratically i.e. turbulence is present, even if, in the average, the convective structure remains ordered. In other words, this kind of turbulence, sometimes called weak turbulence

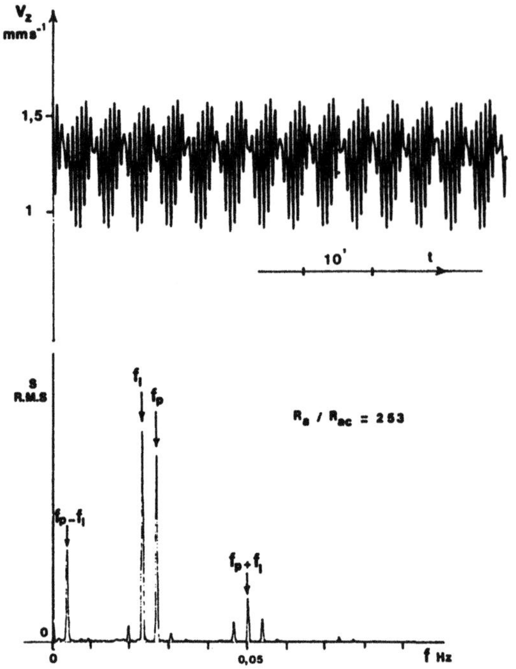

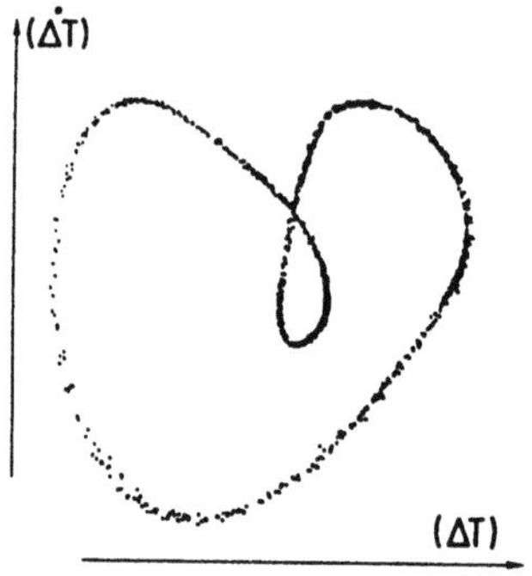

Fig.8 : Time dependence of the convective velocity in a biperiodic regime and the corresponding spectrum.
Clearly, two characteristic frequencies are present. One, f_1, originating from the pulsation of the whole cold boundary layer, the other, f_p, being due to a more localized oscillator, a "plume" located in the same boundary layer. Due to the linear scale of the spectrum, only the more simple combinations mf_1+nf_p can be seen.

Fig.9 : Example of a Poincaré section in a biperiodic regime.
This Poincaré section has been directly obtained through the method schematically drawn in fig.7. This closed loop corresponds to the Poincaré section of a torus, in spite of the presence of the secondary loop due to the presence of harmonics in the spectrum.

by opposition with developed turbulence, is mainly temporal, for it does not imply spatial disorder. One can see in figure 10 the corresponding Poincaré sections in two different cases, and, obviously, the structure has something to do with that of strange attractors (11) (12).

MEASUREMENT OF THE DIMENSION OF AN EXPERIMENTAL ATTRACTOR

Beyond the complex and multilayered aspect of an attractor obtained in a chaotic regime, we have to be more quantitative and to point out the true signature of the deterministic chaos. The more significant characteristic of this one is the dimension of the corresponding attractor, which must be low and non integer. But, in the case of an experiment, the relevant variables X,Y,Z... of the system are not known (except in simple systems in with the oscillators are very well identified); so the first problem is to reconstruct not *the* phase space but an

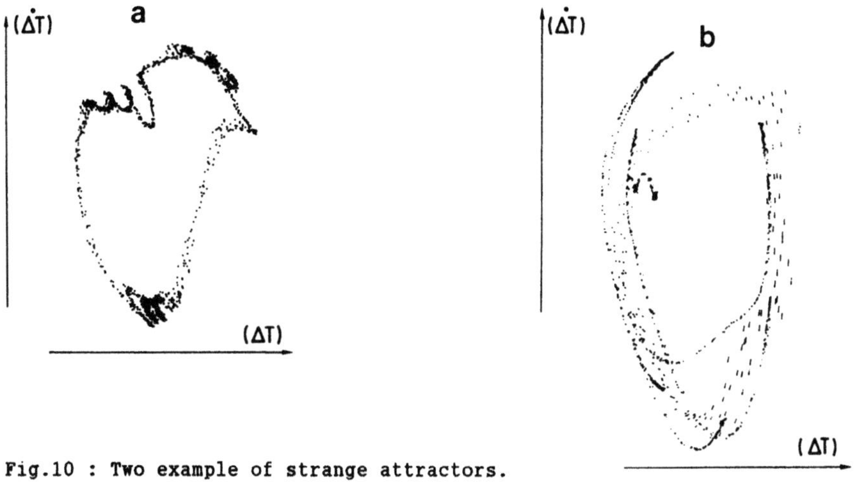

Fig.10 : Two example of strange attractors.
These two strange attractors correspond to chaotic regimes obtained from biperiodic ones.
(a) is obtained from the regime of fig.9 by increasing the Ra number from 520 Ra_c up to 563 Ra_c. Note the foldings typical of strange attractors.
(b) is obtained from another biperiodic regime and has a more complex structure.

equivalent phase space. One can adopt the method of the delays (13) which creates, from a time series obtained by a single measurement V(t), N variables by considering the N successive values of V:

$$V(t),\quad V(t+\tau),\quad V(t+2\tau).....V[t+(N-1)\tau]$$

where τ is a suitable time delay.

The phase space -then the attractor- being reconstructed, we have to use an efficient algorithm to access the geometric characteristic which is the dimension. Such a method has been proposed recently (14); it consists to take an origin on the attractor and to count the number C(r) of its neighbours located at a distance from it lower than r (see figure 11).

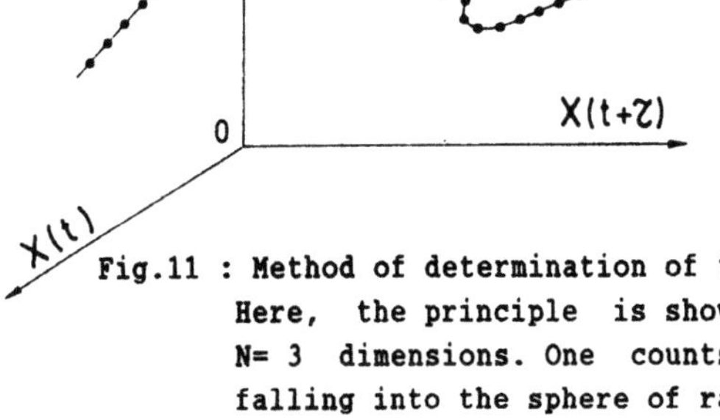

Fig.11 : Method of determination of the fractal dimension.
Here, the principle is shown for a phase space reconstructed with only N= 3 dimensions. One counts the number C(r) of points of the attractor falling into the sphere of radius r.

One can show that, under certain conditions, and by averaging on many different origins,

$$C(r) = K\, r^{\nu}$$

where ν is the dimension of the attractor. So, in a log-log plot, the characteristic C(r) versus r should be a straight line whose slope measures the dimension. Indeed, to have sense, this slope has to be independent of the dimension N of the reconstructed phase space, as soon as this dimension exceeds sufficiently that of the attractor (15) (16). One can see on figure 12 that in a given situation of turbulence in R.B. convection, the measured slope is quite constant as soon as N > 4, the value being a bit lower than 3. This contrasts strongly with the case in which the erratic behaviour originates from the coexistence of a large number of degrees of freedom. For example, such an analysis performed on white noise shows that the (apparent) dimension grows like the dimension of the reconstructed phase space. This behaviour is typical of a system with a very large number of degrees of freedom. Similarly, the same kind of processing applied to developed turbulence seems to indicate that the dimension of the strange attractor, if any, is very large. This gives sense to the very low dimension found in many attractors of chaotic regimes and support strongly the concept of deterministic turbulence and strange attractors (for others examples of strange attractors obtained in the same experiment, see the paper by M. Dubois in the same volume).

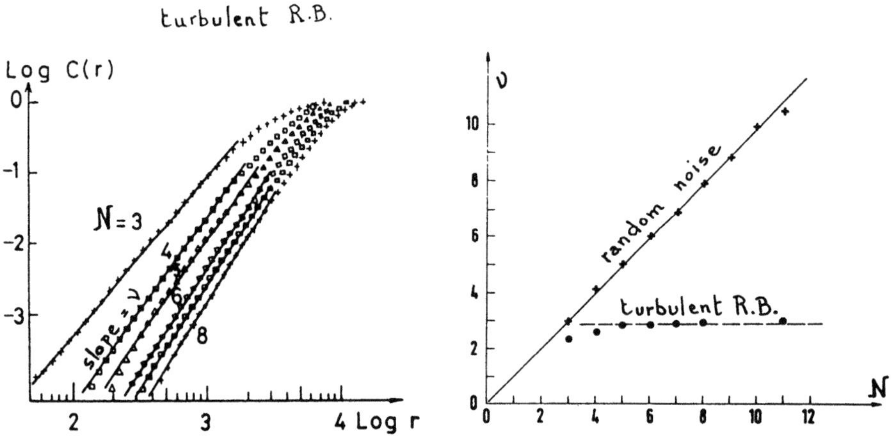

Fig.12 : Characteristics C(r) versus r and variation of the corresponding slope ν with N, the dimension of the reconstructed phase space.
Note the striking difference between the behaviour of erratic signals originating from turbulent convection and erratic signal corresponding to a random noise.

CONCLUSION

In many cases of hydrodynamics where spatial order is maintained, for example by a suitable confining of the fluid or by application of a convenient constraint, the fluid behaves like a dynamical sytem with a very small number of degrees of freedom. In these cases, turbulence occurs through the appearance of a strange attractor of low dimension.

BIBLIOGRAPHY

(1) P. Bergé, Y. Pomeau and C. Vidal, Order within chaos (translated from l'ordre dans le chaos by L. Tuckerman) Wiley interscience Hermann (1986)

(2) E. Ott, Strange attractors and chaotic motions of dynamical systems, Review of Modern Physics, 53, 655 (1981)

(3) M.I. Rabinovich, Stochastic self-oscillations and turbulence, Soviet Physics Uspekhi, 21, 443 (1978)

(4) C. Normand, Y. Pomeau and M.G. Velarde, Convective instability: a physicist's approach Review of Modern Physics 49, 581 (1977)

(5) P. Bergé and M. Dubois, Rayleigh Benard Convection Contemporary Physics, 25, n°6 535 (1984)

(6) J.P. Eckmann, Roads to turbulence in dissipative dynamical systems, Review of Modern Physics, 53, 643 (1981)

(7) J.P. Gollub, S.V. Benson, Many routes to turbulent convection, Journal of Fluid Mechanics, 100, 449 (1980)

(8) M. Dubois and P. Bergé, Experimental evidence for the oscillators in a convective biperiodic regime, Physics Letters 76A, 53 (1980)

(9) M. Dubois and P. Bergé, Instabilité de couche limite dans un fluide en convection. Evolution vers la turbulence, Journal de Physique, 42, 167(1981)

(10) M. Dubois and P. Bergé, V. Croquette, Etude de régimes convectifs instationnaires à l'aide des diagrammes de Poincaré, Comptes Rendus de l'Académie des Sciences de Paris, C293, 409 (1981)

(11) M. Dubois, Experimental aspects of the transition to turbulence in Rayleigh Benard convection in "Stability of Thermodynamic System", Lecture Notes in Physics 164, 177 (1981)

(12) P. Bergé, Study of the phase space diagrams through experimental Poincaré sections in prechaotic and chaotic regimes, Physica Script T1, 71 (1982)

(13) F. Takens, in "Dynamical Systems and Turublence", Lecture Notes in Mathematics 898 Springer Verlag, Berlin (1981)

(14) P. Grassberger, I. Procaccia, Characterization of strange attractors, Physical Review Letters, 50, 346 (1983)

(15) B. Malraison, P. Atten, P. Bergé, M. Dubois, Dimension d'attracteurs étranges: une détermination expérimentale en régime chaotique de deux systèmes convectifs, Comptes Rendus de l' Académie des Sciences de Paris, C297, 209 (1983)

(16) A. Brandstäter, J. Swift, H.L. Swinney, A. Wolf, J.D. Farmer, E. Jen, J.P. Crutchfield, Low dimensional chaos in a system, Physical Review Letters, 51, 1442 (1983)

Chaos Through Biperiodicity

M. Dubois

Service de Physique du Solide et de Résonance Magnétique, CEN-Saclay, F-91191 Gif-sur-Yvette Cedex, France

The general concept of turbulence involves the presence of a great number of degrees of freedom. Nevertheless, it has been demonstrated recently, both theoretically and experimentally, that a small number of degrees of freedom may be sufficient to allow a turbulent behaviour [1], i.e. an unpredictable behaviour ; in this case we will also speak of chaotic states. A characterization of this chaos, apart from the involved number of degrees of freedom, is the fact that chaotic regimes set in continuously, in the sense that the behaviour just after their onset may be deduced from the behaviour just before. So routes to chaos can be defined [2], as the evolution of the dynamics with the increase of a control parameter (or of the amount of non-linear coupling). Here we will discuss of the route through biperiodicity and of related experimental results, obtained on Rayleigh-Bénard convection.

Biperiodicity

Let us take two coupled oscillators, characterized by their own frequency f_1 and f_2. Two parameters control their different states: the frequency ratio $\rho = f_1/f_2$ and the coupling R between the two oscillators. The variables are the respective amplitudes A_1 and A_2 of each oscillator and their phases θ_1 and θ_2; their relative phase θ_{12} plays also an important role. In the plane of the parameters $[R,\rho]$, the first striking feature is the presence of phase-locked tongues (Fig.1) i.e. dynamical regimes where the phase of an oscillator is locked to that of the other oscillator. A very simple model, due to Arnold [3], gives a good example:

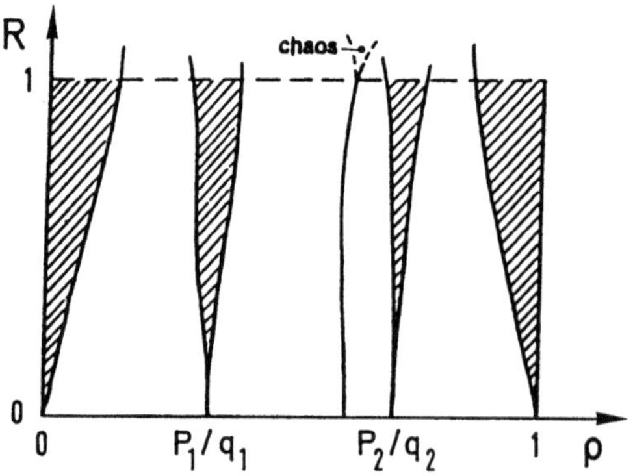

Fig.1 : Schematic representation of the phase-locked tongues in the Arnold model.

$$\theta_{n+1} = \theta_n + \rho - \frac{R}{2\pi}\sin(2\pi\theta_n) \mod 1.$$

where θ_n may be seen as the phase of an oscillator for a given value of the phase of an external oscillator; ρ corresponds to the frequency ratio and R governs the amount of the non-linear coupling.

But does chaos exist? In the Arnold model, the line R=1, (called the critical line), defines the possibility of the appearance of chaotic regimes by increasing R. In realistic models [4] this line is more complicated and depends on the frequency ratio. Now what about experimental observations? Two kinds of situation have been studied: model systems for which the number of degrees of freedom is well controlled [5][6] and physical situations for which it is necessary to prove the presence of chaos by pointing out its specific properties i.e. strange attractors and small dimension of the relevant phase space. In both cases, deterministic chaos has been observed by evolution of biperiodic regimes [7][8][9]. Here, we want to give examples of this in Rayleigh-Bénard convection studied in confined geometry and with a high Prandtl number fluid.

Forced convection

When a horizontal fluid layer is submitted to a destabilizing vertical temperature difference ΔT (generally by heating at the bottom), motion appears in the fluid under the form of rolls [10], as soon as ΔT exceeds a critical value ΔT_c; this is the so-called Rayleigh-Bénard instability. Its properties are controlled by a dimensionless parameter, the Rayleigh number Ra. The experiments of interest here, have been conducted with small containers, so that only a small number of rolls sets in with a spatial arrangement which remains stable in a large domain of Ra values. So periodic behaviours may be observed; they are dire to the presence and the evolution of thermal oscillators, generally located in the thermal boundary layers (high Prandtl number fluid) [11].

The simplest system which may become chaotic is the forced pendulum. The corresponding situation in convection is achieved as follows: keeping the structure fixed, the Rayleigh number is increased until we obtain a periodic regime. Then the thermal oscillator, responsible for this regime, is forced by an artificial oscillator, which is created near the natural one, by a local electrical heating; the electrical power is driven in amplitude Ae and frequency fe by a synthetizor. This set-up allows in principle the exploration of the plane $[R,\rho]$ (R is proportional to Ae).

Results we present in the following have been obtained with Si Oil ($Pr \simeq 38$), confined in a small box. (horizontal extents $\simeq 2$ and 1.2 times the depth). At $Ra/Ra_c = 310$ (Ra_c is the critical value of the convective onset), a natural monoperiodic regime is present, with an oscillator (hot droplet) whose frequency f_1 is $29.5\ 10^{-3}$Hz. When we forced this droplet, phase-locked tongues are observed by varying Ae and fe. But here, we want just to demonstrate the appearance of chaos under the influence of an external oscillator. In Fig.2, the Fourier spectra of a convective signal (semi-local temperature gradient see [1]) show how the dynamical regime previously periodic, becomes progressively chaotic when the amplitude Ae of the external oscillator is increased; the frequency ratio $\rho = fe/f_1$ is kept constant, $\rho = 0.97$.

Three dimensional attractors have been reconstructed in the phase space from the experimental signals. We show on Fig.3 the Poincaré section (cut by a plane) of the attractor with Ae = ± 1.85 Amp.; the typical features of strange attractors

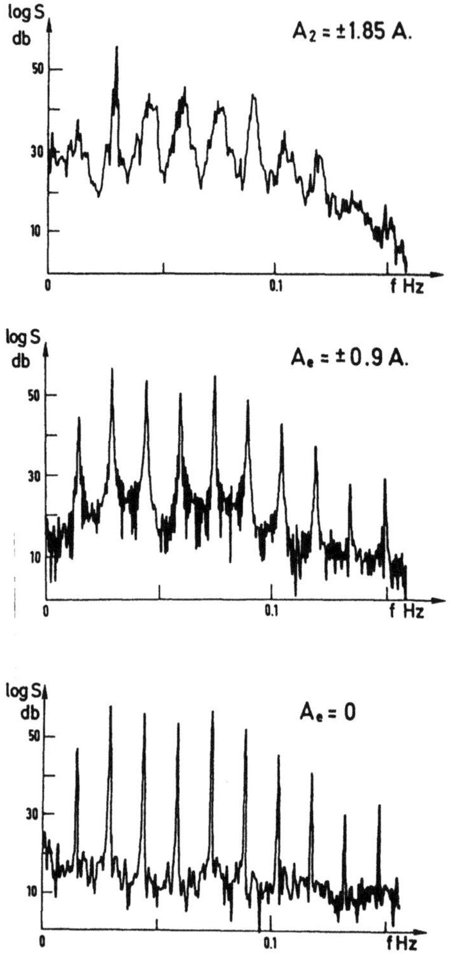

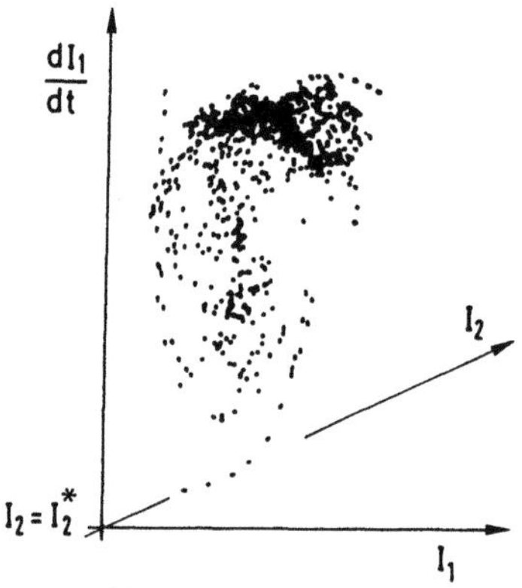

Fig.2 : Fourier spectra of a convective signal for different values of the forcing i.e. of the modulation amplitude Ae of the external oscillator. $\rho = 0.97$; $Ra/Ra_c = 310$.
a) Ae = 0; b) Ae = ±0.9 Amp.; c) Ae = ±1.85 Amp.

Fig.3 : Poincaré section of a 3d attractor in the case of forced convection with Ae = ±1.85A. It has been constructed from two signals, measured in two places in the structure and the time derivative of one of them. $Ra/Ra_c = 310$; $\rho = 0.97$.

(stretching and folding) may be seen, despite the complexity of the distribution of the points. This complexity is partly due to the fact that this section is a projection of an object with three (or perhaps more) dimensions. Indeed, the dimension of the corresponding attractor has been calculated by the method proposed in [12] and discussed in [13] and we found 3.7 ± 0.15. In this regime, the natural oscillator may be described at least by three variables, instead of two, because of the presence of the subharmonic mode. So the calculated dimension is in agreement with what is expected ($\nu < 4$), and furthermore gives a proof of the presence of deterministic chaos, with the insurance that the involved number of degrees of freedom is small.

Free convection

Two natural thermal oscillators may be present in the convective fluid layer, under certain conditions (given structure, range of Ra values...). These oscillators are coupled by the motion of the fluid and the thermal diffusivity. So, by varying the Rayleigh number there is an evolution of the observed dynamical regimes. Opposite to the case of the forced convection, one parameter only can be varied, i.e. the Rayleigh number, which acts on both the coupling intensity and the frequency ratio of the oscillators. Nevertheless, it is interesting to follow the observed dynamical states. In the example presented here, the experimental conditions are similar to those, previously described for forced convection (same fluid, same container), but the forcing has been suppressed and the spatial arrangement is different [14].

Here we focus our interest only on a sequence related to biperiodicity and chaos. From $Ra/Ra_c = 317$ we observe the following evolution

$317 \leq Ra/Ra_c < 320$	chaos	$\rho > \frac{1}{7}$
$320 \leq Ra/Ra_c \leq 322$	phase-locked regime	$\rho = \frac{1}{7}$
$Ra/Ra_c > 322$	biperiodic regime	$\rho < \frac{1}{7}$

The detailed study of the chaotic sequence has been performed through the determination of Fourier spectra of a convective signal, Poincaré sections of 3-d reconstructed phase portraits (see for example Fig.5) and dimension of the corresponding attractors [14]. The main result is that the dynamical behaviour varies dramatically in a small range of Rayleigh numbers and then, that the most relevant parameter is the frequency ratio as we will explain in the following.

At the two ends of the chaotic domain, $Ra/Ra_c \simeq 317$ and $Ra/Ra_c \leq 320$, the amount of chaos is small and the mechanism responsible for it can be clearly identified. For $Ra/Ra_c \simeq 317$, the Fourier spectrum (Fig.4) displays the specific

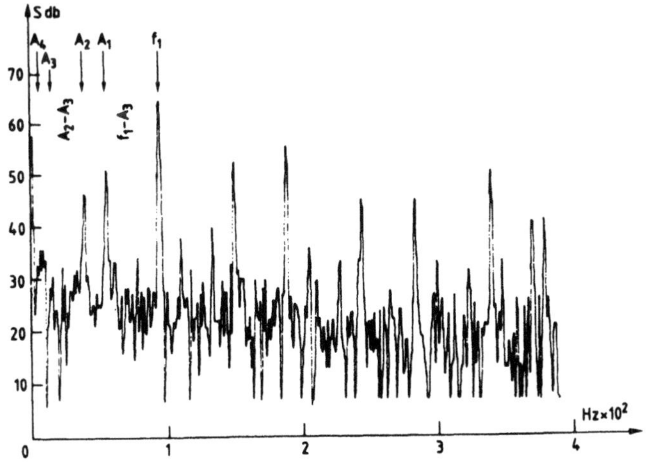

Fig. 4 : Fourier spectrum of the convective signal in natural convection at $Ra/Ra_c = 317$. $\rho = 0.1517$ $(= \frac{1}{6.59})$.

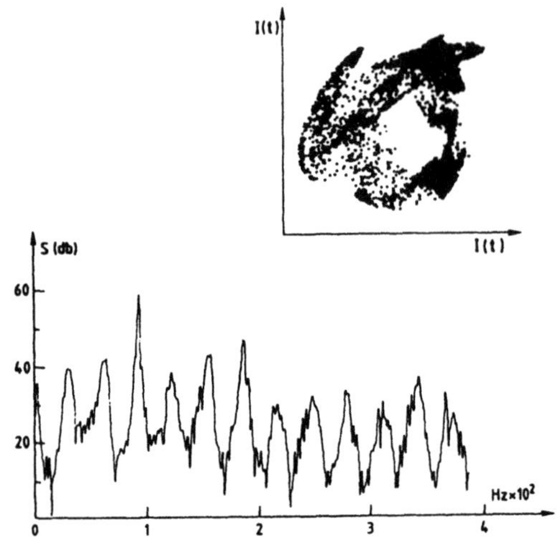

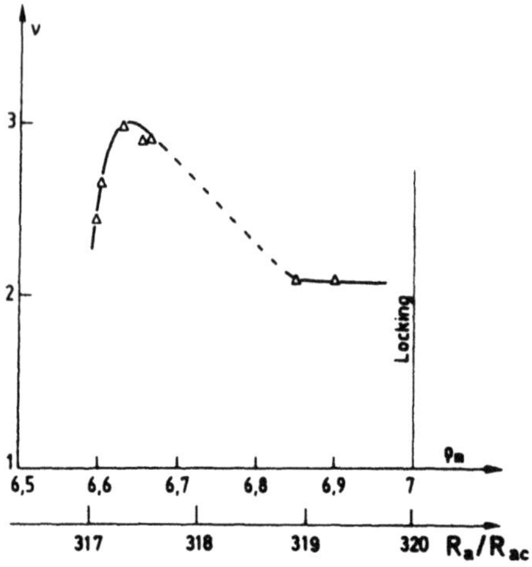

Fig.5 : Poincaré section of a 3d reconstructed attractor for $\rho = 0.151$ (=1/6.62) and the corresponding Fourier spectrum.

Fig.6 : Variation of the calculated fractal dimension ν with the measured frequency ratio in the range of Rayleigh numbers corresponding to the chaotic states.

features of the direct route to chaos from biperiodicity, i.e. chaos is obtained by the overlap of the adjacent phase-locked tongues [15][16][17]; this is revealed by the presence of peaks at low frequencies, given by the distance of the actual frequency ratio ($\rho = 0.1517 = \frac{1}{6.59}$ with $f_1 = 9.52\ 10^{-3}$Hz and $f_2 = 62.74\ 10^{-3}$Hz) to its successive rational approximations, here 1/6, 1/7, 2/13, 5/33. As seen on Fig.4, $A_1 = 6f_1-f_2$, $A_2 = 7f_1-f_2$, $A_3 = 13f_1-f_2$. The peak $A_4 = 33f_1-5f_2$ is present, but broadened as expected near the onset of chaos. For $Ra/Ra_c \leq 320$, $\rho = 0.145 \simeq 1/6.9$, the dynamics is also well understood [18] and the very low amount of chaos is due to the presence of chaotic phase intermittencies; these intermittencies correspond to a succession of quasi phase-locked regimes ($\rho = 1/7$) separated by very brief and fast relative phase (θ_{12}) slipping; this dynamical behaviour is obviously strongly influenced by the vicinity of the phase-locked tongue 1/7.

For $317 < Ra/Ra_c < 320$, the spectra present a relatively high level of chaos (Fig.5). Quantitative information is then obtained by the calculation of the dimension ν of the chaotic attractors, which reflects the amount of chaos. The ν variation is non monotonous and has a maximum value for $\rho = 0.151$ as shown Fig.6. As mentioned previously, the variation of the Rayleigh number induces the variation of the frequency ratio of the two oscillators. In the considered experimental situation, it turns out that an important variation of ρ occurs within a very small range of Ra/Ra_c and that the dynamical properties are then essentially due to the frequency ratio variation, as pointed out by the two end cases. Direct comparison with theoretical and numerical models is difficult but nevertheless results obtained here with natural convection, where the oscillators are free to interact, are in qualitative agreement with theoretical predictions. Note once more that the value of the fractal dimension, calculated for the attractors, confirm that the corresponding states are relevant to deterministic chaos i.e. deterministic turbulence.

Acknowledgment

We would like to thank particularly P. Bergé for constant help and constructive discussions. C. Poitou is acknowledged too for her very efficient technical assistance.

Bibliography

[1] P. Bergé, this book.
[2] See for example: P. Bergé, Y. Pomeau, C. Vidal, "L'ordre dans le chaos" ed. Hermann, Paris 1984; J.P. Eckmann, Rev. of Modern Physics 53, 643 (1981).
[3] V.I. Arnold, Trans. Am. Math. Soc. 2e Sec. 46, 213 (1965).
[4] M. Jensen, P. Back and T. Bohr, Phys. Rev. A30, 1960 (1984).
[5] F.C. Moon, J. of Applied Mechanics 47, 638 (1980).
[6] V. Croquette, Pour la Science 62, 62 (1982).
[7] J.P. Gollub, S.V. Benson, J. of Fluid Mech. 100, 449 (1980).
[8] M. Dubois, P. Bergé, J. de Physique 42, 167 (1981).
[9] J.C. Roux, A. Rossi, S. Bachelart, C. Vidal, Phys. Letters 77A, 391 (1980)
[10] P. Bergé, M. Dubois, Contemporary Physics 25, 535 (1984).
[11] M. Dubois, P. Bergé, Phys. Letters 76A, 53 (1980).
[12] P. Grassberger and I. Procaccia, Phys. Rev. A28, 2591 (1983).
[13] P. Atten, J.P. Caputo, B. Malraison, Y. Gagne, J. de Mécanique, Volume spécial 1984 "Bifurcations et comportements chaotiques".
[14] M. Dubois, P. Bergé, Physica Scripta 33, 159 (1986).
[15] S. Ostlund, D. Rand, J. Sethna, E. Siggia, Physica 8D, 303 (1983)
[16] A. Fein, M.S. Heutmaker, J.P. Gollub, Physica Scripta T9, 79 (1985)
[17] J. Stavens, F. Heslot, A. Libchaber, Phys. Rev. Letters 55, 596 (1985)
[18] P. Bergé, M. Dubois, J. de Phys. Lettres 46, L431 (1985).

A Topological Model for the Onset of Chaotic Motions in the Blasius Boundary Layer

St. N. Savulescu

Max-Planck Institut für Strömungsforschung,
D-3400 Göttingen, Fed. Rep. of Germany

The particle paths equations around H (the point where the u'-phase shifts and where v' attains its maximum v'_{MAX} in a cross-section of a perturbed Blasius boundary layer) are reduced to a quadratic map. A condition of breakdown for the amplified Tollmien-Schlichting (T-S) waves and a new v'/v'_{MAX} distribution are derived. The identification of an invariant hyperbolic set around H and a corresponding similarity transformation group are briefly discussed.

The present topological model uses the theory of nonlinear dynamical systems in order to provide a mathematical description for the observable phenomena of breakdown and spot birth in the transitional flat plate boundary layer (BL).

Let us consider a Blasius BL perturbed by the developing T-S waves. The following notations and approximations for the mean (B) and fluctuating (T-S) quantities are introduced:
$u = U_B + u'$; $v = V_B + v'$ - velocity components ; ν - kinematical viscosity; U_e - free stream velocity; δ, δ^* - BL thicknesses; f_{T-S} - frequency; λ_{T-S} - wave length; T_{T-S} - period; $R = U_e\delta^*/\nu$; $x_a = (x-U_H t)/\lambda_{T-S}$; $\eta_0 = y/\delta$; $y_0 = \eta_0 - \eta_{0_H}$; $t_a = f_{T-S}t$; $F = 2\pi f\nu/U_e^2$; the index H refers to the point H, located by $u'_H = 0$ and $v'_H = v'_{MAX}$. The approximations around H are /1/:
$U_B/U_e = U_{a_B} = [\partial U_a/\partial\eta_{0_H}](\eta_0-\eta_{0_H}) +\dots$; $u'/u'_{MAX} = u_a = [\partial u_a/\partial\eta_{0_H}](\eta_0-\eta_{0_H}) +\dots$; $V_B/U_e = V_{a_B} = (1.5/R)\eta_0 +\dots$; $v'/v'_{MAX} = v_a = 4a\eta_0(1-a\eta_0) +\dots$; where the parameter $a = (1/2)(1+\sqrt{1-(v'_e/v'_{MAX})})$ locates $v_{a_{MAX}}$.

Starting from the general form of the particle path equations, i.e. $dx_i/dt = u_i(x,y,z,t)$ with $u_i = u,v,w$, and using the above notations and approximations, one gets:

$$dx_a/dt_a = \gamma y_0 +\dots ;\ dy_0/dt_a = \lambda_0 + \lambda_1 y_0 - By_0^2 +\dots ;\ dz/dt = 0 \qquad (1)$$

where

$$\gamma = (U_e/f_{T-S}\lambda_{T-S})(dU_{a_B}/d\eta_0-(u'_{MAX}/U_e)(du_a/d\eta_{0_H}))$$

$\lambda_0 = (A-1)\eta_{0_H} - B\eta_{0_H}^2$; $\lambda_1 = A-1-2B\eta_{0_H}$; $B = 4a^2(v'_{MAX}/\delta f_{T-S})$

$A = 1 + (U_e/\delta f_{T-S})(1.5/R) + 4a(v'_{MAX}/\delta f_{T-S}) - (1/\delta)(d\delta/dt_a)$

are intricate functions of x,z,t.

The analytical treatment of (1) is based on the following model assumptions:

(i)- the parameters A,B,λ_0,λ_1 and γ are constants at the moment $t_a^+ \in [0,1]$;

(ii)- during the evolution of the T-S waves the time is parameterized by writing $t_a = f_{T-S}t = n$, with n integer;

(iii)- $\gamma \to 0$ for large amplifications of the T-S waves and (1) reduces to an autonomous system in a cross-section of the boundary layer.

Therefore a sequential amplification of the T-S waves is expressed by the following quadratic, noninvertible map:

$$Y_{0_{n+1}} = A\,Y_{0_n}(1-Y_{0_n}) = f(\,Y_{0_n}) \tag{2}$$

where the linear transformation $Y_{0_n} = (B/A)\eta_{0_n}$ is used. The extensively studied /2/ map (2) undergoes a cascade of accumulated bifurcations when $3 < A < 3.6$. Thus, according to the present model, a condition for the local breakdown reads:

$$2<(U_e/\delta f_{T-S})(1.5/R) + 4a(v'_{MAX}/\delta f_{T-S}) - (1/\delta)(d\delta/dt_a)<2.6 \tag{3}$$

where the second term, due to the increasing v'_{MAX}, brings the major contribution. Examples of this breakdown condition are displayed in Fig.1a,b; the amplification rate $v'_{MAX}/v'_0 = \exp(A_{T-s})$ and $d\delta/dt_a = 0$ were used. The relatively small values of v'_{MAX}/U_e at breakdown (Fig.1b) are physically plausible; in Fig.1a the curve $R_{BK}(F)$ is compared with the neutral stability curve.

According to this model, the period 2 motion occuring after the first bifurcation is physically related to the change of the $v_a(\eta_0)$ distribution from $v_a{}^{(1)} = 4a\eta_0(1-a\eta_0)$ to $v_a^{(2)} \sim f(f(Y_0)\,)$. By using the linear transformation $\eta_0 = (A/B)Y_0$ and by putting $Y_{0_n} - Y_{0_{n-1}} = (dY_0/dt_a)^{(n)} = v'_{MAX}/\delta f_{T-S})v_a^{(n)}$ one obtains:

$$v_a^{(2)}(Y_a;A) = v_a^{(1)}(\,(A-1)Y_a - A\,)(\,(A-1)Y_a - 1) \tag{4}$$

where $Y_a = a\eta_0$, $A = 1 + B/a$ and $d\delta/dt_a = 0$ were used. An exemple, plotted in Fig.2, shows the simple zero of $v_a^{(2)}$ located at $Y < 1$.

The topological structures around H are pointed out by the phase portrait corresponding to (1), when $\gamma \neq 0$. The solution reads:

$$(a^{-1}-\eta_{0_H}-y_0)^{1-a\eta_{0_H}}(\eta_{0_H}+y_0)^{a\eta_{0_H}} = \exp(\,(B/\gamma)(C-x_a)\,) \tag{5}$$

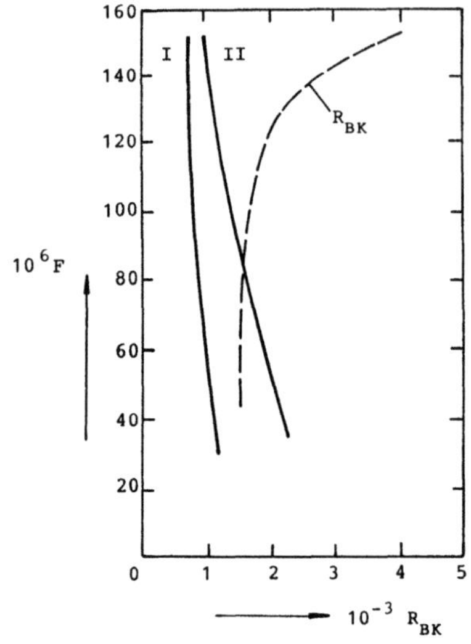

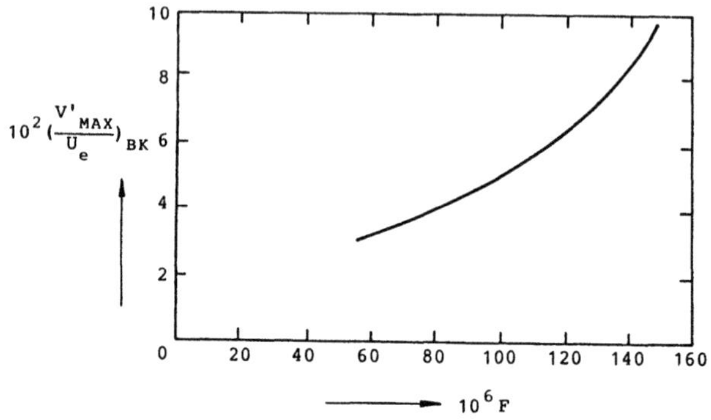

Fig.1a Example of a "breakdown condition": $R_{BK}(F)$
$(4a = 3,\ V'_{o_{MAX}} = 10^{-3} U_e)$

Fig.1b Example of a "breakdown condition": $(V'_{MAX}/U_e)_{BK}(F)$
$(4a = 3,\ V'_{o_{MAX}} = 10^{-3} U_e)$

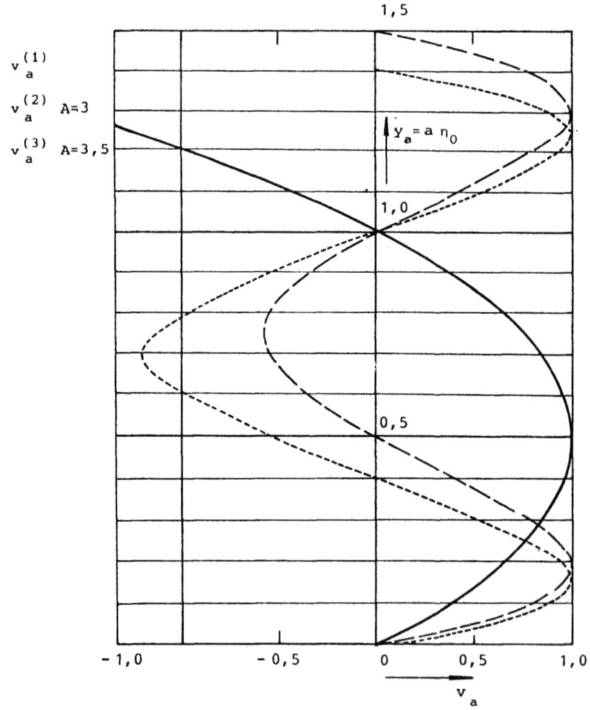

Fig.2 The instantaneous $v_a\ (y_a)$ distribution at the breakdown. Theoretical example

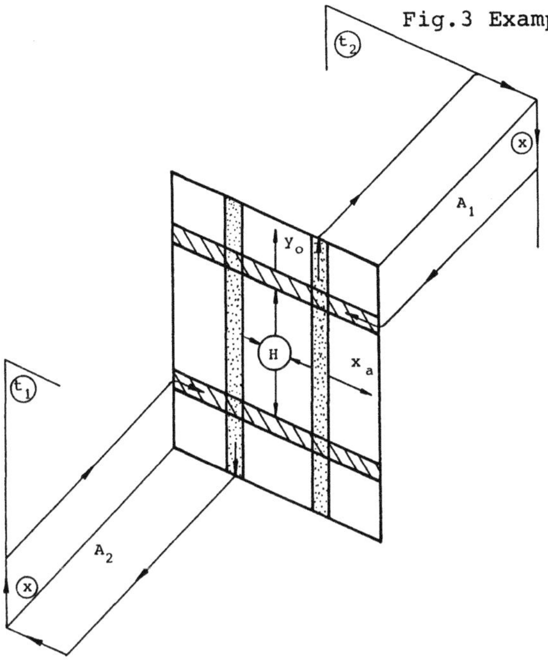

Fig.3 Example of sketched orbits around H

where the constant C marks each trajectory, B = a(A-1) and $v_a = v_a^{(1)}$. Figure 3 is a pictorial view of the homoclinic orbit ensuring the connection between the attractors A_1 and A_2 , via H, in both time directions. Therefore, the set involving H, A_1, A_2 and the trajectories around H can be viewed as a hyperbolic invariant set of mappings /3/, exemplified by the mappings of sqares. The homoclinic orbit taking the union of the x_a-strip to the y_0-strip (see Fig.3) has the essential features of the Smale horseshoe. This complicated set of particle paths, "closed" by a homoclinic orbit, can be related to the onset of a turbulent spot.

The assumed structural stability of the Conley hyperbolic set is strongly connected to a transformation group ensuring its invariance. When the breakdown occurs and the solution of (1) preserves its form, a proper rescaling $\delta \sim \lambda_x \sim U_e t$ and $y_H \sim V_B t$ leads to the conical similarity /4/

$$x_a = (x/U_e t) - (U_H/U_e) \; ; \; y_0 = (y/U_e t) - (V_B/U_e) \qquad (6)$$

and to the normalized /5/ von Mises transformation $\psi/\psi_\delta = \int_0^y u dy / \int_0^\delta u dy$.

Acknowledgments

The support of the Max Planck Institut für Strömungsforschung, Abt. Prof. E.A. Müller, Göttingen, BRD, is gratefully acknowledged.

References

1. W. Tollmien: Nachr. Ges. Wiss. Göttingen, Math.-Phys. K1, 21 (1929)
2. A.J. Lichtenberg, M.A. Lieberman: Regular und Stochastic Motion Appl. Math. Sc., 38, (Springer-Verlag, Berlin, 1983)
3. C. Conley: Lectures Notes in Physics, 38, (Springer Verlag, Berlin, 1975), p.539-550
4. B. Cantwell, D. Coles, P. Dimotakis: J. Fluid Mech. 87, 641, (1978)
5. S.N. Savulescu: Lectures Notes in Physics, 75, (Springer Verlag, Berlin, 1977), p.289-295

One-Dimensional Analog of the Saltzman-Lorenz Problem for Thermal Convection

S. Panchev and V. Roussenov

Department of Meteorology, Faculty of Physics, University of Sofia, 1126-Sofia, Bulgaria

1. Introduction

In a previous paper PANCHEV /1/ proposed the following system of equations

$$W_t + WW_z = \nu W_{zz} + \beta\theta \quad , \quad \theta_t + W\theta_z = \kappa\theta_{zz} + \gamma W \tag{1.1}$$

as one-dimensional analog to the classical system of Rayleigh-Benard or Saltzman-Lorenz, governing the thermal convection in horizontal fluid layer heated from bellow. The fundamental properties of the latter - nonlinearity, diffusivity ($\nu,\kappa>0$) and forcing ($\beta,\gamma>0$) are preserved in (1.1). Moreover, similarly to the Burgers' single equation, it is matematically much more tractable than the complete system.

After appropriate scaling

$$z_* = H/\pi \; , \; W_* = \kappa/z_* \; , \; t_* = z_*^2/\kappa \; , \; \theta_* = \kappa\nu/\beta z_*^3 \tag{1.2}$$

the system (1.1) takes the form

$$W_t + WW_z = \sigma W_{zz} + \sigma\theta \; , \quad \theta_t + W\theta_z = \theta_{zz} + rW \; , \tag{1.3}$$

where

$$r = \beta\gamma z_*^4/\kappa\nu \quad , \quad \sigma = \nu/\kappa \tag{1.4}$$

are the Rayleigh and Prandtl number, while W, θ, z, t now denote dimensionless variables. Besides, $W_t = \partial W/\partial t$, $W_{zz} = \partial^2 W/\partial z^2$ etc.

The functions $W(z,t)$ and $\theta(z,t)$ (velocity and temperature) are defined in the interval (0,H) and are supposed to be of the form

$$f(z,t) = \sum_{n=1}^{\infty} f_n(t)\sin(nz) \quad , \quad f = \theta, W \tag{1.5}$$

Applying standard technique one obtains from (1.3) and (1.5) the spectral equations

$$
\begin{array}{l|l|l|l}
 & (a) & (b) & (c) \\
\dot{W}_1 = \sigma\theta_1 - \sigma W_1 & + \frac{1}{2} W_1 W_2 & + \frac{1}{2} W_2 W_3 & + \dots \\
\dot{\theta}_1 = W_1\theta_2 - \theta_1 + rW_1 & - \frac{1}{2} W_2 \theta_1 & + \frac{3}{2} W_2\theta_3 - W_3\theta_2 & + \dots \\
\dot{\theta}_2 = -\frac{1}{2} W_1\theta_1 - 4\theta_2 & + rW_2 & + \frac{3}{2} W_1\theta_3 - \frac{1}{2} W_3\theta_1 & + \dots \\
\hline
\dot{W}_2 = -\frac{1}{2} W_1^2 - 4\sigma W_2 + \sigma\theta_2 & & + W_1 W_3 & + \dots \\
\hline
\dot{W}_3 = -\frac{3}{2} W_1 W_2 - 9\sigma W_3 + \sigma\theta_3 & & & + \dots \\
\dot{\theta}_3 = - W_1\theta_2 - \frac{1}{2} W_2\theta_1 - 9\theta_3 + rW_3 & & & + \dots \\
\hline
\end{array}
\qquad (1.6)
$$

. . . .

Obviously, (1.6a) is the Lorenz system, derived from (1.3) when the nonlinear term WW_z is neglected. The truncation at n=2 yields the system (1.6b) first derived in /1/. It seems to be the simplest nontrivial generalization of (1.6a) since the contribution of WW_z is now retained.

The comparative investigation of the systems (1.6a) and (1.6b) is a subject of the present paper. Sections 2 and 3 contain analytical results concerning the stationary solutions of the above equations and their stability. Section 4 is devoted to the numerical results and section 5 to their discussion. In the conclusion some possible directions for further studies of the model (1.1) are mentioned .

2. Stationary Solutions. Analytical Results

We denote by $\bar{W}_i$ and $\bar{\theta}_i$ the stationary solutions to (1.6b), i=1,2. Since $\dot{W}_i = \dot{\theta}_i = 0$, then (1.6b) yields

$$\bar{\theta}_1 = \bar{W}_1\left(1 - \frac{1}{2\sigma}\bar{W}_2\right) , \quad \bar{\theta}_2 = 4\bar{W}_2 + \frac{1}{2\sigma}\bar{W}_1^2 . \qquad (2.1)$$

Upon substituting of (2.1) into (1.6b) one obtains

$$\bar{W}_1[2\bar{W}_1^2 + 2(7\sigma+1)\bar{W}_2 + \bar{W}_2^2 + 4\sigma(r-1)] = 0 , \tag{2.2}$$

$$(\bar{W}_2 - 8 - 2\sigma)\bar{W}_1^2 + 4\sigma(r-16)\bar{W}_2 = 0 . \tag{2.3}$$

There exist two kinds of solutions to these equations:

a) Trivial (zero) solution

$$\bar{W}_1 = \bar{W}_2 = 0 , \qquad \bar{\theta}_1 = \bar{\theta}_2 = 0 . \tag{2.4}$$

b) Nonzero solutions. The latter depend on two parameters - $\sigma>0$ and $r>0$. From (2.3)

$$\bar{W}_1^2 = 4\sigma(r-16)\bar{W}_2/(8+2\sigma-\bar{W}_2) \tag{2.5}$$

Upon substituting into (2.2) we derive the fundamental cubic equation for $\bar{W}_2$:

$$\bar{W}_2^3 + 6(2\sigma-1)\bar{W}_2^2 - 4[7\sigma^2 + \sigma(r-2)+4]\bar{W}_2$$

$$- 8\sigma(\sigma+4)(r-1) = 0 . \tag{2.6}$$

The real roots of (2.6) for which $\bar{W}_1^2>0$ will be physically meaningfull solutions to the problem. For $r<16$ all $\bar{W}_2<0$ and $\bar{W}_2>8+2\sigma$ yield $\bar{W}_1^2>0$. If $r>16$, $\bar{W}_1^2>0$ for $0<\bar{W}_2<8+2\sigma$ only.

Exact analytical solutions to (2.2) and (2.3) can be found at $r=1$ and $r=16$. Let $r=1$. Then (2.6) reduces to a quadratic equation

$$\bar{W}_2^2 + 6(2\sigma-1)\bar{W}_2 - 4(7\sigma^2 - \sigma + 4) = 0 . \tag{2.7}$$

with a physically meaningfull solution.

$$\bar{W}_2 = -3(2\sigma-1) - (64\sigma^2 - 40\sigma + 25)^{1/2} \tag{2.8}$$

existing for all $\sigma>0$. At $r=16$ the solution is

$$(\bar{W}_1 = 0, \ \bar{W}_2 \text{ - arbitrary}, \ \bar{\theta}_1 = 0, \ \bar{\theta}_2 = 4\bar{W}_2) . \tag{2.9}$$

Equation (2.6) simplifies considerably at $\sigma = 1/2$:

$$\bar{W}_2^3 - (19 + 2r)\bar{W}_2 - 18(r-1) = 0 . \tag{2.10}$$

The number of the real roots of (2.10) depends on the sign of the function

$$D(r) = 81(r-1)^2 - \frac{1}{27}(19+2r)^3 \qquad (2.11)$$

Evidently D(4) = 0. For r<4 D(r)<0 and (2.10) has two negative real roots and one positive ($\bar{W}_2<9$). The latter root however yields $\bar{W}_1^2<0$ and must be omitted. At r=4 the two negative roots coincides and for r>4 the nonzero stationary solutions do not exist. Similar results can be obtained for arbitrary σ .

3. Stability of the Stationary Solutions. Analytical Results

Following the standard method we linearize (1.6b)

$$\begin{aligned}
\dot{W}_1' &= \left(\frac{1}{2}\bar{W}_2 - \sigma\right)W_1' + \frac{1}{2}\bar{W}_1 W_2' + \sigma\theta_1' , \\
\dot{W}_2' &= -\bar{W}_1 W_1' - 4\sigma W_2' + \sigma\theta_2' , \\
\dot{\theta}_1' &= (\bar{\theta}_2 + r)W_1' - \frac{1}{2}\bar{\theta}_1 W_2' - \left(\frac{1}{2}\bar{W}_2 + 1\right)\theta_1' + \bar{W}_1\theta_2' , \\
\dot{\theta}_2' &= -\frac{1}{2}\bar{\theta}_1 W_1' - rW_2' - \frac{1}{2}\bar{W}_1\theta_1' - 4\theta_2' ,
\end{aligned} \qquad (3.1)$$

where f' = f - $\hat{f}$ and assume f'(t) = $\hat{f}$ exp(λt). The system (3.1) reduces to algebraic one with respect to $\hat{f}$. Hence, the characteristic equation follows

$$\lambda^4 + 5(\sigma+1)\lambda^3 + \alpha_2\lambda^2 + \alpha_1\lambda + \alpha_0 = 0 , \quad \text{where} \qquad (3.2)$$

$$\begin{aligned}
a &= \frac{1}{2}\bar{W}_2 - \sigma , \quad b = \frac{1}{2}\bar{W}_2 + 1 , \quad c = \bar{\theta}_2 + r , \\
\alpha_2 &= 4(\sigma+1)^2 - (ab + \sigma c) + \bar{W}_1^2 + \sigma(16-r) , \\
\alpha_1 &= (\sigma+1)\left[\frac{5}{2}\bar{W}_1^2 - 4(ab+\sigma c) + \sigma(16-r)\right] , \\
\alpha_0 &= \left[\frac{1}{4}\sigma(a+b) + 2\sigma(\sigma-1)\right]\bar{W}_1\bar{\theta}_1 + \left[2(b-a\sigma) + \frac{1}{4}c\sigma + r\sigma\right]\bar{W}_1^2 + \frac{1}{4}\bar{W}_1^4 + \sigma(16-r)(ab+\sigma c) - \frac{1}{4}\sigma^2\bar{\theta}_1^2 .
\end{aligned} \qquad (3.3)$$

For the zero stationary solution (2.4), (3.2) takes the form

$$[\lambda^2 + 4(\sigma+1)\lambda + \sigma(16-r)][\lambda^2 + (\sigma+1)\lambda + \sigma(1-r)] = 0 \tag{3.4}$$

implying that for r<1 this is a stable solution. Equation (3.2) also simplifies for the case (2.9)

$$\lambda(\lambda + 4\sigma + 4)[\lambda^2 + (\sigma+1)\lambda - ab - \sigma c] = 0 \tag{3.5}$$

and can be solved immediately.

4. Numerical Results

Following the tradition, the Rayleigh number r is taken as fundamental bifurcation parameter. The cubic equation (2.6) as well as equations (2.5) and (2.1) on one hand and equation (3.2) on the other, have been solved numerically for various values of the Prandtl number σ . In such a way, the stationary solutions are obtained and their stability investigated. Selected bifurcation diagrams for the most interesting variable $W_1(r,\sigma)$ are presented in Fig.1a - 1c at σ = 1/2, 10, 20 and in Fig.2 at r=12. Various symbols on the figures denote:

$O^{p,q}$ - stationary point in the phase space with dimension p+q=n (= 4 in our case);

p(q) - number of the roots of the characteristic equation (3.2) with negative (positive) real part;

$\Gamma^{a,b}$ - closed periodic trajectories (cycles), appearing at a given value of the bifurcation parameter;

$\emptyset$ - disappearence of stationary points at $r=r_2$.

Table 1 represents the dependences $r_1(\sigma)$ and $r_2(\sigma)$, where r_1 is the critical value of r at which the nonzero stable solutions lose their stability, while r_2 is such that for $r>r_2$ nonzero stationary solutions do not exist.

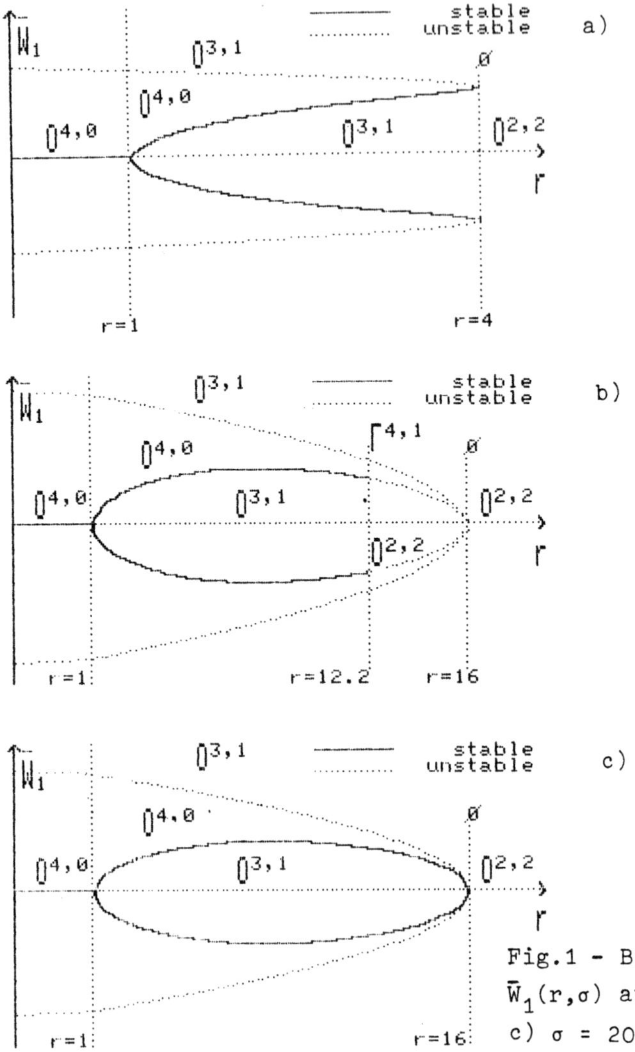

Fig.1 - Bifurcation diagrams for $\bar{W}_1(r,\sigma)$ at: a) σ = 1/2, b) σ = 10, c) σ = 20

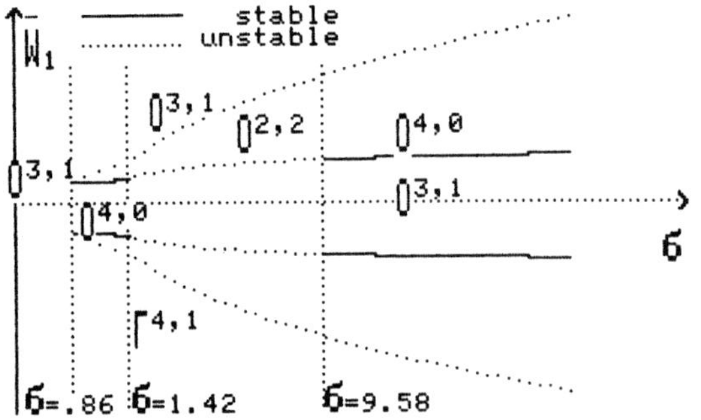

Fig.2 - Bifurcation diagram for $\bar{W}_1(r,\sigma)$ at r = 12

Table 1. Variation of r_1 and r_2 with σ .

σ	r_1	r_2	σ	r_1	r_2
0.05	-	7.95	1.1	15.15	16
0.1	-	4.74	1.2	13.65	16
0.15	-	3.76	1.5	11.65	16
0.2	-	3.35	2	10.48	16
0.3	-	3.16	3	9.78	16
0.4	-	3.40	4	9.76	16
0.5	-	4	5	10	16
0.6	-	5.04	10	12.21	16
0.7	-	6.75	15	14.88	16
0.8	-	9.57	16	15.42	16
0.9	-	14.69	18	15.95	16
1.0	16	16	20	15.98	16

5. Discussion of the Results

It is interesting to compare the properties of the new system (1.6b) with those of the classical Lorenz system (1.6a). The latter has a zero solution

$$\bar{W}_1 = \bar{\theta}_1 = \bar{\theta}_2 = 0 \tag{5.1}$$

stable for r<1 and unstable for r>1. Two stationary nonzero solutions exist for r>1 :

$$\bar{W}_1 = \bar{\theta}_1 = \pm\sqrt{8(r-1)} \quad , \quad \bar{\theta}_2 = 1 - r \quad , \tag{5.2}$$

which are stable until

$$r < r_c = \sigma(7+\sigma)/(\sigma-5) \tag{5.3}$$

and become unstable for $r>r_c$. The solutions (5.2) meet (5.1) at r=1 - Fig.3. If $\sigma=10$ then $r_c=34$. This is a Hopf bifurcation point for (1.6a) where a periodic solution branches from (5.2). For some supercritical values $r>r_c$, a strange attractor appears. Generally, the behaviour of the Lorenz system (1.6a) is now rather well understood.

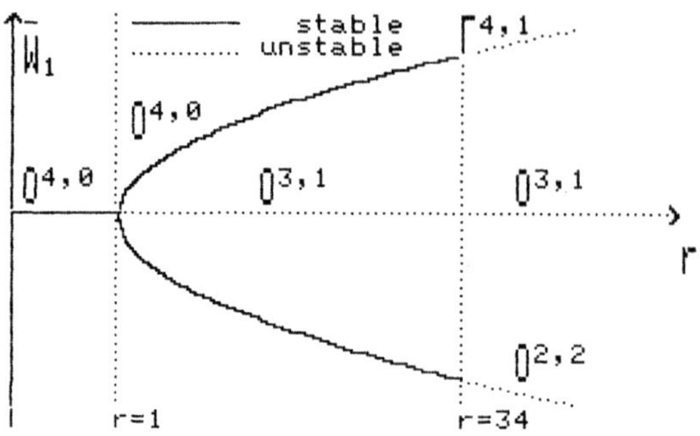

Fig.3 - Bifurcation diagram for $\bar{W}_1(r)$ at $\sigma = 10$, corresponding to the Lorenz system (1.6a)

Our results concerning the new equations (1.6b) can be summarized as follows:

- The nonzero solutions of the latter essentially depend on the Prandtl number σ, while (5.1) do not. This is clearly seen from Fig.1 and Fig.2 .

- Figure 1a: Stationary nonzero solutions exist up to $r_2=4$. Two of them are unstable for all $0<r\leqq 4$, while the zero solution exchanges stability at r=1 with the other two nonzero solutions. At r=4 the stable and unstable nonzero solutions meet and disappear.

- Figure 1b: Stationary nonzero solutions exist up to $r_2=16$. Two of them are unstable for all $0<r<16$, the other two, branching from the primary bifurcation point r=1, lose their stability at $r=r_1$ ($r_1=12.2$ at $\sigma=10$). Therefore, if $r_1 \leqq r \leqq r_2=16$, there will exist four unstable nonzero stationary solutions, which meet the zero solution at r=16.At $r=r_1$ Hopf bifurcation is observed.A characteristic feature of the curves $r_1(\sigma)$ and $r_2(\sigma)$ is the existence of a large minimum around $\sigma = 3$ and $\sigma = 0.3$ accordingly Table 1. Moreover, $r_1 \equiv r_2 \simeq 16$ for $\sigma \geq 20$ and $\sigma \to 1$ (Fig.1c).

- Since the system (1.6b) is two-parametric, one may fix r and vary σ. The result for r=12 is shown in Fig.2. The zero stationary solution is always unstable because r=12. At $\sigma=0.86$ two pairs of nonzero (stable and unstable) solutions arise. At $\sigma=1.42$ the stable solutions become unstable - Hopf bifurcation is observed, while at $\sigma=9.58$ the latter become again stable (inverse Hopf bifurcation) . The range of "total" instability ($1.42< \sigma <9.58$) decreases with decreasing of r (Table 1) and disappears for $r<10$.

6. Conclusion

Our results concerning the stationary solutions of the new system (1.6b) approximating (1.1) and their stability show, that really it is nontrivial generalization to the classical Lorenz system (1.6a). It follows that the nonlinear interaction in the field of velocity (the term WW_z in (1.1)), which is ignored in the Saltzman - Lorenz derivation, increases the number of the equations in the low-order system by one (four instead of three) and changes considerably the properties of the solution. We hope that further studies of the new equations (1.6b) and particularly of their nonstationary solutions will reveal many interesting features. Moreover, the six - component system (1.6c) may also be studied.

REFERENCES

1. S.Panchev : Bulgarian Geophysical Journal, IX, 2 (1983)
2. C.T.Sparrow : The Lorenz Equations: Bifurcation, Chaos and Strange Attractors (Springer, Berlin, Heidelberg 1982)

Part III

Direct and Large Eddy Simulation of Turbulence

On Direct and Large Eddy Simulation of Turbulence

U. Schumann [1] *and R. Friedrich* [2]

[1] DFVLR, Institut für Physik der Atmosphäre, D-8031 Weßling, Fed. Rep. of Germany

[2] Technische Universität München, Lehrstuhl für Strömungsmechanik, D-8000 München, Fed. Rep. of Germany

This review is based on the results of the EUROMECH Colloquium No. 199. It summarizes experiences concerning effects of resolution and subgrid scale models, Galilean transformation, boundary conditions, the impact on statistical modeling, and the analysis of turbulence structures.

1. Introduction

Direct and 'large eddy simulation' are terms which denote two closely connected methods of turbulence research. In a 'direct simulation' (DS), turbulent motion is simulated by numerically integrating the Navier-Stokes equations in three-dimensional space and as a function of time. In a DS, turbulence is simulated purely from first principles without any closure assumptions. All what is needed are proper initial and boundary conditions, an accurate and efficient numerical solution scheme and a large computer. Computer resources limit the resolution in time and space, though simulations with an order of one million discrete points in space are feasible. In principle, direct simulations are valid as long as all the excited scales remain within the band of resolved scales. This means that viscosity must be strong enough to damp out the not resolved scales or the simulation is restricted to a limited integration-time interval only. If these conditions cannot be met, a large eddy simulation would help.

Large eddy simulation (LES) is a combination of direct simulation with statistical modelling. The large 'eddies' are the motion structures at scales which remain after a certain filter has been applied to the turbulent field. Several papers in the recent past have discussed how this filter is related to the discrete numerical representation of the flow field employed in the simulation algorithm, see Mason [10], for example. For a finite difference simulation the large scales must be resolvable by the underlying grid. In contrast, the 'subgrid-scales' are those which are too small for the grid. In LES, these subgrid-scales are taken into account by a subgrid-scale (SGS) model.

Table 1: Differences between Large Eddies and Fine Scale Turbulence under idealized conditions

Large Eddies	Fine Scale Turbulence
* produced by mean flow	* produced by large eddies
* depend on boundaries	* universal
* ordered	* random
* require deterministic description	* can be modelled statistically
* inhomogeneous	* homogeneous
* anisotropic	* isotropic
* long-living	* short-living
* diffusive	* dissipative
* difficult to model	* easier to model

Such a model mainly provides damping of those resolved scales which lie at the border of resolution and which are excited by the large scale motions. Also, the SGS-model has to describe turbulent fluxes not represented by the large scales. Finally one would expect that the SGS-model has to mimic the random forcing of large scale motions induced by the fine scales; however this aspect has usually been neglected and deterministic SGS-models are being employed.

Many of the SGS-models are formally very similar to the models commonly used to describe time- or ensemble-averaged turbulent fields. Thus they have basically the same advantages and disadvantages. Approximation errors in SGS-models are of less importance, however, because the SGS-motion carries only a fraction of total turbulent kinetic energy and fluxes. If the grid (and the related filter) is fine enough, this fraction becomes very small. Moreover, SGS-turbulence has many features, see Table 1, which makes it much easier to construct models of general applicability. The most important feature stems from the fact that the characteristic length scale is controlled by the grid- or filter-scale if this scale is less than the integral-scale of turbulence. The range of applicability of LES is large but not unlimited. Some aspects are mentioned in Table 2.

Direct turbulence simulations were pioneered by Orszag and co-workers since 1969. The first convincing large eddy simulations were reported by Deardorff [3] for a turbulent channel flow and applied to atmospheric boundary layers thereafter. His SGS-model employed a proposal by Smagorinsky [16] for para-

Table 2: Application Domain of Large-Eddy Simulations

easy	difficult
* simple cubic domains	* complicated geometry
* periodic or wall boundaries	* inflow/outflow or open boundaries
* zero mean velocity or suitable transf.	* large mean flow without transf.
* large scale forcing (buoyancy, shear, strain, rotation)	* small scale forcing (viscous friction, point sources, wall roughness)
* simulation of several forces (e.g. shear or buoyancy) at comparable time-scales	* forcing with highly different scales
* interaction of modes with comparable scales	* interaction of modes with highly different scales
* nonstationary turbulence	* stationary turbulence
* homogeneous mean fields	* inhomogeneous mean fields
* evaluation of low order moments and derivatives	* evaluation of high order moments and derivatives

metrization of subgrid-scale motions in meteorology. Lilly [9] showed that this SGS-model is consistent with Kolmogorov's inertial-subrange theory. The state of the art in DS and LES as of 1983 was reported by Rogallo & Moin [13].

In this review lecture, we summarize experiences gained from the EUROMECH-Colloquium No. 199 on the present subject, which we had organized in 1985. Many of the papers presented in this Colloquium have been collected in a volume (Schumann & Friedrich [15]).

The following subjects were investigated in several contributions of the Colloquium:

2. Effect of resolution and SGS-models

MOIN+) reported results of a DS of curved and plane channel flow using a spectral method for the spatial representation of the velocity field. At Reynolds numbers of $Re_{CL} \approx 3000$ (based on centerline mean velocity and channel half width) a grid spacing of 6 wall units in spanwise direction, 18 in mean flow

+) References to papers included in [15] are not quoted in the literature.

direction and between 0.2 and 8.2 in the direction normal to the walls was sufficiently small to reproduce the alternating high and low speed streaks with their mean spacing of 100 wall units. In the region near to the wall Moin observed significantly more solitary vortices than vortex pairs and concluded that the wall layer is not composed of elongated pairs of counter-rotating vortices, as is usually assumed. Finally, he showed that turbulent channel flow consists of a large number of horseshoe (or hairpin) vortices inclined to the wall being the dominant structures in this flow. Using the same spectral DS method as Moin but a computational grid (for $Re_{CL} \approx 3200$) which was refined by a factor of 2 (respectively 4) in the direction normal to the walls KIM studied in great detail the evolution of a horseshoe-shaped vortical structure which is associated with the bursting process in the wall layer of a turbulent channel flow. GILBERT & KLEISER applied their spectral code to investigate the plane channel flow at subcritical Reynolds numbers ($Re_{CL} = 2800$) in a direct simulation. With very coarse resolution in planes parallel to the walls (8 x 8 x 64 grid points) they were able to obtain statistical quantities such as Reynolds stress and rms-values of fluctuating velocity components being surprisingly close to experimental data of Eckelmann [4], cf. figure 1. Of course, such a low resolution is insufficient to study details of turbulent flow structures.

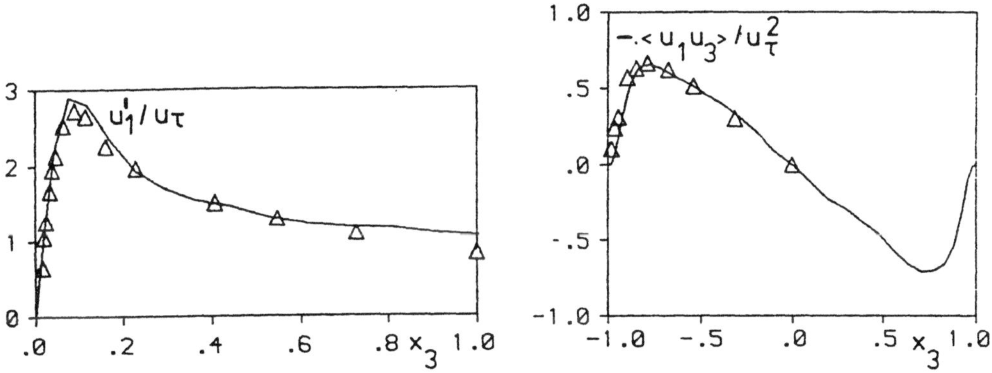

Fig. 1: Rms fluctuation. - calculation, Δ experiment

Reynolds stress. - calculation, Δ experiment

Large-eddy simulations of wall-bounded shear flows by HORIUTI & YOSHIZAWA, by GAVRILAKIS, TSAI, VOKE & LESLIE, by GRÖTZBACH and by SCHMITT, RICHTER & FRIEDRICH confirmed the validity of the simplest Smagorinsky-type SGS-model and of the two-part eddy-viscosity model as proposed by Schumann [14] at least with respect to the first and second-order quantities. Most higher-order quantities are not so well predicted which is due to the fact that these quantities are strongly determined by the small scales. HORIUTI & YOSHIZAWA investigated the turbulent plane channel flow at Reynolds numbers $Re_{\tau} = 640$ (based on channel half width and friction velocity) and compared Smagorinsky's model with a 'one-equation model' in which the SGS kinetic energy determining the eddy-viscosity coefficient is calculated from a transport equation using the Direct Interaction Approach. In the lower resolution cases (32^3 grid points) they found remarkable differences in the turbulence intensities and in spanwise two-point correlation functions, due to the too dissipative nature of the Smagorinsky model. In the higher resolution cases (64^3 grid points) differences between the two models had disappeared, meaning that the choice of the SGS model is crucial only in situations with low resolution. In the case of 64^3 grid points the authors estimated the mean spanwise spacing of streaks as 250 wall units which seems to demand for a still finer grid. Finally they found that the diffusion term in the transport equation for the SGS kinetic energy plays a dominant role besides production and dissipation especially in the vicinity of the walls and there is a remarkable correlation between these three mechanisms.

GAVRILAKIS, TSAI, VOKE & LESLIE presented large-eddy simulations of turbulent plane channel flow at $Re_{\tau} = 194$ obtained with a spectral and a finite-difference method and compared them with the Kreplin-Eckelmann experiment. Their SGS-model is close to that of Schumann [14] and Moin and Kim [11]. With a resolution of 32 x 64 x 32 mesh points for a computational domain of size $2\pi \times \pi \times 2$ (in terms of channel width, streamwise, spanwise and normal to the wall) the agreement between experiment and finite-difference results is satisfactory, whereas no real conclusions can be drawn from the spectral results since the simulation had not achieved a statistical equilibrium. GRÖTZBACH discussed his theory to calculate coefficients of SGS-models which is based on Schumann's two-part eddy-viscosity concept and covers turbulent heat flux. This theory accounts for all details of the finite difference scheme and in fact strongly depends on it. An important feature is that the SGS coefficients vary with local flow parameters which makes the model fully self-adaptive, i.e. it automatically switches the model gradually or totally off in cases where the smallest turbulence scales are of the order of or larger than the grid widths. The power of the method is demonstrated for turbulent low Prandtl number flows in annuli and internally heated horizontal convection layers. The heat source in the latter case is homogeneously distributed over the layer and both walls are kept at constant temperatures. The Rayleigh number chosen is $Ra = 4 \cdot 10^6$. An equidistant grid is used with 64 x 64 x 32 mesh cells in a channel with periodicity lengths of 2.8 channel widths in both horizontal directions. Since the walls are colder than the heated fluid, one finds stably stratified fluid

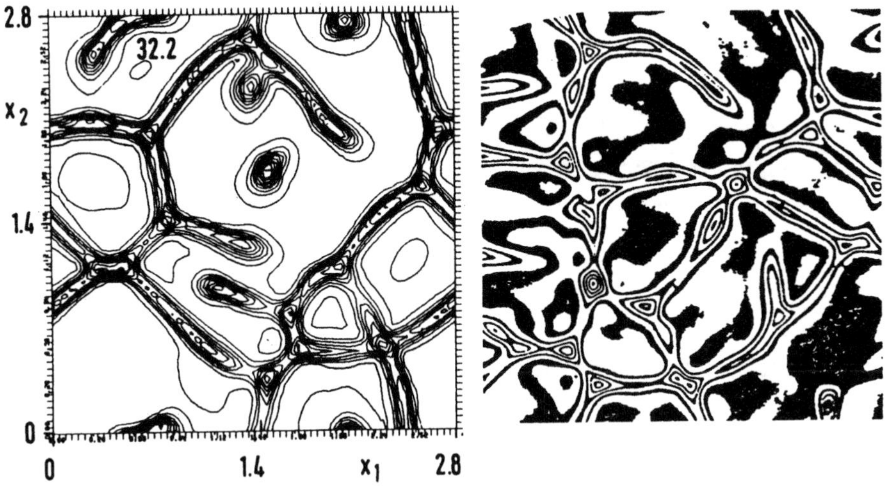

Fig. 2: Calculated and measured temperature fields in a horizontal plane

near the lower wall and an unstable stratification near the upper wall which drives the flow. Figure 2 gives an impression of the convection phenomena on the basis of isotherms in a horizontal cross-section. Most of the scales found in the temperature field are larger than the grid scales and only a few of them are of comparable size. The present theory was used to check whether the chosen grid resolution was sufficient to allow for a DS in this case or not. Since the important SGS coefficient depends on flow variables and on the mean grid width it should tend to zero in the whole field in the DS case and give the proper maximum rms-values of velocity and temperature fluctuations. Figure 3 shows the independence of the temperature fluctuations on the mean grid width for grids which are at least as fine as a 32 x 32 x 16 mesh system.

In a LES of turbulent flat plate boundary layer flow at a Reynolds number of $Re_\tau = 3240$ (based on upstream boundary layer thickness and friction velocity) SCHMITT, RICHTER & FRIEDRICH used Schumann's algebraic two-part SGS model with spatially constant coefficients which however vary with the mesh size. Using 64 x 32 x 32 mesh volumes streamwise, spanwise and normal to the wall they ob-

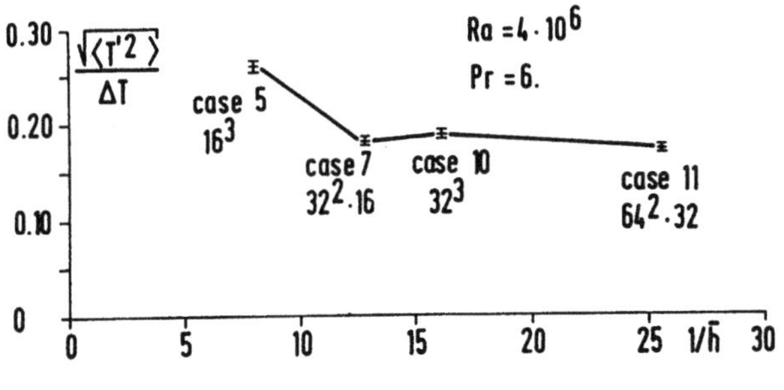

Fig. 3: Influence of the mean grid width $\bar{h}$ on the maxima of temperature fluctuations

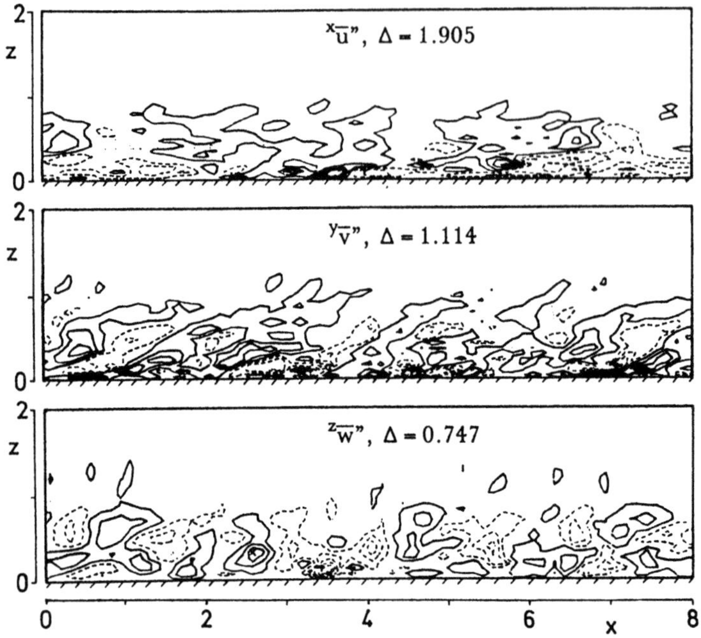

Fig. 4: Contour plots of resolved fluctuating velocity components

tained a resolution of 1/8 x 1/8 x 1/16 boundary layer thicknesses which is of course insufficient to describe the details of the wall layer. In fact, the wall nearest mesh cell extends over more than 200 wall units. Since the authors assume a parameterized wall boundary condition, namely a correlation between the fluctuating wall shear stress and the fluctuating streamwise velocity of the near wall mesh cell, their global treatment of the wall region is reasonable. Figure 4 gives an impression of the computed instantaneous flow field in the zero pressure gradient case. The contour line plots of the spanwise velocity fluctuations, v'', show typical large scale structures inclined in mean flow direction and the intermittent nature of the flow in the outer region. High positive values of u-fluctuations find their equivalent in a spotty occurrence of the instantaneous resolvable shear stress.

It may seem curious for someone who does a hard job in developing sophisticated SGS models that reasonable accuracy in first and second-order mean quantities can be obtained without a specific SGS model if only the numerical scheme contains some high-order numerical damping. This was demonstrated for high Reynolds number flow by KUWAHARA & SHIRAYAMA using a third-order upwind difference scheme for the nonlinear transport terms containing implicit fourth-order damping. The authors have not shown that the results are independent of further grid refinement. This is necessary insofar as the numerical diffusion introduced by this numerical scheme is sensitive to Galilean transformation and its dependence on grid spacing is likely to be different from what one expects on the basis of inertial subrange theory. For high accuracy LESs advanced SGS models are of course inevitable. AUPOIX has discussed SGS modelling for homogeneous, anisotropic turbulence. While standard models try to express the small scale effects in terms of the large eddies, his model not only drains energy from the large scales via an effective viscosity term, but also reinjects the SGS anisotropy into the large scales. It is based on Cambon's spherically averaged EDQNM model [2]. As Aupoix pointed out the linkage of the subgrid anisotropy to the anisotropy of the large scales near the filter cut has some drawbacks as it may introduce numerical noise, mainly for low resolution cases. A test of the SGS model in a LES of Gence's experiment of grid turbulence submitted to two successive plain strains showed that good results can be obtained if the subgrid anisotropy is modelled correctly. A completely new way of deriving a SGS model may be through progressively eliminating the effect of eddies starting with the smallest ones, then the next smallest, until the filter-cut is reached. This is what renormalization group (RG) theory applied to fluid turbulence in principal does. McCOMB in his paper presents the concept of RG and some aspects of a more general and systematic development of the theory in terms of moment expansions which leads to an effective viscosity. Since the theory has not yet been tested in conjunction with a numerical simulation no conclusions can be drawn concerning its efficiency.

3. Effect of Galilean transformation and numerical schemes for mean advection

In different contributions it was pointed out that advection by mean flow has to be treated very accurately. Otherwise the non-invariance of discrete approximations with respect to Galilean transformation causes large discrepancies. Not only the numerical schemes are not Galilean invariant but also SGS models, as was already discussed by Speziale [17]. LAURENCE in his contribution showed how this problem can be circumvented in homogeneous turbulence with a three-level decomposition of the velocity field into a time-dependent supporting field, the usual grid scale and subgrid scale fields. The SGS models have to be formulated in terms of grid scale components and space derivatives of the supporting component only. What the advection of momentum is concerned, LAURENCE noted that his characteristics scheme which naturally distinguishes between advected and advecting velocities, is typically not Galilean invariant. So, in order to find an optimal computation frame for each node point, he chooses the mean velocity of the domain of dependence of the node point as the proper translational velocity. The other drawback of the characteristics scheme, namely the need for high order interpolation (the particle trajectory will

(channel center line)

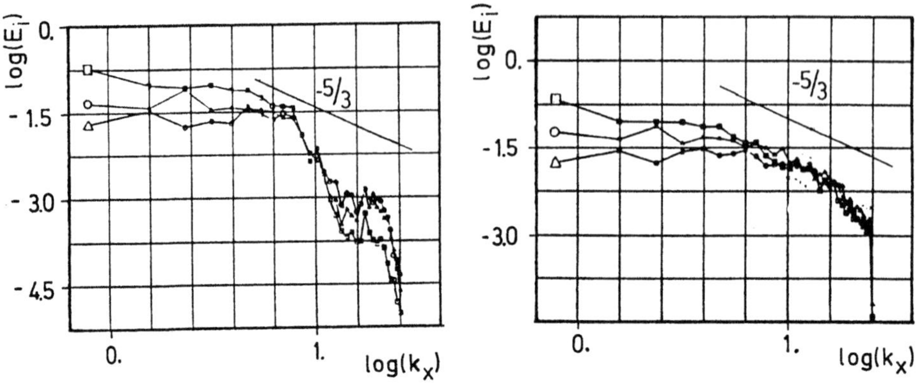

Fig. 5: Standard solution method Fourier interpolation

not in general pass through node points) which leads to numerical filtering, is of minor importance in a weak formulation. SCHMITT, RICHTER & FRIEDRICH found their finite difference scheme (essentially a second-order scheme in time and space on a staggered grid) to produce unphysical effects when applied to turbulent channel flow in a stationary reference frame. Besides a flattening of the mean velocity profile the longitudinal power spectra of the velocity fluctuations were too steep in the high wave number range, figure 5. Translating the coordinate system with an average velocity taken over the channel cross section removed these anomalies. Even better results could be obtained when the standard solution method was applied to the fluctuating part of the advection terms and when the mean advection was taken care of by Fourier interpolation, see fig. 5. The drop off of the highest mode is a consequence of the fact that this mode is not affected by mean advection. In their DS of the effects of stable stratification and shear on homogeneous turbulence in a cubical domain at moderate Reynolds numbers SCHUMANN, ELGHOBASHI & GERZ found, that the inclusion of mean advection in the finite difference scheme (which is second-order energy-conserving in time and space on a staggered grid) resulted in small time steps and moderately large errors in the velocity-derivative skewness. Superimposing a constant advection velocity (Galilean transformation) strongly reduces the accuracy and decreases the skewness. However, if the mean velocity is treated with the Fourier interpolation method, which is nothing but a most accurate characteristics method (feasible only in the case of horizontal homogeneity of the flow), then the results are fairly insensitive to the additional advection.

4. Methods to generate boundary conditions where standard periodicity or wall conditions cannot be applied

In the case of uniform shear studied by SCHUMANN, ELGHOBASHI & GERZ an initially periodic flow field does not remain periodic in the shear direction and demands for modified boundary conditions. In a Lagrangian reference frame which distorts with the mean shear standard periodicity conditions can still be applied (a remapping of the flow field during time-integration is however necessary). With respect to an Eulerian reference frame Baron [1] had proposed "shear-periodic" boundary conditions

$$f(x, y, z + L, t) = f(x - Ut, y, z, t) + (\partial \bar{f}/\partial z)L ,$$

where $\bar{f}$ is the ensemble average, L the periodicity length, $U/L \equiv d\bar{U}/dz$ and $(x - Ut)$ is taken as modulo L. SCHUMANN et al. use Baron's concept with some modification. While Baron restricts the timestep Δt and the grid-spacing Δx to satisfy $\Delta t L \, d\bar{U}/dz = \Delta x$, which eliminates the need for any interpolation between grid points, but may conflict with stability constraints on Δt, Schumann et al. generalize this approach to treat arbitrary Courant-numbers by using a Fourier interpolation to determine $f(x - Ut)$ if the argument does not coincide with discrete grid points. Moreover, to account for shear-periodicity in the pressure field, they have developed an efficient direct Poisson solver. Spatially growing flows, like boundary layers or jets and mixing layers cannot in general be converted into temporally growing flows using periodic boundary conditions unless one tolerates a change in the flow characteristics. For turbulent boundary layer flow SCHMITT, RICHTER & FRIEDRICH have constructed inflow boundary conditions from rescaled data near the outflow plane. A LES needs the grid scale velocity vector in each grid point of the inflow plane for all times. Needless to say that no experiment is able to provide these data. Therefore a possible way out of this difficulty is to take velocity fluctuations from a plane far enough downstream of the inlet and to weight these quantities with factors such that their mean values vanish and their rms-values coincide with experimental data. Rescaling of geometry has yet been avoided because of limited vertical resolution and the difficulty of properly interpolating the grid scale quantities which are defined on a fixed mesh system. Recent investigations seem to indicate that an adjustment of velocity fluctuations mainly with respect to the proper Reynolds stress in the inflow plane is more useful. No doubt, inflow boundary conditions based on large scale structures, such as horseshoe vortices, would produce better flow fields, especially concerning higher order statistics and would allow for shorter computational domains. An enormous amount of work has still to be done in this area.

5. Applications of DS and LES results to check statistical models

The potential of DS and LES to investigate the accuracy of closure models and to evaluate the coefficients in new models is still not used to that extent it should be. Correlations with pressure fluctuations induced by shear, buoyancy or similar forces can be evaluated with great success. A good example to demonstrate the need for LES is the stratus cloud-topped planetary boundary layer studied by MOENG because the application of statistical turbulence models to this problem is still controversial due to the complicated structure and dynamics of the flow and due to a lack of observational data. Moeng's LES model includes buoyancy effects, long wave radiation and condensation processes. The simulation uses 40^3 grid points to cover a domain of 2.5 x 2.5 x 1 km^3. Data generated over 2500 time steps with a time interval of 1.5 s form the basis from which information can be extracted for mixed layer and second-order closure modelling. The first model assumes the conserved mean fields to be always well-mixed within the PBL. The only closure problem consists in specifying the entrainment rate, i.e. to decide which part of the thermally generated turbulent kinetic energy (production) is used in dissipation and which part returns to potential energy through thermally indirect motions (consumption). The answer given by Moeng's LES is that 60 % of the kinetic energy are dissipated and 40 % are converted back to potential energy. Fig. 6a shows a comparison of production and consumption rates for the stratus-topped PBL, w being the vertical velocity component and s_v the virtual dry static energy. Second-order closure modelling of the stratus-topped PBL depends e.g. on realistic

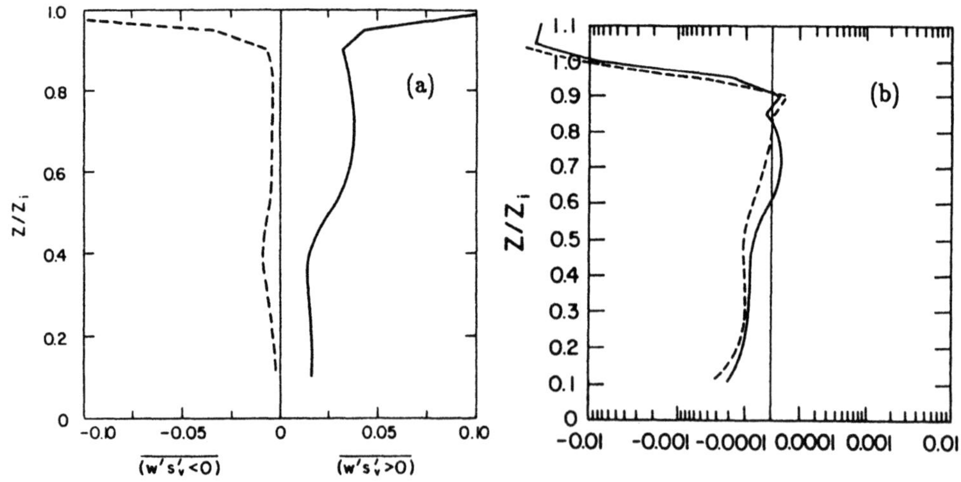

Fig. 6: Stratus-topped PBL. a) buoyant production and consumption b) analyzed and parameterized (dashed) pressure gradient/ liquid water static energy covariance

assumptions for correlations of the pressure gradient with temperature and moisture fields appearing in the transport equations for turbulent heat and mass flux. Moeng has calculated the various contributions to these quantities from turbulence-turbulence interaction, from mean shear, buoyancy and Coriolis effects and concluded that the turbulence-turbulence nonlinear effect can be approximated by Rotta's return-to-isotropy concept while the buoyancy contribution to the pressure-scalar covariance is proportional to the buoyancy production term in the scalar vertical flux equation. Fig. 6b presents a comparison between LES and parametrization for the pressure gradient/liquid water static energy covariance.

The DS of SCHUMANN, ELGHOBASHI & GERZ for stratified homogeneous shear shows in agreement with experimental data of Komori et al. [6] that under strongly stable conditions a counter-gradient heat-flux arises which no longer dampens the turbulent motion but supplies additional turbulent kinetic energy from the reservoir of potential energy, see fig. 7. Along with this effect the authors observed that the magnitudes of the components b_{11} and b_{33} of the Reynolds stress anisotropy tensor increase with Richardson number in the weak stability region and then decrease for higher stability. Their results suggest that this is due to a strong reduction in the pressure strain terms. This is insofar a significant result as it contrasts Launder's model [7] predicting a monotonic increase of b_{11}, b_{33} with Richardson number. It should, however, be said that the usage of this model was not really intended in strong stratification cases. This is another example where DS can be useful in improving statistical models.

The paper by LEE & REYNOLDS [8] presented at the EUROMECH Colloquium, but not included in the Proceedings, seemed to indicate that new developments in statistical turbulence modelling may be necessary in the near future. On the basis of a DS of homogeneous turbulence subjected to irrotational strain and

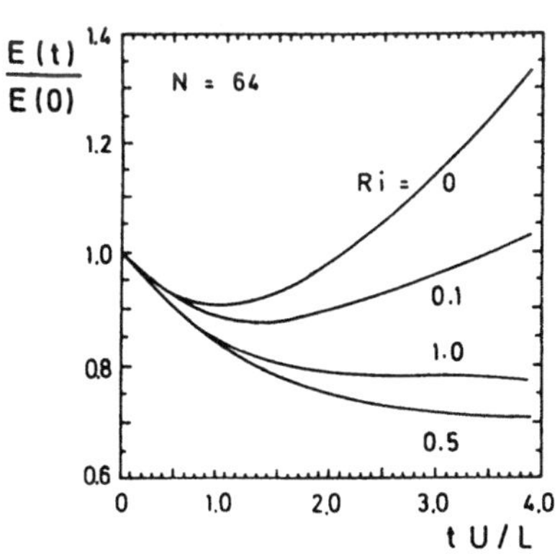

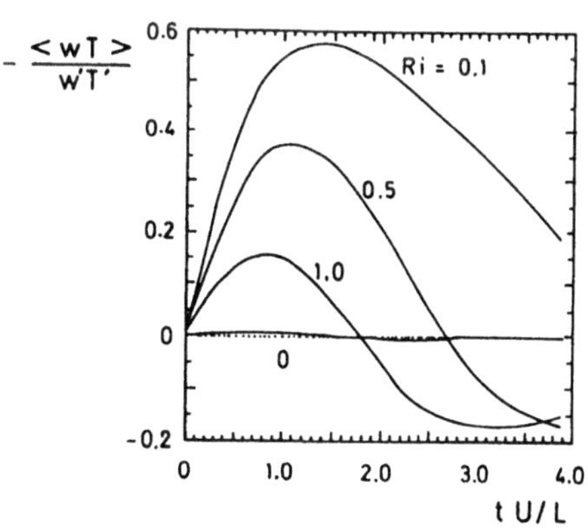

Fig. 7: Time variation of turbulent kinetic energy and vertical heat flux for different Richardson numbers Ri

relaxation after that strain at Reynolds numbers of order 10 (based on Taylor microscale and rms-velocity) they found surprisingly strong anisotropy in the small-scale structure, as reflected by vorticity and dissipation. The Reynolds stress anisotropy did not show up that strong correlation with the pressure-strain term that is a presupposition of Rotta's return-to-isotropy model. But instead the isotropy-restoration term and the anisotropy of the small-scale proved to be well correlated. Moreover, the vorticity anisotropy was, in general, greater than that of the Reynolds stress. From this and further arguments the authors concluded that the Reynolds stress anisotropy can be regarded as arising from vorticity anisotropy.

6. Analysis of turbulence structures from the results of DS and LES

Results are no longer solely presented in terms of mean profiles or contour line plots of the instantaneous fields. Further and sometimes more sophisticated evaluations of simulated data show the structure of the vorticity field and the formation of hairpin-vortices.Under 2) we have already mentioned Moin's findings of streaks in the wall region of turbulent channel flow. MOIN & KIM were also the first to study the three-dimensional structure of hairpin vortices in terms of instantaneous vorticity vectors in planes inclined at 45° to the wall, in terms of vortex filaments (sets of instantaneous vortex lines) and their conditional samples using the spatial analogue of the VITA technique [5,12]. For the first time MOIN has used the Karhunen-Loeve expansion and Lumley's characteristic eddy decomposition to extract organized structures from simulated turbulent channel flow. (In homogeneous turbulence this technique corresponds to a Fourier analysis of the two-point velocity correlation function). This method allows to determine unambiguously the contributions of the individual eddies to turbulence kinetic energy and stresses and to measure their importance to the overall dynamics of the flow. Applying this technique to their LES of a turbulent flat plate boundary layer SCHMITT, RICHTER & FRIEDRICH show the contribution of the dominant eigenfunctions to turbulent kinetic energy and Reynolds stress, fig. 8. As the x- and z-components of the dominant eigenfunctions have opposite sign, they make a negative contribution to the shear stress. The first eigenfunction contributes 23 % to the total resolved turbulence kinetic energy and 58 % to the Reynolds stress. The convergence behaviour indicates a positive contribution to the stress from the higher-order eigenfunctions similar to what Moin found for channel turbulence. Examining again the instantaneous flow field it is apparent from fig. 9 that the boundary layer contains vortical structures of the horseshoe type. Distributions of the inclination angle of the projected vorticity vector demonstrate

Table:
Contribution of the dominant eigenfunction to turbulent kinetic energy and Reynolds shear stress.

$\lambda^{(1)}/E_t$	$P^{(1)}/P_t$	$\lambda^{(1)}/\lambda^{(2)}$
0.236	0.585	1.546

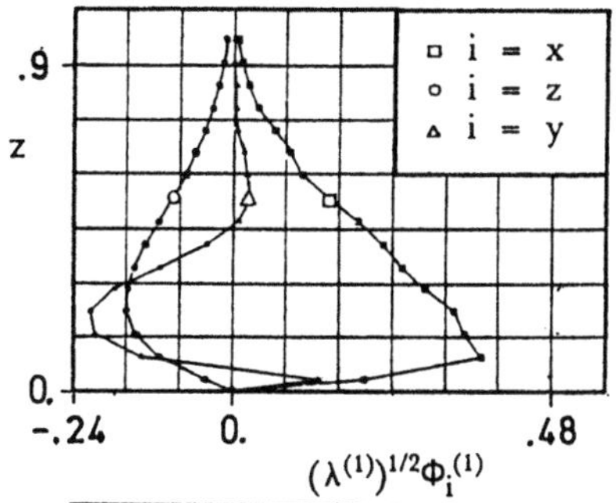

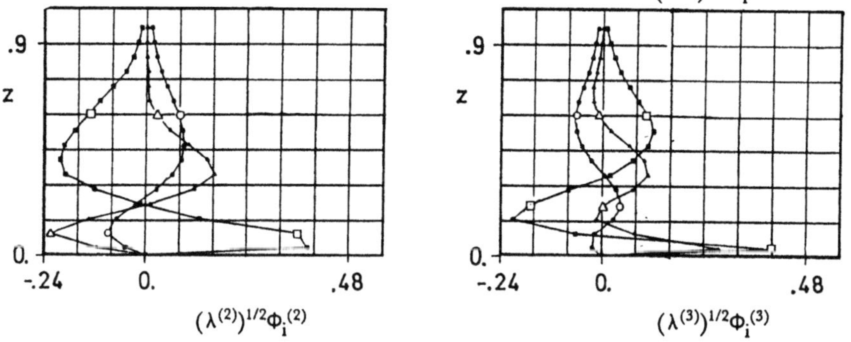

Fig. 8: Components of the first three dominating eigenfunctions of the Karhunen-Loeve expansion

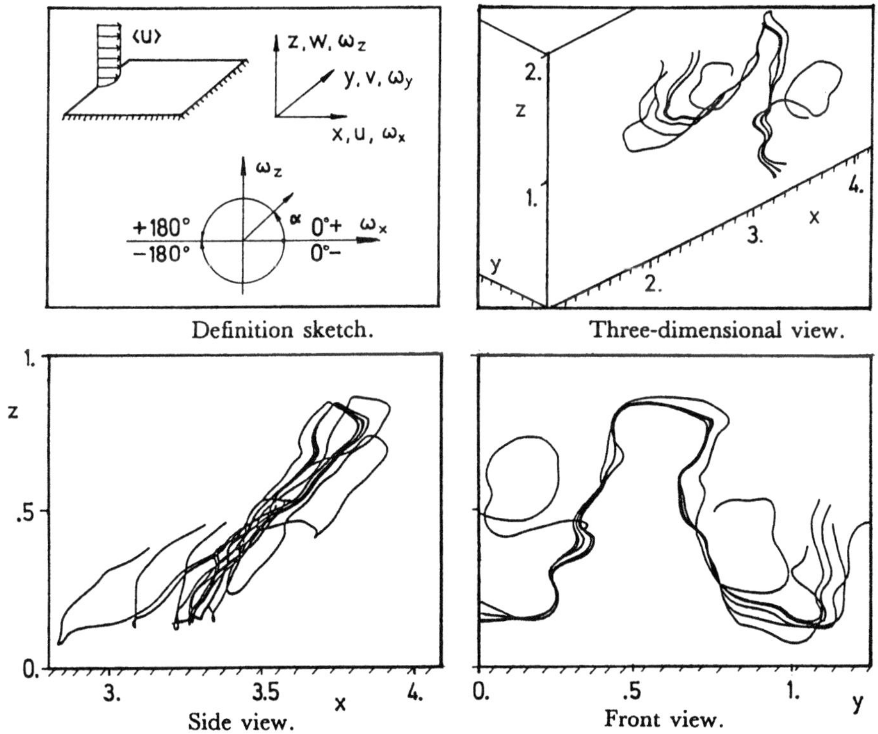

Fig. 9: Instantaneous vorticity lines

the characteristic inclination of these structures with respect to the wall. Since these findings were made in the outer part of the boundary layer without resolving the wall layer, horseshoe-shaped vortices need not be connected with burst and sweep events. They apparently result from the vortex-stretching mechanism which in our opinion gives some support to Theodorsen's conjecture [18] that horseshoe vortices are characteristic of all shear flows.

Coherent structures with a 45° inclination in stratified homogeneous turbulence are obtained in the DS of SCHUMANN et al. Fig. 10 shows contour-line plots of the spanwise velocity component at the initial and at a later time. The initially random field produces orderly structures under the influence of shear. Analysis and interpretation of such structures are the subject of on-going work.

In summary the methods of DS and LES have come to a mature stage with respect to numerical algorithms. However, quite some work is required concerning the

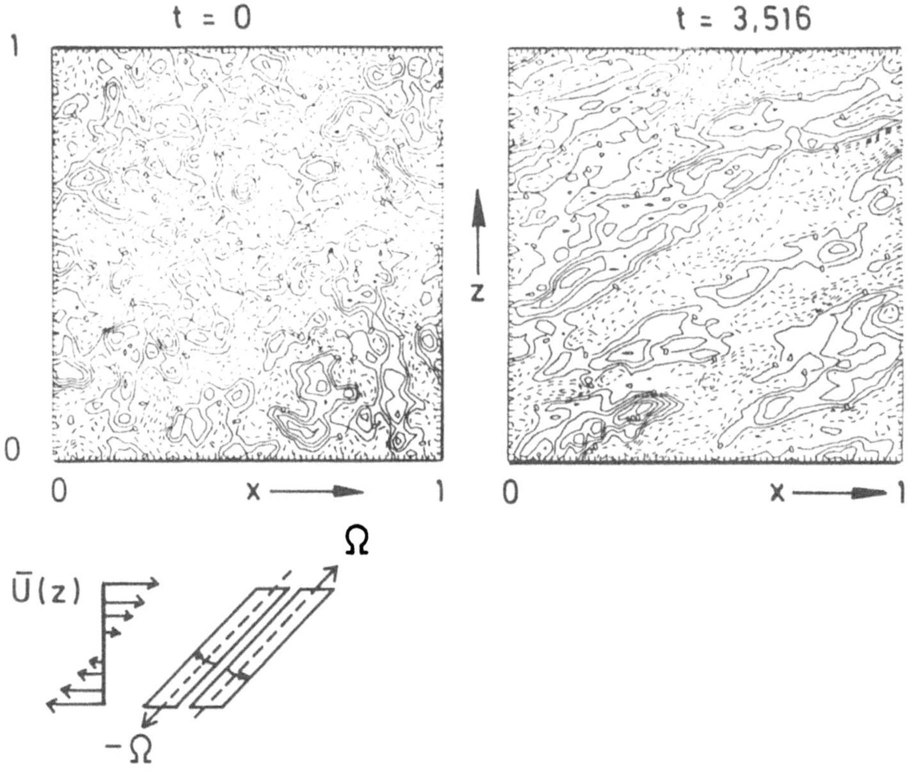

$$\Omega = \frac{\partial v}{\partial x} - \frac{\partial u}{\partial y}$$

Fig. 10: Creation of coherent structures under shear for Ri = 0.1. The contour-line plots show the v-velocity at initial and final time.

construction of proper inflow boundary conditions for spatially developing flows, the further development of suitable SGS models, the illumination of the delicate balance between mesh system, filter, SGS model and numerical schemes and Galilean invariance of the solution methods. Although many applications are known especially with respect to simple geometries, the full potential of usage for turbulence research and for application in geophysics or engineering is by far not yet fully explored.

Literature

1. Baron, F. (1982): Macrosimulation Tridimensionelle d'Ecoulements Turbulents Cisaillés. Thesis and report HE 41/82.21, Electricité de France
2. Cambon, C., Jeandel, D., Mathieu, J. (1981): Spectral modelling of homogeneous nonisotropic turbulence. J. Fluid Mech. 101, pp. 247-262
3. Deardorff, J.W. (1970): A numerical study of three-dimensional turbulent channel flow at large Reynolds numbers. J. Fluid Mech. 41, pp. 453-480
4. Eckelmann, H. (1970): Experimentelle Untersuchungen in einer turbulenten Kanalströmung mit starken viskosen Wandschichten. Mitt. MPI f. Strömungsforschung Nr. 48
5. Kim, J. & Moin, P. (1986): The structure of the vorticity field in turbulent channel flow. Part 2. Study of ensemble-averaged fields. J. Fluid Mech., 162, pp. 339-363
6. Komori, S., Ueda, H., Ogino, F. & Mizushina, T. (1983): Turbulence structure in stably stratified open-channel flow. J.F.M. 130, pp. 13-26
7. Launder, B.E. (1975): On the effects of a gravitational field on the turbulent transport of heat and momentum. J.F.M. 67, pp. 569-581
8. Lee, M.J. & Reynolds, W.C. (1985): Numerical experiments on the structure of homogeneous turbulence. Report No. TF-24. Dept. of Mech. Eng., Stanford University, Stanford, Calif. 94305
9. Lilly, D.K. (1967): The representation of small-scale turbulence in numerical simulation experiments. In: H.H. Goldstine (ed.): Proc. of the IBM Sci. Comp. Symp. Env. Sci., New York, IBM-Form No. 320-951, pp. 195-210
10. Mason, P.J. & Callen, N.S. (1986): On the magnitude of the subgrid-scale eddy coefficient in large-eddy simulations of turbulent channel flow. J. Fluid Mech. 162, pp. 439-462
11. Moin, P. & Kim, J. (1982): Numerical investigation of turbulent channel flow. J. Fluid Mech. 118, pp. 341-377
12. Moin, P. & Kim, J. (1985): The structure of the vorticity field in turbulent channel flow. Part 1. Analysis of instantaneous fields and statistical correlations. J. Fluid Mech. 155, pp. 441-464

13. Rogallo, R.S. & Moin, P. (1984): Numerical simulation of turbulent flows. Ann. Rev. Fluid Mech. 16, pp. 99-137

14. Schumann, U. (1975): Subgrid scale model for finite difference simulations of turbulent flows in plane channels and annuli. J. Comp.Phys., pp. 376-404

15. Schumann, U. & Friedrich, R., eds. (1986): Direct and Large Eddy Simulation of Turbulence. Proc. EUROMECH Coll. No. 199. Notes on Numerical Fluid Mechanics, Vol. 15, Vieweg-Verlag Braunschweig

16. Smagorinsky, J. (1963): General circulation experiments with the primitive equations. I. The basic experiment. Mon. Weather Rev. 91, pp. 99-164

17. Speziale, C.G. (1985): Galilean Invariance of Subgrid-Scale Stress Models. J. Fluid Mech., 156, pp. 55-62

18. Theodorsen, T. (1952): Mechanism of turbulence. In: Proc. 2nd Midwestern Conf. on Fluid Mech. Ohio State Univ., Columbus, Ohio

The Pressure-Strain Correlation of a Turbulent Homogeneous Shear Flow Under Strongly-Stable Stratification

T. Gerz and U. Schumann

DFVLR, Institut für Physik der Atmosphäre,
D-8031 Weßling, Fed. Rep. of Germany

Results are presented concerning the direct numerical solution of the exact time-dependent three-dimensional Navier-Stokes equations and the first law of thermodynamics for an incompressible turbulent flow possessing uniform gradients of the mean velocity and temperature in the vertical direction. The equations are solved in a cubical domain representing a part of the homogeneous flow and divided into 64^3 grid cells. Boundary conditions are periodic in the horizontal and shear-periodic in the vertical (shear) direction. The model is described in detail in ELGHOBASHI et al. [1] and in SCHUMANN et al. [2].

The behaviour of a homogeneous shear-flow with initially fully turbulent, isotropic velocity and temperature fields under strongly-stable stratification is considered. In flows with Richardson-numbers (Ri) larger than 0.25 the turbulent kinetic energy is expected to decrease with time. The trend of increasing conversion of kinetic energy into potential energy for increasing Ri-numbers inverts if the stratification is very stable ($Ri = 1$, see Figure 1). This effect can be explained in studying the dynamical behaviour of the turbulent vertical momentum flux $\overline{uw}$ and the heat flux $\overline{wT}$ for $Ri = 1$. Figure 2 shows that these quantities, here normalized by their rms-values, change from negative to positive fluxes. These counter-gradient fluxes, also observed by KOMORI et al. [3] in a water-tank experiment and discussed in [1], indicate that the initially turbulent flow changes

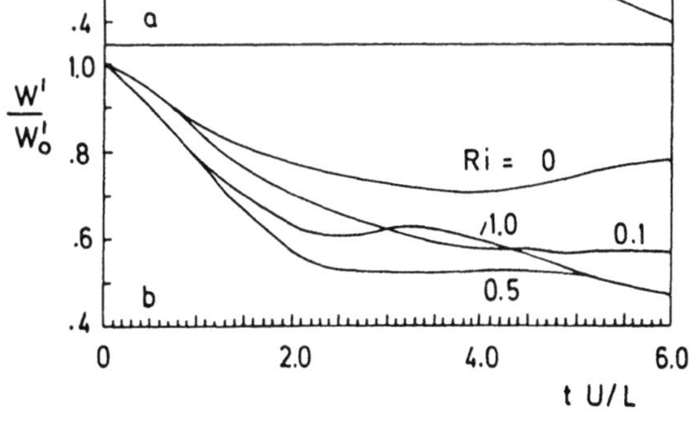

Figure 1. Time variation of the turbulent kinetic energy E (a) and of the rms-value of w (b) normalized by their initial values for different Ri

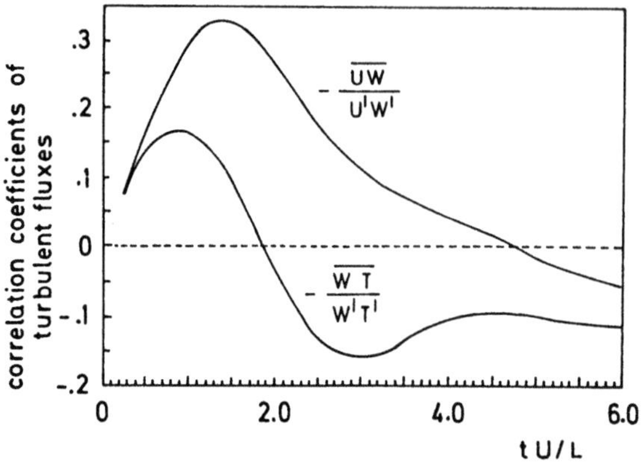

Figure 2. Time variation of the mean one-point correlations $-\overline{uw}/u'w'$ and $-\overline{wT}/w'T'$ normalized by the rms-values for $Ri = 1$

more and more to a wave-like motion. The fluxes no longer damp the turbulent motion rather than enhance the kinetic energy in its vertical component $\overline{ww}$ in the time interval $2.0 < tU/L < 3.5$ (Figure 1).

Interestingly, these dynamics are also reflected in the pressure-strain correlation

$$\phi_{ij} = \overline{p\left(\frac{\partial u_i}{\partial x_j} + \frac{\partial u_j}{\partial x_i}\right)}. \tag{1}$$

Figure 3 shows the dimensionless ϕ_{11}, ϕ_{22}, ϕ_{33} and ϕ_{13} as a function of tU/L obtained by direct simulation. ϕ_{11} and ϕ_{33} change sign in the time-interval $1.8 < tU/L < 2.7$ when the counter-gradient heat-flux occurs and attains its maximum value (Figure 2). Looking at the dynamic equations of the components of the turbulent kinetic energy the interaction of fluxes and strain can be explained even if the linear parts are considered alone. These equations read in non-dimensional

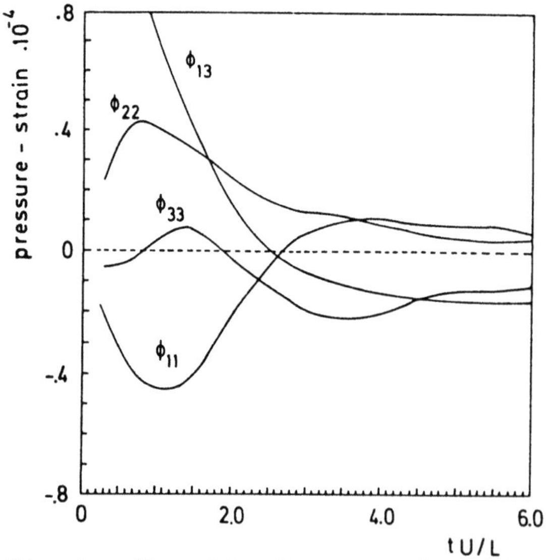

Figure 3. Time variation of the pressure-strain correlations ϕ_{11}, ϕ_{22}, ϕ_{33} and ϕ_{13} obtained by direct simulation for $Ri = 1$

form:

$$\frac{\partial \overline{uu}}{\partial t} = -2S\,\overline{uw} \quad + \phi_{11} - \epsilon_{11} \tag{2}$$

$$\frac{\partial \overline{vv}}{\partial t} = \qquad\qquad + \phi_{22} - \epsilon_{22} \tag{3}$$

$$\frac{\partial \overline{ww}}{\partial t} = 2\,|Ri|\,\overline{wT} + \phi_{33} - \epsilon_{33} \tag{4}$$

where ϵ_{ij} stands for the dissipation rate of momentum fluxes and $S = L/U\, d\overline{U}/dz$ for the non-dimensional shear. L is the scale-length of the domain and U the scale-velocity of the flow. (Note that $\sum \phi_{ii} = 0$.)

Initially, ϕ_{11} is negative and redistributes energy from the streamwise ($\overline{uu}$) component to the lateral ($\overline{vv}$) and the vertical ($\overline{ww}$) parts because $\overline{uu}$ is enhanced strongest due to shear ($\overline{uw}$ is large and negative, (2)) while ϕ_{22} and ϕ_{33}, both positive, act like source terms for $\overline{vv}$ and $\overline{ww}$ (Fig. 2, 3). The former obtains more energy than the latter due to the strong stability of the turbulent flow which prohibits any increase of energy against the buoyancy forces. The heatflux $\overline{wT}$ is negative during this period and thus, acts as a sink term in the $\overline{ww}$-equation (4). The situation, however, is reversed when the turbulent heat-flux changes sign. $\overline{ww}$ is now mostly supported by $\overline{wT}$ ("wave-pumping“) and ϕ_{33} as a negative quantity redistributes energy in equal shares to $\overline{uu}$ and $\overline{vv}$ (ϕ_{11} and ϕ_{22} are positive, see Figure 3). At this time $-\overline{uw}$ is decreasing but still positive. Later on, also a counter-gradient momentum-flux occurs ($-\overline{uw} < 0$, Figure 2) accompanied by a small increase of ϕ_{33} and correspondingly a decrease of ϕ_{11} and ϕ_{22}. The situation has now reached a "quasi-steady“ state ([2]).

After the strong decrease of $-\overline{wT}$ and its sign-change at $tU/L \approx 1.5$ (Figure 2), the flow is no longer purely turbulent but has clearly wave-like characteristics. This is also indicated by the shift of the phase angle between w and T ([3]). With increasing Ri the angle shifts from $\pm\,\pi$ (i.e. turbulent motion) to $\pm\,\pi/2$ (i.e. wave motion) and further on to zero which describes a flow with a wave-like motion intermittent by "hot eddies“ which are responsible for the occurence of the counter-gradient heat-flux.

The reaction of the pressure-strain correlation to these dynamics cannot be described by the second-order parametrization for ϕ_{ij} proposed by LAUNDER [4,5], although all the contributing terms describing the influence of turbulent interaction, shear and buoyancy are included. The equations read in dimensionless form:

$$\phi_{ij} = \phi_{ij_1} + \phi_{ij_2} + \phi_{ij_3} \tag{5}$$

where

$$\phi_{ij_1} = -c_1 \frac{\epsilon}{E}\left(\overline{u_i u_j} - \frac{2}{3} E\, \delta_{ij} \right) \tag{6}$$

$$\phi_{ij_2} = -c_2\left(-\overline{u_i u_k}\, S_{j,k} - \overline{u_j u_k}\, S_{i,k} + [\frac{2}{3}\, \overline{u_i u_k}\, S_{i,k}]\, \delta_{ij} \right) \tag{7}$$

$$\phi_{ij_3} = -c_3\left(|Ri_j|\, \overline{u_i T} + |Ri_i|\, \overline{u_j T} - [\frac{2}{3}\, |Ri_i|\, \overline{u_i T}]\, \delta_{ij} \right) \tag{8}$$

describe the turbulent interaction, the production due to shear and the production due to buoyancy, respectively.

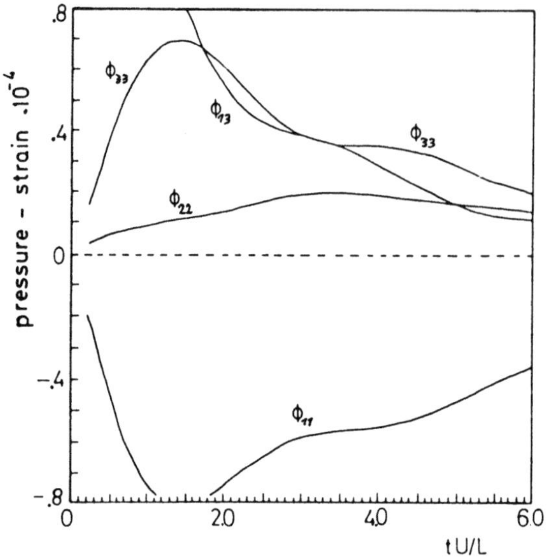

Figure 4. Time variation of the pressure-strain correlations obtained by Launders's model (1975,1976) with the constants $c_1 = 2.0$, $c_2 = 0.6$, $c_3 = 0.3$ for $Ri = 1$

$\epsilon = \nu\overline{(\partial u_i/\partial x_k)^2}$, $E = 1/2\ \overline{u_i u_i}$, $Ri_i = -\alpha g_i L \Delta T/U^2$ $(g_i = (0,0,-g))$, $S_{i,k} = L/U\ \partial \overline{U_i}/\partial x_k = (0,0,S)$ stand for the total rate of dissipation, the kinetic energy, the Richardson-number and the non-dimensional shear, respectively.

Figure 4 shows the time variation of ϕ_{11}, ϕ_{22}, ϕ_{33} and ϕ_{13} obtained by LAUNDER's model for the same flow described above ($Ri = 1$) for the coefficients $c_1 = 2.0$, $c_2 = 0.6$ and $c_3 = 0.3$. Although $-\phi_{11}$ demonstrates the correct development at the beginning qualitatively and decreases when the counter-gradient heat-flux is established, the parametrization with the coefficients mentioned above does not account for the change in the redistribution among the turbulent intensities. The redistribution of energy to the lateral and vertical components is described incorrectly. ϕ_{22} is underestimated and ϕ_{33} is overestimated in comparison with the results of the direct simulation (Figure 3).

The terms ϕ_{ii_1}, ϕ_{ii_2} and ϕ_{ii_3} are shown separately in Figure 5 with the same coefficients as in Figure 4. It can be seen that the turbulent interaction term (6) (dashed lines) dominates and thus, leads to a wrong interpretation of the dynamics of the flow. The value of the coefficient c_1 assumed by LAUNDER which gives good results for a flow under non-buoyant conditions ([5]) is too large for a flow under strongly-stable stratification. On the other hand, the coefficient c_3 which modifies the influence of the buoyancy term (8) seems to be too small for the considered flow.

Looking at the solid and dotted lines of Figure 5 which represent the shear term and the gravitational term the superposition of both describes the time variation of the pressure-strain correlations already in a qualitatively correct manner. Figure 6 now, shows the results of the parametrization obtained after increasing c_3 to a value of 0.6 which was already suggested by LAUNDER [4] and neglecting the share of the turbulent interaction (i.e. $c_1 = 0$). The dynamical behaviour of ϕ_{ij} experiences at least the sign-changes of ϕ_{11}, ϕ_{33} and ϕ_{13} which are predicted by the direct simulation. The situation is fairly good reproduced also quantitatively when the flow has reached a "quasi-steady" state at $tU/L \approx 4.7$.

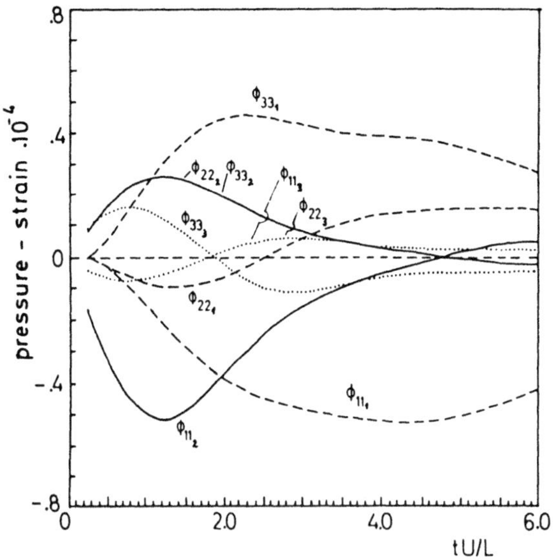

Figure 5. Time variation of the components of the pressure-strain correlations ϕ_{11}, ϕ_{22} and ϕ_{33}; dashed lines: turbulent interaction, solid lines: production due to shear, dotted lines: production due to buoyancy. The coefficients are the same as in Figure 4.

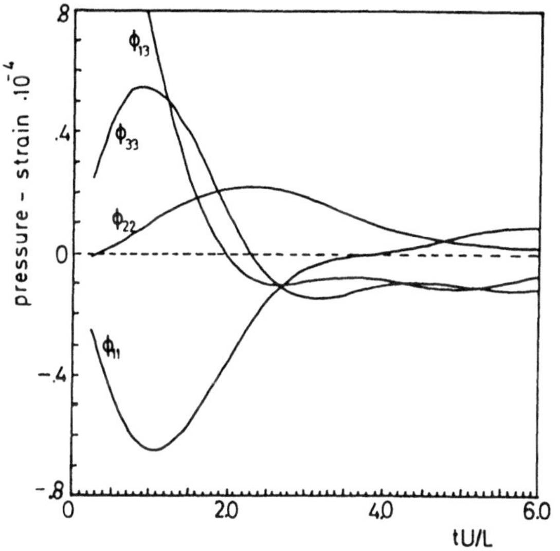

Figure 6. Same as Figure 4 but for the coefficients $c_1 = 0.$, $c_2 = 0.6$, $c_3 = 0.6$.

Not satisfactory, however, is the behaviour of the parametrized quantities ϕ_{22} and ϕ_{33} at the beginning when the vertical momentum flux $\overline{uw}$ dominates the physics of the flow (see Figure 2). A reason might be that almost all energy production due to shear goes from the streamwise to the lateral component because the stratification is too stable to allow an increase of energy in the direction against the buoyancy force. Thus, the energy source of ϕ_{22} due to shear might be larger than the parametrization of ϕ_{22_2} ((7) for $i = j = 2$) suggests. To consider this effect one should examine the redistribution among the flow components for the shear term (7) itself.

In comparison with the results of a direct simulation it could be shown that for flows under strongly-stable stratification ($Ri = 1$) LAUNDER's parametrization of the pressure-strain correlations yields good results when the flow has reached a "quasi-steady" state and if the coefficients are modified to account for the strength of the stability. In the future the dependency of these coefficients on the strength of the stratification will be examined.

Acknowledgement. This work is supported by the Deutsche Forschungsgemeinschaft.

References

1 Elghobashi,S.; Gerz,T.; Schumann,U. 1985: Direct simulation of a turbulent homogeneous shear flow with buoyancy; Proc. 5th Symp. on Turbulent Shear Flows, Aug.7-9,1985, Cornell Univ., Ithaca, New York, pp.22.7-22.12

2 Schumann,U.; Elghobashi,S.; Gerz,T. 1986: Direct simulation of stably stratified turbulent homogeneous shear flows; Proc. EUROMECH 199, Sep.30-Oct.2,1985, Technical University of Munich, Munich (FRG), in press.

3 Komori,S.; Ueda,H.; Ogino,F.; Mizushina,T. 1983: Turbulence structure in stably stratified open-channel flow; J.Fluid Mech., **130**, 13-26.

4 Launder,B. 1975: On the effects of a gravitational field on the turbulent transport of heat and momentum; J.Fluid Mech., **67,3**, 569-581.

5 Launder,B. 1976: Heat and mass transport; in: Topics in Applied Physics, ed. Bradshaw, Vol.12 ("Turbulence"), Springer-Verlag, 232-287.

Normalization of High-Resolution Raster Display Applied to Turbulent Fields

M. Farge

Laboratoire de Météorologie Dynamique du CNRS, 24, rue Lhomond, F-75231 Parix Cedex 5, France

Introduction

With the advent of high-resolution raster graphics, it is essential for their optimal use to be conscious of their possibilities and limitations. Without such an awareness, we run the risk of being mislead by attractive images, but void of information content or very difficult to interpret. This is the reason why it is necessary to precise which representation mode and palette are being used to build a given picture, otherwise the display will be as meaningless as a measure without error bars.

The visual representation and perception process is not neutral, it filters or distorts the information contained in the field we want to analyse. The selection of the best representation mode and color palette to display a given turbulent field is not trivial. It requires most importantly:

- a strong knowledge of the underlying dynamics in order to define the relevant questions we want to ask about this field,
- a certain understanding of the physiology of vision and the psychology of shape and color perception, in order to avoid, or at least be conscious of, different optical illusions or parasite phenomena, for example the problems of *chromostereopsis* /1/ -the shortest wavelength colors tend to focus in front of the retina and thus appear defocused- and *simultaneous contrast* /2/ -a given color is perceived differently depending on the background color-,
- an awareness of the culture dependent symbolism of forms and colors /3/, in order to reduce the risk of misinterpretation due to possible implicit meanings of certain colors -for instance, the color red is usually associated with heat, strength and expansion, while blue is mostly associated with cold, weakness and retraction-.

The ideas discussed in this paper will be applied to the graphic representation of turbulent fields. We will consider, as example, the case of a rotating compressible two-dimensional turbulent flow. We are particularly interested in displaying the following fields: vorticity, divergence, stream function, velocity potential, pressure and potential vorticity. These are all *two-dimensional scalar fields* with constant space integral, *i.e.* they fluctuate around an average value, that we will for convenience adjust to zero. To display such fields, presenting complex structures of very different scales and levels of excitation (*e.g.* coherent vortices, vorticity filaments, modons, gravity waves, inertial waves...), high-resolution raster techniques give a much more detailed representation than the classical vector techniques used with plotters. Using raster techniques a picture is viewed as a set of elementary spots, called *pixels*, having different colors or different grey levels, while with vector techniques it is described as a set of curves satisfying some equation /4/.

We are using the prototype raster display system developed by Jean-François Colonna at LACTAMME, Ecole Polytechnique, Palaiseau. It has a spatial resolution of 512^2 pixels, each of them with one byte memory, which allows $2^8=256$ different values. We therefore scale each field to be displayed by mapping its range into the 0-255 interval, such that the value zero corresponds to 127.

In the first part of the paper we will discuss the *choice of representation modes*, then we will propose a *normalization for the color space* and finally we will define some rules for the *choice of color palettes*.

1. Choice of representation modes

To any particular question raised while analysing a given field, it corresponds a representation mode more adequate than the others. Each distinct representation mode acts as an operator that reveals a given property of the field

For example:

- to study the morphology of the most excited structures in the field, we propose a *frontal perspective representation* with hidden surfaces removal and shadowing (Figure 1a.),
- to see the smoothness of the field, we propose the same *perspective representation*, but *with a vertical section* (Figure 1b.),
- to allow a morphological comparison between the large and small scale structures, we propose a *perspective representation as seen from the top* with shadowing (Figure 1c.),

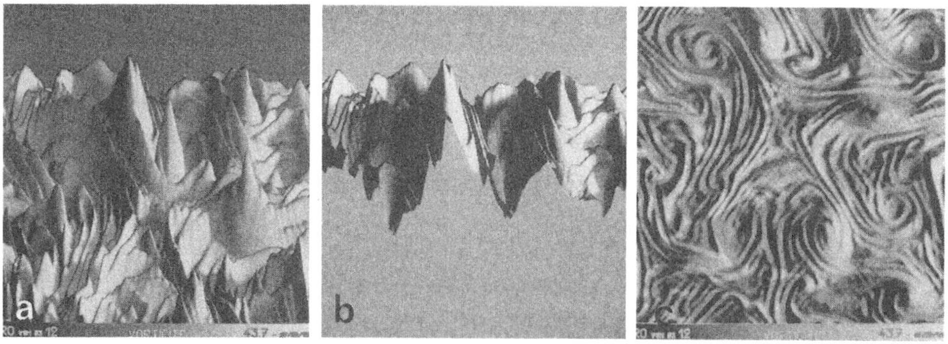

Figure 1a. Figure 1b. Figure 1c.

Perspective representations(+)

(+) The following examples display the *same* vorticity field (Figures 1a. to 2c.)

- to have an integral view of the field, *i.e.* to see only the largest scales, we propose a *cartographic representation with a two-level look-up table* (Figure 2a.), the look-up table defining an equivalence between the different values of the field and several colors or grey levels,
- to have a differential view of the field, *i.e.* to see mostly the smallest scales and gradients, we propose a *cartographic representation with a multi-level look-up table* (Figure 2b.), the number of levels acting as a filter to select a certain scale,
- to analyse the field both morphologically and quantitatively, for instance to compare the excitation of different structures, we propose a *cartographic representation with a continuous look-up table* (Figure 2c.). This representation is continuous in luminance, but presents two thresholds in order to discriminate the most excited structures (the negative and the positive having respectively a dark and a bright center) from the weakly excited ones; it also isolates the value zero as a discrete contour-line with the brightest luminance, in order to clearly distinguish it from the background since the shape of the zero contour-line conveys information on the smoothness of the field.

Figure 2a. Figure 2b. Figure 2c.

Cartographic representations

The choice of the representation mode depends, not only on what we are looking for, but on what we can actually obtain with it. For instance, if we choose a cartographic representation with a continuous look-up table, which varies from dark for the lowest values to bright for the highest values (which seems *a priori* to be the best default solution), the quality of the resulting display will depend critically on the smoothness and small scale isotropy of the field.
To illustrate this idea, let us consider several fields, each of a different nature:

- the *stream function field* which is very smooth, being a space integral of the velocity field, will appear completely blurred (Figure 3a.) because the eye

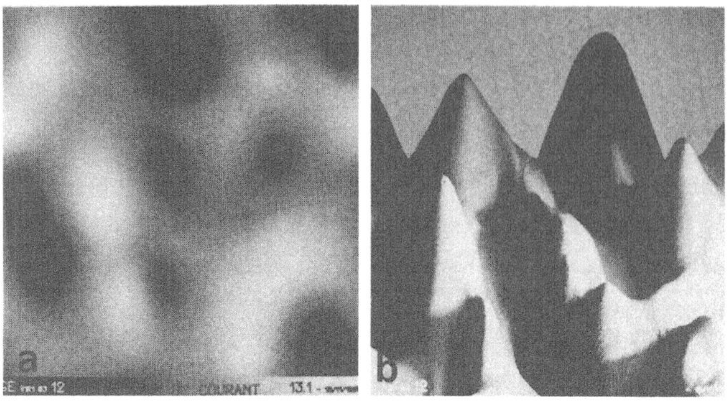

Figure 3a. Stream function field Figure 3b.

is unable to extract a shape when the luminance variation in space is too progressive; therefore only the perspective representation (Figure 3b.) gives a good picture of the stream function field and clearly shows its smooth spatial variation,

. the *divergence field*, which is usually not very smooth (Figure 4b.), being a derivative of the velocity field, presents the same problem because it is isotropic at small scales, and we have again the feeling that the picture is out of focus (Figure 4a.),

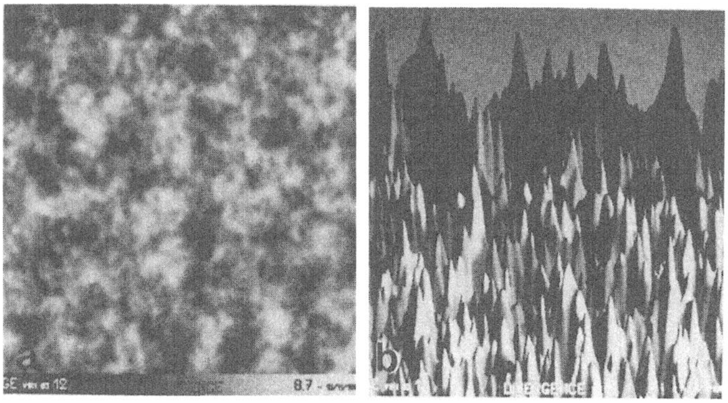

Figure 4a. Divergence field Figure 4b.

. the *vorticity field*, which is at the same time highly irregular, being also a derivative of the velocity field, but anisotropic at small scales (Figure 5b.), will on the contrary appear well focused because the luminance gradients are strong enough for the eye to figure out the shape of the field (Figure 5a).

In practice, it is often necessary to use several representation modes, because each of them reveals complementary informations about the underlying structure of the field. In some cases it may also be interesting to resort to animation. This can be done for two purposes: either to follow the time evolution of a turbulent flow in order to study its dynamics, or to explore a complex coherent structure by moving the

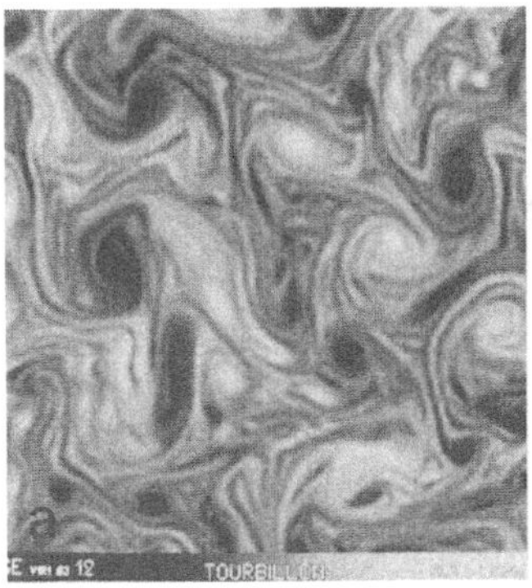
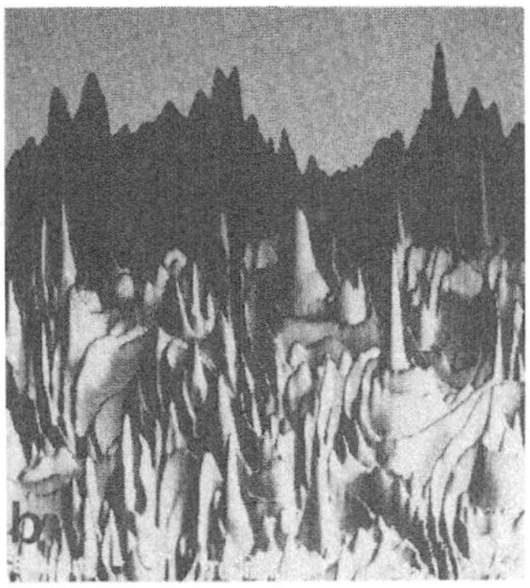

Figure 5a. Vorticity field Figure 5b.

observer around it or closer to enlarge small details. In both cases, the perspective representation is preferable, because the cartographic representation may be difficult to use if, for instance, the value range of the field changes very much with time (as for decaying turbulence), which then drastically reduces the number of levels available at each time step or obliges to redefine the look-up table from time to time.

2. Normalization of the color space

We think that the relevant approach to address the question of color in our case should be essentially pragmatic and bring a simplification to the problem of color choice. This concern leads us first to prefer the color *subtractive synthesis*, and the terminology employed by painters and printers. It is based on the mixture of pigments of primary colors Red-Yellow-Blue which, added together, give the color Black. We have discarded the *additive synthesis*, used with CRT terminals and based on the composition of beams of primary colors Red-Green-Blue which, added together, give the color white, principally because we are less familiar with it, being trained to manipulate colors with paints. The subtractive synthesis also allows us to choose our color palette *a priori*, using only color pencils, without requiring a color display which may be misleading for such tests, due to the present lack of good software tools to manipulate colors.

The second simplification we propose is a *reduction of the dimensionality of color space*. According to the principle of *visual trivariance* for the perception of colors discovered by Thomas Young -the retina contains three kinds of pigments: red, green, blue and the brain analyses any visible color by decoding their signals-, the color space is perceived as tridimensional. Two equivalent representations are then possible: the cube RYB (red-yellow-blue, Figure 6a.) and the cone LHS (luminance-hue-saturation, Figure 6b.). The transformation law between these reference frames has been empirically established by Maxwell from visual tests he performed himself. We adopt the *LHS representation*, because it is more natural from a

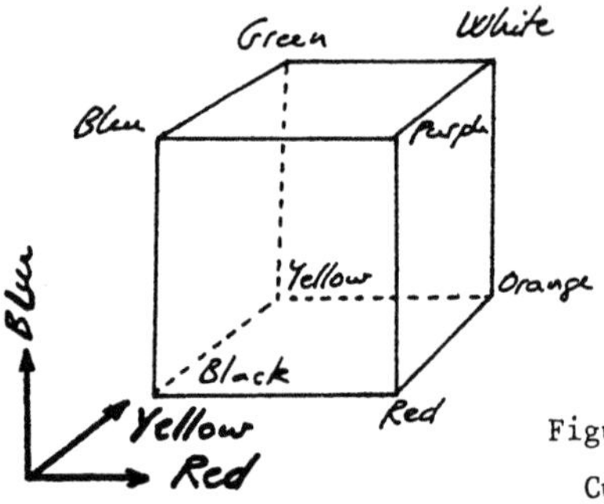

Figure 6a.
Cube RYB

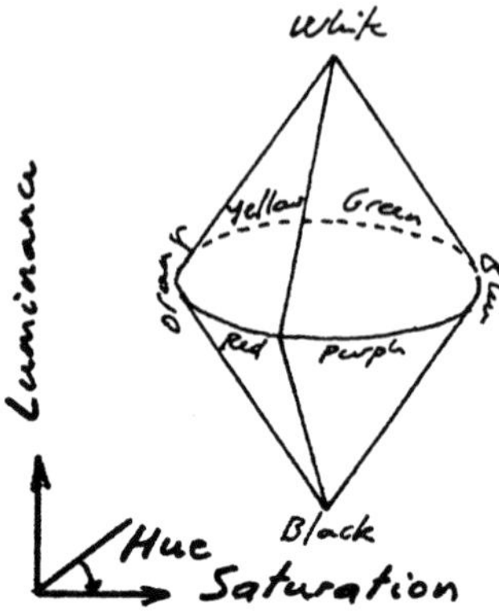

Figure 6b.
Cone LHS

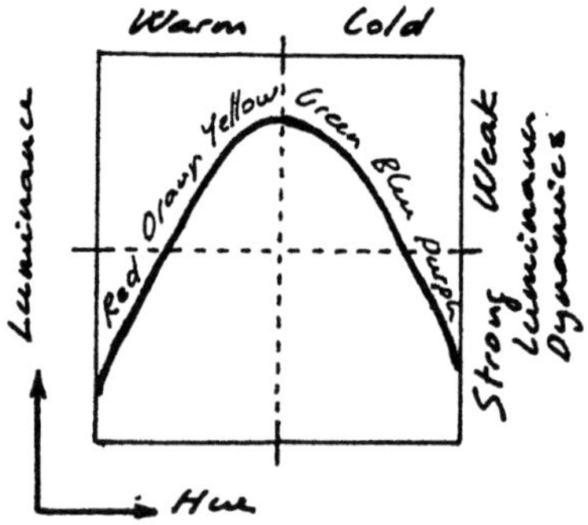

Figure 6c.
Square LH

perceptive point of view: *luminance* corresponds to the intensity of the perceived light, *hue* to the most excited ray in the spectrum of a given color and *saturation* to the width of the excited spectral band, a saturated (or pure) color being monochromatic and a desatured color (*i.e.* a pure color mixed with white or black) being broad band. But in several cases the distinction between luminance and saturation (*i.e.* for instance, the difference between dark light blue and bright obscure blue) is subtle to appreciate, especially with color displays. Therefore we have decided to use only saturated colors, which are the most vivid ones, in order to reduce the dimensionality of color space from three to two, considering only luminance and hue (Figure 6c.).

The third simplification we propose consists of *discretizing and normalizing the hue scale* into twelve basic hues, easy to identify and from which any other hue can then be positioned (Figure 7.). First of all, to guarantee the best possible objectivity in their determination we have defined them in term of the *Pantone standard* /5/, used worldwide by painters, printers and various industries to describe 593 different normalized colors (which has replaced the traditional Munsell Color Notation /6/). It is therefore possible to buy Pantone* papers and markers, that we may use, if necessary, to tune our display in order to obtain by comparison the exact color we want, or to test *a priori* our choice of palettes with those markers. Secondly, in order to simplify the memorization and communication process, we have named these twelve basic hues according to the *Color Naming System*

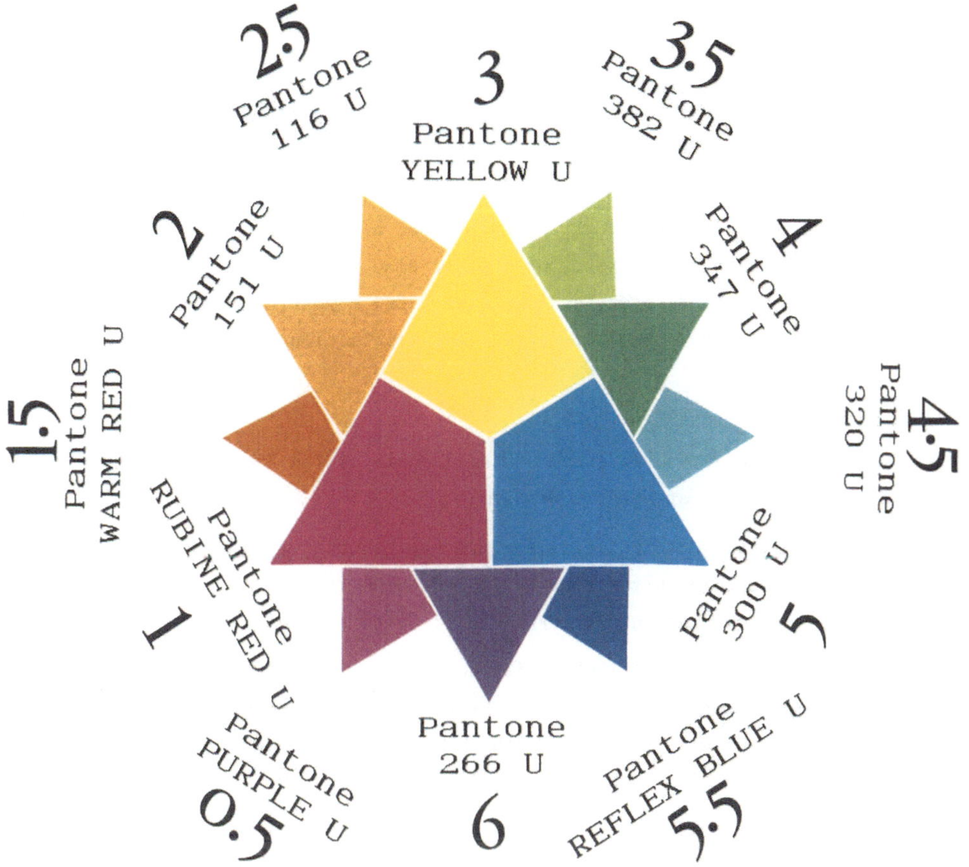

Fig. 7. Farge normalization of the hue scale

(see Appendix) recently designed for computer graphics applications /7/. Thirdly, we have defined a *numbering scheme from 0 to 6* to characterize the organization of hues, such as:

- the three subtractive primary colors red, yellow, blue are designated by the odd integers 1, 3, 5,
- the three subtractive secondary colors orange, green, purple by the even integers 2, 4, 6,
- any other color, obtained by mixing two adjacent colors h_1 and h_2, is determined by a real number intermediate between h_1 and h_2.

We have adopted the *natural order of the light spectrum*, after the popular mnemonic ROY G. BIV, for Red (1), Orange (2), Yellow (3), Green (4), Blue (5), Indigo, Violet, neglecting Indigo, replacing Violet by Purple (6) and adding the non-spectral (*i.e.* non mono-chromatic) color Purple Red (0.5) to give a circular structure to the hue space (solution already proposed by Newton). We can then define the following relations:

- $h < 3$ for warm colors,
- $h > 3$ for cold colors,
- $h \in [2,4]$ for colors having a weak luminance dynamics, *i.e.* which are difficult to brighten or darken without distorting their hue, as it is the case for Yellow (3) which easily tends to white or brown but not for Blue (5) or Purple (6) (Figure 6c.).

This normalization of color space, and the associated numbering scheme we propose, will help us to write algorithms for the automatic selection of color palettes.

3. Choice of color palettes

Following Johannes Itten /2/, one of the leaders of the Bauhaus school, we can define seven types of color contrasts:

- the *cold-warm contrast*, which is maximal between the warmest color Red (1) and the coldest color Green Blue (4.5) and which is minimal between neutral colors as the combination Purple Red (0.5)-Green Yellow (3.5),
- the *complementary contrast*, which is encountered between a given color C and its complementary C_c, defined as the mixture of all other spectral colors besides C and which corresponds to the hue h_c diametrically opposed on the hue circle (Figure 7.) such as $h_c = (h+3) \bmod 6$,
- the *simultaneous contrast*, which comes from the physiological fact stating that, if an association of colors is not in complementary equilibrium (*i.e.* such as if we mix them we obtain the color grey), the eye tends to distort the colors in order to recreate this equilibrium, therefore a given color is perceived differently depending on the surrounding colors (Figure 8.),
- the *quantity contrast*, which is function of the area covered by a given color weighted by its luminance, the smallest quantity contrast being obtained if we respect the following area ratio: yellow 3, orange 4, red 6, green 6, blue 8 and purple 9 (Figure 9.).
- the *luminance contrast*, which exists between saturated colors having different luminance (Figure 10.), the highest luminance contrast corresponding to the combination Yellow (3)-Purple Blue (5.5) and the lowest to the combination Green (4)-Red Orange (1.5),
- the *hue contrast*, which is maximal between primary colors Red (1)-Yellow (3) -Blue (5) and decreases for secondary colors Orange (2)-Green (4)-Purple (6),
- the *saturation contrast*, which is strongest when we associate a saturated (vivid) color with a desaturated (whitened or blackened) color.

Fig. 8. Simultaneous contrast

Fig. 9. Quantity contrast

Fig. 10. Luminance contrast

In practice, any association of different colors combines several kinds of contrasts. Our aim is to be able to recognize them and consciously decide to use or avoid them. For instance, to display turbulent fields we employ the cold-warm contrast in order to easily distinguish *negative values*, which always correspond to a *cold hue*, and *positive values*, which are identified with a *warm hue*. We also

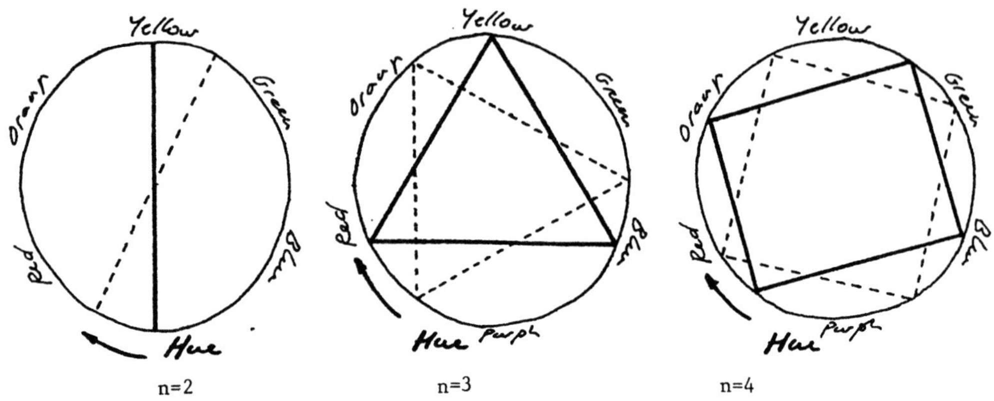

$h_2 = (h-3) \bmod 6$ $h_3 = (h-2) \bmod 6$ $h_4 = (h-1.5) \bmod 6$

i.e. more generally to find the ith hue in complementary equilibrium with n-1 others:

$$h_{n,i+1} = [h_i - (6/n) \bmod 6] \text{ for } i=1,n$$

Figure 11.
Complementary equilibrium between n hues

avoid simultaneous contrast in two ways:

- by separating each color with a *grey background* which will correspond to the values around zero,
- by choosing n different hues to characterize a given field, making sure that they are in *complementary equilibrium*, *i.e.* correspond to the colors associated with the n submits of a polygon inscribed inside the hue circle (Figure 11.).

If we apply these rules to the cartographic representation mode with a continuous look-up table presenting two thresholds (Figure 2c.), we choose n=3 and attribute a cold hue to the values smaller than the first threshold and a warm hue to those larger than the second threshold. Thus, we combine the possibilities of *local analysis*, because the continuous luminance variation allows us to follow small variations (for instance gravity waves), and *global analysis*, because the two thresholds, enhanced by the cold-warm contrast, clearly separate the most excited structures (for instance coherent isolated vortex) from the background turbulent field, where the enstrophy (for two-dimensional flows) or energy (for three-dimensional flows) cascade is acting and producing characteristic vorticity filaments, which are displayed in grey with the zero contour-line characterized by a third hue in complementary equilibrium with the two others (Figure 12.).

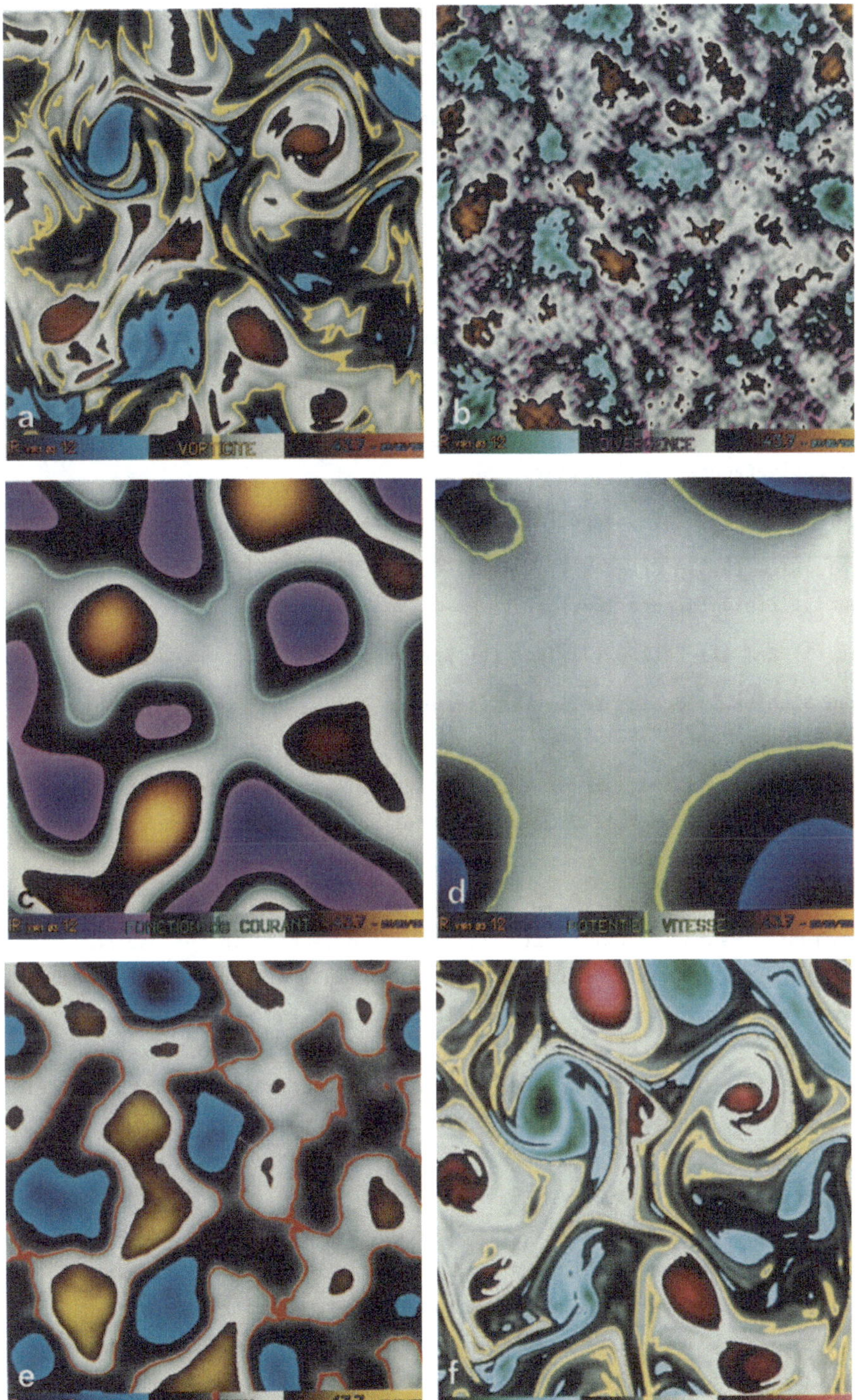

Fig. 12. (a) vorticity; (b) divergence; (c) stream function; (d) velocity potential; (e) pressure; (f) potential vorticity

Therefore, the way we have defined our color palettes for the display of turbulent scalar fields is such that *all information is coded with a continuous variation of luminance*, in order to avoid any loss of information if we have to publish them in black and white, and is enhanced by the *superimposition of discrete hues*, in order to easily distinguish one given field from the others. Each field is identified by three hues *in complementary equilibrium* (*i.e.* such as they correspond to the three submits of an equilateral triangle inscribed in the hue circle, Figure 11.) , with a cold hue ($h>3$) for the negative values, a warm hue ($h<3$) for the positive ones and any complementary hue for the value zero, which gives the following combinations (Negative-Zero-Positive):

- [Blue (5)-Yellow (3)-Red (1)] for the vorticity field (Figure 12a.),
- [Green Blue (4.5)-Purple Red (0.5)-Orange Yellow (2.5)] for the divergence field (Figure 12b.),
- [Purple (6)-Green (4)-Orange (2)] for the stream function field (Figure 12c.).
- [Purple Blue (5.5)- Green Yellow (3.5)-Orange Red (1.5)] for the velocity potential field (Figure 12d.),
- [Blue (5)-Red (1)-Yellow (3)] for the pressure field (Figure 12e.),
- [Green Blue (4.5)-Orange Yellow (2.5)-Purple Red (0.5)] for the potential vorticity field (Figure 12f.).

Now, having adopted this color convention, we are able to identify at the first glance each of these fields, which is very useful if we want to compare them.

Conclusion

High-resolution raster graphics will soon become essential to analyse and communicate the results from numerical experiments. This will especially be true for studying turbulent fields, due to the complexity and the variety of their spatial structure. Such a technique will probably induce new possibilities of investigation, still unknown, concerning the morphological analysis of those fields in term of pattern recognition and local structure decomposition. Therefore, it will bring a new approach, complementary to the spectral analysis which is on the contrary non-local.

We foresee that laboratory experiments, likewise numerical experiments, will tend to use the same digital techniques to display their results. This will then promote and facilitate their intercomparison. But we think that high-resolution raster graphics will not fully play its role, unless a normalization of representation modes and color palettes will be adopted, at least within a given discipline. We hope the normalization proposed here will bring some awareness about this need and could be considered as a first step for further discussions.

If today most people are able to easily understand a road map, it is only because we are accustomed to a certain representation mode and color convention. Imagine the difficulty we would face if we had to adjust to a different coding everytime we use a new map. Now imagine the non sense of having to interpret maps where the legend is not even displayed ! Unfortunately, this is often the case today with high-resolution raster graphics. Without the adoption of a normalization, we run the risk that they will mostly be used as advertising means to attract granting agencies rather than as scientific tools to analyse, compare and communicate meaningful results.

REFERENCES

/1/ G. Murch: Visual accomodation and convergence to multi-chromatic display terminals, Proceedings of the Society for Information Display, 1983

/2/ J. Itten: Art de la couleur, Dessain et Tolra, 1984

/3/ W. Kandinsky: Cours du Bauhaus, Collection Médiations, Denoël, 1975

/4/ J.D. Foley and A. Van Dam: Fundamentals of interactive computer graphics, Addison-Wesley, 1982

/5/ Pantone Color Formula Guide, 18th edition, Pantone Inc., 1985

/6/ A.H. Munsell, A color notation, Munsell Color Inc., 1981

/7/ T. Berk, L. Browston, A. Kaufman: A new color-naming system for graphics languages, Computer Graphics and Applications, May 1982

APPENDIX Correspondance between Farge Numbering Scheme Color Naming System and Pantone* Color Standard

Farge Numbering	Color Naming System	Pantone* Standard
0.5	Purple Red	Purple U
1	Red	Rubine Red U
1.5	Orange Red	Warm Red U
2	Orange	151 U
2.5	Orange Yellow	116 U
3	Yellow	Yellow U
3.5	Green Yellow	382 U
4	Green	347 U
4.5	Green Blue	320 U
5	Blue	300 U
5.5	Purple Blue	Reflex Blue U
6	Purple	266 U

* Pantone is a trademark of Pantone Inc.,
55 Knickerbocker Road, Moonachie, New Jersey 07074 USA.

Acknowledgments

The computing has been done on the Cray 1 of C2VR (Centre de Calcul Vectoriel pour la Recherche), Palaiseau, using as front-end the IBM 3090 of CIRCE (Centre Interdisciplinaire Régional de Calcul Electronique), Orsay. High-resolution raster graphics has been done at LACTAMME, Ecole Polytechnique, in collaboration with Jean-François Colonna. Nüzhet Dalfes helped with the editing. Marie-Christine Cally typed the text. I express them all my gratitude.

Large Eddy Simulations of Isotropic or Stably-Stratified Turbulence

J.-P. Chollet, O. Metais, and M. Lesieur*

Institut de Mécanique de Grenoble[1], B.P. 68,
F-38402 Saint Martin d'Hères Cedex, France

A spectral formulation of a subgrid scale model is used to parameterize the smallest scales of the velocity and temperature fields of a three dimensional turbulence. This model, primarily designed for homogeneous isotropic turbulence, is extended to homogeneous stratified turbulence in order to study the initial stage of the decay of a three dimensional turbulence in a stable stratification density gradient.

1 .INTRODUCTION

Turbulent flows cannot be studied only with experiments since the values of the parameters are severely limited in laboratory or natural phenomenas conditions. Statistical models depend upon arbitrary closure assumptions and do not provide any information about instantaneous and local values of the variables. In order to obtain detailed informations about the velocity fields, the Navier Stokes equations have to be numerically integrated. Direct (or full) numerical simulations, taking a complete account of all the scales up to the molecular disspative scales are limited to values of the Reynolds numbers of 40 to 100. High Reynolds number turbulence cannot be computed completely; fluid dynamicists generally choose a detailed description of the large scales dynamics through a Large Eddy Simulation (LES), and therefore have to be satistified with a rather crude description of the small scales. These small scales are modeled through a subgrid scales (SGS) model, generally formulated in terms of eddy-quantities (eddy viscosity and diffusivity). These quantities are well-defined in isotropic turbulence, their use in more general homogeneous situations (for instance stratified turbulence) can be justified, to some extent.

---

* also National Center for Atmospheric Research, Boulder, Colorado

[1] Institut National Polytechnique et Université Scientifique Technique et Médicale de Grenoble, Unité associée au CNRS

First recalling its formulation in terms of an eddy viscosity and an eddy diffusivity, the SGS model is used in a calculation of freely decaying isotropic turbulence. Then, the Large Eddy Simulation, still with the same SGS modelling, is used to study the initial stage of the decay of an isotropic three dimensional turbulence in a stable stratification density gradient.

2 .DERIVATION OF A SUBGRID SCALE MODEL

Following Kraichnan [5], we have derived an eddy-viscosity in spectral space, using a statistical theory of the turbulence, the Eddy Damped Quasi Markovian approximation (EDQNM). The transfer of energy from large ($k < k_c$) to small (or subgrid) scales ($k > k_c$) is formulated in terms of an eddy viscosity which is shown to be proportional (to the lowest order) to $[E(k_c,t))/k_c]^{1/2}$. k_c is the cutoff wavenumber which corresponds to the smallest scales explicitely computed in the numerical simulation.

Using the same method, we have derived an eddy-diffusivity to model the transfer of passive scalar variance (for instance, temperature or density fluctuations) from large to subgrid scales.

The eddy viscosity is written as :

$$\nu_t(k|k_c,t) = \nu_t^+(k/k_c).[\frac{E(k_c,t)}{k_c}]^{1/2} \tag{1}$$

and the eddy viscosity as :

$$D_t(k|k_c,t) = D_t^+(k/k_c).[\frac{E(k_c,t)}{k_c}]^{1/2} \tag{2}$$

The evolutions of the eddy coefficients ν_t^+ and D_t^+ were computed with the $EDQNM$ statistical theory from $k^{-5/3}$ energy and scalar variance spectra and are plotted on Fig. 1. These coefficients were obtained from considerations about statistical closures and cannot be considered as arbitrary constants to be tuned from numerical tests of Large Eddy Simulations. The eddy Prandtl number obtained from these eddy viscosity and diffusivity has a fairly realistic value of 0.6.

Eddy viscosity and diffusivity of Fig.1 exhibit strong cusps in the neighbourhood of the cutoff k_c; these cusps whose behaviour is discussed in [1] contribute to the SGS model as an high order dissipativity, nevertheless in these three dimensional calculations, this strong dissipativity

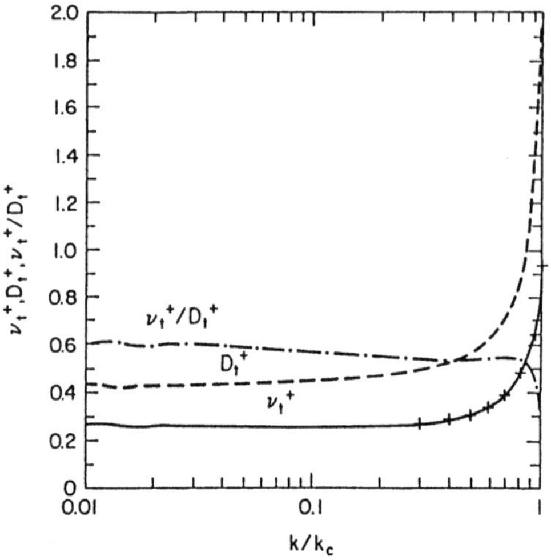

Figure 1: Eddy-viscosity, Eddy-diffusivity and Eddy Prandtl number from a statistical $(EDQNM)$ calculation on inertial and inertio-convective $k^{-5/3}$ spectra (from [2])

is significantly dominated by the eddy viscosity effects contrary to what happens in the approximation of a two dimensional turbulent flow field.

3 .L.E.S. OF ISOTROPIC TURBULENCE

From the formulations (1) and (2), the energy and scalar variance transfers towards small and dissipative scales can adjust themselves to the evolution of the field since the eddy quantities depend upon the value of the energy, $E(k_c, t)$, at the cutoff . Therefore, this parameterization of the small scales effects is especially convenient to model time -evolving flows.

These subgridscale procedures are easily introduced in a pseudo-spectral code to compute the large scales of an isotropic three dimensional turbulence, in a cubical box with periodical boundary conditions. The eddy -viscosity and -diffusivity (1) and (2) are shown to allow a correct evolution in time of the energy and passive scalar spectra for freely evolving turbulence; inertial and inertial- convective range are correctly reproduced, up to the cutoff scale ([2] and [3]). They also give kinetic energy and temperature variance decay laws which are in good agreement with the phenomenological or the stochastic closures predictions, in particular the Saffman $t^{-1.2}$ law, in spite of the fact that, due to the limited size of the computation domain, the large scales cannot increase continuously.

The large eddy simulations, as above mentioned, correspond to an infinite value of the Reynolds number based on the r.m.s. velocity of the most energetic scales and the molecular viscosity. In order to take a more exact account of the Reynolds number, we could use a full coupling procedure as mentioned in [2]. In such a procedure the evolution of the energy and scalar (e.g. temperature, density) variance spectra in the subgrid scale range is calculated with a statistical model ($EDQNM$ for instance); such a calculation is significantly more time -consuming than when using the eddy quantities (1) and (2) and the Reynolds number has to be low enough in order to make a real difference between the two procedures. In a nonhomogeneous flow, low values of a local Reynolds number can appear, for instance near rigid boundaries; the use of a full coupling procedure in such a case would be quite intricate since the turbulence is no more isotropic which makes any statistical ($EDQNM$ for instance) calculations very difficult to handle efficiently and acccurately.

The main assumption which was used in the derivation of the SGS model (just like other models) is the existence of inertial ranges, which means that these models do not hold for rapidly varying turbulence. The rate of subgrid scale transfer is determined by the energy $E(k_c, t)$ at the cutoff wave number number. We can consider that the eddy viscosity and diffusivity are valid as long as the spectra behave like $k^{-5/3}$ in the neighbourhood of the cutoff (around k_c) and in the small scales, in such a way that this subgridscale modelling does not require isotropy for the largest scales. Consequently, these procedures can be used in homogeneous nonisotropic flows such as stably-stratified flows [7] or turbulence with rotation [10]. Extensions can be imagined for free layers (jets, mixed layers) and it must be noticed that our procedures would not be any more identical to Smagorinsky's ones for nonhomogeneous turbulence. Extensions to nonhomogeneous flows such as boundary layers along rigid boundaries are more arguable and would request the derivation of local (in space) values of the eddy quantities.

4 .L.E.S. OF STABLY-STRATIFIED TURBULENCE

Stably stratified density fields influence many natural atmospheric or oceanic flows; they can also be considered as relatively simple cases to study some basic mechanisms involved in industrial chemically reacting flows and combustion processes. As compared to the previous section, the turbulence is still three dimensional, homogeneous and freely-decaying, but it is no more isotropic.

Direct numerical simulations have been carried out by Riley et al. [9] but they were necessarily limited to low values of the Reynolds number with a significant damping effect of the molecular dissipation. In order to reduce the linear transfers due to the molecular viscosity and (scalar) diffusivity and to allow longer periods of time for the evolutions, the calculation is restricted, here, to the largest scales through a Large Eddy Simulations (LES) with the subgrid scale above- mentioned.

The evolutions of velocity and density fields are described in the assumption of the Boussinesq equations

$$\frac{\partial \underline{u}}{\partial t} + \underline{u}\nabla \underline{u} = -\nabla p + \theta \underline{z} + (\nu + \nu_t(k|k_c,t))\nabla^2 \underline{u} \tag{3}$$

$$\nabla . \underline{u} = 0 \tag{4}$$

$$\frac{\partial \theta}{\partial t} + \underline{u}\nabla\theta + N^2 w = (D + D_t(k|k_c,t))\nabla^2\theta \tag{5}$$

where ρ_0, the average density of the fluid across the whole fluid layer, has been taken equal to one (in the pressure gradient), ρ' is the density fluctuation with respect to the mean density profile $\bar{\rho}(z)$, and $\theta = -\rho'/\rho_0$ is a normalized density fluctuation. N, the Brunt Vaisala frequency,is assumed to be constant in a stably stratified situation and given by :

$$N^2 = -\frac{g}{\rho_0}\frac{d\bar{\rho}}{dz} \tag{6}$$

A Froude number is defined as

$$F = \frac{u}{Nl} = (\frac{l_R}{l})^{2/3}$$

where $l = \frac{u^3}{\epsilon}$ is the integral scale of the turbulence and $l_R = \frac{\epsilon}{N^3}^{1/2}$ is a scale characteristic of a balance between inertial and buoyant forces.

Initial conditions are taken isotropic, and the free decay is studied. Let us recall some phenomenology from [9] and [6]; when the time evolves, the integral scale of the turbulence increases while the buoyancy scale decreases, therefore the turbulence should collapse when these two scales meet each other, and this for a value of the Froude number of the order of 1. Since, in isotropic turbulence, $u/l \propto t^{-1}$, the Froude number behaves like $(Nt)^{-1}$ and reaches the value 1 for a time of the order of N^{-1}.

We have developed a Large Eddy Simulation of the Boussinesq equations (3),(4),(5) using a pseudo-spectral code, taking advantage of the homogeneity of the velocity and density fields. The resolutions are of 32^3. The subgrid scales are simulated with the eddy -viscosity and - diffusivity (1) and (2). The initial Froude number is of the order of 3, close to the experimental value of [4]. This parameterization should be valid at least for the preliminary phase of the decay, as long as the integral scale is smaller than the buoyancy scale.

Fig.2 shows the energy decay in the stratified case as compared to the non stratified case; the decay of the total energy (kinetic + potential)

$$\frac{1}{2}(<\bar{u}^2> + N^{-2}\theta^2)$$

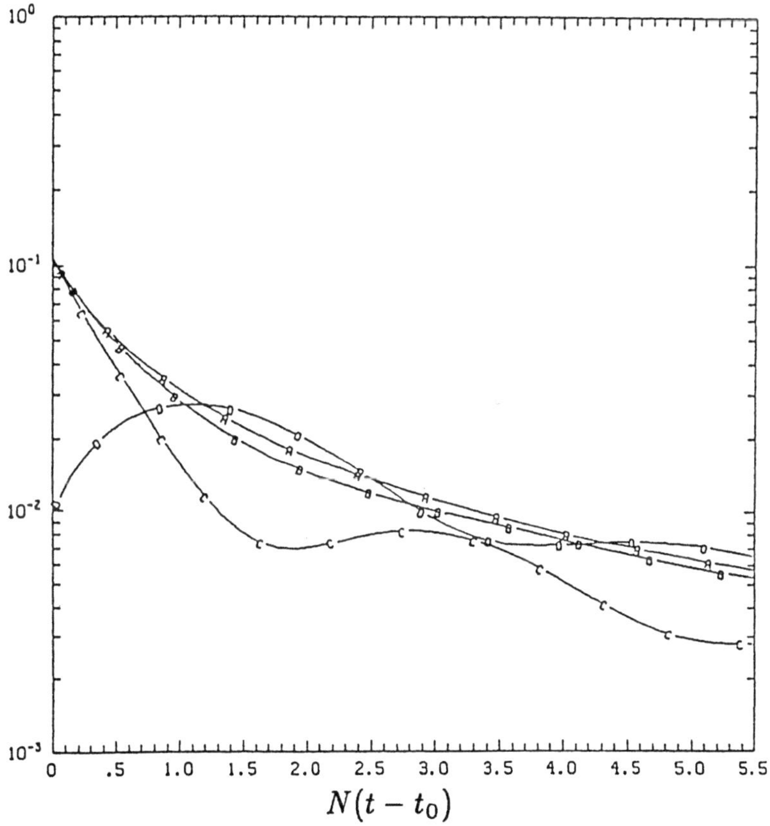

Figure 2:Time-evolutions:

• in the NON STRATIFIED case (A : half-horizontal kinetic energy)

• in the STRATIFIED case (B : vortex kinetic energy, C : wave kinetic energy, D : potential energy) is almost identical in the two cases, approximately like $t^{-1.18}$, giving for this range of time (up to $5N^{-1}$) no evidence of any trend towards two -dimensionality. More precisely the horizontal energy (close to the"vortex" energy in the terminology of [9]) is not affected by the stratification. The vertical kinetic energy (close to the "wave" kinetic energy in the terminology of [9]) exhibits a periodical exchange with the potential energy. The latter is initially zero and grows before undergoing an oscillating decay. These oscillations, of period $\approx \pi N^{-1}$ seem to correspond to gravity waves propagating horizontally: indeed, from the dispersion relation of these waves, the fluctuation $\psi(\underline{k}, \underline{x}, t)$ characterizing the propagation of the wave will be $\propto \sin Nt$ only if the vertical component of $\underline{k}$ is 0; then, the energy of the wave $\propto |\psi|^2$ will be $\propto \sin^2 Nt \propto sin(2Nt + \phi)$, whose period is precisely π/N.

These waves affect essentially the large scales (i.e. low wave numbers $k < k_i$, where $k_i \approx l^{-1}$ characterizes the energy containing eddies): this is clear in particular on the various energy spectra calculated in ref [7]. While the vortex kinetic energy spectra are, in the stratified case, negligibly different from their non stratified counterpart, the wave and potential energy spectra display strong oscillations in their infrared region. All these spectra display however the same Kolomogorov $k^{-5/3}$ cascade close to the ultra violet cutoff wave number k_c.

The calculations were checked by a comparisons with the experiments [4]. The comparison of energy decay, in Fig.3, is quite satisfactory; moreover, it suggests that, as mentioned above, the oscillations of the kinetic vertical energy seems to be artifically enhanced in the numerical calculation.

The same comparison is done in Fig.4 , in terms of characteristic scales : the integral scale l of the turbulence decreases when the buoyancy scale l_R increases. As predicted by the above phenomenology, these two scales collapse at the time N^{-1}. However there is not yet any tendency towards horizontal turbulence, as attests the finite isotropic like energy decay which keeps on up to $5N^{-1}$. This evidence is confirmed by the examination of the spectra of [7] which do not show any significant inverse energy transfer characteristic of two-dimensional turbulence. Laboratory experiments of the collapse of wakes downstream of an obstacle in a stratified flow ([8]) suggests that the collapse could take place at times slightly higher.

We have also written a 64^3 resolution code[1], which is too heavy in

[1] This code, designed for a Cray 1S machine, is run out of core.

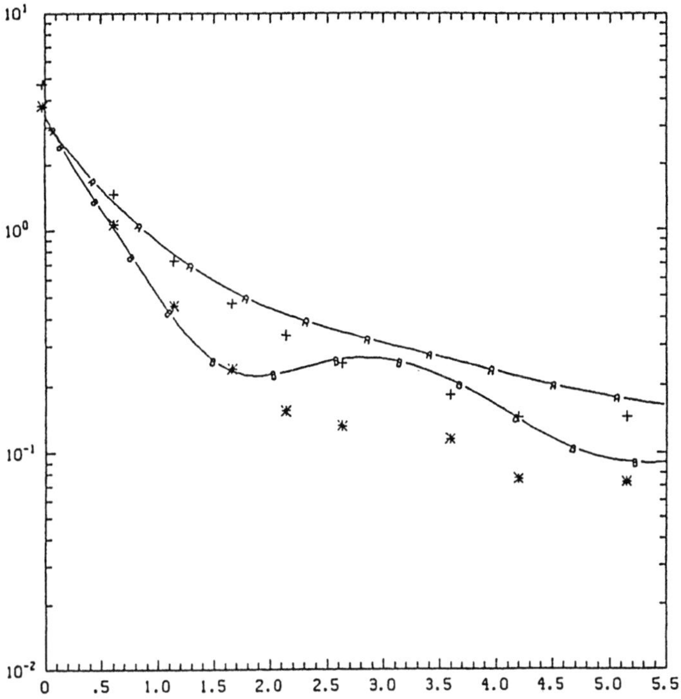

$N(t - t_0)$

Figure 3: Comparisons with experiments by ITSWEIRE, HELLAND and VAN ATTA [4]; time-evolutions of :

- the kinetic energy of one horizontal velocity component (A : computed results ; + : experimental results)
- the kinetic energy of the vertical velocity component (B : computed results ; * : experimental results)

order to allow the same preceeding decay calculations. We have then used it to examine some fundamental dynamical interactions between the vortex and wave components of the flow. The initial state is purely "vortex", that is composed of horizontal slices of fluid vertically decorrelated. This initial state induces strong inflexional vertical gradients of the horizontal velocity. Then, two cases must be considered: if the stratification is too strong, the local Richardson number will be everywhere higher than the critical value 0.25 predicted by the theory of the stratified Kelvin Helmholtz instability, and no significant vertical fluctuations will be able to develop. Then the initial vortex field will not evolve, except for some gravity waves propagating on the isopycnal surfaces. If, on the contrary, the stratification is weaker, Kelvin Helmholtz instability will locally develop, leading to gravity waves which will break

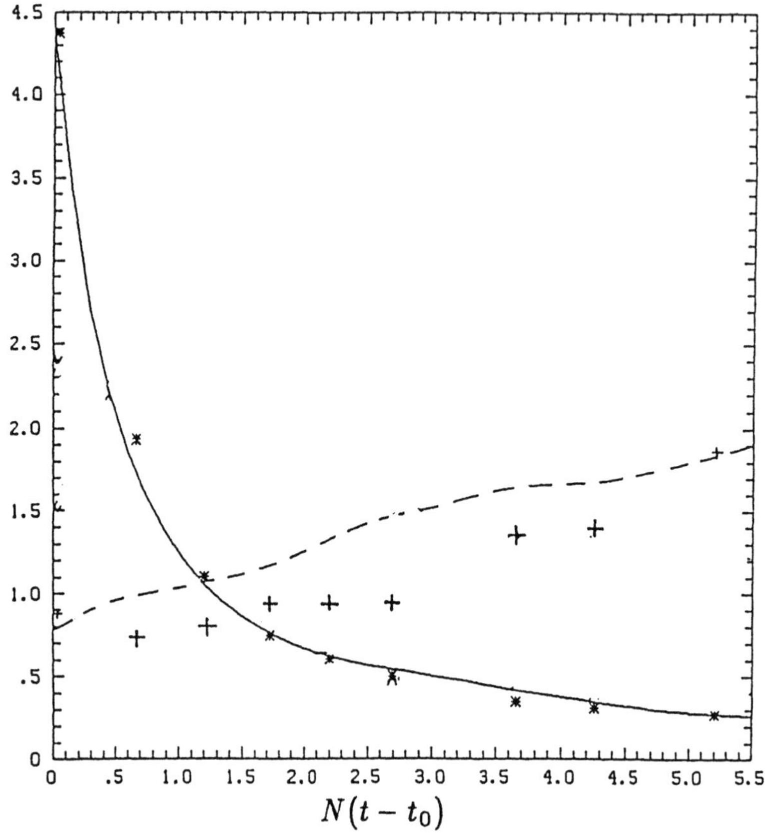

Figure 4: Comparisons with the experiments of ref [4], time evolutions of:
integral scales (+ : experimental results; -- computed results)
buoyancy scales ($*$: experimental results; — computed results)

up, giving rise to small scale three dimensional turbulence. Such a phenomenon was actually obtained in our simulations with the 64^3 code. This is an example of conversion of vortex energy (through the initial vertical shears) into wave energy which, here, is under the form of small scale three dimensional turbulence.

5 .CONCLUSIONS

Computers devoted to scientific numerical simulations are getting more and more powerful in memory size and computing power; new technics enhance the efficiency of these machines (vectorization, multitasking,...). Nevertheless, full direct numerical simulations of high Reynolds numbers three dimensional turbulence remain beyond the capabilities of these computers. Subgridscale modelling procedures have still to be used in order to increase the resolution in the range of the largest scales. These technics of large eddy simulations could also be used on smaller computers (e.g. work stations) even with low space resolution but taking advantage of a greater availability for on-line interactions between the flow calculations and the graphic display of results. Besides, the filtering of the small scales by reducing the amount of informations can make the analysis of some results much easier.

Our isotropic subgrid scale model is conveniently written in spectral formulations and then easily introduced into pseudo-spectral codes. It seems to work satisfatorily, at least for the early stage of the time-evolution, even in nonisotropic flows such as in stably stratified situations when starting from an isotropic initial field. Their extension to nonhomogeneous cases is subject to the possibility of local evaluation, in the physical space, of the spectral eddy quantities introduced above.

From our work, it is clear that the problem of the collapse of three dimensional turbulence in a stably stratified fluid is not yet resolved, even with the increased of the Reynolds number permitted by the subgridscale modelling. Numerical integrations for longer periods of time and higher resolutions are then needed.

ACKNOWLEDGEMENTS

This work has been supported by the D.R.E.T. under contract 85/105 and the ATP-CNRS "Fluides Geophysiques". Computing time was allocated by the Centre de Calcul vectoriel pour la Recherche and by the National Center for Atmospheric Research[2].

REFERENCES

1. J.P. Chollet, in Macroscopic Modelling of Turbulent Flows, proceedings of a Workshop held at INRIA, Sophia-Antipolis, Lecture Notes in Physics 230, Springer -Verlag (1984)

2. J.P. Chollet, in Turbulent Shear Flows 4, Selected Papers from the fourth International Symposium on Turbulent Shear Flows, Karlsruhe, Springer-Verlag (1985)

3. J.P. Chollet, M. Lesieur, La Météorologie, VIe série, N 29 et 30, (1982)

4. E.C. Itsweire, K.N. Helland, C.W. Van Atta, preprint (1985)

---

[2] NCAR is sponsored by the National Science Foundation.

5. R.H. Kraichnan, J. Atmos. Sci., 33 , 1521-1536 (1976)
6. D.K. Lilly, J. Atmos. Sci., 40, 759-761 (1983)
7. O. Métais, Proc. of Turb. Shear Flow Vth, Cornell University,(1985)
8. Y.H. PAO, Boundary-Layer Met., 5, 177 (1973)
9. J.J. Riley, R.W. Metcalfe, M.A. Weissman, M.A. 1981, Proc. of the AIP Conf. n.76: "Nonlinear Properties of Internal Waves", ed B.J. West, 79- 112, (1981)
10. Ph. Roy, these de doctorat es Sciences, Univ. de Nice (1986)

3D Simulation of 2 Length Scales Turbulent Flows by Homogenization

C. Bègue, T. Chacón, F. Ortegón, and O. Pironneau

INRIA, Domaine de Voluceau-Rocquencourt, F-78153 Le Chesnay Cedex, France

A two equations turbulent model is proposed for flows with two length scales involving the mean flow, the mean kinetic energy and the mean helicity. The system is obtained by an asymptotic expansion with respect to the small scales of the velocity and pressure in the Navier-Stokes equations. The performances of the model are compared with a direct 3D simulation of Poiseuille flows and good agreements are obtained.

Notation

Let be u : $R^3 \rightarrow R^3$, R : $R^3 \times R^3 \rightarrow R^3$, A, B $\in R^3 \times R^3$, then we note

$$(\nabla u)_{ij} = \frac{\partial u_j}{\partial x_i} = u_{j,i}, \quad (\nabla .R)_i = \frac{\partial R_{ij}}{\partial x_j}$$

$$A : B = A_{ij} B_{ij}$$

with the summation convention over repeated indices.

$< u >$ denotes the statistical average of u.

O. Introduction

Averaging the Navier-Stokes equations (N-S) is a general technique used in turbulence modelling (see [1] for example); to explain it briefly, consider the dimensionless N-S equations for 3D incompressible flows ; Reynolds' approach to turbulence modelling is to consider the N-S equations with random initial data :

$$\begin{gathered} u_{,t} + (u.\nabla)u + \nabla p - \nu \Delta u = 0, \quad \nabla .u = 0 \\ u(x,0) = u^o(x) + u^{o\prime}(x), \quad u|_\Gamma = u_\Gamma \end{gathered} \tag{1}$$

where ν is the inverse of the Reynolds number, $u^o(x)$, is the mean initial velocity field, and $u^{o\prime}$ the random part, Γ the boundaries.

Splitting u into its mean part **u** and its random oscillating part u' :

$$u = \mathbf{u} + u' \tag{2}$$

and averaging, the N-S equations yield the well-known Reynolds equation

$$\begin{gathered} \mathbf{u}_{,t} + (\mathbf{u}.\nabla)\mathbf{u} + \nabla p - \nu \Delta \mathbf{u} + \nabla . R = 0, \nabla .\mathbf{u} = 0 \\ \mathbf{u}(x,0) = u^o(x). \end{gathered} \tag{3}$$

Here R is the Reynolds' stress tensor i.e. the average of $u'_i u'_j$. In the $k - \epsilon$ model, we have

$$R_{ij} = -\nu_T(\mathbf{u}_{i,j} + \mathbf{u}_{j,i}) \tag{4}$$

(i.e. R is a diffusion tensor). Here $\nu_T \geq 0$ is an eddy viscosity of the form

$$\nu_T = a\frac{k^2}{\epsilon}$$

where a > 0 is a numerical constant, k(x,t) is the turbulent kinetic energy (i.e. the average of 1/2 u'^2) and $\epsilon(x,t)$ is the rate of turbulent viscous energy, (the average of $\nu(u'_{i,j} + u'_{j,i})^2$). The functions k and ϵ are given by the system of advection equations :

$$k_{,t} + \mathbf{u}\nabla k - a\frac{k^2}{\epsilon}\nabla\mathbf{u} : (\nabla\mathbf{u} + \nabla\mathbf{u}^T) + \epsilon - b\nabla.(\frac{k^2}{\epsilon}\nabla k) = 0$$

$$\epsilon_{,t} + \mathbf{u}.\nabla\epsilon + \epsilon ck\nabla\mathbf{u} : (\nabla\mathbf{u} + \nabla\mathbf{u}^T) - \frac{\epsilon^2}{k} - d\nabla.(\frac{k^2}{\epsilon}\nabla\epsilon) = 0, \tag{5}$$

a,b,c, and d are numerical constants. Usual boundary conditions are :

$$k = \frac{u^{*2}}{\sqrt{a}}\epsilon = \frac{u^{*3}}{K\delta}$$

where u* stands for the friction velocity, K for the Von Karman's constant and δ is the distance to the wall.

The model seems to have been derived first for turbulent boundary layer (Rota, Launder, Spalding...see [1] for example) where the mean flow is stationnary. It is now widely accepted in industry as a not too unreasonable model, at least for boundary layers, channel flows and jets i.e. when the flow has a stationnary mean part and a turbulent part in a quasi-equilibrium. However it is not so good and difficult to implement numerically for flows behind obstacles. On the hypothesis for the (heuristic) derivation of the $k - \epsilon$ model, we make the following comments (see [3] for more details):

- the equation for k would be correct if Reynolds' hypothesis is correct and u', $\nabla p'$, $\nabla |u'|^2$ are uncorrolated.

- Reynolds' hypothesis is a consequence of frame invariance in 2D once it is assumed that $< u'_i u'_j >$ is a function of k, ϵ and $\nabla u + \nabla u^T$.

In 3D it is not true ; there one has

$$< u' \otimes u' > = \alpha I + \beta(\nabla u + \nabla u^T) + \gamma(\nabla u + \nabla u^T)^2$$

where α, β, γ are functions of k, ϵ, $tr(\nabla u + \nabla u^T)^2$ and $det(\nabla u + \nabla u^T)$.

- The equation for ϵ is controversial and subject to variations.

- Together with the boundary conditions which come from a matching with the logarithmic sublayer, the $k - \epsilon$ model is numerically stiff.

Turbulence modelling by homogenization

Here we study the turbulence model derived in [2] using homogenization techniques : the existence of two length scales is assumed and an asymptotic expansion with respect to the length scale ratio is done. We report on a comparison between a direct simulations of the problem and the predictions of the model; it is found that the mean velocity field,the kinetic energy of the microstructure, and an oscillatory coupling between the small and large structures are properly predicted.

1. A model based on the homogenization of small eddies

1.1. Description

Consider the initial value problem for the N-S equations in the case of slightly viscous flows :

$$u^\epsilon_{,t} + (u^\epsilon.\nabla)u^\epsilon + \nabla p^\epsilon - \mu\epsilon^2\Delta u^\epsilon = 0, \quad \nabla.u^\epsilon = 0$$

$$u^\epsilon(x,0) = u^o(x) + \epsilon^{\frac{1}{3}} w^o(\frac{x}{\epsilon}, x) \tag{7}$$

$$\mu = 0(1)$$

Assume that $w^o(y,x)$ is periodic in y on Y $=]0,1[^3$and has zero mean

$$< w >= \int_Y w(y,x)dy = 0$$

Our aim is to find the limit u of u^ϵ when ϵ tends to 0.

For that, we look for u^ϵ, p^ϵ in the form

$$u^\epsilon(x,t) = u(x,t) + \epsilon^{\frac{1}{3}} w'(\frac{a(x,t)}{\epsilon}, \frac{t}{\epsilon^{\frac{2}{3}}}; x,t) + \epsilon^{2/3} u'^{(1)}(\frac{a(x,t)}{\epsilon}, \frac{t}{\epsilon^{\frac{2}{3}}}; x,t) + ... \tag{8}$$

$$p^\epsilon(x,t) = p(x,t) + \epsilon^{\frac{2}{3}} \pi(\frac{a(x,t)}{\epsilon}, \frac{t}{\epsilon^{\frac{2}{3}}}; x,t) + \epsilon p^{(1)}(\frac{a(x,t)}{\epsilon}, \frac{t}{\epsilon^{\frac{2}{3}}}; x,t)) + ... \tag{9}$$

where a(x,t) are the inverse of the Lagrangian coordinates of the problem, i.e.

$$a_{,t} + (u.\nabla)a = 0$$

$$a(x,0) = x \tag{10}$$

and the functions $\omega'(y,\tau;x,t)$, $u'^{(1)}(y,\tau;x,t)$, ..., $\pi(y,\tau;x,t)$, $p^{(1)}(y,\tau;x,t)$, ... are $(y-\tau)-$ *periodic* with mean zero.

There are several good reasons for taking such a complicated ansatz. The exponent of $\epsilon^{1/3}$ is taken to agree with the Kolmogorov dimensional analysis ($< u(x)u(x+r) > \cong r^{2/3}$); the variable $a(x,t)/\epsilon$ is chosen because $w^o(x/\epsilon, x)$ is expected to be convected by the mean flow. In fact (10) is not an hypothesis, it is a consequence of the Ansatz.

Substituting these expansions in the N-S equations (7), one can obtain the following canonic equations for the microstructures :

$$\omega_{,\tau} + (\omega.\nabla_y)\omega + C\nabla_y\pi = 0, \quad \nabla_y.\omega = 0 \quad in \quad]0,1[^3 \tag{11}$$

$$u^{(1)}_{,t} + (\omega.\nabla_y)u^{(1)} + (u^{(1)}.\nabla_y)\omega + C\nabla_y p^{(1)} = f, \quad \nabla.u^{(1)} = g \tag{12}$$

with

$$C = \nabla a^T \nabla a, \quad \omega' = \nabla a^{-T}\omega, \quad u'^{(1)} = \nabla a^{-T} u^{(1)}$$

and

$$f = -\nabla a^T[(u+\omega)_{,t} + (u+\omega).\nabla(u+\omega) + \nabla(p+\pi)] \tag{13}$$

$$g = -\nabla.(u+\omega) \tag{14}$$

Equations (11)-(12) cannot be verified simultaneously unless [f,g]verifies certain compatibility conditions. In [4], three sets are found :

$$< f + \omega g >= 0, < g >= 0 \tag{15}$$

$$< \omega C^{-1} f + (\frac{1}{2}|\omega'|^2 + \pi) g >= 0 \tag{16}$$

$$< (\nabla a \nabla_y) \times \omega'.f >= 0 \tag{17}$$

The number of equations like (15)-(17) necessary to makes (12) solvable is strongly connected with the number of invariants of (11). System (11) is a generalized Euler equation for incompressible flows (notice the matrix C in front of the pressure term). We know three invariants for this system :

. the mean velocity

$$< \omega >= 0 \tag{18}$$

. the kinetic energy

$$q = \frac{1}{2} < \omega C^{-1} \omega >= \frac{1}{2} < |\omega'|^2 > \tag{19}$$

. the mean helicity

$$r =< \nabla a^{-T}\omega.(\nabla a \nabla_y) \times (\nabla a^{-T}\omega) >=< \omega'.\nabla' \times \omega' > \tag{20}$$

$$((\nabla' = \nabla a \nabla))$$

1.2. Effective equations

By making use of (13)-(14) in (15)-(17), we obtain a system of nonlinear advection equations for u,q and r:

$$u_{,t} + (u.\nabla)u + \nabla p + \epsilon^{2/3}\nabla.(qR) = 0 \tag{21}$$

$$q_{,t} + u.\nabla q + qR : \nabla u + \epsilon^{\frac{1}{3}}\Delta(q^{\frac{3}{2}}b) + \mu q d = 0 \tag{22}$$

$$r_{,t} + u.\nabla r + qS : \nabla u + \epsilon^{\frac{1}{3}}\nabla(q^{\frac{1}{2}}\nabla r c) + \mu q e = 0 \tag{23}$$

$$u(x,0) = u^o(x), \quad q(x,0) = \frac{1}{2} < |\omega^o(x)|^2 >, \quad r(x,0) =< \omega^o.\nabla_y \times \omega^o > \tag{24}$$

where

$$R_{ij} =< \omega_i' \omega_j' >, \quad S_{ij} = 2 < \omega_i' r_j' >$$

$$b =< (\pi + \frac{|\omega'|^2}{2}) w' >, \quad d =< |\nabla' \omega'|^2 >$$

$$c =< (\pi + \frac{|\omega'|^2}{2}) r >, \quad e = 2 < \nabla' \omega' : \nabla' r' >$$

$$r' = (\nabla a \nabla_y) \times (\nabla a^{-T} \omega).$$

Thus, the system is closed if we are able to resolve the canonic microstructure problem

$$\omega_{,\tau} + (\omega.\nabla)\omega + C\nabla_y \pi = 0, \quad \nabla_y.\omega = 0 \quad in \quad Y$$

$$< \omega >= 0, \quad 1/2 < \omega C^{-1} \omega >= 1, \quad < \nabla a^{-T} \omega (\nabla a \nabla_y) \times (\nabla a^{-T} \omega) >= \frac{r}{q} \tag{25}$$

$$\omega \; y - \tau \text{ periodic in Y } =]0,1[^3$$

which depends on the parameter ∇a, q, r.

We remark that (21)-(23) resembles formally the $k - \epsilon$ model with the substitution of the rate of dissipated energy ϵ by the helicity r. The validity of the model is analyzed in [2] and [3].

1.3. Eddy viscosity terms

It can be shown that the new Reynolds tensor R in this model is not a diffusion tensor (the total energy,the integral on the domain of $\frac{1}{2}u^2 + q$, is preserved). However to find eddy diffusion terms, one has to add to the equations the terms of superior orders in ϵ. In this way, one obtains the system (see [2])

$$u_{,t} + (u.\nabla)u + \nabla p + \nabla.(qR) + \epsilon(\sqrt{q}A : \nabla u) = 0$$

$$\nabla.u = 0$$

$$k_{,t} + u.\nabla k + k(R : \nabla u + \mu\psi) + \epsilon^{\frac{2}{3}} \nabla.(\sqrt{k} V : \nabla k) = 0$$

$$a_{,t} + (u.\nabla)a = 0$$

$$r_{,t} + u.\nabla r + qS : \nabla u + \epsilon^{\frac{1}{3}} \nabla.(q^{\frac{3}{2}} c) + \mu q e = 0$$

$$u(x,0) = u^o(x), \quad k(x,0) = \epsilon^{\frac{2}{3}} q^o(x), \quad a(x,0) = x$$

where A, V are $(3 \times 3)^2$ and 3×3 tensors respectively defined through the canonical linear microstructure problems(11) (see [2]). System (23) is written in terms of the kinetic energy of the microstructure $k = \epsilon^{2/3} q$.

2. A reduced model : numerical experiments

If we consider one dimensional mean flows of the form

$$u(x,t) = (u_1(x_3,t), 0, 0)^T \tag{27}$$

we have

$$\nabla a^T \nabla a = \begin{pmatrix} 1+\alpha^2 & 0 & \alpha \\ 0 & 1 & 0 \\ \alpha & 0 & 1 \end{pmatrix} \tag{28}$$

with

$$\alpha = -\int_0^T u_{1,3}(x_3, s)ds$$

In this case, we can show that all tensors appearing in the helicity equation are proportional to the helicity ; hence if we have r(x,0) = 0, we will continue having r(x,t) = 0 for t > 0. Consequently, the model can be reduced to

$$u_{1,t} + (k\rho)_{,3} - \frac{4}{9}\epsilon(\sqrt{k}u_{1,3})_{,3} = 0$$

$$k_{,t} + k(\rho u_{1,3} + \mu\psi) - \frac{4}{9}\epsilon^{\frac{2}{3}}(\sqrt{k}k_{,3})_{,3} = 0 \tag{29}$$

where $\rho = R_{13}$ and ψ depend only the the parameter α defined in (28). We have compared the solution of (29) with the direct simulation of (7) for 3D Poiseuille flows. We have obtained a reasonable agreement in the behaviour of both the mean velocity field and the kinetic energy of the microstructure (see [2]).

Now, if we neglect all viscous effects in (29), the model reduces to :

$$u_{1,t} + (k\rho)_{,3} = 0, k_{,t} + k\rho u_{1,3} = 0 \tag{30}$$

if we put

$$v(x_3, t) = \int_0^t u_1(x_3, s)ds$$

we can approximate (30) by

$$v_{,tt} - k_o\rho'(0)v_{,33} = 0 \tag{31}$$

Our numerical experiences show us $\rho'(0) > 0$ and then (31) is a wave equation for v. So the effect of small eddies of energy k_o on Poiseuille flows is to create oscillations of periods proportional to $1/\sqrt{k_o}$ (see [2])

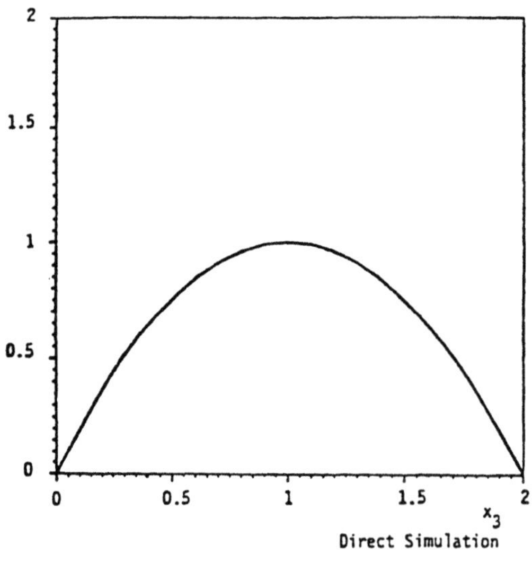

Fig. 1. First component of the mean velocity field. Corresponds to the direct simulation of (7).

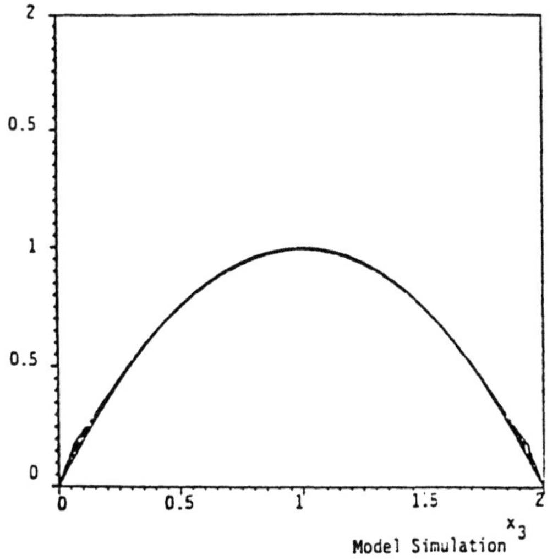

Fig. 2. Function u_1. Corresponds to the reduced model (29).

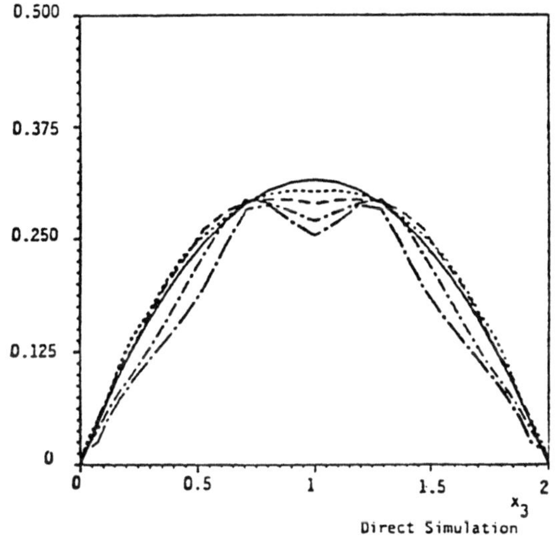

Fig. 3. Mean kinetic energy of the microstructure k. Direct simulation.

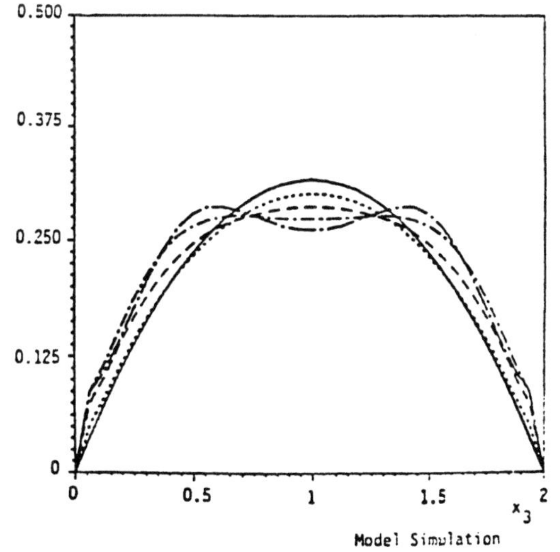

Fig. 4. Mean kinetic energy of the microstructure k. Reduced model.

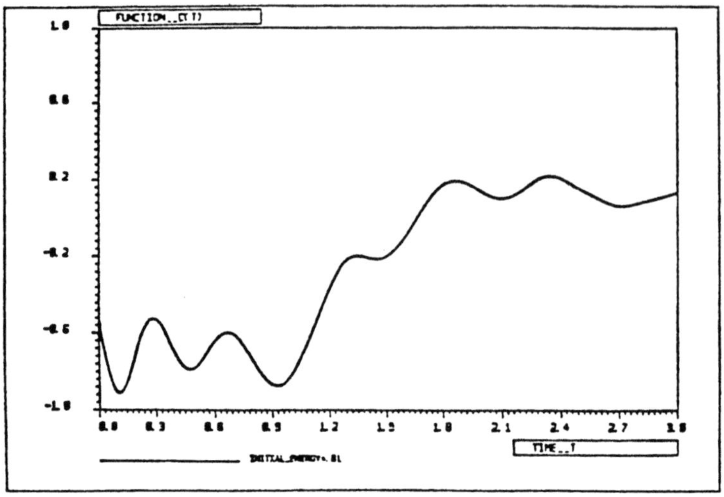

Fig. 5. Rate of exponential growth of the mean microstructure kinetic energy. The oscillations are well demonstrated, with almost constant period for t < 1.

3. Conclusion

Turbulence closure by homogenization is promising. It has not yet reached the level of efficiency of $k-\epsilon$ models but it is likely to be good for flows with several well separated scales ; some effects of small to large interactions were clearly demonstrated to exist in real flows and those cannot be represented by $k-\epsilon$ models in their present forms. Another important aspect is that the present analysis can be extended to compressible flows where the physical intuition to built closure assumptions is difficult to acquire (see [3]).

References

[1] B.E. LAUNDER, D.B. SPALDING, "Mathematical models of turbulence", Academic Press (1972).

[2] T. CHACON, "Oscillations due to the transport of microstructures", to appear.

[3] T. CHACON, O. PIRONNEAU, "On the mathematical foundations of the $k-\epsilon$ turbulent model",in "Vistas in applied Mathematics" A.V. Balakrishnan, A. Dorodnitsyn, J.L. Lions ed. Springer Verlag New York (1986) pp 44-56

[4] D. MCLAUGHIN, G. PAPANICOLAOU, O. PIRONNEAU, "Convection of microstructure", SIAM J. Appl. Math. 45 (1985), 780-793.

The Simulation of Dispersion by Convective Turbulence in the Atmosphere

F.T.M. Nieuwstadt[1] *and J.P.J.M.M. de Valk*[2]
[1]Laboratory for Aero- and Hydrodynamics, NL-2628 AL Delft, The Netherlands
[2]MFO, de Uithof, NL-3584 CC Utrecht, The Netherlands

1. Introduction

A large eddy model gives a numerical simulaton of the large scale motions in a turbulent flow by solving the set of Navier-Stokes equations from which small scale motions are removed by filtering [1]. Here we apply this approach to calculate turbulence in the convective atmospheric boundary layer. This type of turbulence, which generally occurs when the boundary layer is heated from below, is dominated by large scale structures and therefore well suited to simulation by large eddy modeling. These large scale structures are called thermals and they consist of rising plumes of relatively warm air. They start at the surface and they travel towards the top of the boundary layer. The thermals are embedded in regions of relatively cold and slowly subsiding air. These characteristics are well reproduced by a large eddy model as shown in [2].

Our goal is to consider dispersion of pollutants in the convective boundary layer. This is a very complicated process due to the fact that convective turbulence is both highly inhomogeneous and also strongly non-Gaussian. Here we will discuss the results of some preliminary calculations. First we consider the dispersion of passive contaminants. It has already been shown in [3] that in this case large eddy modeling leads to results which are in close agreement with experiments. Our calculations, which use a different numerical procedure, confirm this result.

As a next step we extend our study to the dispersion of non-passive contaminants or in particular buoyant plumes. Here we show some new results.

2. Numerical Procedure

The calculation region is a box of 5 × 5 × 2 km as shown in Fig. 1. In this region the equations of motion are solved by a finite difference technique. For details we refer to [1].
Convective turbulence is generated by heating the surface of our box at a constant rate. When a state of quasi-stationary turbulence has been reached,

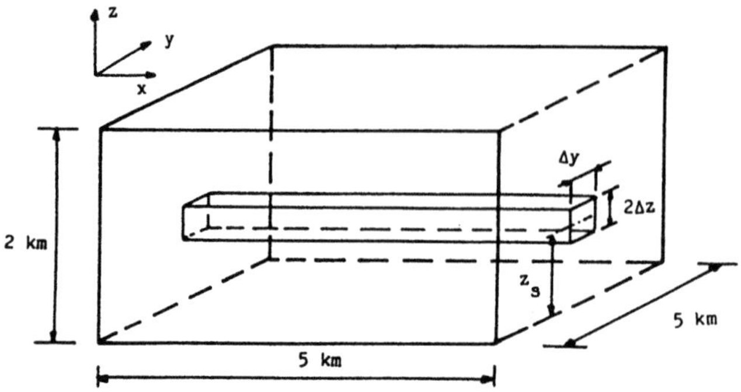

Figure 1. Calculation domain and the initial configuration of the line source of pollutants

we introduce at t = 0 a line source of pollutants along the x-axis, as also shown in Fig. 1. The cross-section of this line source is $2\Delta z \times \Delta y$ where Δz is the vertical and Δy the horizontal grid size. In our calculations $\Delta z = 50$ m and $\Delta y = 125$ m. For the case of a buoyant plume we also increase the temperature within the line source. After t = 0 we continue the calculation of the large eddy model with the same forcing conditions at the surface. However, we carry along an additional equation to describe the evolution of the concentration as a function of time.

The dispersion of the line source as a function of time allows an interpretation in terms of plume dispersion from a continuous point source as a function of distance from the source. In this interpretation we assume that the plume is advected horizontally along the x-axis by a mean wind speed U. In addition we have to assume that the diffusion in the direction of the mean wind speed is negligible with respect to transport. Time t can then be related to distance x by means of the transformation x = Ut. As a consequence line averages of the large eddy results along the x-axis can become equivalent to time averages for the plume.

3. Results

The main conclusion of previous studies such as [2] is that turbulence in the convective boundary layer can be universally scaled in terms of the boundary-layer height h and the convective velocity scale w_*. Moreover, in [4] it is shown that dispersion in this boundary layer can be scaled in a similar way. Therefore, we will present the results of our simulation studies in terms of this scaling approach. From the large eddy simulation we obtain the following values for the scaling parameters: $w_* = 2.4$ ms^{-1} and h = 1500 m.

Two release heights are considered: a near-surface release ($z_s \approx 0.04$h for the non-buoyant release and $z_s \approx 0.16$h for the buoyant release) and a release in the middle of the boundary layer, $z_s \approx 0.49$ h. For the case of a buoyant plume we increase the temperature in the line source by 1.6 °C. The plume buoyancy parameter F/U which is defined as

$$F/U = \frac{g}{\pi T} \iint \Delta T \, ds \ , \tag{1}$$

becomes then equal to 223 $m^3 s^{-2}$. In (1) the integral extends over the cross-section of the plume.

First we consider an example of the instantaneous concentration pattern. We select the case of a buoyant line source in the middle of the boundary layer. In Fig. 2 we show the isolines of concentration on a vertical cross-section along the initial line source. The results are given for two non-

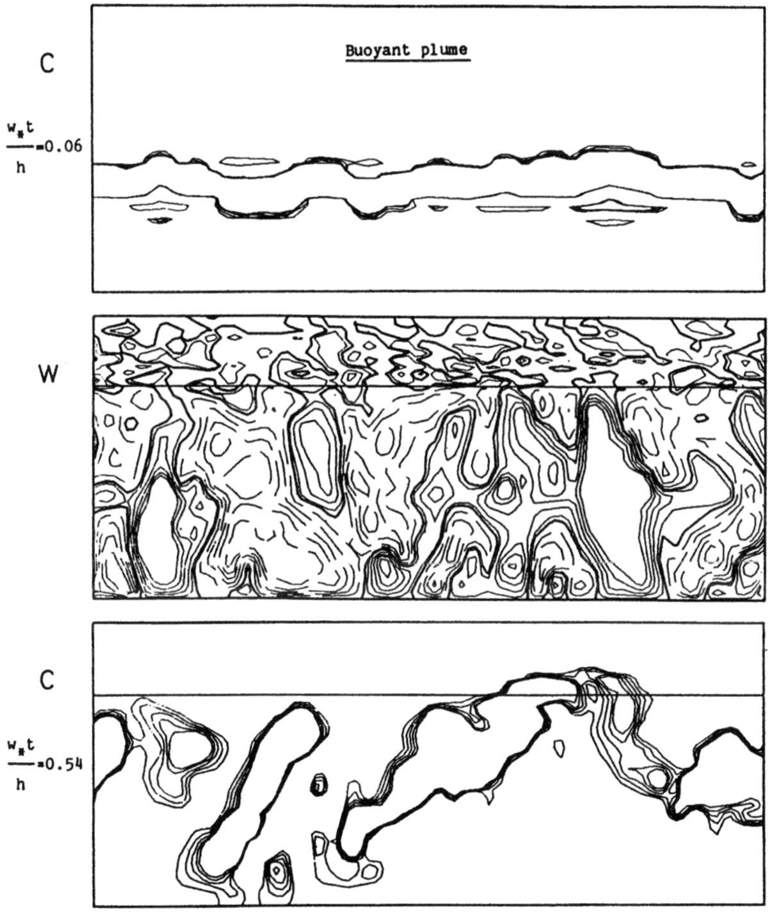

Figure 2. Instantaneous distribution of the concentration and of the vertical velocity fluctuations on a vertical cross-section through the line source. For the vertical velocity fluctuations solid isoline indicate positive (upward) vertical velocity, whereas broken isolines denote negative (downward) velocity. The thick isoline gives the zero value. The horizontal line gives the mean boundary-layer height

dimensional times. In addition we show the isolines of vertical velocity fluctuations, in which we can clearly recognize the structure of the thermals. We find that the shape of the concentration pattern is strongly correlated with the vertical velocity fluctuations: in the thermals the plume is lifted up and even penetrates in the region above the boundary layer, whereas outside thermals the plume follows the subsiding motion of the air. The strong distortion by the turbulent velocity field leads to a plume which at $w_*t/h = 0.54$ is broken up in separate sections. Such a concentration pattern is characteristic for dispersion in the convective boundary layer.

The next result, that we consider, is the cross-wind integrated concentration which in our case is defined as

$$\langle c \rangle = \frac{1}{L} \iint c \; dxdy \tag{2}$$

where c [kg m^{-3}] is the instantaneous concentration and where L is the length of the calculation domain along the x-axis. The integration extends across the whole horizontal domain. A non-dimensional cross-wind integrated concentration is defined as $\langle c \rangle hU/Q$. Q is the strength in kg s^{-1} of the equivalent continuous point source, which is related to the total amount of contaminants q [kg] in the line source by $Q = Uq/L$. With this definition the non-dimensional cross-wind integrated concentration becomes unity in the limit of $t \to \infty$ when the contaminants are uniformily mixed across the depth of the boundary layer. The non-buoyant plume results are shown in Fig. 3 for both release heights in the form of an isoline plot with the non-dimensional height z/h along the vertical axis and the non-dimensional distance $X = w_*x/(Uh)$ along the horizontal axis. In addition to the results of our simulation we give in this figure also experimental data which were obtained from [4] and [5].

Figure 3 shows clearly that the plume of the surface release rises towards the top of the boundary and even forms there a concentration maximum. In contrast the plume of the elevated release descends and impinges on the surface with a relatively high concentration. The explanation of this perhaps surprising behavior lies in the inhomogeneous and non-Gaussian character of convective turbulence. It is also clear from Fig. 3 that our simulation results are in good agreement with the experiments.

As a final result we consider the mean plume height which is defined as

$$\bar{z} = \int_0^h z \frac{\langle c \rangle L}{q} dz \; . \tag{3}$$

Our simulation results are shown in Fig. 4 together with the experimental

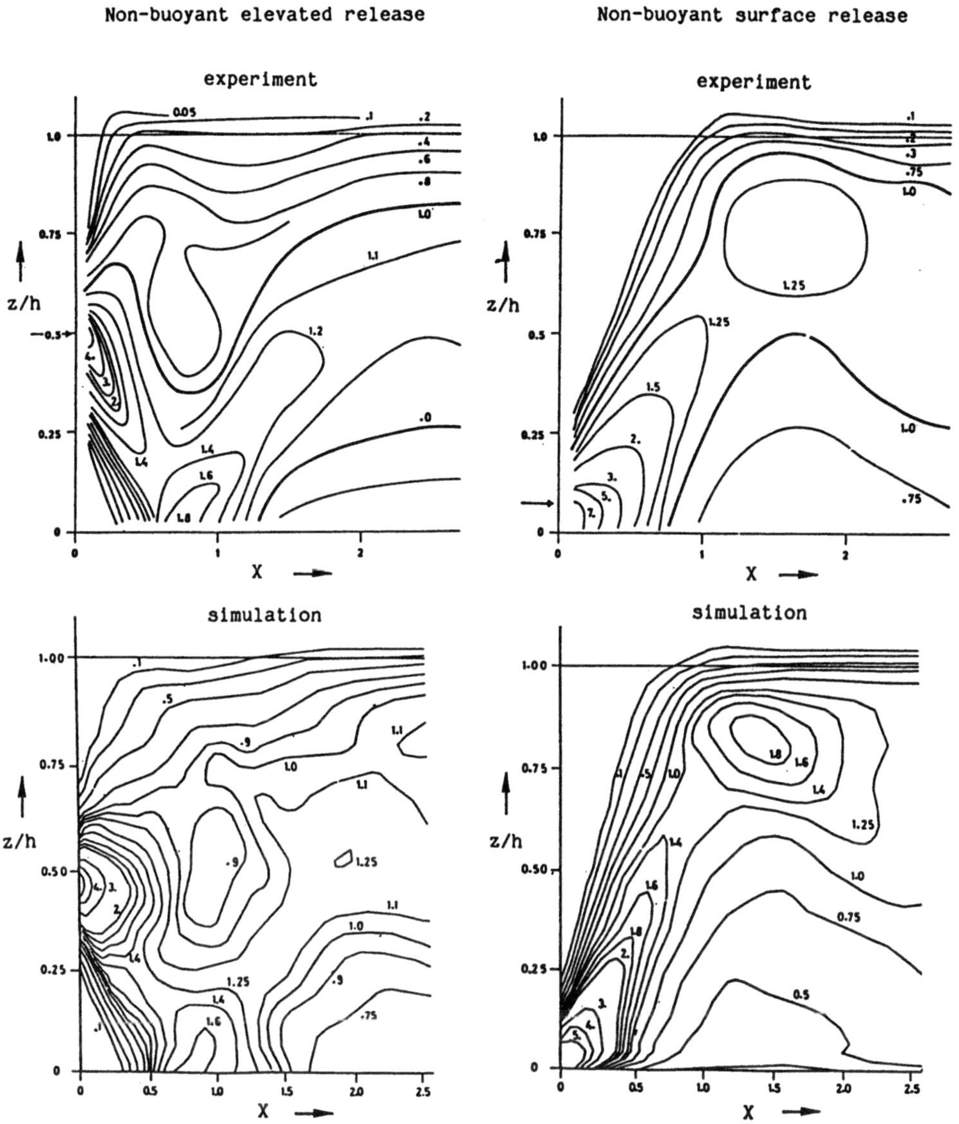

Figure 3. Distribution of the non-dimensional cross-wind integrated concentration as a function of non-dimensional height and of non-dimensional distance. The laboratory data and our simulation results are both drawn on the same scale

data given in [4], [5] and [6]. For the case of a non-buoyant release Fig. 4 summarizes the information already given by the concentration distributions of Fig. 3 and the good agreement with experimental data is confirmed. For the case of the buoyant release at $z_s = 0.49$ h we find an initial rise of the mean plume height. We have compared this result with the well-known plume rise equation of Briggs [7] which we have corrected for the initial dimension of the line source. The plume rise equation then reads

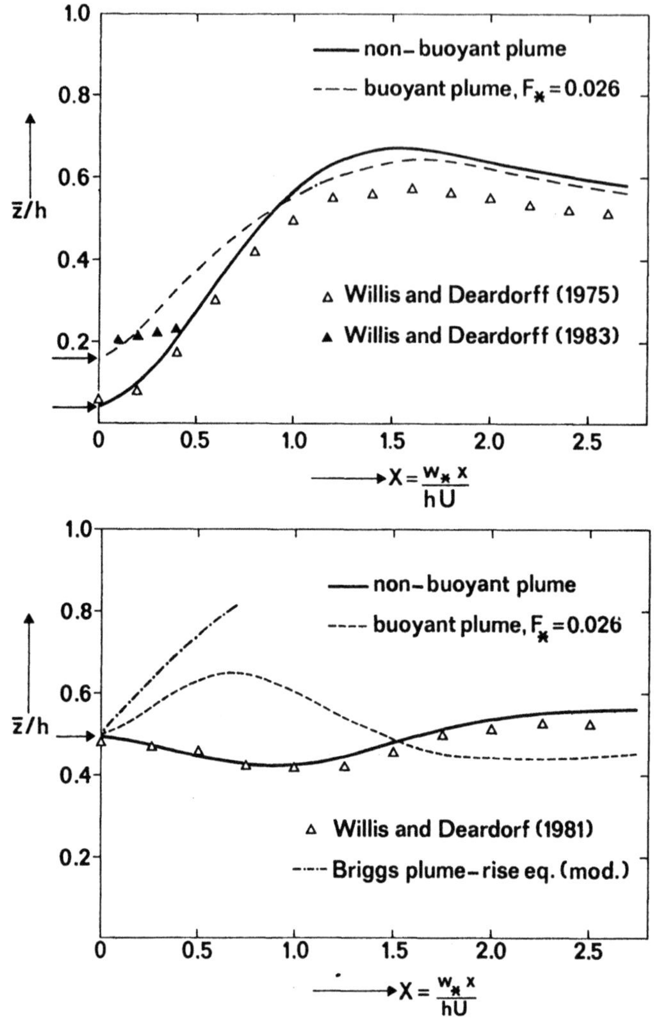

Figure 4. The mean plume height as a function of non-dimensional distance.

$$\frac{z}{h} = (4.2F_*X^2 + a^3)^{1/3} - a\,, \qquad (4)$$

where $F_* = F/(Uw_*^2h)$ and where X is the dimensionless distance. The parameter a is related to the initial cross-section A_o of the line source by $a = 0.94\ A_o^{1/2}/h$. In our case $A_o = 2\Delta y\Delta z = 1.25.10^4\ m^2$. We see that (4) overestimates the rise of the mean plume height, which we attribute to the fact that dispersion by ambient turbulence may already be important at an early stage of plume development.

4. Conclusion

Plume dispersion in the convective boundary layer can be well simulated by a large eddy model. The agreement with experimental data in the case of a non-buoyant release is good. However, the results shown should be interpreted with some care because they are based on a single run with the large eddy model. In general the data of several runs are necessary to calculate a stable and representative average plume behavior.

5. References

1. F.T.M. Nieuwstadt, R.A. Brost and T.L. van Stijn: In Proceedings of the 199th Euromech Meeting of Direct and Large Eddy Simulation of Turbulence, München, 1985
2. J.W. Deardorff: Bound.-Layer Meteorol., 7, 81-106 (1974)
3. R.G. Lamb: In Atmospheric Turbulence and Air Pollution Modelling, F.T.M. Nieuwstadt and H. v. Dop (eds.) (D. Reidel, Publishing Company, 1982)
4. G.E. Willis and J.W. Deardorff: Quart. J. Roy. Meteor. Soc., 102, 427-445 (1976)
5. G.E. Willis and J.W. Deardorff: Atmos. Environ., 12, 1305-1312 (1978)
6. G.E. Willis and J.W. Deardorff: Atmos. Environ., 17, 2435-2447 (1983)
7. G.A. Briggs: Plume rise. U.S.A.E.C. (1969).

Part IV

Coupling of Fourier Modes. Spectral Analysis of Turbulence and Related Problems

Spectral Evolution of the Navier-Stokes Equations for Low Order Couplings of Fourier Modes

J.G. Brasseur[1] *and S. Corrsin*[2]

[1]Department of Mechanical Engineering, Clemson University, Clemson, SC 29634, USA
[2]Department of Chemical Engineering, The Johns Hopkins University, Baltimore, MD 21218, USA

1. Introduction

The complex spacio-temporal development of turbulent flow has its origins in the nonlinear inertial couplings of the Navier-Stokes equations. At high Reynolds numbers these couplings lead to the continuous exchange of energy among a wide range of spacial scales and the creation of new scales. It is this process of scale generation and energy exchange which we discuss in this paper. Space necessarily limits the paper to primary results, and many interesting details must unfortunately await future publication.

Classically, the decomposition of the velocity field into well defined length scales has been accomplished through Fourier decomposition. A structure within the flow is identified as a band of Fourier modes in a narrow region of wave number space, its evolution described by the collective interactions with Fourier modes in all regions of Fourier space. Fourier decomposition of the Navier-Stokes equation leads directly to a triad structure in Fourier space upon which the nonlinear interactions are based. Thus, the triad interaction is the fundamental interaction in the nonlinear process of energy exchange, and this paper shall be devoted to triad evolution. The calculations bear directly on the issue of the correlation, or non-correlation between scales in different regions of the spectrum; and should find application in the interpretation of large scale numerical calculations using spectral methods [2,8].

The triad has been studied previously by Lee [3,4,5] in context with the development of ergodicity and mixing. In his analyses, however, energy is confined to the initial triad and new modes do not appear. Lorenz [7] simplified the dynamic equations to four specialized Fourier modes in two dimensions as a simple model of atmospheric zonal flow.

2. The Structure of the Navier-Stokes Equations in Fourier Space

2.1 Fourier Decomposition

In the Fourier spectral representation, the velocity field is represented as an infinite series of discrete complex Fourier modes with wave vector $\vec{k}$ and complex amplitude $\vec{u}(\underline{k},t)$:

$$\vec{v}(\underline{x},t) = \sum_{\underline{k}} \vec{u}(\underline{k},t)e^{i\underline{k}\cdot\underline{x}} \tag{1}$$

Boundaries are sufficiently distant that the flow is considered unbounded. Equation (1) may therefore be interpreted as the discretized form of a Fourier integral, and periodicity will not be imposed. For incompressible flows, the complex amplitude is constrained by continuity to a plane perpendicular to its wave vector: $\vec{k}\cdot\vec{u}(\underline{k},t) = 0$. The condition that the velocity field be real requires that the wave vector set $[\underline{k}]$ be separable into two sets $[+\underline{k}]$ and $[-\underline{k}]$ and that $\vec{u}(-\underline{k}) = \vec{u}^*(\underline{k})$.

The Fourier transform (1) of the Navier-Stokes equations, with continuity applied to eliminate pressure, leads to a set of coupled differential equations in time for the complex amplitudes $\vec{u}(\underline{k},t)$ [vid.1,6]. These may be written

$$\left(\frac{d}{dt} + \nu k^2\right) \vec{u}(\underline{k},t) = -i \sum_{\underline{k}'} \vec{u}(\underline{k}')_{\perp \underline{k}} \; [\vec{k}\cdot\vec{u}(\underline{k}-\underline{k}')] \tag{2}$$

The second term on the LHS is viscous dissipation, whereas the RHS describes the evolution of mode $\vec{k}$ due to nonlinear interactions with all other modes $\vec{k}'$. $\vec{u}(k')_{\perp \underline{k}}$ is the vectoral projection of the amplitude of mode $\vec{k}'$ onto a plane perpendicular to $\vec{k}$, and incorporates both the inertial exchange of energy between modes and the pressure term. The effect of pressure is to rotate the amplitude vector in a plane perpendicular to $\vec{k}$, causing energy to be exchanged between components.

The physical structure of a complex Fourier mode represents what we call a 'complex shear wave', because it must be specified with two independent phase angles. This is in contrast to a 'simple shear wave' - a sine wave with single phase. Let us write $\vec{u}(\underline{k})$ as a complex vector, $\vec{u}(\underline{k}) = \vec{a}+i\vec{b}$. Since continuity constrains $\vec{a}$ and $\vec{b}$ to a plane perpendicular to $\vec{k}$, $\vec{u}(\underline{k})$ is specified with 4 independent components. Splitting the wave vector set $[\underline{k}]$ into its positive and negative parts and rewriting (1) over the positive set only, one may write the velocity field as a sum of complex shear waves:

$$\vec{v}(\underline{x},t) = 2\sum_{+\underline{k}} [\vec{a}(\underline{k}) \cos \underline{k}\cdot\underline{x} - \vec{b}(\underline{k}) \sin \underline{k}\cdot\underline{x}] \tag{3}$$

The two component sine waves produce a three dimensional velocity field described by a velocity vector which rotates around the wave vector in the spacial direction $\vec{k}$. Changes in 'overall phase' of a complex shear wave arise both from changes in the relative magnitudes and the relative orientations of $\vec{a}$ and $\vec{b}$.

Like the velocity field, the vorticity field of a complex shear wave is in general three dimensional. However, when $\vec{a}$ and $\vec{b}$ are colinear, the two polar phase angles coincide and the complex shear wave reduces to a simple shear wave. It can be shown from (2), however, that if $\vec{a}$ and $\vec{b}$ are initially non-zero and colinear, they will not remain colinear--with the exception that when all wave vectors and amplitudes are in the same plane they remain confined to that plane. This leads to a purely two dimensional evolution with scaler vorticity.

2.2 The Triad Structure of the Inviscid Equation

Since our interest is in the non-linear spectral transfer of energy among modes, we shall eliminate viscous dissipation from (2) and consider inviscid evolution:

$$\frac{d\vec{u}(\underline{k})}{dt} = -i \sum_{\underline{k}'} \vec{u}(\underline{k}')_{\perp_{\underline{k}}} \; [\vec{k}\cdot\vec{u}(\underline{k}-\underline{k}')] \tag{4}$$

The triad structure of the RHS is apparent. The change in mode $\vec{k}$ is given by a superposition of interactions with all other modes, where each interaction involves three modes with wave vectors which form a closed triangle. All nonlinear interactions occur within such triads. The basic triad interaction is represented in Fig. 1a, where $\vec{k}_1+\vec{k}_2+\vec{k}_3 = 0$. The change in mode $\vec{k}_1$ depends on the geometry of the triad, the relative orientations of the amplitude vectors to the triad plane, and the phase relationships among the modes. New modes are created by completing incomplete triads. A triad therefore produces three new modes at the first time step, as shown in Fig. 1b. At the next time step the total is 18, then 60, 219, and so on--an exponential growth in modes. The rate at which energy is delivered to each new set of modes, however, depends on the energy distribution within the current set and the relative geometrical orientations among the modes. Not all modes receive significant energy at any given time step.

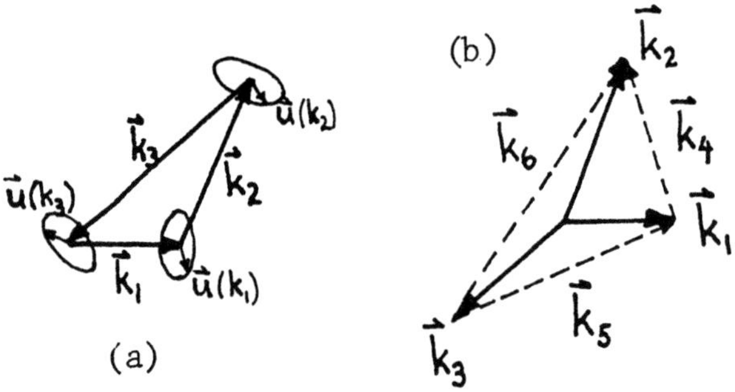

Fig. 1

The structure of the triad interaction. (a) initial condition; (b) the first time step.

It may be shown from the triad equations which describe the interaction in Fig. 1b that the inviscid Navier-Stokes equations conserve energy not only as a whole, but within each triad interaction separately. This has the consequence that when (1) is truncated to a finite number of modes (Galerkin methods), all the energy stays in the existing modes.

3. The Calculations

We begin with triad initial conditions, and search for overall behaviour trends in triad evolution. The dependent variables are non-dimensionalized with velocity scale $\sqrt{E}$, where E is the total energy, and length scale $2\Pi/k_0$, where k_0 is the average of the initial triad wave numbers. The time scale $1/(k_0\sqrt{E})$ decreases with wave number and triad energy, so the rate at which the spectrum evolves in dimensional Fourier space depends on the location of the triad in the spectrum and its energy content. Total energy is normalized to 1.

The initial conditions which must be specified are the wave vector triangle geometry, the energy in each mode, the relative orientations of the complex amplitudes with the triad plane, and the phase relationships. The wave vector triangle geometry ('triad geometry') is systematically varied as shown in Fig. 2a. The long narrow triads of the top row describe interactions between wave numbers widely separated in the spectrum, while the other triads describe more local interactions. Geometrical constraints restrict interactions which spectrally are widely separated to one low wave number and two high wave number modes. In contrast, the interaction between two lower wave number modes and one higher wave number mode (triangles 10, 11, 12) is necessarily local in wave number space. The triad wave vectors to the right in Fig. 2a are near colinear. It may be shown from the triad equations that when all three wave vectors are colinear, no energy is exchanged among modes. In all, 17 triad shapes were computed, including systematic extensions below those shown in Fig. 2a.

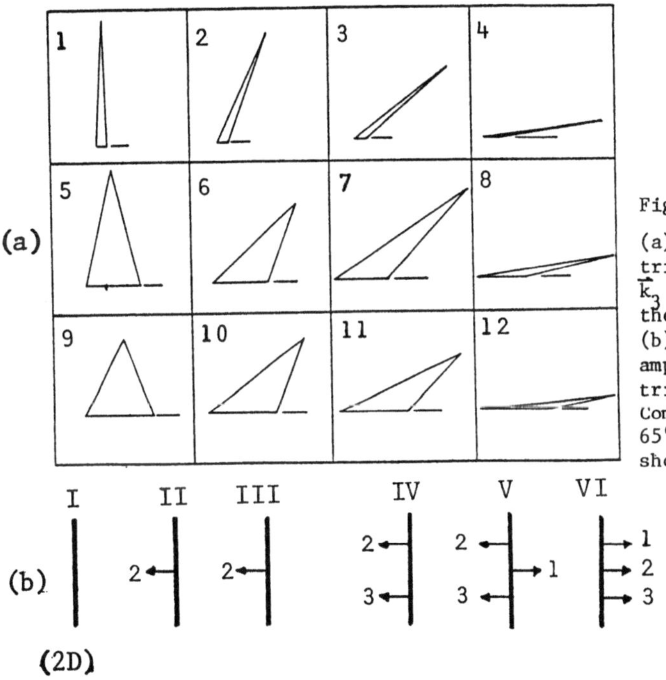

Fig. 2

(a) The initial wave vector triangle geometry. $\vec{k}_1$, $\vec{k}_2$, $\vec{k}_3$ are counterclockwise from the bottom, as in Fig. 1.
(b) The six initial complex amplitude orientations. The triad plane is shown edge on. Complex amplitude vectors are 65° to the triad plane as shown.

For each triad shape the short time evolution was computed for six basic orientations of complex amplitude vectors relative to the triad plane. These are shown schematically in Fig. 2b. The real and imaginary parts of $\vec{u}(\underline{k})$ are initially colinear and either in the triad plane or at an angle of 65° to the triad plane. Except in the two dimensional calculations, $\vec{a}$ and $\vec{b}$ do not remain co-linear at later times. The magnitudes of $\vec{a}$ and $\vec{b}$ and the energies in each of the three modes are initially uniform. In a subsequent series of calculations, the effect of phase is examined by modifying the relative magnitude and orientations of $\vec{a}$ and $\vec{b}$.

A fourth order Runge-Kutta procedure is used to advance in time. Calculations were performed to non-dimensional time $\tau = 2$, for 219 modes. It was found that at the end of the calculations typically no more than 20 modes had significant energy; thus for these short times, the cutoff at 219 provides no major restriction.

4. The Results

4.1 Evolution in Energy

Figure 3 shows the evolution in energy of the 12 triads in orientation III, a three dimensional calculation. The initial triad modes (1, 2, and 3) exchange energy among themselves, and lose energy to new modes (4, 5,...) created through the process of completing incomplete triads. As the initial wave vectors become colinear there is a decrease in the rate of evolution of all modes. Observe also, however, that in the top row the energy in the k_1 mode changes very little relative to the other modes, unlike the other triads where energy changes at comparable rates. This distinction between the $k_1 \ll k_2, k_3$, and the $k_1 \sim k_2 \sim k_3$ classes is observed in all calculations, both two and three dimensional, and is a general feature of the triad interaction. We examine these two classes separately.

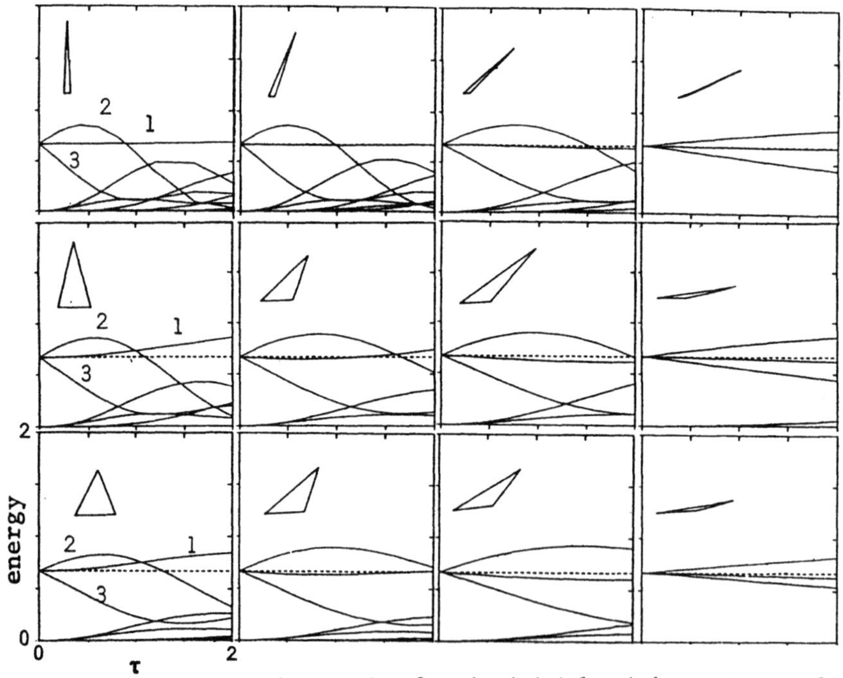

Fig. 3 Evolution of mode energies for the initial triad geometries of Fig. 2a, and the 3D complex amplitude orientation III of Fig. 2b. New modes form triads with previously existing modes.

4.2 The Interaction Between One Large and Two Small-scale Modes

The class $k_1 \ll k_2, k_3$ describes the basic interaction between modes widely spread in wave number. Typical spectral evolution characteristics of this class are shown in Fig. 4, here for triad 2 at three amplitude orientations. The general feature is a spread and redistribution of energy through the the high wave number region of the spectrum, while the energy in the low wave number mode remains relatively unchanged. Since the enstrophy in each mode is given by $k^2E(k)$, the local spread of energy in the vicinity of the k_2, k_3 modes produces only small changes in total vorticity. Modification of phase affects only the details of the energy transfer, and not the overall characteristics shown in Fig. 4.

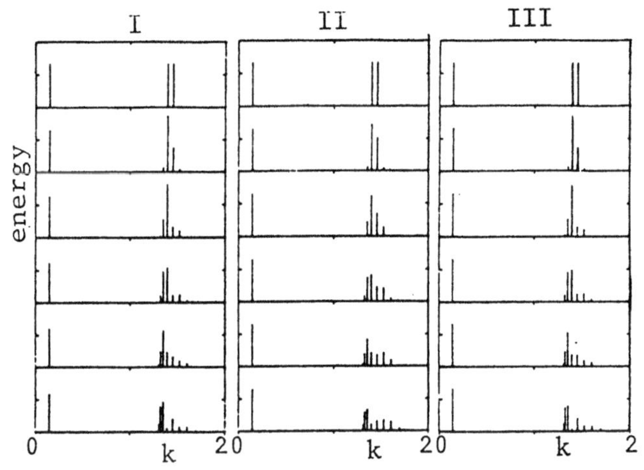

Fig. 4

Evolution of the spectrum of triad 2 of Fig. 2a, and complex amplitudes I, II, III of Fig. 2b. Mode energy is plotted against wave number from $\tau = 0$ to $\tau = 2$, in increments of 0.4.

It is important to recognize that the presence of the low wave number mode is central to the development which takes place; a different geometrical arrangement among the initial wave vectors leads to entirely different evolution characteristics. One may infer from Fig. 4, therefore, that the large scale mode in some way moderates the interaction so as to result in the breakdown of the small scale modes without itself losing energy.

Examining carefully which modes are created, of those modes which can be created, one discovers that only those modes which form a triad with the low wave number mode receive significant amounts of energy. Thus, at each time step the evolution consists of a collection of $k_1 \ll k_2, k_3$ triads with each triad tied to the same low wave number mode. Consequently, new high wave number modes remain in the vicinity of the original triad.

The process can be further understood by defining the small parameter $\varepsilon = k_1/k_2 \sim k_1/k_3$, and reducing the triad equations in the limit $\varepsilon \to 0$. This yields

$$\frac{d\vec{u}(k_1)}{dt} \sim \frac{d\vec{u}(k_6)}{dt} = O(\varepsilon) \qquad [\vec{k}_6 = \vec{k}_2 - \vec{k}_3]$$

$$\frac{d\vec{u}(k_2)}{dt} = -i\ \{\vec{u}^*(k_3)\ [\vec{k}_2\cdot\vec{u}^*(k_1)]\} + O(\varepsilon)$$

$$\frac{d\vec{u}(k_3)}{dt} = i\ \{\vec{u}^*(k_2)\ [\vec{k}_2\cdot\vec{u}^*(k_1)]\} + O(\varepsilon)$$

$$\frac{d\vec{u}(k_4)}{dt} = i\ \{\vec{u}^*(k_2)\ [\vec{k}_2\cdot\vec{u}\ (k_1)]\} + O(\varepsilon) \qquad [\vec{k}_4 = \vec{k}_1-\vec{k}_2]$$

$$\frac{d\vec{u}(k_5)}{dt} = -i\ \{\vec{u}^*(k_3)\ [\vec{k}_2\cdot\vec{u}\ (k_1)]\} + O(\varepsilon) \qquad [\vec{k}_5 = \vec{k}_1-\vec{k}_3] \tag{5}$$

It turns out that changes in all properties associated with the low wave number mode (amplitude, energy, and phase) approach zero with ε. The same is true of the one new mode $\vec{k}_6 = \vec{k}_2-\vec{k}_3$ which does not form a triad with $\vec{k}_1$.

The energy required to produce the new high wave number modes must originate in k_2 and k_3. In the limit $\varepsilon \to 0$ these modes initially change energy at the rate

$$\frac{dE(k_2)}{dt} = -\frac{dE(k_3)}{dt} = i\ \{\vec{u}(k_2)\cdot\vec{u}(k_3)\ [\vec{k}_2\cdot\vec{u}(k_1)]\} + c.c. \tag{6}$$

One mode gains energy at the same rate the other loses it; this is visually evident, for example, in Fig. 3. More importantly, however, because the amplitude of the k_1 mode appears in (6), the energy in the low wave number mode directly affects the rate at which the high wave number modes lose energy. Thus, the large scale mode not only forces certain small scale modes to appear and suppresses the appearance of other small scale modes without itself losing energy, it moderates the rate at which the spectrum evolves at the high wave numbers. This process is illustrated in Fig. 5. In Fig. 5a the k_1 mode has three times the energy of the other two, in Fig. 5c one third, and in Fig. 5b the initial energy is uniformly distributed among the three. The rate at which high wave number modes are produced and evolve clearly increases from right to left.

4.3 Three Interacting Modes of the Same Scale

A triad local in wave number space appears in physical space as eddies of roughly the same scale, as illustrated in Fig. 8; the three modes exchange energy at com-

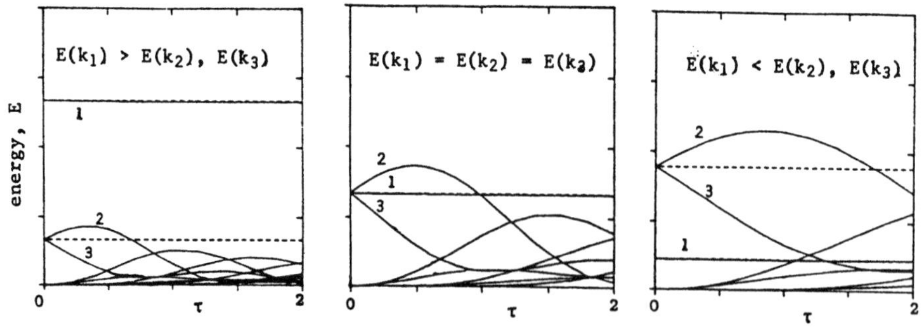

Fig. 5 Effect of relative energies between low and high wave number modes in the $k_1 \ll k_2,k_3$ class. Triad 2; complex amplitude orientation III.

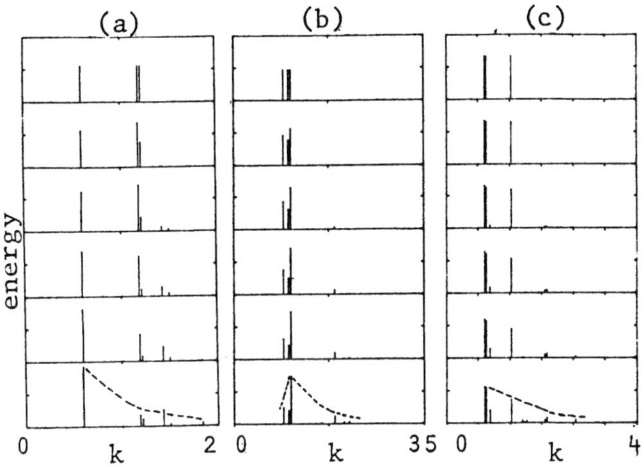

Fig. 6

Examples of spectral energy evolution for the class $k_1 \sim k_2 \sim k_3$. (a) triad 5, orientation I (2D); (b) triad 9, orientation V; (c) triad 10, orientation VI.

parable rates. Examples of energy spectra in this class of triad interactions are shown in Fig. 6. Examination of the energy exchange reveals a short time evolution consistent with our usual notion of energy cascade in homogeneous turbulence. New modes at higher wave numbers are rapidly produced, leading to a spectrum in which energy peaks in the low wave numbers, and rapidly decreases at the high wave numbers.

The process of energy exchange may be described as a series of local interactions in wave number space whereby new triads are created, usually at higher average wave number, connected to earlier triads, usually at lower average wave number. Energy flows through the spectrum via the Fourier modes which connect each pair of triads, thus maintaining energy transfer through local interactions. The generally rapid transfer of energy to higher wave number modes (the rate depends not only on triad shape, but also on the geometrical orientations of wave vectors and complex amplitudes in spectral space) leads to a rapid increase in total vorticity. This is in contrast with the class $k_1 \ll k_2, k_3$ where total vorticity increases slowly. As before, differences in phase produce differences in detail only.

4.4 Dimensionality

Although the cascade of energy is generally toward higher wave numbers, in two dimensions invariance of total vorticity requires that some energy be transferred to lower wave numbers as well. This is readily observed in the total enstrophy spectrum, as shown for example in Fig. 7 for the class $k_1 \sim k_2 \sim k_3$. There is a clear shift in total vorticity, to lower wave numbers in two dimensional calculations, and higher wave numbers in three dimensional calculations. The same shift is observed in the high wave number modes of class $k_1 \ll k_2, k_3$. In 3D, enstrophy increases rapidly at the highest wave numbers, signaling the development of a singularity in the vorticity field [2].

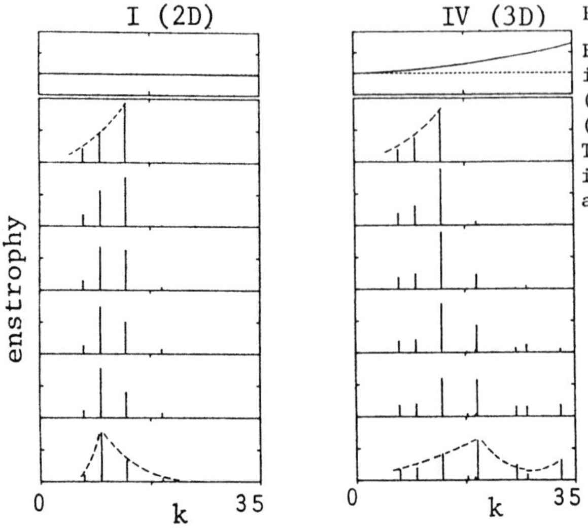

Fig. 7

Evolution of the enstrophy spectrum in two and three dimensions. (a) Triad 6, orientation I (2D); (b) Triad 6, orientation IV (3D). Total enstrophy (vorticity squared) is shown at the top plotted against time.

The evolution of the vorticity field of Fig. 7 is shown in physical space in Fig. 8. Since the initial three wave vectors lie in one plane, all subsequent wave vectors are constrained to that plane, and contours of constant enstrophy may be displayed on any plane parallel to the triad plane (triad evolution is pseudo-three dimensional, in the sense that the complex amplitude vectors evolve three dimensionally while the wave vectors remain in a plane). In purely two dimensional evolution, the smaller scales are produced solely through the convective redistribution of vorticity. As a result the vorticity field remains rather diffuse. In three dimensions, vortex stretching rapidly acts to concentrate vorticity within small eddies and along sheets.

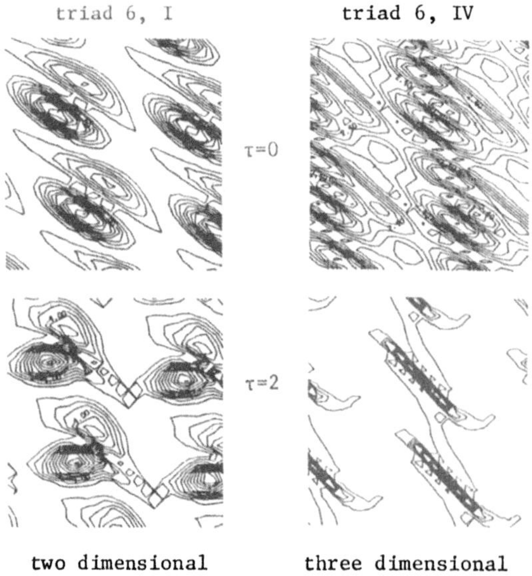

Fig. 8

Contours of constant enstrophy in physical space at non-dimensional time 0 and 2. The initial triads are those of Fig. 7.

5. Concluding Remarks

The analysis is based on the observation that in Fourier spectral space the Navier-Stokes equations evolve via a superposition of triad interactions, where each interaction involves three wave vectors which form a closed triangle. In examining triad evolution, we find that local interactions in wave number space have a fundamentally different character than interactions among wave numbers widely separated in the spectrum.

Three interacting modes, all of roughly the same scale, evolve in a manner consistant with our usual notion of the cascade process. Energy flows through the spectrum as a series of local triad interactions, each new triad formed by completing incomplete triads, and each connected to an existing triad by a common mode through which the energy is transferred. Total vorticity rapidly increases as energy is transfered to higher wave numbers.

The triad interaction between wave numbers which are widely separated in wave number, on the other hand, is quite different. Although there is no direct energy transfer between large and small scale modes, the large scale mode strongly influences the evolution which takes place, first by allowing only new modes connected to it as a triad to receive significant amounts of energy, and second by regulating the rate at which the high wave number modes distribute their energy. The small scales distribute their energy in a rather narrow region of the spectrum, increasing by only small amounts the total vorticity.

We might think of the large scales as overseers of the small scales. When they have a great deal of energy, they force the small scales to adjust to their presence by regulating the process by which the small scales can evolve. However, due to interactions among scales of comparable size, the large scales themselves lose energy to somewhat smaller scales, those to yet smaller scales, and so on (a la Richardson). If there is no source of energy to maintain the strength of the large scales, they eventually decrease in strength, and as they do their command over the small scales weakens.

These conclusions appear to be qualitatively valid regardless of differences in phase, and in both two and three dimensions. The lack of vortex stretching in two dimensions and the consequent invariance of total enstrophy, however, leads to a backward cascade of vorticity to lower wave numbers in two dimensional triad calculations which in three dimensions appears to be primarily forward.

A Final Word from J.G.B.

Stan Corrsin passed away one month before I presented this work to The European Turbulence Conference. We shall miss his knowledge, his insights, and him. I shall miss most of all his wonderful sense of humor.

References

1. G.K. Batchelor: The Theory of Homogeneous Turbulence, (Cambridge University Press, Cambridge 1953)
2. M.E. Brachet, D.I. Meiron, et al.: J. Fluid Mech. 130, 411 (1983)
3. J. Lee: J. Fluid Mech. 120, 155 (1982)
4. J. Lee: Phys. Fluids 22, 40 (1979)
5. J. Lee: J. Math. Physics 16, 1359 (1975)
6. D.C. Leslie: Developments in the Theory of Turbulence, (Clarendon Press, Oxford 1973)
7. E.D. Lorenz: Tellus 12, 243 (1960)
8. S.A. Orszag, G.S. Patterson: Phys. Rev. Letters 28, 76 (1972). Also Lect. Notes in Physics, 12, (Springer, Heidelberg 1972), p.127

Grid Generated Turbulence Exhibiting a Peak in the Spectrum

M. Michard, J. Mathieu, R. Morel, E. Alcaraz, and J.P. Bertoglio

Laboratoire de Mécanique des Fluides et d'Acoustique, Ecole Centrale de Lyon, Unité Associée au CNRS No. 263, B.P. 163, F-69131 Ecully Cedex, France

In order to investigate the behavior of a turbulence associated with a non- equilibrium spectrum, a grid equipped with propellers is used. Reynolds stress profiles clearly show the effects of the mean flow induced by the propellers : a spatially periodic distribution of anisotropy is obtained. The rotation of the propellers introduces a peak in the kinetic energy spectrum. The spectral disturbance, is found to be strongly anisotropic near the grid and to relax towards an axisymmetric state. A large eddy simulation is carried out ; the results show good qualitative agreement with the experimental spatial periodicity of inhomogeneities.

1. INTRODUCTION

Most turbulence models are based on the implicit assumption that the velocity fluctuations have a spectrum of universal form (equilibrium spectrum). Such an assumption is justified for turbulent fields in which any kind of spectral disturbances, due to initial conditions, external effects..., have been removed by nonlinear energy transfer between wavenumbers.

Such flows have been exhaustively studied both experimentally and theoretically. For example, many theories have been developed for homogeneous turbulence and compared to data provided by experimental studies of grid generated turbulence.

Many turbulent flows are, however, more complex. They have spectra departing substantially from equilibrium : several peaks may be present. Such flows often contain two-dimensional organized structures (e.g. the first stages of shear layers and jets) or oriented eddies induced by lift. These structures are associated with high local rotation rates and/or strong inhomogeneities.

In previous studies, several kinds of "active" grids were used to control the initial conditions |1| |2|. The behavior of a high wavenumber spectral peak, generated by a "zither", has also been investigated |3| |4|. Exponential decay of the spectral disturbance was found.

In the present work, an "active" grid equipped with small rotating propellers is used ; this grid generates a turbulent field exhibiting both a spatial periodicity and a peak in the spectrum.

The experimental set-up is described in section 2. Experimental results relative to inhomogeneities and spectral data are reported in section 3. A comparison with large eddy simulation is presented in section 4

2. EXPERIMENTAL FACILITY

A grid equipped with one hundred propellers generates the turbulence. The 3-bladed propellers are driven by five electric motors ; their axes are parallel to the mean flow. In order to avoid instabilities, adjacent propellers counter-rotate. The vortex field issuing from the grid is amplified by an axisymmetric contraction of area ratio 9/25. The grid mesh is 5 cm and the propellers are 4 cm in diameter. A sketch of the experimental set-up is shown in fig. 1 ; the X_1 axis is the centerline of the wind tunnel (fig. 1) . $X_1 = 0$ m corresponds to the end of the distorting duct.

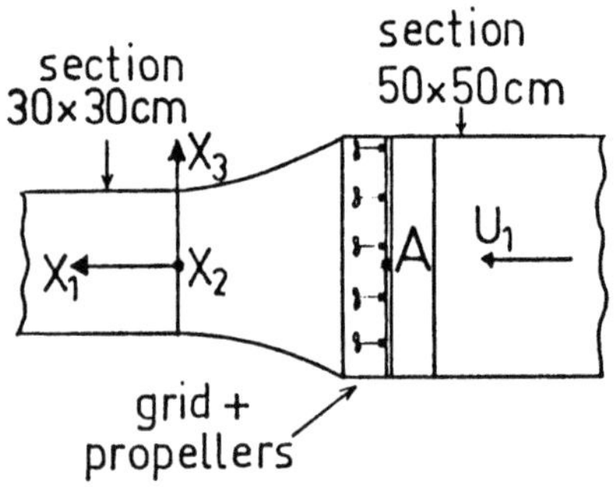

Fig. 1 : Experimental set-up

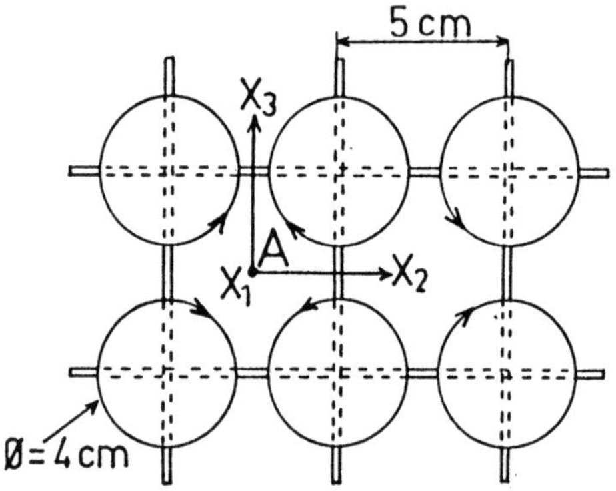

Fig. 2 : Sketch of the grid

The measurements are made with an X-wire anemometer ; the upstream velocity U_1 is 4 ms^{-1} ; the rotation rate of the propellers is N_{ad} = 90 Hz. A typical value of the turbulent Reynolds number $Re = \overline{(q^2)}^2/9\varepsilon\nu$ is 140 at X_1 = .5 m.

3. EXPERIMENTAL RESULTS

With the propellers rotating, spectral disturbances are introduced. Two distinct zones have to be considered. The first zone (A) is located between the grid and the end of the contracting duct. In this zone, three peaks at the blade-passing frequencies (3No)and its first two harmonics (No and 2No) are found.

In the second zone (B), the velocity spectrum exhibits a single peak at frequency No. All the measurements presented later are performed downstream of the section X_1 = 0.5 m. We may vary the basic frequency No as well as the mean velocity U_1 ; we found the following :

- For a given value of the mean velocity U_1, a peak is observed in the spectrum for a limited range of the rotation rate No. The intensity of the peak is maximum for No = Nad.
- The adequate frequency Nad increases as the mean velocity increases ; thus the peak is connected with lift of the blades.

In every case under consideration, the peaks at frequencies 2 No and 3 No vanish in zone (B).

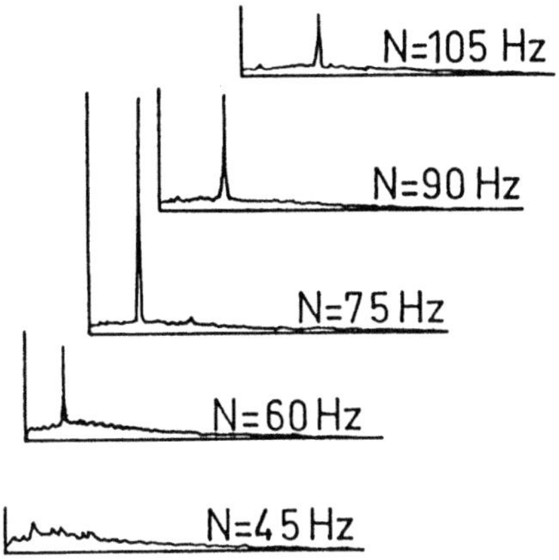

Fig. 3 : Spectra for different values of N_0

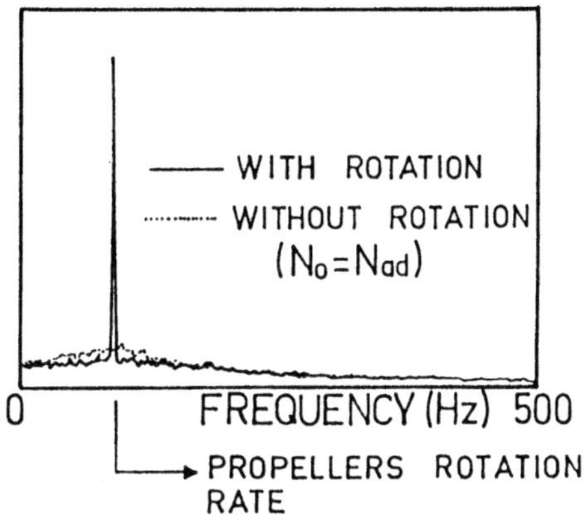

Fig. 4 : Spectra with and without rotation

Velocity spectra obtained for several rotation rates No at a particular location (X_1 = .5 m) on the centerline are given in fig. 3 for a fixed value of U_1. Spectra corresponding to No = 0 and No = N_{ad} at the same location are shown in fig. 4.

In order to investigate the homogeneity of the flow, the diagonal components of the Reynolds stress tensor were measured in the hatched area in fig. 5 at section X_1 = .5 m. This area contains the flow which has passed through four adjacent propellers. The results, presented in fig. 6, show that the axial component $\overline{u_1^2}$ is approximately uniform, whereas the components $\overline{u_2^2}$ and $\overline{u_3^2}$ display a periodic distribution. However, a balance between $\overline{u_2^2}$ and $\overline{u_3^2}$ occurs, so that the turbulent kinetic energy is nearly constant in a given cross

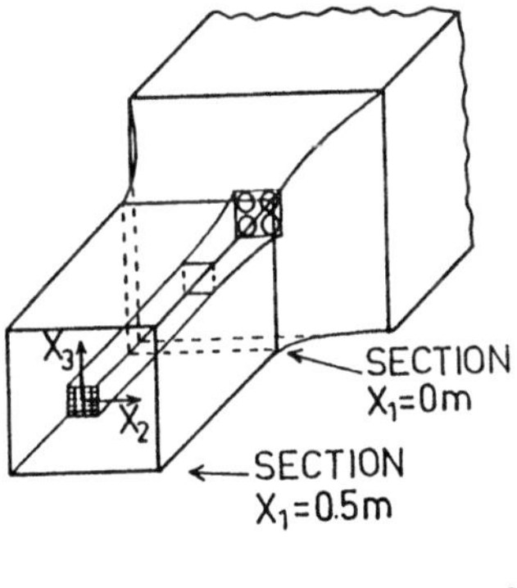

Fig. 5 : Sketch of the test section ; the measurements presented in fig. 6 and 7 are performed in the hatched area

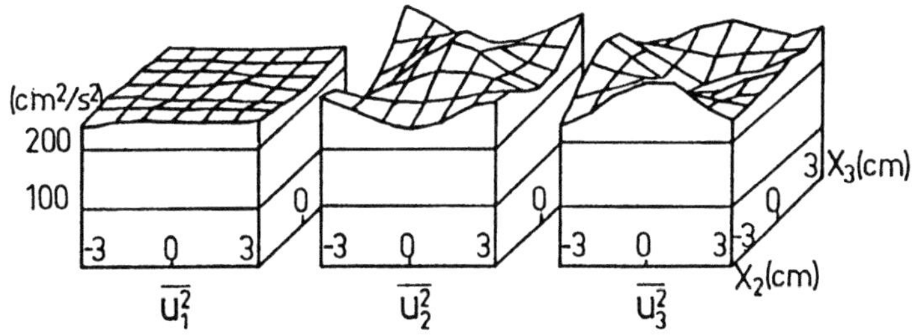

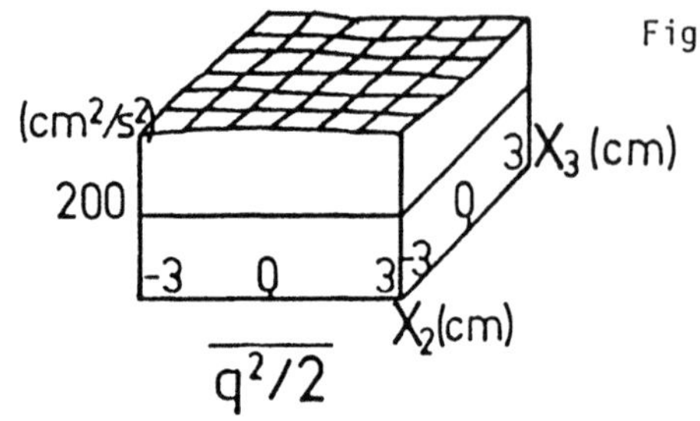

Fig. 7 : Kinetic energy profile

section (fig. 7). This spatial period of the field is 6 cm in both the X_2 and X_3 directions ; this period is exactly twice that of the mesh size of the grid (5 cm), when account is taken of the 3/5 contraction factor imposed by the distorting duct in these directions.

A qualitative explanation of this result can be provided by examining the mean flow induced by the propellers near the grid : at point A, shown in fig. 8, the flow is subjected to a local plane strain whose axes of contraction and dilatation are X_2 and X_3 respectively ; the effect of such a strain is to increase $\overline{u_2^2}$ and to reduce $\overline{u_3^2}$. At point B the sign of the strain is reversed and $\overline{u_2^2}$ decreases and $\overline{u_3^2}$ increases. Therefore, the periodicity of the field is twice that of the grid. The same trends are found when the propellers are at rest, but the phenomenon is not as strong.

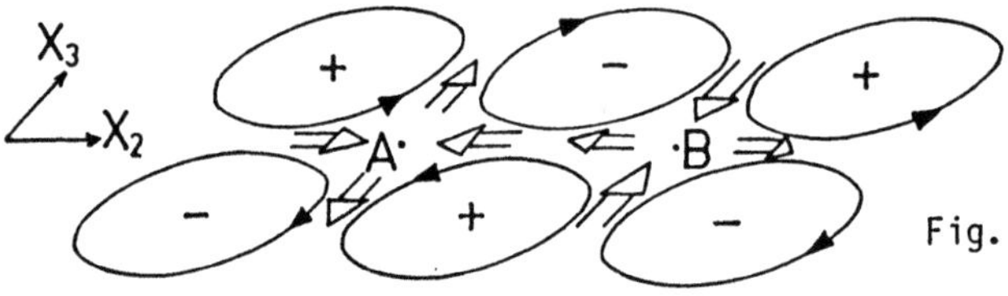

Fig. 8 : Local strains induced by the propellers

The evolution of the peak is measured along two mean streamlines passing through points A and B, respectively (fig. 8) ; along these streamlines, $\overline{u_1u_2}$ and $\overline{u_1u_3}$ have no significant values and $\overline{u_2u_3}$ is expected to be zero. The measurements are made downstream of X_1 = .5 m where only a single peak appears in the spectrum.

The peak values of the three components , $\overline{u_i^2}_p$, are defined by :

$$\overline{u_{ip}^2} = \int_{N_{ad}-\Delta N/2}^{N_{ad}+\Delta N/2} E_{ii}(N)\,dN \qquad i = 1, 2 \text{ or } 3 \qquad (1)$$

where $E_{ii}(N)$ is the one-dimensional spectrum of the u_i velocity component and ΔN is the bandwidth of the spectral disturbance. Results are given in fig. 9. In both cases, the anisotropy of the peak is stronger than that of the Reynolds stress tensor as a whole. The decay of $\overline{u_{1p}^2}$ corresponds to a power law whose exponent is - 1.2. Different behavior is observed for the two other components : in the peak, the energy is redistributed between u_2 and u_3. After the section X_1 = 3 m, $\overline{u_{2p}^2}$ and $\overline{u_{3p}^2}$ are nearly equal and uniformly distributed. It should be noted that at station X_1 = 3 m, the peak in $\overline{u_1^2}$ has nearly disapeared, whereas the two other have still significant amplitudes (fig. 10).

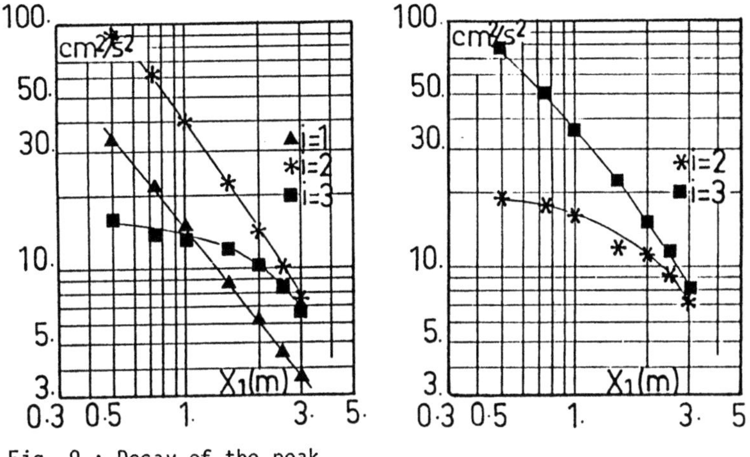

Fig. 9 : Decay of the peak

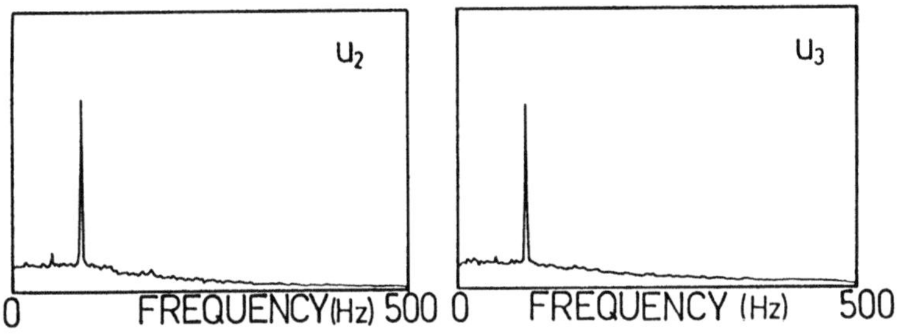

Fig. 10 : Spectra at X_1 = 3 m

In order to get better insight into the anisotropy of the peak, it is convenient to apply the concept of invariants used by LUMLEY and NEWMANN |5|. Let II and III be the two non-dimensional scalars (invariants) defined by :

$$II = b_{ij}\, b_{ji} \qquad \text{and} \qquad III = b_{ik}\, b_{kj}\, b_{ji} \tag{2}$$

where b_{ij} is the deviatoric part of the Reynolds stress tensor :

$$b_{ij} = \frac{\overline{u_i u_j}}{\overline{u_k u_k}} - \frac{1}{3}\delta_{ij} \tag{3}$$

The b_{ij} terms (and hence II and III) are identically zero for an isotropic turbulence. The state of an anisotropic turbulent field can be represented by a point in the (II, III) plane. In the present case, the only non-zero terms in (3) are the diagonal terms (i = j).

In the same way, the anisotropy of the peak may be characterized by the two "spectral invariants" II_s and III_s obtained by substituting $\overline{u_{ip}^2}$ for $\overline{u_i^2}$ in (3). The trajectory of the peak in the (II_s, III_s) plane is shown in fig. 11. It begins on the right side (at X_1 = .5 m) where $III_s > 0$ and moves to the left side where $III_s < 0$ (at X_1 = 3 m). The left boundary corresponds to a state in which $\overline{u_{2p}^2} = \overline{u_{3p}^2}$.

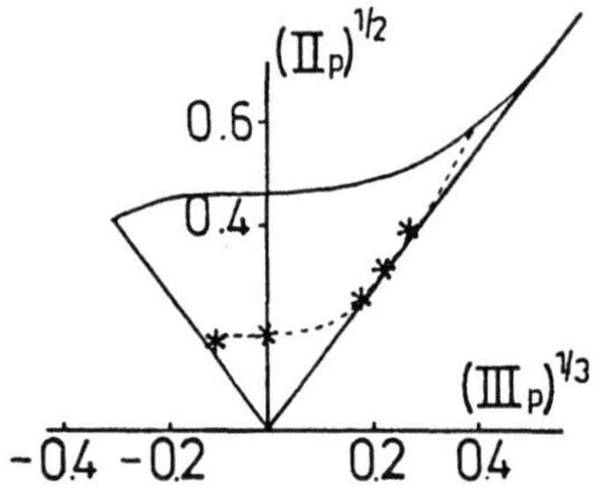

Fig. 11 : Relaxation trajectory in the (II_s, III_s) plane

4. LARGE EDDY SIMULATION

With a view to validating the interpretation of the periodic distribution of anisotropy, a large eddy simulation is made. A pseudo-spectral method is applied on a cubic box with periodic boundary conditions. Four counterrotating vortices with parallel axes are introduced into the initial conditions of the computation (fig. 12). They are embedded in a random fluctuating velocity field. A 16^3 grid is used and an eddy viscosity is used to parameterize the effects of the subgrid scales (SMAGORINSKY |5| or CHOLLET and LESIEUR |6|).

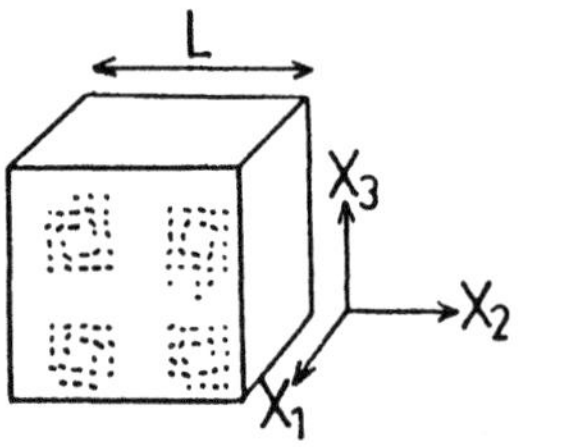

Fig. 12 : Initial field

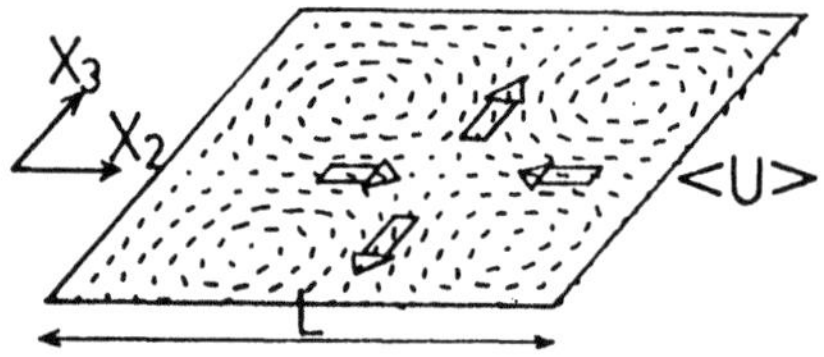

Fig. 13 : Mean field

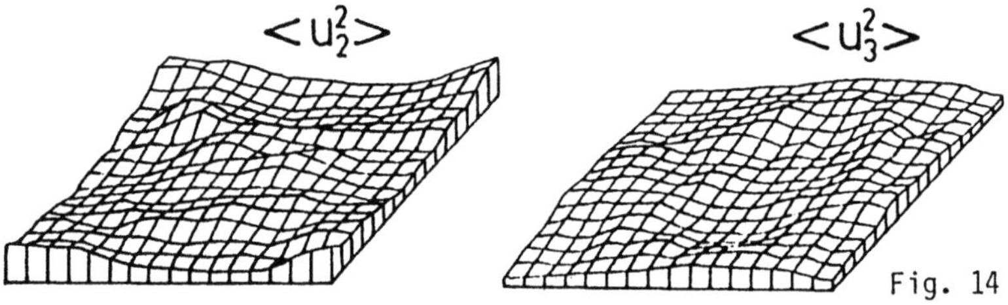

Fig. 14 : Residual field

At each point in the (X_2, X_3) plane, the results are averaged over 16 points in the X_1 direction and over 8 different realizations to produce a mean velocity $<U_i>$ The residual velocity u_i is thus defined :

$$U_i = <U_i> + u_i$$

Fig. 13 and 14 show the mean flow profile $<U_i>$ and the distribution of two components of the Reynolds stress tensor $<u_2^2>$ and $<u_3^2>$ The agreement with the experimental trend is good : the spatial periodicity of the anisotropy distribution clearly appears.

This work was supported by the Direction des Recherches Etudes et Techniques under contract n° 84-355.

REFERENCES

1 M. GAD EL HAK, S. CORRSIN : J. Fluid Mech., Vol. 62, p. 115 (1974)

2. S.C. LING, C.A. WAN : Phys - Fluids, Vol. 15, p. 1363 (1972)

3. R.M. KELLOGG, S. CORRSIN : J. Fluid Mech., Vol. 96, p. 641 (1980)

4. E. ITSWEIRE, VAN ATTA : J. Fluid Mech., Vol. 145, p. 423 (1984)

5. J. LUMLEY, G. NEWMANN : J. Fluid Mech., Vol. 82, p. 161 (1977)

6. J. SMAGORINSKY : Month. Weath Rev., Vol. 91 (1963)

7. J.P. CHOLLET, M. LESIEUR : J. Atm. Sci., Vol. 38, p. 2747 (1981)

Spectral Analysis of a Three Dimensional Homogeneous Turbulence Submitted to a Solid Body Rotation

C. Cambon[1] *and L. Jacquin*[2]

[1]Laboratoire de Mécanique des Fluides et d'Acoustique, Ecole Centrale de Lyon, Unité Associée au CNRS No. 263, B.P. 163, F-69131 Ecully Cedex, France
[2]Office National d'Etudes et de Recherches Aérospatiales, 29 Av. de la Division Leclerc, F-92320 Chatillon, France

1. Anisotropisation mecanisms - In search of a criterion of bidimensionalisation

It is now well admitted that a solid body rotation acting on a three-dimensional and isotropic turbulence leads to the slowing of the kinetic energy decay of the fluctuating velocity field, together with a tendency to its bidimensionalisation. This last effect, which has been experimentally analysed |1|, justifies the theoretical (bi-dimensional) frame in which the turbulent flows submitted to a strong rotation have often been studied up to now. However starting from a three-dimensional state, no clear manifestation of a transition has been brought into light by direct numerical simulation |2|, |3| except in the case of initial conditions that already contained important two-dimensional structures |4|. Nevertheless these simulations reflect a clear change in the structure of the flow, primarily through the anisotropisation of the integral lengthscales, and as we have shown |5|, even in the case of a rigourously isotropic initial state, the behavior of the longitudinal length scales (separations parallel to the rotation axis) can be linked to a strong anisotropic structure of the second order spectral tensor, even if a quasi-spherical Reynolds tensor is maintained. These considerations illustrate the complexity of the problem of the transition, which has often been evaded by calling for the Batchelor-Proudman theorem, and it shows the necessity of a full three-dimensional spectral description in order to take into account the dependence on the orientation of the wave vector.

Because of its dispersive and anisotropic properties, the inertial wave regime associated with the linearised equations can initiate a loss of isotropy which is the prerequisite for the rising of a particular orientation (along the rotation axis) of the vortices in the flow. However in the case of a rigourously isotropic initial state, this influence has no direct effect on the double correlations but only on the triple correlations, which thus are responsible for the generation of orientated structures. This very subtle mechanism has been taken into account in a two-point closure |6| based on the Eddy Damped Quasi Normal Markovian theory |7|.

If one assimilates the non-linear interactions with collisions of modes, a striking analogy can be made between this problem and the formation, from a quasi-uniform distribution of gravitating particles, of rings in the equatorial plane of rotating giant planets. Figure 1 shows a numerical simulation of these collision effects |8|.

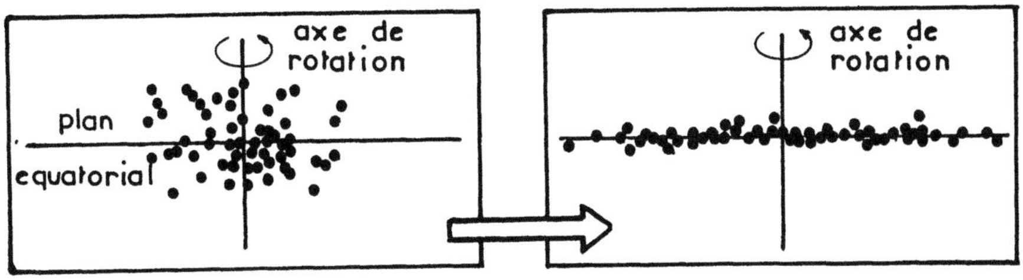

Fig. 1 - The simulation of collisions of lump of matter in the vicinity of the planets shows the rapid formation of a disc in the equatorial plane. From |8|.

In our model, a clear separation between "linear effects" and "non-linear interactions", is obtained by working with the eigen modes of the second order spectral tensor $\hat{U}_{ij}$, which are : the energy $\mathcal{E}(\underline{k}) = \frac{1}{2}\hat{U}_{ii}(\underline{k})$, the complex deviator $Z(\underline{k}) = \frac{1}{2}\hat{U}_{ij}(\underline{k}) N_i^*(\underline{k}) N_j^*(\underline{k})$ and the helicity, $h(\underline{k}) = -\frac{I}{2} k_i \varepsilon_{ijl} \hat{U}_{jl}(\underline{k})$ ($I = \sqrt{-1}$; * denotes the complex conjugate), where $\underline{N}$ and $\underline{N}^*$ are the eigen vectors of the rotation around $\underline{K}$.

In the rotating frame characterized by the rotation vector $\underline{\Omega}$, the rate equations governing $\hat{\underline{\underline{U}}}$ lead to the following system :

$$
\text{(I)}\quad
\begin{aligned}
&\frac{\partial \mathcal{E}}{\partial t} + 2\nu k^2 \mathcal{E} &&= T_{ii} \\
&\frac{\partial Z}{\partial t} + 2\left(\nu k^2 - 2I\,\underline{\Omega}\cdot\frac{\underline{k}}{k}\right) Z &&= T_{ij} N_i^* N_j^* \\
&\frac{\partial h}{\partial t} + 2\nu k^2 h &&= -I\, k_i \varepsilon_{ijl} T_{jl}
\end{aligned}
$$

The modelized transfer tensor $\underline{\underline{T}}$ which characterizes the triple correlations is a quadratic and non-local function of the parameters ($\mathcal{E}$, Z ,h), and it is affected by the rotation Ω through the parameter $\underline{\Omega}\cdot\left(\frac{\underline{k}}{k} \pm \frac{\underline{p}}{p} \pm \frac{\underline{q}}{q}\right)$ which drives the triadic ($\underline{k} + \underline{p} + \underline{q} = \underline{0}$) interactions. Starting from 3D isotropic initial conditions ($\mathcal{E} = \frac{E(k)}{4\pi k^2}, Z = h = 0$) only $\underline{\underline{T}}$ can generate an anisotropy that is a dependence of the energy $\mathcal{E}$ with respect to $\cos\theta = \frac{\underline{k}\cdot\underline{\Omega}}{k\Omega}$ together with the emergence of Z.

The first manifestation of a bidimensionalisation is a relative concentration of the energy $\mathcal{E}$ in the equatorial plane of the sphere of radius k ($\underline{k} \perp \underline{\Omega}$) . A second one, located in the small wave numbers, is the preponderance of the horizontal mode ($\perp \underline{\Omega}$) of the spectral tensor over the vertical one, in this plane. Both effects are visualized on figure 2. The distribution of $\mathcal{E}(k, \cos\theta)$ (at the top) shows the deficit of energy between the polar zone ($\underline{k} /\!/ \underline{\Omega}$, $\cos\theta = 1$) and the equatorial plane. The corresponding distribution of Z (k, $\cos\theta$) is visualized at the bottom ; for each wave number k, and each orientation $\cos\theta$, the segment length is proportional to $|Z|$ that is the difference between the eigen values of the spectral tensor $\hat{\underline{\underline{U}}}(\underline{k})$; its orientation, the same as that of the eigen vector of $\hat{\underline{\underline{U}}}(\underline{k})$ associated with the maximum eigen value, gives the direction of maximum correlation. These particular distributions explain that the increase of the ratio between the longitudinal length scales

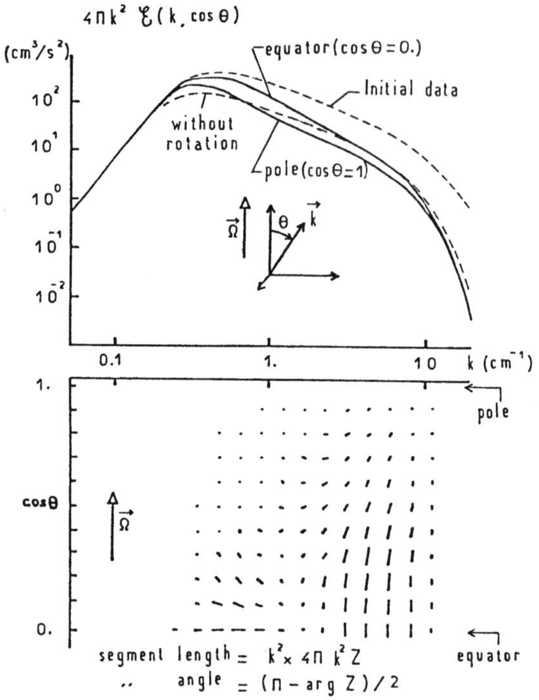

Fig. 2 - Angular distributions of the energy $\mathcal{E}(k,\cos\theta)$ and the complex deviator $Z(k,\cos\theta)$ at $\Omega t/2\pi$ = 0.5 (isotropic initial data, Ω = 10 s^{-1}).

$L_{11,3}/L_{33,3}$, which appears as a quite fine physical consequence of these subtle mechanisms, is not associated with a collapse of the longitudinal component of the Reynolds tensor as it could have been expected. These results, that can be closely related to those obtained in low magnetic Reynolds number MHD |9|, show that the tendency to the alignment of big eddies with the rotation axis direction (which is linked to the rather horizontal orientation of the segments in the small wave number range on the figure) does not exclude vertical velocity fluctuations. Thus, the behavior of the Reynolds stresses and of the length scale is consistant with a trend toward two-dimensionality in the domain of big structures.

2. Rotation effects on an axisymmetric turbulence - Experimental confrontation

From the idea that an experimental isotropic turbulence is obtained by means of a contraction applied to a grid generated turbulence |10|, a realistic spectral configuration ($\mathcal{E}$, Z) for the flow near a grid can then be calculated from the solution of a rapid axisymmetric dilatation applied to an isotropic spectrum $E(k)$. In particular, one can adjust by this way the ratio $\overline{u_3^2}/\overline{u_1^2}$ between the longitudinal (in the mean stream direction) and the transversal components of the Reynolds tensor, in order to obtain the classical value of 1.4. Thus, one can expect that the application of a sudden rotation to such an initial turbulence, will suitably simulate the experimental protocol adopted by Wigeland |11|, for the generation of rotating grid generated turbulence. Coming back to the equations (I) governing ($\mathcal{E}$, Z ,h) pure linear effects are now superposed to the non-linear ones, through the term ($\Omega \cos\theta$), which affects Z by rotating the eigen directions of $\hat{\underline{\underline{U}}}(\underline{k})$ with a rate proportional to ($\Omega \cos\theta$).

During a transitional period (lower than a quarter of a revolution) this scrambling of the orientations of energy would lead to violent changes of the anisotropic structure of the flow, but these changes are partly counterbalanced by not less important and immediat non-linear effects which insure in particular a quasi indifference of the longitudinal component of the Reynolds stress with respect to the rotation as found by Wigeland. After a quarter of a revolution, the linear effects are completely damped and one retrieves the trends brought into light in the case of isotropic initial conditions. Figures 3 and 4 which show the initial distributions of $\mathcal{E}$ and Z and those obtained after one revolution, illustrate the linear action on Z, the spectacular decrease of the initially preponderant energy in the polar zone as well as the clear diminution of the initially vertical structure of Z in the equatorial plane. Figure 5 illustrates the qualitatively good agreement between our results and Wigeland's

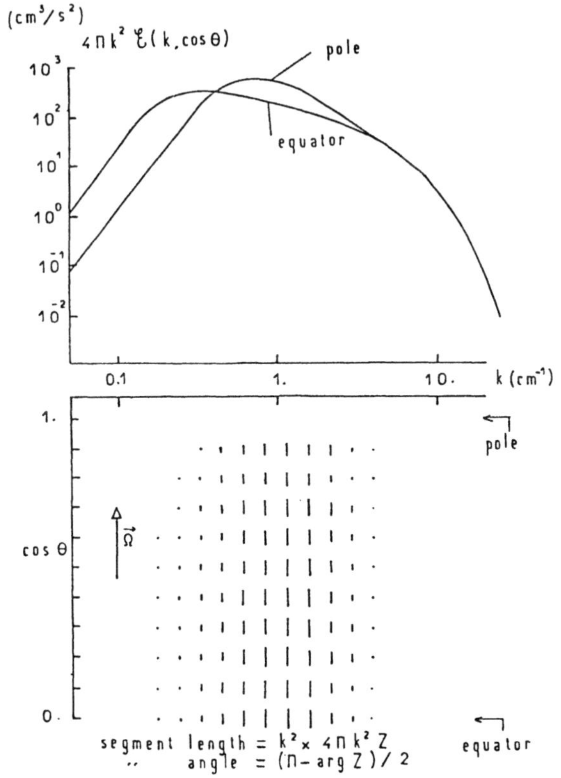

Fig. 3 - Initial angular distributions of the energy $\mathcal{E}$(k, cosθ) and of the complex deviator Z(k, cosθ) corresponding to the dilatation data.

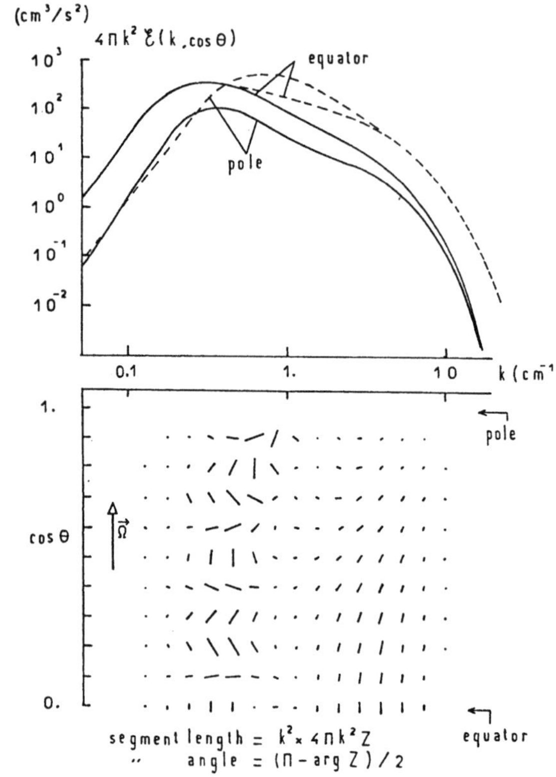

Fig. 4 - Angular distributions of the energy $\mathcal{E}$ (k, $\cos\theta$) and of the complex deviator Z(k, $\cos\theta$) at $\Omega t/2\pi$ = 1 starting from the dilatation data (see figure 3).

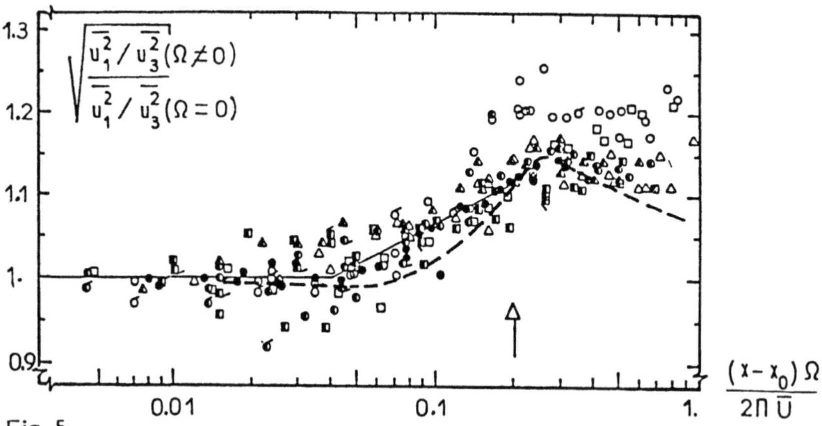

Fig. 5.

experiment. The experimental points relative to the ratio of both $\overline{u_1^2}/\overline{u_3^2}$ with and without rotation versus $\Omega t/2\pi$; $t = (x - x_0)/\bar{U}_3$ corresponds to various rotation rates Ω, grid mesh sizes and mean velocities $\bar{U}_3$. The curve corresponding to the calculation (dashed line) is in quite good qualitative agreement and enables to explain the arrow, drawn by Wigeland for emphasizing the (unexplained) change of behavior, as being the end of the action of the linear terms.

These results show the necessity of considering the grid turbulence as being essentially non-isotropic, at last in the case of fine orientation effects as those of the Coriolis force. Moreover they question the principle of a modelisation separated in "slow" and "rapid" linear and non-linear contributions of the pressure-strain correlations.

References

|1| HOPFINGER E.J., BROWAND F.K. & GAGNE Y., J. Fluid Mech. 125, pp. 505-531 (1982).

|2| DANG K. & ROY Ph., Proceedings of workshop on macroscopic modelling of turbulent flows and fluid mixtures. Springer-Verlag, vol. 230 (1985).

|3| BARDINA J., FERZIGER J.H. & ROGALLO R.S., J. Fluid Mech. 154, 321-336 (1985).

|4| DANG K. & ROY Ph., Proceeding of 5th Symp. on Turb. Shear Flows, Ithaca (NY) (1985).

|5| CAMBON C. & JACQUIN L., 7th A.M.S. Symp. on Turb. and Diff., Boulder (1985).

|6| CAMBON C., State thesis Lyon I (1982).

|7| ORSZAG S.A., J. Fluid Mech. 41, p. 369 (1970).

|8| BRAHIC A., La Recherche, n° 173 (1986).

|9| CAPERAN Ph. & ALEMANY A., J. Mec. Théor. Appl., vol. 4, n° 2, p. 175-200 (1985).

|10| COMTE-BELLOT G., CORRSIN S., J. Fluid Mech., 48, part 2, pp. 273-337 (1971).

|11| WIGELAND R.A., Ph. D. Illinois Institute of Technology (1974).

MIS: An Alternative for the Dissipation Equation

B. Aupoix, J. Cousteix, and J. Liandrat
ONERA-CERT-DERAT, 2 av. E. Belin, F-31055 Toulouse Cedex, France

The evaluation of the turbulence length scale is one of the key problems in turbulence modelling. Transport equations for the length scale or for the dissipation rate can be deduced from the NAVIER equations but each term must be modelled. Standard models are based upon dimensional analysis and tensorial invariance.

1. Spectrum Shape Assumptions

The Spectral Integral Method is an extension of a model previously introduced by COMTE-BELLOT and CORRSIN /3/ and REYNOLDS /6/. A priori shapes, based upon physical grounds, are proposed for the energy spectrum. The energy spectrum for the very large eddies can be described by a power law $E(k) = Ak^s$ while, between the energy containing range and the dissipative range, at sufficiently high REYNOLDS number, there exists an inertial range following KOLMOGOROV's law $E(k) = K_o\,\varepsilon^{2/3}k^{-5/3}$.

The energy spectrum can be mimicked with these two simple laws :

$$k \leq k_{max} \qquad E(k) = Ak^s$$

$$k \geq k_{max} \qquad E(k) = K_o\,\varepsilon^{2/3}k^{-5/3}$$

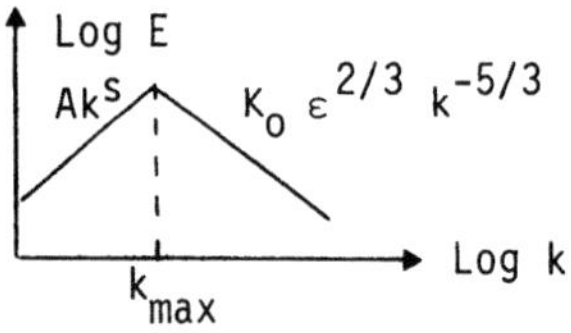

so that the turbulent kinetic energy reads :

$$\frac{1}{2}\,q^2 = \int_0^\infty E(k)dk = \frac{3s+5}{2(s+1)}\,K_o^{\frac{3s/3}{3s+5}}\,A^{\frac{2}{3s+5}}\,\varepsilon^{\frac{2(s+1)}{3s+5}} .$$

More physical spectrum shapes can be assumed in the vicinity of the energy maximum. If the spectrum shape is governed by the same parameters, the kinetic energy reads :

$$\frac{1}{2}\,q^2 = F\,A^{\frac{2}{3s+5}}\,\varepsilon^{\frac{2(s+1)}{3s+1}} . \tag{1}$$

2. Basic Assumptions

Extra hypotheses are needed to describe the evolution of the energy spectrum :

* As concerns the very large eddies ($k \ll k_{max}$, $E(k) = Ak^s$), their behaviour can be modelled with the help of inviscid rapid distortion theory as these very large eddies have large turnover times, thus are quite insensitive to turbulence/turbu-

lence interactions and are mainly governed by the mean velocity gradients. Inviscid rapid distortion theory then shows that the exponent s remains constant and gives the evolution of the level A.

* A one scale hypothesis is needed to identify the energy flux ε cascading along the inertial range with the rate at which energy is dissipated by viscous effects at the end of the inertial range.

In the following, we shall restrict ourselves to homogeneous (i.e. translation invariant) turbulence. No diffusion occurs and the problem can be reduced to a time-evolution problem in a reference frame conveyed by the mean flow. The evolution equation of the turbulent kinetic energy then reads :

$$\frac{d}{dt}\frac{1}{2}q^2 = P - \varepsilon$$

where $P = -\langle u_i'u_j'\rangle \frac{\partial U_i}{\partial x_j}$ is the production due to the action of the mean velocity gradient upon the REYNOLDS stresses.

The formula (1) can then be differenciated to obtain :

$$\frac{1}{\varepsilon}\frac{d\varepsilon}{dt} = -\frac{1}{s+1}\frac{1}{A}\frac{dA}{dt} + \frac{3s+5}{2(s+1)}\left(\frac{P-\varepsilon}{\frac{1}{2}q^2} - \frac{1}{F}\frac{\partial F}{\partial t}\right) \qquad (2)$$

i.e. there is a linear relation between the inverses of the evolution time-scales of the dissipation, the very large scale level A, the turbulent kinetic energy and the coefficient F linked to the shape of the spectrum in the energy containing range. The coefficients in this relation are constants, linked to the exponent s of the spectrum in the very large eddies.

3. Decaying Turbulence

For purely decaying turbulence, in the absence of velocity gradient, there is no energy production (P = 0). Moreover, the rapid distortion theory shows that the very large eddies are invariant ($\frac{dA}{dt} = 0.$). At last, experiments and simulations show that there is a self-similar decay, i.e. the spectrum shape in the energy containing range does not change if the REYNOLDS number is high enough. The equation then reduces to :

$$\frac{1}{\varepsilon}\frac{d\varepsilon}{dt} = -C_{\varepsilon 2}\frac{\varepsilon}{\frac{1}{2}q^2} \qquad C_{\varepsilon 2} = \frac{3s+5}{2(s+1)}$$

as previously proposed by COMTE-BELLOT and CORRSIN /3/ and REYNOLDS /6/.

For turbulence decaying in presence of a solid body rotation, there is still no energy production and the rapid distortion theory still shows that the large eddies spectrum does not change. It has been shown that the role of rotation is to inhibit energy transfers so that the energy is trapped in the energy containing range. The spectrum shape will thus change to store the extra trapped energy ; the dissipation equation now reads :

$$\frac{1}{\varepsilon}\frac{d\varepsilon}{dt} = - C_{\varepsilon 2}\left(\frac{\varepsilon}{\frac{1}{2}q^2} + \frac{1}{F}\frac{dF}{dt}\right) .$$

Moreover, we have already /2/ modelled this equation as :

$$\frac{1}{\varepsilon}\frac{d\varepsilon}{dt} = - (C_{\varepsilon 2} + C^*_\varepsilon)\frac{\varepsilon}{\frac{1}{2}q^2}$$

where C^*_ε is a function of the ratio of the rotation and turbulence time scales :

$$C^*_\varepsilon = \frac{0.2236\,\omega^{*2} + 0.0303\,\omega^*}{0.2540\,\omega^{*2} + 0.1567\,\omega^* + 1} \qquad \omega^* = \frac{\omega\frac{1}{2}q^2}{\varepsilon}$$

The evolution of the shape of the spectrum in the energy containing range can then be modelled as :

$$\frac{1}{F}\frac{dF}{dt} = \frac{C^*_\varepsilon}{C_{\varepsilon_2}}\frac{\varepsilon}{\frac{1}{2}q^2}$$

4. Flows with Energy Production

We shall assume that the spectrum shape F is only changed by energy trapping due to transfer inhibition by rotation. On the one hand, this transfer inhibition will still be governed by the ratio between rotation and turbulence characteristic time scales. On the other hand, the evolution of the spectrum shape is linked to the evolution of the turbulent kinetic energy so that the above expression now reads :

$$\frac{1}{F}\frac{dF}{dt} = -\frac{C^*_\varepsilon(\omega^*)}{C_{\varepsilon 2}}\frac{P - \varepsilon}{\frac{1}{2}q^2} \qquad \omega^* = \frac{\omega\frac{1}{2}q^2}{\varepsilon}$$

The time evolution of the very large scales can be obtained with the help of rapid distortion theory. However, rapid distortion computations are cumbersome and lengthy ; a simplified rapid distortion theory is needed to obtain a tractable model.

A solution is the use of the tensorial volumes of turbulence introduced by LIN and WOLFSHTEIN /4/. These volumes are defined as the volume integrals of two-point correlations :

$$\frac{1}{2}q^2 V_{ij}(\underline{x}) = \iiint \langle u'_i(\underline{x})\, u'_j(\underline{x} + \underline{r})\rangle\, d^3\underline{r}$$

and we have shown /1/ that they make sense only when the spectra of all REYNOLDS stresses behave like k^2 for wavenumbers close to zero. The spectra $\phi_{ij}(k)$ of the REYNOLDS stresses $\langle u'_i u'_j\rangle$ are then expressed as :

$$\phi_{ij}(k) = A_{ij}\,k^2 \qquad k \ll k_{max}$$

and the coefficients A_{ij} satisfy the relations :

$$\frac{1}{2}A_{ii} = A \qquad A_{ij} = 4\pi\, q^2\, V_{ij}$$

An evolution equation for the tensorial volumes of turbulence can be obtained from the NAVIER equation. For homogeneous flows, this equation reduces to :

$$\frac{d}{dt} A_{ij} = - (A_{ik} \frac{\partial U_j}{\partial x_k} + A_{jk} \frac{\partial U_i}{\partial x_k})$$

i.e. a simple linear model with pressure effects discarded. The omission of the pressure terms and the failure to build a better model are discussed in /1/. Figure 1 shows the evolution of the very large eddies time scale for isotropic turbulence submitted to a plane strain ($\frac{\partial U}{\partial x} = - \frac{\partial V}{\partial y} = D$) or a shear ($\frac{\partial U}{\partial y} = S$). Rapid distortion theory (RDT) and the simple model deduced from the tensorial volume of turbulence approach (LW) give the same results for strained turbulence but slightly differ for sheared turbulence. The difference of behaviour of the very large eddies time scale according as the flow is subjected to strain of shear must be stressed. For strained turbulence, the time scale rapidly reaches an asymptotic value while for sheared turbulence, it first increases and then decreases towards zero.

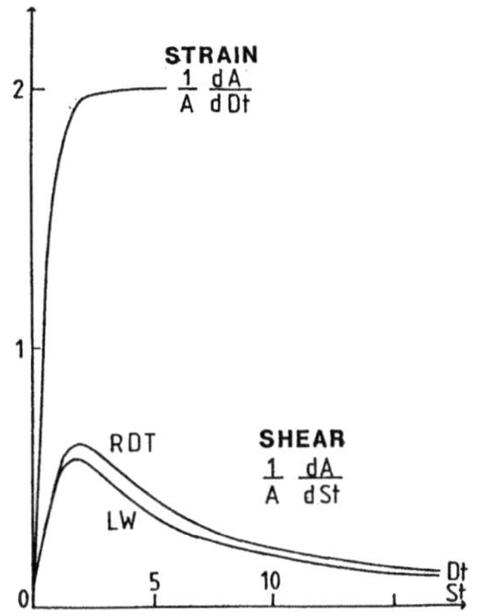

Fig.1

The use of the tensorial volumes approach to build a simple rapid distortion model now imposes a value of two for the exponent s of the very large eddies spectrum.

7. Low REYNOLDS Number Flows

The proposed spectrum shapes made with a power law in the very large eddies and a KOLMOGOROV law in the inertial range with a smooth transition in the energy containing range can mimic realistic spectra at high REYNOLDS numbers. At low REYNOLDS numbers, the presence of the dissipative range must be accounted for. The KOLMOGOROV spectrum is then modified as :

$$E(k) = K_o \, \varepsilon^{2/3} \, k^{-5/3} \, f(k/k_D) \qquad k_D = (\frac{\varepsilon}{\nu^3})^{1/4}$$

where several damping functions f have been proposed. Instead of dealing with this form of spectrum, we shall use truncated KOLMOGOROV spectrum having the same energy i.e. :

$$\int_0^{\infty} K_o \, \varepsilon^{2/3} \, k^{-5/3} \, f(k/k_D) dk = \int_0^{\alpha k_D} K_o \, \varepsilon^{2/3} \, k^{-5/3} \, dk$$

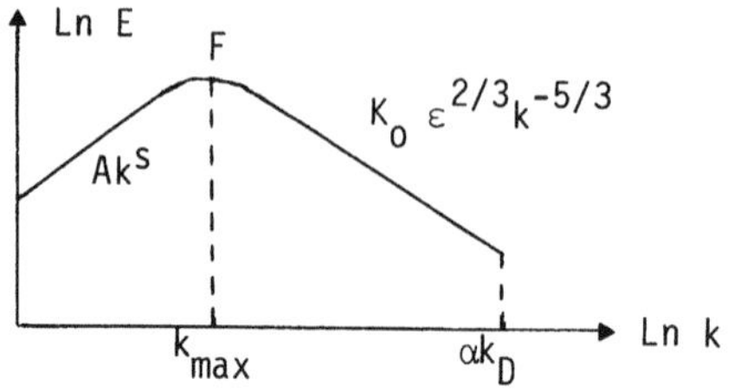

The dissipation equation can then be obtained after a similar algebra. It reads :

$$\left[1 + (2 - C_{\varepsilon 2}) \frac{\beta}{3\sqrt{R}}\right] \frac{1}{\varepsilon}\frac{d\varepsilon}{dt} = - \frac{1}{s+1} (1 + \frac{2\beta}{3\sqrt{R}}) \frac{1}{A}\frac{dA}{dt} + C_{\varepsilon 2} \left[\frac{P - \varepsilon}{\frac{1}{2} q^2} - (1 + \frac{2\beta}{3\sqrt{R}})\frac{1}{F}\frac{dF}{dt}\right]$$

where $R = \frac{q^4}{9\nu\varepsilon}$ is the turbulence REYNOLDS number and β a coefficient linked to α. Numerical computations with viscous damping laws f given by PAO, SAFFMAN or obtained as a solution of EDQNM equations /1/ gave values of α close to one and for β, respectively 2.065, 2.047 and 2.079. We choose a mean value $\beta = 2.06$.

We have no theoretical tool to predict the possible changes of the spectrum shape coefficient F at very low REYNOLDS number. For purely decaying turbulence, without mean velocity gradients, if we assume no variation of the spectrum shape as for high REYNOLDS numbers, the model gives :

$$- \left[1 + (2 - C_{\varepsilon 2}) \frac{\beta}{3\sqrt{R}}\right] \frac{\varepsilon^2}{\frac{1}{2} q^2} \frac{d\varepsilon}{dt} = C_{\varepsilon 2} = \frac{11}{6} (s = 2) \qquad (s = 2)$$

On figure 2, this equation is compared with numerical studies of decaying turbulence obtained with the EDQNM closure. The agreement is fairly good and superior to any low REYNOLDS number model compiled by PATEL et al /5/. This result leads us to neglect, as a first approximation, variations of the spectrum shape in the energy containing range due to low REYNOLDS number effects.

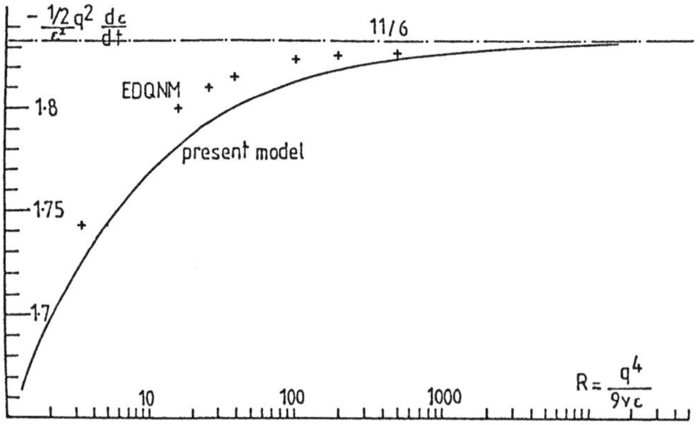

Figure 2

6. Comparison with Experiments

The MIS dissipation equation has been checked by comparison with experiments. Details about the computation and the initial data can be found in /1/. On fig. 3 to 8, the time evolutions of the kinetic energy obtained with various dissipation equations are compared. The continuous line is obtained with the standard dissipation equation, including rotation effects to improve sheared flow predictions as presented in /2/. The dissipation equation used reads :

$$\frac{d\varepsilon}{dt} = (C_{\varepsilon 1} + C_\varepsilon^*) \frac{P}{\frac{1}{2} q^2} \varepsilon - (C_{\varepsilon 2} + C_\varepsilon^*) \frac{\varepsilon^2}{\frac{1}{2} q^2}$$

where $C_{\varepsilon 1} = 1.44$, $C_{\varepsilon 2} = 1.92$.

The three other lines are results obtained with the MIS dissipation equation, including one after another rotation effects (i.e. F-variations) and low REYNOLDS number effects.

For strained experiments such as TOWNSEND experiment (fig. 3) and GENCE and MATHIEU experiment (fig. 4), all the models gave similar results. It must be pointed out that, in the standard dissipation equation, both coefficients $C_{\varepsilon 1}$ and $C_{\varepsilon 2}$ are tuned with reference to experiment while in the MIS approach, the only free parameter s is assumed equal to two to simplify the rapid distortion model.

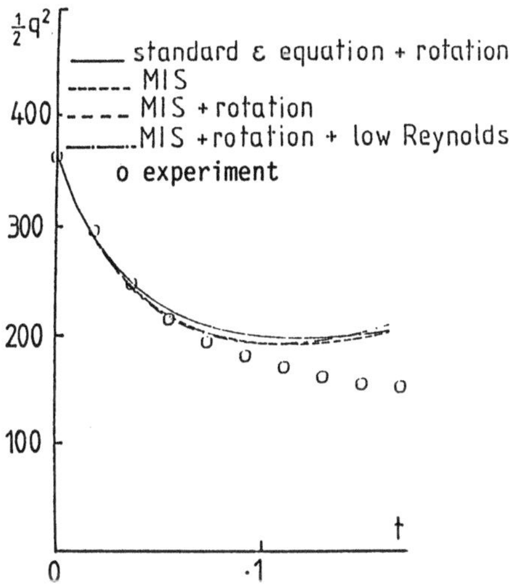

Fig. 3 - Strained flow - TOWNSEND experiment

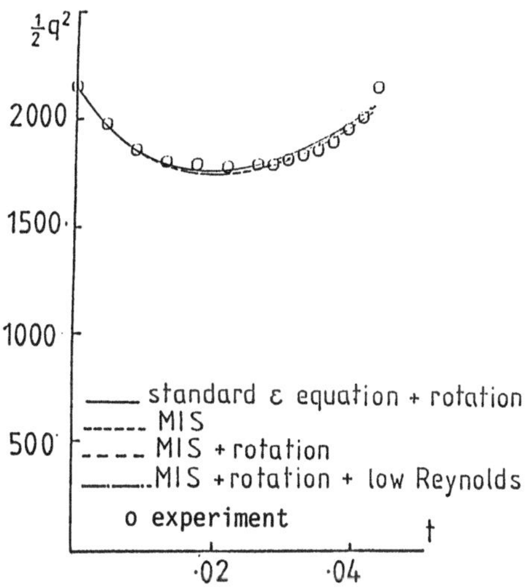

Fig. 4 - Strained flow - GENCE and MATHIEU experiment

For weakly sheared flows such as ROSE experiment (fig. 5) and CHAMPAGNE et al experiment (fig. 6), the discrepancy between the predictions is small. One can note that the low REYNOLDS number effects correction slightly improve the prediction.

For strongly sheared flows such as HARRIS et al experiment (fig. 7) and TAVOULARIS et al experiment (fig. 8), the prediction is highly improved with the MIS approach. On the one hand, the rotation effects improve the prediction as can be seen by comparing MIS predictions with and without rotation effects. This effect has been already brought into evidence in /2/. The new improvement due to the MIS approach is the introduction of the very large scale time-scale which has a different behaviour for strained turbulence, weakly sheared turbulence (small S.t) and highly sheared turbulence (large S.t) as shown on fig. 1. This effect, which is not accounted for in the standard dissipation equation, highly improves the prediction.

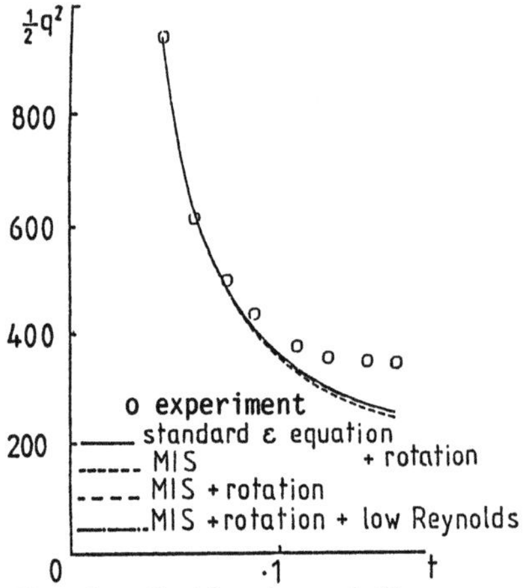

Fig. 5 - Weakly sheared flow ROSE experiment

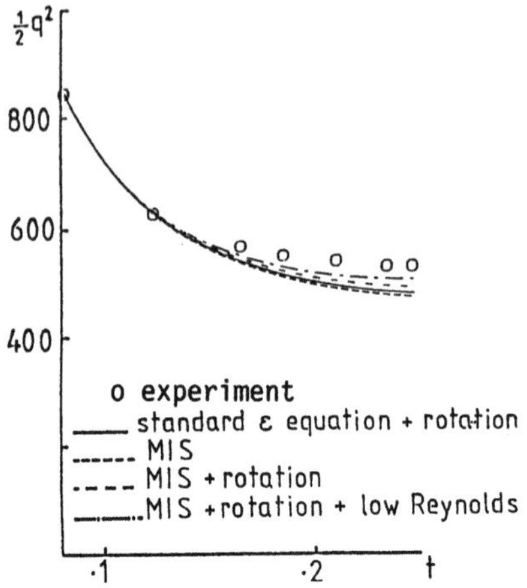

Fig. 6 - Weakly sheared flow CHAMPAGNE et al experiment

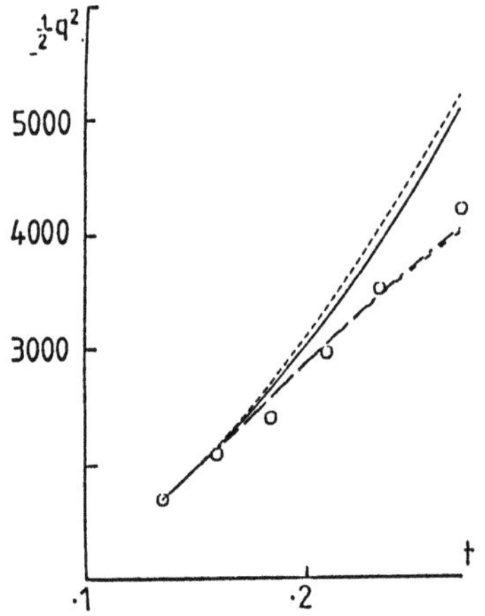

Fig. 7 - Strongly sheared flow HARRIS et al experiment

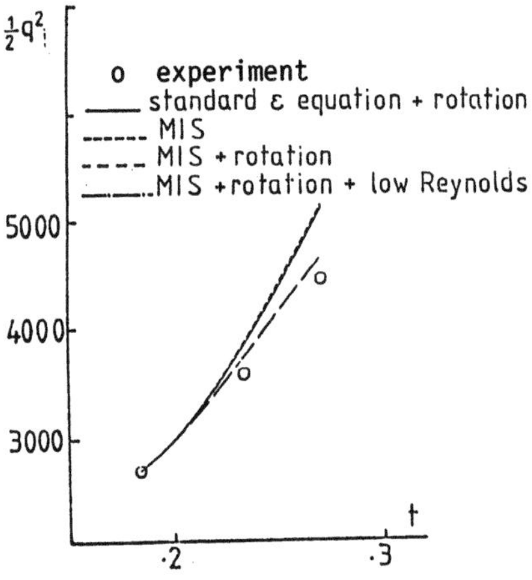

Fig. 8 - Strongly sheared flow TAVOULARIS et al experiment

7. Conclusion

The MIS approach is based upon simplified energy spectrum shapes. It leads to a new dissipation equation which brings into evidence new time-scales. The very large eddies time-scale can be obtained with the help of a simplified rapid distortion model. The spectrum shape time-scale is linked to the inhibition of energy transfer by rotation and to the kinetic energy time-scale.

Without any tuned coefficient, the MIS model gives predictions as good as standard, tuned, dissipation equation for homogeneous strained or weakly sheared flows and highly improves the prediction of strongly sheared flows.

1. B. AUPOIX : Thèse d'Etat, Université de LYON (1986)
2. B. AUPOIX, J. COUSTEIX, J. LIANDRAT : 4th Turbulent Shear Flow Symposium, KARLSRUHE (1983)

3. G. COMTE-BELLOT, S. CORRSIN : JFM, Vol. 25, Part 4, pp. 657-682 (1966)
4. A. LIN, M. WOLFSHTEIN : Physics of Fluids, N° 23-3, pp. 644-646 (1980)
5. V.C. PATEL, W. RODI, G.SCHEUERER : 3rd Turbulent Shear Flow Symposium, DAVIS (1981)
6. W.C. REYNOLDS : AIAA Paper, N° 74-556 (1974)

Transient Behavior of a Stably Stratified Homogeneous Turbulent Flow

R.C. Sanderson[1], *J.C. Hill*[1], *and J.R. Herring*[2]
[1]Iowa State University, Ames, IA 50011, USA
[2]National Center for Atmospheric Research, Boulder, CO 80307, USA

1. Abstract

The direct interaction approximation (KRAICHNAN [1]) is used to study the problem of scalar transport by a uniform mean scalar gradient in a decaying homogeneous turbulent flow. Cases that are studied include a passive scalar, such as temperature, and an active scalar, such as density, in a stably stratified flow. The direct interaction approximation is shown, from these preliminary results, to correctly predict the qualitative dynamics associated with these two problems.

2. Introduction

The problem of scalar transport by a uniform mean gradient in a homogeneous, incompressible turbulent velocity field was proposed by CORRSIN [2] as a physically simple test of closure methods. His prediction that the scalar gradient would remain constant received experimental confirmation through the studies of WISKIND [3], SIRIVAT and WARHAFT [4], and BUDWIG, TAVOULARIS, and CORRSIN [5] among others. These studies yielded asymptotic values for the scalar transfer correlation coefficient ($\rho_{v\theta} = \overline{v\theta}/v'\theta'$) in the range of -0.7 to -0.8 and for the ratio of the eddy diffusivity to molecular diffusivity ($\varepsilon_H/\alpha = -\overline{v\theta}/a\alpha$, where a is the magnitude of the scalar gradient) in the range 15 to 22 at R_λ of 30. Sirivat and Warhaft found that asymptotic results for such quantities as the scalar transport correlation coefficient and scalar to velocity length scale ratio appeared to be relatively insensitive to the initial conditions. HILL [6] made an abridged direct interaction approximation (DIA) calculation for Wiskind's problem and obtained (corrected) values for ε_H/α between 14 and 18 and values for $\rho_{v\theta}$ of about -0.35.

The case of stably stratified flow described by the Boussinesq approximation is a simple generalization of Corrsin's problem. GIBSON [7] proposed three dynamical regimes delimited by the relationship between the characteristic length scales of the density (or buoyancy) and turbulent velocity fields. The density field could convect the turbulence, preferentially suppress the larger scales of the turbulence while convecting the smaller scales, or convert the turbulent field into an internal gravity field when the density length scales are larger than, equal to, or smaller than the turbulent length scales, respectively. Gibson further proposed parameteric limits for these regimes and defined a critical dissipation rate necessary to maintain active turbulence. STILLINGER, HELLAND and VAN ATTA [8] performed experiments in a salt water tunnel which gave experimental limits for these three regimes in reasonably good agreement with Gibson's predictions.

RILEY, METCALFE, and WEISSMAN [9] and HERRING and MÉTAIS [10] have used direct numerical simulation to study stably stratified flows and saw no evidence of a transition to waves, even at very high values of Brunt-Väisälä frequency (the Brunt-Väisälä frequency, N, is defined as $N^2 = a\beta g$ where β is the bulk coefficient of expansion).

The present study applies the full direct interaction approximation (DIA) both to a stably stratified turbulent flow and to passive scalar transport in decaying homogeneous turbulence with uniform mean gradients. This will complement the DIA study done by DANNEVIK [11] for the case of Rayleigh-Benard convection, the unstably stratified problem.

3. Method

The DIA equations for this problem were derived by KRAICHNAN [1], who showed they conserve kinetic and thermal energies and satisfy certain realizability conditions. These equations are then rewritten in terms of their Fourier transforms, resulting in a set of first order ordinary differential equations for the evolution of tensor quantities with vector arguments, and then reduced to equations for scalar quantities of scalar arguments using the Herring-Craya representation (see HERRING [12]). This amounts to transforming all quantities to eigenfunction form and characterizing the angular dependence by use of Legendre polynomials. For example the representation of the velocity correlation tensor has the form

$$\langle u_i(\underset{\sim}{k},t)u_j(-\underset{\sim}{k},t')\rangle = U_{ij}(\underset{\sim}{k},t,t') = \sum_{\ell=0}^{\infty}\sum_{\lambda=1}^{2} \phi^{\lambda}_{uu_\ell}(k,t,t')e^{\lambda}_i(\underset{\sim}{k})e^{\lambda}_j(\underset{\sim}{k})P_\ell(\mu_k) \tag{1}$$

where the eigenvectors are defined in Fig. 1 and $P_\ell(\mu_k)$ is the ℓth Legendre polynomial of μ_k, where μ_k is defined as $\cos(\underset{\sim}{k},\underset{\sim}{n})$. The DIA equations for $\phi^{\lambda}_{uu_\ell}(k,t,t')$ are of the form

$$\left(\frac{d}{dt} + \nu k^2\right)\phi^{\lambda}_{uu_{\ell'}}(k,t,t') + \sum_{\ell} Bou^{\lambda}_{\ell\ell'}(k)\phi^{\lambda}_{\theta u_\ell}(k,t,t') = \{\text{terms like}$$

$$\sum_{\mu,\eta}\sum_{\ell mn}\iint_{\substack{\Delta \\ \underline{p}+\underline{q}=\underline{k}}} A^{\lambda\mu\nu}_{\ell'\ell mn}(k,p,q)\int_0^{t'} g^{\lambda}_{uu_\ell}(k,t',s,)\phi^{\mu}_{uu_m}(p,t,s)\phi^{\eta}_{uu_n}(q,t,s)ds\,dp\,dq\}$$

$-\{$terms like

$$\sum_{\mu,\eta}\sum_{\ell mn}\iint_{\substack{\Delta \\ \underline{p}+\underline{q}=\underline{k}}} B^{\lambda\mu\nu}_{\ell'\ell mn}(k,p,q)\int_0^{t} g^{\mu}_{uu_m}(p,t,s)\phi^{\eta}_{uu_n}(q,t,s)\phi^{\lambda}_{uu_\ell}(k,s,t')ds\,dp\,dq\}. \tag{2}$$

Equation (2) is for the two point velocity correlation while similar equations are

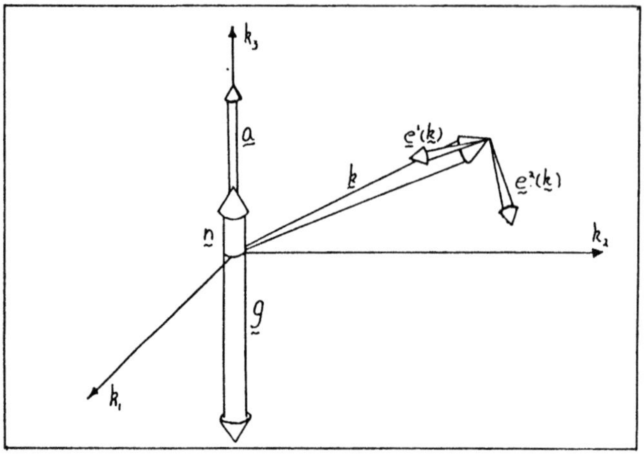

Fig. 1. Schematic of the problem geometry: $\underset{\sim}{a}=\nabla T$, $a = |\underset{\sim}{a}|$, $\underset{\sim}{n}=\underset{\sim}{a}/a$, $\underset{\sim}{e}^1(\underset{\sim}{k})=\underset{\sim}{k}\times\underset{\sim}{n}/|\underset{\sim}{k}\times\underset{\sim}{n}|$, $\underset{\sim}{e}^2(\underset{\sim}{k})=\underset{\sim}{k}\times\underset{\sim}{e}^1(\underset{\sim}{k})/|\underset{\sim}{k}\times\underset{\sim}{e}^1(\underset{\sim}{k})|$, $\underset{\sim}{g}$=gravity vector

needed for the scalar, velocity-scalar and scalar-velocity correlations and their corresponding response functions. The coefficients A and B in (2) are calculated using the procedure outlined in SCHUMANN and HERRING [13].

In this study, twelve logarithmically spaced wavenumbers between one and sixty-four were used. Due to the relatively large size of the system, these preliminary calculations employed only the zeroth Legendre polynomial (the spherical average) which cannot account for all the anisotropies to be expected in these problems. The initial energy spectra were of the form $Ak^3\exp(-Bk)$. The remaining conditions are summarized in Table I.

Table I. Conditions for the computer runs

Run	I	II	III	IV	V
$R_{\lambda o}$	35	35	35	35	35
a	1.5	1.5	2	9.825	1.965
βg	0	0	0.29	2.545	12.72
N	0	0	0.76	5	5
θ'_o	1.0	0	0	1.414	1.414
$\overline{v\theta}_o$	-1.0	0	0	-0.707	-0.707

$v'_o = 1.0$ and $Pr = 0.7$ for all runs.

The numerical scheme makes use of a predictor-corrector time stepping method which treats the viscous term exactly and is discussed in more detail by KRAICHNAN [14]. The time history integrals are calculated using Simpson's rule while the p-q integrals were effected by integration of the cubic spline representations for the integrands.

4. Results

Because the computer runs were limited to just over one eddy circulation time it was suspected that initial conditions may have a significant effect on the results. We confirmed that the qualitative dynamics were those of the DIA and not merely those of the initial conditions or linear effects by making separate rapid distortion calculations in which the nonlinear terms are set equal to zero.

The runs for the passive scalar transport were begun using the initial conditions discussed above and allowed to decay to conditions similar to those of Wiskind's experiment; the results are summarized in Table II. The difference in behavior for the two sets of initial conditions for these runs is noticeable for these short time results. At the time at which the eddy diffusivities are sampled for Table II, they are still evolving rapidly and each of the two cases appear to be approaching values intermediate between those given in the table.

Table II. Passive scalar transport results (* estimated)

	Wiskind	Budwig	Sirivat	Hill	This Study I	This Study II
R_λ	29.3	28	26	29.3	29.9	28.9
$\rho_{v\theta}$	-0.48	-0.7 to -0.8	-0.7	-0.35	-0.85	-0.86
ε_H/α	17.4	18*	19*	14-18	53	6.46

The time evolution of various statistical quantities for the case of stable stratification is shown in Figs. 2 and 3. Figure 2 shows that the intensity of scalar fluctuations increases for a time, similar to the increase obtained in the passive scalar case and as predicted by Corrsin. This figure illustrates that the

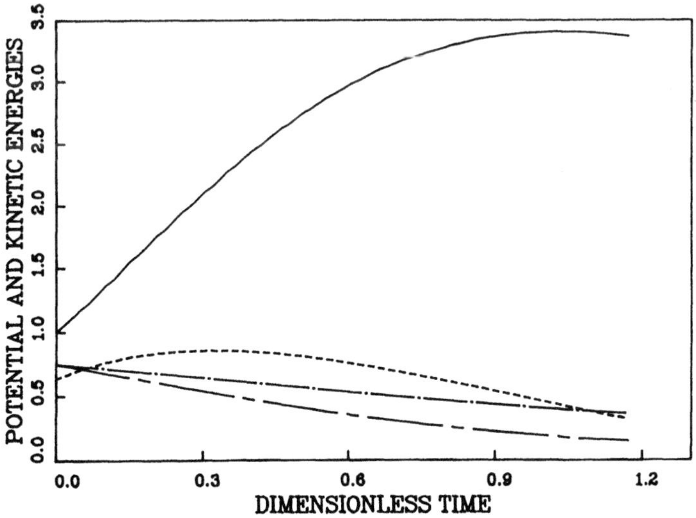

Fig. 2. Evolution of stably stratified turbulence "energies" for Run IV: —— scalar "energy", $\frac{1}{2}\theta'^2$, — · — horizontal kinetic energy; — — — vertical kinetic energy; --- $\frac{1}{2}$ $v\theta$. ("Potential energy" = scalar energy × $2N^2/a^2$).

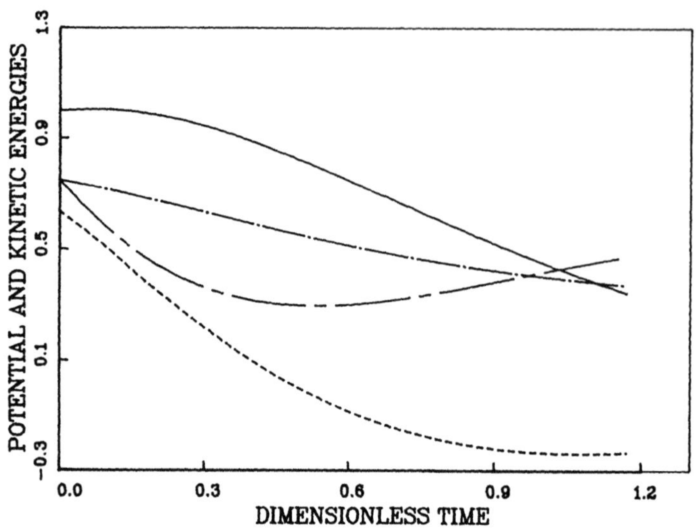

Fig. 3. Evolution of stably stratified turbulence "energies" for Run V: symbols same as in Fig. 2

scalar intensity (or "potential" energy) passes through a maximum value and that the vertical components of the turbulence are preferentially suppressed. In this run the scalar gradient is large relative to the value of βg (9.825 and 2.454 respectively). Figure 3 shows the results of another run where this situation is reversed (1.965 and 12.72), and where there seems to be an exchange between the vertical kinetic energy and the potential energy. The scalar transport changes sign when the conversion of potential energy to kinetic energy exceeds the gradient

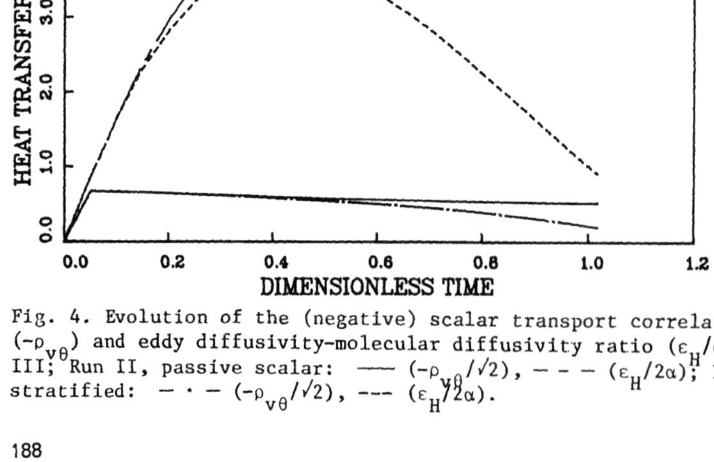

Fig. 4. Evolution of the (negative) scalar transport correlation coefficient $(-\rho_{v\theta})$ and eddy diffusivity-molecular diffusivity ratio (ε_H/α) for Runs II and III; Run II, passive scalar: —— $(-\rho_{v\theta}/\sqrt{2})$, – – – $(\varepsilon_H/2\alpha)$; Run III, stably stratified: — · — $(-\rho_{v\theta}/\sqrt{2})$, --- $(\varepsilon_H/2\alpha)$.

transfer, as in the studies by Stillinger et al., Riley et al., and Herring and Métais.

Figure 4 illustrates the effect of stable stratification upon the scalar transpor characteristics compared with the passive scalar case. Although the scalar transpor grows initially, for both $\rho_{v\theta}$ and ε_H/α the effect of the stable stratification is to suppress the vertical kinetic energy and, in turn, the scalar transport.

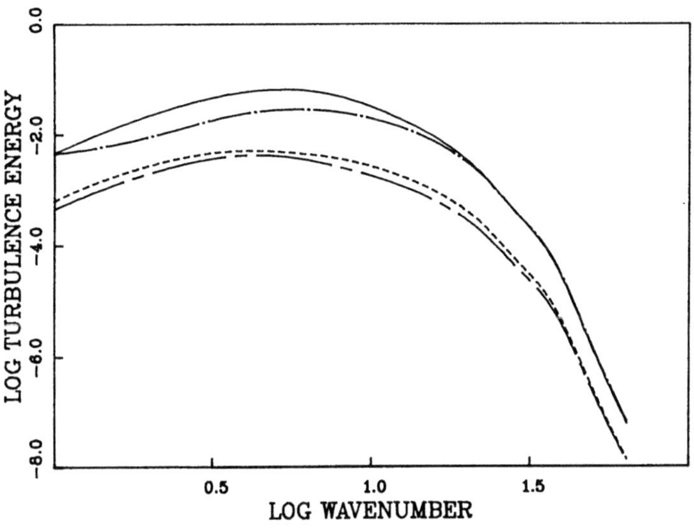

Fig. 5. Spectral evolution of the horizontal and vertical components of kinetic energy for Run V: for t=0.4; —— horizontal; -·- vertical: for t=1.2; — — — horizontal; - - - vertical

Figure 5 shows the horizontal and vertical kinetic energy spectra corresponding to Fig. 3 at two different times. The vertical spectrum shows that the buoyancy affects the lower wavenumbers and this effect reaches higher wavenumbers as time passes, which is consistent with the model proposed by Gibson.

5. Conclusions

In this preliminary study the DIA seems to correctly predict the qualitative dynamic for both the passive scalar and the stably stratified flow problems. Because of truncation of all but the zeroth Legendre polynomials, these results should only be considered qualitative and preliminary to a more complete study now in progress.

6. Acknowledgements

The authors wish to acknowledge the support and facilities provided by the Iowa State University Computation Center. The National Center for Atmospheric Research is sponsored by the National Science Foundation.

7. References

1. R. H. Kraichnan: Phys. Fluids 7, 1048 (1964)
2. S. Corrsin: J. Appl. Phys. 23, 113 (1952)
3. H. K. Wiskind: J. Geophys. Res. 67, 3033 (1962)
4. A. Sirivat, Z. Warhaft: J. Fluid Mech. 128, 323 (1983)
5. R. Budwig, S. Tavoularis, S. Corrsin: J. Fluid Mech. 153, 441 (1985)
6. J. C. Hill: Chem. Eng. Comm. 12, 69 (1981)
7. C. H. Gibson: In Nonlinear Properties of Internal Waves, ed. by B. West, A.I.P. Conf. Proc. no. 76 (A.I.P., New York 1981) p.159
8. D. C. Stillinger, K. N. Helland, C. W. van Atta: J. Fluid Mech. 131, 91 (1983)
9. J. J. Riley, R. W. Metcalfe, M. A. Weissman: In Nonlinear Properties of Internal Waves, ed. by B. West, A.I.P. Conf. Proc. no. 76 (A.I.P., New York 1981) p.79
10. J. R. Herring, O. Metais: "Numerical and Theoretical Calculations of Stratified Turbulence", presented at the Tenth Symposium on Turbulence, (University of Missouri-Rolla 1986)
11. W. P. Dannevik: Two-Point Closure Study of Covariance Budgets for Turbulent Rayleigh-Benard Convection, Ph.D. Thesis (Saint Louis University, 1984)
12. J. R. Herring: Phys. Fluids 17, 859 (1974)
13. U. Schumann, J. R. Herring: J. Fluid Mech. 76, 755 (1976)
14. R. H. Kraichnan: Phys. Fluids 7, 1030 (1964)

A Stochastic Analysis of the Displacements of Fluid Elements in Inhomogeneous Turbulence Using Kraichnan's Method of Random Modes

C. Turfus and J.C.R. Hunt

Department of Applied Mathematics and Theoretical Physics,
University of Cambridge, Silver Street, Cambridge CB3 9EW, UK

At an instant in time a homogeneous turbulent flow field can be represented as a summation of random Fourier components; this can be extended to describe turbulence near an interface or a boundary in shear-free flow, by adding an irrotational velocity field. The time evolution of the flow can be simulated by assuming that the modes are oscillatory functions of time with random frequencies and random amplitudes: i) our model simulates the energy containing eddies and the inertial subrange eddies with a (5/3) energy spectrum; ii) the time scale for the decorrelation of each eddy is made proportional to its length-scale corresponding to an Eulerian power spectrum proportional to $\omega^{-5/3}$.

Near a rigid surface, the Eulerian variance $\overline{w^2}$ and correlation function ($R_{ww}(\tau,z)$) for vertical velocity fluctuations are calculated analytically; the former gives $\overline{w^2} \propto z^{2/3}$ and the latter a $(\tau/z)^{2/3}$ decay as $z \to o$ and $(\tau/z) \to o$. A Lagrangian time scale is found by statistically averaging over a large number of numerically-simulated trajectories of particles released from a fixed height. The computations indicate that the Lagrangian time scale T_L for vertical fluctuations near the surface is proportional to L_x^{ww}/σ_u in contrast to the generally assumed estimate L_x^{ww}/σ_w; i.e. $T_L \propto z$, rather than $T_L \propto z^{2/3}$. (This may help to explain the well-known discrepancy between surface similarity theory where $K_z \approx L_x^{22} . \sigma_w \propto z^{4/3}$ and observations which suggest $K_z \propto z^{3/2}$. Our simulation suggests $K_z \propto z^{5/3}$.)

The time development of the probability distribution function for a particle released at a fixed height is estimated by similar methods. These results agree broadly both with a random-flight simulation and with a statistical method applicable for small t (van Dop, Nieuwstadt & Hunt [1]).

1. Introduction

Very little is known, even now, about the Lagrangian statistics of the velocity and displacements of fluid elements (or particles) in turbulent flows, despite the importance of such information for calculating the dispersion and mixing of matter or heat released into turbulent flow. This approach has become quite practical with the availability of large computers and stochastic computer models (reviewed by Durbin [1], van Dop et al. [2], Sawford & Hunt [3].

But in these stochastic models assumptions have to be made about the variables ($\sigma_u^2, \sigma_v^2, \sigma_w^2$) and time scales $T_L^{(u)}$, of the different components of movement of the particles in inhomogeneous flows, despite the lack of either experimental or computational information. Most models are based on the physical scaling argument of Corrsin [4] that relates the Lagrangian time scale to the Eulerian length scale

L_x (e.g. of the velocity fluctuations measured at fixed points separated by distances in the x-direction), and by a relevant velocity scale. So for the z-component (the vertical component in horizontal shear layers or boundary layers), it is assumed that

$$T_L^{(w)} \approx L_x^{(w)}/\sigma_w \ . \qquad (1.1)$$

This estimate has been justified by several laboratory experiments for homogeneous grid turbulence (e.g. by Snyder & Lumley [5]). However it is also widely used without much justification in inhomogeneous turbulence. If the variances of velocity components are proportional to each other throughout a flow, e.g. a turbulent boundary layer without buoyancy forces, (1.1) appears to be an adequate estimate. But in a shear-free boundary layer, e.g. one with strong thermal convection, or a region in a river below the free surface, σ_u is approximately constant, while σ_w tends to zero at the surface or interface. Then the choice of one or other component in (1.1) has an important effect on the estimation of T_L. From the computations presented we suggest which is the correct choice and what the physical reasons are.

Even if the direct computation of Lagrangian statistics in turbulent flows is performed with the largest computers in the world, they are still not representative of high Reynolds number flows of interest in geophysical applications. A much simpler method is possible, based on the model of Kraichnan [6]. Construct a velocity field made up of random uncorrelated Fourier modes. The energy of the modes is specified so that the spectrum of the velocity field agrees with the known Eulerian spectrum. At high wave numbers this should be the Kolmogorov spectrum for the inertial subrange. Kraichnan [6] and Drummond et al. [7] only considered a delta function spectrum. There is no dynamical interaction between the Fourier modes. This raises a subsidiary question of how important is this interaction in affecting the statistics of particle motions? Our hypothesis is that it is a weak effect. (Our argument is too weak and too lengthy to be worth developing here!)

Our other new development to the Kraichnan simulation reported here is to model in as simple a way as possible the effects of inhomogeneity of turbulence. We consider a 'shear-free' turbulent boundary layer near a rigid surface at z=o, which blocks vertical velocity fluctuations at z=o. But if say the Reynolds number is very large, at the surface there are still horizontal velocity fluctuations, except within a very small distance of the surface (which is $O(Re^{-1/2}L_x)$ of the surface . These shear-free layers are quite widespread in engineering and geophysical flows (e.g. in thermal convection, at gas-liquid interfaces, at density interfaces, etc. (Hunt [8]). In modelling such layers we make the second hypothesis that the effect of the surface on the initially homogeneous turbulence is, to the first approximation, to add an irrotational velocity field such that w=o at z=o.

Simulating particle motions and understanding Lagrangian and Eulerian statistics in such layers has many applications. We show how the simulated velocity field can be used to simulate the statistics of displacements of particles released from a source (and hence the mean concentration field). We show that these results are similar to those obtained from a random-flight simulation of particle displacements, using Lagrangian statistics obtained from our flow-field simulation. Another application of this flow-field model is to understand how internal waves are generated by turbulence at density interfaces (Carruthers et al. [9]).

2. Modelling the Homogeneous Flow Field by Random Fourier Methods

In a homogeneous flow field the velocity field $\underline{u}^{(H)}(\underline{x},t)$ can be represented by a three-dimensional Fourier transform in space. It is convenient also to assume that the turbulence is stationary in time, by ignoring its slow decay. Then

$$\underline{u}^{(H)}(\underline{x},t) = \iiiint_{-\infty}^{\infty} e^{i(\underline{\kappa}.\underline{x} + \omega t)} \underline{S}(\underline{\kappa},\omega) d\underline{\kappa}\, d\omega \tag{2.1}$$

where $\underline{\kappa}$ is the three-dimensional wavenumber and ω the frequency. If there is a mean flow, we take co-ordinates moving with the mean flow.

The variance of the transform is defined by the three-dimensional space and time spectrum

$$\chi_{ij}^{(H)}(\underline{\kappa},\omega) = \left[\overline{S_i^*(\underline{\kappa},\omega)\, S_j(\underline{\kappa},\omega)} \right] \frac{(2\pi)^4}{TXYZ}$$

where the turbulence is defined within a box of sides X,Y,Z and over a time T. $(\,)^*$ denotes the complex conjugate. Because it is difficult to measure, there are still no definitive measurements of $\chi_{ij}^{(H)}(\kappa,\omega)$, but there is information about the three-dimensional energy spectrum

$$\Phi_{ij}^{(H)}(\underline{\kappa}) = \frac{1}{(2\pi)^3} \iiint \overline{u_j^{(H)}(\underline{x})u_j^{(H)}(\underline{x}+\underline{r})}\; e^{-i\underline{\kappa}.\underline{r}} d\underline{r}\ .$$

For isotropic turbulence (Batchelor [19]), Φ_{ij} is related to the energy spectrum $E(\underline{\kappa})$ by

$$\Phi_{ij}^{(H)}(\underline{\kappa}) = \frac{E(k)}{4\pi k^4} (k^2\delta_{ij} - \kappa_i\kappa_j) \tag{2.2}$$

where $k^2 = \kappa_i\kappa_i$ and

$$\int_0^\infty E(k)dk = \frac{3}{2}\left[\overline{u_1^2}\right]^{(H)}, \text{ and } \left[\overline{u_1^2}\right]^{(H)} = \left[\overline{u_2^2}\right]^{(H)} = \left[\overline{u_3^2}\right]^{(H)} . \tag{2.3}$$

Tennekes [11] and Townsend [12] argued that the turbulent energy at a frequency ω, for an observer moving with the mean flow, is primarily associated with energy-containing eddies, with variance u_H, advecting eddies of scale k^{-1} past the observer where $k \approx \omega/u_H$ (Fig.1.). Therefore we make the hypothesis

$$\text{(H1)} \quad \chi_{ij}^{(H)}(\underline{\kappa},\omega) = \delta(\omega - k\,u_H)\, \Phi_{ij}^{(H)}(\underline{\kappa})\ . \tag{2.4}$$

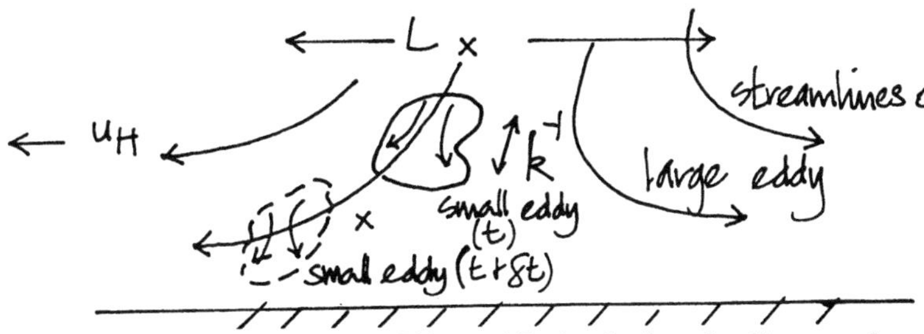

Fig.1. Showing how large eddies with typical velocity u_H advect small eddies with length scale k^{-1} and produce velocity fluctuations with frequency $\omega \approx u_H k$. (The local velocity within the eddy, u_k, contributes to much lower frequencies of order $u_k k$.)

For homogeneous turbulence a form of the spectrum E(k), that corresponds closely to many operations (e.g. in grid turbulence and boundary layers) and reduces at high wave number to the form of the Kolmogorov inertial subrange, is the von Karman form

$$E(k) = c_1 k^4/(c_2^2 + k^2)^{17/6} \tag{2.5}$$

where $c_1 = 0.97\, c_2^{2/3} (3/2\, \overline{u_1^2}^{(H)})$ and $c_2^2 = 0.56$. Note that

$$L_x^{(u)} = \int_0^\infty \overline{u_1(x) u_1(x+r_1)}\, dr_1 = 1 \ .$$

We now generalise the methods of discretisation of Kraichnan [6] and Drummond et al.[7] who assumed that all the energy is concentrated in a shell of radius $k = |\underline{\kappa}|$ in wave-number space. We express $\underline{u}(\underline{x},t)$ as a Fourier series, with properties of the coefficients defined below:

$$\text{(H2)} \quad \underline{u}^{(H)}(\underline{x},t) = \sum_{n=1}^{N_k} \sum_{m=1}^{N_\omega} \left[\begin{array}{l} \underline{a}_n \times \hat{\underline{\kappa}}_n \cos(\underline{\kappa}_n \cdot \underline{x} + \omega_m t) \\ + \underline{b}_n \times \hat{\underline{k}}_n \cos(\underline{\kappa}_n \cdot \underline{x} + \omega_m t) \end{array} \right] \tag{2.6}$$

where $\hat{\underline{\kappa}}_n = \underline{\kappa}_n / k_n$, and $\underline{a}_n$, $\underline{b}_n$ are independent random vectors uniformly distributed over all directions. For isotropic turbulence satisfying (2.4), their magnitudes are specified by

$$\overline{(\underline{a}_n \times \hat{\underline{\kappa}}_n)^2} = \overline{(\underline{b}_n \times \hat{\underline{\kappa}}_n)^2} = \gamma \int_{k_n}^{1.088\, k_n} p(k) dk \ .$$

which implies

$$2/3\ \overline{|\underline{a}_n|^2} = 2/3\ \overline{|\underline{b}_n|^2} \approx \gamma\,(0.088) k_n p(k_n) \tag{2.7}$$

where γ is a constant (≈ 1.0) chosen so that $\overline{|\underline{u}|^2}^{(H)} = 3$ and $p(k) = 2/3\, E(k)$, where E(k) is the energy spectrum, defined by (2.5). The 64 values of k_n were specified in the range 0.5 to 101.5 with a distribution

$$k_n = (1.088)^{n-1} \cdot 0.5 \qquad n = 1,2,\ldots 64 \ . \tag{2.8}$$

Following the assumption (2.4), only the sum over n is required, and

$$\omega_m = k_n u_H \delta_{nm} \ . \tag{2.9}$$

In Kraichnan's formulation ω_m is independent of k_n, which does not appear to be physically correct for a full spectrum. Note that (2.6) is an approximate statistical representation of the second-order moments of the Eulerian velocity field. It does not correctly represent the interaction between the modes, defined by third-order moments of $\underline{S}(\underline{\kappa})$

$$\text{e.g.} \quad \overline{\underline{S}^*(\underline{\kappa}_1) \underline{S}(\underline{\kappa}_2) \underline{S}(\underline{\kappa}_1 + \kappa_2)} \ .$$

The representation in (2.6) does not include scales of turbulence smaller than about 1/100. For some computations of trajectories it is important to represent these smaller scales, in which case we add a small-scale velocity field $\Delta\underline{u}(\underline{x},t)$ to (2.6). The energy from this component follows from (2.3);

$$\overline{(\Delta\underline{u})^2} = \frac{2}{3}\int_{k_{64}}^{\infty} E(k)\,dk = 0.055 \; . \tag{2.10}$$

This component accounts for 5% of the energy in homogeneous turbulence.

Thence, using (2.9) and on normalising the r.m.s. velocity $u_H = 1$, the representation of the homogeneous velocity field becomes:

$$\underline{u}^{(H)}(\underline{x},t) = \sum_{n=1}^{64}\left\{\underline{a}_n \times \hat{\underline{\kappa}}_n \cos\phi_n + \underline{b}_n \times \hat{\underline{\kappa}}_n \sin\phi_n\right\} + \Delta\underline{u}(\underline{x},t) \tag{2.11}$$

where $\phi_n = \underline{\kappa}\,.\,\underline{x} + k_n t$.

3. Inhomogeneous Turbulence in a Shear-Free Boundary Layer

3.1 The Model Velocity Field

In turbulent flow near rigid boundaries in the absence of large velocity gradients the rate of dissipation of turbulent energy per unit mass ϵ is approximately constant with height. Such shear-free boundary layers are quite different to boundary layers in which there is mean flow along a rigid surface when the shear and the energy dissipation are very large and vary rapidly near the surface, i.e. $dU/dz \propto z^{-1}$ and $\epsilon \propto z^{-1}$. In fully developed turbulent flows at high Reynolds number ϵ is proportional to the mean square vorticity $\underline{\omega}$, i.e. $\epsilon = \nu\,\overline{|\underline{\omega}|^2}$. Therefore if $\partial\epsilon/\partial z \approx 0$, $\partial\overline{|\omega|^2}/\partial z \cong 0$.

In a number of such shear-free turbulent flows it has now been shown that the variances, cross correlations and spectra (especially of the normal component) are rather well described by making the hypothesis (H3) that the effect of the surface is to add an irrotational velocity field to a homogeneous flow field (Fig.2) [8],[13]:

$$\text{(H3)} \quad \underline{u} = \underline{u}^H + \nabla\phi \; , \tag{3.1}$$

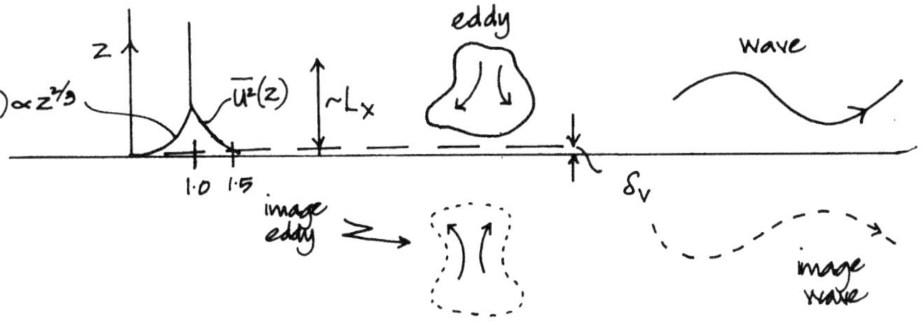

Fig.2. A diagram to illustrate how in a shear-free boundary layer, the effect of a rigid surface on high Reynolds number turbulence is (i) to ensure no vertical fluctuations; (ii) amplified horizontal fluctuations (except within a very thin viscous layer of thickness δv - neglected in this paper), and (iii) that these changes are to a good approximation the irrotational effects produced by image eddies (or waves).

since $\nabla . \underline{u} = 0$, ϕ satisfies

$$\nabla^2 \phi = 0 \qquad (3.2)$$

and to satisfy $u_3 = o$ on $z = o$, the boundary condition on ϕ is

$$\partial\phi/\partial z = - u_3^{(H)}(x,y,o,t) \ . \qquad (3.3)$$

When turbulent flow suddenly meets a surface parallel to the mean flow (e.g. free-stream turbulence on a flat plate), equation (3.1) is a rigorous result for a mean travel time T along the plate given by $T \leqslant T_L$ [14].

For a homogeneous flow field $\underline{u}^{(H)}$ defined by (2.6), the solution to (3.1) for $\underline{u}$ near the surface (but at a distance that is larger than the smallest 'resolved' scale, i.e. $z > 1/100$) are for the horizontal components, u_1, u_2, or(u,v)

$$\underline{u} = \sum_1^{64} \left\{ (\underline{a}_n \times \hat{\underline{\kappa}}_n) \cos\phi_n + (\underline{b}_n \times \hat{\underline{\kappa}}_n) \sin\phi_n - \left[(\underline{a}_n \times \hat{\underline{\kappa}}_n)_3 \sin\psi_n + (\underline{b}_n \times \hat{\underline{\kappa}}_n)_3 \cos\psi_n \right] \frac{\underline{\kappa}_n}{k_{12n}} e^{-k_{12}z} \right\} + \Delta\underline{u} \ . \qquad (3.4a)$$

and for the vertical (or w) component

$$w = u_3 = \sum_1^{64} \left\{ (\underline{a}_n \times \hat{\underline{\kappa}}_n)_3 (\cos\phi_n - \cos\psi_n \ . \ e^{-k_{12n}z}) + (\underline{b}_n \times \hat{\underline{\kappa}}_n)_3 (\sin\phi_n - \sin\psi_n \ . \ e^{-k_{12n}z}) \right\} + \Delta u_3 \qquad (3.4b)$$

where $\phi_n = \underline{\kappa}_n . \underline{x} + k_n t$, $\psi_n = \underline{\kappa}_{12n} . \underline{x} + k_n t$, $\underline{\kappa}_{12n} = (k_{1n}, k_{2n}, o)$, $k_{12n} = |\underline{\kappa}_{12n}|$.

Solutions to (3.1) for a continuous velocity field represented by Fourier transforms were given in [14].

3.2 Eulerian Statistics of the Model Velocity Field

From (3.4b) and the properties of $\underline{a}_n$ and $\underline{b}_n$ the variance and length scales of the different velocity components were computed and shown to agree well with the computed and analytical results in [8] and [13] for a continuous spectrum.

In particular for the variances:

$$\text{when} \quad \left. \begin{array}{ll} z \geqslant 2, & \overline{u^2} = \overline{v^2} = \overline{w^2} \\ z \to 0, & \overline{u^2} \to 1.5 u_H^2 \ , \overline{w^2} \to 1.4 u_H^2 z^{2/3} . \end{array} \right\} \qquad (3.5)$$

(The 2/3 law dependence is explained in [8] in terms of the -5/3 spectrum being cut off at a wavenumber $k \approx z^{-1}$.)

For the length scales:

$$\text{when} \quad \begin{aligned} z \geqslant 2, \quad & L_x^{(w)} \to 1/2.\ L_x^{(u)} \to 1 \\ z \to o, \quad & L_x^{(w)} \to 2z,\ L_x^{(u)} \to 0.7 \end{aligned} \quad . \tag{3.6}$$

The new results that can be obtained from this model are the Eulerian time cross-correlations near the surface that could not be obtained from the purely spatial analysis in [14]. If $R_{ij}(\underline{x},\tau) = \overline{u_i(\underline{x},t)u_j(\underline{x},t+\tau)}/(\overline{u_i^2}\,\overline{u_j^2}(\underline{x}))^{1/2}$ then far from the surface

$$R_{33}(\tau, z\to\infty) = \frac{2}{27\,\Gamma(1/3)}\left[\frac{\tau^*}{2}\right]\left[55\,K_{1/3}(\tau^*) - 66\tau^* K_{4/3}(\tau^*) + 9(\tau^*)^2 K_{7/3}(\tau^*)\right] \tag{3.7}$$

where $\tau^* = \tau/c_2$, showing an exponential decrease when $\tau\to\infty$. But when $\tau \ll 1$,

$$R_{33}(\tau,\infty) = 1 - 1.95\,(\tau^*)^{2/3} + O(\tau^{*2}) \tag{3.8}$$

which is equivalent to the (-5/3) Eulerian time spectrum discussed by Tennekes (1975). Near the surface, when $z \ll 1$,. but $\tau/z \ll 1$.

$$R_{33}(\tau,z) = 1 - 1.14\,(\tau/z)^{2/3} + O(\tau^2/z^2) + O((z\tau)^{2/3})\ . \tag{3.9}$$

Thus the correlation peak at $\tau = o$ becomes sharper near the surface, i.e. the vertical turbulence is dominated by the short time scale eddies near the surface.

4. Computation of Particle Trajectories and Lagrangian Statistics

4.1 Method

The computation of the displacement of particles moving in the inhomogeneous turbulent flow requires a careful evaluation of the effects of the random small-scale velocity field $\Delta\underline{u}(\underline{x},t)$. Although it accounts for only 5% of the total energy, it has a much larger effect on the vertical displacements near the surface z=o, where $w\to o$.

The fluid displacements X are computed over many small time steps δt for particles released from particular points in the flow. The time steps chosen are about 10% of the integral time scale T_L (defined from the surface). This is about equal to the time scale of $\Delta\underline{u}$. However, we assume that the displacements produced by the small-scale velocity field are uncorrelated between time steps, and can be represented as a white-noise process with a diffusivity b_i in the ith direction. So if the turbulence is inhomogeneous in the z (or x_3) direction, the displacement in the ith direction is given by

$$X_i(t+\delta t) = X_i(t) + u_i(\underline{X}(t),t)\,\delta t + b_i\,\delta W + b_j\,\frac{\partial u_i}{\partial x_j}\,W_s + F_3\,\delta_{i3} \tag{4.1}$$

where

$$b_i = \left[2\,\overline{\Delta u_i^2}\;\Delta T_L^{(i)}\right]^{1/2} \tag{4.2}$$

$\Delta T_L^{(i)}$ being the time scale of the ith component, $\delta W, W_s, F_s$ being white-noise processes. The term in W_s is caused by gradients in $\underline{u}_i$ and the F_3 is caused by gradients in b_3. The analysis (given in detail by Turfus [15], which is based on earlier work by Rao et al.[16]), shows that

$$\text{(i)}\qquad \Delta T_L^{(i)}(z) = T_L^{(i),(H)}\,\frac{L_x^{(i)}(z)}{L_x^{(i)(H)}}\left[\frac{\overline{u_i^2}}{\overline{\Delta u_i^2}}\right]^{(H)} \tag{4.3}$$

(ii) δW and W_s are normal random variables such that

$$\overline{\delta W} = 0,\quad \overline{\delta W^2} = 0, \overline{W_s} = 0,\quad \overline{W_s^2} = \delta t^2/3\,,$$

$$\text{and}\quad \overline{\delta W W_s} = \frac{1}{2}\,\Delta t^2\quad , \tag{4.4}$$

$$\text{(iii)}\ F_3 = \partial b_3/\partial z\ w\{\delta t\ \delta W - W_s\}$$

$$+\frac{1}{2}\,b_3\,\frac{\partial b_3}{\partial z}\,(\delta W^2 - \delta t) + \frac{1}{2}\,b_3\left[\frac{db_3}{dz}\right]^2\left\{\frac{1}{3}(\delta W)^3 - \delta W\,.\,\delta t\right\}$$

$$+\frac{1}{2}\,b_3\left[\frac{db_3}{dz}\right]^2\left\{\frac{1}{3}(\delta W)^3 - W_s\right\}\,. \tag{4.5}$$

4.2 Results

Particle trajectories were calculated with time steps δt of 0.01 for release height Z_o = 0.05, 0.1, 0.5 and ∞. For each value of Z_o and realisation of the flow (i.e. values of the random coefficients $\underline{a}_n, \underline{b}_n, \underline{\kappa}_n$), 64 particles were released from 64 different source positions $\underline{X}_o$, in the x,y plane. ($\underline{X}_o = (3m, 3n, Z_o)$, m = 0,1...7; n = 0.1...7.) The spacing was large enough that the trajectories were uncorrelated. Trajectories were run for $o < t < 1$, (or $t \leqslant 100\,\delta t$). Typically 60 different flow realisations were computed for each value of Z_o (40 minutes CPU time on the IBM 3081).

Reducing δt to 0.005 or 0.0025 did not produce significantly different results in the statistics (to within about 15%). But increasing δt to 0.02 did.

When $Z_o = \infty$, the mean square displacement $\frac{1}{3}\overline{(X_i X_{oi})^2}\,(t)$ was computed. From its asymptotic form $2/3\left[\sum_i^3 \overline{u_i^2}^{(H)}\ \ T_L^{(H)}\right]. t$, we derived

$$T_L^{(H)} = 0.30\ . \tag{4.6}$$

(This can be compared with the measured value in [5] of $T_L^{(H)(w)} \approx (L_x^{(w)}/\sqrt{(\overline{w^2})}) = 0.5$, in our normalisation.)

From the displacement of the particles $\underline{X}(t)$ released from a source position $\underline{X}_o$, the cross correlation between their velocity at different times after release can be computed. We have computed the cross correlation $R_L^{(w)}(\tau;\underline{X}_o)$ between the vertical velocity at release and a time t latter. The suffix L denotes Lagrangian, i.e. following particles

$$R_L^{(w)}(\tau;X_o) = \frac{\overline{w_L(t+\tau)w_L(t)}}{\left[\overline{w_L^2}(t)\ \overline{w_L^2}(t+\tau)\right]^{1/2}}, \text{ where } \underline{X}(t) = \underline{X}_o , \tag{4.7}$$

where $w_L(t) = w(\underline{x} = \underline{X}(t),t)$.

Typical Lagrangian autocorrelation curves of $R_L^{(w)}(\tau)$ are presented in Fig.3 for particles released from $Z_o = 0.3$, and $Z_o = 0.15$. Note how the $w_L(t)$ decorrelates faster in the latter case, and note how the curves display 'noisy' oscillations around $R_L = o$ for $\tau \geqslant 0.4$. Consequently it is sufficiently accurate to estimate the integral time scale $T_L^{(w)}(Z_o)$, for particles released from Z_o, by the first integral

$$T_L^{(w)}(Z_o) = \int_0^1 R_L^{(w)}(\tau;Z_o)d\tau . \tag{4.8}$$

Profiles of $T_L^{(w)}(Z_o)$ for different Z_o are plotted in Fig.4. When $Z_o \leqslant 0.1$, the contribution by the random small scale becomes relatively important. As mentioned

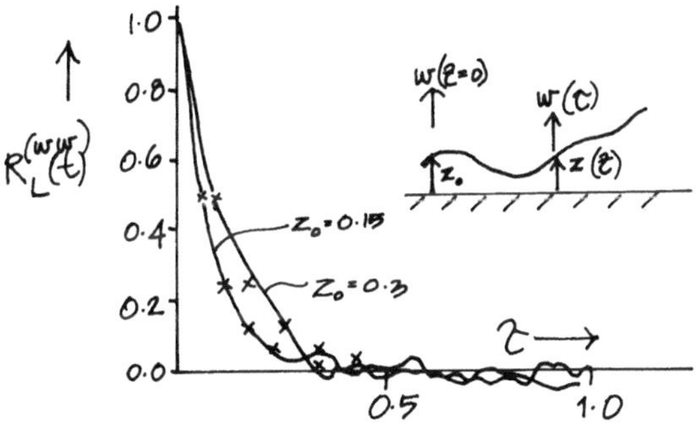

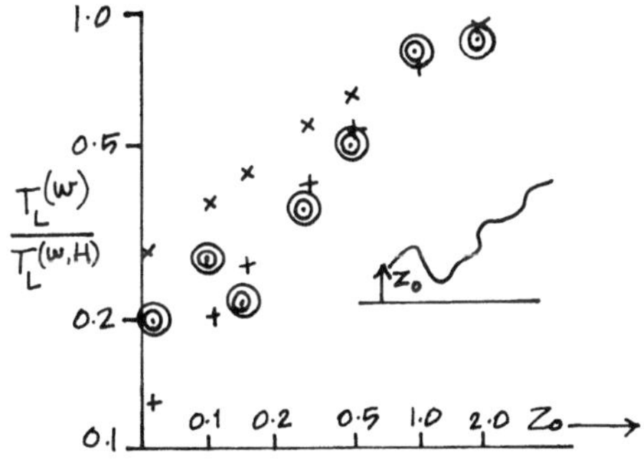

Fig.3. The Lagrangian decorrelation for the vertical velocity of particles released from two source heights Z_o=0.3 and Z_o=0.15. Computations from the random modes velocity field. **x** denotes values of e^{-t/T_L} where T_L is chosen for the best fit.

Fig.4. Lagrangian Integral Timescale $T_L^{(w)}$ for particles released at different heights Z_o above the surface in a shear-free turbulent boundary layer, using the random modes computation. $T_L^{(w)}$ is normalised on its value for homogeneous turbulence $T_L^{(w,H)}$. Comparisons are made with different estimates. (Note the log-log scales.) o computation: $T_L^{(w)}/T_L^{(H)}$; + $T_L^{(w)} \propto L_x^{(w)}$ - see eq.(4.9). (This implies $T_L^{(w)} \propto z$ as $z \to o$). x $T_L^{(w)} \propto L_x^{(w2)}/\sqrt{\overline{w^2}}$ - conventional estimate - (probably wrong for inhomogeneous turbulence).

in the introduction, it has been suggested (on similarity grounds) that even in inhomogeneous flows, $T_L^{(w)}$ should depend on the local values of $L_x^{(w)}$ and $\overline{w^2}$. So in Fig.4 we have plotted $L_x^{(w)}/\sqrt{\overline{w^2}}$ normalised on its homogeneous value. It clearly does not follow the simulated Lagrangian time scale. However if we develop another hypothesis that $T_L^{(w)}$ should depend on the length scale of w and the r.m.s. value of <u>all</u> velocity components, then $T_L^{(w)} \approx L_x^{(w)}/\sqrt{\overline{|u|^2}}$. Since $\overline{|u|^2}$ does not change significantly through the shear-free layer, it follows that

$$T_L^{(w)}(Z_o)/T_L^{(w)(H)} \approx L_x^{(w)}(z = Z_o)/L_x^{(w)(H)} \ . \tag{4.9}$$

this formula agrees well with the simulation.

4.3 Interpretation

The usual concept of particle motions in eddies at a surface is that the particles move around <u>within</u> an eddy with a vertical velocity $\sqrt{\overline{w^2}}$ and over a scale $L_x^{(w)}$ and therefore the time scale is of order $L_x^{(w)}/\sqrt{\overline{w^2}}$. However particles are also being moved around in the horizontal plane much faster with a velocity of order $\sqrt{\overline{u^2}}$. (Much of this motion is driven by eddies far from the surface.) As they move around, the particles are deflected vertically by eddies with a velocity of order $\sqrt{\overline{w^2}}$; the spacing of these eddies is $L_x^{(w)}$. So the time scale between these vertical motions is of order $L_x^{(w)}/\sqrt{\overline{u^2}}$. This is a much <u>smaller</u> time scale as Fig.4 showed.

Carruthers & Hunt [13] showed that this latter time scale is the critical time scale for the generation of internal waves in density interfaces by turbulence.

Another implication of this result is for the variation of the vertical diffusivity K_z for heat or matter at a height z in a shear-free turbulent boundary layer. Based on the usual estimate $K_z(z) \approx \overline{w^2}(z)T_L^{(w)}(z)$. Since as $z \to o$, $\overline{w^2} \propto z^{2/3}$, and from (4.9) $T_L \propto z$, it follows that

$$K_z(z) \propto z^{5/3} \ . \tag{4.10}$$

The usual estimate is $T_L^{(w)} \propto L_x/\sqrt{\overline{w^2}} \propto z^{2/3}$, which leads to $K_z(z) \propto z^{4/3}$. However field experiments and large eddy simulations (Wyngaard & Brost [17]) have established that this exponent

$$K_z(z) \propto z^{3/2} \ . \tag{4.11}$$

This exponent is well enough established to be different to 4/3, yet there has been no previous explanation for this difference. Our models does not lead to (4.11), but it indicates why $z^{4/3}$ is too slow a variation with height.

5. <u>Comparison of particle displacement statistics derived from random-flight simulation and random modes</u>

The advantage of computing particle displacements in complex flows by random <u>flight</u> simulations is that it is easy to programme, to incorporate local measurements or computations of the velocity field, and to use the same velocity statistics for different concentration distributions. However the methods are based on numerous assumptions (see [2] including the appendix), which can only be tested by comparison with computations of Lagrangian velocity and displacement statistics.

In the random-flight model, displacements are modelled by a sequences of straight lines over small periods of time dt. These displacements dz are a discrete random process but are correlated with previous displacements. The eventual problem is finding a method of relating the Lagrangian velocity w_L of these displacements to the statistics of the Eulerian velocity field. The following procedure can be justified in the limit of small times from the release of particles from the source and by appeal to the general but approximate, arguments based on the equation for the probability distribution or their movements [2,18]. For simplicity we consider here one component of motion. For particles starting at height z_o, with mean square vertical velocity $\overline{w^2}(z_o)$, dz is given by:

$$dz = w_L(t)\,\delta t \quad \text{where}$$

$$d\,w_L = -\frac{w_L}{T_L(z(t))}\,\delta t + d\,W_t\,, \quad \text{and}$$

$$z(t = o) = z_o, \quad \text{and} \quad w_L(t = o) = \overline{w^2(z_o)}^{1/2}\;N$$

where N is a Gaussian random variable with zero mean and unit variance, and $T_L(z)$ is the Lagrangian time scale of the vertical velocity fluctuations. W_t is a stochastic process which is related to the vertical component of the Eulerian velocity field $\underline{w}(\underline{x},t)$ by

$$\overline{d\,W_t} = \partial\,\overline{w^2}(z)/\partial z\;.\;\delta t$$

$$d\,\overline{W_t^2} = (2\,\overline{w^2}(z)/T_L(z) + \partial\overline{w^3}/\partial z)\;\delta t\;.$$

In most applications of the random-flight method, $\overline{w^2}$, T_L and $\overline{w^3}$ are measured or estimated from measurements (eg. [18]). Note that $T_L(z)$, $\overline{w^2}$, $\overline{w^3}$ are taken to be <u>independent</u> of the position where the particles are released. This is a critical assumption, but one that greatly simplifies the model.

In this investigation we used the values obtained from the random modes simulation. Since we assumed that the amplitude of the random modes had a Gaussian distribution, it follows that in our simulation $\overline{w^3} = o$.

In Fig.5 we present the comparison of the particle displacement results obtained from the random-mode computation, the random-flight model (using random-mode parameters) and from the diffusion equation. We show the mean height $\overline{Z}(t)$ and variance $(\overline{Z^2}\,t - \overline{Z}^2(t))^{1/2}$ for 2000 trajectories of particles reduced at three initial source positions z_o = 0.5, 0.1 and 0.5 (normalised as with all dimensions on the scale of turbulence).

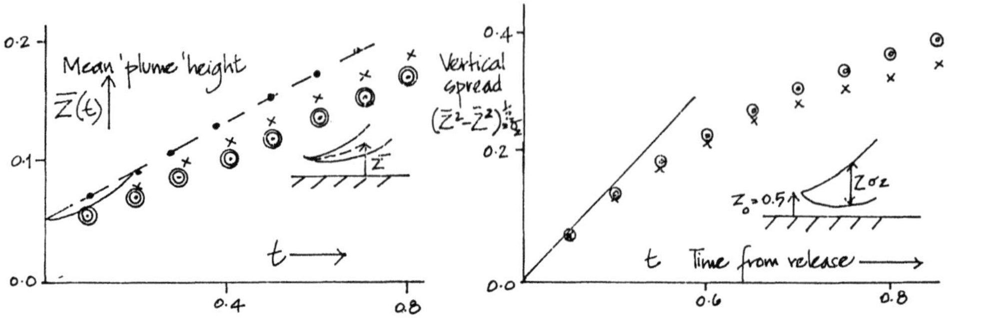

Fig.5. Moments of displacements of particles reduced from the surface at different source height Z_o. Comparison of computations based on o particle trajectories using random mode velocity field; x particle trajectories using random flight model (with velocity statistics obtained from the random mode model); . diffusion equation.

Note the close but not complete agreement between the random modes and random-flight computation, and their significant difference with the results of the diffusion equation (where $K_z = \overline{w^2}(z)T_L(z)$). This latter difference is to be expected especially near the source. Since the diffusion equation leads to $d\overline{Z}/dt = dK_z/dz$, it overestimates $\overline{Z}(t)$ near the source [19], which explains the latter difference. The former differences in $\overline{Z}(t)$ are probably caused by the fact that generally the rate of increase of z of the variance of marked particles $\partial\overline{w_L^2}/\partial z$ is less than the $\partial\overline{w^2}/dz$ for all particles (i.e. particles rise on an energetic upward eddy!). This leads to $\partial\overline{Z}/\partial t$ being slightly too great in the random-flight method. The differences in $\sigma_z(t)$ between the random flight and random modes computation are probably caused by the definition of $T_L(Z)$ in the random flight computation. It is derived in terms of all particles released at z, whereas it is defined in the model for particles at height z released at z_o, i.e. $T_L(z;z_o)$. Inevitably $T_L(z;z_o) > T_L(z)$, and therefore $\sigma_z(t)$ for the random flight model is likely to be less than for the random modes computation.

These questions need further exploration. But there is enough agreement between the methods to indicate that the random flight method is sufficiently accurate provided the correct values of $\overline{w^2}$ and $\overline{w^3}$ and $T_L(z)$ are used. (The usual criteria for $T_L(z) \approx L_x/\sqrt{\overline{w^2}}$ is not generally correct and can give poor answers!) Lagrangian methods are very useful for providing these data and for indicating where and why Eulerian methods fail.

Acknowledgement

The research of CJT was sponsored by the British Gas Corporation. JCRH is grateful to the Ecole Central de Lyon for their hospitality during the preparation of this paper.

[1] Durbin, P.A. (1983) Stochastic differential equations and turbulent dispersion. NASA Reference Publication 1103.

[2] van Dop, H., Nieuwstadt. F.T.M & Hunt, J.C.R. (1985) Random walk models for particle displacements in inhomogeneous unsteady turbulent flows. Phys. Fluids. 28, 1639-1653.

[3] Sawford, B. & Hunt, J.C.R. (1986) Effects of turbulence structure, molecular diffusion and source size on fluctuations of concentration in homogeneous turbulence. J. Fluid Mech. 165, 373-400. J. Fluid Mech.

[4] Corrsin, S. (1963) J. Atmos. Sci. 20, 115-119.

[5] Snyder, W.H. & Lumley, J.L. (1981) Some measurements of particle velocity autocorrelation functions in a turbulent flow. J. Fluid Mech. 48, 41-71.

[6] Kraichnan, H. (1970) Diffusion by a random velocity field. Phys. Fluids. 13, 22-31.

[7] Drummond, I.T. Duane, S. & Horgan, R.R. (1984) Scalar diffusion in simulated helical turbulence with molecular diffusivity. J. Fluid Mech. 138, 75-91.

[8] Hunt, J.C.R. (1984) Turbulent structure in thermalconvection and shear-free boundary layers. J. Fluid Mech. 138, 161-184.

[9] Carruthers, D.J., Hunt, J.C.R. & Turfus, C. (1986) Turbulent flow near density inversion layers. Proc. of Euromech. 199, 'Direct and Large Eddy Simulation of Turbulent Flows'. Ed. U. Schumann.

[10] Batchelor, G.K. (1953) Theory of Homogeneous Turbulence. C.U.P.

[11] Tennekes, H. (1975) Eulerian and Lagrangian time microscales in isotropic turbulence. J. Fluid Mech. 67, 561-567.

[12] Townsend, A.A. (1966). Internal waves produced by a convective layer. J. Fluid Mech. 24, 307-319.

[13] Carruthers, D.J. & Hunt, J.C.R. (1986) Velocity fluctuations near an interface between a turbulent region and a stably stratified layer. J. Fluid Mech. 165, 475-501.

[14] Hunt, J.C.R. & Graham, J.M.R. (1978) Free-stream turbulence near plane bounaries. J. Fluid Mech. 84, 209-235.

[15] Turfus, C. (1985) Stochastic Modelling of Turbulent Dispersion near Surfaces. Ph.D. Dissertation, University of Cambridge.

[16] Rao, N.J., Borwankar, J.D., Ramkrishna, D. (1974) Numerical solution of Ito integral equations. SIAM J. Control 12, 124-139.

[17] Wyngaard, J.C. & Brost, R.A. (1984) Top-down and bottom-up diffusion in the convective boundary layer. J. Atmos. Sci. 41, 102-112.

[18] Thomson, D.J. (1984) Random walk modelling of diffusion in inhomogeneous turbulence. Quart. J. Roy. Met. Soc. 110, 1107-1120.

[19] Hunt, J.C.R. (1985) Turbulent diffusion from sources in complex flows. Ann. Rev. Fluid Mech. 17 447-485.

Fully Developed Turbulence and Complex Time Singularities

T. Dombre[1], *Y. Gagne*[2], *and E. Hopfinger*[2]

[1]Groupe de Physique du Solide de l'E.N.S., 24 Rue Lhomond, F-75231 Paris Cedex 05, France
[2]Institut de Mécanique de Grenoble, B.P. 68, F-38402 St. Martin d'Hères Cedex, France

1. Introduction

In 1981 Frisch and Morf (1) proposed to relate the intermittent bursts shown by a high pass filtered turbulent flow to complex time singularities of the solution of Navier Stokes equations. Their arguments may be summarized as follows : let v(t) be a velocity signal measured at one point in a turbulent flow. If there is dissipation, v(t) will in general never become singular on the real time axis but it is free to develop singularities as soon as one goes into the complex time domain. These spontaneous singularities are a salient feature of non linear differential equations. Their positions cannot be easily predicted but their exponent as well as their amplitude (to be defined below) are intimately linked to the structure of the equations governing the flow.

Let now v(t) be singular at some point to = $z + i\delta$ as $a.(t - t_0)^{\alpha}$ (α is by definition the exponent of the singularity, a its amplitude), its Fourier transform $\hat{v}(\omega)$ will contain at high frequency ω (rd/sec) a contribution of the form :

$$\hat{v}(\omega) \underset{\omega\to\infty}{\sim} a/\omega^{\alpha+1} .\exp(i\omega z).\exp(-\omega\delta) \tag{1}$$

Each singularity gives rise in the high pass filtered signal at frequency ω to a burst centered at t_0, whose amplitude is essentially given by formula (1). Due to the exponential factor, the closer the singularity to the real axis, the stronger the corresponding burst in the signal. This furnishes an elegant explanation of the intermittency observed experimentally : as one increases the filter frequency, only the bursts corresponding to values of δ^*, the minimum of δ, will be left behind, which events become rarer and rarer. Furthermore, if δ^* is finite, it follows from eq. (1) that the energy spectrum decreases at high frequency exponentially rather than algebraically as :

$$E(\omega)= \langle\hat{v}(\omega).\hat{v}(-\omega)\rangle \sim \exp(-2\omega\delta^*) \tag{2}$$

δ^* is a quantitative measure of the smallest scales present in the flow. At finite Reynolds number it is not completely obvious that δ^* should be finite, although physically quite appealing (some numerical evidence may be found in ref. (2)).

The purpose of this work was to see if experience corroborates this picture in terms of singularities for the dissipation range in fully developped turbulence.

We stress that for Navier Stokes equations, contrarily to the situation in the area of dynamical systems[3], nothing is known, whether analytically or numerically, about the nature of such singularities. Therefore this study must be taken as a prospective one which, we hope, will encourage further work in this direction.

2. Experiments and results

In order to test predictions (1) and (2) one needs to probe somewhat deeper as usual into the dissipation range. This is a difficult task, mainly because of the rapid deterioration of the signal to noise ratio, accompanying the eventual exponential decrease of the turbulent spectrum that we are looking for. On the other hand, low Reynolds numbers are required to get high values of δ^* : indeed, if δ^* becomes too small, one can show that singularities are strongly distorted by the experimental measure device. We have chosen a grid turbulence experiment in the following configuration : mean velocity of the flow U = 6,1 m/s, mesh-length d = 7,5 cm. At the point of observation $\frac{x}{d} = 40$, $R_\lambda = 67$, the dissipation length $\eta = 0{,}5$ mm, which gives for the Kolmogorov frequency $f_K = U/2\pi\eta \simeq 2$kHz. The results to be presented below have been obtained from a filtered velocity signal, recorded at a sampling frequency of 20 kHz during a total time of 7 mn. The high and low=pass frequencies were set respectively to 1 kHz and 6 kHz.

We sketch now our main results before discussing them briefly in the following.:

a) the energy spectrum displays a typical exponential behaviour which settles well before the Kolmogorov frequency, indicating for δ^* a finite value of the order of the inverse of f_K.

b) the individual spectra of isolated bursts reveal also exponential behaviour in agreement with eq. (1). Our technique of signal processing, while failing up to now to give the algebraic exponents α of singularities, is sensitive to variations of δ. The values of δ, which we found by this analysis were of the order of our previous estimation of δ^* or higher.

c) In order to gain insight into the organization of singularities in the complex plane, we studied the statistical distribution of the burst amplitudes and their recurrences in time. We find that the probability $P(\delta)$ for a singularity to be at distance δ from the real axis, vanishes as δ, approaches its minimal value δ^* like $(\delta - \delta^*)^\beta$, where $\beta \simeq 11/3$. As concerning $P(t)$, the probability of finding two sucessive bursts of given amplitude, distant from t, it seems to be well described by a law in $t^{-1/2}.\exp(-t/2t_m)$, where t_m is the average time between the bursts. This kind of law is reminiscent of what is found for the so-called type II or III intermittencies [4] in the context of transition to chaos.

a) Figure 1.a and 1.b show the energy spectrum of the high pass velocity signal in log-log and lin-log coordinates. Note that four decades in energy are covered, for frequencies ranging from 1 kHz to 3kHz, before entering the noise level (somewhat obscured here by the accidental intrusion of harmonics of the regulation

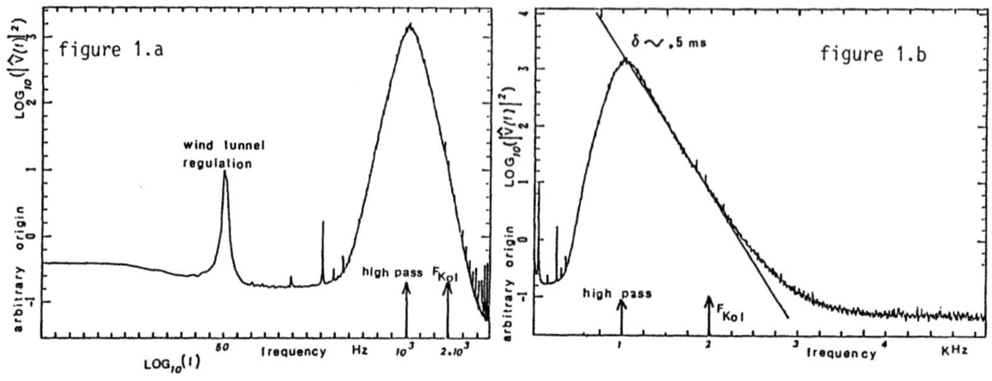

device). The pseudo-inertial range of the non filtered signal (not shown here) lies about four decades still higher in energy, for frequencies ranging from 10 Hz to 200 Hz. In figure 1.a, the spectrum is continuously curved, discording any fit to a given power law. On the other hand, the quasi linear behaviour observed in figure 1.b gives strong evidence for a finite value of δ^*, which we estimate of the order of 0.5 ms equal to the Kolmogorov time scale. Two comments, however, are in order. Firstly a slight change of slope is observable in figure 1.b around 1,5 kHz : we get $\delta^* = 0.5$ ms for 1 kHz $< f <$ 1,5 kHz, whereas for 1,5 kHz $< f <$ 2,5 kHz we rather find 0.4 ms. We do not know whether this effect is intensic or is a precursor of the subsequent rounding of the spectrum by noise. We note that this result agrees quantitatively with this deduced by Sreenivasan [5] from the Comte-Bellot and Corrsin measurements. Secondly, expression (2) gives only the leading behaviour at high frequency of the energy spectrum. We expect, in front of the exponential, algebraic prefactors, $E(\omega) \sim \omega^{-n}.\exp(-\,2\delta^*\omega)$, where n may be linked to the exponents α of singularities or at least to some average on them. In our relatively small range of experimentally accessible frequencies, this will only result in a deviation of the measured value of δ^* from its real one by an amount which varies in proportion with n. As we cannot hope to improve significantly in the next future the signal to noise ratio, experiments at lower high-pass frequencies would certainly be helpful in order to clarify these two points.

b) To submitt the theory of Frisch and Morf to a more stringent test, we selected from our high-pass filtered signal, well-characterized dissipation bursts and calculated their individual spectra. Figure 2.a shows for example the strongest burst of the whole recording and its associated spectrum is presented in figure 2.b in lin-log coordinates (before Fourier transforming, we multiplied the signal by a Gaussian window centered around the burst). From the slope of the spectrum around 2 kHz we extract the value $\delta = 0.51$ ms, which is in a surprising agreement with our previous estimation of δ^*. The shape of the spectrum is very analogous to the ones that we obtain by simulating fictitious singularities in an equivalent noise level. The comments, made at the end of the last section, are still perti-

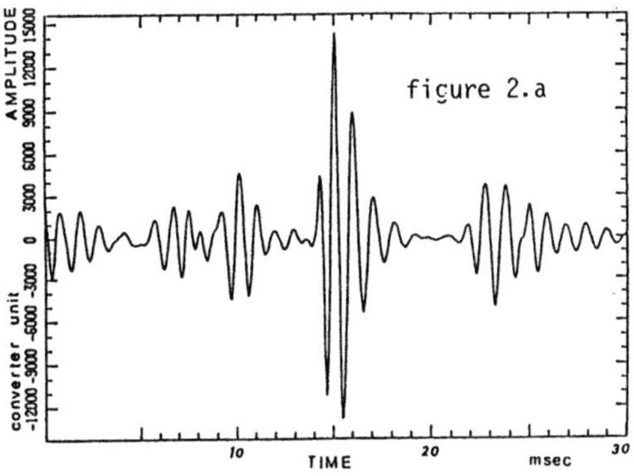

figure 2.a

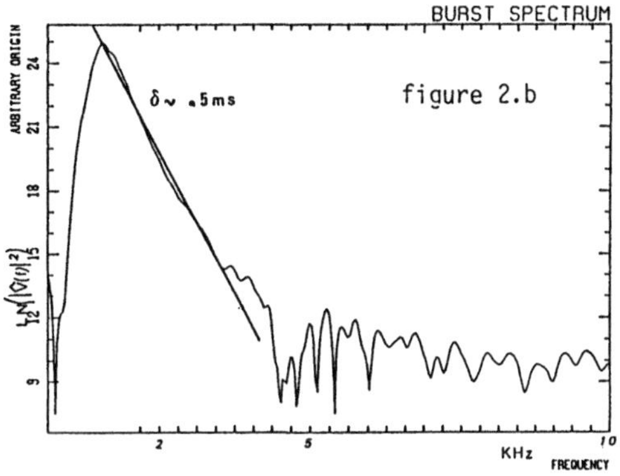

figure 2.b

nent : as we cannot give any precision on the value of α, our determination of δ is only a relative one. This analysis yielded systematically higher values of δ for bursts of lower amplitude. Were a and α the same for all singularities, then, according to (1), the logarithm of the burst amplitude should depend lineari-ly on δ. This prediction seemed well satisfied for events of high amplitudes but we observed strong departures at small amplitudes, explaniable only by assuming different values of α or a (a not too surprising result in view of the complexi-ty of Navier-Stokes equations).

c) For the last part of our study we took as granted that the logarithms of burst amplitudes are essentially proportionnal to the distance δ of the correspon-ding singularities in complex t-plane. In figure 3.a we have plotted the histo-gram N(δ) of these logarithms for amplitudes between 6000 and 32768 (the satura-tion level of our converter). In this way we obtain the shape of the repartition of singularities along the imaginary axis P(δ). The power-law behaviour is most easily seen in a log-log plot where the zero of P(δ) is considered as an adjusta-ble parameter. The best linear fit, shown in figure 3.b, was obtained for a maxi-mal logarithm of amplitude slightly higher than the last case of the histogram, a supplementary argument in favour of a finite δ*.

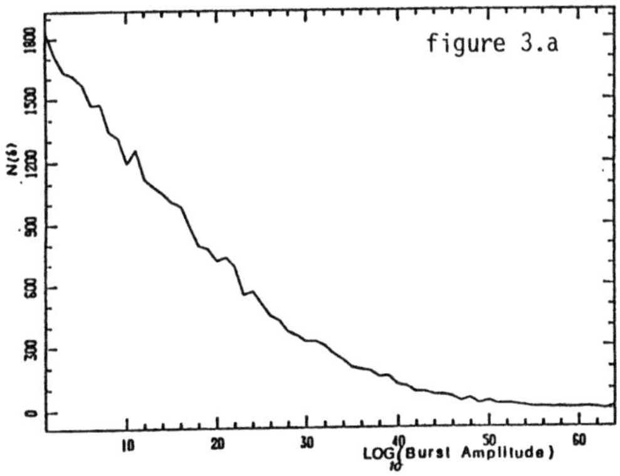

figure 3.a

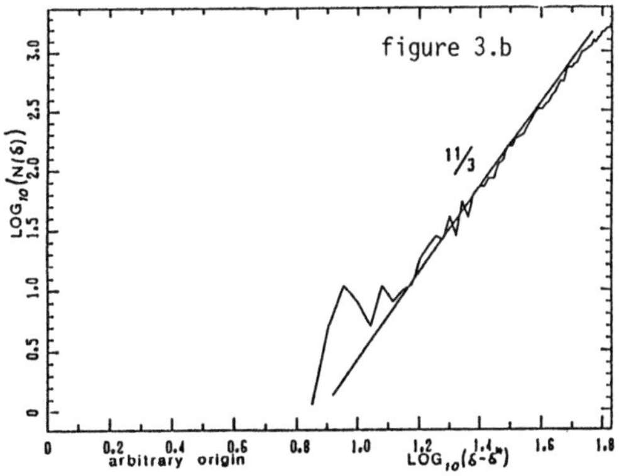

figure 3.b

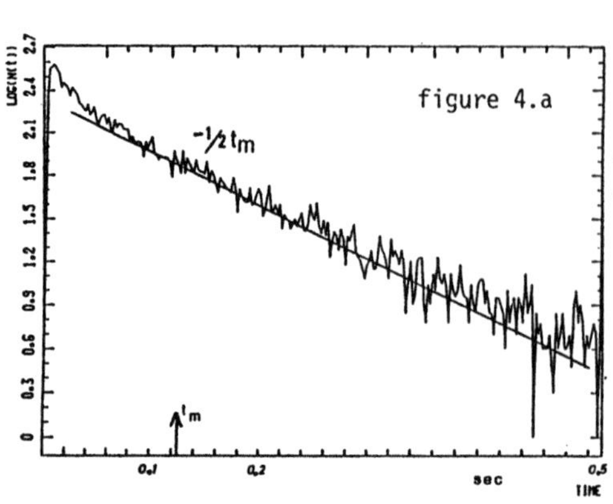

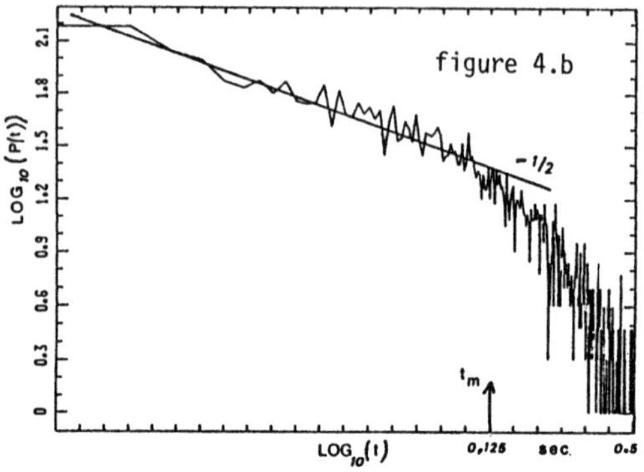

Figure 4.a is concerned with temporal aspects : we fixed to 8000 the threshold in amplitude (converter unit) and picked up the times between successive bursts whose amplitude exceeded this threshold. This time distribution P(t) exhibits at long times an exponential tail characteristic of random processes with loss of memory. At short times on the contrary, P(t) diverges as $t^{-1/2}$, as revealed by figure 4.b in log-log coordinates. The cross-over time as well as the inverse logarithmic decrement in the exponential regime scale roughly as the mean time between the bursts. The similar results were observed for different values of the threshold in amplitude.

References

1. U. Frisch and R. Morf : Phys. Rev. A 23, 2673 (1981)
2. M.E. Brachet et al. : J. Fluid Mech. 130, 411 (1983)
3. See for example Y.F. Chang, M. Tabor and J. Weiss : J. Math. Phys. 23, 531 (1982)
4. P. Bergé, Y. Pomeau, C. Vidal : in "L'ordre dans le chaos", edited by Hermann (1984)
5. K.R. Sreenivasan : J. Fluid Mech. 151, 81 (1985)
6. G. Comte-Bellot and S. Corrsin : J. Fluid Mech. 48, 273 (1971).

Reynolds Number and Prandtl Number Influence on the Determination of Isotropic Velocity and Temperature Turbulent Length Scales

C. Rey and J.-M. Rosant

Laboratoire de Mécanique des Transferts Turbulents et Diphasiques,
G.S. 410005 du CNRS, ENSM, 1 rue de la Noë, F-44072 Nantes Cedex, France

The spectral cascading for transfer of turbulent kinetic energy, heat and mass is directly related with the appropriate representation of the inertial subrange of the energy spectra.

In the case of small Reynolds numbers, with significant spectral transfer, these mechanisms interact in large and small wave number ranges and no variation using -5/3 power law in the inertial zone can be found. Therefore, the determination of the integral length scales for both turbulent fields is complicated, whereas dissipation length scales are well defined.

1. Case of large Reynolds number

On the experimental point of view, small structures are not affected by unavoidable deviation to isotropy in wind tunnel, whereas the largest one are. Measurements of Taylor scale λ_g and Corrsin scale λ_θ by means of correlations or spectra give very precisely the virtual origin x_o^u and x_o^θ of the fields, and the power law exponent of turbulence decreases : then, we write [1,2,3,4] :

$$\frac{\lambda_u^2}{M^2} Re_M = \frac{20}{n} \left(\frac{x-x_o^u}{M}\right) \longrightarrow n \text{ and } x_o^u \longrightarrow \frac{u'^2}{\bar{U}^2} = A_u \left(\frac{x-x_o^u}{M_u}\right)^{-n}$$

$$\frac{\lambda_\theta^2}{M^2} Re_M \, Pr = \frac{12}{m} \left(\frac{x-x_o^\theta}{M}\right) \longrightarrow m \text{ and } x_o^\theta \longrightarrow \frac{\theta'^2}{\Delta T^2} = A_\theta \left(\frac{x-x_o^\theta}{M_\theta}\right)^{-m}$$

where u'^2 and θ'^2 (respectively the variance of velocity and temperature fluctuations) are independently measured ; $Re_M = \bar{U} M/\nu$, $Pr = \nu/\gamma$, ν and γ respectively are the viscous diffusivity and the thermal diffusivity.

These relations give the dissipation ε, thermal dissipation ε_θ, Kolmogorov length scale η and the Batchelor length scale η_θ. We note that this determination of the turbulence decrease power law corresponds to exact isotropic turbulence, and not always give better fitting of experimental data of u'^2 and θ'^2 dowstream near the grid (Fig. 1).

The integral length scale can be calculated in isotropic turbulence, using von Kārmān spectrum which corresponds to the asymptotic solution for very large Reynolds number ; in this case :

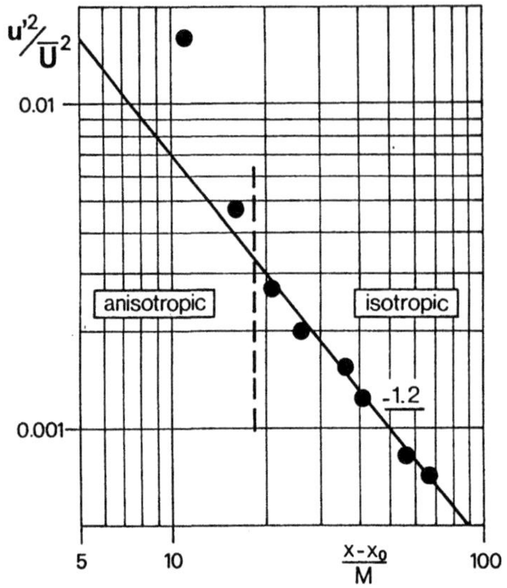

Fig. 1 Decay of velocity fluctuations with the distance from the grid
- Hydrodynamic grid turbulence, LDA measurements ; $Re_M = 6000$ -

$$\left.\begin{aligned} E(k,t) &= \alpha\, \varepsilon^{2/3}\, \ell_u^{5/3} \frac{(k\ell_u)^4}{[1+(k\ell_u)^2]^{17/6}} \\ L_u &= \frac{\pi}{2u'^2}\int_0^\infty \frac{E(k,t)}{k}\, dk \end{aligned}\right\} \quad \ell_u = \left(\frac{55\, L_u}{9\pi\alpha\ell_u}\right)^{3/2} \frac{u'^3}{\varepsilon}$$

$$\left.\begin{aligned} E_\theta(k,t) &= \beta\, \varepsilon_\theta\, \varepsilon^{-1/3}\, \ell_\theta^{5/3} \frac{(k\ell_\theta)^2}{[1+(k\ell_\theta)^2]^{11/6}} \\ L_\theta &= \frac{\pi}{2\theta'^2}\int_0^\infty \frac{E_\theta(k,t)}{k}\, dk \end{aligned}\right\} \quad \ell_\theta = \left(\frac{10\, L_\theta}{3\pi\beta\ell_\theta}\right)^{3/2} \frac{\theta'^3 \varepsilon^{1/2}}{\varepsilon_\theta^{3/2}}$$

with $L_u = \ell_u$ and $L_\theta = \ell_\theta$, $\alpha = \frac{55}{9\pi}$ and $\beta = \frac{10}{3\pi}$, these relations lead to the usual estimation of integral turbulent length scales :

$$\ell_u = \frac{u'^3}{\varepsilon} \quad \text{and} \quad \ell_\theta = \frac{\theta'^3 \varepsilon^{1/2}}{\varepsilon_\theta^{3/2}}$$

2. Case of small Reynolds number, influence of the Prandtl number

In the present work we propose relations giving integrals scales of kinetic energy and heat turbulence in the case of small Reynolds number and very large variations of Prandtl number (or Schmidt number for mass concentration turbulence). These relations use a modelisation of the energy spectrum distribution for velocity and temperature, called Kàrmàn-Pao spectrum, which can be written for velocity spectrum [4,5,6,7]:

$$E(k,t) = \alpha\, \varepsilon^{2/3}\, \ell_u^{5/3} \frac{(k\ell_u)^2}{[1+(k\ell_u)^2]^{11/6}} \exp\{-\tfrac{3}{2}\alpha(k\eta)^{4/3}\}$$

and for temperature spectrum [8,9,10,11]:

$$E_\theta(k,t) = \beta\, \varepsilon_\theta\, \varepsilon^{-1/3}\, \ell_\theta^{5/3} \frac{(k\ell_\theta)^2}{[1+(k\ell_\theta)^2]^{11/6}} [1+Pr^{1/3}(k\eta_\theta)^{2/3}]$$

$$\exp\{-\beta(k\eta_\theta)^2 - \frac{3}{2}\beta\, Pr^{-1/3}(k\eta_\theta)^{4/3}\}$$

A preliminary experimental study of the grid turbulence has been developed, with large variations of the flow velocity, for validation of the velocity spectrum relation, when the Reynolds number is small ($8600 < Re_M < 76000$). Then, for a better fitting of experimental data, we selected rather k^2 than k^4 as the increasing of the spectrum in the small wave numbers range (Fig. 2).

The evolution of the temperature spectrum, with the Prandtl number and for a fixed Reynolds number, calculated by the model, is presented on Fig. 3. We note that it corresponds to the Batchelor's model for large wave numbers (viscous convective range).

The definition of the integral length scales L_u and L_θ together with these models for $E(k,t)$ and $E_\theta(k,t)$ gives:

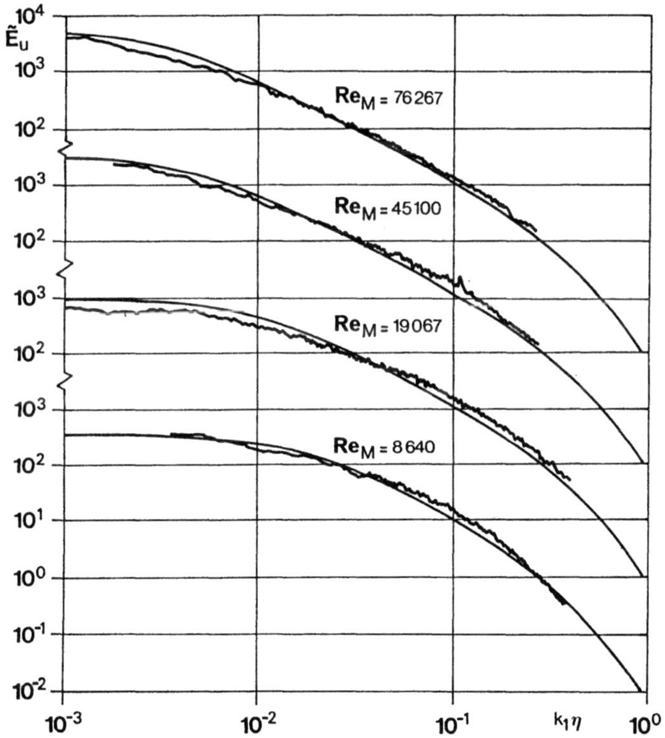

Fig. 2 Normalised kinetic turbulence spectrum $\tilde{E}_u = E_u/\varepsilon^{2/3}\, \eta^{5/3}$. Comparisons with experimental data in a wind tunnel ; distance from the grid : $x/M = 15$, mesh size : $M = 5.5$ cm

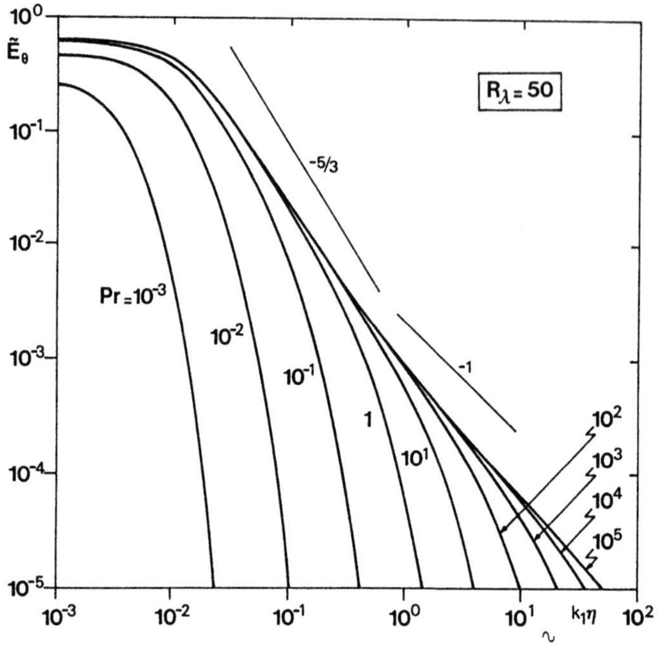

Fig. 3 Normalised thermal turbulence spectrum $\tilde{E}_\theta = 2E_\theta \, \varepsilon^{1/3}/\beta \, \varepsilon_\theta \, \ell_\theta^{5/3}$ variations with various values of Prandtl number for $R_\lambda = 50$

$$\ell_u = \frac{u'^3}{\varepsilon}\left[\frac{10L_u/3\pi\alpha\ell_u}{1+I_u(\frac{\eta}{\ell_u})}\right]^{3/2} \quad \text{with} \quad I_u(\frac{\eta}{\ell_u}) = (\frac{\eta}{\ell_u})^{5/3}\int_0^\infty \frac{-2\alpha \exp\{-\frac{3}{2}\alpha x^{4/3}\}}{[(\frac{\eta}{\ell_u})^2 + x^2]^{5/6}} x^{1/3}\, dx$$

and

$$\ell_\theta = \frac{\theta'^3 \varepsilon^{1/2}}{\varepsilon_\theta^{3/2}}\left[\frac{10L_\theta/3\pi\alpha\ell_\theta}{1+I_\theta(\frac{\eta_\theta}{\ell_\theta},Pr)}\right]^{3/2}$$

where the integral I_θ is expressed in two asymptotic situations, when Prandtl number is large (Pr >> 1) :

$$I_\theta(\frac{\eta_\theta}{\ell_\theta},Pr) = (\frac{\eta_\theta}{\ell_\theta})^{5/3}\left[\int_0^\infty \frac{-2\beta\, x \exp\{-\beta x^2\}}{[(\frac{\eta_\theta}{\ell_\theta})^2 + x^2]^{5/6}}\, dx + \frac{5}{3} Pr^{1/3}\int_0^\infty \frac{x^{5/3}\exp\{-\beta x^2\}}{[(\frac{\eta_\theta}{\ell_\theta})^2 + x^2]^{11/6}}\, dx\right]$$

and, when Prandtl number is small (Pr << 1) :

$$I_\theta(\frac{\eta_\theta}{\ell_\theta}, Pr) = -(\frac{\eta_\theta}{\ell_\theta Pr^{1/4}})^{5/3}\int_0^\infty \frac{2\beta \exp\{-\frac{3}{2}\beta x^{4/3}\}}{[(\frac{\eta_\theta}{\ell_\theta Pr^{1/4}})^2 + x^2]^{5/6}} x^{1/3}\, dx$$

corresponding evolutions for ℓ_u and ℓ_θ are illustrated on Fig. 4 in non dimensional form, using numerical calculations, with restrictive hypothesis :

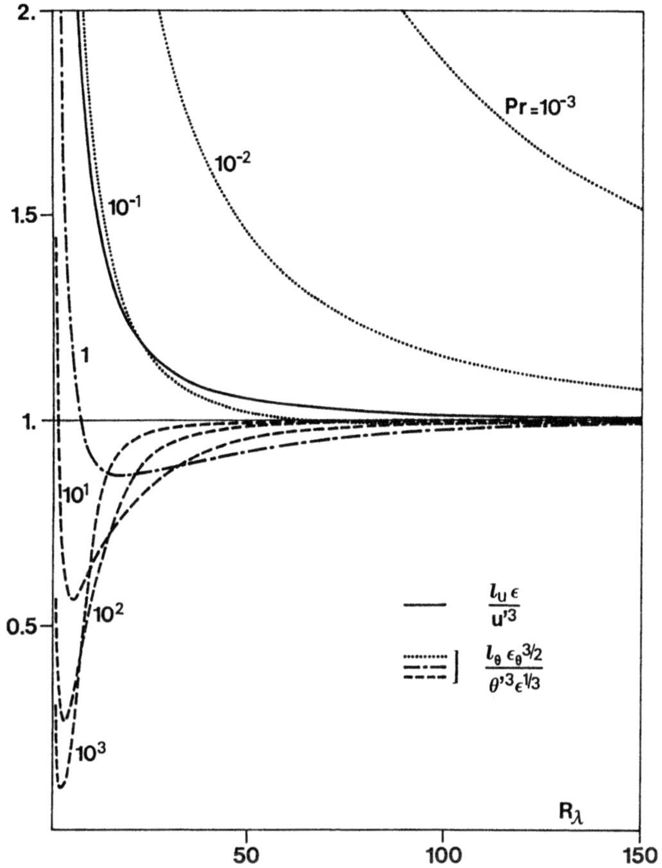

Fig. 4 Numerical calculations of relations giving integral scales and taking into account the influence of Reynolds and Prandtl numbers.

$$\frac{10L_u}{3\pi\alpha\ell_u} = \frac{10L_\theta}{3\pi\beta\ell_\theta} = 1$$

For small Reynolds numbers and small Prandtl numbers these hypothesis is not justified, that shows the difficulty of exact calculation of L_u and L_θ.

3. Consequences

The study of the integrals $I_u\ (\frac{\eta}{\ell_u})$ and $I_\theta\ (\frac{\eta_\theta}{\ell_\theta},\ Pr)$ gives new informations for discussion on grid turbulence experimental data [12,13]. Precisely, the difference which can be observed below the grid inside evolutions of velocity and temperature integral length scales is represented by the ratio L_θ/L_u versus x/M and corresponds to :

$$\frac{L_\theta}{L_u} = \left(\frac{3}{2}\,\frac{n}{m}\,\frac{x-x_o^\theta}{x-x_o^u}\;\frac{1+I_u(\frac{\eta}{\ell_u})}{1+I_\theta(\frac{\eta_{\hat\theta}}{\ell_\theta},\ Pr)}\right)^{3/2}$$

this relation uses the previous restrictive hypothesis between L_u and ℓ_u, L_θ and ℓ_θ. We note here that the size of the thermal mesh M_θ, which can be different from the kinematic one M_u, has no influence in the phenomenon.

In the flow downstream the magnitude of the gap between the virtual origins decreases when the distance from the grid increases whereas I_u and I_θ integrals may not be negligible comparing to unity, according to the values taken by the Reynolds number and Prandtl number.

References

1. S. Corrsin : Journal of Aeronautic Sciences, Vol. 18, p 417 (1951)
2. S. Corrsin : Journal of Applied Physics, Vol. 22, n° 4, p 469 (1951)
3. M.S. Uberoi & S. Wallis : The Physics of Fluids, Vol. 10, p 1216 (1967)
4. G. Comte Bellot & S. Corrsin : Journal of Fluid Mechanics, Vol. 48, p 273 (1971)
5. R.L. Schapker : AIAA Journal, Vol. 9, n° 5, p 952 (1971)
6. J.T. Lin : The Physics of Fluids, Vol. 15, n° 1, p 205 (1972)
7. J. Qian : The Physics of Fluids, Vol. 27, n° 9, p 2229 (1984)
8. Y.H. Pao : The Physics of Fluids, Vol. 8, n° 6, p 1063 (1965)
9. Y.H. Pao : The Physics of Fluids, Vol. 11, n° 6, p 1371 (1968)
10. R.J. Hill : Journal of Fluid Mechanics, Vol. 88, part. 3, p 541 (1978)
11. J. Qian : The Physics of Fluids, Vol. 28, n° 5, p 1299 (1985)
12. Z. Warnaft & J.L. Lumley : Journal of Fluid Mechanics, Vol. 88, p 659 (1978)
13. K.R. Sreenivasan, S. Tavoularis, R. Henry & S. Corrsin : Journal of Fluid Mechanics, Vol. 100, part. 3, p 597 (1980)

On the Detection of Dissipation Events in a High Reynolds Number Turbulent Boundary Layer

L. Jacquin [1] *and P. Mestayer* [2]

[1] O.N.E.R.A., 29 Av. de la Division Leclerc, F-92320 Chatillon, France
[2] I.M.S.T., 12 Av. du Général Leclerc, F-13003 Marseille, France

An original conditional sampling method is proposed, to investigate the intermittency of temperature dissipation. It is based on a combined threshold analysis of amplitude and duration of the squared derivative of the temperature signal and applied to measurements realized in the large boundary layer of I.M.S.T. Air-Sea Interaction Simulation Tunnel. An extensive investigation of the method brings into light the existence of a continuous organisation of particular scales in the intermittent structure of the signal. However, with a convenient arbitrary choice of detection parameters, this method enables to isolate the events containing the major part of the dissipation $\overline{(\partial\theta/\partial t)^2}$. This analysis has been coupled with a quadrant analysis of the shear stress measured at the same location. These analyses, as well as visual observations of the conditionned signals show, to a certain extent, how dissipation can be related to the 'ramp structures' of the temperature signal and to the movements of fluid associated to the quadrants.

1. Introduction

Since Landau's connection between the small scale intermittency [1] and the phenomenological statistical description of homogeneous turbulence, internal intermittency has been mainly investigated in the context of Kolmogorov's local self-similarity theory [2], [3]. Therefore it was considered within the framework of Kolmogorov's first hypothesis of small scale statistical independency with respect to macroscopic scales and geometry of the flow. These investigations essentially concern refinements of Kolmogorov's original theory [3], experimental determinations of the 'power law' exponents (see reviews of references in [4]), the development of the fractal cascade concept (see most reference in [5]), and the band-pass-filter analyses of the high frequencies of longitudinal velocity u' and temperature θ'([5],[6],[7]).

Yet there are some reasons to question these purely statistical approaches.

Some experimental results tend to contradict or to strongly limit the domain of validity of the hypothesis of local isotropy, a cornerstone of the universal self-similarity theories (see references in [8] and [9]). Direct numerical simulations [10] and three-dimensional E.D.Q.N.M. modelisations [11] have put into evidence (in relatively low Reynolds number flows) strong dependencies of small scale vorticity structure with respect to mean velocity gradients. A possible explanation could be found in the role of 'elongated non-local spectral interactions' [12] i.e. direct interactions between structures having very different characteristic scales. From KRAICHNAN's [13] analysis of intermittency of band passed velocity, FRISCH & MORF [14] have shown that internal intermittency can depend on production scales and that the dissipation range can be nonuniversal for finite Reynolds numbers.

On the other hand, the existence of intermittent organized large scale motions has been demonstrated, even in relatively large Reynolds number shear flows [15] [16]. They have been clearly identified by various conditional sampling methods, usually associated to arrays of probes (see e.g. [16]). To a lesser extent, the large scale ramp-like patterns put into evidence in the temperature signals obtained in shear flows ([17], [18]) have been shown to be related to these intermittent organized motions [19].

These reasons have already led several authors to develop new methods of investigation of the possible interactions between large scale intermittent structures and internal intermittency ([19] to [23]). Yet, the first method ([19] [20]) is based on the conditional sampling of the relevant signals from the detection of the temperature ramp structure signature i.e. their sharp front. Therefore, this a priori method can, and do effectively, only investigate the influence of these fronts on other conditionned functions, specifically the dissipation statistics.

The other method ([21],[22],[23]) proposes a more open way of investigation by isolating a signal representing the intermittent small scales, the streamwise velocity filtered by a narrow-band-pass frequency-filter. Its relations with large scale functions are then investigated by the classical methods. Unfortunately, the results appear too dependant on the precise characteristics of the frequency filter to provide objective interpretations.

We propose here an alternate method minimizing the arbitrary choices of conditionning parameters and allowing the construction of signals representing the intermittent small scales to investigate the possible interactions, even those of semi-deterministic nature, between large scale structures and internal intermittency.

2. Conditional Detection of Temperature Dissipation

The first problem to solve is the choice of a signal which, if properly conditionned, can represent by itself the dissipation intermittency, i.e. have high and low values about in the same time than dissipation.

The average rate of kinetic energy dissipation is defined as :

$$\overline{\varepsilon} = \nu \sum_{i=1}^{3} \sum_{j=1}^{3} \overline{(\partial u_i/\partial x_j)(\partial u_j/\partial x_i)} \qquad (1)$$

where u_i, u_j represent the components of velocity fluctuation in the directions of space x_i, x_j, ν is kinematic viscosity and the overbar indicates ensemble averaging, or time averaging in stationnary flows. The determination of ε so implies the measurement of 9 gradient-product functions, except when the local isotropy assumption holds, in which case the average rate can be obtained from the average value of only one of these functions. As for the instantaneous kinetic energy dissipation it seems very unlikely that it can be correctly represented by any one of these 9 gradient-product functions alone and it is very difficult to measure correctly and simultaneously more than one [5].

Because temperature is a scalar, the average rate of dissipation of temperature variance :

$$\bar{\varepsilon}_\theta = D_\theta \sum_{i=1}^{3} \overline{(\partial\theta/\partial x_i)^2} \qquad (2)$$

where D_θ is temperature diffusivity, depends on only 3 gradient functions. Although ANTONIA et al [20] have shown that large excursions in these gradient functions do not seem to coincide well, we expect the signal $S(t) = (\partial\theta(t)/\partial t)^2$ to have an intermittent structure quite similar to that of the instantaneous dissipation. Although this assumption is supported by some experimental results [25], [26] it should be the subject for an independant study. The strong intermittent behaviour of S(t) is illustrated on the time traces of Fig. 1, by comparison to the squared derivative of a white noise.

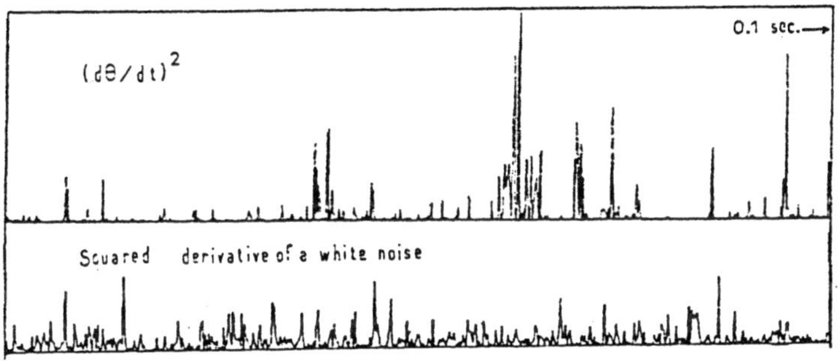

Fig. 1 - Compared structures of squared derivatives of temperature signal and white noise signal

To localize in S(t) the 'events' containing the large contributions to dissipation, the method combines an amplitude threshold analysis and a duration threshold analysis (Fig. 2). The amplitude threshold analysis isolates all parts of the signal, actually spikes, larger than the threshold H. The duration threshold analysis, applied to the durations Δt between the detected spikes, musters the spikes and the accompanying low amplitudes belonging to the same 'event'. The intermittency function thus obtained is defined by :

$$I(t) = \begin{cases} 1 & \text{if } S(t) \geqslant H \text{ .or. if } S(t) < H \text{ during } \Delta t \leqslant \tau \\ 0 & \text{if } S(t) < H \text{ during } \Delta t > \tau , \end{cases} \qquad (3)$$

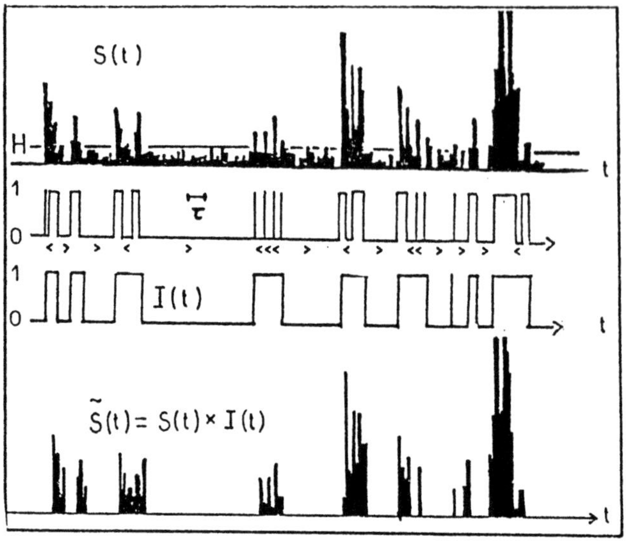

Fig.2 - Synoptic of conditional sampling method

where τ is the duration threshold. The smaller τ, the more acute the detection ; if $\tau = 0$ only the spikes remain. The larger τ, the looser the detection but the larger the probability to isolate the complete duration of real events.

Could the intermittent structure of S(t) be characterized by one specific time scale τ' ? In that case, in the domain $[\tau \in V(\tau')]$ the probability density functions of the durations of the events and of the separations between events would be nearly independant of the amplitude threshold H, at least in a defined range $[H \in V(H')]$. The systematic observation of monotonous evolution of these functions show that it is not the case [24].

3. Characteristics of the Conditionned Dissipation

The detection method can be considered as a time-filtering. In a way similar to the analysis of KUO and CORRSIN's [6] frequency-filtering method, it is interesting to determine the influence of the filtering parameters, the thresholds, on the statistical characteristics of the intermittency function I(t) and of the filtered signal :

$$S^{\circ}(t) = S(t) \cdot I(t). \tag{4}$$

The signal studied here is the squared derivative of the temperature fluctuation measured with a cold wire resistance thermometer in the boundary layer of I.M.S.T. Large Air-Sea Interaction Simulation Tunnel. The characteristic parameters of the flow are the following : $z/\delta = 0.33$, $\delta = 0.75$m, $U_\infty = 7.5$ m/s, $\lambda = 0.027$m, $\eta = 0.27 \cdot 10^{-3}$m, $R_\lambda = 610$, $R_\theta = 36\ 000$ and $\theta(z) - \theta_{surface} = -10$K.

Figure 3 presents the evolutions of the average number N of 'events' per unit of time and their mean length $\langle L \rangle = \overline{U.I(t)}/N$ normalized by the Kolmogorov microscale η, when H varies, for different values of τ. For comparison, Fig. 4 shows

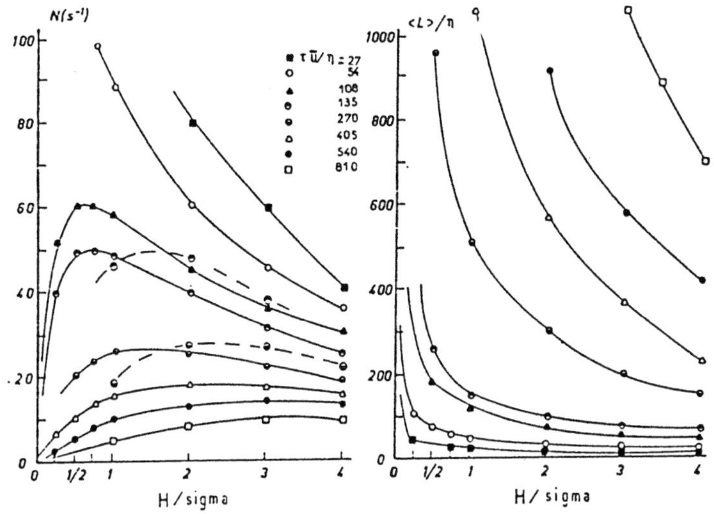

Fig.3 - Average number N and mean width L of the pulses detected, as functions of detection thresholds H and τ. Dashed lines are results obtained with $(\partial u/\partial t)^2$

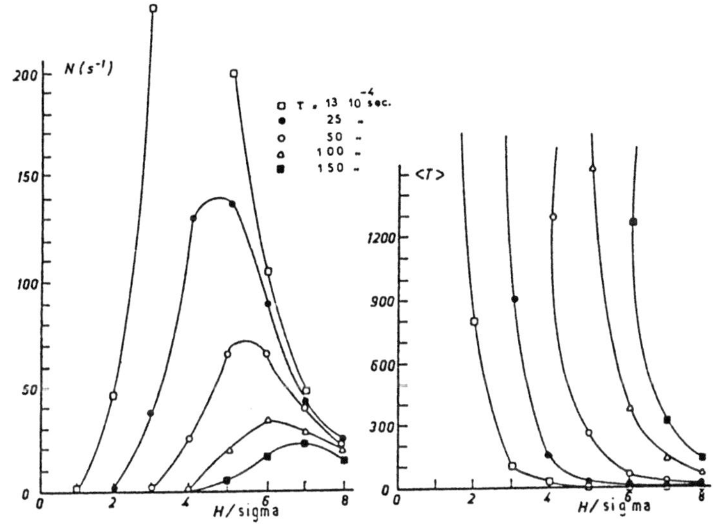

Fig.4 - Same as Fig.3 when detection is applied to white noise squared derivative

the results obtained when the detection is applied to the squared derivative of a white noise signal. In Fig. 4, the mean duration of the 'events' detected is $\langle T \rangle = I(\tau)/N$. The amplitude threshold is normalized by sigma = $\overline{S(t)^2}^{\frac{1}{2}}$ and the duration threshold by Kolmogorov time scale $\eta/\overline{U}$.

For both signals the average number of events N reaches a maximum for a given range of H that depends on τ. Visual observations indicate that for S(t) this range corresponds somewhat to an equilibrium between the rapid fragmentation of large 'events' due to the loss of small amplitude spikes, and the progressive elimination of generally short events that do not contain sufficiently high amplitudes. The first process prevails for small values of H whereas the second one prevails in the decrease of N for larger values of H. For those values one can observe a damping in the decrease of the mean width $<L>$ of the 'events'. However, for the larger duration thresholds τ it becomes more difficult to localize the maximum of N and the damping of $<L>$ decrease disappears.

For the white noise intermittency function the more symetric shape of the curves of N associated to the abrupt decrease of $<T>$ correspond to a process of rapid apparitions of very narrow pulses that are then rapidely eliminated. The shape of the curves do not change with τ : this process of pulse generation-elimination exists for all values of τ and is shifted toward the larger values of H when τ increases.

By opposition, the organized character of the intermittency of $(\partial\theta/\partial t)^2$ appears in the following features :

- For a given value of the duration threshold τ, N is an asymetric function of the amplitude threshold H : the presence of more or less coherent and localized zones of high activity in S(t) is the cause of the damping in the decrease of N and $<L>$ for large values of H. As a comparison, the behaviour of the intermittency function of $(\partial u/\partial t)^2$ is illustrated in Figure 3 (dash lines). The already observed less-intermittent structure of $(\partial u/\partial t)^2$ appears in the curves of N by their weaker asymetry and by the shift of the maximum toward the larger values of H.
- For the high values of H, the shape of the curves varies progressively with τ. When τ decreases, the large 'events' detected are very progressively subdivized into smaller ones down to individual spikes.

Of course, the acuteness of the detection performed by this method depends on H and τ. If one agrees that the internal intermittency can be characterized by the curves of Fig. 3 and that they do present features that are characteristic of its organization, the observation of Fig. 5 then becomes appealing. This figure is an enlargment of Fig. 3 for the small values of and H. The whole sets of curves in Fig. 3 and in Fig. 5 superimpose nearly perfectly : this is a characteristic feature of a self-similar process.

This detection method is a time-filtering and it must be emphasized how different it is from the narrow-band-pass frequency-filtering method used by KUO and CORRSIN [6] and others ([5],[7],[21],[22]). This is well illustrated by Fig. 6 that presents the frequency spectra of S(t), the filtered signal S°(t) and the signal in the zones rejected by the detection (the quiet zones). This Fourier analysis in zones has been carefully realized to avoid any window effects [24]. The choice for the filtering parameters, H/sigma = 2 and $\tau.\overline{U}/\eta$ = 810 is especialy interesting as it produces an intermittency factor of 50%. The integrals under the curves in Fig. 6 are respectively proportional to the mean dissipation rate ε_θ and

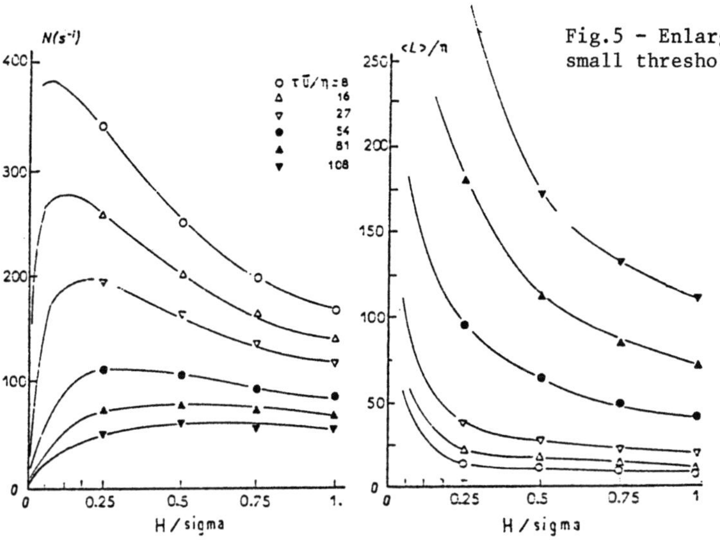

Fig.5 - Enlargement of Fig.3, for small thresholds H and τ

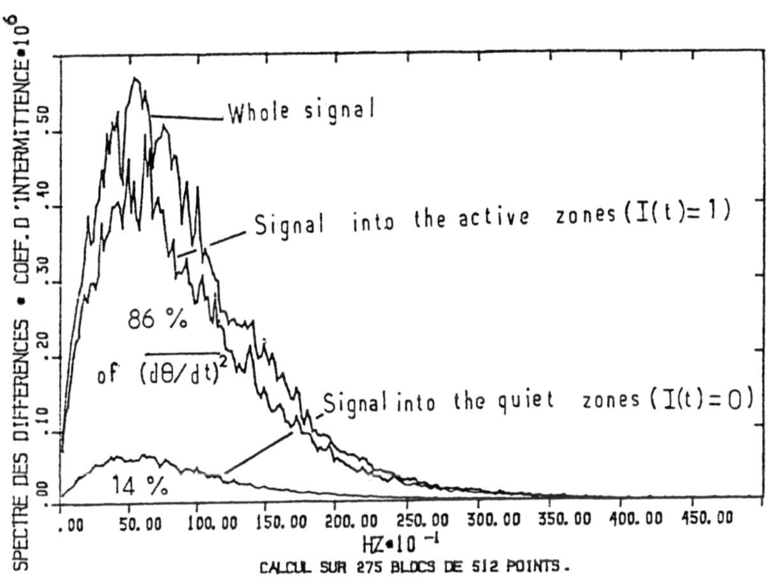

Fig.6 - Variance spectra in the quiet and active zones (H = 2 sigma; $\tau = 810\ \eta/\overline{U}$; $\langle I \rangle$ = 50%) compared to the spectrum of the whole signal S(t)

to its contributions from the 'dissipative events' and the quiet zones. It appears that this time-filtering method allows to select 50% of the signal that contains 86% of the dissipation. It also appears that the variance spectra of the original signal and of the filtered signals have nearly the same shape.

4. On Search of a Relation Between Dissipation Intermittency and Large-Scale Events

A large number of techniques have been proposed to put into evidence the large movements of fluids in a boundary layer or to characterize them. Most of these techniques are efficient only in low to moderate Reynolds number flows, and most of them rely on large arrays of sensors. As an illustration of the possibilities offered by the new filtering technique proposed here, two types of large scale events have been considered, that can be detected easily from measurements in one point, the temperature 'ramp-like' structures and the 'ejections-sweeps' movements normal to the wall. The signals available are the fluctuations of temperature θ' and those of the velocity components along the stream wise direction u' and along the vertical direction, normal to the water surface, w'.

The sweeps and ejections from the wall have been characterized by the 'Reynolds-stress quadrant-analysis' ([15],[27]). This technique separates the signal u'w'(t) in four signals according to the signs of u' and w'. The quadrants I and III (u'w' > 0) correspond to 'weak-interaction zones' when the quadrant II (u' < 0, w' > 0) corresponds to slower 'ejections' of fluids away from the wall and the quadrant IV (u' > 0, w' < 0) to 'sweep' motions back toward the wall. The pertinency of this method for a high Reynolds number turbulent boundary layer the wall of which is a water surface, has never been demonstrated. Especially, the role of pressure in the wall region over water wavy surface is a study that is still to be entreprized [29]. Yet the statistical structure of this flow, in the central and outer regions [28], seems, in first approximation, sufficiently similar to that of a boundary layer over a rough plate to allow use of this technique. The data in Fig. 7 confirm this estimation. The normalized quadrant contributions to the Reynolds stress and the ratio of the two most energetics one, obtained in the central region of this boundary layer (z/δ = 0.33) are in

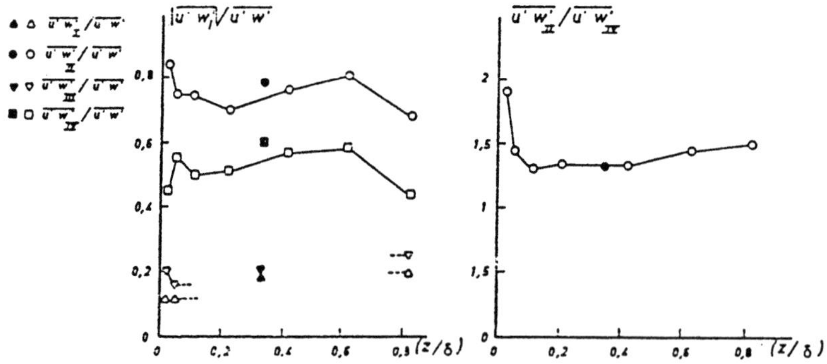

Fig. 7 - Quadrants contributions to Reynolds stress (Δ = I; O = II; ∇ = III; □ = IV), open symbols, LU & WILLMARTH [15], solid symbols, this study

excellent agreement with the results of LU and WILLMARTH [15]. The momentum Reynolds number is one order of magnitude larger here ($R_\theta \simeq 38000$) than in WILLMARTH's flow ($R_\theta = 4320$). This implies an important extension of the similitude of these large scale motions and probably of the universality of the bursting process near the wall.

The existence of the ramp-like structures in the temperature signals obtained in shear flows has been demonstrated since long [17], [18]. If their signature, a sharp front the sign of which depends on the mean temperature gradient sign, is detectable quite unambiguously, the whole structure seems more difficult to isolate, especially in a boundary layer (see [18]). In Fig. 8, the simultaneous time traces of u'w'(t) in quadrants II and IV, of $\theta'(t)$ and of the dissipation intermittency function (same parameters as for Fig. 6) show that the dissipative 'events' have a large factor of coincidence with the ramp fronts, visualy detected and indicated by arrows. A further study of these coincidences should be of very high interest but, up to now, no method has been found to determine without ambiguity the length of the ramp structures in a boundary layer.

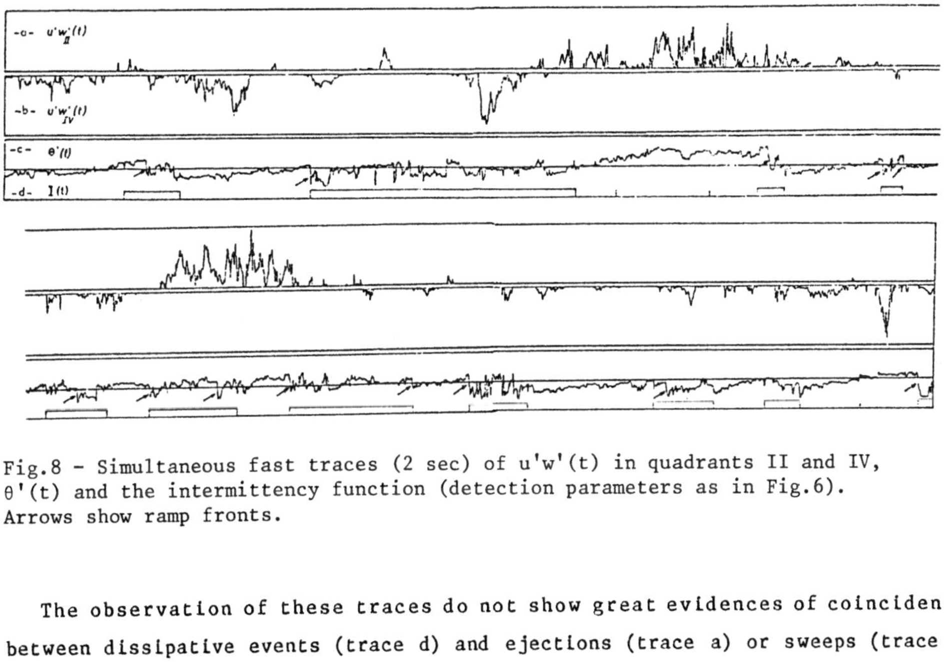

Fig.8 - Simultaneous fast traces (2 sec) of u'w'(t) in quadrants II and IV, $\theta'(t)$ and the intermittency function (detection parameters as in Fig.6). Arrows show ramp fronts.

The observation of these traces do not show great evidences of coincidences between dissipative events (trace d) and ejections (trace a) or sweeps (trace b). To confirm this observation of full quadrant analysis has been entreprized [23] by introducing an amplitude threshold allowing to filter out the ambiguous parts of the quadrant signals [27]. Some of the results are presented in Figure 9. The absolute values of the quadrant intermittency factors and contributions to the Reynolds stress depend on the value of the threshold, but not their relative values. As for the relation between quadrants and dissipation, it is clear that the ejections correspond to lower contributions to $\overline{\varepsilon_\theta}$.

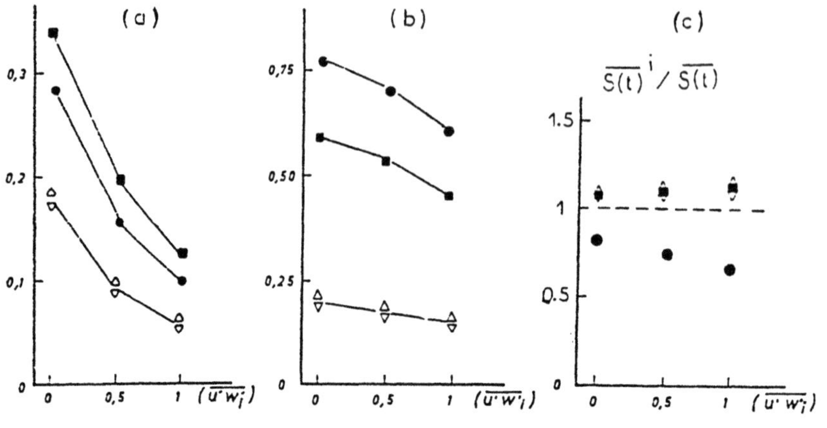

Fig.9 - Quadrant (qud) analysis versus qud threshold H_i (in fraction of $\overline{u'w'}_i$)(Δ = I; • = II; ∇ = III; ■ = IV); a) qud intermittency factor; b) qud contribution to $\overline{u'w'}$; c) qud normalized contribution to $\overline{\varepsilon}_\theta$

5. Conclusions

A new detection method is proposed to investigate the relations of statistical and semi-deterministic natures between the internal intermittency of dissipative scales and the observable intermittent macro-scale events. It is based on a time-filtering of $(\partial\theta/\partial t)^2(t)$ by means of a combination of amplitude and duration threshold analyses. It seems to produce results consistent with the hypothesis of self-similarity of internal intermittency. It separates the zones of the signal that contain the major part of the dissipation, with the same spectral repartition that the unfiltered dissipation signal. This technique allows to detect coincidences and to compute statistical correlations with other signals representative of large scale events of interest, by means of the techniques usually applied to intermittent functions.

6. Acknowledgements

This study has been realized at I.M.S.T. with support from the A.T.P. Recherches Atmosphériques of I.N.S.U.

7. References

1. G.K. Batchelor & A.A. Townsend : Proc. Roy. Soc. A129, 238 (1949)
2. A.N. Kolmogorov : Dokl. Akad. Nauk. S.S.S.R. 30, 301 (1941)
3. A.N. Kolmogorov : J. Fluid Mech. 13, 82 (1962)
4. P. Mestayer : Bulletin Dir. Et. & Recherches EDF, série A, N°4 (1981)
5. Y. Gagne, Thèse Doct. Ing. I.N.P. Grenoble (1980)
6. A.Y.S. Kuo & S. Corrsin : J. Fluid Mech. 50, 285 (1971)
7. S. Tavoularis & S. Corrsin : J. Fluid Mech. 104, 349 (1981)

8. P. Mestayer : J. Fluid Mech. 125, 475 (1982)
9. R.A. Antonia & L.W.B. Browne : J. Fluid Mech.163, 393 (1986)
10. R.S. Rogalo : NASA TM 81315 (1981)
11. C. Cambon & L. Jacquin : 7th A.M.S. Symp. on Turbulence and Diffusion (Boulder, 1985)
12. P. Mestayer, J.P. Chollet & M. Lesieur : in Turbulence and Chaotic Phenomena, Ed. by T. Tatsumi, Elsevier Sci. Publi. B.V. (North-Holland, 1984) p.285
13. R.H. Kraichnan : Phys. Fluids 10, 2080 (1967)
14. U. Frisch & R. Morf : Physical Rev. A23, 2673 (1981)
15. S.S. Lu & W.W. Willmarth : J. Fluid Mech. 60, 489 (1973)
16. Z. Zaric, R.E. Falco & R.F. Blackwelder : in Turbulence and Chaotic Phenomena, Ed. by T. Tatsumi, Elsevier Sci.Publi. B.V. (North-Holland, 1984) p. 439
17. P.G. Mestayer, C.H. Gibson, M.F. Coantic & A.S. Patel : Phys. Fluids 19, 1279 (1976)
18. C.H. Gibson, C.A. Friehe & S.O. McConnel : Phys. Fluids 20, S156 (1977)
19. K.R. Sreenivasan, R.A. Antonia & D. Britz : J. Fluid Mech. 94, 745 (1979)
20. R.A. Antonia, F. Anselmet & A.J. Chambers : J. Fluid Mech. 163, 365 (1986)
21. K.N. Rao, R. Narasimha & M.A. Badri Narayanan : J. Fluid Mech. 48, 767 (1976)
22. M.A. Badri Narayana, S. Rajogopalan & R. Narasimha : J. Fluid Mech. 80, 237 (1977).
23. S. Rajagopalan & R.A. Antonia : Phys. Fluids 27, 1966 (1984)
24. L. Jacquin : Thèse Doctorat 3° Cycle, IMST, Univ. Aix-Marseille II (1983)
25. P. Mestayer, M. Coantic, J.P. Pages & J. Saissac : in Structure of Turbulence in Heat and Mass Transfer, Ed. by Z. Zaric, Hemisphere Publ. Corp. (Washington, 1982)
26. L. Fulachier & R.A. Antonia : Intl. Comm. Heat Mass Transfer 10, 435 (1983)
27. J. Sabot & G. Comte-Bellot : J. Fluid Mech. 74, 767 (1976)
28. M. Coantic, A. Ramamonjiarisoa, P. Mestayer, F. Resch & A. Favre : J. Geophys. Res. 86, 6607 (1981)
29. J.P. Giovanangeli & A. Memponteil : J. Fluid Mech. 159, 69 (1985)

Part V

Two-Dimensional Velocity Fields. Geophysical and Astrophysical Turbulence

Geophysical and Astrophysical Turbulence

H.K. Moffatt

Department of Applied Mathematics and Theoretical Physics,
University of Cambridge, Silver Street, Cambridge CB3 9EW, UK

1. Introduction

Turbulence in large-scale geophysical and astrophysical contexts is dominated on the one hand by Coriolis forces associated with global rotation, and on the other by Lorentz forces associated with the currents that flow in the electrically conducting medium of astrophysical bodies and the magnetic fields to which these currents give rise.

Coriolis forces in the upper atmosphere of the Earth constrain the turbulence to be approximately two-dimensional in horizontal planes. Direct numerical simulation (e.g. McWilliams 1984) has proved effective in revealing the characteristic structures that emerge under evolution that is constrained to remain two-dimensional. There are two apparently conflicting theories of two-dimensional turbulence, the 'vortex patch' theory of Saffman (1971) which yields a k^{-4} energy spectrum, and the enstrophy cascade theory of Kraichnan (1967) and Batchelor (1969) which yields a k^{-3} spectrum. We shall argue that these theories are not irreconcilable, and that a k^{-4} spectrum will gradually evolve towards a k^{-3} spectrum through interaction of vortex patches leading to tightly wound-up spiral structures.

In section 3, we briefly review the theory developed in Moffatt (1985, 1986 a, b) which regards fully three-dimensional turbulence as an agglomeration of 'random vortex sheets and coherent helical structures'. The starting point is the Euler equation, regarded as a dynamical system evolving in an infinite-dimensional function space. The fixed points of this system are the steady solutions of the Euler equations (or 'Euler flows'), and by consideration of the problem of magnetic relaxation towards analogous magnetostatic equilibria, it may be shown that there exist Euler flows of arbitrarily complex topology. These flows are characterised by toroidal blobs within which the streamlines are ergodic and the flow has maximal helicity, separated by families of stream-vortex surfaces on which vortex sheets may be located. We tentatively identify the toroidal blobs as the 'coherent structures' of the turbulent flow that evolves from the Euler flow through Kelvin-Helmholtz instability of the vortex sheets. These coherent structures are persistent simply because the vorticity remains nearly parallel to the velocity so that nonlinear energy transfer to smaller scales remains weak. Viscous dissipation is concentrated in the vortex sheets; and the inertial range, with its associated Kolmogorov spectrum is to be found in the spiral structures arising from the Kelvin-Helmholtz instability (cf Lundgren 1982, Moffatt 1984).

In section 4, we consider briefly the process by which current-sheet discontinuities may appear during the magnetic relaxation problem mentioned above. This can occur even in two-dimensional situations, in which the topological constraints are more easily apprehended. The formation of current sheets has been noted in a number of direct numerical simulations of 2-dimensional MHD turbulence (e.g. Frisch et al 1983) and is known to be associated with the presence of X-type (or hyperbolic) neutral points of the magnetic field, which of course occur randomly when the field is random.

In section 5, we consider what still appears to be the central problem of dynamo theory, at least in astrophysical contexts, namely determination of the parameters α (generation coefficient) and β turbulent diffusivity) in mean-field electrodynamics (see, for example, Moffatt 1978, Krause & Radler 1980) for turbulence with helicity in the limit of large magnetic Reynolds number R_m. If, as frequently maintained in the astrophysical literature,

$$\alpha \sim u_o \quad , \quad \beta \sim u_o \ell \quad \text{as} \quad R_m \to \infty \quad ,$$

where u_o is the rms turbulent velocity, and ℓ is a characteristic turbulent length-scale, then the mean field dynamo equations imply exponential growth of large-scale magnetic field with a growth rate p independent of R_m in the limit $R_m \to \infty$. The dynamo is then a 'fast' dynamo, in the sense of Vainshtein & Zel'dovich (1972). It has been argued by Moffatt & Proctor (1985) that, for the case of a steady velocity field, a fast dynamo has the property that the field becomes non-differentiable nearly everywhere as $R_m \to \infty$, and it seems probable that this property will carry over to unsteady (and turbulent) velocity fields also. The solar dynamo (for which $R_m \sim 10^4$) must inevitably be of this type. It is therefore of vital importance to clarify the behaviour of turbulent dynamos in the large R_m limit. The renormalisation-group procedure described by Moffatt (1983, section 11) still appears to be the best approach to adopt, although difficult to justify with any degree of rigour; this procedure is reviewed in section 5.

Finally, in Section 6, we consider certain features of what may be described as 'chromospheric turbulence', i.e. turbulence in the solar atmosphere outside the photosphere. Here, the density is low, and the dynamics is controlled by the magnetic field, whose field lines are 'rooted' on the solar surface. The footpoints move randomly with the sub-photospheric turbulent motion, and the field in the chromosphere is then instantaneously determined. Particular interest attaches to the question of formation of discontinuities (current sheets) in this region; such sheets are the site of strong Ohmic heating, and may provide an explanation for solar flares and similar intense bursts of activity. One mechanism by which discontinuities can appear is presented in section 6.

2. Spiral Structures in Two-dimensional Turbulence

It is well-known (see, for example, Aref 1983) that when 4 or more point vortices move under their mutual influence, the corresponding dynamical system is in general non-integrable, and each point vortex moves on a chaotic trajectory. It may therefore be readily appreciated that fluid particles in a two-dimensional turbulent field $\underline{u} = (u(x,y,t), v(x,y,t), o)$ will follow an equally chaotic path, and that any small 'patch' of fluid marked by dye will rapidly spread under the random distorting effect.

There are two apparently conflicting theories of two-dimensional turbulence; the first is the 'vortex patch' theory of Saffman (1971) which conceives of turbulence as the mutual interaction of a large number of patches in each of which the vorticity takes a constant value, a situation that persists since the vorticity equation reduces to $D\omega/Dt = 0$. The vorticity on any straight-line transversal then has a finite number of discontinuities per unit length, and so the vorticity spectrum $k^2E(k)$ falls off as k^{-2}. Viscous effects will of course provide an exponential cut-off for wave-numbers $k_v = \ell_v^{-1}$ characterising the real width of the 'discontinuities', but we suppose that viscosity is so weak that this effect may be ignored for all times of interest.

The second theory is the 'enstrophy cascade' theory of Kraichnan (1967) and Batchelor (1969). This starts from the enstrophy (or mean-square vorticity) equation,

$$\frac{d}{dt}\langle\omega^2\rangle = \eta - \nu\langle(\nabla\omega)^2\rangle , \qquad (2.1)$$

where η is the rate of supply of enstrophy at low wave-numbers of order k_0 (and equally the rate of dissipation of enstrophy when a statistically steady state is attained). It is then argued, in parallel with the Kolmogorov theory of 3-dimensional turbulence, that enstrophy cascades through the spectrum at a rate η towards the viscous sink at high k. On dimensional grounds, the energy spectrum in the inertial range $k_o \ll k \ll k_v$ $(\sim(\eta^{1/3}/\nu)^{1/2})$ is given by

$$E(k) = C_2\, \eta^{2/3} k^{-3}, \tag{2.2}$$

where C_2 is a universal constant (the suffix 2 referring to the two-dimensional situation). Kraichnan (1967) has indicated that refinements of the above argument indicate the need for logarithmic corrections to this result, but this can be ignored for the purpose of the present discussion

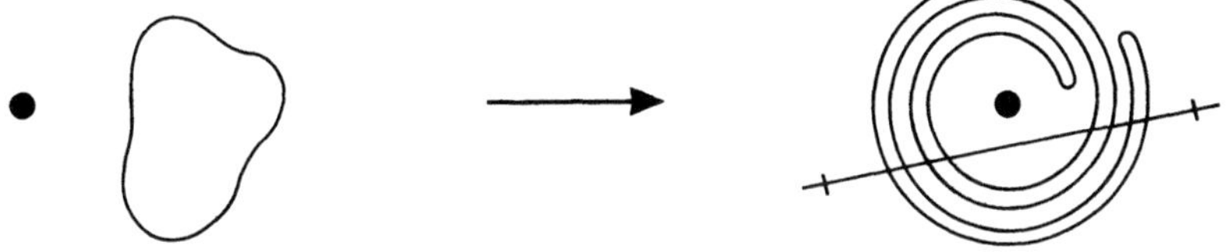

Fig. 1. Interaction of a small strong patch of vorticity and a weak large patch: the large patch is wound up into a tight spiral structure which makes many intersections with any straight line transversal.

We then have two theories, one giving a k^{-4} spectrum, the other a k^{-3} spectrum, in apparent conflict. How are we to reconcile these results?

Let us consider more closely the interaction of vortex patches. Suppose in particular that a small strong patch strays into the neighbourhood of a large weak patch (Fig. 1). The large patch is then distorted like a passive scalar in the r^{-1} velocity field of the small patch (which may be treated like a point vortex). The large patch is thus wound up into a spiral of increasingly fine-structure, its area being conserved. A transversal passing near the centre of the patch now 'sees' a large number N of discontinuities of vorticity per unit length, and $N \to \infty$ as $t \to \infty$. In calculating the spectrum, we are thus dealing with the Fourier transform of a function having an accumulation point of discontinuities, a situation similar to that considered in the context of 3-dimensional turbulence by Moffatt(1984). The resulting spectrum falls off as $k^{-\lambda}$ where $\lambda < 4$ over a range of wave-numbers (the 'spiral range') characterising this type of spiral structure which may be supposed to be randomly distributed. Detailed calculations (A.D. Gilbert 1986, in preparation) have shown that $\lambda = {}^{11}/_{3}$, account being taken of all possible positions of the transversal. The spectrum is then as indicated in Fig. 2.

This of course only brings us part way towards reconciling the k^{-4} and k^{-3} predictions. Observe now however that the same effect may

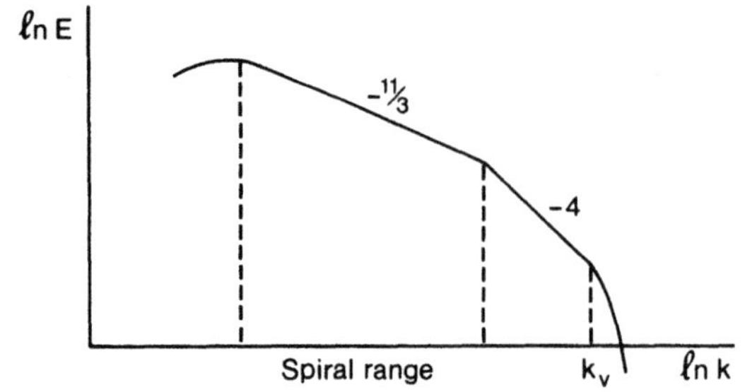

Fig. 2. Form of energy spectrum resulting from random distribution of spiral vortex structures; a spiral range with spectral exponent -11/3 appears (A.D. Gilbert 1986).

be repeated, as indicated in Fig 3: if another strong vortex patch should wander into the neighbourhood of an already formed spiral structure, then it will induce a secondary spiral deformation, thus increasing further the spatial complexity of the vorticity field and leading to a decrease in the relevant index λ in the spiral range. Repeated spiral distortion of this kind will lead to a vorticity field structure that is self-similar on all scales, and it is this property that characterises a k^{-3}-spectrum. We thus conjecture that a k^{-3}-spectrum will emerge from repeated spiral distortions of the kind indicated above. A possible reconciliation between the k^{-4} and k^{-3} theories may thus be recognised.

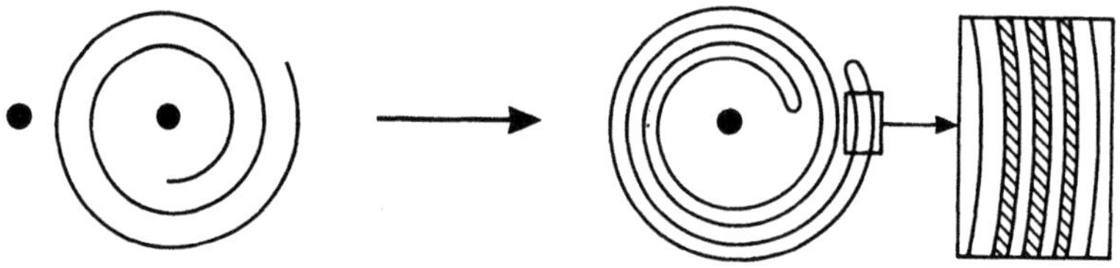

Fig. 3. Interaction of a second strong vortex with a pre-existing spiral structure: a secondary spiral is formed in which the beginnings of a fractal structure may be discerned.

We note that substantial evidence for the 'stacking' of vorticity discontinuities is provided by the numerical simulations of Brachet et al (1986), presented in the poster session at this meeting. The formation of spirals is one mechanism by which such stacking may be understood.

3. Fixed Points of the Euler Equations

Turning now to three-dimensional turbulence, it is natural to consider the general structure of solutions of the Euler equations ($\nu = 0$), recognizing that vortex sheets may be present in such solutions, and that viscosity will eventually resolve the structure of such sheets (cf. Townsend 1951). The Euler equations may be written in the form

$$\frac{\partial \underline{u}}{\partial t} = \underline{u} \times \underline{\omega} - \nabla h, \tag{3.1}$$

where $\underline{\omega}$ = curl $\underline{u}$ and $h = {}^{p}/_{\rho} + \frac{1}{2}\underline{u}^2$, and we regard this as an infinite-dimensional dynamical system, analogous to a finite dimensional system of the form

$$\frac{d\underline{u}}{dt} = \underline{F}(\underline{u})$$

where $\underline{u} = (u_1(t), u_2(t), \ldots u_n(t))$. From such a viewpoint, it is natural to locate first the fixed points of the system (where $\underline{F}(\underline{u}) = 0$), and then to consider the stability of these fixed points. Even if they are all unstable, and therefore in a sense unphysical, their location provides valuable information about the evolution of the system in its phase-space, as reference to such well-known examples as the Lorenz system will make clear.

The fixed points $\underline{u}^E(\underline{x})$ of the Euler equations, described as Euler flows, are solutions of the steady equations

$$\underline{u} \times \underline{\omega} = \nabla h. \tag{3.2}$$

An approach, suggested first by Arnold (1974), has been developed by Moffatt (1985, 1986 a, b), and is remarkably powerful in revealing an incredible richness in the structure of Euler flows. We first note the well-known exact analogy with the problem of magnetostatics of a perfectly conducting fluid

$$\underline{j} \times \underline{B} = \nabla p \tag{3.3}$$

with $\underline{j}$ = curl $\underline{B}$, $\nabla \cdot \underline{B} = 0$. Here, $\underline{B}$ represents magnetic field, and $\underline{j}$ electric current. The analogy is between the variables

$$\underline{u} \longleftrightarrow \underline{B}\,,\ \underline{\omega} \longleftrightarrow \underline{j},\ \ h \longleftrightarrow -p. \tag{3.4}$$

Location of solutions of (3.3) lends itself to a method of 'magnetic relaxation' that is not available (without introducing very artificial physics) for the problem (3.2). Nevertheless, to every solution of (3.3) found in this way, there corresponds an Euler flow, via the exact analogy (3.4). The analogy does not extend to questions of stability of the steady states thus determined.

The magnetic relaxation argument proceeds as follows. We suppose that $\underline{B}_o(x)$ is an arbitrary initial field (at t = o) in a domain D (which may be infinite) and that the fluid is perfectly conducting (so that $\underline{B}$-lines are frozen in the fluid) but viscous (so that energy is dissipated when the fluid moves). Such assumptions are physically artificial, but no matter: they are merely an artifice, a means towards an end. In general the Lorentz force $\underline{j} \times \underline{B}$ is rotational at time t = o, so thefluid will move. The total energy (magnetic plus kinetic) then decreases, but subject to the vital constraint that the topology of the $\underline{B}$-field is conserved. As $t \to \infty$, the viscous dissipation must tend to zero, and so the velocity field must everywhere tend to zero. In the final stages of this process, discontinuities of $\underline{B}$ may form through a 'squeeze-film' mechanism in which flux tubes embrace each other in increasingly close contact. The final field $\underline{B}^E(\underline{x})$ is not for this reason strictly topologically equivalent (or homeomorphic) to the initial field $\underline{B}_0(\underline{x})$, but it is nevertheless 'topologically accessible' from it, being obtained by the convective action of a smooth velocity field $\underline{u}(\underline{x},t)$ $(o < t < \infty)$ of finite total dissipation.

This argument implies that to every field $\underline{B}_0(\underline{x})$, there corresponds at least one magnetostatic equilibrium $\underline{B}^E(\underline{x})$ topologically accessible from it, the lines of force, magnetic surfaces, and ergodic regions of $\underline{B}_0$ mapping faithfully to those of $\underline{B}^E(\underline{x})$. By the analogy referred to above, we may then make a similar statement about Euler flows: to every solenoidal velocity field $\underline{U}(\underline{x})$, there corresponds at least one Euler flow $\underline{u}^E(\underline{x})$, whose streamlines may be obtained by distortion of the streamlines of $\underline{U}(\underline{x})$ by an (imagined) velocity field $\underline{v}(\underline{x},t)$ $(0<t<\infty)$ of finite total 'dissipation integral', i.e.

$$\int_0^\infty \int_D \left(\frac{\partial v_i}{\partial x_j}\right)^2 dV\,dt < \infty . \tag{3.5}$$

We emphasise that the Euler flow $\underline{u}^E(\underline{x})$ will in general contain vortex sheets (which may well be stacked on top of each other, or exhibit other curious features) even although the 'reference' field $\underline{U}(\underline{x})$ is everywhere differentiable. With this qualification, it may be confidently asserted that there is an uncountable infinity of topologically distinct Euler flows in any domain D.

Consider now the 'generic' structure of such a flow, satisfying (3.2). Obviously

$$\underline{u}.\nabla h = 0, \quad \underline{\omega}.\nabla h = 0 , \tag{3.6}$$

so that both streamlines and vortex lines lie in surfaces h = cst. It may however happen that in some subdomain D' of D, h is identically constant, so that $\underline{u} \times \underline{\omega} \equiv 0$ in D', i.e.

$$\underline{\omega} = \alpha(\underline{x})\, \underline{u} \quad \text{in } D' \tag{3.7}$$

for some scalar field $\alpha(\underline{x})$, satisfying

$$\underline{u}.\nabla\alpha = 0 \tag{3.8}$$

since both $\underline{u}$ and $\underline{\omega}$ are solenoidal. Hence now the streamlines (coinciding with the vortex lines) lie on surfaces α = cst. Yet again, it may happen that, in some subdomain D_1 of D', α is identically constant, so that

$$\underline{\omega} = \alpha_1\, \underline{u} \quad \text{in } D_1 \;, \; \alpha_1 = cst. \tag{3.9}$$

Only in this circumstance can the streamlines be ergodic in D_1 (or in some subdomain of D_1). Conversely, if we know that the streamlines of $\underline{u}$ are ergodic in a subdomain D_1, then the Beltrami property (3.9) must be satisfied. This is important, because the arbitrary reference field $\underline{U}(\underline{x})$ will in general have ergodic regions which will map into corresponding regions D_1, D_2, - - -, in which the associated Euler flow is likewise ergodic. The picture is indicated schematically in Fig. 4: blobs of maximal helicity ($\underline{\omega} = \alpha_n \underline{u}$) in which the streamlines are ergodic are separated by families of stream-vortex surfaces, on which, through confluence of stream-surfaces, vortex sheets may be located.

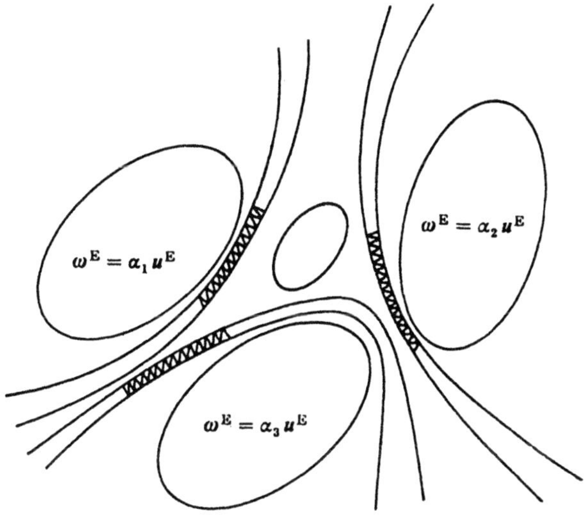

Fig. 4. Generic structure of Euler flow: ergodic blobs are separated by families of stream-vortex surfaces on which vortex sheets may be located.

We have observed above that the analogy between Euler flows and magnetostatic equilibria does not extend to questions of stability. The magnetostatic equilibria, arrived at through a relaxation process, may be expected to be stable within the perfect conductivity framework, the magnetic energy being minimal in the subspace of topologically accessible fields. The analogous Euler flow $\underline{u}^E$ may however be unstable in the different subspace of velocity fields whose vorticity fields $\underline{\omega}$ are topologically accessible from $\underline{\omega}^E$. This dichotomy has been explored in detail (Moffatt 1986a) where it is shown, by way of example, that the ABC-flow studied by Dombré et al (1986) is unstable to large-scale helical perturbations, whereas the analogous magnetostatic equilibrium is stable.

When vortex sheets are present in the Euler flow, these are of course subject to the fundamental Kelvin-Helmholtz instability leading to intrinsic and inescapable wind-up into the familiar double spiral structures. When the wave-crests of the instability are inclined to the vorticity in the sheet as is inevitable for a general toroidal topology (Moffatt 1984), the instability stretches the vorticity perpendicular to the crests, so that enstrophy increases (unlike the two-dimensional situation); this of course is an essential feature of 3-dimensional turbulence.

The elemental picture of random vortex sheets (Townsend 1951) gives a k^{-2} energy spectrum, modified by an exponential cut-off on the <u>inner</u> (Kolmogorov) scale. We conjecture that the double spiral wind-up associated with the Kelvin-Helmholtz instability, possibly occurring on a succession of decreasing length-scales, will convert this spectrum to the slower $k^{-5/3}$ fall-off required by the cascade (similarity) theory of Kolmogorov (see, for example, Batchelor 1953) in parallel with the corresponding 2-dimensional problem described in §2 above. Here, the picture is close to that of Lundgren (1982) who has succeeded in extracting a $k^{-5/3}$-spectrum from a 'strained spiral vortex' model of turbulence.

As regards the ergodic blobs of maximal helicity, we may tentatively associate these with the 'coherent structures' that have been identified in a variety of turbulent flows. To be sure, the condition $\underline{\omega} = \alpha_n \underline{u}$ is no longer exactly satisfied in the blob D_n when perturbed by the influence of Kelvin-Helmholz instabilities in adjacent layers; but if, in some sense, the real turbulent flow follows a trajectory (in function space) around an unstable fixed point (as suggested by the behaviour of simpler finite-dimensional systems) then ergodic blobs in which $\underline{\omega}$ is <u>nearly</u> parallel to $\underline{u}$ are to be expected; within these blobs, nonlinear energy transfer represented by the term $\underline{u} \times \underline{\omega}$ of the Euler equation (3.1) is then weak, and so they will be relatively persistent, and therefore identifiable. Independent evidence has been provided by Tsinober & Levich (1983) that coherent structures in real turbulent flows do indeed generally exhibit non-zero helicity in the appropriate co-moving frame of reference; and (Shtilman et al 1985) that coherent structures of non-zero helicity (positive and negative) do emerge in direct numerical simulations of certain space-periodic flows, of which the Taylor-Green vortex may be regarded as the prototype (see also Levich 1985).

4. Appearance of Discontinuities in the Magnetic Relaxation Problem

The appearance of discontinuities is a very important aspect of the magnetic relaxation problem, foreshadowed by the discussion of Arnold (1974) who simply pointed out that the problem would in general have no solution 'within the class of smooth vector fields'. Discontinuities form even in two-dimensional situations, by a mechanism that we shall now indicate.

Fig. 5. Relaxation of field with one elliptic neutral point: each line of force becomes a circle.

Consider first the very simple topology indicated in Fig. 5, in which the initial field $\underline{B}_0(\underline{x})$ within a cylinder of circular cross-section has a single elliptic (or 'O-type') neutral point in the interior of the domain. Magnetic relaxation proceeds through contraction of $\underline{B}$-lines, but here subject to the essentially topological constraint that the area within each convected field-line remains constant. Equilibrium is attained when the length of each $\underline{B}$-line is minimised subject to this constraint, i.e. when each $\underline{B}$-line is circular; moreover these circles must be concentric with the cylinder in order to equalise the magnetic pressure along $\underline{B}$-lines. There is thus a unique equilibrium towards which the field relaxes, and this has no discontinuities.

Consider now the more complex topology of figure 6 in which the initial field $\underline{B}_0(\underline{x})$ has two elliptic neutral points and one hyperbolic (or 'X-type') neutral point in the interior of the domain. Again, each field line would like to become circular in order to minimise length subject to constant enclosed area, but this is no longer compatible with the topology. The system must seek an alternative route to equilibrium. The disposition of Lorentz forces in the neighbourhood of the X-type neutral point is such as to eject fluid from the acute angles, and so to decrease these angles. This process continues until (as $t \to \infty$) a field discontinuity (or current sheet) must form. This effect has been confirmed numerically by Bajer (1986, in preparation) who finds an asymptotic configuration with cusped structure as indicated in Fig. 6.

Fig. 6. Relaxation of field with two elliptic neutral points and one hyperbolic point: a current sheet discontinuity forms with a cusped structure at its end-points (K. Bajer 1986).

For a more general topology, and in particular for a random magnetic field, such behaviour is to be expected in the neighbourhood of every X-type neutral point. This indeed is the behaviour that has been revealed in the direct numerical simulations of Frisch et al (1983). These simulations were carried out with non-zero magnetic diffusivity η (in fact with $\eta = \nu$), but the tendency to form near-discontinuities was nevertheless apparent.

Formation of current sheet discontinuities is an important process in the solar atmosphere, as persuasively argued by Parker (1979, chap. 14) who considers the possible structure of force-free fields and of magnetostatic equilibria of prescribed non-trivial topology, with the conclusion that in general these must exhibit discontinuities. This is entirely consistent with the point of view developed in the present paper. We return to the solar atmosphere context in section 6 below.

5. The Fast Turbulent Dynamo

As indicated in the introduction, a turbulent dynamo acting in the convection zone of any late-type star (e.g. the Sun), and in which the magnetic Reynolds number is of order 10^4 or greater, must presumably be of 'fast' type in the sense of Vainshtein & Zel'dovich (1972). The growth rate of such a dynamo (neglecting back-reaction effects) is independent of R_m in the limit $R_m \to \infty$, i.e. it exhibits a 'magnetic Reynolds number similarity' analogous to the Reynolds number similarity that is familiar in conventional turbulence. It has been shown by Moffatt & Proctor (1985) that if the magnetic diffusivity η is zero (i.e. $R_m = \infty$), then dynamo action is impossible with a steady velocity field $\underline{u}(\underline{x})$, even if this is a random function of position. It seems likely that this is true also for unsteady (turbulent) fields $\underline{u}(\underline{x},t)$, although opinions are divided on this matter.

It seems intuitively clear that, as $R_m \to \infty$, the magnetic field fluctuations will have progressively finer scale, and will inevitably be larger (by some positive power of R_m) than the mean field. This has been used by Piddington (1972) as an argument against the applicability of the methods of mean-field electrodynamics developed by Steenbeck, Krause & Rådler (1966) (for a comprehensive account of this and subsequent papers, see Krause & Rädler 1980); the criticism is however not sound: the mean-field method does not require that the fluctuation be small compared with the mean. This is a requirement of the frequently used 'first-order smoothing' approximation (see Moffatt 1978, Chap.7) which is certainly not valid when $R_m >> 1$. What is needed when $R_m >> 1$ is an alternative approach, still within the general framework of the mean-field method. This means essentially that we need a good theory for the determination of the 'generation coefficient α (in the 'α-effect') and the turbulent magnetic diffusivity β , which remains valid no matter how large R_m may be.

One possible approach, which seems to incorporate the right physics, involves the renormalisation group technique, as described by Moffatt (1983). Here we simply review the elements of this procedure. The energy spectrum $E(k)$ and the helicity spectrum $H(k)$ (on which the α-effect depends) are first discretised,and first order smoothing is applied to the smallest scales (or largest wave-numbers), whose associated magnetic Reynolds number $R_m(k)$ is small. This yields a 'mean-field' equation involving an α-effect and a (weak) turbulent diffusivity which supplements the (weak) molecular diffusivity. This process is then repeated: another 'bite' is taken of the energy and helicity spectra, these being progressively devoured from the '$k = \infty$' end. The mean-field equation is modified only through change in the values of α and β (i.e. through renormalisation). After n similar steps, we obtain values α_n, β_n in terms of α_{n-1}, β_{n-1}, $E(k_n)$ and $H(k_n)$. Now, we make the essential assumption, which is hard to justify with any rigour, that cumulative errors are small, and that we may return to continuum (rather than discrete) spectra. This yields differential equations for $\alpha(k)$ and $\beta(k)$, the effective α and β associated with all scales of motion smaller than k^{-1}.
These equations are

$$\left.\begin{aligned} \frac{d\alpha}{dk} &= \frac{1}{3}\;\frac{2\alpha E(k) + (\eta+\beta)H(k)}{k^2(\eta+\beta)^2 - \alpha^2}\;, \\ \frac{d\beta}{dk} &= -\frac{1}{3}\;\frac{2(\eta+\beta)E(k) + \alpha k^{-2}H(k)}{k^2(\eta+\beta)^2 - \alpha^2}\;, \end{aligned}\right\} \tag{5.1}$$

and these are to be integrated in from $k = \infty$, with 'initial' conditions $\alpha(\infty) = 0$, $\beta(\infty) = 0$.

The interesting thing now is that a little molecular diffusivity η is needed to, as it were, get the system started: if $\eta = 0$, the first step of the smoothing process cannot be justified. At the same time, if $\eta > 0$ (no matter how small), then $|\alpha(k)|$ increases as k decreases to a maximum value $|\alpha(k)|_{max}$ independent of η. In fact, $\eta+\beta$ may be eliminated from (5.1) giving

$$\frac{d}{dk} \alpha^2 = \frac{2 \alpha^2 (2 \alpha^2 E - h H)}{3 (kh - \alpha^2)(kh + \alpha^2)} \tag{5.2}$$

where $$h(k) = \frac{1}{3} \int_k^\infty k^{-2} H(k)\, dk. \tag{5.3}$$

When k is large (and $|\alpha|$ small), (5.2) becomes

$$\frac{d}{dk} \alpha^2 = \frac{-2\alpha^2 H}{3k^2 h} . \tag{5.4}$$

Suppose, for example, that $H(k) \sim C k^{-m}$ for large k, so that $h(k) \sim \frac{1}{3}C (m+1)^{-1} k^{-(m+1)}$, and

$$\frac{d}{dk} \alpha^2 \sim -2 k^{-1} (m + 1) \alpha^2 . \tag{5.5}$$

Then $$\alpha^2 \sim k^{-2(m+1)} \quad \text{as } k \to \infty. \tag{5.6}$$

Thus, as k decreases from ∞, α^2 will 'latch on' to this solution. The typical behaviour of α^2 is indicated in figure 7.

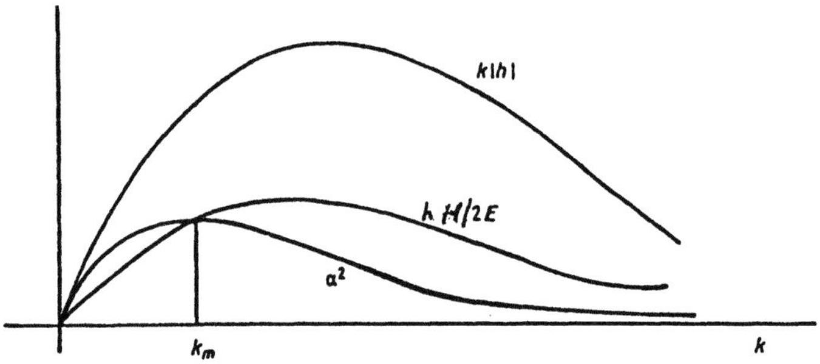

Fig. 7. Form of the function $\alpha^2(k)$ as determined by integration of equation (5.2).

In a real system such as the Sun, it is of course unrealistic to pursue the integration all the way to $k = 0$. Rather, one should stop around the wave-number k_m where α^2 is maximal and use equations (5.1) to determine $\alpha(k_m)$, $\beta(k_m)$ (this is just a form of sub-grid modelling). Scales larger than k_m^{-1} should ideally be treated by full numerical simulation, in the manner of Gilman & Miller (1981).

6. Chromospheric Turbulence, and the Formation of Discontinuities

Outside the visible surface of the Sun (the photosphere) the density of the solar atmosphere falls off very rapidly, and the dynamics is totally dominated by the magnetic field emanating from the surface.

This field presumably adjusts itself more or less instantaneously to be force-free ($\underline{j} \times \underline{B} = 0$), since otherwise infinite accelerations would occur. Actually, the process by which this adjustment occurs in response to a changing photospheric field is not at all clear; let us assume however, following Low (1977), that it does occur and that the field in the chromosphere is essentially force-free at all times. The field lines emerge from the photosphere, where they are rooted to the sub-photospheric plasma which may be assumed perfectly conducting on the time-scale characteristic of solar magnetic activity. This field has a complex random structure, which is constantly changing as the footpoints of the field-lines move with the plasma.

Adopting, for simplicity, a two-dimensional model of this field, we may expect a random distribution of O-type and X-type neutral points below the photosphere. The field topology in the neighbourhood of any X-type neutral point is as indicated in figure 8. Clearly one field line ABC above the neutral point must touch the photosphere (at B). Consider now two field lines A'C' and $A_1B_1B_2C_2$ just above and below this 'critical' field line. As the footpoints move due to sub-photospheric turbulence, the field instantaneously adjusts itself to remain force-free. This means that the field in the chromosphere is instantaneously determined by the motion of the footpoints, i.e. by the motion of A' and C' for the field line A'C', and by A_1 and B_1 for the field line A_1B_1. Since the motions of B_1 and C' are quite independent, we must therefore expect in general that a tangential discontinuity (or current sheet) will form on AB and likewise on BC. Detailed calculation confirms this conclusion.

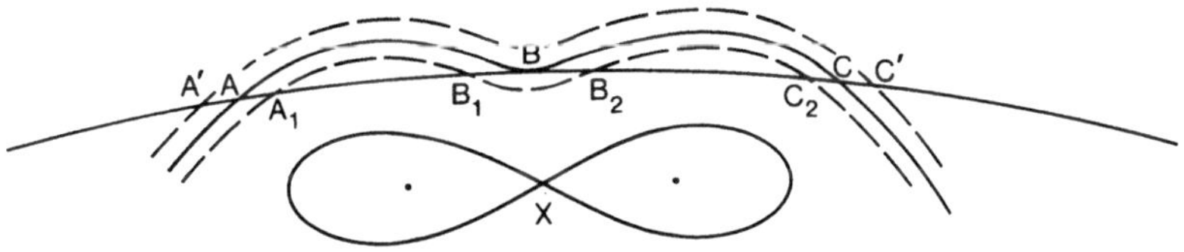

Fig. 8. Field structure above a hyperbolic neutral point just below the photosphere: when the footpoints move, a current sheet appears on the critical line of force ABC.

The importance of X-type neutral points in the chromospheric context was first pointed out by Sweet (1969) and Syrovatsky (1966) — see also Priest (1982, chap.10). The fact that such neutral points below the photosphere may equally have an important influence on chromospheric activity has not however been previously recognized. Current sheet discontinuities, and the means by which they may appear, have an important bearing on the phenomenon of solar flares and similar bursts of magnetic activity on the Sun.

8. Summary

This review of geophysical and astrophysical turbulence has necessarily been very selective, and we have focussed mainly on aspects of both two-dimensional and three-dimensional turbulence that may fairly be described as topological in character. When a magnetic field is present (as it almost inevitably is in a high conductivity plasma) the fact that $\underline{B}$ - lines are frozen in the fluid on time-scales small compared with the (long) diffusion time makes topological (as opposed to purely analytic) arguments even more appealing. This point of view has been urged by Parker (1979) and others; and the fact that helicity, already known to be of vital importance in the dynamo context, has a topological interpretation, lends further force to this standpoint. The method described in §3, involving topological ideas in an essential way, is remarkably powerful in providing new insights both for the problem of turbulence, and for certain classical problems of laminar flow theory (Moffatt 1986b). It seems that we may be on the threshold of major advances in basic understanding, which will have a profound influence on further analytical and numerical work.

REFERENCES

Aref, H. 1983 Integrable, chaotic, and turbulent vortex motion in two-dimensional flows. Ann.Rev.Fluid Mech. 15, 345-389.

Arnold, V. 1974 The asymptotic Hopf invariant and its applications. In Proc.Summer School in Differential Equations. Armenian SSR Acad.Sci. (in Russian).

Batchelor,G.K. 1953 Homogeneous Turbulence. Cambridge University Press.

Batchelor,G.K. 1969 Computation of the energy spectrum in homogeneous two-dimensional turbulence. Phys.Fluids Suppl. 12, II-233 - II-239.

Brachet, M.E., Meneguzzi, M. & Sulem P.L. 1986 Small scale dynamics of high Reynolds number two-dimensional turbulence. Submitted to Phys. Rev.Letters.

Dombré, T., Frisch, U., Green, J.M., Henon, M., Mehr, A. & Soward, A.M. 1986 Chaotic streamlines in the ABC flows. J.Fluid Mech. 167, 353-391.

Frisch, U., Pouquet A., Sulem, P.L. & Meneguzzi, M. 1983 The dynamics of two-dimensional ideal magnetohydrodynamics. J.de Mécanique, Numéro Spéciale, 191-216.

Gilman, P.A. & Miller J. 1981 Dynamically consistent nonlinear dynamos driven by convection in a rotating spherical shell. Astrophys. J.Suppl. 46, 211-238.

Kraichnan, R.H. 1967 Inertial ranges in two-dimensional turbulence Phys.Fluids 10, 1417-1423.

Krause, F. & Rädler K-H. 1980 Mean-field Magnetohydrodynamics and Dynamo Theory (Oxford: Pergamon)

Levich, E. & Tzvetkov, E. 1985 Helical inverse cascade in three-dimensional turbulence as a fundamental dominant mechanism in meso-scale phenomena. Phys.Rep. 128, 1-37.

Low, B.C., 1977 Evolving force-free magnetic fields. I. The development of the pre-flare stage. Astrophys.J. 212, 234-242.

Lundgren, T.S. 1982 Strained spiral vortex model for turbulent fine structure. Phys.Fluids 25, 2193-2203.

McWilliams, J.C., 1984 The emergence of isolated coherent vortices in turbulent flow. J.Fluid Mech. 146, 21-43

Moffatt, H.K., 1969 The degree of knottedness of tangled vortex lines. J.Fluid.Mech. 35, 117-129.

Moffatt, H.K., 1978 Magnetic field generation in electrically conducting fluids. Cambridge Univ. Press.

Moffatt, H.K., 1981 Some developments in the theory of turbulence. J.Fluid Mech. 106, 27-47.

Moffatt, H.K., 1983 Transport effects associated with turbulence with particular attention to the influence of helicity. Rep.Prog. Phys. 46, 621-664.

Moffatt, H.K., 1984 Simple topological aspects of turbulent vorticity dynamics. In Proc.IUTAM Symp. on Turbulence and Chaotic Phenomena in Fluids. (ed. T. Tatsumi) Elsevier, 223-230.

Moffatt, H.K., 1985 Magnetostatic equilibria and analogous Euler flows of arbitrarily complex topology. Part I. Fundamentals. J.Fluid Mech. 159, 359-378.

Moffatt, H.K., 1986a Magnetostatic equilibria and analogous Euler flows of arbitrarily complex topology. Part 2: Stability considerations. J.Fluid Mech. 166 359-378.

Moffatt, H.K., 1986b On the existence of localized rotational disturbances which propagate without change of structure in an inviscid fluid. J.Fluid Mech. 173 (to appear).

Moffatt, H.K. & Proctor,M.R.E. 1985 Topological constraints associated with fast dynamo action. J.Fluid Mech. 154, 493-507.

Parker, E.N., 1983 Absence of equilibrium among close-packed twisted flux tubes. Geoph.Astrophys.Fluid Dyn. 23, 85-102.

Parker, E.N., 1979 Cosmical magnetic fields, Oxford: Clarendon.

Piddington, J.H., 1982 The origin and form of the galactic magnetic field, I. Parker's dynamo model. Cosmic Electrodynamics 3, 60-70.
Priest, E.R., 1982 Solar Magnetohydrodynamics, Reidel Publishing Co.
Saffman, P.G., 1971 Studies in Appl.Math. 50, 377-
Shtilman, L., Levich, E., Orszag, S., Peltz, R., and Tsinober, A., 1985 On the role of helicity in turbulent flows. Phys.Lett.A.
Steenbeck, M., Krause, F., & Rädler, K-H. 1966 Berechnung der mittleren Lorentz Feldstarke fur ein leitendes Medium in turbulenter, durch Coriolis-Krafte beeinfluster Bewegnung. Z.Naturforsch 21a, 369-376.
Sweet, P.A., 1969 Mechanisms of solar flares. Ann.Rev.Astron. Astrophys. 7, 149-176.
Syrovatsky, S.I., 1966 Dynamic dissipation of a magnetic field and particle acceleration. Soviet Astron. 10, 270-280.
Townsend, A.A., 1951 On the fine-scale structure of turbulence. Proc.Roy.Soc. A 208, 534-542
Tsinober, A., & Levich, E., 1983 On the helical nature of three-dimensional coherent structures in turbulent flows. Phys.Lett. 99a, 321-324.
Vainshtein, S.I., & Zel'dovich, Ya.B., 1972 Origin of magnetic fields in astrophysics. Sov.Phys.Usp. 15, 159-172.

Computer Simulation of Decaying Two-Dimensional Turbulence

M.E. Brachet[1], *M. Meneguzzi*[2], *H. Politano*[1], *and P.L. Sulem*[1;3]

[1]C.N.R.S., Observatoire de Nice, B.P. 239, F-06007 Nice Cedex, France
[2]Service d'Astrophysique, C.E.N.-Saclay and C.N.R.S., F-91191 Saclay, France
[3]School of Mathematical Sciences, Tel-Aviv University, Tel-Aviv 69978, Israel

1. Introduction

It has early been realized that very high resolutions are required to simulate large Reynolds number turbulence, by direct numerical integration of the Navier-Stokes equation

$$\partial_t v + v.\nabla v = -\nabla p + \nu\Delta v$$
$$\nabla.v = 0. \qquad (1)$$

Focussing on the internal dynamics rather than on the effects of special geometries, we assume periodic boundary conditions. We thus use spectral methods for the space variables. We concentrate here on the simulation of a turbulent flow at a resolution of $(2048)^2$ collocation points, and especially on visualizations of the small scales of the flow. Computations at lower resolutions have been reported previously[1].

To reach such a resolution on a CRAY1S computer with a central memory of one megaword, we have resorted to consider a flow with large scale symmetries as in [1]. The code is out of core but the waiting time due to uncovered input/output is only 30% of the total computer time. The flow is characterized by a stream function

$$\Psi(x,y,t) = \sum a_{l,m} \sin lx \sin my \qquad (2)$$

where the $a_{l,m}$ Fourier components do not vanish only when l and m are both even or both odd. This representation which is compatible with the Navier-Stokes equation, corresponds in physical space to the following flow symmetries: i) invariance of rotation of π around the point $x = \pi/2$, $y = \pi/2$; ii) reflexional symmetry on the side of an impermeable box given by $x = 0$ and π, $y = 0$ and π.

Initially, the $a_{l,m}$ coefficient are zero-mean gaussian random variable such that the initial velocity has an energy spectrum

$$E(k,0) \sim k \exp\{-(k/k_0)^2\} \qquad (3)$$

with $k_0 = 5$ .For the particular realization we have used, the energy is 0.61 and the enstrophy 3.6 . We used a viscosity $\nu = 2.33\ 10^{-5}$, which insures a high precision at all scales of motion.

2. Visualization of the small scales

Fig. 1 displays the vorticity field as visualized on a VICOM image processor. This apparatus uses a Raster technique where the field to be visualized is represented with a continuous palette from black to white. It thus provides a visualization of the function itself while the usual contour line representation emphasizes regions of strong gradients, by close iso-field lines.

Vorticity is shown at times from 1 to 12 by steps of 1 . Characteristic time scales for the flow evolution are provided by the skewness (a non-dimensional measure of the enstrophy transfer) and the enstrophy dissipation. The maximum of the skewness corresponds to the end of the inviscid phase while the maximum of the enstrophy dissipation indicates roughly the moment when turbulence reaches maturity. For the run presented here, they are respectively equal to $t_s \simeq 1$ and $t_d \simeq 6$.

Various dynamical phenomena are observed in the simulation. The initial conditions can be viewed as a collection of a few vortices with different signs and intensities. At early times, some vortices are distorted by pairs of vortices rotating either in the same or in the opposite direction . This leadsto the formation of vorticity gradient sheets. Latter, some of these sheets become aligned and are packed together; others condensate into isotropic vortices the importance of which in the long time behaviour of the two-dimensional flows was stressed by McWilliams[3]. We also observe strong vortices which are first distorted and whose isotropy is latter restaured, spinning around them vorticity gradient sheets or even layers resulting from the packing of such sheets.

Fig.1 (next page): Vorticity field visualized on VICOM for times from 1 to 12 by steps of 1 . The left hand side figures correspond to the domain $0<x<\pi/2$, $0<y<\pi/2$, the right hand side to $\pi/2<x<\pi$, $0<y<\pi/2$. The complete field can be reconstructed by symmetries.

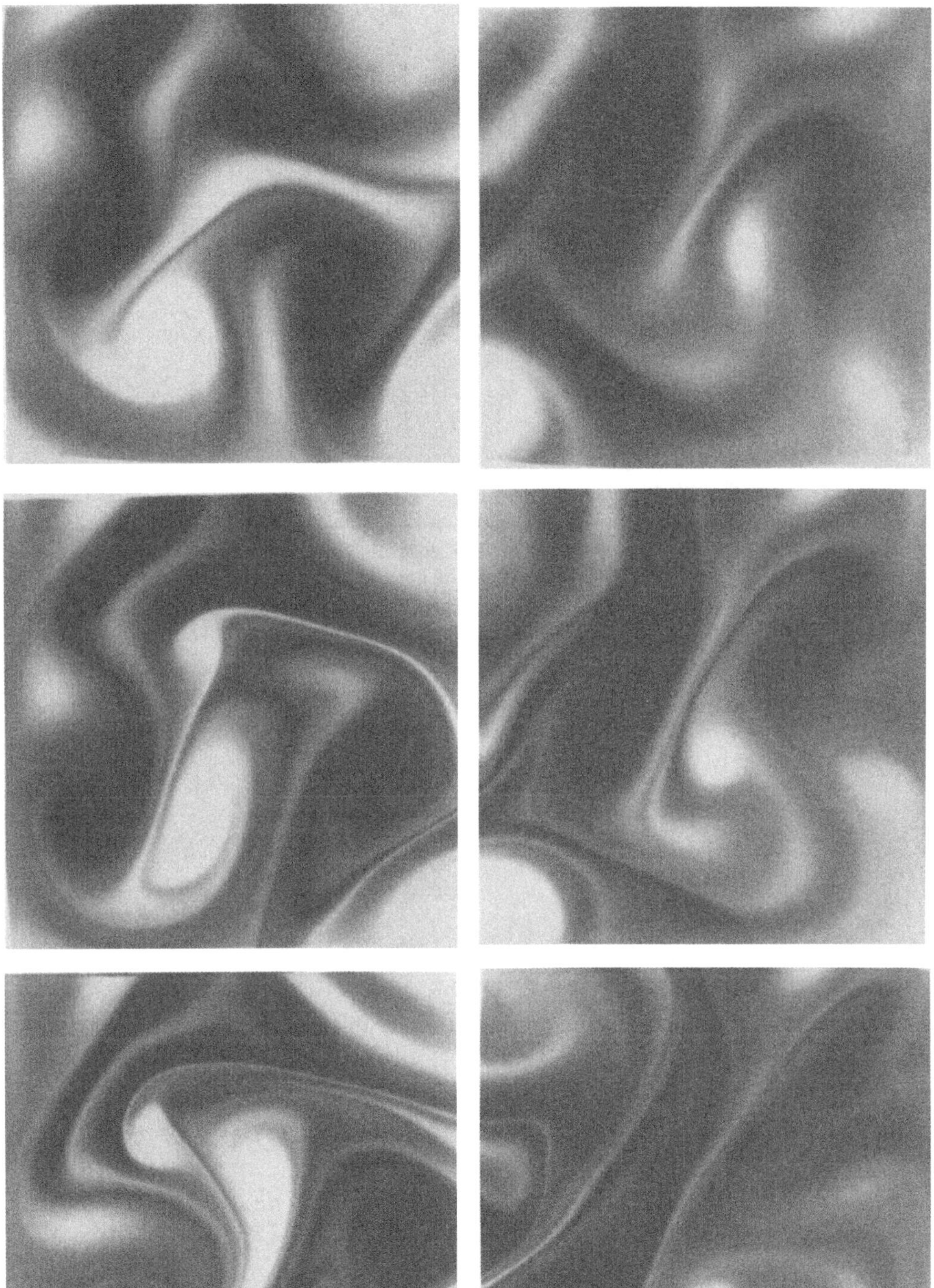

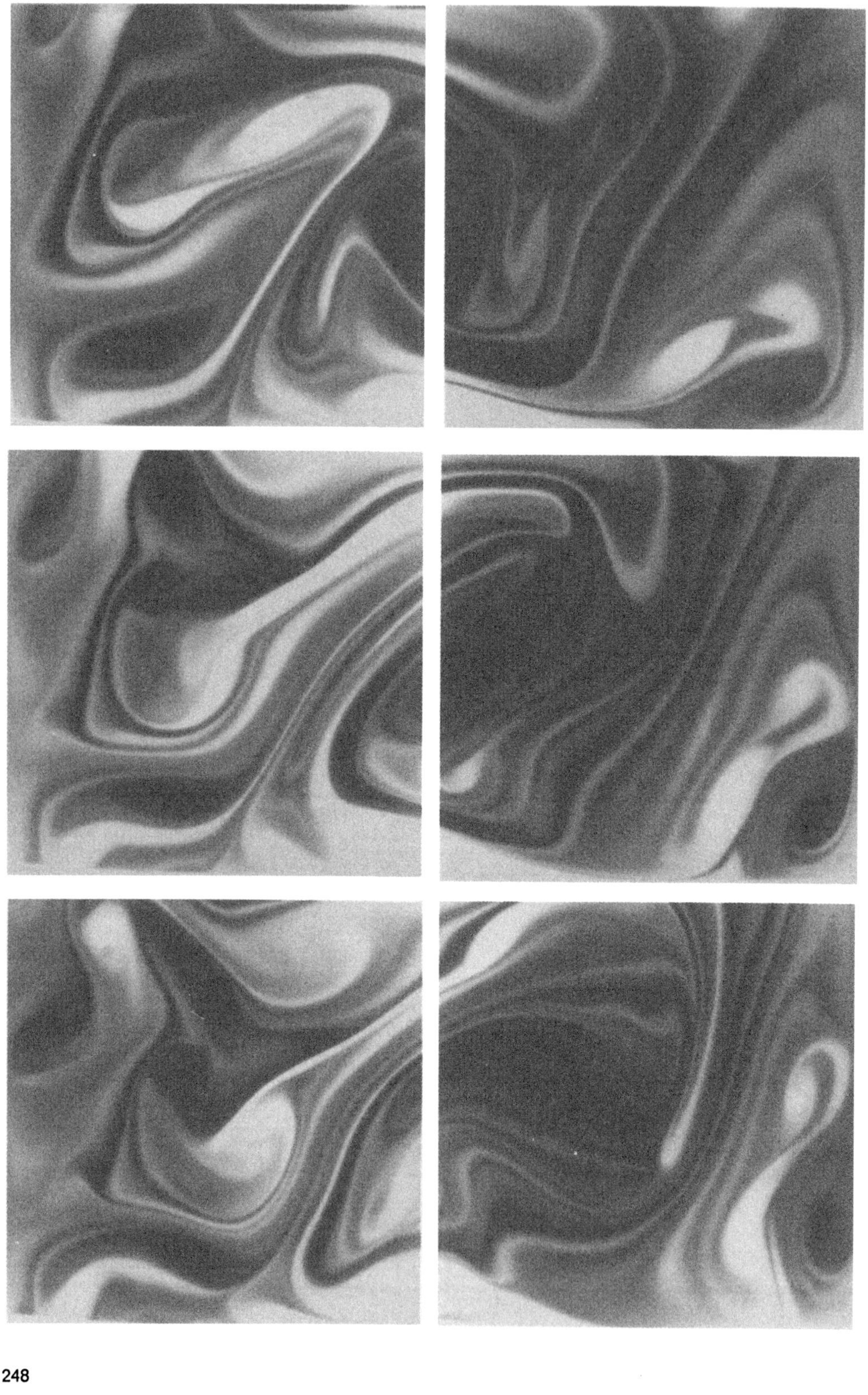

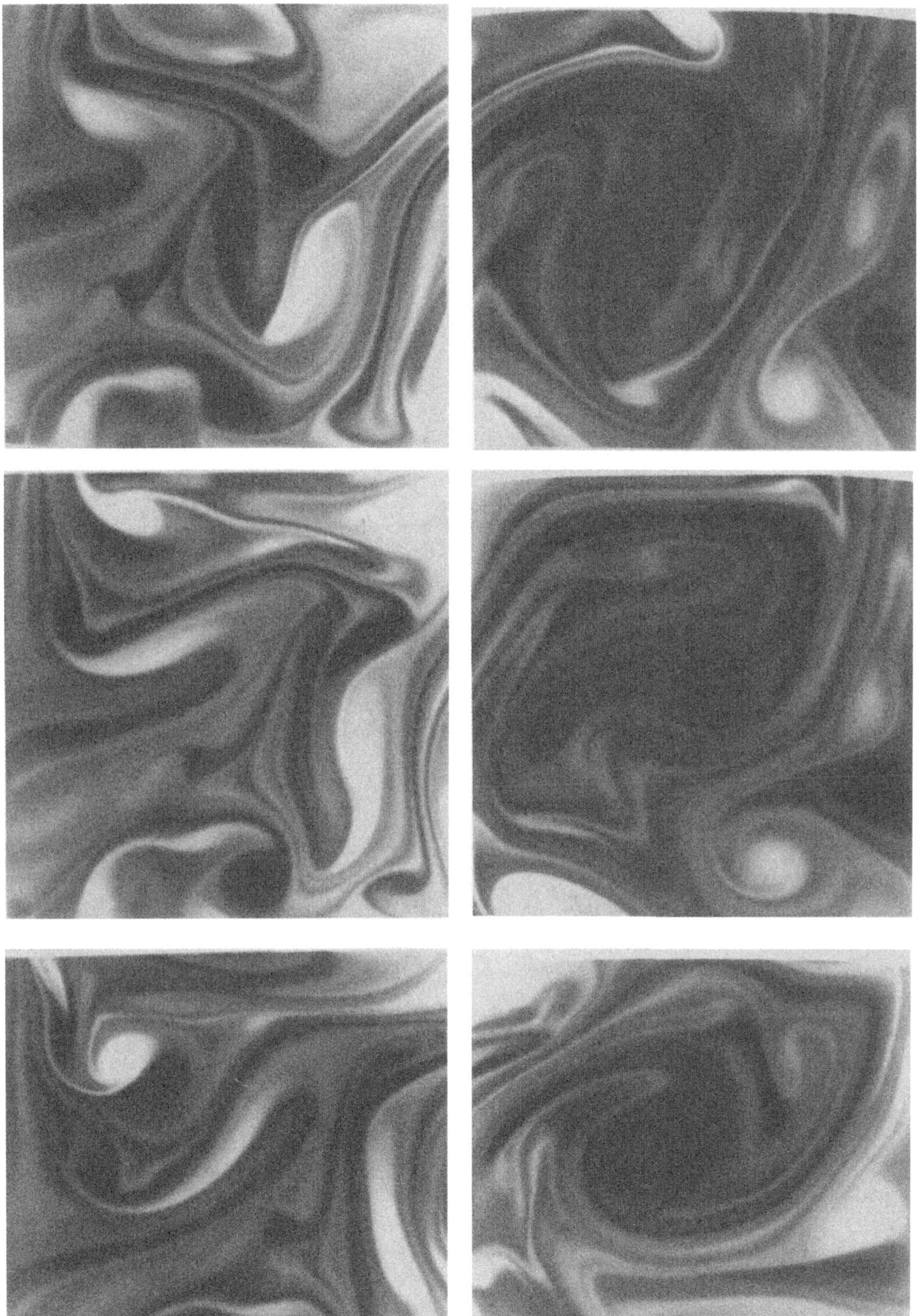

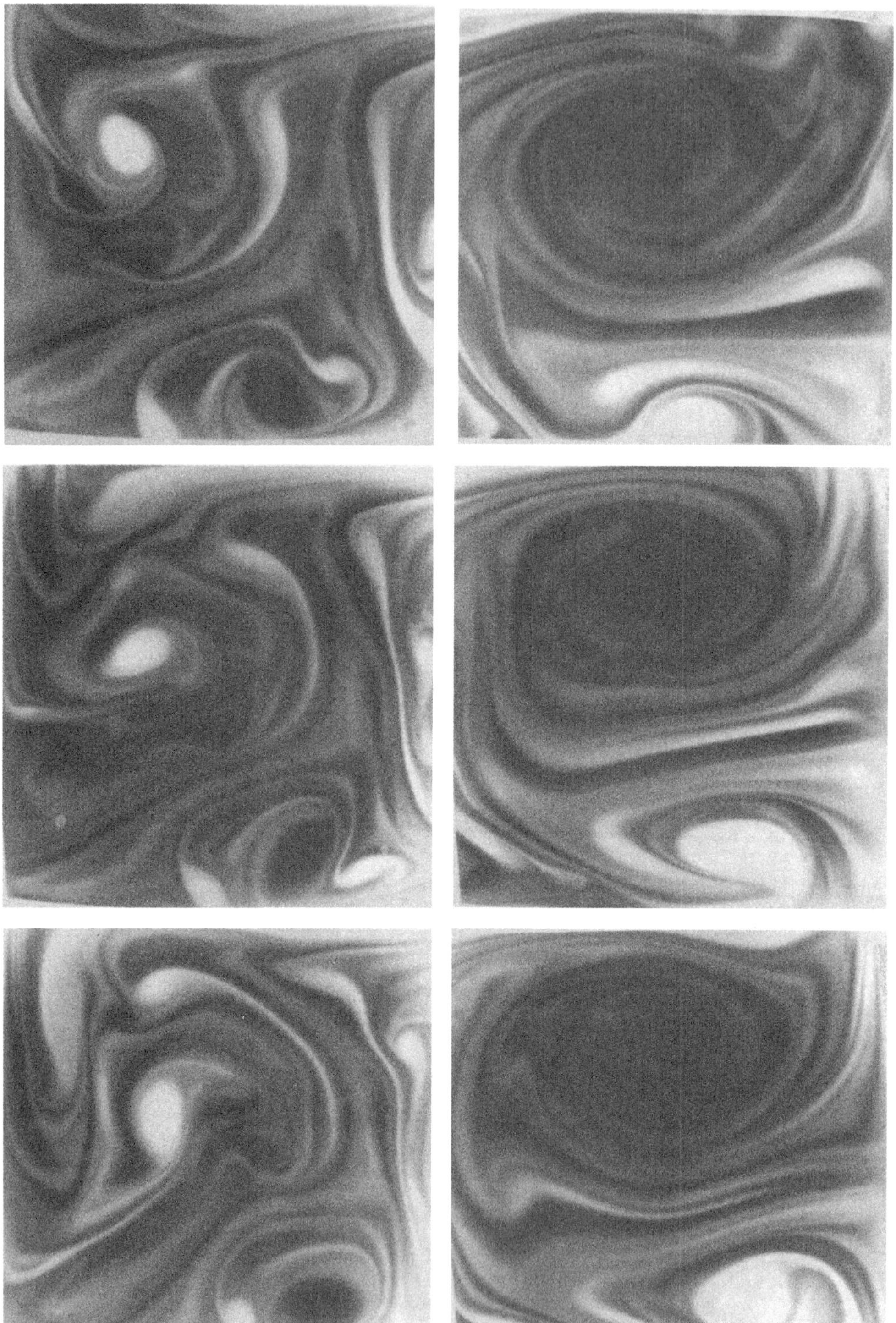

Most of the dynamics can be interpreted in terms of a simple model, first considered by Weiss[4]. In two-dimensions, when viscosity is neglected, the curl of vorticity (which is deduced from the vorticity gradient by a $\pi/2$-rotation) satisfies

$$D_t \,\mathrm{Curl}\, \omega = \mathrm{Curl}\, \omega . \nabla v \qquad (4)$$

where D_t denotes the Lagragian derivative. This equation looks similar to that obeyed by the vorticity in three dimensions. There is in fact a significant difference: Curlω is a second derivative of the velocity field. As a consequence, one may resonably assume that ∇v varies much more slowly than Curlω both in space and time. Eq.(4) is thus essentially linear in Lagrangian coordinates. The dynamics of Curlω is thus locally prescribed by the eigenvalues of ∇v. In the case of a two-dimensional incompressible flow, these are given by $\pm\sqrt{-\det(\nabla v)}$. One easily checks that

$$-\det(\nabla v) = S_{11}^2 + S_{12}^2 - \omega^2 \qquad (5)$$

When strain dominates vorticity, the eigenvalues are real and of opposite sign. Curlω is thus exponentially elongated in the positive direction and shrinked in the negative one until viscosity comes into play. This mechanism generates vorticity gradient sheets observed in the simulation. Conversely, when vorticity dominates, the structures are essentially rotating.

The produced vorticity gradient sheets are then transported by the flow. In the region where strain dominates, they will tend to pack up, and when coming close enough, reconnection takes place by dissipation, leading to the formation of complex structures. When a segment of an isolated sheet moves in a region where vorticity dominates, the sheet is locally distorted to an s-shape and a localized blob of vorticity is formed. We also observe strong vortices which are first stretched and then move to rotation regions where they recover their isotropy. In their neighbourhood, structures resulting from packing and reconnection of vorticity gradient sheets are also subject to rotating motion and spin up around the coherent vortices.

To conclude this section, we stress that the zero-vorticity lines play a special role in the dynamics. First, these lines are intrinsic. Indeed, a rigid rotation has no effect on a two-dimensional incompressible flow, since all its contributions can be absorbed in a modified pressure. Second, near the zero-vorticity lines, the eigenvalues of ∇v are real, Curl ω is stretched and vorticity gradient sheets are created. This leads to a distinction between the dynamical influence of pairs of vortices with opposite or with the same signs. In the former case, a line of zero-vorticity is present and in its neighbourhood vorticity gradient sheets are generated. In contrast, two vortices with the same sign may either act as rolling mill and generate vorticity gradient sheets if strain dominates vorticity, or rotate one around the other and eventually coalesce when the strain is weak.

3. Spectral analysis

in order to analyze the flow in Fourier space, we have fitted the energy spectrum

$$E(k) = \sum_{k \leqslant |\mathbf{k}'| < k+1} |v(\mathbf{k}')|^2 \qquad (6)$$

with a function

$$E(k) = C\, k^{-n} \exp(-\beta k). \qquad (7)$$

At early times, namely up to the maximum of the skewness, the logarithmic decrement β decreases exponentially in time. This correspondsto the inviscid phase during which the thickness of the gradient sheets shrinks exponentially. In the viscous regime, β characterizes the dissipation scale and the algebraic prefactor k^{-n} reflects the inertial range. Fig. 2 shows the inertial exponent versus time. We see that the regime of isolated vortex sheet gradient, predicted by Saffman[5] corresponds to a k^{-4} inertial range

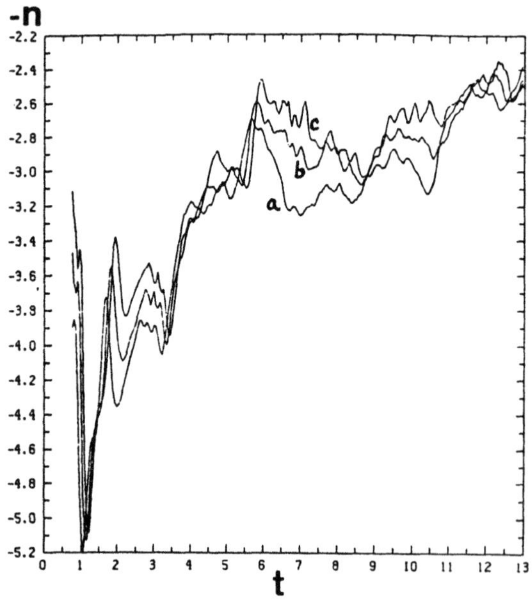

Fig.2: Spectral index n vs. time for the symmetric flow defined in Sect.1. The 3 curves correspond to fits in the ranges: 10<k<300 (curve a), 10<k<400 (curve b), 10<k<500 (curve c).

while the more mature regime where the sheets are packed together is characterized by a k^{-3} inertial range, associated to the enstrophy cascade predicted by Kraichnan[6] and Batchelor[7].

The isolated sheets can indeed be interpreted as resulting from the action of pairs of vortices which, acting either as a rolling mill or as a shear, bring together different values of the vorticity.

On the other hand, the packing of the vorticity gradient sheets subject to a large scale shear is efficiently modeled by the Batchelor theory of the passive scalar[8], applied to vorticity, as suggested by Kraichnan[9].

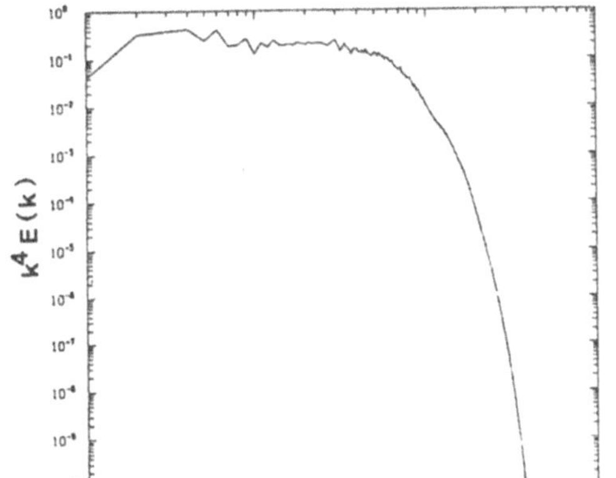

Fig.4: $k^4E(k)$ vs. k at t = 13.

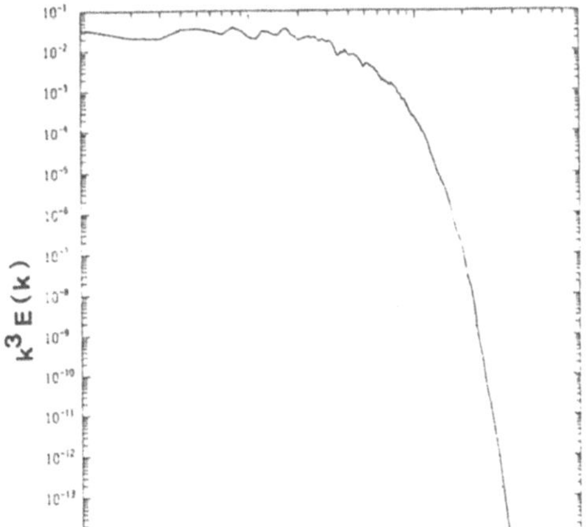

Fig.6: $k^3E(k)$ vs. k at t = 40.

4. General periodic flow

The results presented in the previous chapter are not specific of symmetric flows. A similar dynamics is obtained in the case of general periodic flows. Results of a simulation at a resolution of $(800)^2$ collocation points is presented in Figs. 3-6. The initial velocity field has an energy spectrum of the form (3) with $k_0 = 2$. The energy and enstrophy of the realization we have used are $4.\ 10^{-2}$ and $2.27\ 10^{-1}$ respectively. The viscosity is $7.5\ 10^{-5}$. The vorticity field and the corresponding energy spectrum are presented at times $t = 13$ and $t = 40$, characteristic of the k^{-4} and the k^{-3} regimes respectively.

Acknowlegements

We acknowledge useful discussions with C. Bsdevant, U. Frisch, J. Herring, R.H. Kraichnan, S.A. Orszag and R. Sadourny. The computations were done on the Cray1S of the CCVR (Palaiseau), using Fast Fourier Transform of S.A. Orszag and C. Temperton and the NCAR graphics software. We benefited of partial support from a DRET contract and from the CNRS-ATP: "Dynamique des fluides astro and géophysiques".

[1] Brachet M.E. and Sulem P.L. , in "Single and Multi Phase Flows in an Electric Field (Proc. 4th-Beer Sheva Seminar on MHD Flows and Turbulence,1984), eds. H. Branover, P.S. Lykoudis, M. Mond., "Progress in Astronautics and Aeronautics", **100**, p. 100 (1985); Brachet M.E., Meneguzzi M. and Sulem P.L., in "Macroscopic Modelling of Turbulent Flows",Lecture Notes in Physics, **230**, p. 347, 1985, Springer Verlag; Brachet M.E., Meneguzzi M. and Sulem P.L., Phys. Rev. Lett,**57**, 683 (1986)

[2]Brachet M.E, Meiron D.I., Orszag S.A., Nickel B.G., Morf R.H. and Frisch U., J.Fluid Mech. **130**, 411 (1983)

[3]McWilliams J.C., J. Fluid Mech. ,**146**, 21 (1984)

[4]Weiss J. "The dynamics of enstrophy transfer in two-dimensional hydrodynamics", La Jolla Inst. , La Jolla, California, LJI-TN-81-121 (1981).

[5]Saffman R.G., Studies in Applied Math., **50**, 377 (1971)

[6] Kraichnan H.R., J. Fluid Mech.,**10**, 1417 (1967)

[7]Batchelor G.K., Phys. Fluids, **12**, 233 (1969)

[8]Batchelor G.K., J. Fluid Mech., **5**, 113 (1959)

[9]Kraichnan H.R., J. Fluid Mech., **67**, 155 (1975)

Energy Cascades of Kelvin-Helmholtz Instabilities in the Stably Stratified Atmospheric Boundary Layer

M.S. Darby, S.D. Mobbs, and J. Brindley

Department of Applied Mathematical Studies, University of Leeds, Leeds LS29JT, UK

1. Introduction

A problem of fundamental importance in atmospheric boundary layer theory, as in many other fields of fluid mechanics in which waves can exist, is that of the non-linear evolution in space and time of an assemblage of waves. In the boundary layer, these waves occur under conditions of stable stratification, and may be thought of as internal waves, influenced to a greater or less extent by shear.

We are concerned in this paper with an analysis of data obtained from the very strongly stratified atmospheric boundary layer which can develop in the Antarctic winter night, where several months of darkness allow observations to be made in the absence of the complicating diurnal cycle. Our objectives are twofold; firstly to indentify the mechanisms by which the observed waves are generated, and secondly, to discover how the energy associated with these mechanisms is redistributed in the domains of wave-number and frequency.

At the lower end of the length scale spanned by atmospheric gravity waves (from a few metres to hundreds of kilometres) the important question of generation of or interaction with turbulence is encountered. In the middle to upper range the possible exchange of energy between disturbances on the scale of the boundary layer and larger scale gravity shear modes has important consequences for the redistribution of momentum in the lower atmosphere.

In addition to the usual spectral analysis of the data we have carried out bispectral analyses, which use phase information to throw light on wave-wave and mean-wave flow interactions.

2. Observational data

The observations were made at Halley Bay, a base operated by the British Antarctic Survey. It is situated near the seaward edge of the Brunt Ice Shelf, which is about 50 km wide and devoid of large scale topographic features which might locally influence wave activity.

The primary evidence for such gravity wave activity may be seen in temperature records from platinum resistance thermometers on a 10m mast. The sensors are situated at 2, 4 and 10 metres above the ground and the 2 metre temperature, together with the difference between this and the 4 and 10 metre temperatures, have been recorded for a number of years. During the winter night, strong temperature inversion conditions exist near the surface, with a temperature gradient as large as 10°C per metre, for approximately half the time. The inversion conditions are <u>always</u> accompanied by strong, irregular, wave-like activity which can be seen as fluctuations in the 2 metre temperature and 4 and 10 metre temperature differences with amplitudes of several degrees Celsius. Figures 1 and 2 show the temperature data for 12 hour periods at the beginning and end of separate periods of wave activity (or "wave events") observed in June 1981. It is noticeable how the wave activity builds up as soon as the inversion conditions are established and dies away as the inversion disappears. Since the wave events are so well defined, it is possible to measure their duration, which typically ranges from a few hours to a few days. Further evidence of wave activity is contained in measurements of wind speed and direction. These records are, however, dominated by high frequency fluctuations, and we have not analysed them in detail.

Information on atmospheric conditions above the lowest 10 metres is available from daily radiosonde ascents, in which data was recorded at standard pressure intervals equivalent to several hundred metres in height. Several high resolution

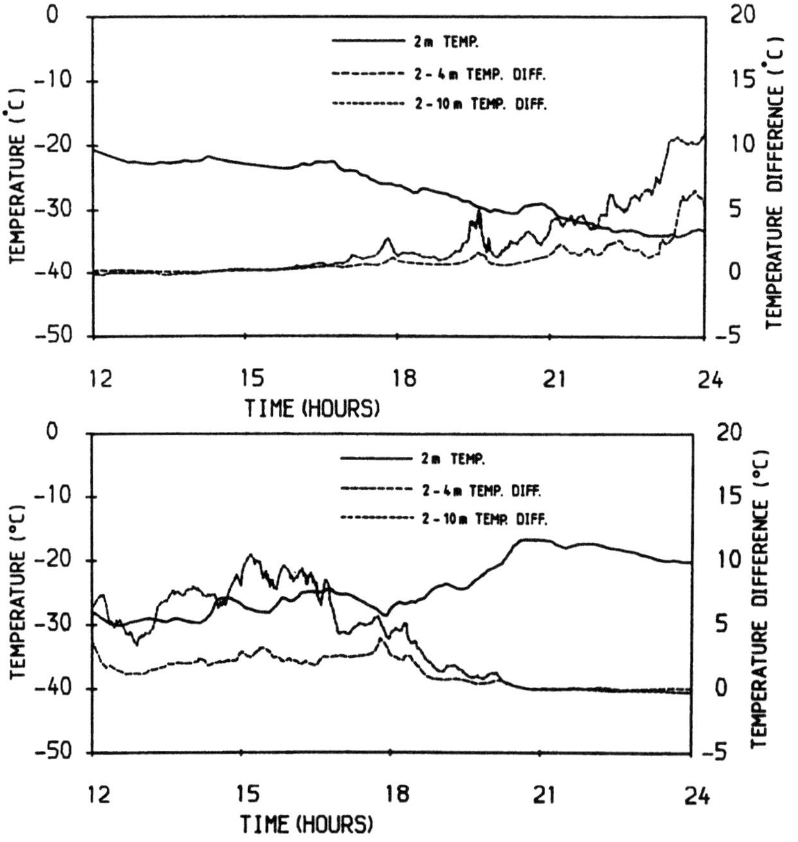

Figs. 1 and 2. Temperature plots at the beginning and end of a "wave event"

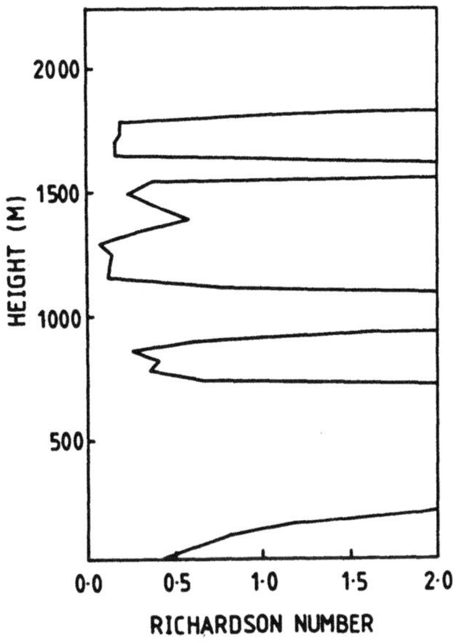

Fig. 3. Variation of Richardson number with height: radiosonde ascent of 1/2/86.

radiosonde ascent has revealed much more detail, on a scale of tens of metres (e.g. 1/2/86 see Fig. 3), which is of some significance in recognising the mechanism of generation of the waves. As the title of the paper implies, this data suggests the importance of shear instabilities (Kelvin-Helmholtz) as an energy source for wave generation.

3. Analysis of the Data

Examination of spectra and bispectra of several "wave events" suggests a characteristic pattern of behaviour as the event progresses, and the results we present here come from a rather more thorough analysis of a single, apparently typical, event, which lasted roughly from 5 a.m. on 8/6/81 to 5 a.m. on 12/6/81.

Spectra of Temperature Time-Series

The 2 to 10 metres temperature difference shows the most easily observed fluctuations so spectra have been calculated for this time series by the following method:

(i) Interpolotation used to give 1000 evenly spaced points.

(ii) A simple prewhitening used:

$$\Theta(t) = \Theta(t) - 0.95*\Theta(t - \delta t),$$

where $\Theta(t)$ is temperature difference at time t.

(iii) First and last 5% of data sample tapered using a split-cosine bell function.

(iv) Mean corrected.

(v) Autocorrelation taken to give correlogram.

(vi) Correlogram multiplied by a Parzen window with lag of 300.

(vii) Fast Fourier Transform taken to give smoothed spectrum.

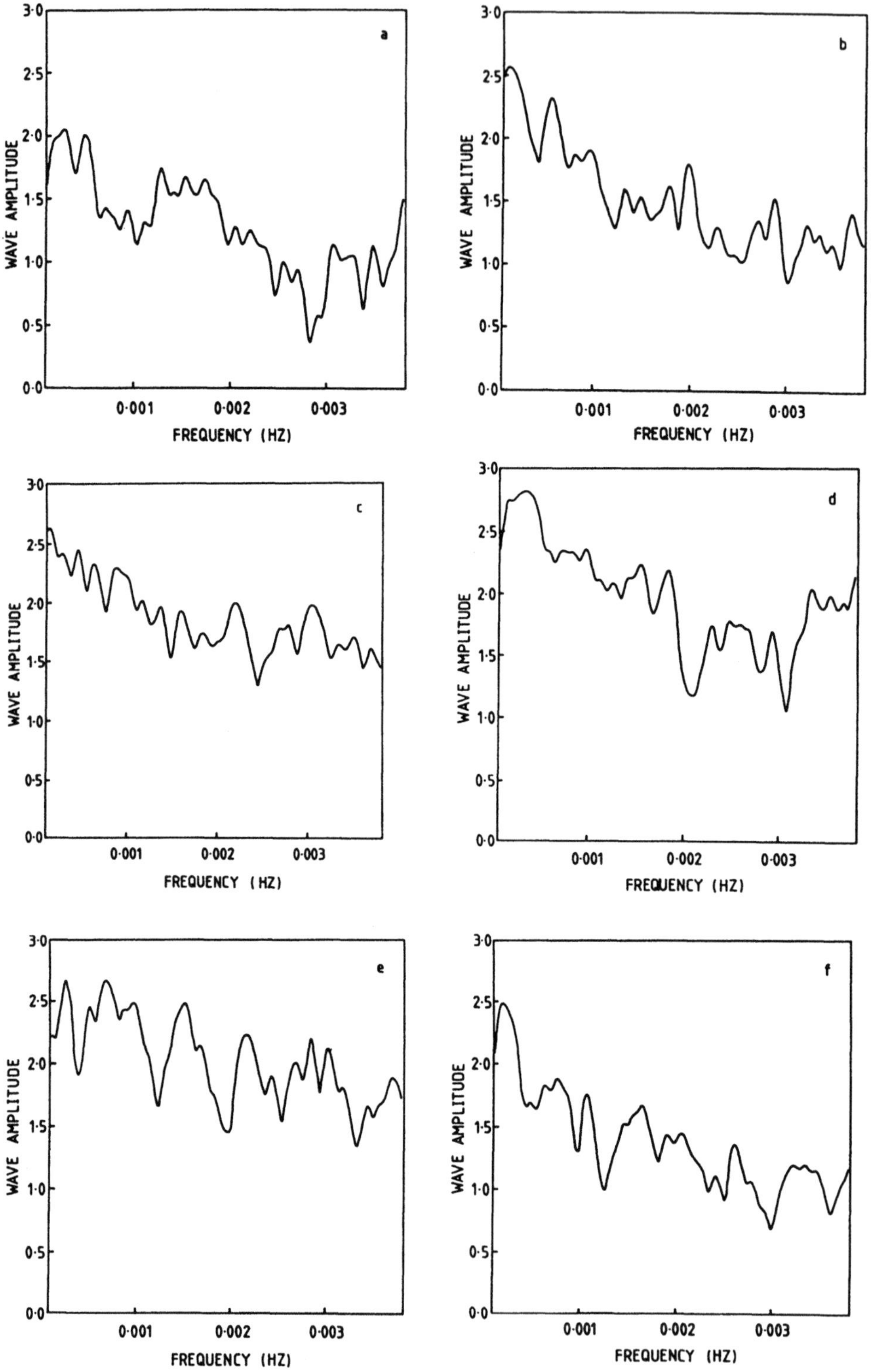

Fig.4a-f. Caption see opposite page

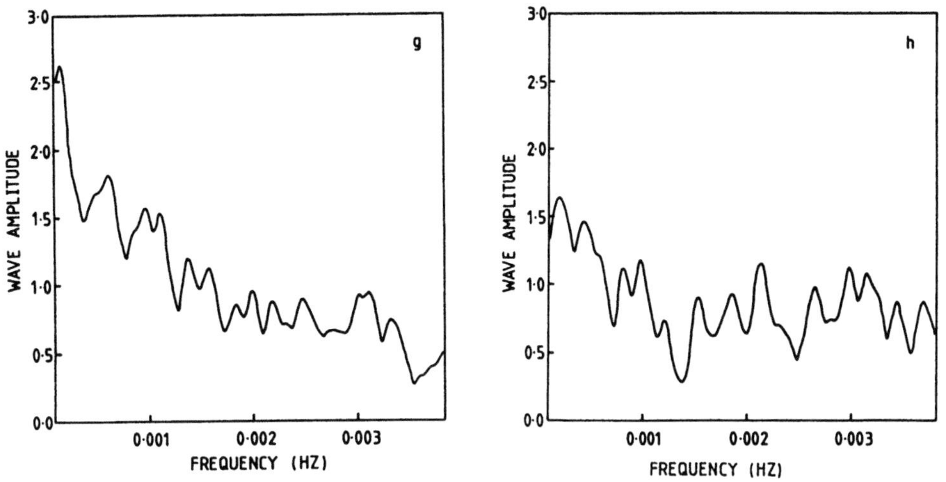

Fig. 4 (a)-(h)

Spectra of 2-10m temperature differences for successive 12 hour periods for the whole of one "wave event" (8-11/6/81).

Figures 4(a) to 4(h) show the spectra of the 2-10m temperature difference for successive 12 hour periods from 8/6/81 a.m. to 11/6/81 p.m., the whole of one wave event. There is no tendency for particular peaks in the spectrum to persist from one 12 hour period to the next. However, it is noticeable that whilst the energy at the lower end of the spectrum changes little throughout the event, there is a build-up of energy at the higher frequencies during the first half of the event, followed by a decay at these frequencies during the second half; this pattern of evolution has been seen in several wave events.

A spectrum merely shows which frequencies are present in a time-series and gives no information about the interactions occuring between waves. It is interesting to speculate as to whether the build-up of energy at the higher frequency end of the spectrum is due to instability at those frequencies or whether it is due to a transfer of energy from lower frequencies. This is one of the reasons for calculating the bispectrum of the time series, as described in the next section.

Bispectra

We define the bispectrum $\mathcal{B}(\omega_1,\omega_2)$ associated with a time series as

$$\mathcal{B}(\omega_1,\omega_2) = \mathcal{R}[\mathcal{F}(\omega_1)\mathcal{F}^*(\omega_2)\mathcal{F}^*(\omega_1-\omega_2)\exp(-i\alpha)] \tag{3.1}$$

where $\mathcal{R}$ denotes real part, $\mathcal{F}(\omega)$ is the complex Fourier transform of the time-series and an asterisk denotes complex conjugation. Physically, we can interpret a peak in $\mathcal{B}(\omega_1,\omega_2)$ as meaning that waves ω_1 and ω_2 are interacting to give the wave with frequency $\omega_1-\omega_2$ and relative phase α.

Thus the spectrum $\mathcal{B}(\omega_1,\omega_2)$ can be used to determine whether triads of waves are present in the time-series with the chosen value of α. The technique has been used most commonly in the study of ocean surface waves [1,2].

If sufficient stationary data is available then it is possible to calculate the bicoherence, $b(\omega_1,\omega_2)$ and biphase $\beta(\omega_1,\omega_2)$ thus avoiding the need to specify α [2]. These are defined as

$$b(\omega_1,\omega_2) = \frac{|B(\omega_1\ \omega_2)|^2}{<|F(\omega_1)|^2|F^*(\omega_2)|^2> <|F^*(\omega_1-\omega_2)|^2>} \tag{3.2}$$

and $$\beta(\omega_1,\omega_2) = \tan^{-1}\left(\frac{I[B(\omega_1,\omega_2)]}{R[B(\omega_1,\omega_2)]}\right) \tag{3.3}$$

where $$B(\omega_1,\omega_2) = <F(\omega_1)F^*(\omega_2)\ F^*(\omega_1-\omega_2)>. \tag{3.4}$$

Angle brackets denote ensemble averages, i.e. averages over many (repeatable) experiments and I denotes the imaginary part.

Unfortunately the Halley Bay data to hand was insufficient to obtain reliable averages for b and β. Instead we used equation (3.1) to calculate bispectra for particular values of α.

The values chosen are suggested by the form of the equations of fluid motion. The nonlinear terms in these equations imply nonlinear wave-wave and wave-mean flow interactions, and a simple qualitative argument enables us to determine the relative phases of the interacting waves in two simple cases. For simplicity, we restrict attention to the self-interaction of a wave, giving rise to its harmonic. The same arguments apply, however, to the interactions of waves of different frequencies.

In the absense of mean velocity shear, the linearised, Boussinesq approximation to the temperature equation is:

$$\frac{D\theta'}{Dt} + \bar{\underline{u}}.\nabla\theta' + w\frac{\bar{\theta}}{g}N^2 = 0 \tag{3.5}$$

where $\underline{u}$ is the velocity vector, w is vertical component of the velocity, θ is the potential temperature, N is the Brunt-Väisälä frequency and g is the gravitational acceleration. Wave quantities are denoted by a prime and mean quantities by an overbar. The first case we consider is a freely propagating, two dimensional wave with propagation components in the x and z directions. Then it is easily shown that in the Boussinesq approximation, if

$$u \sim \sin(\Phi_1 + m_1 z)$$

then $$\theta \sim \cos(\Phi_1 + m_1 z),$$

where $\Phi_1 = k_1 x - \omega_1 t$. u_1' and θ_1' denote the horizontal velocity and potential temperature perturbations respectively, due to wave 1. k_1 is the horizontal wave number, m_1 ia the vertical wave number and ω_1 is the wave frequency for wave 1. One of the nonlinear terms in equation (3.5) is:

$$u_1' \frac{\partial \theta_1'}{\partial x} \sim \cos[2(\Phi_1 + m_1 z)] + \text{mean term} . \qquad (3.6)$$

Equating this to the time derivative in (3.1) gives an expression for the time evolution of the first wave harmonic contribution to the potential temperature:

$$\theta_2' \sim \sin[2(\Phi_1 + m_1 z)] = \cos[2(\Phi_1 + m_1 z) - \pi/2] \qquad (3.7)$$

Therefore, if the phase of the primary wave is taken to be zero, then the first harmonic has a phase $\pi/2$.

The second simple case we consider is that of waves which are propagating in the horizontal (x) direction but have a modal structure in the vertical (z) direction. Again it is easily shown that if

$$u_1' \sim \sin(m_1 z)\cos(\Phi_1) \text{ then } \theta_1' \sim \cos(m_1 z)\cos(\Phi_1)$$

and hence

$$u_1' \frac{\partial \theta_1'}{\partial x} \sim \text{ain}(2m_1 z)\cos(2\Phi_1). \qquad (3.8)$$

In this case, equating the nonlinear contribution (3.4) to the time derivative of the first harmonic in (3.5) gives

$$\theta_2' \sim \sin(2m_1 z)\cos(2\Phi_1). \qquad (3.9)$$

Thus the phase of the first harmonic is the same as that of the primary wave.

In the more general case where waves of different frequencies interact to give a third wave, the conclusions are similar. If we represent the periodic parts of the expression for potential temperature corresponding to the two waves by $\cos(\phi_1)$ and $\cos(\phi_1)$, then $\cos(\phi_1)$ interacts with $\cos(\phi_2)$ to give $\cos(\phi_1 + \phi_2 + \alpha)$ and we find that:

(a) for free waves, the phase $\alpha = \pi/2$ or $3\pi/2$, depending on the directions of propagation and

(b) for trapped waves $\alpha = 0$ or π.

Figures 5(a) and 5(b) show the bispectra $\mathcal{B}(\omega_1,\omega_2)$ plotted as grey shades for the case $\alpha = 0$ two 12 hour periods during a wave event (lasting from 8/6/81 p.m. to 11/6/81 p.m.). Only values between the lines $\omega_1 = \omega_2$ and $\omega_1 = 2\omega_2$ (shown as dotted lines) are plotted, since all other values are repeats of these obtained by exchanging ω_1 and ω_2 or one of ω_1 and ω_2 with $\omega_1 - \omega_2$. (Note that the grey levels are scaled differently in each figure, but that comparisons between the values of $\mathcal{B}$ in different figures are meaningful.) The diagrams have been produced by retaining only those points where the phase is within 10° of the specified value. No values of $\mathcal{B}(\omega_1,\omega_2)$ have been calculated close to the line $\omega_1 = \omega_2$ since

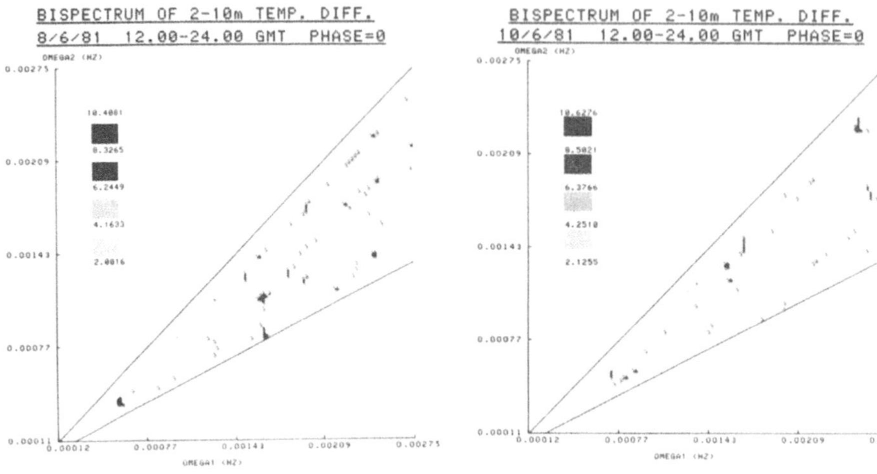

Fig. 5. (a) and (b)
Bispectra for the case $\alpha = 0$ for two 12 hour period in the "wave event" of 8-11/6/81

these points imply a difference frequency $\omega_1 - \omega_2$ which is too low to be resolved by the length of the data records.

Note that in Fig. 5(a) there is a tendency for peaks to lie along lines of constant $\omega_1 - \omega_2$, indicating that several interactions are transferring energy into or out of particular modes. A study of the bispectra from several wave events suggests a tendency for the main peaks to move away from the line $\omega_1 = \omega_2$ (indicating generation of very low frequency waves or interaction with the mean flow) as the event progresses. This is not particular clear from one event alone. The most striking feature of the case $\alpha = 0$ is that at the end of the wave event there are virtually no interactions occurring, even though there are waves with significant amplitude remaining.

No plots of $\mathcal{B}$ are shown for the last 24 hours of the event since there are no significant peaks in these bispectra. This contrasts markedly with the case $\alpha = \pi/2$ shown in Figs. 6(a) to 6(d). Here the most striking feature is the dominance of interactions with this phase during the last 24 hours of the event, suggesting that freely propagating (or radiating) waves are dominant during this part of the event.

4. Discussion and Conclusions

We have presented results, which we regard as preliminary, concerning gravity wave activity in the nocturnal boundary layer in the Antarctic. The data available were far from ideal, and our conclusions at this stage are tentative, but they have been useful in planning a range of much more comprehensive and detailed observations in the Antarctic winter of 1986, as part of our ongoing research collaboration with the British Antarctic Survey. Nevertheless we can draw a number of conclusions from the results summarised in section 3.

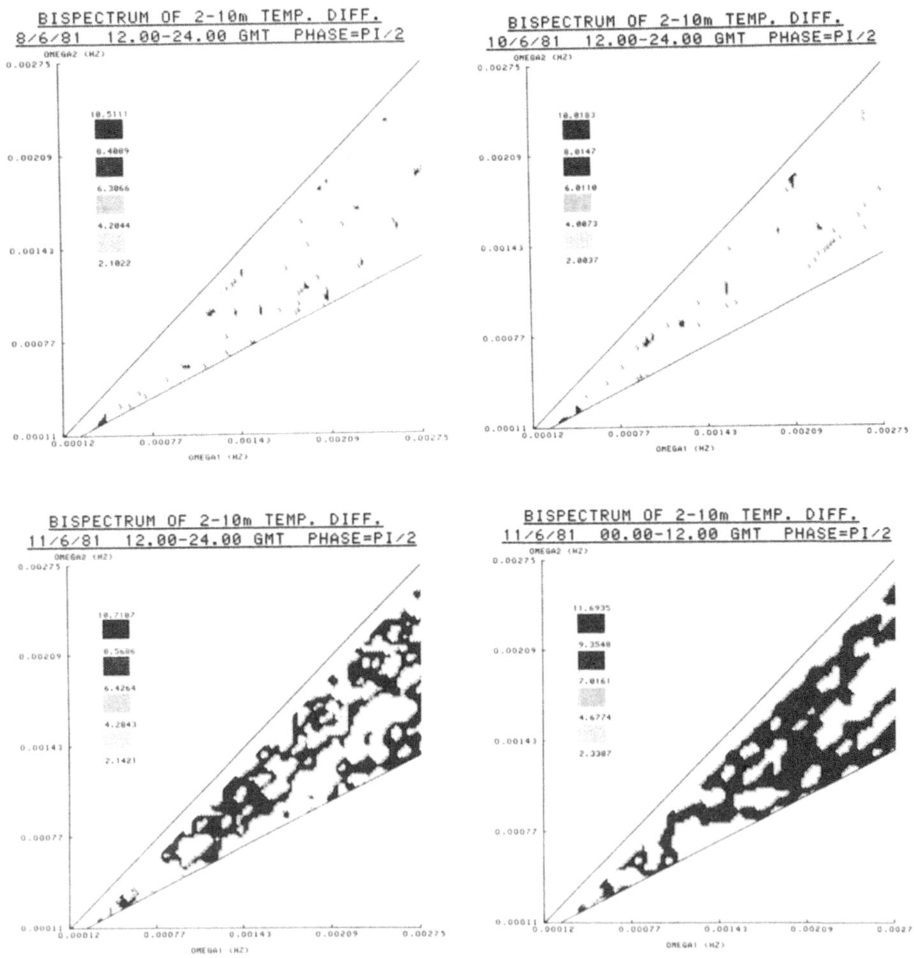

Fig. 6 (a)-(d)

Bispectra for the case $\alpha = \frac{\pi}{2}$ for four 12 hour periods in the "wave event" of 8-11/6/81.

(i) Generation Mechanism

The single radiosonde ascent (Fig. 3) shows thin layers of low Richardson number defined as $Ri \equiv \frac{g}{\theta}\frac{d\theta}{dz} \Big/ [(\frac{du}{dz})^2 + (\frac{dv}{dz})^2]$. Values of $Ri < \frac{1}{4}$ suggest possible instabilities of Kelvin-Helmholtz type.

A linear stability analysis of a very recent radiosonde profile (from February 1986), containing similar thin layers of low Ri, has revealed instabilities at frequencies compatible with those prominent in the boundary layer spectra. This leads us to conjecture (and much more detailed work is presently in hand [3], King et al. 1986) that such shear instabilities are an important generation mechanism for the observed waves in the boundary layer.

(ii) Energetics of the Waves

(a) Active waves occur in well defined "events", lasting from a few hours to several days.

(b) During the course of a wave event, there is a build-up of energy in waves with periods of 5 to 10 minutes, followed by a decay of energy associated with these waves during the second half of the event.

(c) Bispectra indicate that towards the end of an event, there is a strong tendency for all nonlinear interactions between waves to involve freely propagating waves rather than trapped waves.

Acknowledgements

The data for this study was provided by the British Antarctic Survey. We wish to thank both Mr. D. W. S. Limbert for bringing the phenomena described here to our attention (and for allowing us to use the data), and the British Antarctic Survey staff at Halley Bay who collected the data. We also wish to thank Dr. J. C. King for many helpful discussions. MSD is grateful to the Natural Environment Research Council for the provision of a CASE research studentship at the British Antarctic Survey funded under the Antarctic Special Topic scheme.

Reference

1. Elgar, S. and Guza, R. T. 1985. Shoaling gravity waves: comparisons between field observations, linear theory and a nonlinear model. J. Fluid Mech. 158, 47-70.

2. Hasselman, K., Munk, W. and MacDonald, G. 1963. Bispectra of ocean waves. In Time Series Analysis (ed. M. Rosenblatt). 125-129. Wiley.

3. King, J. C. Darby, M. S., Mobbs, S. D. and Rees, J. M. Observations of an internal gravity wave in the lower troposphere at Halley, Antarctica, to be submitted to Boundary Layer Meteorology.

An Experimental Study of the Inverse Cascade of Energy in Two-Dimensional Turbulence

J.M. Nguyen Duc, Ph. Caperan, and J. Sommeria
MADYLAM, E.N.S.H.M.G., BP 95, F-38402 St. Martin d'Héres Cedex, France

A quasi two dimensional turbulence is studied, which is created in a horizontal layer of mercury submitted to a vertical, homogeneous and permanent magnetic field B (of which effect is to stabilize the flow to a bidimensional state [1]). A well controlled injection of electrical currents from a periodic network of electrodes allows us to drive the flow. In Fourier space, this corresponds to injection of enstrophy and energy on the discrete two dimensional set of wavenumbers corresponding to the spacial periodic repartition of exciting electrodes [2]. The flow is studied by streak photography or using measuring electrodes (Fig. 1) which have been arranged at the bottom of the box. In theses types of flows the measured potential, when removed the part generated by injection currents, is directly proportional to the stream function. A good model of the evolution of this flow is given by the two dimensional Navier Stokes equations with a steady forcing term and a linear friction one due to the Hartmann layer, of which effect is estimated by the parameter R_h[3]. Two different shapes of boxes have been used in this study.

The first one is a square with a set of 6 x 6 exciting electrodes [2]. The behaviour of the flow has been studied at different increasing parameter R_h. When

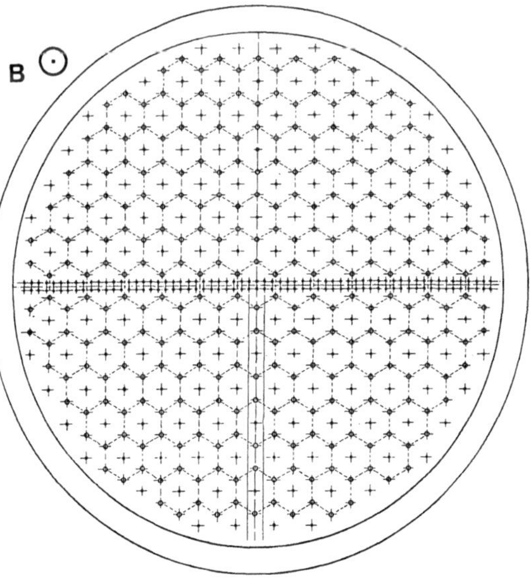

Fig. 1 : The experimental box ; hexagonal network of injection electrodes. Note the measuring electrodes at the center of each cell, and on one diameter. (internal diameter is 26 cm)

R_h is less than 1.78 a stable network of 36 alternated vortices is observed, while above 1.78 a first instability occurs. This transition could be described as occuring in the elementary cell of four vortices of the network, where two diagonally opposite vortices will merge, while the two other ones are stretched. This basic mechanism quickly leads as R_h is increased to a fully turbulent regime. An inverse energy cascade is observed, which is limited at large scales either by linear friction ($4 < R_h < 30$) or by the finite size of the box ($R_h > 30$). In the case $R_h < 30$ we have measured monodimensional energy spectra which exhibit a $k^{-5/3}$ law on the range of the lowest modes than the injection one, with a Kolmogorov constant in the range 3 to 7. In the second configuration, $R_h > 30$, a condensation of the turbulent energy in the lowest mode is observed.

We recently made new experiments in a circular box in which the flow is created by injection of current from an hexagonal network of 282 electrodes. This new apparatus allows a better homogeneity and isotropy of the flow, as well as an increased resolution in wave number (Fig. 1). We first explore the general behaviour of the flow by means of streak photography while signal processing provides us with quantitative informations at some chosen values of R_h.

The photo a illustrates the stable state pertaining when $R_h \sim 1.15$. The maximum velocity is about 0.25 cm/s and is observed at the boundary between each contrarotating triangular vortices. During this regime, a constant electrical potential is observed on each of the measuring electrodes of the diameter of the box. Small perturbations of these potential signals are occuring, which could be related to oscillations of the network. But these observations need to be confirmed since we are at the limit of resolution of our measuring system, where thermal drift is of the same order as the possible oscillation.

When R_h is increased to 1.35 the flow endures some unstability (Photo b) which seems to correspond to the association of triad of corrotating vortices. Those triads correspond to enlarged structures, which grow at the expense of the opposite signed vortices on their boundaries. We can see that as soon as such a structure is formed, it will favour the formation of identical ones along its three axii of symetry (Fig. 2). From this, two ways to destabilisation will compete : negative or positive vortices interaction.

At $R_h \sim 11.3$ the orthoradial velocity as measured from two central electrodes (Fig. 3) presents high frequency fluctuations. This is explained by the turbulent behaviour of the flow, as it is observed on photo c. At this stage and for higher R_h visualisations reveal a broad band of scales bigger than the basic injection one. These flows are organised in a set of vortices separated by large streams of potential (or weak vorticity) flows. The characteristic scales of the flow increase with R_h, as can be deduced from visualisation or from lateral correlation measurements (Fig. 4). These lateral correlation functions correspond to statistical averaging of the homogeneous correlation function obtained from the 62 electrodes disposed on a diameter of the box. This is related to the inverse transfer of

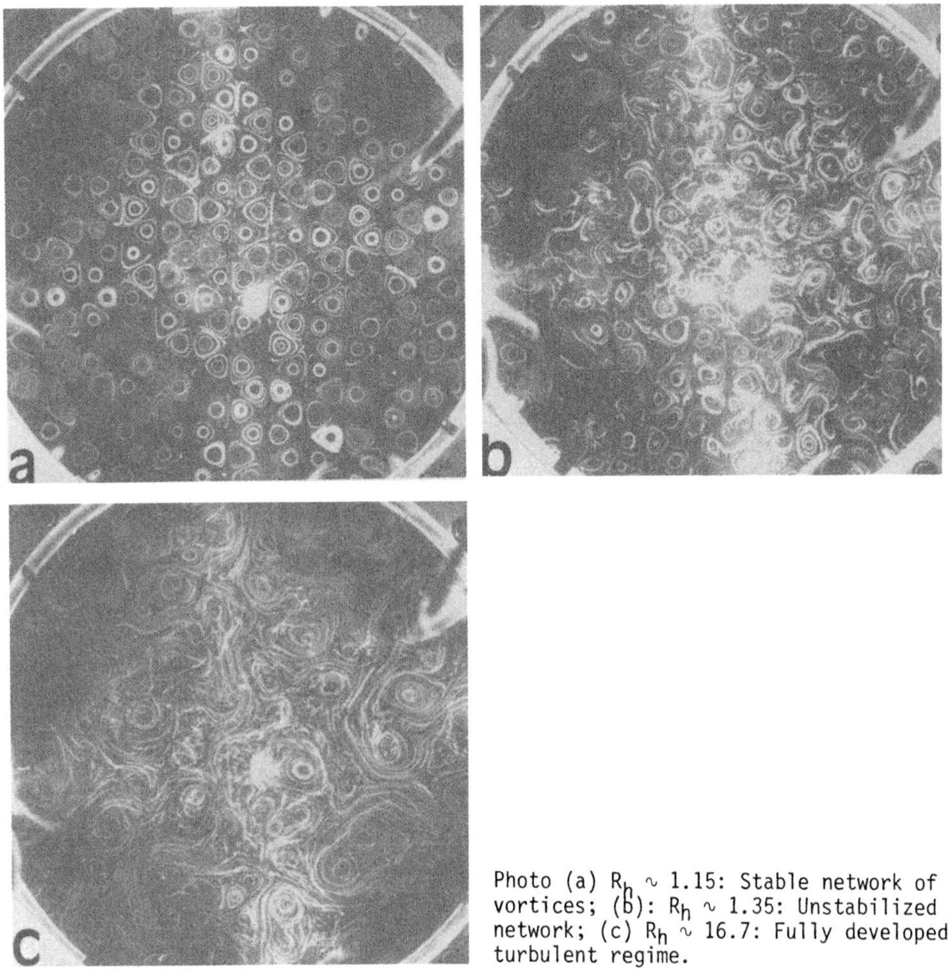

Photo (a) $R_h \sim 1.15$: Stable network of vortices; (b): $R_h \sim 1.35$: Unstabilized network; (c) $R_h \sim 16.7$: Fully developed turbulent regime.

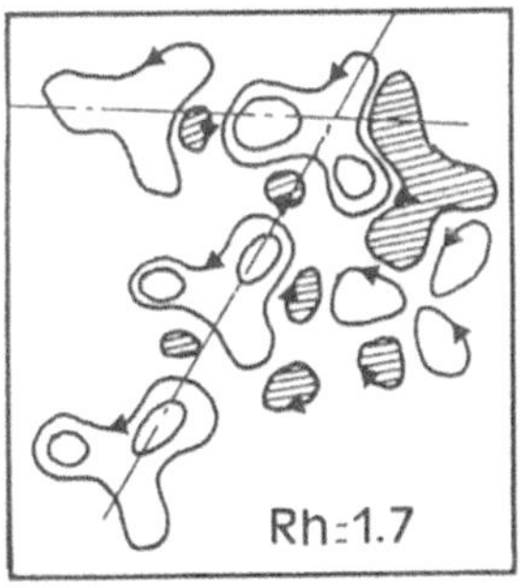

Figure 2 : Mechanism of unstability as observed from photo c.

energy occuring in two dimensional flows, which has been first experimentally quantified by Sommeria [2].

These results compose the first set of observations that we have been able to make on the new circular box. They illustrate the wide range of behaviours

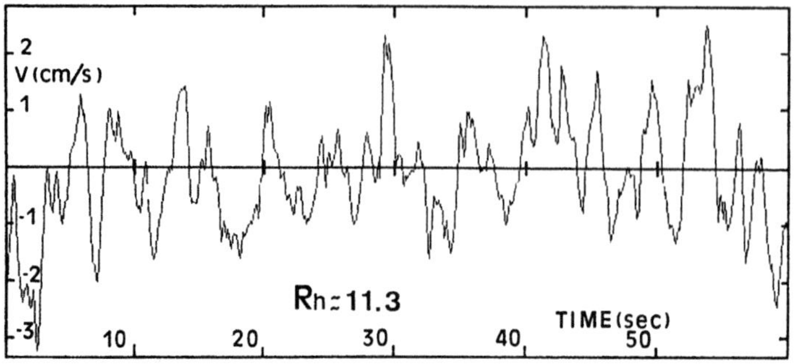

Fig. 3 : Temporal evolution of orthoradial velocity as observed from a central doublet of electrodes.

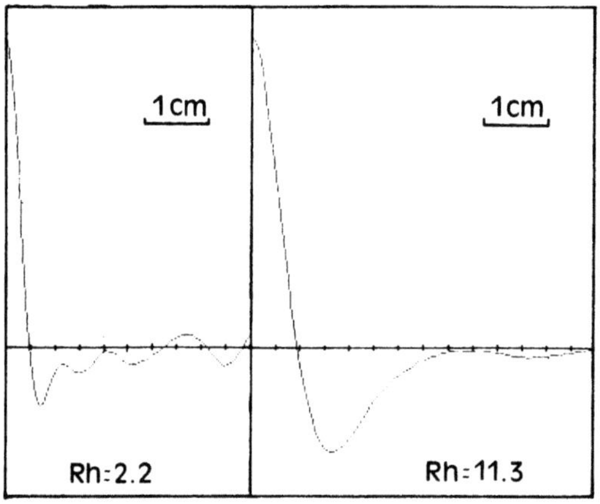

Fig. 4 : lateral correlation ; function bars corresponds to measuring electrodes.

reachable with this installation. We expect to make a detailed analysis of inertial transfers by bispectral measurements, and characterize in a more precise way the different regimes described above.

1. J. SOMMERIA, R. MOREAU, J. Fluid Mech., Vol. 18, pp 507-518
2. J. SOMMERIA, J. Fluid Mech., Vol. 170, pp 139-168
3. $R_h = (eI/B\sigma\nu)^{1/2}/L$ with ν,σ viscosity and conductivity of the fluid, e thickness of the layer and L characteristic size of the box.

Inhomogeneous Two-Dimensional Turbulence in the Atmosphere

T.G. Shepherd

Department of Applied Mathematics and Theoretical Physics,
University of Cambridge, Silver Street, Cambridge CB3 9EW, UK

1. Introduction

Faced with the strongly nonlinear and apparently random behaviour of the energy-containing scales in the atmosphere, geophysical fluid dynamicists have attempted to understand the synoptic-scale atmospheric flow within the context of two-dimensional homogeneous turbulence theory (e.g. FJØRTOFT [1]; LEITH [2]). However atmospheric observations (BOER and SHEPHERD [3] and Fig.1) show that the synoptic-scale transient flow evolves in the presence of a planetary-scale, quasi-stationary background flow which is approximately zonal (east-west). Classical homogeneous 2-D turbulence theory is therefore not strictly applicable to the transient flow. One is led instead to study 2-D turbulence in the presence of a large-scale (barotropically stable) zonal jet inhomogeneity.

2. Strongly Nonlinear Wave, Mean-Zonal-Flow Interaction

The model equation to be studied is

$$\nabla^2\psi_t + (\beta - U_{yy})\psi_x + U\nabla^2\psi_x + J(\psi,\nabla^2\psi) = \tilde{S}(x,y,t), \qquad (1)$$

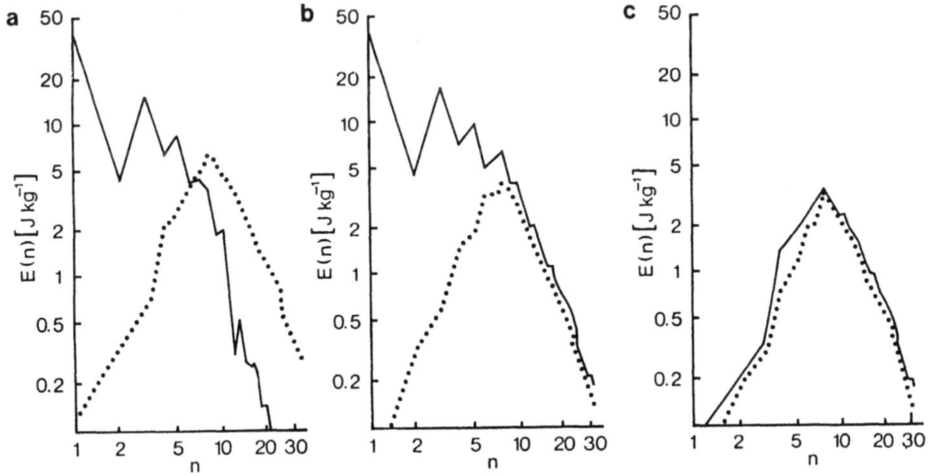

Figure 1. Spectra of (a) stationary (solid) and transient (dotted) kinetic energy; (b) zonal (solid) and meridional (dotted) kinetic energy; and (c) zonal (solid) and meridional (dotted) components of transient kinetic energy; as functions of total spherical harmonic index n. Global atmospheric FGGE data from January 1979. From SHEPHERD [4]

where $U(y)$ is the prescribed zonal basic flow velocity, $\psi(x,y,t)$ the disturbance streamfunction, β the northward gradient of the Coriolis parameter, and $\tilde{S}(x,y,t)$ a zonally homogeneous source-sink term. The zonal coordinate x increases eastward and the meridional coordinate y northward. This is the traditional beta-plane approximation (e.g. PEDLOSKY [5]). The disturbance velocities are obtained from the streamfunction by $u = -\psi_y$, $v = \psi_x$ (the usual geophysical sign convention).

There are two important features to note about (1) which provide the key to the analysis presented here (and which are explained in greater detail in SHEPHERD [6]). The first is that, because of the presumed scale separation between the basic zonal flow and the disturbance, $|U_{yy}\psi_x| \ll |U\nabla^2\psi_x|$ and hence the global integral of disturbance enstrophy $\frac{1}{2}(\nabla^2\psi)^2$ is approximately conserved (apart from $\tilde{S}$). The second feature is a simple geometrical constraint on the wave, mean-zonal-flow interaction. Because the basic flow is zonal, wavevector triads involving the basic flow and the disturbance must couple disturbance components which have the same zonal wavenumber k (Fig.2a). (This property remains true in spherical geometry [4] although the details are more complicated.)

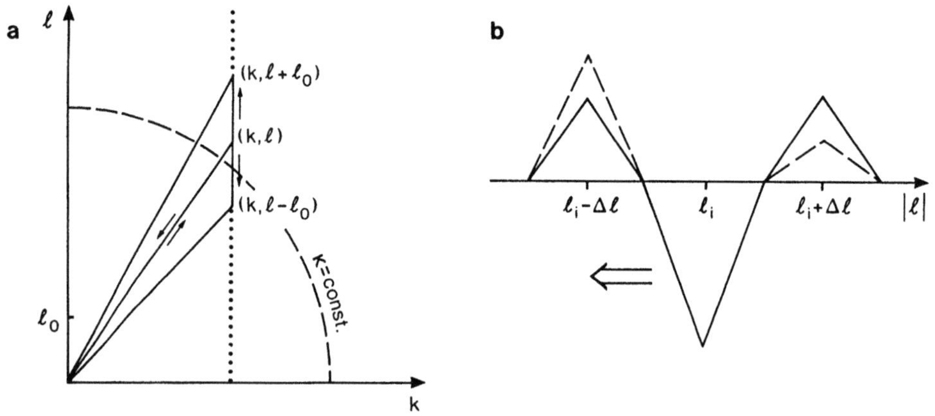

Figure 2. (a) Possible triad interactions between a basic flow $(0,\ell_0)$ and an arbitrary disturbance component (k,ℓ). Arrows indicate direction of energy transfers if the disturbance is to spread. (b) Schematic of enstrophy (solid) and energy (dashed) transfer terms, scaled by ℓ_i^2 one to another, for interaction between a disturbance of meridional scale ℓ_i and a zonal flow, assuming spectral symmetry. Arrow indicates the sense of the implied energy "cascade"; the energy loss at $\ell = \ell_i$ is covered over by the enstrophy loss. From [6]

The two above points together imply that the interaction between the basic flow and the disturbance may be characterized as a "shear-induced spectral transfer" of disturbance enstrophy along lines of constant zonal wavenumber k. In this transfer the disturbance energy $\frac{1}{2}|\nabla\psi|^2$ is generally not conserved. The sense of the transfer is dependent on relative phase relationships, and can be calculated from ray-tracing theory.

For example, consider a spectrally symmetric initial disturbance with given k centred about meridional wavenumber $|\ell| = \ell_i$; this might be just two wave packets

(k, ℓ_1) and $(k, -\ell_1)$. The shear-induced enstrophy transfer is then symmetric in $|\ell|$, but there is an associated up-scale (i.e. to smaller $|\ell|$) transfer and amplification of disturbance energy (Fig.2b). This latter is essentially just the "Orr effect" of temporary amplification in a shear flow, and is a simple consequence of the fact that enstrophy equals $k^2+\ell^2$ times energy for a wave packet. It is easy to see that this process leads to meridional anisotropy ($v^2 > u^2$) at large scales and to zonal anisotropy ($u^2 > v^2$) at small scales. According to linear theory, the large-scale development is subsequently reversed and the disturbance is eventually absorbed by the basic flow. But nonlinear effects act to isotropize the flow and can make the initial up-scale transfer irreversible (see [6] and section 4 below).

3. Stationary-Transient Interaction Diagnostics

We wish to consider (1) in the context of forced-dissipative turbulence in a state of statistical equilibrium. Then a time-mean flow (consisting of the basic flow $U(y)$ plus the stationary part of the disturbance) can be defined and the stationary-transient interaction investigated. Since the stationary flow will be zonally symmetric the discussion of section 2 carries over directly.

In classical 2-D homogeneous turbulence theory (e.g. KRAICHNAN [7]; LEITH [8]) one usually considers a spectral energy equation of the form

$$\frac{\partial}{\partial t} E(\underset{\sim}{\kappa}) = I(\underset{\sim}{\kappa}) + S(\underset{\sim}{\kappa}) \tag{2}$$

for each wavenumber $\underset{\sim}{\kappa} \equiv (k, \ell)$; here $E(\underset{\sim}{\kappa})$ is the total flow energy in wavenumber $\underset{\sim}{\kappa}$, $I(\underset{\sim}{\kappa})$ is the energy transfer into $\underset{\sim}{\kappa}$ arising from nonlinear interactions with all other scales, and $S(\underset{\sim}{\kappa})$ is the source-sink term. If the turbulence is isotropic then everything is a function of $\kappa \equiv |\underset{\sim}{\kappa}|$ alone.

In the present case, however, we must consider instead the pair of equations

$$\frac{\partial}{\partial t} E_S(\underset{\sim}{\kappa}) = I_S(\underset{\sim}{\kappa}) + C_S(\underset{\sim}{\kappa}) + S_S(\underset{\sim}{\kappa}), \quad \frac{\partial}{\partial t} E_T(\underset{\sim}{\kappa}) = I_T(\underset{\sim}{\kappa}) + C_T(\underset{\sim}{\kappa}) + S_T(\underset{\sim}{\kappa}), \tag{3}$$

corresponding respectively to the stationary and transient flow components. The nonlinear interaction terms $I_S(\underset{\sim}{\kappa})$ and $I_T(\underset{\sim}{\kappa})$ are defined analogously to $I(\underset{\sim}{\kappa})$ in (2), and represent respectively the interactions involving stationary motions alone and transient motions alone. They act to redistribute stationary or transient energy and enstrophy among the various spectral components; hence

$$\iint I_S(\underset{\sim}{\kappa})\, d\underset{\sim}{\kappa} = \iint \kappa^2 I_S(\underset{\sim}{\kappa})\, d\underset{\sim}{\kappa} = \iint I_T(\underset{\sim}{\kappa})\, d\underset{\sim}{\kappa} = \iint \kappa^2 I_T(\underset{\sim}{\kappa})\, d\underset{\sim}{\kappa} = 0 . \tag{4}$$

The stationary-transient conversion terms $C_S(\underset{\sim}{\kappa})$ and $C_T(\underset{\sim}{\kappa})$ are given by

$$C_S(\underset{\sim}{\kappa}) \equiv \frac{1}{2}\hat{\bar{\psi}}^*(\underset{\sim}{\kappa})\, \overline{\{J(\psi', \nabla^2\psi')\}(\underset{\sim}{\kappa})} + \text{c.c.} \tag{5a}$$

$$C_T(\underset{\sim}{\kappa}) \equiv \frac{1}{2}\overline{\hat{\psi}'^*(\underset{\sim}{\kappa})\left[\{J(\bar{\psi}, \nabla^2\psi')\}(\underset{\sim}{\kappa}) + \{J(\psi', \nabla^2\bar{\psi})\}(\underset{\sim}{\kappa})\right]} + \text{c.c.} , \tag{5b}$$

where the cap and the brace denote Fourier transforms, the overbar a time average, and the prime the deviation therefrom.

4. Numerical Simulation: Cosine Jet

The foregoing ideas are applied first to a numerical investigation of an idealized situation where the basic flow is the sinusoidal jet $U(y) = -U_0 \cos y$ in doubly-periodic geometry ($0 \leqslant x,y < 2\pi$). This case is examined in great detail in [6]. Here it is sufficient to point out that $\tilde{S}$ of (1) consists of: an isotropic and homogeneous "negative-viscosity" forcing in the spectral band $9 \leqslant \kappa < 12$; a high-order diffusion $\nu\nabla^8\psi$; a scale-independent damping $-r\nabla^2\psi$ chosen to (energetically) balance the intermediate-scale forcing; and a relaxation tendency to the basic flow $-(r+\nu)U_0 \sin y$ (this ensures that the basic flow is a steady solution of (1)). The spectral truncation is $0 \leqslant |k|,|\ell| \leqslant 32$, and $u_{rms} \simeq 1$. Geophysical considerations suggest the choices $\beta = 25$ and $U_0 = 2\sqrt{2}$. The "cascade arrest" scale of RHINES [9], characterizing the cross-over between Rossby waves and 2-D turbulence in the classical homogeneous problem, is given by $\kappa_\beta \simeq 5$; the linear damping coefficient r is 0.5, corresponding to a dimensional spin-down time of 5 days; and the turbulent eddy-turnaround time is 0.1.

The differences between classical beta-plane turbulence and the present situation are apparent in Figs.3a,b, which show the time-averaged energy spectra, decomposed into zonal and meridional components (u^2 and v^2), in both cases. One may note the following effects of the zonal-flow inhomogeneity:

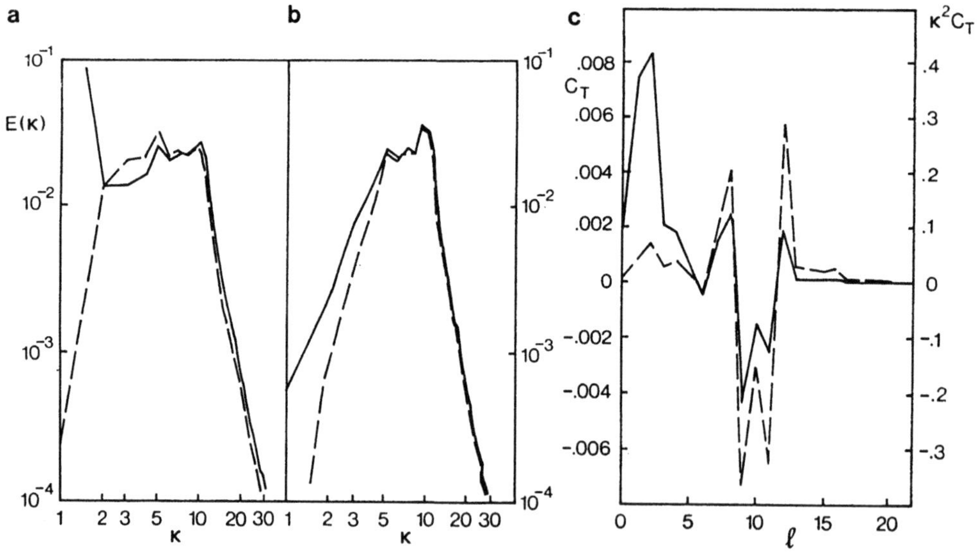

Figure 3. Time-averaged energy spectrum $E(\kappa)$, decomposed into zonal (solid) and meridional (dashed) components, for (a) beta-plane turbulence in a jet, as described in section 4, and (b) homogeneous beta-plane turbulence. (c) Conversion terms of energy $C_T(k;\ell)$ (solid) and of enstrophy $\kappa^2 C_T(k;\ell)$ (dashed), shown for fixed $k = 2$ as functions of ℓ, corresponding to the numerical simulation of (a). From [6]

(i) While the turbulent dynamics are still confined to $\kappa > \kappa_\beta$, the disturbance field penetrates to the largest scales of motion;

(ii) The larger scales exhibit a tendency to meridional rather than zonal anisotropy, namely towards $v^2 > u^2$ rather than vice-versa;

(iii) The spectral transfer rate away from an isotropic intermediate-scale source is significantly enhanced by the shear-induced transfer associated with straining by the zonal flow.

These conclusions are all robust to changes in the external parameters (see [6]), as expected on theoretical grounds. Point (iii), and the general importance of the process of shear-induced spectral transfer of disturbance enstrophy, are highlighted in Fig.3c. This shows the induced transfer as a function of ℓ for $k = 2$; one sees a symmetric transfer of enstrophy associated with an up-scale transfer and amplification of energy, much as in Fig.2b.

5. Atmospheric Observations

The aim is now to try to interpret atmospheric observations in the light of what has been discussed. This analysis is discussed in greater detail in [4]. The observations used are global FGGE data from January 1979 (see [3] and [4]), with samples taken twice daily; stationary in this case thus refers to the monthly mean. The diagnostics are compiled on each pressure level and averaged over the 12 levels. The total spherical harmonic index n is analogous to κ in planar geometry, and the resolution extends to $n = 40$.

First the kinetic energy and enstrophy conversion terms are shown as functions of meridional wavenumber ($n-m$ in spherical geometry) for zonal wavenumbers $m = 2$ and $m = 7$ (Fig.4). In the first case one sees a pattern much like that of Figs.2b and 3c, corresponding to a net conversion of kinetic energy from the stationary to the transient flow. In the case of $m = 7$, however, the sense of the conversion is the opposite; the shear-induced enstrophy transfer is entirely to smaller scales, implying a loss of transient energy.

It must be emphasized that the validity of the concept of shear-induced enstrophy transfer presented here is not self-evident: it depends on the separation in scale between the stationary and transient flows, and on the nearly zonal nature of the former. Figures 1a,b show that these conditions are only weakly satisfied in the atmosphere. Nevertheless the transient enstrophy is very nearly conserved along lines of constant m in the stationary-transient interaction; that is, the sum of $n(n+1)\,C_T(n,m)$ over all n for fixed m is very small. For example, in Figs.4b,d the average residual per mode of such a sum is just over 5% of the maximum transfer.

From the interaction term $I(\underset{\sim}{\kappa})$ of (2) it is conventional to construct nonlinear spectral fluxes of energy and enstrophy following a projection onto a single spectral index. In spherical geometry the spectrum is of course discrete; the

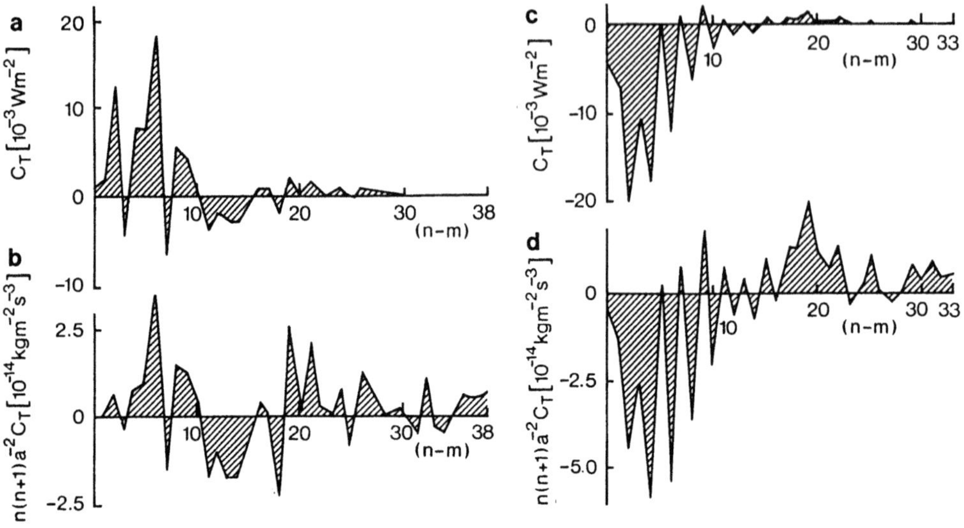

Figure 4. Atmospheric stationary-transient conversion terms: (a) of energy, $C_T(n;m=2)$, and (b) of enstrophy, $n(n+1)a^{-2}C_T(n;m=2)$, for fixed $m = 2$ as functions of $n-m$; (c) of energy, and (d) of enstrophy, for fixed $m = 7$. From [4]

energy and enstrophy fluxes as functions of n are defined respectively by

$$F(n) = -a^{-1} \sum_{\nu=0}^{n} \sum_{m=-\nu}^{\nu} I(\nu,m) \quad \text{and} \quad H(n) = -a^{-3} \sum_{\nu=0}^{n} \sum_{m=-\nu}^{\nu} \nu(\nu+1) I(\nu,m), \quad (6)$$

a being the radius of the earth. $F(n)$ and $H(n)$ are given by the solid curves in Fig.5. Since a positive value of the flux denotes net transfer to smaller scales, Fig.5 shows energy and enstrophy cascades respectively to larger and to smaller scales of motion, in accord with classical 2-D turbulence theory ([1]; [7]; [8]).

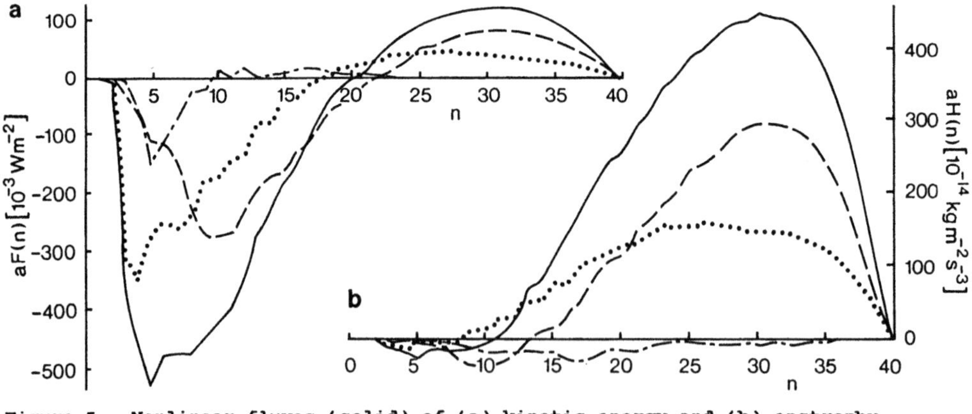

Figure 5. Nonlinear fluxes (solid) of (a) kinetic energy and (b) enstrophy, decomposed respectively into pure stationary (dash-dot), $F_S(n)$ and $H_S(n)$, pure transient (dashed), $F_T(n)$ and $H_T(n)$, and mixed stationary-transient (dotted), $F_{ST}(n)$ and $H_{ST}(n)$, components. From [4]

Our purpose here is to understand the nature of these apparent cascades. Since $I(n,m)$ is a triple correlation term, its time average can be decomposed into three distinct parts; schematically one has

$$\overline{I(\psi,\psi,\psi)} = I(\bar{\psi},\bar{\psi},\bar{\psi}) + \overline{I(\psi',\psi',\psi')} + \overline{I(\bar{\psi},\psi',\psi')} \,, \tag{7}$$

the last term including also the cyclic permutations. Equation (7) takes the precise form

$$\bar{I}(n,m) = I_S(n,m) + I_T(n,m) + I_{ST}(n,m) \,, \tag{8}$$

where $I_{ST}(n,m) \equiv C_S(n,m) + C_T(n,m)$ can be interpreted as the transfer induced by the stationary-transient interaction.

Figure 5 exhibits the breakdown of the fluxes (6) into their components corresponding to (8). The pure transient fluxes $F_T(n)$ and $H_T(n)$ are arguably explainable by 2-D turbulence theory. However the mixed fluxes $F_{ST}(n)$ and $H_{ST}(n)$ cannot be so explained; rather, they reflect the nature of the stationary-transient interaction. From Fig.6a, showing the net energetic interaction for each m, it is evident that this interaction takes one form (like Figs.4a,b) for small m and another (like Figs.4c,d) for large m; in fact, the net effect is dominated by the latter. Hence one may characterize the stationary-transient interaction as in Figs.6b,c: transient enstrophy is transferred to high n along lines of constant m, by the action of the shear in the stationary flow (Fig.6b); this implies a decrease in transient energy, which shows up as an induced energy flux $F_{ST}(n)$ to large scales (Fig.6c). However this flux involves highly nonlocal interactions, and should not be thought of as a classical reverse energy cascade.

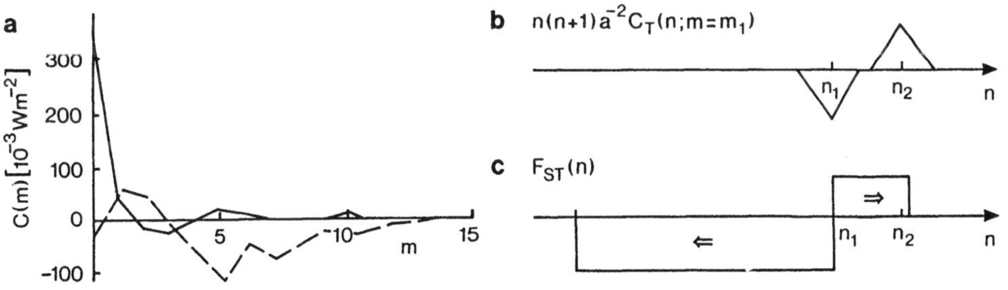

Figure 6. (a) Stationary-transient conversion terms $C_S(m)$ (solid) and $C_T(m)$ (dashed) as functions of zonal wavenumber m. (b) Schematic of induced down-scale enstrophy transfer, with (c) the associated energy flux. From [4]

6. Numerical Simulation: Multimode Jet

Compared to the complexity of the real atmosphere, the numerical simulation presented in section 4 is highly idealized. Here another simulation is presented which, though still idealized, relaxes one of the simplifications of the previous case and produces more realistic results.

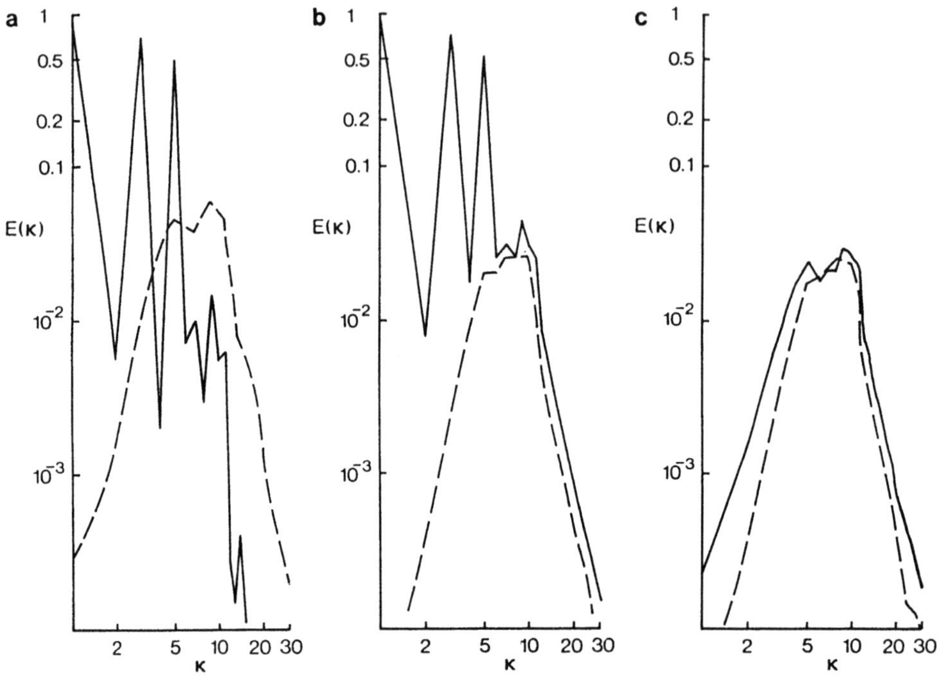

Figure 7. Spectra of (a) stationary (solid) and transient (dashed) energy; (b) zonal (solid) and meridional (dashed) energy; and (c) zonal (solid) and meridional (dashed) components of transient energy; as functions of total wavenumber κ, for the numerical simulation described in section 6

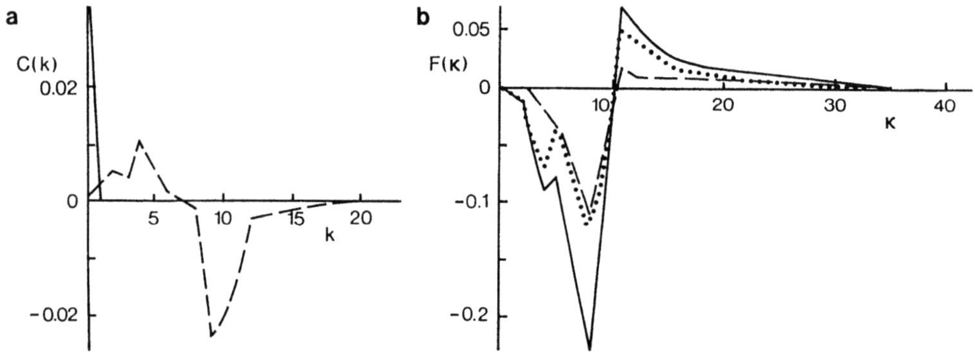

Figure 8. (a) Stationary-transient conversion terms $C_S(k)$ (solid) and $C_T(k)$ (dashed); and (b) energy flux $F(\kappa)$ (solid), decomposed into pure transient (dashed), $F_T(\kappa)$, and mixed stationary-transient (dotted), $F_{ST}(\kappa)$, components; corresponding to Fig.7. $F_S(\kappa)$ is indistinguishable from zero

Everything from section 4 carries over except that the basic flow is now given by $U(y) = -2\cos y - \sqrt{2}(\cos 3y + \cos 5y)$, and the relaxation tendency is altered accordingly; this choice is made simply because the atmospheric stationary flow has strong peaks at wavenumbers 1, 3 and 5. The stationary and transient spectra obtained in the model are shown in Fig.7, and may be compared with Fig.1; the similarity is striking, to say the least. The similarity extends to the stationary-transient interactions (Fig.8a) and to the nonlinear fluxes (Fig.8b). It is remarkable that such a simple model is able to capture so much of the nonlinear dynamics of the atmosphere. Of course one would not want to make quantitative comparisons; among other things, the numerical model is barotropic and planar. Nevertheless the processes studied here do seem to be important to the understanding of stationary-transient interaction in the atmosphere.

7. An Enstrophy-Cascading Inertial Subrange in the Atmosphere?

It has sometimes been argued (e.g.[2]) that the atmosphere might possess an enstrophy-cascading inertial subrange in the smaller synoptic scales. Such a theoretical solution of the homogeneous and isotropic form of (2) (see [7] and [8]) would be characterized by a constant positive flux of enstrophy, a vanishing flux of energy, and an n^{-3} power law dependence for the energy, within the relevant subrange of the spectrum. While the spectra are not inconsistent with this, the evidence from the fluxes has so far been considered inconclusive [3].

The work presented here suggests that the effects of spatial inhomogeneity are likely to reduce the possibility of an inertial-subrange solution, as in addition to the pure transient (read turbulent) interactions considered in the homogeneous theory, there will be systematic, phase-dependent straining effects associated with the stationary-transient interaction. To test for an inertial subrange, the fluxes of kinetic energy and of enstrophy were computed under spectral truncations of N = 20, 32, and 40 (Fig.9). If an enstrophy-cascading inertial subrange did exist then one would presumably expect to see the maximum enstrophy flux levelling off as the resolution increased; however it increases dramatically! Of course there is little reason to believe that the smaller synoptic scales are unforced, as assumed by the theoretical solution.

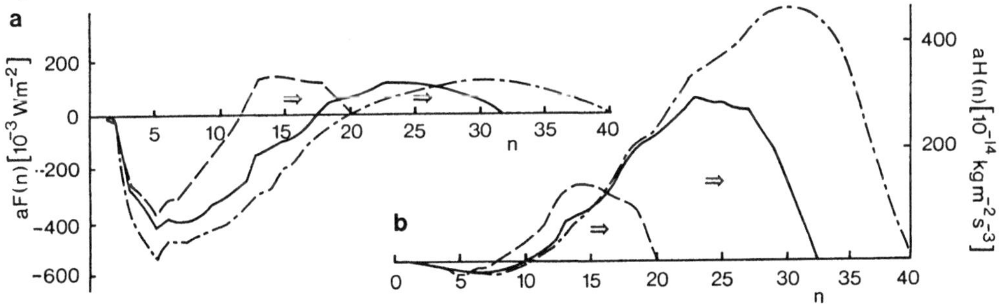

Figure 9. Atmospheric nonlinear fluxes of (a) kinetic energy and (b) enstrophy, for spectral truncations of N = 20, N = 32, and N = 40

It may be noted that changes in resolution have non-negligible effects on the maximum reverse energy flux; for example, when interactions with waves $32 < n \leqslant 40$ are included, the maximum reverse flux at n = 5 increases by over 25% (Fig.9a). This reflects the importance of nonlocal interactions, and suggests that any quantitative theory of the atmospheric stationary flow must account for the effects of a wide range of transient scales of motion.

References

1. R. Fjørtoft: Tellus 5, 225 (1953)
2. C.E. Leith: J.Atmos.Sci. 28, 145 (1971)
3. G.J. Boer, T.G. Shepherd: J.Atmos.Sci. 40, 164 (1983)
4. T.G. Shepherd: J.Atmos.Sci., submitted
5. J. Pedlosky: Geophysical Fluid Dynamics (Springer-Verlag, New York, 1979)
6. T.G. Shepherd: J.Fluid Mech., to appear
7. R.H. Kraichnan: Phys.Fluids 10, 1417 (1967)
8. C.E. Leith: Phys.Fluids 11, 671 (1968)
9. P.B. Rhines: J.Fluid Mech. 69, 417 (1975)

Laminarization by a Strong Magnetic Field

C. Bardos[1], *C. Sulem*[2], *and P.L. Sulem*[3]

[1]Ecole Normale Supérieure, Paris and Université Paris XIII, Paris, France
[2]C.N.R.S., Ecole Normale Supérieure, Paris and
Ben-Gurion University of the Negev, Beer-Sheva 84105, Israel
[3]C.N.R.S., Observatoire de Nice and Tel-Aviv University, Tel-Aviv 69978, Israel

1. Introduction.

A fundamental question in mathematical physics is whether the problem is well posed. In other words, does it have a unique solution? Concerning the dynamics of incompressible fluids, the answer is very incomplete, especially in three-dimensions. For example, one does not know whether the 3D Navier-Stokes equation is well posed for all time. Only the existence of (possibly non unique) weak solutions, i.e. solutions in the sense of distribution, is established[1].

In two dimensions, the problem is well posed for all time (at least in bounded domain or for solutions decaying at infinity) even at zero viscosity. In the latter case, the result is based on the conservation of the vorticity (curl of velocity)[2].

In the case of electro-conducting fluids governed by the MHD equations

$$\partial_t v + v.\nabla v - \nu \Delta v = -\nabla p + B.\nabla B$$
$$\partial_t B + v.\nabla B - \lambda \Delta B = B.\nabla v$$
$$\nabla.v = 0 \; ; \quad \nabla.B = 0 \, ,$$

the above results are also valid, except the well-posedness of the non dissipative two-dimensional flow, because the Lorentz force breaks the vorticity conservation [3]. Numerical simulations of this flow, performed with periodic boundary conditions, indicate nevertheless that the solutions remain globally smooth. When neutral points of X-type are present, the velocity gradients grow roughly exponentially in time and the width of analyticity strip, which measures of the smallest appreciably excited scale, decreases exponentially. In contrast, when the magnetic field does not vanish, the width of the analyticity is bounded from below[4]. The inhibition of small scale formation is associated to the presence of Alfven waves. Defining the Elsasser variables

$$Z^+ = v + b$$
$$Z^- = v - b$$

the MHD equations are written

$$\partial_t Z^+ + Z^-.\nabla Z^+ = -\nabla p$$
$$\partial_t Z^- + Z^+.\nabla Z^- = -\nabla p$$
$$\nabla.Z^+ = 0\ ;\ \nabla.Z^- = 0$$

Separating in $Z^\pm$, the contribution of a background uniform magnetic field B_0, from that of fluctuations $z^\pm$ and linearizing the equations, one gets $z^\pm$–Alfven waves which propagate in opposite directions along the magnetic field lines, with a velocity B_0. As a consequence, the interaction time of the $z^\pm$–eddies is reduced by the background magnetic field [5]. Proving that this may lead to the regularity of the flow as observed in the numerical simulations is still an open problem. The phenomenon has however been illustrated in the case of a simple one-dimensional model which is an extension of the Burgers equation to conducting fluids[6]. In that case, the existence of smooth solutions is obtained when initially, the absolute value of the magnetic field is larger than one half the excursion of the initial velocity. This condition excludes neutral points.

Rigorous results concerning the the existence and the long time behaviour of conducting flows are established in [7]. They concern localized solutions in the entire two or three dimensional space in the presence of a strong enough magnetic field. They are briefly reported in Section 2.

2. Effect of a strong magnetic field on a localized flow.

Let us consider initial fluctuations $z_0^\pm$ which are localized ,i.e. which decay fast enough at infinity, for example

$$z_0^\pm(x) = \varphi^\pm(x)/(1 + |x|^2)$$

with $\varphi^\pm(x)$ uniformly bounded.

Two families of characteristics are assiociated to the MHD equations

$$\dot{x}^-(t) = Z^-(x^-(t),t) \qquad \dot{x}^+(t) = Z^+(x^+(t),t)$$
$$x^-(0) = a^- \qquad x^+(0) = a^+$$

We define the operators A^+ and A^- by

$$A^+z^+(x,t) = f(x+B_0t)\ z^+(x,t)$$
$$A^-z^-(x,t) = f(x-B_0t)\ z^-(x,t)$$

where, in the case of the above initial conditions

$$f(x)= 1 + |x|^2.$$

Similar weights were introduced by Klainerman[8] in the context of non-linear wave equations.

To justify the above definitions, we remind that, in the linear problem, $z^{\pm}$ are only transported along the characteristics $x^{\pm}(t) = \pm B_0$. It follows that, the quantities $A^{\pm} z_{/}^{\pm}(x,t)$ associated to the linear solutions $z_{/}^{\pm}$ are bounded uniformly in time. For the nonlinear problem, we have

$$A^+z^+(x^-(t),t) = A^+z_0^+(a^-) - \int_0^t f(x^-(\tau)+B_0\tau)\, \nabla p(x^-(\tau),\tau)\, d\tau \; +$$
$$\int_0^t \frac{A^-z^-(x^-(\tau),\tau)}{f(x^-(\tau)-B_0\tau)} \; \frac{\nabla f(x^-(\tau)+B_0\tau)}{f(x^-(\tau)+B_0\tau)}\, A^+z^+(x^+(\tau),\tau)\, d\tau$$

and ($j^+=$ Curl z^+)

$$A^+j^+(x^-(t),t) = A^+j_0^+(a^-) - \int_0^t A^-\nabla z^- \wedge A^+\partial_k z^+(x^-(\tau),\tau)/f(x^-(\tau)-B_0\tau)\, d\tau \; +$$
$$\int_0^t \frac{A^-z^-(x^-(\tau),\tau)}{f(x^-(\tau)-B_n\tau)} \; \frac{\nabla f(x^-(\tau)+B_0\tau)}{f(x^-(\tau)+B_0\tau)}\, A^+j^+(x^+(\tau),\tau)\, d\tau$$

Analogous equations hold for $A^-z^-(x^+(t),t)$ and $A^-j^-(x^+(t),t)$.

The method consists of showing that for the nonlinear problem also, $A^{\pm}z^{\pm}$ and $A^{\pm}j^{\pm}$ remain uniformly bounded in $C^{0,\alpha}(\mathbb{R}^n)$ under some smallness hypothesis on the initial fluctuations.

The space $C^{0,\alpha}(\mathbb{R}^n)$ $(0<\alpha<1)$ of (vectorial) functions which are Hölder continuous is equipped with the norm

$$|u|_{0,\alpha} = |u|_0 + \sup_{x \neq y} \{|u(x)-u(y)|/|x-y|^{\alpha}\}$$

where

$$|u|_0 = \sup_x |u(x)|$$

$C^{1,\alpha}(\mathbb{R}^n)$ which is the space of functions which are Hölder continous together with their first derivatives, is equipped with the norm

$$|u|_{1,\alpha} = |u|_{0,\alpha} + \Sigma_i\, |\partial u/\partial x_i|_{0,\alpha}\,.$$

One first establishes the following estimates:

(i) For $z^{\pm}$ satisfying $\nabla . z^{\pm} = 0$, and $j^{\pm} = \text{Curl } z^{\pm}$, we have

$$|A^{\pm}\nabla z^{\pm}|_{0,\alpha} \leqslant C \; (|A^{\pm}z^{\pm}|_{0,\alpha} + |A^{\pm}j^{\pm}|_{0,\alpha}) .$$

(ii)

$$|A^{+}z^{+}(.,t)|_{0,\alpha} \leqslant (|fz_o^{+}|_{0,\alpha} + |fj_o^{+}|_{0,\alpha}) \exp [M^{-}(t)/(2B_o - M^{-}(t))]$$

$$|A^{-}z^{-}(.,t)|_{0,\alpha} \leqslant (|fz_o^{-}|_{0,\alpha} + |fj_o^{-}|_{0,\alpha}) \exp [M^{+}(t)/(2B_o - M^{+}(t))]$$

$$|A^{+}j^{+}(.,t)|_{0,\alpha} \leqslant (|fz_o^{+}|_{0,\alpha} + |fj_o^{+}|_{0,\alpha}) \exp [M^{-}(t)/(2B_o - M^{-}(t))]$$

$$|A^{-}j^{-}(.,t)|_{0,\alpha} \leqslant (|fz_o^{-}|_{0,\alpha} + |fj_o^{-}|_{0,\alpha}) \exp [M^{+}(t)/(2B_o - M^{+}(t))]$$

where

$$M^{\pm}(t) = \sup_{0 \leq \tau \leq t} \; (|A^{\pm}z^{\pm}(.,\tau)|_{0,\alpha} + |A^{\pm}j^{\pm}(.,\tau)|_{0,\alpha}).$$

They lead to the

Proposition: For sufficiently small fluctuations , namely

$$\delta = |fz_o^{+}|_{1,\alpha} + |fz_o^{-}|_{1,\alpha} < \beta \, |B_o| ,$$

the quantities $M^{\pm}(t)$ remain bounded independently of time.

Theorem: Under conditions of the Proposition, there exits for all time a unique solution (z^{+},z^{-}) which is a continuous function of time with values in $C^{1,\alpha}(\mathbb{R}^n)$. This solution is such that $|A^{\pm}z^{\pm}(.,t)|_{0,\alpha} + |A^{\pm}\nabla z^{\pm}(.,t)|_{0,\alpha}$ is uniformly bounded. Moreover, it tends to a solution of the linear problem

$$\partial_t w^{\pm} \mp B_o . \nabla w^{\pm} = 0$$

when t tends to $+\infty$.

References

[1] Leray, J., Acta Mathematica, 63,1934, 193.

[2] Wolibner, W., Math. Z., 37, 1933.

[3] Sulem, C., C.R.Ac.Sc. Paris, 285, Série A, 1977,365.

[4] Frisch, U., Pouquet, A., Sulem, P.L., Meneguzzi, M., J. Méca Théor. Appl., Special issue on two dimensional turbulence, 1983, 191.

[5] Kraichnan, R.H., Phys. Fluids 8, 1965, 1365.

[6] Sulem,C., Fournier,J.D., Frisch, U., Sulem, P.L., C.R.Ac.Sc., Paris, A288, 1979, 571.

[7] Bardos, C., Sulem, C., Sulem, P.L., Long time dynamics of a conductive fluid in the presence of a strong magnetic field, Submitted to Trans. Amer. Math. Soc.

[8] Klainerman, S., Comm. Pure Appl. Math., 36, 1983, 325.

Transition of Magnetohydrodynamical Waves in the Solar Atmosphere

L. Nocera

IFAM-CNR, Via del Giardino 7, I-56100 Pisa, Italy

Transition to irregular behaviour of Magnetohydrodynamical (MHD) waves is investigated using a simple four-mode dynamical system worked out from the basic MHD equations by means of a Fourier and a multiple-scale analysis. In the initial stage of the transition, when most of the energy in concentrated in the low-wavenumber Fourier component, the system behaves like a van dr Pol oscillator, exhibiting saddlenode bifurcations. Hopf bifurcations are found numerically when the low- and high-wavenumber components have comparable magnitude. Two Hopf bifurcations leading to a two-D torus are followed by bifurcation to a nonperiodic attractor.

1. Introduction

The relevance of the turbulent behaviour of plasmas for the dynamics of the atmosphere and convective layer of the Sun has been pointed out in the review by Moffatt in this same volume. In more detail, some authors have suggested that turbulent features of the velocity field are observed in the solar photosphere (e. g. MATTIG and NESIS [1]) and is also desirable, from a theoretical standpoint, as a possible coronal heating mechanism (e.g. HOLLWEG [2]).

MHD waves are sustained against dissipation by some sort of forcing acting at photospheric level and which, ultimately, takes energy out of the convective layer: shear motions at the photospheric boundary generate perturbations which then propagate away as MHD waves. The fast and slow magnetoacustic modes are soon reflected and filtered out by the transition layer at the chromosphere-corona boundary, so that we may assume that only the Alfvén mode makes a substantial contribution to the energy flux into the solar corona (e.g. HOLLWEG [2]). Thus, away from the photospheric boundary, we may think of a "sea" of body Alfvén waves pervading the whole coronal plasma and providing a body acceleration field denoted by $\underline{g}$ in (2.2). For the sake of simplicity, we assume these waves to be linearly polarised (say along y) and propagating along z.

At this stage, advective, inductive and anisotropic viscous nonlinear effects couple the three MHD modes again. This interaction is responsible for a turbulent cascade which keeps transferring energy from low-to high-wavenumber components of the Fourier decomposition of, say, the velocity field in the Alfvén wave and leads to an irregular temporal behaviour possibly in one of the popular ways described by ECKMANN [3]. We refer to these two features as "transition" of MHD waves.

In this work, we study transition of MHD waves by means of a simple four-dimensional dynamical system.

2. The Dynamical System.

Following LIFSHITZ and PITAEVSKII[4], we describe the solar coronal plasma by means of the basic equations

$$\partial\rho_{\alpha} / \partial t = - \operatorname{div} (\rho \underline{v}) \tag{2.1}$$

$$\rho\{\partial v_{\alpha} / \partial t + (\underline{v}.\nabla) v_{\alpha}\} + \partial p/\partial x_{\alpha}$$

$$- [(\operatorname{rot} \underline{B}) \wedge \underline{B}]_{\alpha} / 4\pi - \partial\sigma'_{\alpha\beta} / \partial x_{\beta} = g_{\alpha} \tag{2.2}$$

$$\operatorname{div} \underline{B} = 0 \tag{2.3}$$

$$\rho T \; ds/dt = - \sigma_{\alpha\beta} \; \partial v_{\alpha} / \partial x_{\beta} \tag{2.4}$$

where the following definitions

ρ = mass density
p = gas pressure
T = gas temperature
s = entropy per unit mass
$\underline{v}$ = fluid velocity
$\underline{B}$ = magnetic field
σ = viscous tensor
$\underline{g}$ = body wave-field acceleration

Equations above are now analysed by a multiple-scale (averaging) technique starting from the assumptions

i. Perturbations (denoted by δ)entrained by the external wave-field are small: $\epsilon \simeq [\delta v_y / v_A]^2 \ll 1$, $v_A = B_0 / (4\pi\rho_0)^{\frac{1}{2}}$ being the Alfvén speed and "o" denoting equilibrium values.

ii. the plasma is locally homogeneous

iii. the plasma is highly magnetized, i.e. the proton-proton collision time τ_p and the proton gyrofrequency ω_p are such that $\omega_p\tau_p \ll 1$

and from the following perturbation scheme

$$\delta v_y = \epsilon^{\frac{1}{2}} v_A\{V_1 (\tau)\exp (i\theta)+V_3(\tau)\exp(3i\theta)\}+ \text{c.c.} \tag{2.5}$$

Here, $\theta = (z-t)/2\pi$, $\tau = \epsilon t$; space and time coordinates are referred to the wavelenght and the period of the linear Alfvén wave respectively

Perturbation analysis, using similar expansions for δv_x, δv_z, δp, $\delta\rho$, and δs and the Bogoliubov-Mitropolskii theorem leads (NOCERA and PRIEST [5]) to the dynamical system

$$\frac{dV_1}{d\tau} = (p|V_1|^2 + q|V_3|^2)\,V_1 + r(V_1^\star)^2 V_3 + g_1(\tau)/2, \qquad (2.6)$$

$$\frac{dV_3}{d\tau} = (s|V_1|^2 + \sigma|V_3|^2)V_3 + \phi\, V_1^3 + g_3(\tau)/2, \qquad (2.7)$$

where g_1 and g_3 are the Fourier amplitudes of the y-component of the external forcing $\underline{g}$ with respect to the eigenmodes $\exp(i\theta)$ and $\exp(3i\theta)$; their dependence on the slow time τ accounts for a modulation of the forcing due to supergranular motions in the solar convective region;p, q, r, s, σ, ϕ, are functions of the two parameters

$$Re_0 = v_A\,\ell\,\rho_0/\eta_0,\ \eta_0 \cong 10^{-16}[T(°K)]^{5/2}\ \ g/(cm.s) \qquad (2.8)$$

and

$$\beta = (v_S/v_A)^2, \quad v_S = (\Gamma\, p_0/\rho_0)^{\frac{1}{2}}, \qquad (2.9)$$

Γ being the adiabatic exponent.

3. Local Bifurcation for $|V_3| \ll |V_1|$ and $g_1 = G_1 \exp(i\omega_0\tau)$, $g_3 = 0$

NOCERA et al. [6] showed that the anisotropic nature of viscous dissipation in magnetized plasmas gives rise to a cubic wave damping, which in fact, we retrieve in (2.6) - (2.7). It is not surprising, therefore, that, under suitable rescaling, the forced system (2.6) - (2.7) reduces to the van der Pol equations, describing forced, cubically damped oscillations

$$\dot{\rho} = (\mu + \lambda\rho^2)\rho - F\sin\psi, \quad \rho\dot{\psi} = (\omega - \rho^2)\,\rho - F\cos\psi, \qquad (3.1)$$

where

$$\rho = |V_3|, \qquad \psi = \arg(V_3) - 3i\omega_0\tau, \qquad (3.2)$$

ω_0 is the supergnanular forcing frequency and μ, ω, λ, F are again functions of β and Re_0. The relevant bifurcations for the system (3.1) are shown in Figure 1. Bifurcations occur when (F, ω) is on the curve B-0-A. 1B. Phase portraits in the different regions of Fig. 1A: notice the seddlenodes on the bifurcation sets 0-A and 0-B. 3C. Bifurcation diagram:ρ_0 is the modulus of the fixed point V_3 = const.; $\omega = -0.1$. Turning points correspond to saddlenode bifurcations. Notice the appearance of an hysteresys cycle (broken lines).

4. Numerical Results for $|V_3| \approx |V_1|$ and $g_1(\tau) = 4\exp(5i\tau)$, $g_3(\tau) = 0$.

Bifurcations of the full system (2.6) - (2.7) are studied numerically, using a predictor-corrector algorithm with the trapezoidal rule as a corrector, and with an absolute error test with an accurancy of 10^{-6}. The trapezoidal rule, which has the smallest error constant amongst the A-stable algorithms was iterated to conve-

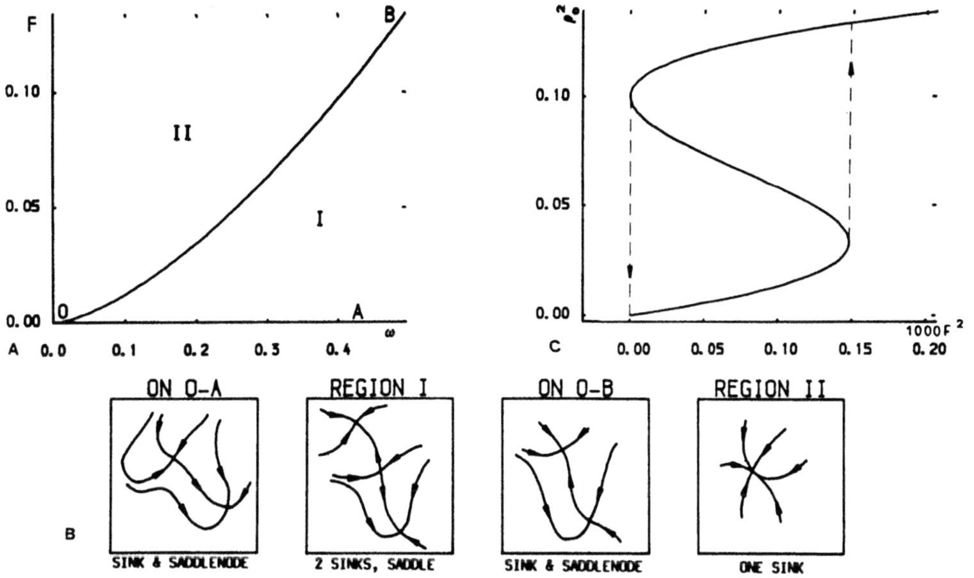

Fig. 1. 1A. Saddlenode bifurcations for (2.6)-(2.7) in the limit $|V_3| \ll |V_1|$. $F = |\phi|/|\mathrm{Im}\,\sigma|$, $\omega = (\mathrm{Im}\ s - 3\omega_0/|V_1|^2)/|\ \mathrm{Im}\sigma\ |$.

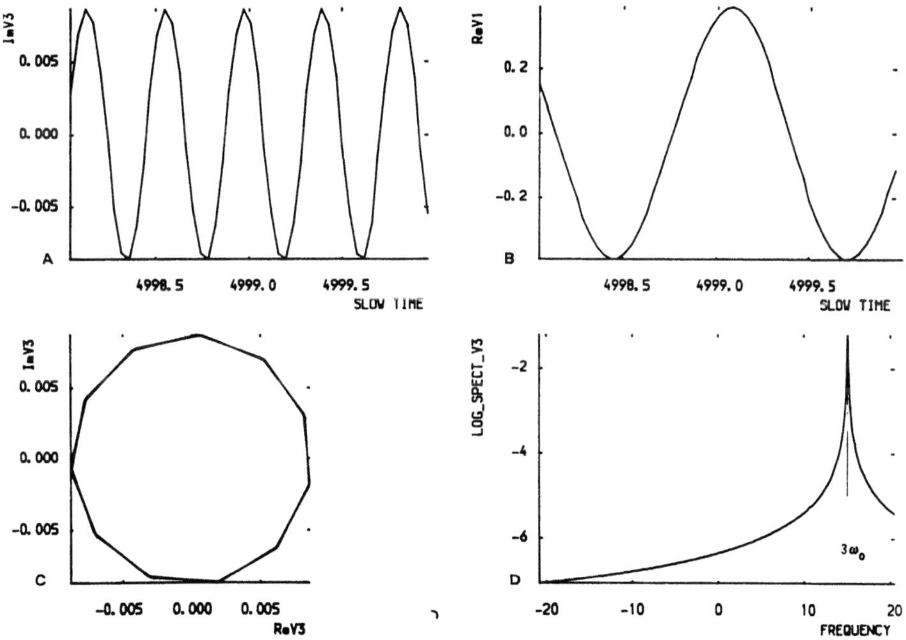

Fig. 2. Numerical solution of (2.6) - (2.7) and its temporal spectrum. $\beta = 0.7$, $Re_0 = 128$: $\omega_0 = 5$ is the supergranular forcing frequency. 2A. The imaginary part of V_3 vs. the slow time $\tau = \epsilon t$. 2B. Same for the real part of V_1. 2C. Projection of the phase trajectory onto the ReV_3-ImV_3 plane. Temporal spectrum of V_3 about $\omega = 0$. Notice the capture of the orbit by a limit cycle.

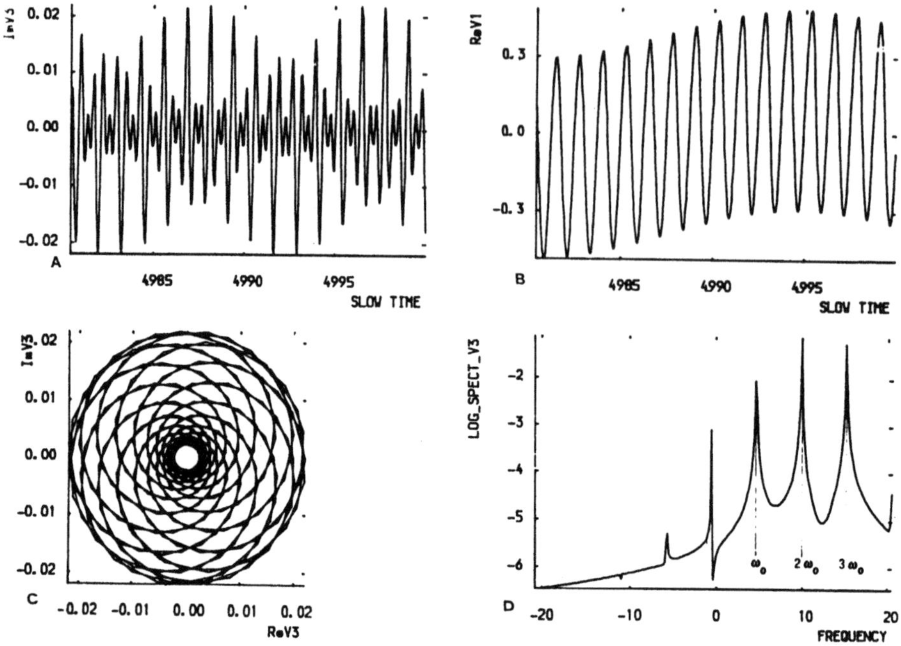

Fig.3. Same as Fig. 2 but for Re_0 = 2048. Two rationally related frequencies have appeared as a result of Hopf bifurcations. A third not rationally related frequency ($\omega \cong 0$) is also present, making the orbit in C no longer periodic. Energy is being transferred from the forced component V_1 to the unforced component V_3 which has an initial amplitude $|V3(\tau = 0)| = 10^{-4}$ (energy cascade).

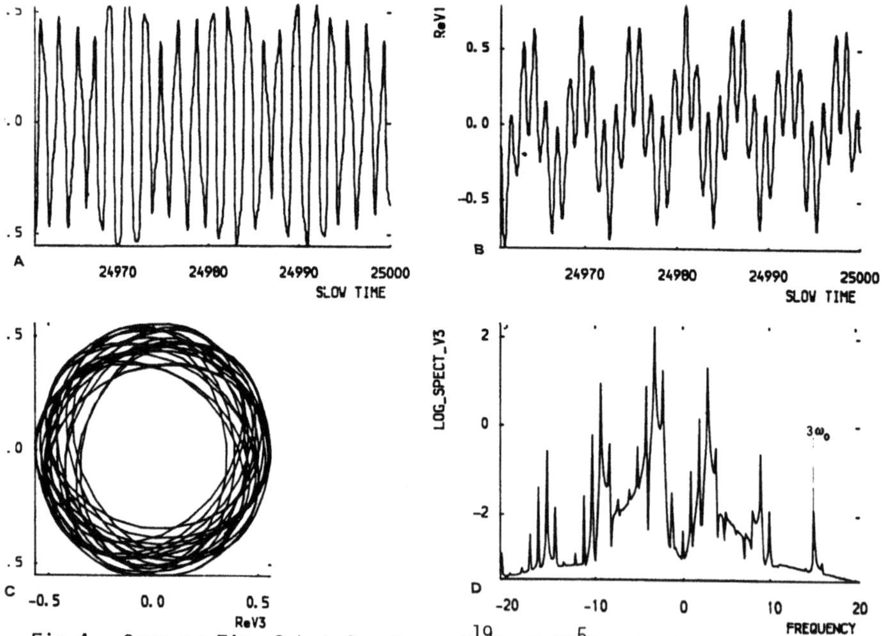

Fig.4. Same as Fig. 2 but for $Re_0 = 2^{19} (\cong 5.10^5)$. Notice the appearance of broad-band features.

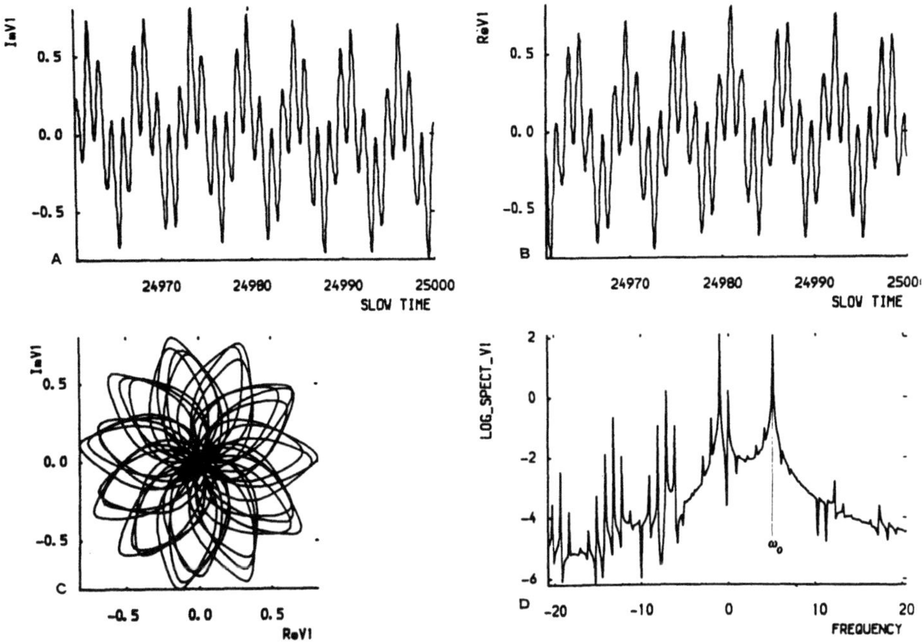

Fig. 5. Same as in Fig. 4 but for the real and imaginary part of V_1. Notice again the appearance of broad bands.

gence. Numerical experiments of Figs. 2-3 were repeated using different time steps, whereas experiments of figs 4-5 were prohibitively long to advice any further tests. We allowed integration to run for a long enough time for convergence on the attractor to be obtained. The last part of the evolution is plotted in Figs. 2-5 for several values of the Reynolds number Re_0, keeping $\beta = 0.7$.

5. Conclusions

We reproduced two basic features of transition in solar coronal plasmas described by the MHD approximation, by means of a simple dynamical system. These are:

i a nonperiodic behaviour apparent from the Fourier spectra of the velocity field (Figs. 4-5). This is attained through a sequence of two Hopf bifurcations and a third bifurcation to an apparently nonperiodic attractor (Fig.3). This suggests that the 'Ruelle-Takens scenario' could be at work in our case.

ii Energy transfer from a forced low-wavenumber component to an unforced high-wavenumber component which has initially a small amplitude (Figs.2-5). We argue that this may represent the first stage of a turbulent cascade.

As a main indication stemming out of our study, we may conclude that nonlinear propagation of MHD waves in plasmas is structurally unstable. Structural instability leads to a nonperiodic behaviour and the investigation of the nature of the attractor of Figs. 4-5 would be higly worthwhile.

Acknowledgements

The author wishes to thank the staff of the Applied Maths. Department of St.Andrews University, where the main part of this work was carried out.

References

1. W.Mattig, A.Nesis: "Studies of Granular Velocities IV: Statistical Analysis of Granular Dopper Shifts" . Solar Phys. 36, 3-9 (1974).
2. J.V.Hollweg: "Coronal Heating by Waves" In Solar wind V, ed. by E.M.Naugebauer, NASA Conference Publication 2280, p.5 (1983).
3. J.P.Eckmann : "Roads to Turbulence in Dissipative Dynamical Systems" Rev. Mod. Phys. 53 , 643-668 (1981).
4. E.M.Lifshiftz, L.P.Pitaevskii : "Physical Kinetics" (Pergamon Press Oxford 1981).
5. L.Nocera, E.R.Priest : "Onset of an Energy cascade and Nonperiodic Behaviour in Nonlinear Propagation of MHD Waves" Submitted to Geophys. and Astrophys. Fluid Dynamics (1986) . In press.
6. L.Nocera, E.R.Priest, J.V. Hollweg : "Nonlinear Development of Phase-mixed Alfvén Waves" Geophys. and Astrophys. Fluid Dynamics, 35, 111-129 (1986)

Sensitivity of Two-Dimensional MHD Turbulence to Alfven Waves and Correlations

P.L. Sulem[1,2], *A. Pouquet*[1], *and M. Meneguzzi*[3]

[1]C.N.R.S., Observatoire de Nice, B.P. 139, F-06003 Nice Cedex, France
[2]School of Mathematical Sciences, Tel Aviv University, Tel-Aviv 69978, Israel
[3]Service d'Astrophysique, CEN-Saclay and C.N.R.S., F-91191 Gif-sur-Yvette Cedex, France

Direct numerical simulations of decaying, two-dimensional incompressible MHD turbulence are presented. The exchanges of kinetic and magnetic energy lead to their oscillating with an amplitude and a characteristic frequency which are larger when the Alfvenic time is shorter. These oscillations together with the energy cascade to small scale are sensibly reduced when the velocity and magnetic field are strongly correlated. Quantitative results are also presented on the influence of the rate of correlation on the inertial range spectral exponents .

1. Introduction

Planar incompressible conducting fluids governed by the MHD equations:

$$\begin{aligned} &\partial_t \mathbf{v} + \mathbf{v}.\nabla\mathbf{v} = -\nabla p + \mathbf{b}.\nabla\mathbf{b} + \nu\, \nabla^2\mathbf{v} \\ &\partial_t \mathbf{b} + \mathbf{v}.\nabla\mathbf{b} = \mathbf{b}.\nabla\mathbf{v} + \lambda\, \nabla^2\mathbf{b} \qquad (1) \\ &\nabla.\mathbf{v} = 0 \qquad \nabla.\mathbf{b} = 0 \end{aligned}$$

have often been used in modelling various problems in plasma physics , astrophysics , geophysics , and controlled fusion . In eq. (1), the magnetic field **b** has been dimensionalized to an Alfven velocity and the magnetic pressure has been included to the total pressure p. In this paper, the magnetic Prandtl number ($P_m=\nu/\lambda$) is taken equal to unity. From the velocity and the magnetic field, one defines the following quadratic quantities: the kinetic energy $E^V=1/2\langle v^2\rangle$, the magnetic energy $E^M=1/2\langle b^2\rangle$, the total energy $E^T=E^V+E^M$, and the cross-correlation $E^C=1/2\langle \mathbf{v}.\mathbf{b}\rangle$, where $\langle\,\rangle$ denotes spatial averaging. The total energy and the cross-correlation are invariant in the non-dissipative limit, together with, in dimension two, the squared magnetic potential $E^A=1/2\langle a^2\rangle$, where **b**=Curl a. It has been shown [1] that the long time dynamics is often governed by the relative importance of E^C and E^A. Two simple dimensionless parameters can be associated with the initial conditions, the rate of correlation $\rho = E^C/E^T$ and the energy ratio $\rho' = E^M/E^V$. In this paper we start an investigation of the influence of these parameters on the dynamics of the flow. The method we use is that of direct numerical integration of the MHD equations. In order to stress the internal dynamics, instead of effects due to special geometries and external constraints, we have used periodic boundary conditions. In this case, spectral methods, based on Fourier series expansions are particularly useful since they are both economical (thanks to extensive vectorization and machine langage FFT routines), and very precise: for analytic solutions, the truncation error is exponentially small. The temporal scheme is second order (periodically stabilized) leap-frog for the non-linear terms and Crank-Nicolson for the dissipative terms.

2. Kinetic and magnetic energy exchanges.

In contrast with a uniform velocity field, a uniform magnetic field has significant dynamical effects It produces Alfvén waves which are known to inhibit small scale formation [2], [3]. Alfvén waves are also responsible for exchanges of energy between velocity and magnetic field leading to oscillations of E^V and E^M which are particularly regular in the absence of neutral X-points [4]. This effect is less visible when neutral X-points are present, because of the development of small scale excitation. Indeed, not only is the dynamics more chaotic, but the enhanced dissipation they procure drastically dampens such oscillations. However these are recovered when we consider the ratio E^M/E^V [5]. A simple prototype of MHD flow with neutral X-point is

$$\psi_0(x,y) = 2\,(\cos x + \cos y)$$
$$a_0(x,y) = 2s(\cos x+\cos 2y),$$

where ψ is the stream function. When $s=1$, $E^M=E^V=2$, corresponding to the standard Orszag-Tang (O.T.) vortex [6]. We report here on computations in which s takes respectively the values 2 (run A), 1 (run B) and 1/2 (run C). In these experiments, the Alfvenic time is thus changed by an overall factor of four, whereas the large scale eddy turnover time remains constant. All three runs were performed with a resolution of $(256)^2$ points and a

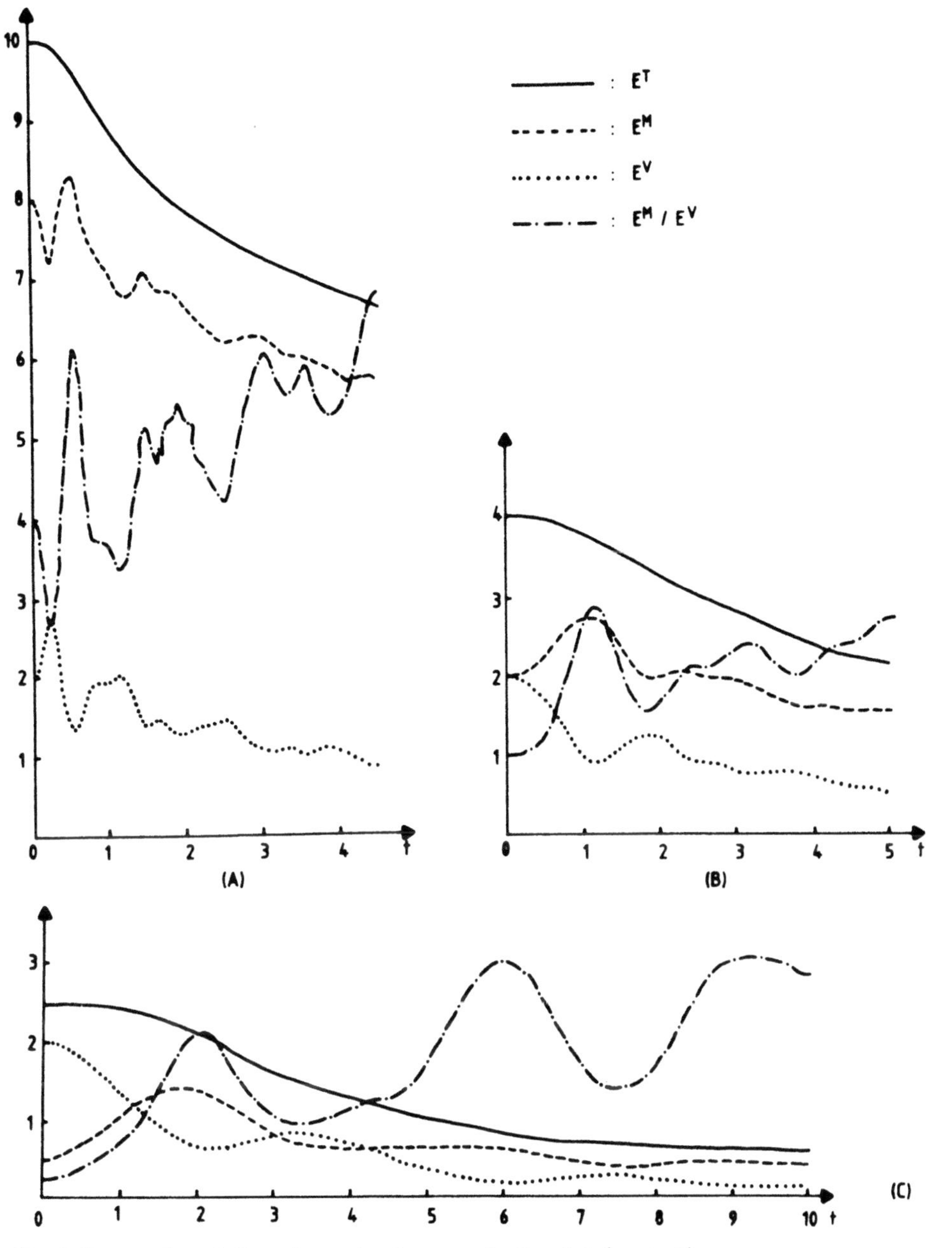

Fig. 1: Temporal variation of energies for runs A, B and C (see text).

viscosity $\nu = 5.\ 10^{-3}$. We also consider random initial conditions. In Fig. 1, we show the time evolution of the total, kinetic and magnetic energies together with the ratio E^M/E^V for runs A, B and C. We observe that the amplitude of the oscillations and their characteristic frequency is

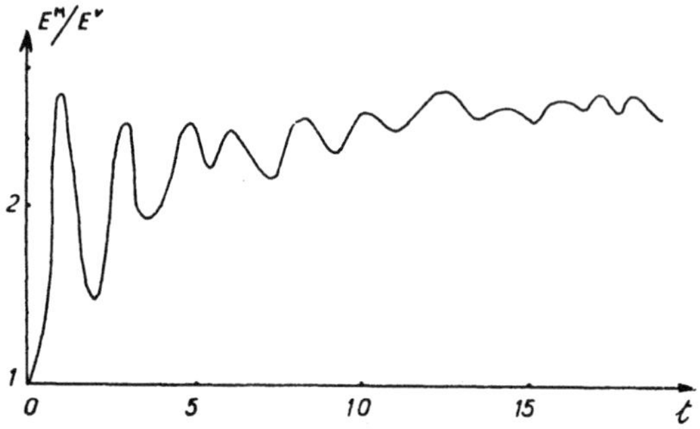

Fig. 2. Persistence of Alfvenic oscillations for long times on the O.T vortex.

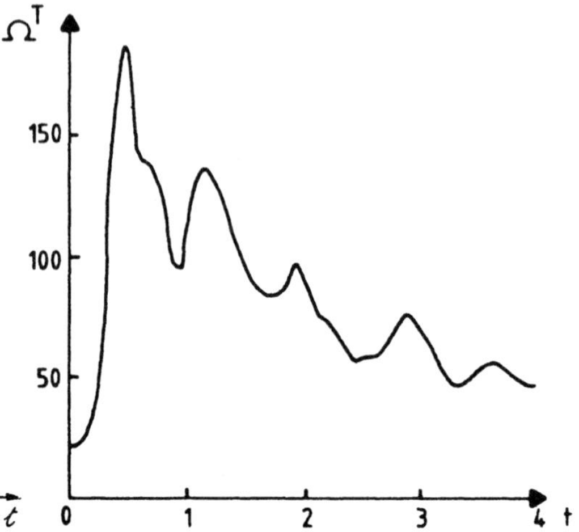

Fig. 3: Total enstrophy vs time for run C with strong initial magnetic field.

enhanced when the Alfven time is shorter. The oscillations of E^M/E^V persist for long times as seen on Fig. 2 for the same initial conditions than in run B and a viscosity $\nu = 1.25\ 10^{-2}$. These oscillations are also visible on the total enstrophy (proportional to total dissipation) of the run A, displayed in Fig. 3 , as well as on its kinetic and magnetic components (not shown). This leads to inflection points in the evolution curves of the energies when plotted versus time. For the run B with the same viscosity, the oscillations of the total enstrophy are less prominent. They are enhanced when the Reynolds number is increased, as seen in Fig. 4 which displays the O.T vortex (s=1) total enstrophy for viscosities $\nu = 1.25\ 10^{-2}$, $\nu = 5\ 10^{-3}$, and $\nu = 2.5\ 10^{-3}$; a resolution of $(512)^2$ has been used for the latter case. When separating the total enstrophy in its kinetic and magnetic components (see Fig.5), we always observe in this type of run that both oscillate, and that the magnetic enstrophy is dominant. This result is for a random initial conditions run with a correlation of 10%.

The oscillations of E^M/E^V shown in Fig. 1 integrate contributions from all the scales of motion. In order to discriminate between them, the time behaviour of the ratio $E^M(k)/E^V(k)$ of the magnetic and kinetic energy spectra for varying wavenumber k was considered in [5]. Although time intervals much longer than those available in our simulations would be required to make the analysis conclusive, we noticed that for sufficiently correlated flows, the oscillations at the lowest wavenumbers of the energy

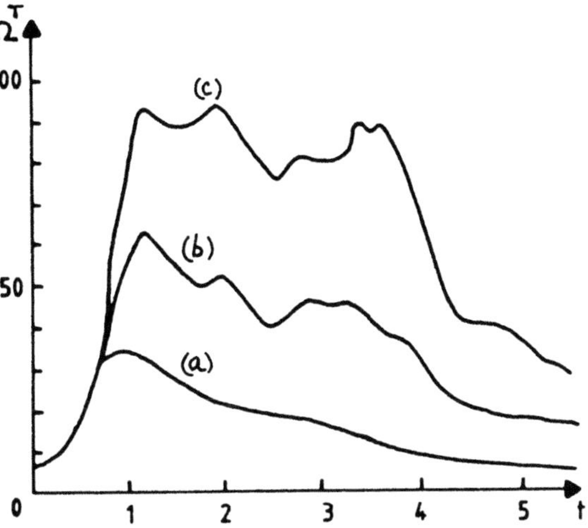

Fig. 4: Total enstrophy vs time for run B for three different viscosities: .0125 (a), .005 (b) and .0025 (c).

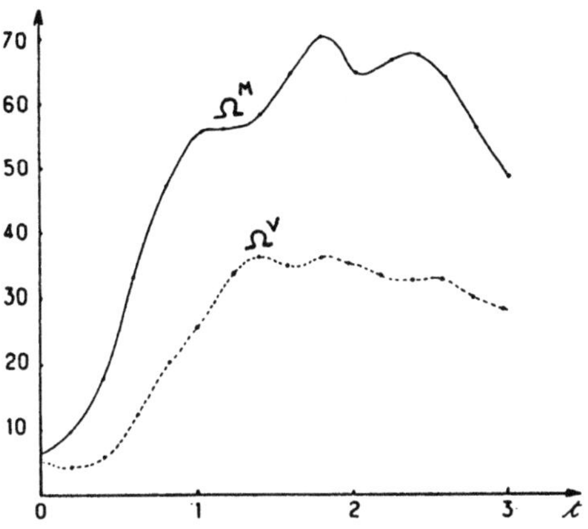

Fig. 5: Kinetic and magnetic enstrophies versus time for a random flow with initial correlation of 10%.

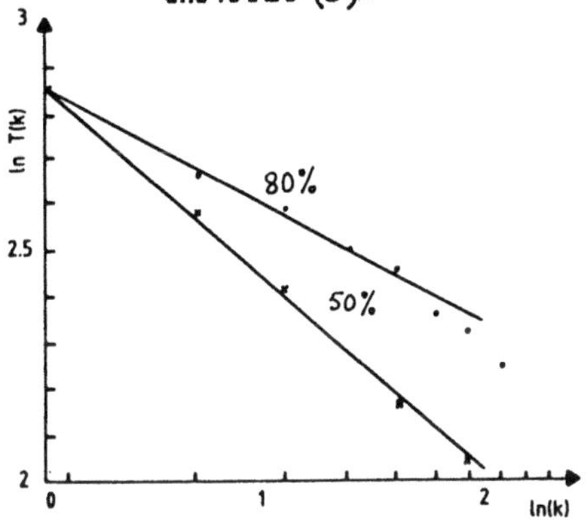

Fig. 6: Log-log plot of the period of oscillation versus wavenumber k of the ratio of magnetic to kinetic energy at wavenumber k in the energy range for 50% and 80% correlated flows.

range look regular, possibly periodic. The oscillations at higher wavenumbers are more complex. If we try to measure a period of oscillation T(k) for $E^M(k)/E^V(k)$ in this range, we observe that it scales approximately like a power of the wave-number. Fig. 6 shows a log-log plot of T(k) versus k for run B and for a 80%-correlated random initial condition run which has similar initial kinetic and magnetic energy. We observe that these curves are sensitive to the correlation. The straight line that fits the experimental points is shallower when the flow is more correlated. The convergence of the two curves at the same point reflects the independence of the characteristic time of oscillation of the total energies from the rate of correlation. If we were to interpret these data as representing the variation of the local eddy turn-over time as a function of scale, giving a dependence $\tau(k) \sim (k^3 E(k))^{-1/2}$, the difference of behaviour between the two runs may be indicating different decay laws for the energy spectra for varying correlations. However, at lower correlation (10%), the dynamics takes a more chaotic aspect and no single frequency can be deduced from the data.

3. Influence of cross-correlation.

As briefly observed in Section 2, another global parameter which influences the dynamics is the rate of correlations between the velocity and the magnetic field. This point has already been discussed in [4], [5] and [7]. We shall here briefly report on computations presented in [5], some of them being obtained with a 512^2 points code. This code makes systematic use of peripheral memory of the Cray 1S with one million words of central memory, and requires 2 seconds of CPU per time step. In order to vary the cross-correlation, without affecting sensibly the kinetic and magnetic energy, random initial conditions have been considered. The main observation is that cross-correlation inhibits both the Alfvénic oscillations and the direct energy cascade, leading to a weaker energy dissipation. Furthermore, cross-correlation influences the energy spectra. Consider for example the spectra $E^+(k)$ and $E^-(k)$ of the Elsässer variables $\mathbf{z}^+ = \mathbf{v}+\mathbf{b}$ and $\mathbf{z}^- = \mathbf{v}-\mathbf{b}$. These spectra are fitted with functions of the form $E(k) = C\, k^{-m}\, e^{-\beta k}$, where β measures the dissipation scale, while m estimates the inertial exponent. We obtain that the inertial exponents m^+ and m^- of the two spectra oscillate in time and depend on the rate of correlation. Quantitative information are presented in Fig. 7 which displays the spectral exponents as a function of initial correlation; the extended spots represent the amplitude of their oscillations and take into account the growth in time of the rate of correlation. Both exponents vary monotonically with the correlation coefficient. For the initial conditions used in the runs for which the results are presented, the exponent m^+ increases with the rate of correlation, whereas m^- decreases. The reverse situation can also occur, depending on the sign of the initial correlation. The values of the exponents are close at low correlation, but significantly different at large correlation where m^+ is nearly 3; this point

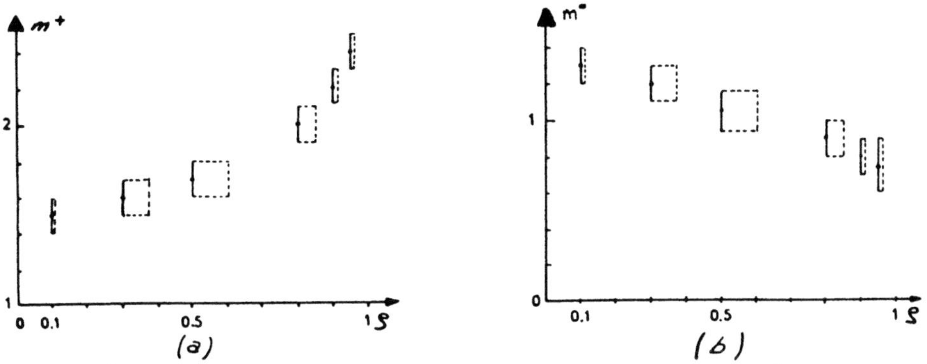

Fig. 7: Inertial range exponents m^+ (a) and m^- (b) of the spectra of the $z^{\pm} = v \pm b$ fields as a function of correlation. Note the strong increase of m^+ for high correlations.

corroborates previous results [4] obtained for highly correlated flows without neutral X-points. Furthermore, their sum is approximately constant, varying from 2.8 to 3.1. This is in agreement with the value 3 obtained with two-point closures and with the phenomenology of [8], which extends the Kraichnan analysis[9] to correlated MHD turbulence.

4. Three-dimensional MHD turbulence.

Carrying over the above numerical experiments to three dimensional turbulence is difficult because of the lack of resolution. We integrated the three-dimensional MHD equations at a resolution of $(64)^3$ points, with $\nu = \mu = 2.\ 10^{-2}$, and initial conditions given in [10]. We observe on Fig. 8 that the oscillations of the energy ratio are also present in three dimensions, although they are rapidly damped because of the moderate Reynolds numbers achieved on present day computers. Concerning the influence of the correlations, we note that the phenomenology presented in [8] is independent of the space dimension and that the closure calculations of [8] were made for the three dimensional case. Indeed, the influence of the dimension is probably less stringent in MHD than in hydrodynamic turbulence because the Lorentz force breaks the vorticity conservation in 2D, allowing an energy cascade to small scales in both 2D and 3D. The self-similarity of the energy spectrum, as postulated by Kolmogorov, thus probably does not apply strictly to the MHD case. Our numerical results, as well as previous closure results, strongly suggest a dependence on the rate of correlation. It is also plausible that some dependence on other initial parameters, such as E^M/E^T, influence the inertial range. This kind of

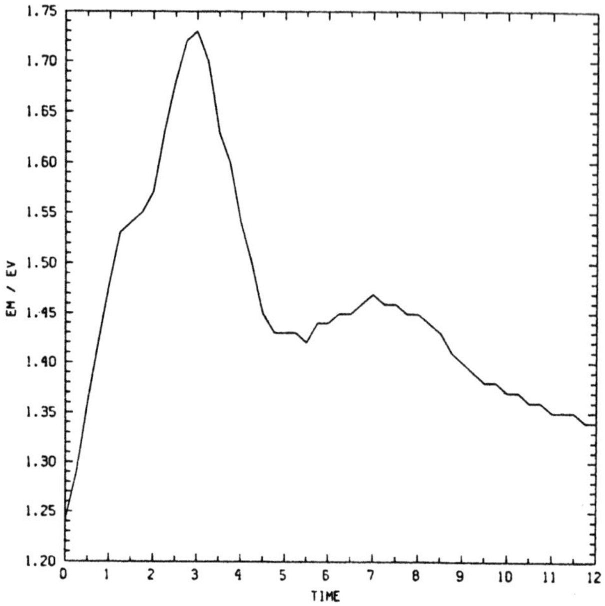

Fig. 8: Temporal oscillations of E^M/E^V for a three-dimensional MHD flow (64^3 resolution).

non-universality has already been noticed in the context of compressible flows, where the Mach number was shown to affect the spectral index in the acoustic regime [11]. Extension of this work to MHD is now in progress. Finally, we would like to note that one way to simulate flows with larger Reynolds numbers, is to implement special symmetries [12], [13] to reduce both the storage and the computing time . Such computations are under way.

Acknowlegements

We are thankful to R. Grappin and T. Passot for useful discussions. The Fast Fourier Transform algorithms we used are due to S.A. Orszag and C. Temperton. The graphics routines originate from the National Center for Atmospheric Research. Computations were performed on the CRAY1S of the Centre de Calcul vectoriel pour la Recherche. We benefit from partial support from the A.T.P. " Dynamique des Fluides Astrophysiques et Géophysiques" of the C.N.R.S.

References

[1] A.C. Ting, W.H. Matthaeus and D. Montgomery, preprint, Dartmouth College (1986).

[2] U. Frisch, A. Pouquet, P.L. Sulem and M. Meneguzzi, J. Méca. Théor. Appl., numéro spécial, 191 (1983).

[3] C. Bardos, C. Sulem, and P.L. Sulem, "Long-time Dynamics of Conducting Fluid in the Presence of a Strong Magnetic Field", submitted to Trans. Amer. Math. Soc. (1986).

[4] R. Grappin, Phys. Fluids, in press (1986).

[5] A. Pouquet, P.L. Sulem and M. Meneguzzi, "Influence of velocity magnetic field correlation on decaying MHD turbulence with neutral X-points", submitted to Phys. Fluids (1986).

[6] S.A. Orszag and C.M. Tang, J. Fluid Mech. **90**, 129 (1979).

[7] W.H. Matthaeus, M.L. Goldstein and D. Montgomery, Phys. Rev. Lett. **51**, 16 (1983).

[8] R. Grappin, A.Pouquet and J. Léorat, Astron. Astrophys. **126**, (1983).

[9] R.H.Kraichnan, Phys. Fluids **8**, 1385 (1965).

[10] A. Pouquet, M. Meneguzzi and U. Frisch, Phys. Rev. **A 33**, 4266 (1986).

[11] S.S. Moiseev, V.I. Petviashvily, A.V. Toor, V.V. Yanovsky, Physica **2D**, 218 (1981).

[12] M.E. Brachet, D.I. Meiron, S.A. Orszag, B.G. Nickel, R.H. Morf and U. Frisch, J. Fluid Mech. **130**, 411 (1983).

[13] M.E. Brachet, M. Meneguzzi and P.L. Sulem, Phys. Rev. Lett. **57**, 683 (1986).

The Structure of the Turbulent Temperature Field Above Heated Planes

T.J. Tsirikoglou * *and D.D. Papailiou* **

Laboratory of Applied Thermodynamics, University of Patras, Patras, Greece

Abstract. The results of an experiment are presented, conserning the structure of the free convection turbulent temperature field,created above a heated plane, in relation to large scale structures. Two configurations were examined namely, the horizontal and that inclined at 45 degrees. Since the flow above the horizontal plate is nearly homogeneous this configuration was also used as a reference case. The obtained measurements consisted of mean and statistical turbulent quantities. In the horizontal case the turbulent field was found double structured and departing significantly from homogenity. The detected rising columns of heated air were originated from instabilities created within the "convective"layer. In the case of the inclined plate the flow field was found to be complicated exhibiting the combined characteristics of a thermal boundary layer and a free convection flow above a horizontal plate.

1. Introduction

Research on turbulent free convection continues for almost a century, the interest on the subject stemming from the numerous applications these flows find in many areas of science and technology. Regarding free convection over plane surfaces, two limiting configurations namely, free convection along a vertical plane(1,2, 3) and that above a horizontal plane surface (4,5,6,7) have been subjected to extensive investigation as representing typical cases of these flows.

The two flows exhibit foundamental differences regarding the structure of their turbulent velocity and temperature fields as well as their respective turbulent transport mechanisms (2).

In the vertical plate configuration the velocity distribution resembles to that of a wall-jet. As a consequence of this velocity distribution,characterized by two layers of opposite vorticity, an organized vortex motion exists in the thermal layer (9) which is absent in the case of a heated horizontal plate. In the latter case the turbulent field above the heated surface is nearly homogeneous with rising columns of hot fluid originating from a "conduction"thin layer adjacent to it (5).

The assumption has been made by Townsend (4,5) that near the horizontal surface, convection is independent of distant boundaries and that it is determined by the heat flux Q, the kinematic viscosity v,and the thermal conductivity λ,of the fluid.

An unsuccessful attempt was made in (2) to apply the same scales in free convection along a vertical heated plate. This failure underlines the existing differences in the turbulent structure and heat transfer mechanism between the two configurations. It was suggested that among the factors contributing to these differences is the large scale vortex motion present in the vertical free convection layer (2,8).

Work on free convection above inclined plane surfaces is limited to the study of overall characteristics of the flow such as transition from laminar to turbulent regimes and the development of empirical heat transfer rate relations (10,11). The question of flow stability has been also investigated. It was found that for inclination angles less than 14° from the vertical position a wave instability exists while for angles over 17° the instability takes the form of longitudinal vortices (12,13).

On the basis of the discussed evidence the present experiments aim at studuing the structure of the turbulent field and transfer mechanisms related to free convection flows above heated planes of various orientation. Particular emphasis is given to the study of the nature and involvement in the structure of turbulence of large scale eddies which, as suggested from previous experiments (2,8), contribute to the differences in the transfer mechanisms existing between the two limiting free convection configurations. Notwithstanding the fact that in the present experimental work a specific turbulent flow was considered, it is hoped that the obtained results would be helpful in improving the understanding of the more general question of large scale structures participation in turbulent processes.

2. The Experiment

The experimental set-up is shown in Fig. 1. It consists of a stainless steel, non-magnetic plate of dimensions 1000x500x12 mm, heated by ten uniformly spaced resistance elements. These are connected independently to an equal number of variacs which are fed from a voltage regulated source. Temperature uniformity on the plate surface is achieved by controlling the power delivered by each heating element. To further improve temperature uniformity, a 20 mm thick layer of heat-resisting cement is placed between the back surface of the plate and the heating elements. Plate temperature distribution was measured by a number of thermocouples located as shown in the figure. The experimental set-up was encased in a wooden box filled with fiberglass insulating material with only the surface of the heated plate exposed to the atmospheric air. The case could rotate around an axis allowing the plate to attain any orientation between the horizontal and vertical positions. To protect the flow from possible disturbances due to draughts existing in the room, the experimental set-up was surrounded by four partitions with suitable glass windows for flow visualization studies.

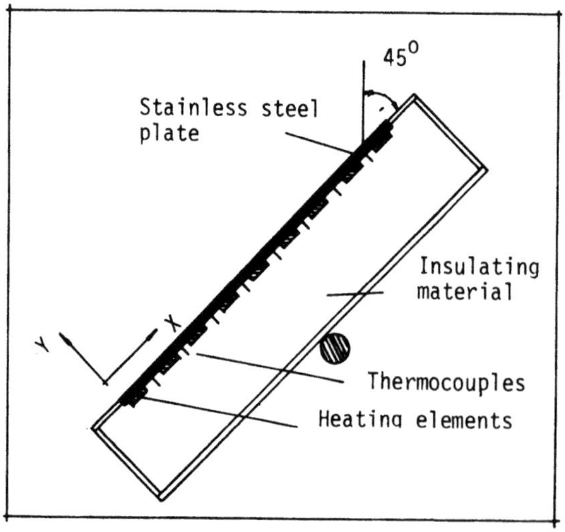

Fig.1. Experimental set-up

Two configurations corresponding to the horizontal and 45^0 plate positions were investigated. The two experiments were conducted at the same Grashof number (based on the length,L, of the plate), equal to $Gr_L = 6.45 x 10^9$. Under those conditions the flow above the plate was found turbulent in both cases. Temperature uniformity over the plate's heated area was found satisfactory,the variation in the temperature distribution not exceeding the two degrees centigrade (2^oC).

For temperature measurements a hot film anemometer (TSI,Model 1040) was used, with the probe operating as a platinum thermometer. The response of the sensor to temperature fluctuations was less than 1 ms corresponding to turbulent frequencies well above their upper limit detected in the flow (approximatelly 100c/sec).

The voltage analog signal obtained from the anemometer's output,corresponding to measured temperature, was digitised and consequently stored to a microcomputer for further processing. The sampling frequency was 200 Hz which was at least twice higher than the indicated above upper turbulent frequency limit,fullfiling therefore the condition for accuracy demanded by Nyquist's theorem (14).

In the case of the inclined plate, measurements were conducted at three locations namely, at distances 70,450 and 710 mm from its leading edge while,in the case of the horizontal plate position only at its center.

The obtained measurements consisted of mean and statistical quantities related to the turbulent thermal field. More specifically,they consisted of mean temperature distributions and heat transfer coefficients as well as turbulent temperature intesities θ', probability distributions, autocorrelations and turbulence spectra, Skewnes factor, S, and Flatness factor, F. Schlieren optical methods were also applied for the study of the free convection flow patterns above the heated plate.

3. Results and Conclusions

Since the free convection turbulent field above the horizontal plate was expected to be nearly homogeneous (5) this flow was initially planned to serve mainly as a reference case for the study of structure changes occuring in the turbulent flow field due to changes in plate orientation. However, the obtained detailed measure-

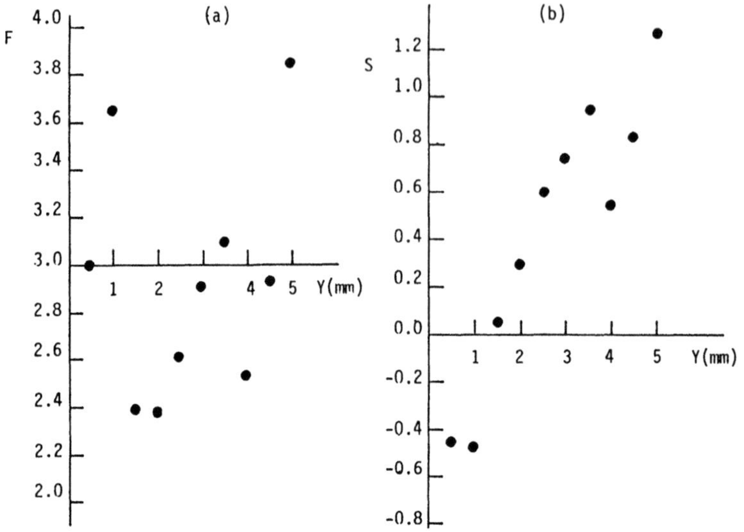

Fig.2. (a) Flatness and (b) Skewness distribution (horizontal plate)

ments of the flow statistical characteristics revealed some new aspects regarding its overall structure and, more important, the nature of the "conduction" layer as well as certain physical processes occuring in it. Mean and fluctuating temperature data were collected at several locations along the vertical direction to the plate,starting from a distance of o,5 mm from the surface.

The thickness of the "conduction"layer, as deduced from Schlieren photographs and obtained measurements, was approximatelly 4,0 mm.

Skewness and Flatness distributions in the layer (Fig.2) proved the lack of homogeneity in the turbulent flow. Regarding the origin and development of the thermals,which are seen in the Schlieren photographs to depart from the edge of the layer in the form of rising "eroptions", the obtained measurements suggest that, the initial instabilities as well as the subsequent thermal formation processes, take place within the layer. This becomes evident from the study of the form of the spectra and Skewness and Flatness distributions. All the turbulent spectra obtained within the layer exhibit distinctive peaks in their lower frequency part (up to about 10 c/sec). The form of the Flatness factor distribution and espessially the part negative and part positive Skewness factor distribution in the layer, further corroborates the suggested physical mechanism of thermal formation (2).

Inspection of the spectra obtained in the layer shows the existence of a -5/3 "inertial"subrange. This finding is not in agreement with the early stage turbulent structure of the flow expected to exist near the heated plate. Indeed, as Schlieren visual observation confirms, the flow in the layer is influenced by a large circulatory system dominating the entire flow field, contrary to the convection pattern described in (5). Therefore, the flow in the layer has already a history as being part of the mentioned system thus, justifying its well developed turbulent character. In conclusion, the "conduction"layer appears to form a significant flow "entity"within which a number of physical processes, decissive to the development of the convection heat transfer scheme occurs.

The flow outside the layer appears double-structured and also lacking of homogeneity as shown in the spectra and Skewness and Flatness distributions (Figs.2, 3). It should be noted that the mean temperature and turbulent intensity distributions were found similar indicating, that similarity is not drastically affected in this case by the lack of homogeneity.

The results of the inclined plate experiment are discussed in the following: The obtained Schlieren photographs reveal in this case, a complicated flow pattern consisting of three distinctive regions. As in the horizontal plate the "conduction"layer appears, adjacent to the heated surface. Next to it, a boundary layer like flow is observed followed by a third region of upward flow motion similar to

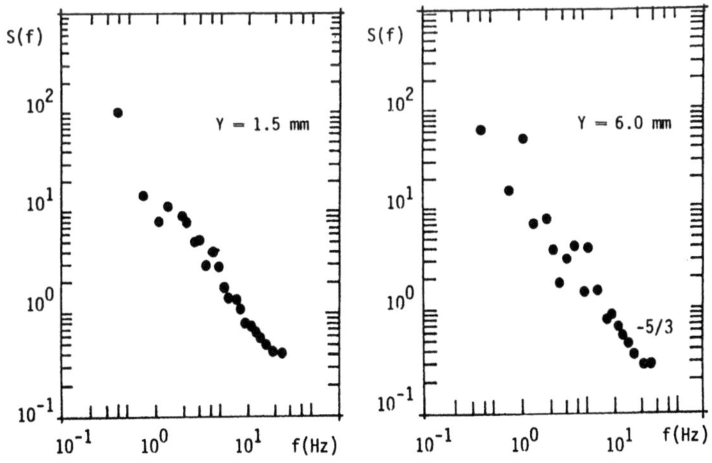

Fig.3. Temperature spectra (horizontal plate)

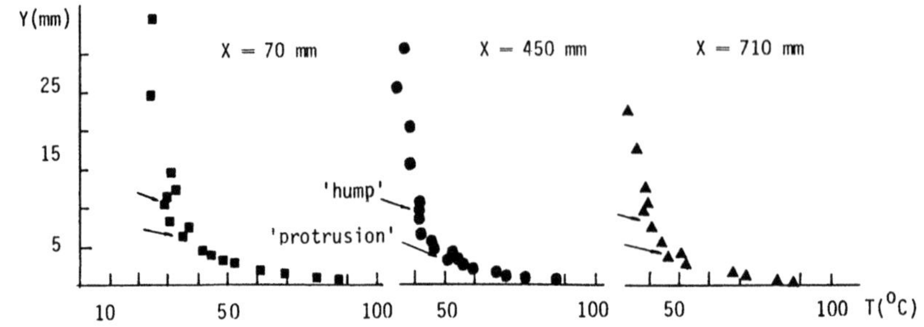

Fig.4. Mean temperature (T) distribution (inclined position)

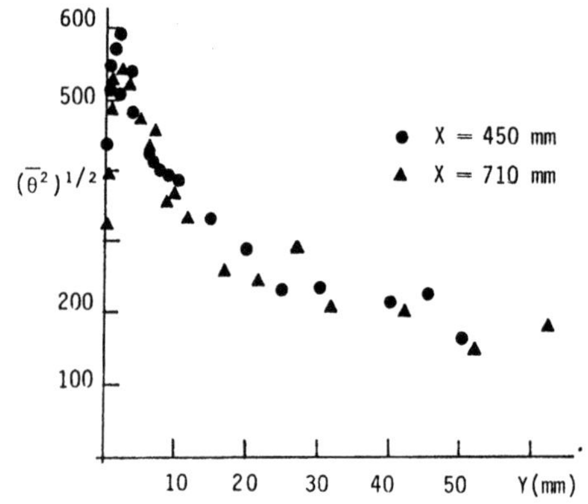

Fig.5. Turbulent temperature intensity distribution

that far above a heated horizontal surface. Thus, the flow field above the inclined plate appears to be a combination of the horizontal and vertical free convection flows.

At the edge of the "conduction"layer large vortices are formed a number of which travel along the plate, within the boundary layer, in what visually appeared as a swirling motion. The remainder of the vortices escaped through the boundary layer, to the third flow region in an upward motion.

The obtained mean temperature and turbulent intensity distributions were not similar in this case (Figs. 4,5).

The shape of the middle part of the mean temperature profiles appeared irregular which, initially was attributed to existing data scatter. However, more carefull inspection showed that the same irregular form, consisting of a "protrusion" followed by a "hump",persisted in all the measured profiles. Since at the last two measuring locations (x=450mm,x=710mm),where the flow is well developed,the"protrusion"temperature discontinuity occurs at a distance from the plate equal to the"conduction"layer thickness these irregularities, in consistence with the visual observations, could be associated with the observed flow regions.

Heat transfer rate measurements, based on the obtained temperature profiles showed that the heat transfer coefficient remains constant along the plate while the local Nusselt number, Nu_x, is expressed by the following relation

$$Nu_x = 0.27\ (Gr_x Pr)^{1/3}$$

Regarding the structure of the "conduction" layer,the measured Skewness and Flatness distributions (Fig.6) as well as the turbulent spectra (Fig.7) showed that it resembles to that found in the horizontal configuration. However, as already mentioned, the eddies departing from the layer's edge are of different structure in the two cases.

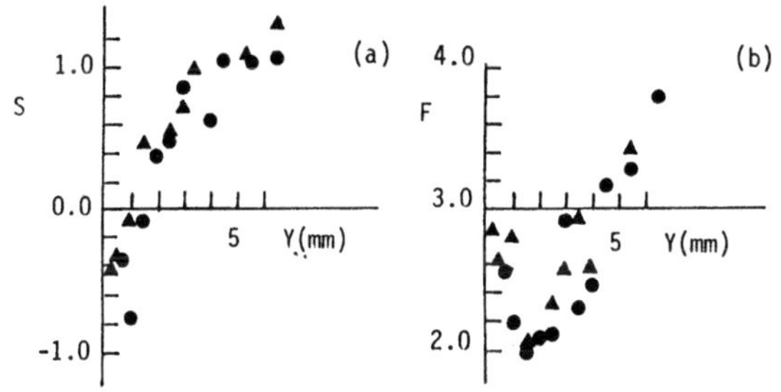

Fig.6. (a) Skewness(S) and (b) Flatness(F) distribution

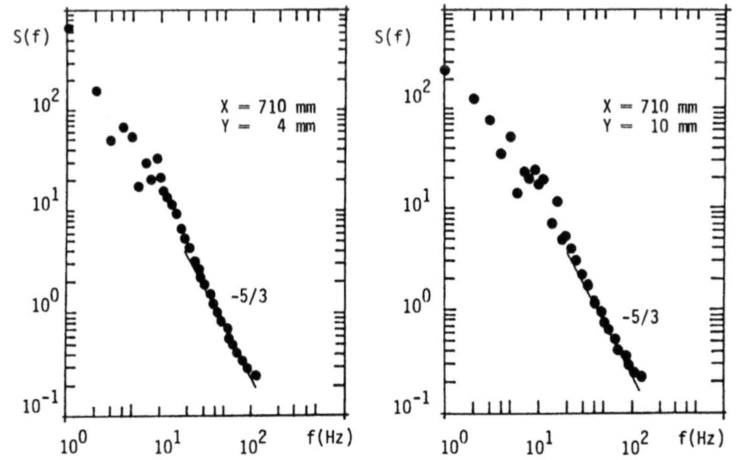

Fig.7. Temperature spectra(inclined position)

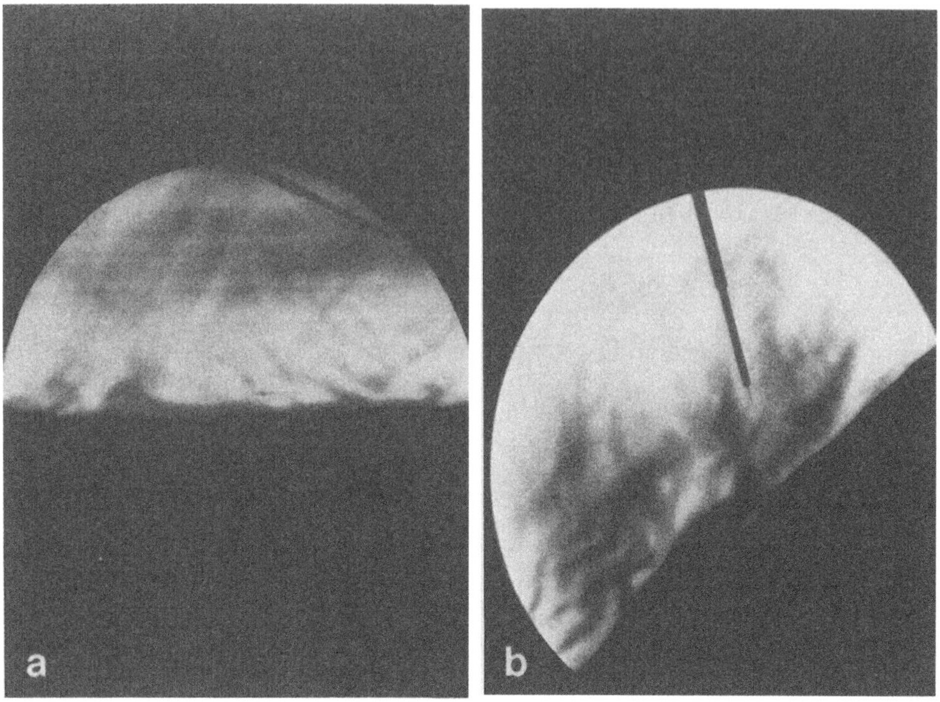

Schlieren technique flow photographs.(a) Horizontal position (b) inclined position.

The above discussed experimental results lead to the following conclusions: In the case of the horizontal plate the turbulent flow field was found to be inhomogeneous although, this has no drastic effect on similarity. The "conduction" layer appears to dominate heat transfer processes as well as the structure of the turbulent field above the plate.In the inclined plate configuration the structure of the turbulent flow field contains combined characteristics of both the vertical and horizontal convection flows. A physically sound theoretical approach to the problem appears difficult at the present state of understanding, due to the complicated nature of the flow. The presented evidence indicate the predominent role of large scale structures in the transfer processes however, complete understanding of their participation in these processes needs further investigation.

1. E.R.G. Eckert, T.W. Jackson: NACA TN 2207 (1950)

2. D.D. Papailiou, P.S. Lykoudis: Int. J. Heat Mass Transfer vol.17 pp 161-172 (1974)

3. G.C. Vliet, C.K. Liu: Journal of Heat Transfer,Nov.1969 pp517-531

4. D.B. Thomas, A.A. Townsend: J. Fluid Mech. 2, pp 473-492 (1957)

5. A.A. Townsend: J. Fluid Mech. 5, pp 209-241 (1959)

6. W.V.R. Malkus: Proc. Roy. Soc. A,225 pp 196-212 (1954b)

7. J.W. Deardorff, G.E. Willis: J. Fluid Mech. vol.28,part 4,pp 675-704 (1967)

8. D.D. Papailiou: Int. J. Heat Mass Transfer vol.23,pp 889-895 (1980)

9. T. Fujii: Bull. Japan Soc. Mech. Eng. 2(8),p 555 (1959)

10. G.C. Vliet: J. Heat Transfer,pp 511-516,Nov.1969

11. B.R. Rich: Trans. ASME vol.75,p 489 (1953)

12. J.R. LLoyd, E.M. Sparrow: J. Fluid Mech. vol.42,part 3,pp 465-470 (1970)

13. K.L. Tzuoo, T.S. Chen, B.F. Armaly: J. Heat Transfer vol.107,pp 107-132 Febr.1985

14. M. Klerer, G.A. Korn: Digital computer user's handbook, Mc Graw-Hill

On the Equations Governing the Propagation of Disturbances in Turbulent Shear Flow

V.I. Vasanta Ram

Institut für Thermo- und Fluiddynamik, Ruhr-Universität Bochum, D-4630 Bochum, Fed. Rep. of Germany

1. Introduction and scope of the work

Research work of the past two decades in turbulent shear flows has suggested the important role propagation characteristics of disturbances might play in the formation and maintenance of coherent structures. Studies on the behaviour of periodic disturbances introduced into a well defined turbulent flow have therefore been persued with much vigour, the channel flow occupying a pivotal role among these configurations. The early studies by HUSSAIN and REYNOLDS [1] have provided much insight into the subject. However, later experiments of KISKE [2] have, while supporting some findings, uncovered features not noticed earlier. A prominent new feature in [2] is: The rate of evolution of the imposed disturbance is characterized by different length scales in the regions nearer to and farther away from the disturbance source. Figure 1 reproduced from experiments in the channel flow by KISKE [2] illustrates this point. The relation between such an observed behaviour of the disturbance and the theoretical framework in [1] is not very clear. There is, therefore, a need for a closer reexamination of the mechanisms underlying disturbance propagation in turbulent shear flows. The present work is part of an effort towards this end. The focal point in this paper is the observed two-scale behaviour of the disturbance. It is an attempt to explain such an observed feature in terms of the dynamics of a disturbance in a basic turbulent flow. Ap-

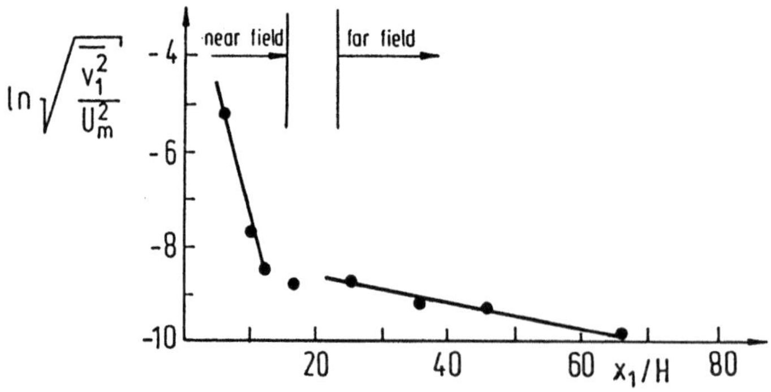

Fig. 1: Two scaled decay of disturbance Reynolds stress in channel flow from Kiske [2]. Two-dimensional disturbance introduced at $x_1 = C$.

proximations that describe the evolution of the disturbance in the two regions are set up and their meaning discussed.

2. Outline of work

The main source of difficulty in the analysis of disturbances in turbulent flow - a difficulty not present in its laminar flow counterpart - is the effect of the disturbance on the Reynolds stresses. Since Reynolds stresses arise from averageing it is only appropriate that propagation characteristics of disturbances in a turbulent flow ought to be studied prior to averaging. This is the approach taken in this paper.

In this work the velocity field of the basic flow is itself the sum of a mean part $\overline{U}_{bl}$ and a fluctuating part u'_{bl} as follows (index "b" denotes basic flow):

$$u_{bl} = \overline{U}_{bl} + u'_{bl} \quad , \qquad l = 1,2,3 \quad , \tag{1a}$$

with

$$\overline{U}_{bl} = U_o U_{bl} \quad \text{and} \quad u'_{bl} = u_\tau \tilde{u}_{bl} \quad , \tag{1b}$$

where (U_o, u_τ) are characteristic velocities of the mean and the fluctuating parts and $(U_{bl}, \tilde{u}_{bl})$ are nondimensional. The equations for a small amplitude disturbance u_{sl} (index "s" stands for the disturbance) on this basic flow are derived through straightforward linearisation of the equations of motion prior to any sort of averaging. It is then obvious that the evolution of the disturbance depends on the characteristics of both the mean and the fluctuating parts of the basic flow. Since the turbulent motion of the basic flow in its entirety is too complex to permit analysis of a disturbance a more tractable model problem is posed for this purpose. The fluctuating part of the basic flow is represented herein by a simpler wave-like form with only two length scales. Essential elements of turbulent motion, viz. time-dependence and three-dimensionality are retained in this formulation. The model problem posed in this manner draws attention to some salient differences between the propagation characteristics of long-wave and short-wave disturbances, thus suggesting a scheme for the classification of disturbances in turbulent flow. With this simpler two-scale wave representation of the basic flow turbulence the evolution of an initially two dimensional short-wave disturbance introduced into the steady fully developed channel flow is analysed in some detail. The study shows that the mechanisms governing disturbance behaviour are different in the near and the far fields, which in turn is associated with different laws of evolution in these regions.

3. The equations for the disturbance

The starting point is the non-averaged equations of motion which, in the customary notation (see e.g. TENNEKES and LUMLEY [3]), may be written as:

$$\frac{\partial u_l}{\partial x_l} = 0 \quad , \tag{2a}$$

$$\frac{\partial u_l}{\partial t} + u_m \frac{\partial u_l}{\partial x_m} + \frac{1}{\rho}\frac{\partial p}{\partial x_l} - \nu \frac{\partial^2 u_l}{\partial x_m \partial x_m} = 0 \quad . \tag{2b}$$

The velocity u_{bl} in (1) and the pressure field p_b of the basic flow obey the above set of equations. The equations governing a small amplitude velocity and pressure disturbance (u_{sl}, p_s), derivable by straightforward linearisation around the basic flow (1), are:

$$\frac{\partial u_{sl}}{\partial x_l} = 0 \quad , \tag{3a}$$

$$\frac{\partial u_{sl}}{\partial t} + (\bar{U}_{bm} + u'_{bm}) \frac{\partial u_{sl}}{\partial x_m} + u_{sm} \frac{\partial}{\partial x_m} (\bar{U}_{bl} + u'_{bl}) + \frac{1}{\rho}\frac{\partial p}{\partial x_l} - \nu \frac{\partial^2 u_{sl}}{\partial x_m \partial x_m} = 0 \quad . \tag{3b}$$

Solving (3) for the disturbance in a real turbulent flow would call for more detailed knowledge of the fluctuating part of the basic flow than is feasible at present. The author has therefore chosen an alternative approach of analysing disturbances in a more tractable model problem in which the fluctuating part is of a simpler structure than in a real turbulent flow but which nevertheless preserves important gross features of the same.

4. The model problem

In the model problem instead of solving the non-averaged equations of motions (2) for the basic flow, the fluctuating part u'_{bl} is regarded as given. The mean part $\bar{U}_{bl}$ then follows from the equations of motion suitably averaged. The dimensionless fluctuating velocity $\tilde{u}_{bl}$ in (1) is prescribed to be of *wave-like form* with only two length scales as follows:

$$\tilde{u}_{bl} = \mathrm{Re}\left\{ \tilde{A}_{bl}(X_m) \cdot \exp\{ i [\tilde{\Phi} + \tilde{\varphi}_{bl}(X_m)]\} \right\} \quad , \tag{4a}$$

with

$$\tilde{\Phi} = 2\pi f \tilde{t} - \sum_{k=1}^{3} \lambda_k \tilde{x}_k \quad , \tag{4b}$$

where X_m and $\tilde{x}_m$ denote the long and short-scale variables respectively and $\tilde{A}_{bl}$ and $\tilde{\varphi}_{bl}$ are real functions of the long-scale variable X_m only. The above form is taken from WHITHAM [4]. In writing the velocity in this form the idea of multiple scales (see e.g KEVORKIAN and COLE [5], NAYFEH [6]) has been used which implies that the derivative $\partial u'_{bl}/\partial x_m$ is expressible as a sum of quantities of differing orders of magnitude as follows:

$$\frac{\partial u'_{bl}}{\partial x_m} = \frac{u_\tau}{L}\frac{\partial \tilde{u}_{bl}}{\partial X_m} + \frac{u_\tau}{l_\tau}\frac{\partial \tilde{u}_{bl}}{\partial \tilde{x}_m} , \tag{5}$$

where L and l_τ are the two characteristic lengths ($l_\tau \ll L$). The fluctuating velocity should satisfy the equation of continuity (2a) which for (4) assumes the following form:

$$\mathrm{Re}\left\{ -\frac{u_\tau}{l_\tau}\tilde{A}_{bl}\,\lambda_l \exp\{i(\tilde{\Phi}+\tilde{\varphi}_{bl})\} + \frac{u_\tau}{l_\tau}\exp(i\tilde{\Phi})\frac{\partial}{\partial X_l}\{\tilde{A}_{bl}\exp(i\tilde{\varphi}_{bl})\}\right\} = 0 . \tag{6}$$

Using (1) and (4) in the equations of motion (2) and averaging over a phase (i.e. over the short-scale variables) yields the equation that determines the mean part of the basic flow $\overline{U}_{bl}$. This is:

$$\frac{\partial U_{bl}}{\partial X_l} = 0 , \tag{7a}$$

$$U_{bm}\frac{\partial U_{bl}}{\partial X_m} + \frac{\partial P}{\partial X_l} + \frac{1}{4}\frac{u_\tau^2}{U_o^2}\frac{\partial}{\partial X_m}\left\{\tilde{A}_{bl}\,\tilde{A}_{bm}\left[\exp\{i(\tilde{\varphi}_{bl}-\tilde{\varphi}_{bm})\} + \exp\{-i(\tilde{\varphi}_{bl}-\tilde{\varphi}_{bm})\}\right]\right\}$$
$$+ \frac{\nu}{U_o L}\frac{\partial^2 U_{bl}}{\partial X_m \partial X_m} = 0 , \tag{7b}$$

where $\partial P/\partial x_l$ stands for the suitably nondimensionalised averaged pressure gradient.

The above equation bears a strong resemblance to well known Reynolds' averaged equation for turbulent flow in the limited sense that the mean motion is subject to the influence of Reynolds' stresses that are ascribable to the fluctuating part of the motion alone. In the model problem the Reynolds' stresses are:

$$\tau_{lm} = \frac{1}{4}\rho\, u_\tau^2\left\{\tilde{A}_{bl}\,\tilde{A}_{bm}\left[\exp\{i(\tilde{\varphi}_{bl}-\tilde{\varphi}_{bm})\} + \exp\{-i(\tilde{\varphi}_{bl}-\tilde{\varphi}_{bm})\}\right]\right\} . \tag{8}$$

While (8) shows the important role of the phase difference between the velocity components for the Reynolds' stresses, (7) draws attention to the need for at least two length scales to characterize the fluctuating motion. For, if the amplitude and the phase of the fluctuating motion were independent of the longer-scale variable X_m, as in a more conventional wave, the averaged motion would be uninfluenced by Reynolds' stresses thus losing the probably most outstanding feature of turbulent shear flows.

To complete the wave description of the fluctuating velocity field it is necessary to specify its dispersion relation. Proposals for the dispersion relation valid for a turbulent flow have been put forth in literature (see eg. LANDAHL and

MOLLO-CHRISTENSEN [7]) but their appropriateness for the model problem as at present is questionable. Fortunately, important qualitative features of the disturbance can be inferred from the equations without explicitly invoking the dispersion relation, and this is the course the author has persued in the present paper. We presume that a dispersion relation of the kind

$$G(f, \lambda_l, \tilde{A}_{bl}) = 0 \tag{9}$$

which may be amplitude dependent (see WHITHAM [4]), can be specified and leave its precise form open.

5. General classification scheme for disturbance

At this stage it is meaningful to classify disturbances according to whether their wavelength is small, large or comparable to that of the basic flow wave. Table 1 shows the relative orders of the terms in (3) where l_s denotes the wavelengths of the disturbance of the order $O(u_{\tau s})$. It is readily seen that the dominant terms for short- and long-wave disturbances are different. For the *short-wave disturbance* $l_s \ll l_\tau$, and (3) gets simplified to:

$$\frac{\partial u_{sl}}{\partial x_l} = 0 \ , \tag{10a}$$

$$\frac{\partial u_{sl}}{\partial t} + (\overline{U}_{bm} + u'_{bm}) \frac{\partial u_{sl}}{\partial x_m} + \frac{1}{\rho}\frac{\partial p}{\partial x_l} = \frac{l_s}{l_\tau}\left(u_{sm}\frac{\partial u'_{bl}}{\partial x_m}\right) + \frac{l_s}{L}\left(u_{sm}\frac{\partial \overline{U}_{bl}}{\partial x_m}\right) . \tag{10b}$$

For the *long-wave disturbance* $l_s \gg l_\tau$ and the appropriate simplification is:

$$\frac{\partial u_{sl}}{\partial x_l} = 0 \ , \tag{11a}$$

$$\frac{\partial u_{sl}}{\partial t} + \overline{U}_{bm}\frac{\partial u_{sl}}{\partial x_m} + u_{sm}\left(\frac{\partial u'_{bl}}{\partial x_m} + \frac{l_\tau}{L}\frac{\partial \overline{U}_{bl}}{\partial x_m}\right) + \frac{1}{\rho}\frac{\partial p}{\partial x_l} = \frac{l_\tau}{l_s}\left(u'_{bm}\frac{\partial u_{sl}}{\partial x_m}\right) . \tag{11b}$$

In both (10) and (11) viscous terms have been dropped.

We wish to draw the reader's attention to the important differences between the two extreme cases. Whereas for the short-wave disturbance the convective term due

Table 1:

Term in (3b)	$\frac{\partial u_{sl}}{\partial t}$	$U_{bm}\frac{\partial u_{sl}}{\partial x_m}$	$u'_{bl}\frac{\partial u_{sl}}{\partial x_m}$	$u_{sm}\frac{\partial U_{bl}}{\partial x_m}$	$u_{sm}\frac{\partial u'_{bl}}{\partial x_m}$	$\nu\frac{\partial^2 u_{sl}}{\partial x_m \partial x_m}$
Order O(.)	$\left(\frac{u_{\tau s}}{t_s}\right)$	$\left(\frac{U_o u_{\tau s}}{l_s}\right)$	$\left(\frac{u_\tau u_{\tau s}}{l_s}\right)$	$\left(\frac{U_o u_{\tau s}}{L}\right)$	$\left(\frac{u_\tau u_{\tau s}}{l_\tau}\right)$	$\left(\frac{\nu u_{\tau s}}{l_s^2}\right)$

the basic flow "turbulence" $u'_{bm}\,(\partial u_{sl}/\partial x_m)$ is more dominant than the vortex-stretching term $u_{sm}(\partial u'_{bl}/\partial x_m)$ the opposite is true for the long-wave disturbance. Particularly noteworthy is also the formal correspondence in respect of the convection term between the long-wave approximation (11) and the equations derived in literature for the "phase-averaged" quantity in the triple-decomposed velocity field since this is the form often used (see eg. [1]) as the equations for the "coherent structure" in turbulent flow. This correspondence is not present in the short-wave approximation (12).

A closer inspection of (3) shows that the broad classification scheme above needs refinement when the disturbance wavelengths in the three directions are not all of comparable order. This is the case for a two-dimensional disturbance.

6. Two-dimensional short-wave disturbances in the channel flow

For the steady fully developed turbulent channel flow the basic flow (1),(4) has a simpler structure in which U_1, $\tilde{A}_{bl}$ and $\tilde{\varphi}_{bl}$ depend only on X_2 and $U_2 = U_3 = 0$. Three-dimensionality of the fluctuating part of the basic flow is thus preserved on the small scale. Equation (7) then reduces to the familiar relation between the pressure gradient and Reynolds stress in channel flow with the latter as in (8).

With these channel flow properties we may proceed along lines of Sec. 5 and assess the relative importance of the various terms in (3) when the wavelength in x_3 direction is much larger than in the other two directions. To this end, we characterize the derivatives of the disturbance through two wavelengths of different order retaining the notation l_s for the order of the wavelengths in the x_1 and x_2 directions, and denoting by l_{s3} this quantity in the x_3 direction. Examining the limit $l_s \ll l_\tau \ll l_{s3}$ we arrive at the *near-field approximation*. Dropping viscous terms again these are:

$$\frac{\partial u_{s1}}{\partial x_1} + \frac{\partial u_{s2}}{\partial x_2} = 0 \quad , \tag{12a}$$

$$\frac{D_{bn}u_{sl}}{Dt} + \frac{1}{\rho}\frac{\partial p_s}{\partial x_l} = \frac{l_s}{l_\tau}\sum_{m=1}^{2} u_{sm}\frac{\partial u'_{bl}}{\partial x_m} + \delta_{1l}\frac{l_s}{L}u_{s2}\frac{d\overline{U}_{b1}}{dx_2} + \frac{l_s}{l_{s3}}u'_{b3}\frac{\partial u_{sl}}{\partial x_3} \qquad l=1,2 \;, \tag{12b}$$

$$\frac{D_{bn}u_{s3}}{Dt} = \frac{l_s}{l_\tau}\sum_{m=1}^{2} u_{sm}\frac{\partial u'_{b3}}{\partial x_m} + \frac{l_s}{l_{s3}}\left(u'_{b3}\frac{\partial u_{s3}}{\partial x_3} + \frac{1}{\rho}\frac{\partial p_s}{\partial x_3}\right) \;, \tag{12c}$$

where the operator D_{bn}/Dt is defined as

$$\frac{D_{bn}}{Dt} = \frac{\partial}{\partial t} + (\overline{U}_{b1} + u'_{b1})\frac{\partial}{\partial x_1} + u'_{b2}\frac{\partial}{\partial x_2} \quad . \tag{12d}$$

The reasons for the validity of the above approximation being limited to the near field region will be evident on examining the characteristics of a disturbance governed by (12) more closely. Solving these equations explicitly is not necessary for this purspoe. The decisive features can be summarized in physical terms as follows:

1. To the leading order the disturbance stays two-dimensional. The spanwise component u_{s3} is small of the order $O(l_s/l_\tau)$. As a consequence the evolution of the other two components u_{s1} and u_{s2} is unaffected by u_{s3}.

2. The equation for the spanwise component u_{s3} contains the nontrivial forcing term $\Sigma u_{sm} (\partial u'_{b3}/\partial x_m)$. This causes a growth of the spanwise component, the result of which is that the intially two-dimensional disturbance does not remain two-dimensional. Equation (12) is therefore inadequate to describe the disturbance evolution in the far field.

The disturbance equations for the far field may be obtained following the same procedure as before but with setting wavelengths in the different directions to be of the same order. We omit details and give below the resulting *far field equation* .

$$\frac{\partial u_{s1}}{\partial x_1} = 0 \quad , \tag{13a}$$

$$\frac{D_{bf} u_{s1}}{Dt} + \frac{1}{\rho}\frac{\partial p_s}{\partial x_1} = \frac{l_s}{l_\tau} u_{sm} \frac{\partial u'_{b1}}{\partial x_m} + \delta_{11} \frac{l_s}{L} \frac{d\bar{U}_{b1}}{dx_2} \quad , \tag{13b}$$

where the operator D_{bf}/Dt is defined as

$$\frac{D_{bf}}{Dt} = \frac{\partial}{\partial t} + (\bar{U}_{b1} + u'_{b1}) \frac{\partial}{\partial x_1} + u'_{b2} \frac{\partial}{\partial x_2} + u'_{b3} \frac{\partial}{\partial x_3} \quad . \tag{13c}$$

Comparison of (12) and (13) shows clearly the differences between the evolution rates in the near and the far fields. The evolution rates are transparent in a frame of reference that may be termed "Lagrangian with respect to the basic flow". In this frame of reference the evolution rates of u_{s1} and u_{s2} in the near field are governed only by the derivatives $\left[(\partial u'_{b1}/\partial x_1),(\partial u'_{b1}/\partial x_2)\right]$ and $\left[(\partial u'_{b2}/\partial x_1),(\partial u'_{b2}/\partial x_2)\right]$, whereas in the far field derivatives in the third direction also contribute to changes. The difference in evolution rate is associated with the change in the physical character of the disturbance. The two-dimensional character of the imposed disturbance is retained in the near field but is destroyed in the far field. The mechanism responsible for this change is *vortex stretching by the three-dimensional fluctuating basic flow.*

7. Conclusions

Different physical mechanisms govern the behaviour of long- and short-wave disturbances in a single-wave basic flow. These differences may be expected to be present in a real turbulent shear flow too.

The evolution of two-dimensional disturbances introduced into a single-wave basic flow is characterized by a two-scaled nature, caused by different mechanisms being effective in the near and the far field. The conversion of the two-dimensional disturbance into a three-dimensional one by the wave in the basic flow is the feature that might be expected to be present in the real turbulent shear flow, which would explain the two-scaled behaviour observed in experiments.

<u>List of symbols</u>

t, x_1, x_2, x_3	time and length coordinates
u_1, u_2, u_3	velocity components
p	pressure
U_o	characteristic velocity of the mean part of the basic flow, vide (1b)
u_τ	characteristic velocity of the fluctuating part of the basic flow, vide (1b)
$u_{\tau s}$	characteristic velocity of the disturbance
L	characteristic length of the mean part of the basic flow
l_τ	characteristic length of the fluctuating part of the basic flow
l_s	characteristic length of the disturbance
τ_{lm}	Reynolds stress
ρ	density
ν	kinematic viscosity
X_1	long-scale variable, vide (5)
$\tilde{x}_1$	short-scale variable, vide (5)

Quantities in the model problem, vide (4)

f	frequency
λ_k	wave number
$\tilde{A}_{bl}$	amplitude
$\tilde{\varphi}_{bl}$	phase angle

Φ phase, vide (4b)

Indices

b refers to basic flow

s refers to disturbance

Achnowledgement

Research work reported herein has been supported by the Deutsche Forschungsgemeinschaft.

References

1. A.K.M.F. Hussain and W.C. Reynolds: The Mechanisms of an Organized Wave in Turbulent Shear Flow. Journal of Fluid Mechanics Vol.41 (1970), pp. 241-255.
2. S. Kiske: Ein Beitrag zur Beschreibung turbulenter Scherschichten mit gestörter Turblenzstruktur. Dr.-Ing. Dissertation, 1983, Abteilung für Maschinenbau, Ruhr-Universität Bochum, F.R.G.
3. H. Tennekes and J.L. Lumley: A First Course in Turbulence (MIT Press 1972).
4. G.B. Whitham: Linear and Nonlinear Waves (John Wiley & Sons 1974).
5. J. Kevorkian and J.D. Cole: Perturbation Methods in Applied Mathematics (Springer Verlag 1981).
6. A.H. Nayfeh: Introduction to Perturbation Techniques (John Wiley & Sons 1981).
7. M.T. Landahl and E. Mollo-Christensen: Turbulence and Random Processes in Fluid Mechanics (Cambridge University Press 1986).

Part VI

Coherent Structures in Turbulent Flows. Conditional Averaging. Pattern Recognition

Coherent Structures

H.E. Fiedler

Hermann-Föttinger-Institut für Thermo- und Fluiddynamik,
Technische Universität Berlin, D-1000 Berlin 12, Germany

Everything should be as simple as possible
- but not too simple ... A. Einstein

1. Introduction

The concept of coherent structures (CS) in turbulent flows became popular only some 15 years ago since BROWN & ROSHKO's [1] spectacular observation in the mixing layer, although the phenomenon was not new: PRANDTL [2] proposed concepts of turbulent exchange arrived at from experimentally studying characteristic motions. What became known of this was the mixing length hypothesis based on the postulation of "Flüssigkeitsballen", which were considered responsible for transport of momentum across the flow. They appear to be the first CS in the history of turbulence. No particular characterization of those structures was proposed by Prandtl and it was only much later for BRADSHAW [3] to explain and for LANDAHL [4] to justify his ideas. Beginning around 1950 we find again increasing interest in visualization of turbulence. The more important observations and concepts of those and later years were:

TOWNSENDS's [5] big eddy hypothesis, postulating the existence of a "double structure": big eddies in energetic equilibrium and the stoachstic (sub-)structure, containing most of the turbulent energy (see also [6]).

A decade later KLINE [7] reported about longitudinal streaks in the wall proximity of turbulent boundary layers. This observation - which had some less noticed predecessors ([8], [9]) - initiated a renaissance of interest in boundary layer structure (e.g. [10], [11]).

Why then did it take so long until the already perceived phenomenon achieved adequate recognition? Surely our preconditioned view of turbulence being a purely stochastic phenomenon provided a serious obstacle for this kind of concept, and the specific tools used to describe turbulence in the classical sense, namely the Reynolds equation and, in this context, the Reynolds-averaging procedure, which indiscriminately irons out all structural peculiarities, were not helpful for discovering anything beyond the stochastic.

In the present context the notion of coherence is - in contradistinction to its classical meaning - used in a more general sense,describing a flow with discernible correlation - i.e. an element of turbulent motion, set off against the stochastic background, which by its specific properties and composition characterizes a specific flow. Fig. 1 shows some examples

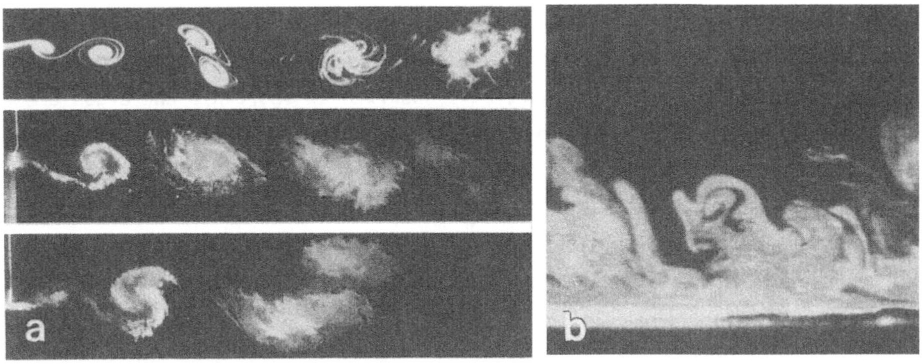

Fig. 1: a) Transitional and turbulent shearlayer, b) boundary layer

These are "young" and(possibly) not fully developed flows of relatively small Reynolds-numbers. The structures observed in the turbulent flow may therefore be only relicts of the transitional structures (which they closely resemble) and thus no true reflections of turbulent flow at all! Observations in nature gives an answer to this objection advanced in [12]: Meteorological structures, oil - wakes of damaged tanker-ships and - as an extreme case - Jupiter's great red spot are all characterized by Reynolds-numbers orders of magnitudes larger than those in laboratory flow. Another argument supporting the true nature of CS is their (re)development in situations where L/T-transition is suppressed (turbulent upstream boundary conditions) or after their deliberate destruction ([13],[14]). The above examples may serve as a basis for a "describing"

2. DEFINITION - APPEARANCE/BEHAVIOUR/QUALITY -

Thus CS are spontaneously formed, non-stationary motional systems of correlated vorticity Ω [15] -. They are

- in most cases of large scale, comparable to the lateral flow dimension, being flow-specific in shape and composition.
- recurrent, having a characteristic life-span - typically of the average passage time of a structure.
- they exhibit a high measure of organization in structure as well as in dynamics yet their appearance is at best quasi-periodic.
- CS in a specific turbulent flow are similar to the corresponding L/T-transitional structures.

This definition is probably - inspite of fashionable criticisms - for most cases sufficient. It is descriptive, distinctive, exclusive and consistent. Indeed what is seen from the photographs of, say, mixing layer flow is sometimes misleading, as it suggests single structures instead of the characteristic complexity found at closer inspection. The latter has given rise also to other conceptions, e.g. where a multitude of CS in a certain flow is suggested ([16]) - see also [17].

3. ORIGINS and CAUSES - KOSMOS/CHAOS -

This question touches on a phenomenon of greater generality: Spontaneous formation of organized structures out of the chaotic is observed in unorganic and organic nature. Known as Synergetics it has received strong attention during the last, say, ten years ([18]). Chaos Theory, a specific approach of the above, has also recently become popular as a possible basis for the description of flow turbulence. Thus a CS might be interpreted as a physical manifestation of a strange attractor. Whether this concept will - for the time being - further our understanding is, however, an open question.

Encouraged by the similarity of transitional - and turbulent structures formation of the latter is generally treated as a phenomenon of instability of the mean turbulent velocity profile. The ambiguity of this procedure is in the fact that this profile has no physical reality. Still the results seem to justify the method: For a plane shearlayer (e.g.) the characteristic line vortices are obtained. They are - as we know from visualization - unstable along their axes, developing - as do ring vortices - secondary instabilities ([19],[20],[21]). In addition there are phenomena of feedback and of resonance which may have bearing on formation and the quasi-periodic recurrence of the structures ([22],[23],[24]). Arguments for helicity ($\underline{c} \times \Omega$) being intrinsic of three-dimensional turbulence and as such the basic element of coherent structures have been presented in [25],[26],[27]. While it is clearly understood that helicity is vital for turbulence it seems, however, (to this author) to be neither necessary nor sufficient for creation or existence of coherence.

4. 1st DESCRIPTION - MORPHOLOGY/TOPOLOGY

For a morphological taxonomy CS as identified in "simple" flows can be reduced to a small number of elements. These are

- Line-vortex, Hairpin-vortex, Ring-vortex, Helical-vortex -

For topological description critical point theory was introduced by [28] (also [29],[30],[31]). Structures are characterized by Nodes, Saddles and Foci. Topological and morphological elements are shown in Fig. 2. While we find a high degree of organization in the structures themselves their temporal sequence remains, however, stochastic, showing the characteristic phenomenon of structural intermittency ([32]).

5. METHODS of ANALYSIS - METHODOLOGY/ANATOMY -

To obtain information about structural characteristics in a turbulent flow recurrent events of certain similarity are collected, from which a typical structure is obtained by ensemble averaging. Necessary techniques are of almost unlimited complexity and expenditure. One of the basic requirements therefore is the digital computer, the leading role of which is unquestioned in modern turbulence research.

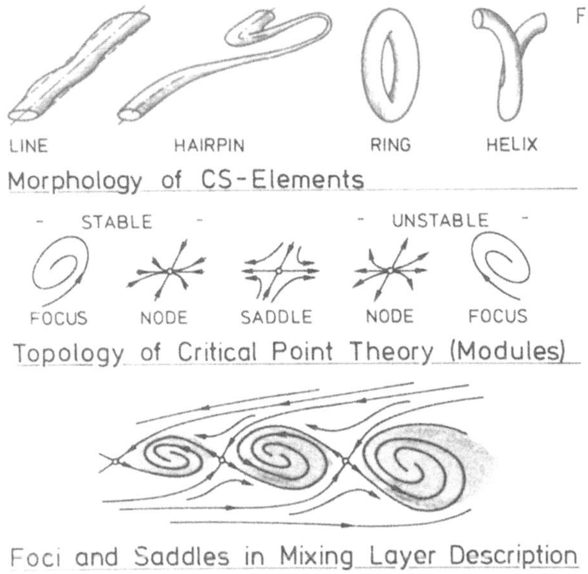

Fig. 2: Elements of CS description

Apart from the again popular visualization-methods which are advantageously used in combination with synchronous hot-wire measurements [100] - two ways to obtain hard data are followed:

1. Conditional averaging and
2. Pattern recognition

In addition there are methods of signal manipulation, by which a coherent content is amplified against (extracted from) the stochastic background (e.g. adaptive filtering ([33]), minimum entropy method ([34]). Introduction of passive scalar tracers is useful to identify CS. While simple correlation-techniques appear to be of limited use for problems of this kind, mode analysis ([35],[36]) (see Fig. 5) serves well for spacial interpretation of structures for which (in most cases) the above procedures give but a two-dimensional cross-section. Conditional averaging necessitates a trigger signal of sufficient stringency. Difficulties arise when applying this method to fully turbulent flow, primarily caused by:

- Phasejitter (Fluctuation around the mean phase-position)
- varying size and shape of single structures
- one-dimensional, locally undefined cut through essentially three-dimensional structure

To minimize these problems stabilization by periodic excitation of the basic flow at natural frequency in the dominant mode is often applied ([37],[38],[39],[40],[41],[42]). It is, however, not fully clarified as to what extent and detail the "frozen" structures represent the natural ones. In excited flow secondary modes as well as secondary instabilities are suppressed and the structure obtained is too "clean" to be natural.

Comprehensive discussions of problems in conditional averaging techniques is found in [43],[44],[45],[46].

Pattern recognition is - in comparison - even more extravagant. What makes it principally different from the conditional averaging methods is its independence of an external reference. Pattern recognition methods work only with the basic signal, out of which - on the basis of certain criteria - the characteristic structure is educed. They are therefore more versatile than conditional averaging schemes yet give not always unambigous results. A discussion of some better known methods (VITA [47],[48] and TPAV [49]) is presented in [33] together with a new scheme (see also [50] and [51]). For the sake of completeness the quadrant analysis ([52],[53]), and the search for CS via vortex models ([54]) are mentioned. Both methods have been successfully employed in wall-bounded flows.

6. 2nd DESCRIPTION - DYNAMICS -

Refined analysis of a turbulent flow requires refined description as compared to the classical Reynolds-decomposition. Splitting of turbulent shear stress into an organized and a stochastic component was already proposed in [55]. Similar procedures are common in dealing with CS. Comprehensive discussions of the two different possibilities of decomposition and their consequences is given in [15]. Figure 3 shows vectorplots of a shear layer with periodic excitation at a particular phase position. (a) is obtained by triple-decomposition, (b) by double-decomposition (quasi-Lagrange view): in triple-d. the coherent and the stochastic fluctuations are considered additional motions superimposed on the mean motion which is thus treated as a physical reality. In double-d. on the other hand the coherent vortices (moving with celerity u_c and their velocity field are the basic flow as such. Consequences of either choice become obvious from comparison with flow visualization -(c)-: What is seen from streakline pictures is depicted by double-d. We find counter-rotating vortices in a mixing layer between the visible ones, which have no physical reality.

7. SCENARIOS

Turbulent flows undergo metamorphoses before reaching a final state and every intermediate state is characterized by typical flow structures. While transient states generally lead to qualitative changes in structure it appears that L/T-transitional structures are largely similar to the subsequently developing turbulent ones. Consequently transitional structures are taken as models for CS in the corresponding flow, some of which are discussed below.

A. FREE FLOWS

The plane mixing layer:

This particular flow has not only lead the CS chase, but also remained probably the most often and most closely investigated "simple" configuration ([56]). Still there is a notable number of poorly understood items, most of which have been discussed

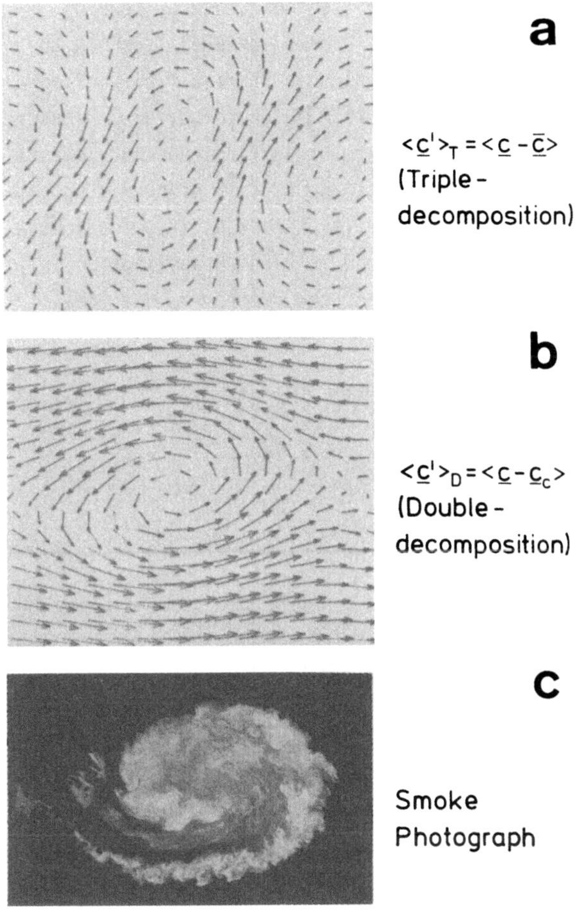

Fig. 3: Decompositions

in the workshop "Puzzles in free shear layers" ([57]). One of the most controversial issues appeared to be the question of whether pairing is central to spread. Preferred modes - a phenomenon of all free shear flows - is again - inspite of some strong statements - not fully understood ([58]) and the same is true for the coupling of large (coherent) scales and the small or medium (stochastic) scales ([59], [60]). Typical Strouhal-values for the preferred mode ("column-frequency") are ~.45 ([61],[62], [33]). Another important aspect is mixing as a consequence of CS ([63]). Influences of density differences of the two streams on the strength of CS as investigated by [1] has been taken up again and is revisited in our laboratory by numerical as well as by experimental route.

Now to the structure: basically two-dimensional line-vortices are formed in transitional ([64]) and turbulent flow alike. They develop three-dimensionalities, i.e. transverse wavyness. Longitudinal vortices (helices) of dimension typically an order of magnitude smaller than the primary structures appear on the interconnecting braids ([65]). It is there that the turbulence energy is produced ([66],[67]). Primary structures either amalgamate via double- or multiple pairing - which in itself is mostly a three-dimensional (helical) process - or lose coherence by tearing ([68]). A conceptual sketch of the typical structure - based on [69] is shown in Figure 4. Particular aspects of CS-behaviour are studied in special flow configurations in our laboratory: the accelerated shearlayer ([70]), whose taylored pressure gradient provides a situation where spread exists without pairing. Another investigation is concerned with the influence of axial strain on CS in a plane mixing layer at zero pressure gradient. Main aspect of this study is the stability of a vortex under longitudinal strain, as numerically investigated in [71]: Stabilization is expected for the expanded vortex and randomization for the compressed one (see also [72]).

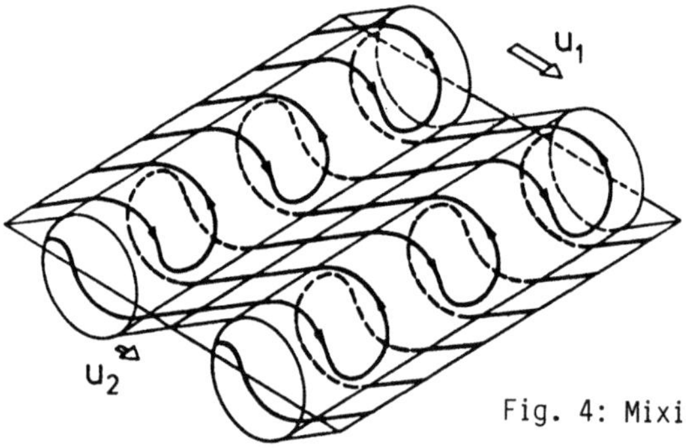

Fig. 4: Mixing layer structure

Free jets

are structurally related to shear layers in having CS of comparable simplicity: basic modules are again line vortices for two-dimensional configurations and ring- or helical vortices in axially symmetric cases (Fig.5). Both kinds of configurations have been investigated although not to the extent as the mixing layer. Substantial investigations into the far region of the round jet are found in [73],[74], [75],[33]. In [33] an average of 10% coherent energy content of the u'-fluctuation was evaluated. In [73] on the other hand the single helix is found to be the strongest structure (characteristic Strouhal-Number based on max. velocity and half-velocity-diameter is ~,5). Beginning with [76] we find a much larger number of papers on the intermediate and near jet of round and plane configurations at neutral and excited flow condition ([77],[78],[79],[80],[81]). In the excited round jet the particularly interesting phenomenon of jet splitting was observed ([82],[19]). For the plane far-jet CS were studied in [83]. "Strainwise rollers" are proposed which are also found in the far wake [84]. Interesting structural developments are also observed in three-dimensional jets ([85]), where elliptical ring vortices with their interesting dynamic evolution ([86],[87]) or the elliptical helix are basic structural moduli - as well as in jet flames with its double row of vortices ([88], [89]).

The Turbulent Wake:

Here CS were always known but taken as a specific feature. Papers on this flow are

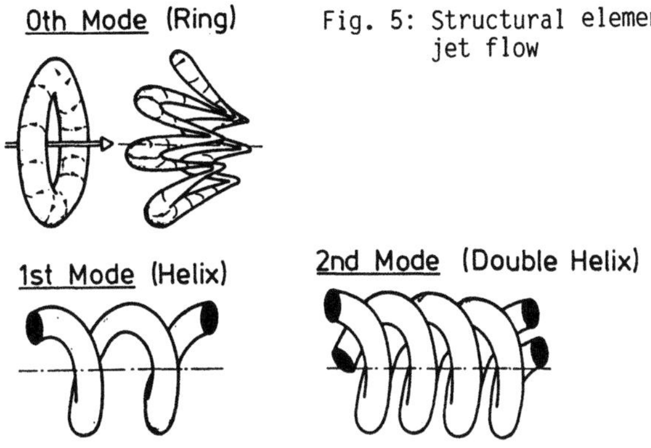

Fig. 5: Structural elements of jet flow

numerous and stabilization of the shed vortices was a well established technique long before the advent of the CS concept. There is little point in commenting on any of the well known results. Some recent investigations are, however, worth reporting: In [90] it was found that CS in the two-dimensional far wake appear in staggered as well as in varicose mode. Another interesting phenomenon is the "unforgetfulness" of the flow for upstream boundary conditions. Different spread rates as well as turbulent intensity distributions are found for different bodies of equal resistance - which touches upon the frequently disputed question of the uniqueness of self-preserving states. A recent investigation into the near wake structure of a stationary-, an axially vibrating- and a tumbling circular disc was reported in [91]. Dominating helical mode was found by inference from the perfect lock-in behaviour of the tumbling case. Comparative measurements with a sphere showed little difference other than the expected Reynolds-dependence, which in fact is also observed for the analogous situation of a jet issuing from a venturi nozzle.

B. WALL-BOUNDED FLOWS

The wall-boundary layer:
The turbulent boundary layer is the final result of transitional spots having eventually grown together. Consequently the idea nearest at hand is to take the spot as the basic structure of the fully developed flow. This was particularly fostered by Coles who investigated the "synthetic boundary layer" as a superposition of evenly spaced, artificially created spots in a laminar surrounding ([92],[93]). Prime objective of these studies was to find typical signal traces, then to be taken for the characteristic "footprint" of the turbulent structure. The spot is - as many investigations ([94],[95],[96]) have demonstrated, an agglomeration of hairpin vortices, the number of which increases with increasing Reynolds-number. So it seems reasonable to assume that not the spot as such but its constituents, i.e. the vortices are the basic structural elements of the boundary layer, The majority of observations support, however, the concept of a more complex scenario of structures and their "cyclic" interplay rather than simple CS, which indeed in boundary layers (as in all wall bounded flows) appear to be indistinct as compared to, say,

structures in the mixing layer. Thus when studying the literature on boundary layer flow one is soon lost in a zoo of structures, e.g. horseshoe- and hairpin - eddies ([95],[97]), pancake- and surfboard - eddies ([4]), typical eddies, vortex- rings, mushroom eddies ([97],[98]), arrowhead - eddies ([99]), etc. .. . The "official" picture of boundary layer structure is made up of a considerable number of elements ([100]):

The viscosity-dominated wall region is characterized by low-speed-**Streaks** ([7], [101]) and **Bursts** ([102],[103]). A process connected with the burst is the **Ejection.** Other elements are **Pockets** ([104]) and **Large Scale Sweeps.** - It is suggested by all observations that it is the structures and the structural events in the wall region that are responsible for the production of turbulent energy.

The inviscid outer region is characterized by **Large Scale Motions** and **Mushroom Eddies** ("Falco-Eddies" or "Typical eddies"). Comparing this complex picture with the well-documented transitional spot we find that resemblance is but superficial. In [105] eduction of the "characteristic eddy" in a turbulent boundary layer and comparison with the transitional spot was undertaken, where indeed significant differences were found. It was on the other hand convincingly shown in [97] that some of the structural varieties as reported above seem to be mainly in the eye of the beholder and that the CS seems to be essentially the simple Re-depending "hairpin". This attractive concept - sketched in Fig. 6 - finds support in [106],[107],[108], [109] thereby supporting our hope for a limited and manageable system.

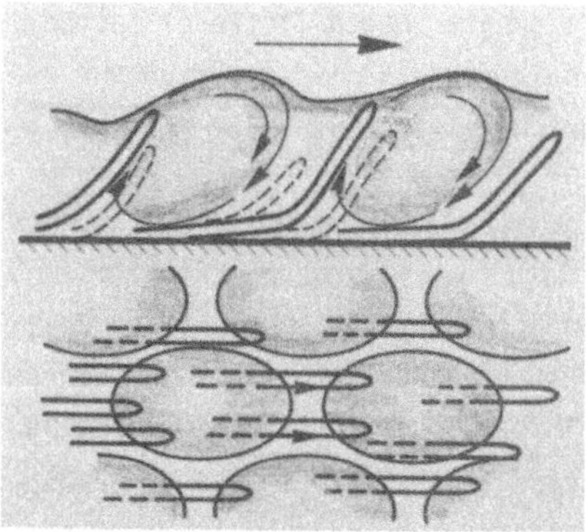

Fig. 6: Boundary layer structure

"characteristic eddy" in a turbulent boundary layer and comparison with the transitional spot was undertaken, where indeed significant differences were found. It was on the other hand convincingly shown in [97] that some of the structural varieties as reported above seem to be mainly in the eye of the beholder and that the CS seems to be essentially the simple Re-depending "hairpin". This attractive concept - sketched in Fig. 6 - finds support in [106],[107],[108],[109] thereby supporting our hope for a limited and manageable system.

Other wall flows:

For Poiseuille flow COLES [100] assumed the typical structure to be the ring vortex - as in a turbulent puff. In [111], however, helicity was reported to exist in the core region of pipe flow. From a recent investigation ([112]) we know that in pipe flow - as in most other axisymmetric configurations - the 1st (helical) mode has approximately the same probability as the 0th (and in some cases also the 2nd) mode. This corroborates the existence of helicity without substantially contradicting Coles' model, since it shows co-existence of different modes. Helical structures were also recently discussed in ([113]).

Channel flow has been looked into in [114]. Structures and characteristic time time scales of retardation- and acceleration-events were obtained by VITA-method. Open channel flow was studied in [54], where oblique vortical structrues (with positive and negative (!) tilt) were identified. They are considered responsible for production of turbulent shear stress.

8. ENTRAINMENT

was one of the conceptually controversial phenomena in turbulent shear flows for long time: The original understanding was based on the idea of "small scale nibbling" at the interface ([115]). Much work has since then been dedicated to its consistent physical explanation and discussion would be incomplete without mention of TOWNSENDS's [116] work - see also [117]. Actual proof for the connection between entrainment and large scale (coherent) motions corrugating the free surface of a turbulent flow and entraining fluid from the potential flow region by "engulfment" or "folding" was finally given in [118]. This concept was corroborated by visualization ([119]).

Entrainment is one of the main features of lateral growth of a turbulent flow and it has been conjectured that in a shear layer spread is by vortex pairing alone. It was, however, shown in [120] that only about 20% of the total entrainment flux goes to the credit of pairing and indeed even the heretical suggestion of pairing being altogether a chimaera was put forward in [41], being possibly a wrong interpretation of streak-line patterns. In any case it seems undisputed by now that the phenomenon of pairing - be it real or not - is of less significance for the flow development than was believed hitherto (see [38] and [70]).

Another issue of dispute was the asymmetry of the mixing layer flow and its consequence with regard to the entrainment-flux ratio being different from unity. A typical ratio of $\dot{V}_{HIGH}/\dot{V}_{LOW}$ ~1.4 for the one-stream layer was already reported in [121].

9. INFLUENCES - MANIPULATION -

This is one of the most interesting and complex aspects of the coherence-phenomenon. Knowledge of dependence of CS on boundary conditions, fluid-characteristics or any other condition opens ways to manipulating turbulent flows via influencing its CS.

Meditations about this question are, however, only useful once it is established that the structures to be influenced have sufficient bearing on the dynamics of the flow considered. We know that

- Mixing and combustion (Entrainment)
- Noise production and -emission and
- Wall-shear (frictional drag)

are essentially determined by large scale structures. Exact assessment of the fraction E_{CS} of the total turbulent energy E is practically impossible. Estimates will, however, suffice. We find for

plane mixing layer	:	~ 20% ([38],[40])
far jet	:	~ 10% ([33],[83])
near jet	:	~ 50% ([122])
far wake	:	~ 20%
near wake	:	~ 25% ([91])

From these data (for wall bounded flow we may assume fractions of order 10%) improvements of turbulence models by adequately modelling the coherent constituents seem possible and manipulation of turbulent flows via influencing CS promising.

Some of the more important activities and obvious results pertinent to this concept are the following:

- coherent structures are recognized as principal sources of flow-noise. Their suppression or destruction significantly reduces the radiated noise level ([123], [124],[125],[126]). The acoustic sources are, however, not convected but are confined to the locations of vortex pairing ([127]). Thus it is not the CS per se that are responsible for the noise but their formation.
- effects of CS in chemical engineering processes, in mixing and combustion are well known and have been a central object of research in recent years ([128], [88],[89]).
- manipulation of CS was shown to be relevant for problems of airplane design: high lift is achieved by delay of separation on a stalled airfoil via excitation of CS ([42],[129]). Possibilities of drag reduction on the other hand via CS-manipulation in boundary layers appears already within reach.

How then can these structures be efficatiously handled: We shall sketch some of the better understood ways. Thus enhancement of coherent structures is possible:

- by periodic excitation ([40],[41],[42])
- polymere-additives produce significant intensification of CS-constituents and at the same time attenuation of the incoherent small-scale structure in free shear flows ([130],[131]). Reduction of wall shear stress in turbulent wall flows by polymeres is known for some time.Correlations of this effect are, however, only recent: significant increase of frequency and of lateral dimension of the pockets in a boundary layer as a result of polymere (Sapran AP273) additives was found in [132].

Suppression of CS is achieved by

- periodic excitation at higher frequencies ([133],[134]).
- special boundary conditions or additional manipulators to destroy or disrupt the coherent structures, e.g. winglets (Large Eddy BreakUp devices) - [135], [136]-, straighteners, screens- [137],[33]) -
- A different concept is derived from a detailed study of sharkskin, the simplest abstraction of which appears to be a surface with longitudinal parallel grooves. Cross-sectional shape and - in particular - lateral spacing of the grooves is of essence, since the most evident explanation for its effect is its interference with the sublayer streaks ([138]).

10. SOME FINAL REMARKS

Basic understanding (and benefit) of coherent structures becomes questionable or limited in view of the multitude and apparent complexity of structural shapes and dynamics as found in most wall-bounded flows as well as in jet flow. Looking at some smoke pictures we are easily convinced of the existence of "characteristic structures" as well as of "characteristic (cyclic) motions", which may serve as a basis for physical understanding and mathematical modelling; but it needs imagination to see coherence in the simple sense suggested by pictures of mixing layer flow. For one the various structures and sub-structures seem ambiguous - which is understood in view of the experimental difficulties described - and the picture becomes progressively vaguer the more differentiated and sophisticated it is painted. One must ask then: is it worth trying to improve on it in the sense of further differentiation? The answer is less simple than suggested by the seemingly trivial Leitmotif: From the predictor's point of view, who searches for a sensible and manageable model of maximum efficiency it would be unwise to simulate every detail of sub- and sub-sub-structures. For the researcher seeking understanding the limit may be further away but it is there, since the CS-concept will otherwise inevitably metamorphose into the statistical. One of the many differences in structural dynamics between wall- and free flows is its genealogy or development: while in free flows we observe creation of new structures preferrably by some kind of amalgamation, the corresponding process in wall flows seems better described as agglomeration and/or decay and reformation. On the other hand there is an important piece of mutuality and thus universality: In all kinds of shearflows we find at least two kinds of structures co-existing:

1. primary structures of large dimension, comparable to the overall (lateral) flow dimension and
2. secondary (longitudinal) structures of Reynolds-No.-depending dimensions, typically one order of magnitude smaller than the primary structures.

Secondary structures are responsible for energy production via vortex stretching ([68],[69]);the primary structures control the (energy-, momentum-, ...) exchange across the flow.

Because of space limitations discussion of modelling CS had to be omitted. Also nothing has been said about CS in complex or disturbed flows, as I believe, that despite all the knowledge collected so far on "simple" flows we are not yet in the position to extrapolate that (only partly understood) knowledge beyond the known limits. For this we need to know the general principles and the "law of order" for the CS by much more than the bits and pieces we have been able to untangle so far. There is, on the other hand, a need for investigations into the more complex and more than just a brave few ought to find the moral strength to seriously doing it. One must keep in mind that, at least in a technical sense, the CS-concept will prove itself useful only as long as the number of CS-moduli remains small and thus allows for a useful description of an arbitrary number of flow configurations. This touches on the question of research strategies: The majority of experiments done so far were specifically "clean" ones in - preferrably - self preserving flow regions etc., which in a scientific sense should yield the best, i.e. least ambiguous information. Flows of practical relevance are, however, almost invariably unclean, disturbed, and/or - at least for free flows - transient (i.e. "young") and it is particularly in these regions that CS are dominant. Maybe the systematic approach via clean and simple scenarios is too much of a detour to come to an understanding and thus to a basis of description and manipulation of more complex cases. It seems that those ought to be taken on by way of a second, more direct route.

12. LITERATURE

1. G.L. Brown, A. Roshko: J. Fluid Mech. 64,4, 775 (1974)
2. L. Prandtl: ZAMM, Bd 5, 136 (1925)
3. P. Bradshaw: Nature 249, 135 (1974)
4. M.T. Landahl: Z. Flugwiss. Weltraumforsch. 8, 4, 233 (1984)
5. A.A. Townsend: Turbulent Shear Flow, (Cambr. Univ. Press 1956)
6. I.S. Gartshore: J. Fluid Mech. 24,1, 89 (1966)
7. S.J. Kline, W.C. Reynolds, F.A. Schraub, P.W. Rundstadler: J. Fluid Mech. 30,4 741 (1967)
8. F. Hama: In S. Corrsin: Publication 515, Natl. Academy of Sciences - Natl. Res. Council 1957
9. J.R. Ruetenik: Rep. No. I-19, The Johns Hopkins University 1954
10. E.R. Corino, R.S. Brodkey: J. Fluid Mech. 37, 1, 1 (1969)
11. H. Eckelmann: MPI Göttingen, Nr. 48 1970
12. C. Chandrsuda, R.D. Mehta, A.D. Weir, P. Bradshaw: J. Fluid Mech. 85, 693 (1978)
13. R.E. Breidenthal: Ph.D. Thesis CALTECH 1978
14. I. Wygnanski, D. Oster, H. Fiedler, B. Dziomba: J. Fluid Mech. 93,2, 325 (1979)
15. A.K.M.F. Hussain: Phys. Fluids, 26 , 10, 2816 (1983)
16. H. Tenneckes: In Turbul. and Predictability in Geophys. Fluid Dynamics and Climate Dynamics, ed. M Ghil (North Holland 1985)

17. B.J. Cantwell: Ann. Rev. Fluid Mech. 13, 457 (1981)
18. H. Haken: Synergetik - eine Einführung (Springer 1983)
19. A. Glezer: PhD Thesis, CALTECH 1981
20. E.M. Schneider: Z. Flugwiss. Weltraumforsch. 5, 307 (1980)
21. S.E. Widnall, J.P. Sullivan: Proc. R. Soc. A 322, 335
22. C.-M. Ho: In Numerical and Physical Aspects of Aerodynamic Flows, ed. T. Cebeci (Springer 1981) 521
23. J. Laufer, P.A. Monkewitz: AIAA Paper No. 80-0962 (1980)
24. W. Koch: J. Sound Vibr. 99, 53 (1985)
25. E. Levich, A. Tsinober: Physics Letters, 93A 6, 293 (1983)
26. A. Tsinober, E. Levich: Physics Letters, 99A, 6,7, 321 (1983)
27. R.Z. Sagdeev, S.S. Moiseev, A.V. Tur, G.A. Khomenko, V.V. Yanovsky: In Self-Organization, ed V.I. Krinsky,(Springer 1984) p. 74
28. A.E. Perry, B.D. Fairly: Geophysics 18B, 299 (1974)
29. H. Poincaré: In Oeuvres de Henri Poincaré, Tôme I (Gauthier-Villars, Paris 1951) XLIX-CXXIX
30. D. Coles: AIAA 23rd Aerospace Sciences Meeting 1985
31. A.E. Perry, M.S. Chong: In Turbulence and Chaotic Phenomena in Fluids, ed. T. Tatsumi, North Holland 1984) p. 351
32. M. Lesieur: In Turbulence and Chaotic Phenomena in Fluids, ed. T. Tatsumi (North Holland 1984) p. 339
33. R. Nieberle: Dissertation TU Berlin 1985
34. T.C. Corke, Y.G. Guezennec: AIAA Journal 24,3 377 (1986)
35. A. Michalke, H.V. Fuchs: J. Fluid Mech. 70 179 (1975)
36. H.V. Fuchs, E. Merker, U. Michel: In Turbulent Shear Flow 2, ed. Bradbury, Durst, Launder, Schmidt, Whitelaw (Springer 1980) p. 282
37. D. Oster, I. Wygnanski: J. Fluid Mech. 123 91 (1982)
38. H.E. Fiedler, P. Mensing: J. Fluid Mech. 150 281 (1985)
39. A.K.M.F. Hussain, H.S. Fusain, K.B.M.Q. Zaman, J. Tso, M. Hayakawa, R. Takaki, M.A.Z. Hasan: AIAA 24th Aerospace Meeting 1986
40. C.-M. Ho, P. Huerre: Ann. Rev. Fluid Mech. 16 365 (1984)
41. I. Wygnanski, R.A. Petersen: AIAA Shear Flow Contr. Conf. 1985
42. J.P. Stone, J. McKinzie Jr.: NASA Tech. Memo 83772 1984
43. R.A. Antonia: Ann. Rev. Fluid Mech. 13 131 (1981)
44. R.F. Blackwelder: Phys. Fluids 20, 10 S232 (1977)
45. A. Yule: In Turbulent Shear Flows 2 , ed. Bradbury, Durst, Launder, Schmidt, Whitelaw (Springer 1980)
46. A.K.M.F. Hussain: AIAA 24th Aerospace Sciences Meeting 1986
47. R.F. Blackwelder, R.E. Kaplan: J. Fluid Mech. 76, 89 89 (1976)
48. A.J. Chambers, R.A. Antonia, L.W.B. Browne: In Turb. Shear Flows 4, ed. Bradbury, Durst, Launder, Schmidt, Whitelaw (Springer 1983)
49. J.M. Wallace, R.S. Brodkey, H. Eckelmann: J. Fluid Mech. 83 673 (1977)

50. T.Y. Young: Handbook of Pattern Recognition, Acad. Press 1986
51. J.M.G. Kunen: PhD Thesis, Delft University of Technology 1984
52. S.S. Lu, W.W. Willmarth: J. Fluid Mech. 60 481 (1973)
53. J.M. Wallace, H. Eckelmann, R.S. Brodkey: J. Fluid Mech. 54 39 (1972)
54. Th. Dracos: Inst. für Hydromech. und Wasserw., ETH Zürich 1984
55. S.-I. Pai: NACA TN 892 (1943)
56. F.K. Browand, C.-M. Ho: J. de Mêchanique (1983)
57. Workshop "Puzzles in Turbulent Mixing Layers", C.-M. Ho, P. Huerre, Providence, R.I., 1984
58. A. Michalke: Progress Aerospace Sci. 21 159 (1984)
59. J.T.C. Liu: In Transition and Turbulence,(Acad. Press 1981) p. 167
60. J.G. Brasseur, S. Corrsin - this volume
61. A.K.M.F. Hussain, K.B.M.Q. Zaman: In Lecture Notes in Physics 75, ed. H. Fiedler, (Springer 1978) p. 31
62. C.-M. Ho, F.B. Hsiao: In Structure of Complex Turbulent Shear Flows, ed. Dumas, Foulachier (Springer 1983) p. 121
63. M.M. Koochesfahani: PhD Thesis CALTECH 1984
64. P. Freymuth: J. Fluid Mech. 25 683 (1966)
65. S.J. Lin, G.M. Corcos: J. Fluid Mech. 141 139 (1984)
66. A.K.M.F. Hussain: In Turbulence and Chaotic Phenomena in Fluids, ed. T. Tasumi (North Holland 1984) p. 453
67. D. Coles: In Turbulence and Chaotic Phenomena in Fluids, ed. T. Tatsumi (North Holland 1984) p. 397
68. P.G. Saffman: In Lecture Notes in Physics 136, ed. J. Jimenez (Springer 1981)
69. L.P. Bernal: PhD Thesis CALTECH 1981
70. H.E. Fiedler: In Turbulence and Chaotic Phenomena in Fluids, ed. T. Tasumi (North Holland 1984) p. 411
71. R. Takaki: J. Phys. Soc. Japan 38,5 1530 (1975)
72. F.F. Keffer, J.G. Kawall, J.C.R. Hunt, M.R. Maxey: J. Fluid Mech. 86,3, 465 (1978)
73. J. Tso: PhD Thesis, The Johns Hopkins University, 1983
74. J. Tso, L.S.G. Kovasznay, A.K.M.F. Hussain: AIAA-80-1355, 1980
75. S. Kumori, H. Ueda: J. Fluid Mech. 152, 337 (1985)
76. S.C. Crow, F.H. Champagne: J. Fluid Mech. 48, 547 (1971)
77. H. Fiedler, D. Korschelt: Proc. Turbulent Shear Flow 2, London 1979
78. L. Bernal, V. Sarohia: JPL Publication 84-91, 1984
79. M. Favre-Marinet, G. Binder, Hak, V. Te: J. Fluid Engg. 103, 609-614 (1981)
80. G. Binder, M. Favre-Marinet: Unsteady Turbulent Shear Flows (Springer 1981)
81. R.A. Peterson: J. Fluid Mech. 89,3, 469 (1978)
82. A. Leonhard: Annual Rev. Fluid Mech. 17, 523 (1985)
83. J.C. Mumford: J. Fluid Mech. 118, 241 (1982)
84. J.C. Mumford: J. Fluid Mech. 137, 447 (1983)

85. H.S. Husain, A.K.M.F. Hussain: Phys. Fluids 26(10), 2763 (1983)
86. H.J. Lugt: Wirbelströmungen in Natur und Technik, G. Braun (Wissenschaft und Technik, Karlsruhe 1979)
87. M.R. Dhanak, De Bernardinis: J. Fluid Mech. 109, 189 (1981)
88. H. Eickhoff: Prog. Energy Combust. Sci. Vol. 8, 159 (1982)
89. H. Eickhoff: In Recent Contribution to Fluid Mechanics, ed. W. Haase (Springer 1982)
90. I. Wygnanski, F. Champagne, B. Marasli: Rep. Dept. of Aerospace and Mech. Engg. Univ. of Arizona, 1983
91. D. Scholz: Dissertation TU Berlin 1985
92. Ö. Savas, D. Coles: J. Fluid Mech. 160, 441 (1985)
93. M. Zilbermann: PhD Thesis, Tel Aviv University 1981
94. I. Wygnanski: In Lecture Notes in Physics, ed. J. Jimenez, 136 (Springer 1981) p. 304
95. A.E. Perry, T.T. Lim, E.W. Teh: J. Fluid Mech. 104, 387 (1981)
96. J.J. Riley, M. Gad-el-Hak: In Frontiers in Fluid Mechanics, ed. Davis & Lumley (Springer 1985) p. 123
97. P.S. Jang, D.J. Benney, Y.M. Chen: Bull. Am. Phys. Soc. II, 29, No. 9, Paper BB8, p. 1528 (1984)
98. J. Laufer, M.A. Badri Narayanan: Phys. Fluids 14 , 182 (1971)
99. R.F. Blackwelder, J.H. Haritonidis: J. Fluid Mech. 132, 87 (1983)
100. R.E. Falco: Phys. Fluids 20 (10), 5124 (1977)
101. M.R. Head, P. Bandyopadhyay: J. Fluid Mech. 107, 297 (1981)
102. R.E. Falco: Report TSL 82-2, MSU 1982
103. R.E. Falco: In Summary of the AFOSR/MSU Res. Spec. Workshop on Coher. Struct. in Turb. Boundary Layers,AFOSR-TR-80-0290, 1979
104. V. Zakkay, V. Barra: ASME 82-GT-78
105. M. Zilbermann, I. Wygnanski, R. Kaplan: Res. Rept., Tel Aviv Univ., School of Engg.
106. A.C.M. Beljaars: PhD. Thesis, Techn. Hogeschool Eindhoven 1979
107. J.M. Wallace: In Turbulence and Chaotic Phenomena in Fluids, ed. T. Tatsumi, (North Holland 1984) p. 447
108. T.J. Black: AIAA-Paper No. 68-42 (1969)
109. R.E. Britter, D.D. Stretch: this volume
110. D. Coles: Proc. Indian Acad. Sci. (Engg. Sci.) 4,2, 111 (1981)
111. J. Sabot, G. Comte-Bellot: J. Fluid Mech., 74,4, 767 (1976)
112. B. Stahl: Dissertation TU Berlin 1986
113. P.R. Bandyopadhyay: J. Fluid Mech. 163, 439 (1986)
114. A.V. Johannson, P.H. Alfredsson: J. Fluid Mech. 122, 295 (1982)
115. S. Corrsin, A. Kistler: NACA T.N. No. 3133 (1954)
116. A.A. Townsend: J. Fluid Mech.41,1, 13 (1970)

117. R. Chevray: Proc. Energy Combust. Sci. 8, 303 (1982)
118. P.M. Bevilaqua, P.S. Lykudis: AIAA Journal 15,8, 1194 (1977)
119. P.E. Dimotakis, R.C. Miake-Lye, D.A. Papantoniou: GALCIT Report FM82-01 CALTECH (1982)
120. M.A. Hernan, J. Jimenez: J. Fluid Mech. 119, 323 (1982)
121. H.E. Fiedler: Advances in Geophysics 18A, 93 (1974)
122. V. Kibens: AIAA Journal 18, 434
123. A. Michalke: In Lecture Notes in Physics 76, ed. H. Fiedler (Springer 1978)
124. D.G. Crighton: In Lecture Notes in Physics 136, ed. J. Jimenez (Springer 1981)
125. R.M. Armstrong, A. Michalke, H.V. Fuchs: AIAA J. 15,7, 1011 (1977)
126. D. Bechert, E. Pfizenmaier: AIAA Paper No. 76-489 (1976)
127. J. Laufer, Yen Ta-Chun: J. Fluid Mech. 134, 1 (1983)
128. N. Peters, F.A. Williams: In Lecture Notes in Physics 136, ed. J. Jimenez (Springer 1981) p. 364
129. I. Wygnanski: Lecture APS-Meeting 1985 (Tucson)
130. M. Kwade: Rheologica Acta 21, 120 (1982)
131. M.F. Hibbert: this volume
132. H. Yoda: Thesis, Michigan State University 1981
133. E. Berger: Phys. Fluids Supplement 1967, p. 191
134. K.B.M.Q. Zaman, A.K.M.F. Hussain: J. Fluid Mech. 103, 133-159 (1981)
135. T.C. Corke: PhD Thesis, Illinois Institute of Technology 1981
136. A.P. Dowling: J. Fluid Mech. 160, 447 (1985)
137. K.S. Yajnik, M. Acharya: In Lecture Notes in Physics, ed. H. Fiedler, (Springer 1978) p. 249
138. D. Bechert, M. Bartenwerfer, G. Hoppe, W.-E. Reif: ICAS, Paper No. 86-1.8.3, London 1986

Topology of Organised Structures in a Turbulent Plane Wake

R.A. Antonia, L.W.B. Browne, and D.K. Bisset

Department of Mechanical Engineering, University of Newcastle, N.S.W., 2308, Australia

1. Abstract

The velocity vector field associated with the organised motion in the turbulent far-wake of a circular cylinder has been determined by first identifying spatially coherent temperature fronts. The emerging topological picture underlines the importance of the diverging separatrix and the saddle point, especially with regard to the entrainment of ambient temperature irrotational fluid. It is also concluded that the organised motion is more efficient at transporting heat than momentum. This conclusion supports that obtained in the self-preserving region of a turbulent plane jet by ANTONIA et al [1].

2. Introduction

The topology of the turbulent Karman vortex street in the near-wake of a circular cylinder has been determined using different conditional sampling and averaging techniques [2-4]. The Karman street eventually disappears but as the distance from the cylinder increases, a relatively ordered larger scale motion can be observed, albeit intermittently with respect to time and space. Several interesting aspects of this motion have already been highlighted by either flow visualisations [5-8] or hot wire measurements [9-11]. Topological information similar to that obtained in the near-wake for the plane of main shear is not yet available for the far-wake.

In a previous paper [1], the topology of the self-preserving region of a far jet was obtained by using temperature as a passive marker of the flow and by conditioning the velocity field on the detection of spatially coherent temperature fronts. The present paper, which considers the self-preserving region of a turbulent wake, uses the same experimental strategy. The velocity vector field as well as the temperature field, both conditioned on the detection of temperature fronts, are used to gain some insight into the physics of the large scale motion, its contribution to the averaged momentum and heat transfers, and the part this motion plays in entraining irrotational fluid.

3. Experimental Details

Only a brief description of the experiments is given here; a more detailed description was given in [12]. Wind tunnel measurements were made with zero pressure gradient at a distance of 420 diameters downstream of a heated circular cylinder (d = 2.67 mm) and a Reynolds number $U_1 d/\nu$ of about 1200 (U_1, the free stream velocity, is 6.7 m/s) with two experimental arrangements. In the first (Fig. 1), a spanwise array of cold wires and an X-probe/cold wire combination, centrally positioned in the rake, were traversed across the wake. In the second, the spanwise array of cold wires was fixed at one lateral position while the X-probe/cold wire combination was traversed.

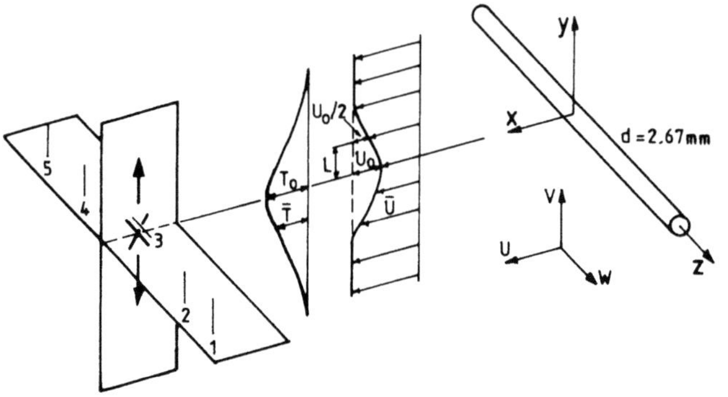

Fig. 1 Definition sketch and experimental arrangement

All cold wires were made of 0.63 μm diameter Wollaston (Pt-10% Rh) with an active length of 1.0 mm. These wires were operated with in-house constant current circuits, the magnitude of the current (0.1 mA) being small enough for the wires to respond essentially to the temperature fluctuation. In the arrangement used (Fig. 1), cold wires 1,2,4,5 are aligned in the y direction while wire 3 is in the z direction, perpendicular to the plane of the X-probe and located 0.5 mm upstream of the crossing point of the X-wires. The X-wires were made of 5 μm diameter Wollaston (Pt-10% Rh) with an active length of 0.85 mm. These wires were operated with constant temperature circuits. Signals from the constant current and constant temperature circuits were digitised, after suitable amplification and low-pass filtering, using an 11 bit + sign A/D converter and processed on a PDP 11/34 computer.

At the measurement station, the maximum velocity defect U_0 (Fig. 1) was equal to 0.36 m/s while the centreline temperature excess T_0 (Fig. 1) was 0.82°[C]. This latter value is sufficiently small for temperature to be an effective passive marker of the flow and, in particular, the large scale organised motion of the flow.

Normalised flow quantities are denoted by an asterisk : the scales U_0 and T_0 are used for velocity and temperature while L (= 12.3 mm, the half-mean-velocity-defect width) is the normalising length scale. Times have been converted to lengths by multiplying by U_c = 6.52 m/s, an appropriate convection velocity of the large scale motion [12].

4. Definition of Conditional Averages

Velocity and temperature fluctuations were conditioned on the occurrence of spatially coherent temperature fronts (such as shown in Fig. 2) detected from the cold wire

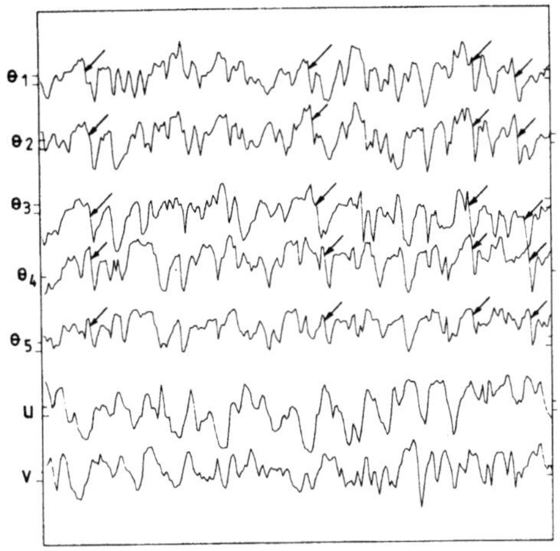

Fig. 2 Simultaneous traces of temperature and velocity fluctuations obtained with the array of wires (Fig. 1) at y* = 0.81. Time increases left to right. The duration of the record shown is 0.127 s. Temperature fronts are indicated by arrows.

data. The detection was determined using computer algorithms [13] that took into account temperature data from all wires in the array.

In previous studies of coherent structures in turbulent flows, instantaneous variables have been decomposed using either the double or the triple decompositions. The advantages and disadvantages of such decompositions have been discussed by [14] and others (e.g. [2]). Although neither decomposition is fully satisfactory, the double decomposition views the large scale coherent motion at a lower level than the triple decomposition since, in the double decomposition, this motion is essentially the mean flow. On the other hand, the triple decomposition allows the contribution from the organised motion to be more easily distinguished from that of the Reynolds averaged mean flow. Since one of the present aims is to evaluate the momentum and heat transport efficiencies of the organised motion, the triple decomposition was used. The instantaneous variables U, V, T are given by

$$\left.\begin{aligned} U &= \overline{U} + \langle u \rangle + u_s \\ V &= \overline{V} + \langle v \rangle + v_s \\ T &= \overline{T} + \langle \theta \rangle + \theta_s \end{aligned}\right\} \qquad (1)$$

where the overbar denotes Reynolds averaging, the angular brackets denotes the coherent large scale component and the subscript s denotes the small scale motion. In the extended abstract prepared for this conference [15] we described the third term on the right side of (1) as the "random" or "incoherent" motion. Neither term seems particularly appropriate in view of the significant degree of spatial organisation exhibited by this motion (e.g. [2,3]). The quantities $\langle u \rangle$, $\langle v \rangle$, $\langle \theta \rangle$ are associated with the coherent large scale motion although the dichotomy between large scale and small scale motions, as expressed in (1), may be an oversimplification.

To form $\langle u \rangle$, $\langle v \rangle$, $\langle \theta \rangle$, about 2000 pairs of detections were selected on the basis that the distance ($\simeq$ 3.4L) between successive temperature fronts coincided approximately with the average wavelength of structures. Further selection gave 500 sets of three detections arranged strictly asymmetrically about the centreline. These structures have been recognised in photographs of the far-wake (e.g. [5,6,8]) and in pattern recognition experiments based on hot wire data (e.g. [10,11]). Although the asymmetric arrangement is not the only one that is observed, there is sufficient evidence to indicate that it is the most prevalent flow mode. Computational details of the formation of the conditional averages can be found in [12].

5. Velocity Vector Pattern for the Large Scale Motion

Once $\langle u \rangle$ and $\langle v \rangle$ are determined, the velocity vector whose components are $\overline{U} + \langle u \rangle$ and $\overline{V} + \langle v \rangle$ can be constructed (here, $\overline{V}$, as obtained from the continuity equation, is very small and has been neglected). The next step in studies of coherent structures usually consists in seeking a frame of reference in which the mean flow appears to be stationary or quasi-stationary. CANTWELL [16] has pointed out that, in such a frame of reference, the integrated particle path equations coincide with the streamlines. There is inevitable arbitrariness in selecting the convection velocity U_c of the frame of reference. One would expect however the dispersion in U_c to be smaller in the far-wake than in the near-wake as a result of the smaller mean velocity defect in the far-wake. In the present work, a value of $0.97U_1$ was chosen for U_c (further details are given in [12]).

The velocity vector pattern that survives the conditional averaging procedure is characterised (Fig. 3) by readily distinguishable critical points or points at which the slope of a streamline is indeterminate : centres (marked by the letter C) and saddles (marked by the letter S). The spatial disposition of these points is consistent with the existence of a vortex street. Saddles lie at the intersection of converging and diverging separatrices (Fig. 3), the diverging separatrices coinciding approximately with the directions of the temperature fronts.

In Fig. 3 (and in subsequent figures) the co-ordinates Δx^* and y^* are used, where $\Delta x^* = -\tau\, U_c/L$ (τ is the time measured from the instant at which detection occurs). As the flow direction is right to left, positive and negative values of Δx^* indicate locations which are downstream and upstream respectively of the detection (or temperature front) location.

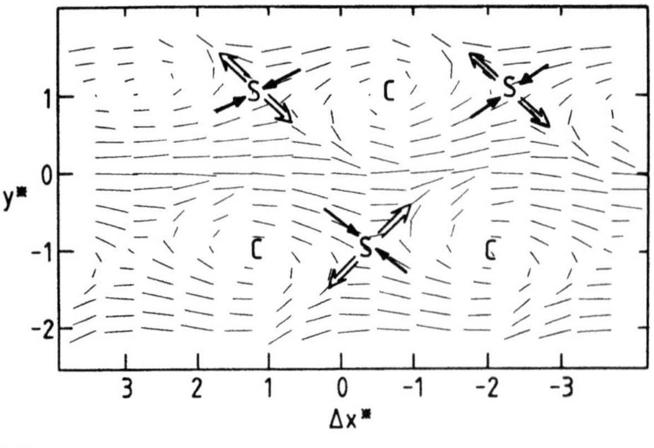

Fig. 3 Velocity vectors relative to an observer moving right to left at the convection velocity U_c. Centres and saddles are denoted by C and S respectively. Diverging separatrices are indicated by arrows directed away from S. Converging separatrices are denoted by arrows pointing to S

6. Momentum and Heat Transports of the Large Scale Motion

The information in this section relates only to the large scale motion. Information on the small scale motion will be presented elsewhere. Isocontours of $\langle u^* \rangle$, $\langle v^* \rangle$ and $\langle \theta^* \rangle$ are presented in Fig. 4 in the Δx^* vs y^* plane. These contours are generally elliptical in shape with a tendency to be flattened near the centreline. One universal feature exhibited by all contours is that the largest values occur on either side of the saddle points along the diverging separatrix. Contours of $\langle u^* \rangle$ and $\langle \theta^* \rangle$ on opposite sides of the centreline have opposite sign. Contours of $\langle v^* \rangle$ on opposite sides of the centreline are more symmetric than those of $\langle u^* \rangle$ or $\langle \theta^* \rangle$ and do not change sign. Isocontours (not shown here) of the Reynolds averaged correlation between the lateral velocity fluctuations v measured at a fixed point ($y^* = -2$) and

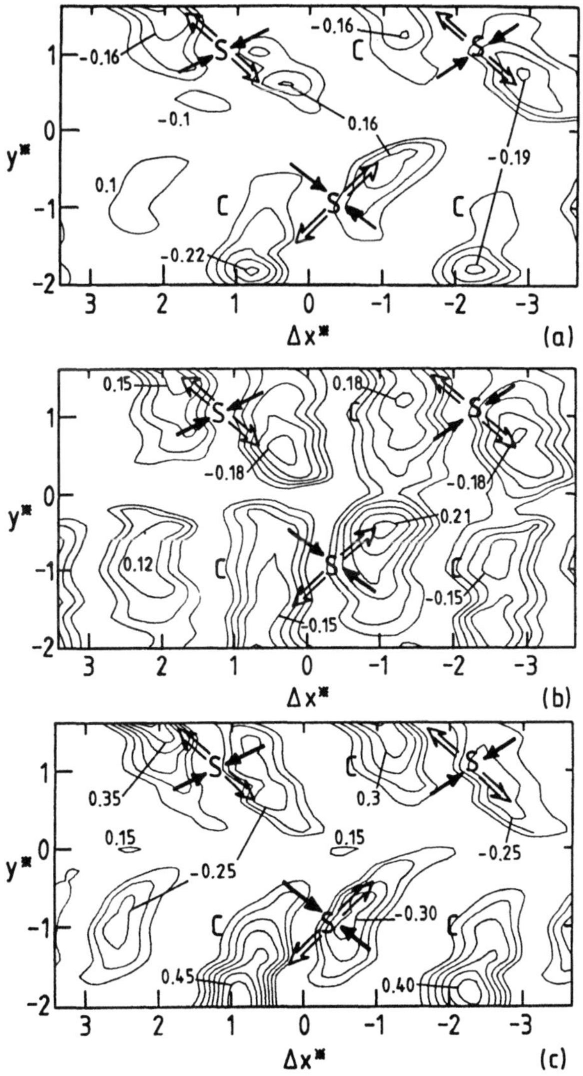

Fig. 4 Contours of constant values of $\langle u^* \rangle$, $\langle v^* \rangle$, $\langle \theta^* \rangle$. Contours are shown at intervals of ±0.03 for $\langle u^* \rangle$ (a), ±0.03 for $\langle v^* \rangle$ (b) and ±0.05 for $\langle \theta^* \rangle$ (c)

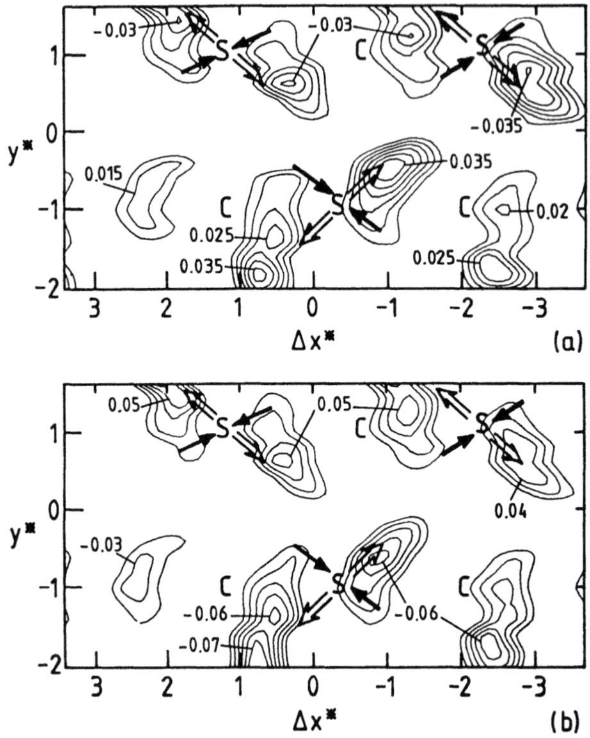

Fig. 5 Contours of constant values of $\langle u^* \rangle \langle v^* \rangle$ and $\langle v^* \rangle \langle \theta^* \rangle$. Contours are shown at intervals of ± 0.005 for $\langle u^* \rangle \langle v^* \rangle$ (a) and ± 0.01 for $\langle v^* \rangle \langle \theta^* \rangle$ (b)

at a second point which varied across the wake remained positive. Lateral velocity fluctuations have also been found to be positively correlated across the entire width of a turbulent plane jet (e.g. [1,17,18]). This result is not surprising when viewed in the context of the topological similarity that exists between the self-preserving regions of the plane jet and the plane wake.

Contours of constant $\langle u^* \rangle \langle v^* \rangle$ (Fig. 5) and constant $\langle v^* \rangle \langle \theta^* \rangle$ are very similar with respect to their geometrical appearance in the Δx^* vs y^* plane. The locations of the peak magnitudes follow from the distributions in Fig. 4 : they occur along the diverging separatrices on either side of saddle points. For $\langle u^* \rangle \langle v^* \rangle$, the magnitude of the peak close to the centreline is slightly larger than that at larger $|y^*|$. This trend is reversed for $\langle v^* \rangle \langle \theta^* \rangle$. The different trends do not seem surprising in the light of the different shapes (Fig. 6) of the distributions for $\overline{u'^* v'^*}$ and $\overline{v'^* \theta'^*}$ (here a

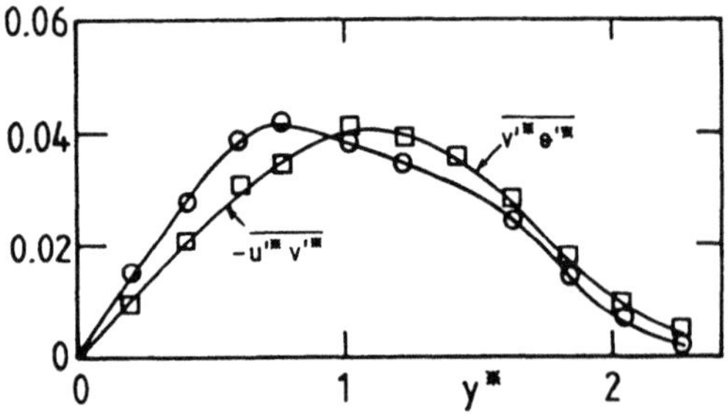

Fig. 6 Variation across one side of the wake, of the Reynolds averaged shear stress and lateral heat flux

prime denotes a fluctuation in the Reynolds decomposition, viz. $U = \overline{U} + u'$, $V = \overline{V} + v'$, $T = \overline{T} + \theta'$ where the overbar represents a time average). It is worth noting however that although the maximum value of $|\overline{u'^* v'^*}|$ is comparable to that of $|\overline{v'^* \theta'^*}|$, the contours of $\langle v^* \rangle \langle \theta^* \rangle$ attain magnitudes that are almost twice as large as those attained by the contours of $\langle u^* \rangle \langle v^* \rangle$. This suggests that the coherent large scale motion makes a larger contribution to the conventionally averaged heat transfer than to the momentum transfer. Calculations, made by [12], indicate that the contribution to the averaged lateral heat transfer is of the order of 40% whereas, for the averaged momentum transfer, it is only about 20%. Slightly larger contributions from the large scale motion to both momentum and heat transfer were estimated for the plane jet [1]. The relative magnitudes of these contributions were however such as to indicate that the large scale motion is more efficient at transporting heat than momentum.

7. Entrainment of External Fluid

Some insight into the manner in which irrotational fluid is entrained in the wake can be gained by considering contours of the quantity $\overline{T} + \langle \theta \rangle$. The lines in Fig. 7 are stretched at approximately 45° to the centreline, in the direction of the diverging separatrix. The spacing between consecutive lines is also reduced in these regions, reflecting the large temperature gradients associated with the fronts. If we focus our attention on these regions, temperatures commonly encountered near the centreline are observed near $|y^*| \simeq 1$ on the downstream side of the diverging separatrix. Conversely, temperatures, more likely to be associated with external cooler fluid, can be observed near the centreline on the upstream side of the diverging separatrix. These observations are consistent with a scenario in which ambient temperature fluid is swept in towards the centreline on the downstream side of a rotational bulge while relatively warm fluid is swept out towards the edge of the wake on the upstream side of the previous bulge. The saddle point plays an important part in this scenario if it is conjectured, e.g. [2,3], that significant vortex stretching occurs, about the saddle, in the direction of the diverging separatrix. Such vortex stretching would draw ambient temperature fluid towards the centreline

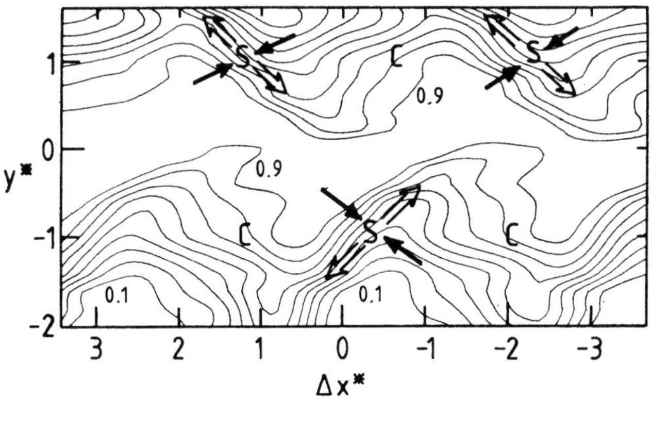

Fig. 7 Isotherms ($\overline{T}^* + \langle \theta^* \rangle$ = constant) associated with the large scale motion. (Contours are shown at intervals of 0.1)

and, at the same time, cause warm fluid to be swept out towards the external stream. BEVILAQUA and LYKOUDIS' [19] visualisation experiment showed a marked droplet of ambient fluid being drawn into the wake of a sphere on the downstream side of a rotational bulge with mixing occurring only in the final phase of entrainment. These authors emphasised the importance of the enfolding action of the large eddies. Although the present ensemble averaged topology tends to show entrainment as a continuous rather than time-varying process, it identifies the regions of space where the momentum and heat transfers, due to the large scale motion, are especially important.

8. A Summary of the Flow Topology

The flow topology, based on results presented here, is shown schematically in Fig. 8. It is very similar to the topology determined in the self-preserving region of a plane jet [1], both topologies exhibiting all characteristics associated with a vortex street. It seems likely that temperature fronts, which separate adjacent masses of fluid with high spanwise vorticity, are formed as a result of the strong vortex stretching directed away from the saddle point. It should be recalled that, like the schematic topology proposed for the plane jet [1], the picture in Fig. 8 is an oversimplified cut through a motion which is three-dimensional with the possibility that vortical structures may interact in a complex manner.

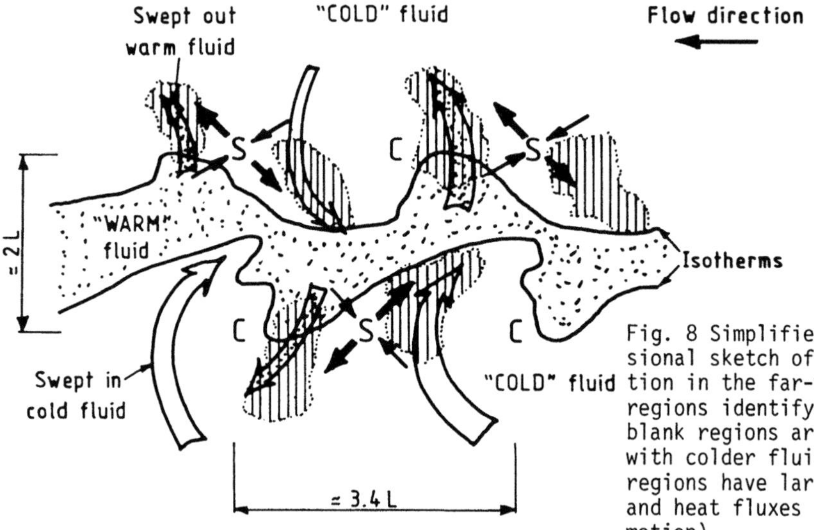

Fig. 8 Simplified two-dimensional sketch of organised motion in the far-wake. Dotted regions identify warmer fluid; blank regions are associated with colder fluid. Cross-hatched regions have large shear stresses and heat fluxes (large scale motion)

9. Acknowledgement

The support of the Australian Research Grants Scheme is gratefully acknowledged.

10. References

1. R. A. Antonia, A. J. Chambers, D. Britz, L. W. B. Browne: J. Fluid Mech. (to appear)

2. B. Cantwell, D. Coles: J. Fluid Mech., 136, 321 (1983)
3. M. Hayakawa, A. K. M. F. Hussain: Proc. Fifth Symposium on Turbulent Shear Flows, Cornell University, 4.33 (1985)
4. M. Kiya, M. Matsumura: Bulletin of JSME, 28, 2617 (1985)
5. S. Taneda: J. Phys. Soc. Japan, 14, 843 (1959)
6. J. F. Keffer: J. Fluid Mech., 22, 135 (1965)
7. D. D. Papailiou, P. S. Lykoudis: J. Fluid Mech., 62, 11 (1974)
8. J. M. Cimbala: Proc. Fifth Symposium on Turbulent Shear Flows, Cornell University, 4.1 (1985)
9. H. L. Grant: J. Fluid Mech., 4, 149 (1958)
10. A. A. Townsend: J. Fluid Mech., 95, 515 (1979)
11. J. C. Mumford: J. Fluid Mech., 137, 447 (1983)
12. L. W. B. Browne, R. A. Antonia, D. K. Bisset: Phys. Fluids (to appear)
13. D. K. Bisset, L. W. B. Browne, R. A. Antonia: Report T.N. FM 86/1, Department of Mechanical Engineering, University of Newcastle
14. A. K. M. F. Hussain: Phys. Fluids, 26, 2816 (1983)
15. R. A. Antonia, L. W. B. Browne: Abstracts European Turbulence Conference, Lyon, 43 (1986)
16. B. J. Cantwell: Arch. Mech., 31, 707 (1979)
17. J. C. Mumford: J. Fluid Mech., 118, 241 (1982)
18. F. O. Thomas, E. G. Brehob: Phys. Fluids, 29, 1788 (1986)
19. P. M. Bevilaqua, P. S. Lykoudis: AIAA Jnl., 9, 1657 (1971)

Models for Coherent Structures in the Wall Layer

N. Aubry[1], *P. Holmes*[2,3], *J.L. Lumley*[1], *and E. Stone*[2]

[1]Sibley School of Mechanical and Aerospace Engineering, Cornell University, Ithaca, NY 14853, USA

[2]Department of Theoretical and Applied Mechanics, Cornell University, Ithaca, NY 14853, USA

[3]Department of Mathematics and Center for Applied Mathematics, Cornell University, Ithaca, NY 14853, USA

1. Introduction

There has been much interest in recent years regarding the application of methods, ideas and models from dynamical systems and bifurcation theory to the study of turbulent fluid flow. The first paper in which an explicit link was made between strange attractors and turbulence is that of RUELLE and TAKENS [1], although, unknown to those authors, LORENZ [2] (considerably earlier) had proposed a three dimensional set of ordinary differential equations (ODEs) as a minimal model for two dimensional Bénard convection.

Recent work in this area, of which accounts can be found in collections such as [3], [4] and [5], has concentrated on "closed" flow systems such as the Rayleigh-Bénard or Taylor-Couette problems. In such problems, simultaneous variation of two or more parameters (Reynolds or Rayleigh numbers, Prandtl number, aspect ratios, etc.) gives rise to multiple bifurcations in which two or more normal modes become simultaneously unstable. Center manifold and normal form methods then permit the analysis of such modal interactions in terms of low dimensional dynamical systems: relatively small sets of ODEs. For background see [6] or [7]. These methods are closely related to the classical amplitude expansions of nonlinear hydrodynamic stability theory [8]. (In fact, Galerkin and other discretization techniques are not always easy to apply to such problems, since, even if the steady (bifurcating) solution is known analytically, the normal modes are not usually elementary functions and linear stability analyses are not straightforward.) Even in this situation, however, rigorous results have not gone beyond the first preturbulent bifurcations from steady, laminar flows. There is nothing close to a proof of Ruelle and Takens' conjecture that the Navier Stokes equations have a strange attractor for a realistic flow geometry. On the other hand, sophisticated analyses of experimental data have lent much support to the thesis that low dimensional dynamics do control certain aspects of preturbulent flow in closed fluid systems (see [9] and references therein).

In view of this, the present study of an open flow: the fully developed boundary layer on a flat plate, might seem hopeless. As MONIN [10] has pointed out, low dimensional models cannot hope to account for the detailed, high wavenumber, spatially chaotic aspects of fully developed turbulence. Moreover in open flow systems, natural modal decompositions such as the Fourier modes of linear theory are lacking or of limited use in the inhomogeneous directions. Thus another expansion is necessary and a coherent structure approach seems to be worth investigating for a judicious determination of relevant modes when large eddies are experimentally seen to play an important role. Since the early work of THEODORSEN [11], which suggested the importance of large scale horseshoe shaped vortices in turbulent transport in a turbulent boundary layer, flow visualization and conditional sampling techniques have revealed the existence of organized structures in many shear flows.

Nevertheless the extraction of deterministic features from a random, fine grained turbulent flow has been a challenging problem. ZILBERMAN et. al. [12] write: "there are no consistent methods for identification which are independent of the techniques and the observer" and "we cannot unambiguously define the signature of an eddy without *a priori* knowledge of its shape and its location relative to the observation station and cannot map such an eddy because we do not have a proper criterion for pattern recognition". More recently, KIM [13] has used a conditional sampling technique based on numerical data but he also admits the subjectivity of the method due to the arbitrary choice of detection functions and threshold values.

In contrast, LUMLEY [14] proposed an unbiased technique for identifying such structures. The method consists of extracting the candidate which is the best correlated, in a statistical sense, with the background velocity field. Such a quantity is called, by definition, a *structure*. The different structures are identified with the orthogonal eigenfunctions of the proper orthogonal or Karhunen-Loève decomposition theorem of probability theory [15]. This is thus a systematic way to find organized motions in a given set of realizations of a random field. The question remains of how energetic and characteristic of the full turbulence of the flow such structures are. The method applied here is optimal in the sense that the series of eigenmodes converges optimally fast; faster than any other representation.

Nevertheless the use of these modes for a low dimension dynamical system study requires a very fast convergence of the series. The method we propose here is therefore limited in application to certain types of flows in which coherent structures contain a large fraction of the energy. It has been demonstrated that axisymmetric turbulent jet mixing layers [16] and wall regions of turbulent boundary layers [17, 18] belong to this group. In the wall region of a turbulent boundary layer, large eddies, which are thought to be characteristic organized structures, are experimentally observed to be intermittent in time and space (in the streamwise direction). They are associated with a succession of events in which their sudden breakup corresponds to peaks in the turbulent energy production. These violent motions of the streaks, called *bursting events*, are also thought to be the basic phenomenon for turbulent energy transfer between the inner and outer regions of the layer, and hence they contribute significantly to the entire turbulence process.

Despite the contribution of many researchers, the description of these eddies is still vague. THEODORSEN [11] was probably the first to suggest the formation of these vortices inclined at an angle of 45 degrees to the wall. HEAD and BANDYOPADHYAY's [19] experiments convincingly established the existence of horseshoe or hairpin vortices - depending on the value of Reynolds number - in the wall region of a boundary layer (cf. [20]). More recently, from a large eddy simulation, KIM and MOIN [21] have used conditional sampling criteria to produce spatial structures and detect the presence of three-dimensional hairpin vortical structures inclined at 45 degrees to the wall. A second group argues for the existence of streamwise elongated patterns, streaks and vortices. The experiment of HAMA [22] first revealed streamwise streaks which were also observed by KLINE et al. [23] who were able to determine the average streak spacing. BAKEWELL and LUMLEY [24] find streamwise counter-rotating eddy pairs as the main structure of the wall region (see also [10]). WILLMARTH [25, 26] and BLACKWELDER [27] give a similar picture of the wall region. It is not rare to find people belonging to both schools. For example, the investigations of [19] extending to the immediate vicinity of the wall provided visual evidence of streamwise filaments. They conclude that "the hairpins have their origin in longitudinal vortex motions very close to the wall". Nevertheless they do not feel able to provide "firmly established connection(s)". KIM's results [13, 28] obtained from a sampling technique (producing a time structure applied to numerical data) indicate that the bursting event is associated with a pair of counter-rotating streamwise vortices, since it survives in an ensemble-averaging technique. Whether or not these streamwise vortices are related to hairpins is not clear and although many connections have been proposed, nothing has been proved. For example, [29] states that these streamwise vortices are the legs of a horseshoe vortex structure.

The time-dependent behavior of these structures has also remained unclear. Experimental results obtained from flow visualization provide a qualitative description in terms of a quasi-cyclic succession of events which consists of a gradual lift up of the longitudinal, streamwise streaks, a sudden oscillation, a breakup or bursting event, a downward motion of high speed fluid towards the wall, called a *sweep event*, and a violent upward motion of low velocity fluid called an *ejection event*. This intense updraft is postulated to be closely related to the high local production of turbulent energy associated to the breakup of the streaks. For more details see [23]. Quantitative estimates have been given for the bursting period, i.e. the average time between two consecutive bursting events at a given spatial location [23, 30]. Many assumptions have been made to explain the bursting phenomenon. The most common is a secondary instability of an inflexional velocity profile in the updraft between the counterrotating eddies, but nothing has really been proved.

2. Outline of approach

A major purpose of this study is to understand better the patterns of the wall region structures, their dynamics, the role they play in the bursting phenomenon and their importance in the turbulence production mechanism. Specifically, we will develop a model for the wall region of the boundary layer (from $y_+ = 0$ to $y_+ = 40$ in wall units [31]), using the proper orthogonal decomposition [14, 32] in the direction normal to the wall, in which the flow is strongly non-homogeneous. In the streamwise and spanwise directions the flow is more homogeneous, and Fourier modes will suffice. Used in conjunction with Galerkin projection, the proper orthogonal decomposition yields an optimal set of basis functions in the sense that the resulting truncated system of ODEs captures the maximum amount of kinetic energy among all possible truncations of the same order. The method has obvious advantages over "a priori" decompositions, based on linear normal modes, but it does not appear to have been used before due to the difficulty of computing the proper orthogonal modes. For this one requires three dimensional autocorrelation tensors averaged over many realizations of the flow in question, data only obtainable from lengthy experiments and analyses or from detailed numerical simulations. In our case complete data is only available from experimental work in a glycerine tunnel [18], although [17] has derived two-dimensional orthogonal modes from large eddy simulations. However, as we shall see, knowledge of the autocorrelation tensor, and use of the Navier Stokes equations, does allow one to uniquely determine the unsteady flow, in contrast to CANTWELL's [33] expectation.

Using this modal decomposition, we have derived and studied minimal truncations of various orders of the Navier Stokes equations in the wall layer. The loss of energy to modes neglected in the truncation are accounted for by an eddy viscosity model (reminiscent of Heisenberg's model for the energy spectrum) and we include a feedback effect to account for the stabilizing role of the interaction between the mean (driving) velocity profile and the fluctuating velocity field in the wall region as disturbances grow. That is, we use cross-stream averaging; the quasi-steady mean velocity equation relates the mean velocity gradient to the Reynolds stress produced by the disturbances. We use this expression for the mean velocity gradient in the disturbance equation. The final models were sets of 3-6 ODEs for the complex modal amplitudes. We have studied these equations analytically and numerically, and interpreted their behavior in the light of dynamical systems theory. This study should be seen as preliminary and substantial modelling and mathematical issues remain to be resolved, but the initial indications are very encouraging.

For high eddy viscosities our model displays a unique, globally attracting fixed point which corresponds to (steady) streamwise vortices or rolls similar to those which have been observed in many experiments [19, 23]. This is not surprising, since the orthogonal and Fourier modes used in the Galerkin projection encode this velocity field structure. However, as the eddy viscosity is reduced to about 1.6, the fixed point destabilizes and intermittent bursting behavior occurs in which the modal content, amplitudes and geometry of the rolls changes quasiperiodically. Starting with almost steady rolls, a period of slow oscillatory growth occurs, culminating in a "burst" after which the velocity field relaxes quickly to the steady state. This agrees remarkably well with the bursts observed in experiments [23] and it is even possible to match typical burst periods by judicious choice of eddy viscosity. Thus our model appears to capture not only the static behavior of the coherent structures in the wall layer, but also the major features of their dynamics. Moreover, it suggests that the bursts originate very close to the wall as suggested in [23].

In the simplified model results presented here we retain a single eigenmode, a single (zero) streamwise wavenumber, and six cross-stream wavenumbers; the lowest (zero) cross stream wavenumber is inactive - if it has a non-zero initial value, it simply decays to zero. We space the cross-stream wavenumbers so that they cover the wavenumber range in which there is significant energy as measured by [18]. In the wall region, the first eigenfunction contains about half the energy, and the first three about 90%. The first eigenfunction has close to 80% of the Reynolds stress . We use the eigenfunctions deduced from measurement by [18], and obtain equations for the time-varying coefficients by Galerkin projection. The lowest order eigenfunction has the form of two streamwise rolls forming a strong updraft between them, with a weaker downdraft on either side.

3. Description of the results

In Fig. 1, we display the form of the solution when the viscosity parameter has a value of 2, so that the solutions consists of simple, stable rolls. There are two pairs of rolls in the computational domain, one in the left half, and one in the right half. The spacing corresponds roughly to that observed in [18]. This should be compared with Fig. 2, reproduced from [17], showing the dominant eigenfunction obtained from exact simulation of channel flow. We may note that the second order modal decomposition cannot distinguish between figure 1 as displayed, and the same figure with the arrows reversed. This is a disadvantage of second order decompositions, which must be supplemented by third order information to make this distinction. Here, however, we substitute these structures into the Navier Stokes equations; these equations add sufficient information to remove the ambiguity.

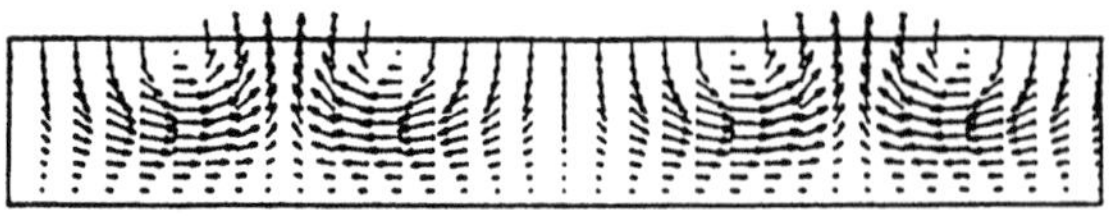

Fig. 1: Solution for viscosity parameter = 2; stable rolls

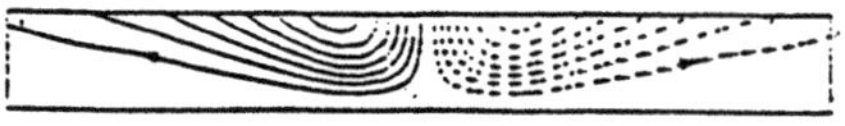

Fig. 2: Principal eigenfunction in the wall region from exact simulation [17]

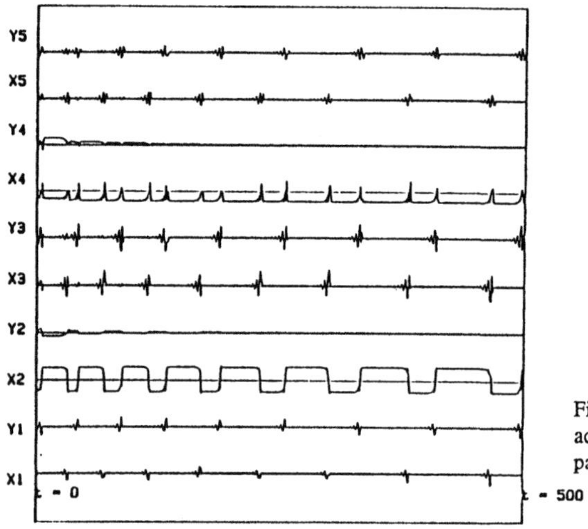

Fig. 3: Real and imaginary parts of coefficients for five active modes in intermittent regime. Viscosity parameter = 1.5

In Fig. 3, we display the values of the real (Xi) and imaginary (Yi) parts of the coefficients of the five active modes, for a value of the viscosity parameter producing intermittent behavior. We have two solutions here, coresponding essentially to two values of X2, one positive and one negative. The system is alternating between these two solutions. In the intervals between transitions the system is close to, but not quite in, an invariant set, corresponding to zero values of all odd coefficients. The odd coefficients are undergoing very small, but growing oscillations, which finally become visible just before the transition. In Fig. 4, we show the cross section of the eddies during one of these transitions. Time runs from top to bottom. It is clear that in the first picture there is an updraft in the middle of the computational domain, and one at each edge; thus, an eddy pair at the center, and half an eddy pair on each edge. After the system has undergone the transition to the other solution (bottom picture) there is a down draft in the middle and at each edge of the computational domain, with an eddy pair in the right half and in the left half of the computational domain. There has simply been a phase change; the form of the eddy pair is the same, but the pair has been displaced a quarter period. In Fig. 5, we show the same situation for a slightly larger value of the viscosity parameter. It is clear, by comparison with figure 3, that the period between transitions is somewhat longer. In both figures, the period between transitions is lengthening gradually. As explained below, introduction of the pressure term would completely dominate this effect, controlling the times at which transitions occur. In Fig. 6, we have reduced the value of the viscosity parameter below that of Fig. 3, and the system now ungergoes continuous almost periodic fluctuations. Finally, in Fig. 7, we show the Reynolds stress calculated at a distance of $y^+ =$

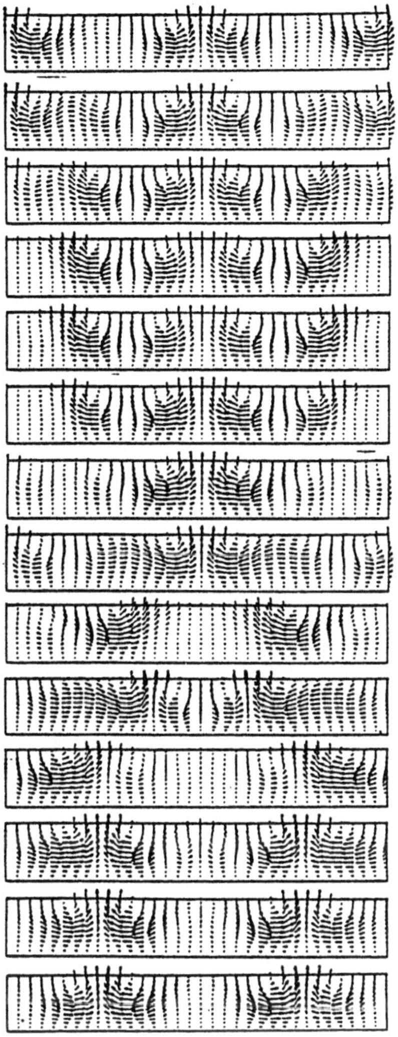

Fig. 4: Solution during a burst; time from top to bottom

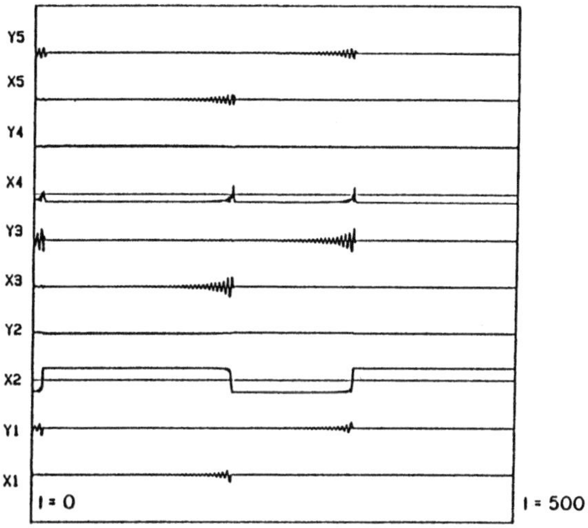

Fig. 5: As in Fig. 3 for viscosity parameter = 1.6

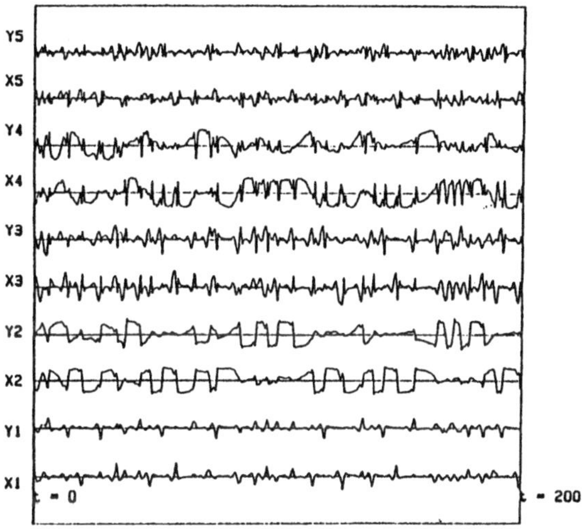

Fig. 6: As in Fig. 3 for viscosity parameter = 1.3, producing chaotic behavior

35 during two successive bursts, with the viscosity parameter equal to 1.4. It is clear that there is a momentary increase in the Reynolds stress, but not by a large factor. Closer to the wall, the Reynolds stress drops. The significance of this is discussed below.

4. Discussion

Our model is rudimentary in more ways than the simple sparsity of modes retained. Many non-essential mechanisms have been suppressed in the interests of simplicity. For example, the structure of the eigenmodes normal to the wall is held fixed as the modes evolve. In nature, the eddies in the wall layer are observed to be approximately round in cross section. To mimic this behavior we could allow the scale of the modes normal to the wall to grow and shrink ; this could be done either by tying the scales to the slope of the mean velocity profile at the wall (so that the eddies always fill approximately

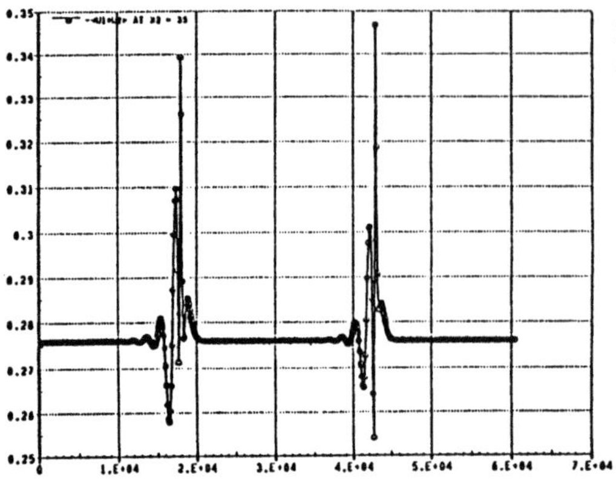

Fig. 7: Reynolds stress at $y^+ = 35$ during two successive bursts for viscosity parameter = 1.4

the same fraction of the mean profile as the shape of that profile changes), or the scales could be made to change proportional to the centroid of the energy in the cross-stream modes. That is, if the energy were concentrated in 2k rather than k, the scale normal to the wall would be reduced. Either of these would produce a more realistic behavior.

We do not have present in our modal population a mechanism to represent the production of higher wavenumber energy when an intense updraft is formed, presumably as a result of a secondary instability. We quote, for example, the description of the measurements of [34] from [35]: "When the rising filament of low-speed fluid reached a height in the range $8 < y^+ < 12$, an oscillatory motion of marked fluid lines (time lines) within the parcel was observed. The oscillatory motions, which were of various types, appeared to be associated with a swirling motion of the fluid. As the amplitude of the swirling motion of the rising fluid became larger, the pattern "broke up" at a distance from the wall in the range $10 < y^+ < 40$. During the breakup process, a significantly more random chaotic motion occurred in which marked lines of fluid were obliterated owing to the sudden increase in turbulent mixing." [35] also quote [36]: "When the ejected low-speed fluid encountered the interface between high- and low-speed fluid, at the high shear layer, a violent interaction occurred with intense, abrupt and chaotic movements. The intense interaction continued as more fluid was ejected. The end result was the creation of a relatively large-scale region of turbulent motion reaching into the sublayer as the violent interaction region spread out in all directions." It is clear that the higher wavenumber turbulence produced in this "violent interaction region" has strong transport properties, since it transports both the marked lines, obliterating them, and itself, spreading the region of higher wavenumber turbulence. To the extent that there are inequalities of momentum of larger scale in this region (and there are) it will transport them also. Thus, the higher wavenumber component will make a substantial contribution to the Reynolds stress. Unfortunately, the various measurements (see particularly [37]) do not appear to have discriminated between the high wavenumber and low wavenumber contributions to the Reynolds stress; they are not separated by conditioning (or, at least, not by the conditioning of [37]) since they occur at more-or-less the same time. The only indication is given by the calculations of [36], who have calculated by hand the contributions to the Reynolds stress of the large-scale motions in the ejection phase, finding a value between 50% and 70% of the long time average value, the percentage falling with Reynolds number. Since [37] and [35] measure values of the Reynolds stress during a burst several times the long term average, a considerable fraction of the Reynolds stress must be produced by the higher wavenumber component. The higher wavenumber component will also be expected to make a contribution to the streamwise and cross-stream energy.

Although our eddies are capable of exhibiting the basic bursting and ejection process, the labor is in vain - there is no sequel, no production of intense higher wavenumber turbulence. A contribution is made only to the low wavenumber part of the streamwise fluctuating velocity and the Reynolds stress. We could easily simulate the production of this high wavenumber turbulence, although we do not expect its inclusion to change the dynamics of our system qualitatively. We are holding the value of our eddy viscosity parameter constant . However, its value should rise and fall with the level of this intense higher wavenumber turbulence in the vicinity of the wall. The transport effectiveness of the intense higher wavenumber turbulence would be expected to damp the system, suppressing the interesting dynamics until the higher levels are either blown downstream or lifted and diffused to the outer edge of the boundary layer. The production of higher values of the eddy viscosity parameter could easily be parameterized by a single time constant first order equation tied to a measure of the amplitude of the coefficients. The net effect of this would simply be to cut off each burst more rapidly than at present.

The absence of this component of the Reynolds stress has other effects. At present each bursting event produces a considerably change in the Reynolds stress, but the magnitude is too small, and the direction is positive over too much of the layer, although it has the right sign in the outer part of the layer. The contribution of the higher wavenumber Reynolds stress would change this situation, since the latter Reynolds stress would be negative and large, and would occur in the outer part of the wall region, near the tops of the large eddies.

In this connection we may mention the convergence of the representation of the velocity field. Our representation converges in quadratic mean. This means that it is a complete representation of all the second order statistics, including the Reynolds stress. It is, of course, quite possible (even likely) that there are dynamically significant events that make negligible contribution to the second order statistics, but contribute importantly to the fourth order statistics, say. We are thus probably excluding in a fundamental way something important in our representation, something that cannot be recaptured by simply keeping more eigenmodes or Fourier modes. This is, however, not the case with the higher wavenumber Reynolds stress; inclusion of higher wavenumbers in the streamwise and cross-stream directions, and of higher eigenmodes (we may, in fact, already have high enough wavenumbers in the cross-stream direction) will recapture this effect. So far as effects that cannot be recaptured are concerned, the way in which our simple system mimics reality encourages us to believe that we have captured much of the important dynamics.

The occurrence of the bursts for values of the eddy viscosity parameter between 1.4 and 1.6 appears to be pseudo-random. This is essentially due to round-off error in the calculations. The transitions from one solution to the other and back are incredibly sensitive to the precise solution trajectory, and a minute change can make a considerable change in the time at which the next transition occurs. In our model we have not exercised the pressure term. The pressure term appeared due to the finite domain of integration. It represents the interaction of the part of the eddy that we have resolved with the part above the domain of integration, which is unresolved. The order of magnitude that we have estimated for this term is small, and for that reason we have neglected it. It would, however, have an important effect, while not changing the qualitative nature of the solution. The term would have the form of a random function of time, with a small amplitude. This would have the effect of slightly perturbing the solution trajectory constantly, resulting in a thorough randomization of the transition time from one solution to the other, while having little effect on the structure of the solution during a burst.

Probably the most significant finding of this work is that the bursting phenomenon appears to be determined by wall region parameters. That is, the bursts appear to be produced autonomously by the wall region, without the necessity of feedback from the outer part of the layer. Whether the bursting period scales with inner or outer scales has been a controversy in the turbulence literature for a number of years. The matter has been obscured by the fact that the experimental evidence has been measured in boundary layers with fairly low Reynolds numbers lying in a narrow range, so that it is not really possible to distinguish between the two types of scaling. The turbulent polymer drag reduction literature is particularly instructive, however, since the sizes of the large eddies, and the bursting period, all change scale with the introduction of the polymer (see for example [38, 39]). [39] tentatively conclude that inner scales are relevant. The results of the present work seem to us a very important indication that the wall region is capable of producing bursts autonomously, which is consistent with inner scaling, and confirms the tentative findings from the polymer literature. We must introduce a caveat, however. The pressure signal felt in the wall region originates much farther from the wall. The fluctuating pressure signal will perturb the system, causing transitions at random times. There is a possibility, however, that some characteristic of the pressure signal (say, an unusually high or low value) will trigger the transition, if it occurs at the right time in the life cycle of the wall layer eddies; if this proves to be true, it would provide a mechanism for the outer layer to lock events in the wall layer into phase, at least some of the time. If this happened all the time, the scaling would be with outer variables. This matter must be investigated with great care.

Finally, we can mention some of the possible applications of this approach. Many drag reduction schemes appear to depend on a change of boundary conditions or of distribution of effective viscosity in the wall region. For example, it has been suggested that polymer drag reduction is due to an increase in the effective viscosity in the turbulent part of the flow (effectively, a suppression of the intense higher wavenumber turbulence), while it remains the same in the viscous sublayer. Microbubbles have been said to have their effect due to reduction of the mean density, with an increase in effective viscosity, in the turbulent part of the wall layer, with essentially no change in the viscous sublayer. It has been suggested that riblets have their effect due to an increase in the viscous losses at the boundary for the transverse velocities in the large eddies. It would be relatively simple to introduce changes in the distribution of the effective viscosity with distance from the wall, or in the boundary conditions, and see the effect of this on the dynamics of the wall region. One could imagine introduction of a compliant boundary. One could consider the pressure footprint of the eddy system on the wall (although much of the pressure signal sensed at the wall is transmitted from the outer part of the boundary layer [40]). It seems that our approach to the wall region has many interesting possibilities that could be explored.

5. *Acknowledgements*

Supported in part by the U. S. Office of Naval Research under the initiative Select Research Opportunity IV, Contract Number N00014-85-K-0172. A more complete report of this work, with all mathematical details, may be found in [41].

6. Bibliography.

1. Ruelle, D. and Takens, F.: Comm. Math. Phys. *82*, 137 (1971)
2. Lorenz, E.N.: J. Atmos. Sciences *20*, 130 (1963)
3. Rand, D.A. and Young, L.S. (eds): *Dynamical Systems and Turbulence*. Springer Lecture Notes in Mathematics 898, (Springer-Verlag, New York, Heidelberg, Berlin, 1981)
4. Swinney, H.L. and Gollub, J.P. (eds): *Hydrodynamic Instabilities and the Transition to Turbulence*, (Springer-Verlag, New York, 1981)
5. Campbell, D. and Rose, H. (eds.): *Order in Chaos*. (North Holland, Amsterdam, 1983)
6. Guckenheimer, J. and Holmes, P.J.: *Nonlinear Oscillations, Dynamical Systems and Bifurcations of Vector Fields*, (Springer-Verlag, New York, 1983). Corrected second printing, 1986.
7. Golubitsky, M. & Guckenheimer, J. (eds): *Multiparameter Bifurcation Theory*. A.M.S. Contemporary Mathematics Series, No. 56. (American Mathematical Society, Providence, RI, 1986)
8. Drazin, P.G. and Reid, W.H.: *Hydrodynamic Stability*, (Cambridge University Press, Cambridge UK, 1981)
9. Swinney, H.: Physica *7D*, 3 (1983) Also in [5]
10. Monin, A. S.: Sov. Phys. Usp. *21*, 429 (1978)
11. Theodorsen, T.: In *Proc. 2nd Midwestern Conference on Fluid Mechanics*. (Ohio State U., Columbus, OH, 1952)
12. Zilberman, M., Wygnanski, I. and Kaplan, R. E.: The Physics of Fluids *20*, 5258 (1977)
13. Kim, J.: Phys. Fluids *26*, 2088 (1983)
14. Lumley, J. L: .In *Atmospheric turbulence and radio wave propagation*, ed. by A. M. Yaglom and V. I. Tatarski, (Nauka, Moscow, 1967) p. 166
15. Loève, M.: *Probability Theory*. (Van Nostrand, New York, 1955)
16. Glauser, M. N., Leib, S. T. and George, W. K.: In *Turbulent Shear Flows 5*, ed. by F. Durst et al (Springer, Heidelberg, 1985) . In press.
17. Moin, P.: Preprint No. *AIAA-84-0174*, 22nd Aerospace Sciences Meeting
18. Herzog, S.: *The large scale structure in the near wall region of turbulent pipe flow*. (Ph. D. Thesis, Cornell University, 1986)
19. Head, M. R. and Bandyopadhyay, P.: J. Fluid Mech. *107*, 297 (1981)
20. Willmarth, W. W. and Tu, B.-J.: Physics of Fluids Supplement. *10*, 5134 (1967)
21. Kim, J. and Moin, P.: J. Fluid Mech. *162*, 339 (1986)
22. Corrsin, S.: In *Symposium on Naval Hydrodynamics*, ed. by F. S. Sherman, Publ. 515, (Nat. Acad. Sci./Nat. Res. Counc., Wash., D.C., 1957) p. 373
23. Kline, S. J., Reynolds, W. C., Schraub, F. A. and Rundstadler, P. W.: J. Fluid Mech. *330*, 741 (1967)
24. Bakewell, H. P. and Lumley, J. L.: Physics of Fluids *10*, 1880 (1967)
25. Willmarth, W. W.: In *Advances in Applied Mechanics*, ed. by C.-S. Yih, Vol. 15 (Academic, New York, 1975). p. 159
26. Willmarth, W. W. and Bogar, T. J. : The Physics of Fluids *20*, 59 (1977)
27. Blackwelder, R. F. and Eckelmann, H.: J. Fluid Mech. *94*, 577 (1979)
28. Kim, J.: Phys. Fluids. *28*, 52 (1985)

29. Smith, C. R.: In *Proc. Eighth Symposium on Turbulence* , ed. by G. K. Patterson and J. L. Zakin (Dept. Chem. Eng., Univ. of Missouri-Rolla)
30. Rao, K. N., Narasimha, R. and Narayanan, M. A.: J. Fluid Mech. *48*, 339 (1971)
31. Tennekes, H. and Lumley, J. L.: *A First Course in Turbulence.* (MIT Press, Cambridge, MA, 1972)
32. Lumley, J. L.: In *Transition and Turbulence,* ed. by R. E. Meyer (Academic, New York, 1981) p. 215
33. Cantwell, B. J.: Ann. Rev. Fluid Mech. *13*, 457 (1981)
34. Kim, H. T., Kline, S. J. and Reynolds, W. C.: J. Fluid Mech. *50*, 133 (1971)
35. Willmarth, W. W. and Lu, S. S.: J. Fluid Mech. *55*, 65 (1972)
36. Corino, E. R., and Brodkey, Robert S.: *J. Fluid Mech. 37*, 1 (1969)
37. Lu, S. S. and Willmarth, W. W.: J. Fluid Mech *60*, 481 (1973)
38. Kubo, I. and Lumley, J. L.: *A study to assess the potential for using long chain polymers dissolved in water to study turbulence.* Annual Report, NASA-Ames Grant No. NSG-2382. (Cornell, Ithaca, NY, 1980)
39. Lumley, J. L. and Kubo, I.: In *The Influence of Polymer Additives on Velocity and Temperature Fields.* IUTAM Symposium Essen 1984, ed. by B. Gampert.. (Springer, Berlin/Heidelberg, 1984) p. 3
40. Farabee, T. M.: *An experimental investigation of wall pressure fluctuations beneath non-equilibrium turbulent flows.* DTNSRDC-86-047. (David W. Taylor Naval Ship Research and Development Center , Bethesda, MD,, 1986)
41. Aubry, N., Holmes, P. Lumley, J. L., and Stone, E.: J. Fluid Mech. Submitted.

Orthogonal Decomposition of the Axisymmetric Jet Mixing Layer Including Azimuthal Dependence

M.N. Glauser and W.K. George

Turbulence Research Laboratory,
Department of Mechanical and Aerospace Engineering, University at Buffalo,
SUNY, Buffalo, NY 14260, USA

In 1967 LUMLEY [4] proposed an approach to the objective determination of coherent structures. The method uses an orthogonal decomposition to extract eigenvectors from two point velocity measurements, the lowest order eigenvector representing the largest structure. If the flow is homogenous, stationary or periodic the orthogonal decomposition reduces to a harmonic decomposition. These directions therefore are fit with the harmonic eigenfunctions before applying Lumley's orthogonal decomposition to the inhomogeneous directions.

In the current work experiments were performed in the mixing region of an axisymmetric turbulent jet at x/D = 3. These experiments involve adding the azimuthal variation to the problem reported in [2] where the decomposition was utilized for variations in time and in the radial direction. The motivation for these additional experiments is to shed further light on the turbulent structure in the jet mixing layer and to examine the effect on the orthogonal decomposition of adding a dimension to the problem. Results from this current work are compared to those in [2].

Nomenclature

A_{ij}	Azimuthal Complex Coefficients
B_{11}	A_{11} summed over frequency
m	Azimuthal Mode Number
R_{ij}	Velocity Correlation Tensor
r	Radial Distance Across Jet Mixing Layer
t	Time
$\underset{\sim}{u}$	Velocity Vector
$\underset{\sim}{x}$	Inhomogeneous Spatial Vector
θ	Azimuthal Angle in Jet
λ	Eigenvalues
Φ_{ij}	Cross-Spectral Tensor
ϕ	Eigenvectors
ψ	Eigenvectors in Transformed Domain
ω	Frequency

Experiment and Apparatus

The experiment was performed in a circular air jet facility. The jet nozzle was of fifth order polynomial design and has an exit diameter of 0.098m with a contraction ratio of 10:1. The exit conditions of the jet were as follows: The turbulence intensity was 0.35%, the Reynolds number was 110,000 and the boundary layer thickness at the exit ($U = .99\ U_e$) was 0.0012. Two specially-designed probes containing seven single hot wire sensors each were positioned at x/D = 3. The individual sensors were spaced 0.0127m apart radially with the center sensors in the middle of the shear layer. The azimuthal variation was realized by keeping one 7-wire probe stationary at an arbitrarily chosen $\theta = 0$, while the other 7-wire probe was moved azimuthally in 12° increments. 500,000 cross-spectral estimates were obtained which corresponds to approximately 500 mbytes of original time series data. The duration of the experiment was about 40 hours.

The Orthogonal Decomposition

In 1967 LUMLEY [4] proposed that the coherent structure should be that structure which has the largest mean square projection on the velocity field. If $\underset{\sim}{\phi}(\underset{\sim}{x},t)$ is our candidate structure, then $\underset{\sim}{\phi}$ should be chosen to maximize

$$\overline{|\underset{\sim}{u} \cdot \underset{\sim}{\phi}|^2} = \overline{|\alpha|^2} \tag{1}$$

where $\underset{\sim}{u} = \underset{\sim}{u}(\underset{\sim}{x},t)$ is the instantaneous velocity. The above expression is assumed to be normalized by the modulus of $\underset{\sim}{\phi}$ so that (1) depends only on the degree of projection and not the magnitude. $\underset{\sim}{\phi}$ is then, in the averaged mean square sense, the most likely occurrence of $\underset{\sim}{u}$.

Maximizing $\overline{|\underset{\sim}{u} \cdot \underset{\sim}{\phi}|^2}$ leads to the following integral eigenvalue problem:

$$\int\int R_{ij}(\underset{\sim}{x}, \underset{\sim}{x}', t, t')\, \phi^{(n)}_j(\underset{\sim}{x}', t')\, d\underset{\sim}{x}'dt' = \lambda^{(n)} \phi_i(\underset{\sim}{x}, t) \tag{2}$$

where $R_{ij}(\underset{\sim}{x}, \underset{\sim}{x}', t, t') = \overline{u_i(\underset{\sim}{x}, t)\, u_j(\underset{\sim}{x}', t')}$

and $\lambda = |\alpha|^2$.

The Orthogonal Decomposition Theorem reduces to the Harmonic Decomposition Theorem if the direction is stationary, homogeneous or periodic. This implies that these directions must first be removed by Fourier analysis.

In the axisymmetric jet being studied the time 'direction' is stationary and the azimuthal direction is periodic as well as symmetric so that R_{ij} in equation (2) breaks down in the following manner. First the time direction is removed by using the Fourier transform resulting in

$$R_{ij}(\underset{\sim}{x}, \underset{\sim}{x}', \theta, t) \longrightarrow \Phi_{ij}(\underset{\sim}{x}, \underset{\sim}{x}', \theta, \omega) \tag{3}$$

where Φ_{ij} is the cross-spectral tensor.

A further decomposition (of the cross-spectral tensor now) over the periodic and symmetric direction gives

$$\Phi_{ij}(\underset{\sim}{x}, \underset{\sim}{x}', \theta, \omega) = \sum_{m=0}^{\infty} \Lambda_{ij}(\underset{\sim}{x}, \underset{\sim}{x}', \omega, m) \cos m\theta \tag{4}$$

where m is the azimuthal mode number and the complex coefficients of the above series are defined by

$$A_{ij}(\underset{\sim}{x}, \underset{\sim}{x}', \omega, 0) = \frac{1}{\pi}\int_0^\pi \Phi_{ij}(\underset{\sim}{x}, \underset{\sim}{x}', \theta, \omega)\, d\theta \tag{5}$$

and by $$A_{ij}(\underset{\sim}{x}, \underset{\sim}{x}', \omega, m) = 2/\pi \int_0^\pi \Phi_{ij}(\underset{\sim}{x}, \underset{\sim}{x}', \theta, \omega) \cos m\theta d\theta. \tag{6}$$

Now the inhomogeneous direction must be decomposed by using A_{ij} in equation (2),

$$\int A_{ij}(\underset{\sim}{x}, \underset{\sim}{x}', \omega, m)\, \psi_j^{(n)}(\underset{\sim}{x}', \omega, m)\, d\underset{\sim}{x}' = \lambda^{(n)}(\omega, m)\, \psi_i^{(n)}(\underset{\sim}{x}, \omega, m). \tag{7}$$

The eigenvalues have now become azimuthal mode dependent eigenspectra

$$\lambda^{(n)} \longrightarrow \lambda^{(n)}(\omega, m) \tag{8}$$

and the eigenvectors are frequency and mode number dependent

$$\phi_i^{(n)}(\underset{\sim}{x}, t) \longrightarrow \psi_i^{(n)}(\underset{\sim}{x}, \omega, m). \tag{9}$$

From the general Hilbert-Schmidt theory it can be shown that A_{ij} can be reconstructed as

$$A_{ij}(\underset{\sim}{x}, \underset{\sim}{x}', \omega, m) = \sum_{n=1}^{\infty} \lambda^{(n)}(\omega, m)\, \psi_i^{(n)}(\underset{\sim}{x}, \omega, m)\, \psi_j^{(n)*}(\underset{\sim}{x}', \omega, m). \tag{10}$$

A Simpler version

A scaler version of the orthogonal decomposition which utilizes the streamwise velocity measurements from the jet can be derived and formulated as,

$$\int A_{11}(r, r', \omega, m)\, \psi^{(n)}(r', \omega, m)\, dr' = \lambda^{(n)}(\omega, m)\, \psi^{(n)}(r, \omega, m) \tag{11}$$

where r is the radius in the jet mixing layer. Using the measured values of A_{11} equation (11) can be solved numerically for the eigenvalues and eigenfunctions.

Results of the Azimuthal Decomposition Analysis

Before examining the results of (11), the breakdown into azimuthal modes warrants examination. From (5) and (6), setting $i = j = 1$,

$$A_{11}(r, r', \omega, 0) = \frac{1}{\pi}\int_0^\pi \Phi_{11}(r, r', \omega, \theta)\, d\theta \tag{12}$$

and,

$$A_{11}(r, r', \omega, m) = \frac{2}{\pi}\int_0^\pi \Phi_{11}(r, r', \omega, \theta) \cos m\theta d\theta, \tag{13}$$

if both sides of (12) and (13) are summed over ω they become respectively

$$B_{11}(r, r', 0) = \frac{1}{\pi}\int_0^{\pi} R_{11}(r, r', \theta)\, d\theta$$

and
$$B_{11}(r, r', m) = \frac{2}{\pi}\int_0^{\pi} R_{11}(r, r', \theta) \cos m\theta d\theta$$

where
$$R_{11}(r, r', \theta) = \sum_{\omega} \Phi_{11}(r, r', \theta, \omega) = \overline{u(r, \theta_o)\, u(r', \theta_o + \Delta\theta)}$$

and
$$B_{11}(r, r', m) = \sum_{\omega} A_{11}(r, r', m, \omega).$$

$R_{11}(r, r', \theta)$ can be reconstructed from the $B_{11}(r, r', m)$ as,

$$R_{11}(r, r', \theta) = \sum_{m=0}^{N-1} B_{11}(r, r', m) \cos m\theta.$$

It is advantageous to set $r = r'$, so that B_{11} and R_{11} can be plotted for each radial position in the jet mixing layer. A sampling of these are plotted in figures 1 - 4.

Figure 1(a) shows measurements of $R_{11}(\theta)$ at a position near the potential core at $r/D = 0.13$. There is clearly a strong correlation over the entire 180° span. The corresponding $B_{11}(m)$ is plotted in Fig. 1(b). The 0th mode (axisymmetric mode) can be seen to contain most of the energy indicating a strong ring-like coherent structure at this position.

Figure 2(a) shows measurements of $R_{11}(\theta)$ at a position on the high speed side of the mixing layer at $r/D = .26$. Again there is a strong correlation over the entire 180° span. The corresponding $B_{11}(m)$ is plotted in Fig. 2(b). The first few modes again contain most of the energy indicating that there is still a ring-like structure at this position.

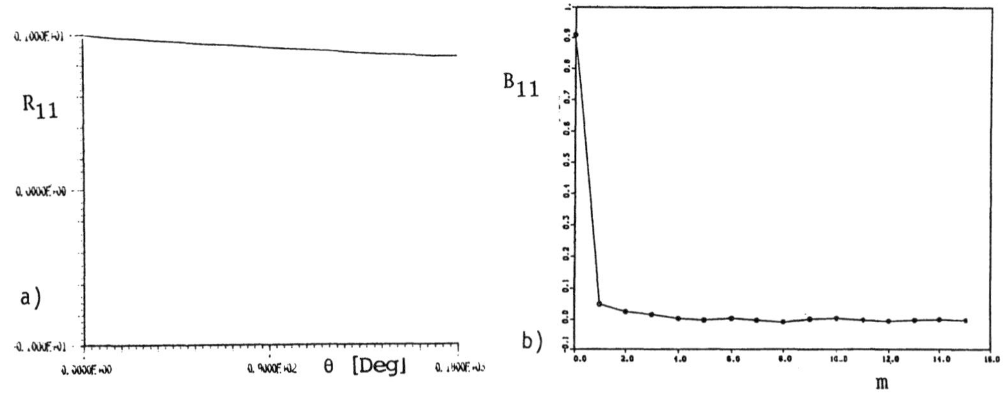

FIGURE 1. (a) $R_{11}(\theta)$ at $x/D = 3$, $r/D = 0.13$, (b) $B_{11}(m)$ at $x/D = 3$ $r/D = 0.13$

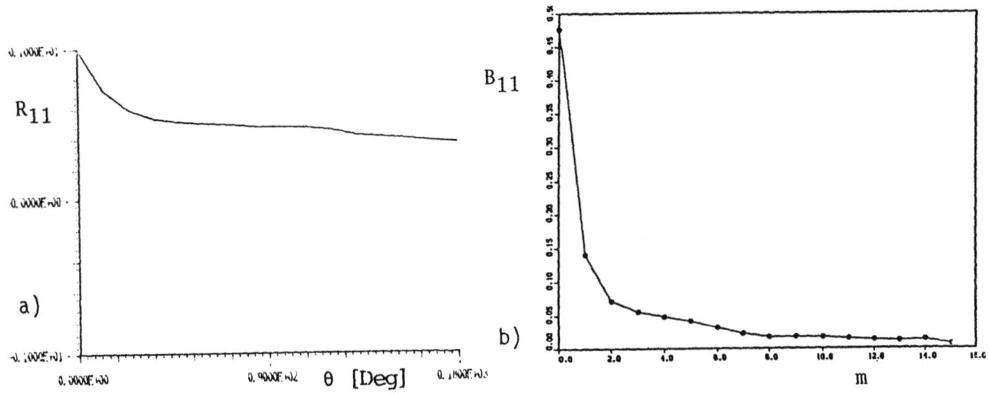

FIGURE 2. (a) R_{11} (θ) at x/D = 3 r/D = 0.26, (b) B_{11} (m) at x/D = 3, r/D = 0.26

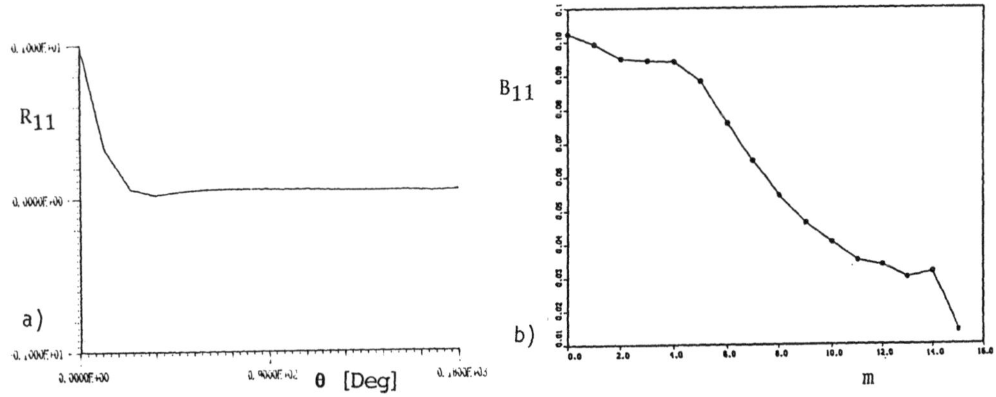

FIGURE 3. (a) R_{11} (θ) at x/D = 3, r/D = 0.39, (b) B_{11} (m) at x/D = 3, r/D = 0.39

Figure 3(a) shows measurements of $R_{11}(\theta)$ at a position on the high speed side of the mixing layer at r/D = .39. The correlation falls off very rapidly indicating a breakdown of the ring-like structure at this position. This can be seen by examining $B_{11}(m)$ in Fig. 3(b). The energy is now spread out over many of the modes when compared to Figs. 1(b) and 2(b).

Figure 4(a) shows measurements of $R_{11}(\theta)$ at a position in the center of the mixing layer at r/D = 0.52. Again the correlation falls off very rapidly. The corresponding $B_{11}(m)$ is plotted in Fig. 4(b). An interesting phenomena occurs at this position. Instead of the first few modes dominating, as in the afore mentioned cases, there is a shift here to the fourth, fifth and sixth modes.

The other three positions in the mixing layer, r/D = 0.65, r/D = 0.78, and r/D = 0.91 where measurements where taken are not shown because R_{11} and B_{11} at these positions exhibit the same basic characteristics as those in Figs. 4(a) and 4(b).

A suggestion for the breakdown of this azimuthal coherent structure may be that high shear across the mixing layer destorts it resulting in bifurcations into many more modes. This would suggest that these structures appear to be more stable toward their inside edges than their outside edges as suggested by

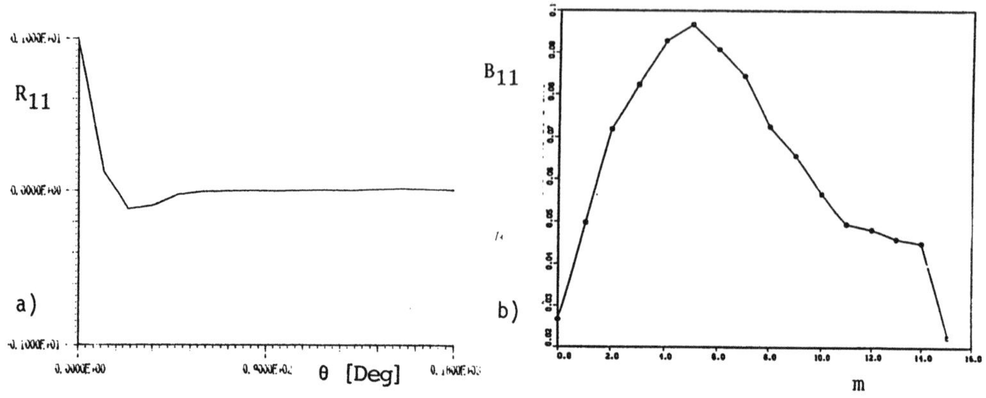

FIGURE 4. (a) R_{11} (θ) at x/D = 3, r/D = 0.52, (b) B_{11} (m) at x/D = 3, r/D = 0.52

[5]. The preference for the fourth, fifth and sixth-lobe configuration as seen in Fig. 4 is intriguing. A sixth-lobe preference (although weak) was seen by [5] at a position closer to the potential core at x/D = 1. This may indicate that as the structure progresses downstream the mean shear smears this sixth lode preference over more modes. A detailed stability analysis for vortex rings has been carried out by [6]. From this analysis the number of preferential lobes is seen to depend on the circulation as well as on the ratio of the vortex core radius to the ring radius. Neither of these terms are known in the present jet with any confidence but the sixth-lobe preference noticed here is not inconsistant with the results from [6].

Results of the Orthogonal Decomposition

The inhomogenous problem was solved using Lumley's Orthogonal Decomposition. For each mode number frequency combination (11) was solved numerically using the measured values $A_{11}(r, r', \omega, m)$. The numerical approximation to (11) was discussed by Leib et.al. [3].

The first 3 eigenspectra for modes 0 and 1 are shown plotted in Figs. 5 and 6. The first eigenspectrum for each of the two modes is seen to dominate.

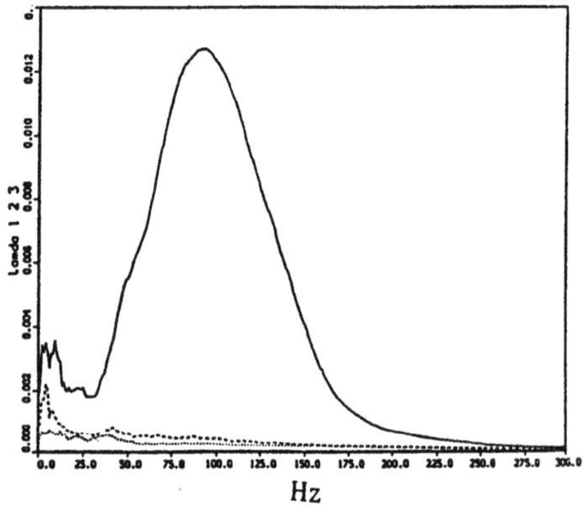

FIGURE 5. Eigenspectra mode 0

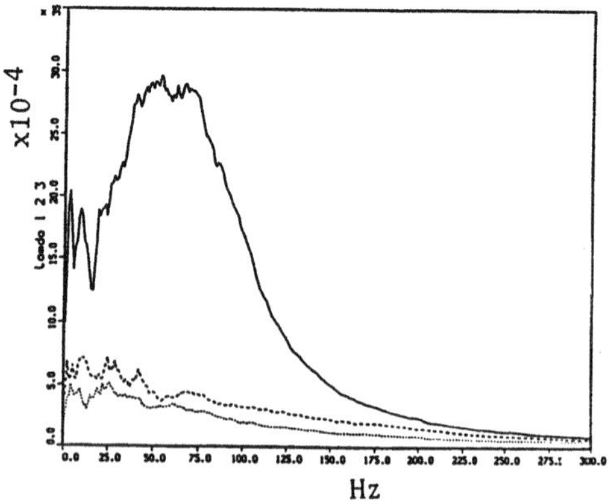

FIGURE 6. Eigenspectra mode 1

This is a desirable result considering that the first eigenspectrum (in particular of the 0th mode) is related to the large eddy. It is useful then to look at the amount of energy contained in $\lambda^{(1)}(\omega)$ of the 0th mode compared to $\lambda^{(1)}(\omega)$ of all the modes combined to see if it is dominant. The ratio of interest is,

$$\sum_{\omega} \lambda^{(1)}(\omega, 0) / \sum_{\omega} \sum_{m} \lambda^{(1)}(\omega, m). \tag{14}$$

This ratio is approximately equal to 0.23.

An additional interesting result can be seen by examining Figs. 5 and 6. The peak of $\lambda^{(1)}(\omega, 0)$ in Fig. 5 is seen to be approximately 95 Hz which corresponds to an exit Strouhal number equal to 0.45. This is higher than the value of 0.3 which is given by some authors [1]. When examining Fig. 6, however, $\lambda^{(1)}(\omega, 1)$ is seen to peak around 50 Hz which corresponds to an exit Strouhal number of 0.25. This would suggest that the Strouhal number of 0.3 represents an average over the first few azimuthal modes. If so, the orthogonal decomposition used here would seem to have a real advantage over the earlier techniques in that it can distinguish clearly these phenomenon.

For the sake of comparison to the initial experiments reported in [2] a brief review of that problem is given below. The eigenvalue problem that was solved is

$$\int_0^R \Phi_{11}(r, r', \omega) \psi^{(n)}(r', \omega) dr' = \lambda^{(n)}(\omega) \psi^{(n)}(r, \omega) \tag{15}$$

where Φ_{11} is the cross-spectral tensor.

From the general Hilbert-Schmidt theory Φ_{11} can be reconstructed from the eigenvalues and eigenfunctions solved for in (15) by

$$\Phi_{11}(r, r', \omega) = \sum_{n} \lambda^{(n)}(\omega) \psi^{(n)}(r, \omega) \psi^{*(n)}(r', \omega). \tag{16}$$

Note the difference between (11) and (15). In (11) the integral eigenvalue problem is performed using the azimuthal coefficients A_{11} obtained from the breakdown of $\Phi_{11}(r, r', \omega, \Theta)$ versus its azimuthal modes. The problem in (15), however, is performed over $\Phi_{11}(r, r', \omega)$ since the azimuthal dependence had not been obtained at that time.

The eigenspectra computed from (15) are shown plotted in Fig. 7. A similar dominance of the first eigenspectrum (compared to Figs. 5 and 6) is observed here also. Although no obvious analytical relationship exists between $\lambda^{(1)}(\omega)$ and $\lambda^{(1)}(\omega, m)$ it is still instructive to compare them by looking at the following ratio,

$$\sum_{\omega} \lambda^{(1)}(\omega) / \sum_{\omega} \sum_{m} \lambda^{(1)}(\omega, m). \tag{17}$$

In other words, by adding the azimuthal dependence, has more of the total energy been gained or did the initial experiment lump all the energy into $\lambda^{(1)}(\omega)$? In this case the ratio in (17) is approximately equal to 0.9. This

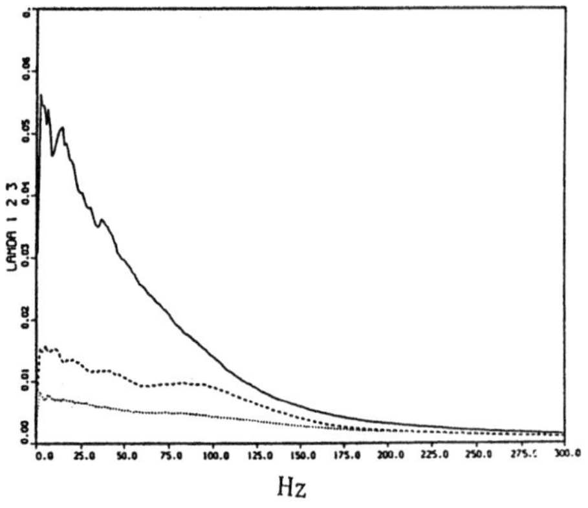

FIGURE 7. First 3 eigenspectra from the original problem (radius and time only)

would suggest that the initial problem did contain most of the azimuthal variation, only that it was mapped somehow, to the lower dimensional problem. These results are consistant with those of Moin (private communication).

A further comparison can be made between the two experiments. Taking (10), setting $i = j = 1$, $x = x' = r$ it reduces to

$$A_{11}(r, r, \omega, m) = \sum_{n=1}^{N} \lambda^{(n)}(\omega, m)\, \psi^{(n)}(r, \omega, m)\, \psi^{(n)*}(r, \omega, m) \tag{18}$$

and, also, examining (16), setting $r = r'$ it reduces to

$$\Phi_{11}(r, r, \omega) = \sum_{n=1}^{N} \lambda^{(n)}(\omega)\, \psi^{(n)}(r, \omega)\, \psi^{*(n)}(r, \omega). \tag{19}$$

If both sides of (18) and (19) are summed over ω, and n is set equal to 1, the following ratios can be derived,

$$\sum_{\omega} \lambda^{(1)}(\omega, m)\, \psi^{(1)}(r, \omega, m)\, \psi^{(1)*}(r, \omega, m) \Big/ \sum_{\omega} A_{11}(r, r, \omega, m) \tag{20}$$

and

$$\sum_{\omega} \lambda^{(1)}(\omega)\, \psi^{(1)}(r, \omega)\, \psi^{(1)*}(r, \omega) \Big/ \sum_{\omega} \Phi_{11}(r, r, \omega). \tag{21}$$

For the 0th mode the ratio in (20) gives an indication of the contribution of the 1st term (obtained now from the inhomogeneous problem) to the total energy in the 0th mode. From Fig. 8 an idea of the value of (20) for the 0th mode can be obtained. The solid line represents the denominator and the dashed line the numerator. Approximately 85% of the total is seen to be contained in the first term. It must be remembered, however, that the results of (20) must be interpreted along with those of (14) for the azimuthal problem. This combination gives a total contribution to the energy of at most 20% for the large eddy. From Fig. 9 the contribution of the large eddy (the ratio in 21) can be seen to be about 40% for the initial problem. This discrepency in the two problems would suggest that the results of the initial experiments (radial

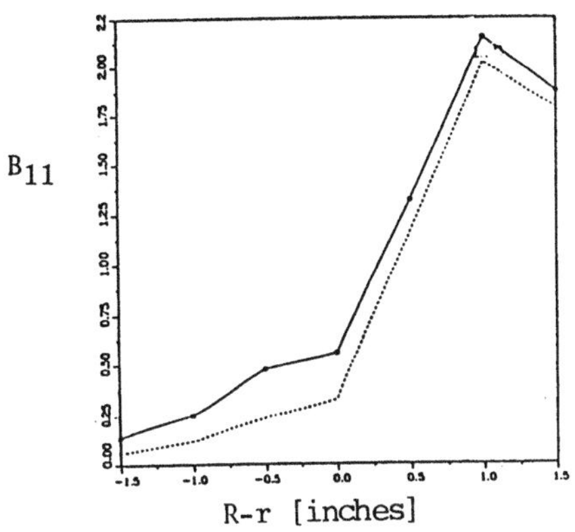

FIGURE 8. (•—•) B_{11} for mode 0, (••••) contribution from first term to B_{11} for mode 0

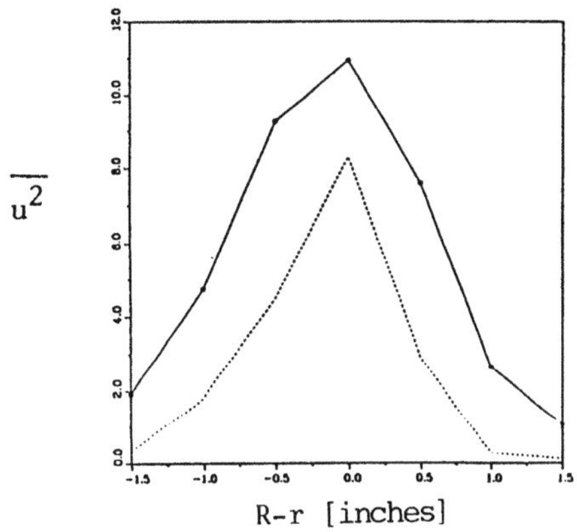

FIGURE 9. (•—•) $\overline{u^2}$, (••••) contribution from first term to $\overline{u^2}$

and time only) are somewhat misleading in that the large eddy in that case contains 40% as opposed to less than 20% when the azimuthal dimension is added.

From the results of (20) for the 0th mode some additional insight can be gained. The fact that the dominate term contained 85% of the energy in the 0th mode would suggest that the large eddy is almost axisymmetric. This result was used in [3] to calculate the radial component of velocity of the large eddy using the continuity equation. The particle paths of the large eddy were then computed. This implies that the results reported earlier were realistic.

Conclusion

The orthogonal decomposition is seen to work more efficiently in the inhomogenous direction (across the shear layer) than the harmonic decomposition applied in the azimuthal direction. This would suggest that the prevailing view of the large eddy as an axisymmetric vortex be modified somewhat. Instead of only using the 0th azimuthal mode, at least the 4th, 5th and 6th modes should be included in the definition. This would increase the energy content of the big eddy, and give a more realistic physical view of it.

Acknowledgements

This research was initiated under support from the National Science Foundation under Grant No. ENG7617466 and the Air Force Office of Scientific Research under Contract Nos. F4962078C0047 and F4962080C0053. Work has continued under NSF Grant No. 8316833 and with the support of the NASA trainee program for Mark Glauser at NASA/Ames. We gratefully acknowledge the technical support and continuing interest of Parvis Moin at NASA/Ames.

References

1.) Crow, S.C., Champagne, F.H. (1971) J. Fluid Mech. 48, 547.

2.) Glauser, M.N., Leib, S.J., George, W.K. (1985) Coherent structures in the axisymmetric jet mixing layer, Proc. of the 5th symposium turbulent shear flow conf. Cornell. (also Springer, selected papers from TSF5).

3.) Leib, S.J., Glauser, M.N., George, W.K. (1984) An Application of Lumley's orthogonal decomposition to the axisymmetric turbulent jet mixing layer. Proceedings 9th Rolla Symposium.

4.) Lumley, J.L. (1967) The structure of inhomogenous turbulent flows. In Atm. Turb. and Radio Wave. Prop., (Yaglom and Tatarsky, ed.) Nauka, Moscow, 166-178.

5.) Sreenivasan, K.R. (1984) The azimuthal correlations of velocity and temperature fluctuations in an axisymmetric jet. Phys. Fluids 27 (4).

6.) Widnall, S.E., Sullivan, J.P. (1973) Proc. R. Soc. London Ser. A 332, 335.

Analysis of Cut-and-Connect of Vortex Filaments

R. Takaki[1] *and A.K.M.F. Hussain*[2]

[1]Tokyo University of Agriculture and Technology, Fuchu, Tokyo 183, Japan
[2]Department of Mechanical Engineering, University of Houston, Houston, TX 77004, USA

1. Introduction

Vortex dynamics is a powerful tool for theoretical investigation of flows in which vorticity is confined within narrow regions and viscosity plays no essential role. Two cases have been investigated extensively: an ensemble of two-dimensional point vortices, motions of which are governed by the Hamiltonian dynamics [1], and a single or several vortex filaments in three-dimensional motion which is determined uniquely by Biot-Savart's law. The motion of a single vortex filament is analysed by Hama [2], Betchov [3], Hasimoto [4] and Takaki [5]. Interactions of two vortex filaments are treated by Crow [6], Moffatt [7], Hopfinger et al. [8] and Takaki & Hussain [9]. The short wave instability on a single vortex ring, investigated by Widnall et al. [10] and Saffman [11], also causes interaction between parts of a vortex filament.

Among many phenomena in vortex dynamics certain types of motion are specially interesting, where vortices approach each other and undergo strong interactions. A typical example of the strong interaction in two-dimensional flows is the vortex merger, investigated by many workers. An example of strong interaction in the three-dimensional flow is the recombination of two vortex filaments (Hama [12], Crow [6], Kambe & Takao [13], Oshima & Asaka [14], Fohl & Turner [15]), where local elements of two filaments are cut and then connected after being switched. This situation of topologically inaccessible transformation appears to be central to the enstrophy cascade in turbulent flows. Except for the MHD flows (Yeh & Axford [16]) this singular interaction mechanism has not been studied theoretically and remains an outstanding challenge for theoretists, as Saffman & Baker [17] point out. This phenomenon is quite complicated, because the fluid motion varies rapidly in an essentially three-dimensional manner. What is unknown even qualitatively is the behavior of the flow field in the localized region where two vortex filaments make contact with each other and the vorticity lines are cut and reconnected.

The purpose of the present analysis is to deal with this phenomenon by a simple model of the time-dependent flow field which simulates the real flow with in this region. In the proposed model the flow must satisfy the incompressible Navier-Stokes equation or the vorticity equation (with the viscous term retained), because viscosity plays an essential role in the mechanism (a vortex line can not be cut in an Euler flow).

The present analysis is motivated not only by the interest in fundamental vortex dynamics but also by our claims that this process is one of the main aspects of development of three-dimensional turbulence from two-dimensional, non chaotic flows and that mixing and noise production in the near field of a jet are primarily caused by the breakdown process of the initial toroidal structures through the recombination process (Hussain [18]).

2. Basic Assumptions

Consider that two vortex filaments with opposite sense of rotation approach each other. Then, parts of the filaments close to each other will deform strongly and are pushed against each other and upwards as shown in Fig. 1(a). After a recombination process the new filaments are arranged as shown in Fig. 1(b), and the adjacent parts then move downwards. It is reasonable to assume that this motion is confined to a local part of the filaments and that the distant parts away from the interaction region remain unaffected (for a short time) during the interaction. This local fluid region is called an interaction region.

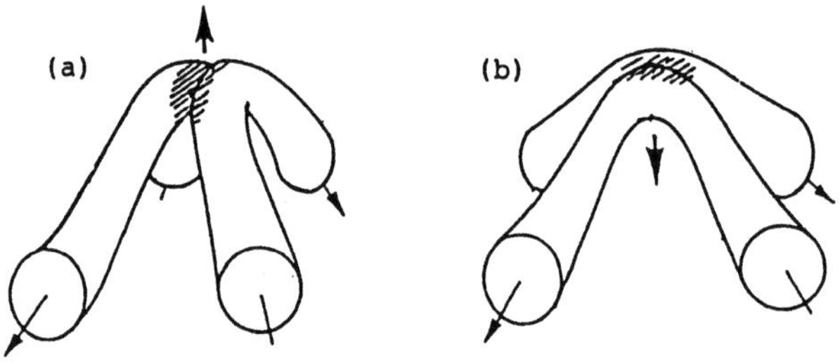

Fig. 1. Vortex configurations before (a) and after (b) the recombination

Now, we consider a change of the fluid impulse between the initial and the final states. The impulse is defined by

$$P = \frac{1}{2} \rho \int r \times \omega \, dV, \tag{1}$$

where ρ is the fluid density, r is the position vector, ω is the vorticity, and the volume integral is taken over the whole fluid region. From the above assumption the change in P comes only from the integral in the interaction region. Contributions from the interaction region give an upward- and a downward-directed impulses for the initial and the final states, respectively, hence violating impulse conservation [19]. To compensate for this impulse jump, we need another fluid motion (i.e. a compensating flow field) with an upward-directed impulse.

An example of such a motion is a recoil fluid element ejected in the upward direction. This complicated process containing the reconnection and the recoil can be looked upon as superposing a pair of collinear, small ring vortices on the initial vortex filaments as shown in Fig. 2. These rings have equal sizes and opposite senses of rotation and of equal circulations, which are also equal to that of either of the initial filaments. The lower ring overlaps with the interaction region and produces a new vortex configuration, while the upper one plays the role of recoil (see Fig. 2(b)). There may be other mechanisms satisfying the fluid impulse conservation, but do not appear likely to be amendable to simple analytical treatment. The following analysis does not depend much on the precise of recoil mechanism.

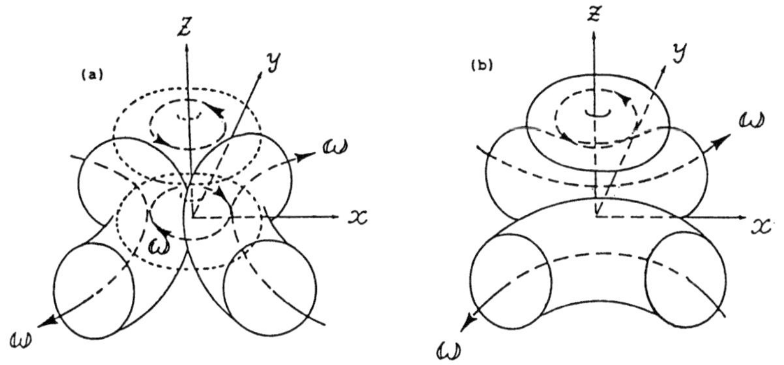

Fig. 2. Assumed mechanism of the process.

This process is expressed quantitatively by asymptotic conditions of the velocity and the vorticity fields. Let the velocity and the vorticity in the initial and the final states and the ring pair be denoted by u_i, ω_i, u_f, ω_f, u_r, ω_r, respectively. The origin of time t is chosen so that this process occurs at $t \cong 0$. Since the process is very rapid, instants before and after the process, to which the asymptotic conditions of the intermediate flow functions should refer, correspond to t=0- and 0+ on the time scales of the initial and the final states, respectively. Then, the asymptotic conditions are

$$u_r(-\infty)=0 \quad \text{and} \quad u_i(0-)+u_r(t) \to u_f(0+) \qquad \text{for } t \to \infty, \tag{2a}$$

$$\omega_r(-\infty)=0 \quad \text{and} \quad \omega_i(0-)+\omega_r(t) \to \omega_f(0+) \qquad \text{for } t \to \infty, \tag{2b}$$

The coordinate axes are introduced as shown in Fig. 2. It is assumed that the two filaments have a configuration with two planes of inversion symmetry: the xz- and yz-planes. The initial and final states differ in dependence on x and y, because the vortex filaments are aligned nearly with the y- and the x-axes and the velocity vectors are nearly confined in the xz- and yz-planes in these two states, respectively. Next, the velocity and the vorticity fields are expanded as polynomials of coordinates up to $O(r^2)$, where $O(r^m)$ denotes the m-th order polynomial.

Now, since the x-component of the velocity (u) is an odd function of x and an even function of y, and is independent of y in the initial state (see Fig. 2), only polynimials allowed for u in the initial state are x and xz up to $O(r^2)$. The y-component of velocity (v) vanishes in the initial state. The z-component of velocity (w) is an even function of both x and y, and allowed polynomial terms are 1, z, x^2, y^2. Polynomials for the vorticity and also the flow fields in the final state are obtained in the same way. The quantities $\Delta\omega_x$ and $\Delta\omega_y$ are expanded independently.

Expansion coefficients in the initial, the intermediate and the final states are related to each other because of the asymptotic conditions (2a,b). It is assumed here that the coefficients contain only one function of time T(t) which stands for the change of flow field with respect to time. Then, after reducing a number of independent coefficients by the use of (2a,b), the flow field during the whole process of recombination up to $O(r^2)$ is

$$\begin{aligned} u &= -m(1-T)x + k(1-T)xz, \\ v &= mTy - kTyz, \\ w &= w_0(0-) - \delta T + m(1-2T)z - \frac{l}{2}(1-T)x^2 + \frac{l}{2}Ty^2 - \frac{k}{2}(1-2T)z^2, \end{aligned} \tag{3}$$

$$\begin{aligned} \omega_x &= (k+l)Ty - nTyz, \\ \omega_y &= (k+l)(1-T)x - n(1-T)xz, \\ \omega_z &= 0. \end{aligned} \tag{4a}$$

$$\omega_x = g_x Ty, \qquad \omega_y = g_y(1-T)x, \qquad \omega_z = 0, \tag{4b}$$

where T(t) satisfies

$$T(-\infty) = 0, \qquad T(\infty) = 1. \tag{5}$$

Order estimations of these coefficients are made in terms of the core size σ, the circulation Γ and a curvature L of the filament, as follows:

$$\begin{aligned} &k = O(\Gamma/\sigma^3); \quad n = O(\Gamma/\sigma^4); \quad w_0(t) = O(\Gamma/\sigma), \\ &m = O(\Gamma/L^2). \end{aligned} \tag{6}$$

3. Solution of the Vorticity Equation

The equation which governs variations of these coefficients is derived from the vorticity equation

$$\frac{\partial}{\partial t}\omega + (u\cdot\nabla)\omega - (\omega\cdot\nabla)u - \nu\Delta\omega = 0, \qquad (7)$$

where ν is the kinematic viscosity. Substitution of Eqs.(4) and (5a,b) into this equation yields equations for T(t), while the parameters m, k, l, z and $w_0(0-)$ are treated as given. From the x- and y-components of the vorticity equation, we have, up to O(r),

$$x: \quad (k+l)dT/dt + m(k+l)T - (w_0(-0)-\delta T)nT + (k+l)Tm(1-T) = \nu g_x T, \qquad (8a)$$

$$y: \quad -(k+l)dT/dt - m(1-T)(k+l)(1-T) - (w_0(0-)-\delta T)n(1-T) - (k+l)(1-T)mT = \nu g_x(1-T). \qquad (8b)$$

From these equations and condition (5), values of g_x and g_y are determined as follows:

$$\nu g_x = m(k+l) - (w_0(0-)-\delta)n, \qquad \nu g_y = -m(k+l) - w_0(0-)n. \qquad (9)$$

Then, two equations (8a,b) are reduced to the same simple form

$$dT/dt - t^{*-1}T + t^{*-1}T^2 = 0, \quad \text{where } t^* = \delta n/(k+l) = O(\Gamma/\sigma^2). \qquad (10)$$

This equation does not contain viscosity explicitly, but, since the parameter t* is related to the viscosity through (9), the viscosity is certainly playing an essential role. The two components of the vorticity equation (8a,b) contradict each other, even in the lowest order, if the viscous terms are neglected.

Solution of the differential equation (10) satisfying (5) is

$$T = e^{\tau}/(1 + e^{\tau}), \quad \text{where } \tau = t/t^*. \qquad (11)$$

This solution has an expected behavior that it varies from 0 to 1 during a short time of O(t*).

4. Discussion

The approximate solution obtained here for the recombination process explains well this rapid fluid motion observed by flow visualization. The role of the viscosity can be discussed better on the basis of Eq.(7), and the vorticity in the final state is proved not to be obtained from the initial state without viscosity. The present analysis requires a recoil fluid motion. More experimental investigation of the detailed flow structure in the interacting region is necessary and worthwhile.

The only solution obtained in the past for a recombination process is given by Yeh & Axford [16] for MHD flow, where The magnetic field lines are recombined at the center of a stagnation flow. The fluid is assumed inviscid, but to have a finite conductivity. This mechanism is similar to our mechanism in the sense that the dissipative effect occurs in a relatively narrow region, while the mechanism differs in several points. Their MHD flow is steady and two-dimensional, and the three forces on the fluid, i.e. inertia, pressure and Lorentz forces balance with each other. On the other hand, in our problem inertia, pressure and viscous forces balance. However, it can be shown that a certain limiting case for the values of our expansion coefficients corresponds to the inner solution of Yeh & Axford [16].

The rapid process of recombination is also considered as an efficient source of the aerodynamic noise. For example, jet noise can be predicted in terms of accumulation of sounds from many recombination processes which are considered to occur at the stage of breakdown to turbulence in the jet flow. The present authors are going to pursue this possibility.

The present authors would like to express their thanks to Drs. K. Oshima, Y. Oshima and J. Bridges for their interest in this work and also for valuable discussion. This work is supported by the Japan-U.S. Cooperative Science Program of Japan Society for the Promotion of Science and US National Science Foundation.

1. H. Aref: Ann. Rev. Fluid Mech. 15, 345 (1983)
2. F. R. Hama: Phys. Fluids 6, 526 (1963)
3. R. Betchov: J. Fluid Mech. 22, 471 (1965)
4. H. Hasimoto: J. Fluid Mech. 51, 447 (1971)
5. R. Takaki: J. Phys. Soc. Jpn. 38, 1530 (1975)
6. S. C. Crow: AIAA J. 8, 2172 (1970)
7. H. K. Moffatt: J. Fluid Mech. 106, 27 (1981)
8. E. J. Hopfinger, F. K. Browand and Y. Gagne: J. Fluid Mech. 125, 505 (1982)
9. R. Takaki and A. K. M. F. Hussain: Phys. Fluids 27, 761 (1984)
10. S. E. Widnall, D. B. Bliss and C.-Y. Tsai:J. Fluid Mech. 66, 35 (1974)
11. P. G. Saffman: J. Fluid Mech. 84, 625 (1978)
12. F. R. Hama: Proc. Heat Tranc. & Fluid Mech. Inst., 92 (1960)
13. T. Kambe and T. Takao: J. Phys. Soc. Jpn. 31, 591 (1971)
14. Y. Oshima and S. Asaka: J. Phys. Soc. Jpn. 42, 708 (1977)
15. T. Fohl and J. S. Turner: Phys. Fluids 18, 433 (1975)
16. T. Yeh and W. I. Axford: J. Plasma Phys. 4, 207 (1970)
17. P. G. Saffman and G. R. Baker: Ann. Rev. Fluid Mech. 11, 95 (1979)
18. A. K. M. F. Hussain: Phys. Fluids 26, 2816 (1983)
19. G. K. Batchelor: An Introduction to Fluid Dynamics (Cambridge at the University Press 1967)

A New Look at the Near-Wall Turbulence Structure

Kwing-So Choi

Applied Fluid Mechanics Division, British Maritime Technology Ltd., Teddington, Middlesex TW11 0JJ, UK

An experiment was carried out in the BMT environmental wind tunnel (4.8m x 2.4m x 15m) in order to study the near-wall structure of the turbulent boundary layer, particular attention being given to the dynamics of the "near-wall bursts" [1]. Conditional samping of the wall-shear stress fluctuations was extensively used along with a simultaneous application of flow visualisation using a streak-smoke wire and a sheet of laser light. The results suggested that a "near-wall burst" was taking place between a pair of smoke tubes, which was interpreted as a pair of stretched legs of neighbouring hairpin loops. The spanwise spacing of the "near-wall bursts" determined from a conditional space correlation of skin-friction signals was found to be a function of the threshold value used in burst detection.

1. INTRODUCTION

A large number of studies have been conducted in the last two decades on the coherent structure of turbulent shear flows since a group at Stanford University published a series of reports of their work [2, 3] in the sixties. Among the studies on coherent structures, investigations into the structure of the turbulent boundary layer have been particularly in the interest of aerospace engineers who expected that they would be useful for the management of turbulence, e.g. a reduction of turbulent skin friction.

Roughly speaking, the study of turbulent boundary layers can be classified into two categories: one for the outer layer, another for the inner layer. It has been conjectured, only with limited experimental data however, that these two layers were interrelated with each other like a closed feed-back loop. One of the proposed conceptual models [2] was based on the sequence of "bursts", i.e. the "sweep" and the "ejection". In this model the near-wall activities were assumed to play rather passive roles being a consequence of the initiative taken by the "sweep", where the "sweep" was an event associated with an inrush of higher-momentum fluid from the outer layer into the inner layer, followed by the "ejection" of lower-momentum fluid away from the wall.

However, there have been extensive studies of the near-wall turbulence structure of the boundary layer which seemed to suggest that the near-wall activities in the inner layer might be playing more active roles than was thought originally in overall turbulence dynamics [4, 5]. They hinted at the existence of longitudinally stretched vortices near the viscous sublayer.

Recently, Choi [1] suggested from his flow visualisation study that there seemed to exist pairs of counter-rotating vortices very close to the wall near the sublayer. The study also indicated that the "near-wall burst", which was an event associated with an instantaneously large skin friction, was taking place between the pair of vortices.

The purpose of this paper is to examine further the near-wall structure of the turbulence boundary layer by using newly acquired experimental data. Particular attention is given to the dynamics of the "near-wall bursts".

2. EXPERIMENTAL SETUP

All the experiments were carried out in the BMT environmental wind tunnel (4.8m x 2.4m x 15m) with a free-stream velocity of 1.5m/s except for the experiment in Section 6 where a higher velocity of 3.0m/s was used for a measurement of the conditional space correlation. The Reynolds number based on the momentum thickness was 2.7×10^3 and 4.6×10^3, respectively. TSI model 1471 miniature hot-film sensors of 1.5mm diameter were used along with a DISA 55M constant temperature anemometer system to measure the fluctuating signal of skin-friction. Eight sensors in total were mounted flush with the wall surface in the spanwise direction.

Simultaneous application of flow visualisation was carried out in order to correlate the flow visualisation pictures with the skin-friction signals. The streak-smoke wire technique [1] and a slice of laser light were employed for the visualisation of the near-wall turbulence structure, whose sequence was recorded on video tape. Both the wire and the laser light were placed parallel to the wall, the former at $y^+=4$ and the latter at a higher position of $y^+=8$.

3. DATA ACQUISITION and REDUCTION

Data acquisition was carried out using a PDP 11/44 mini computer with an A-D board having a capability of sample-and-hold up to 16 channels. Signals from all of the hot-film sensors were sampled simultaneously with anti-aliasing filters set at the Nyquist frequency. The "near-wall bursts" were detected using the VITA technique [6] and signatures during the bursts were ensemble averaged. Typically, a few hundred realisations of the burst event were used for the average.

4. SIMULTANEOUS MEASUREMENTS

Figure 1 shows a typical sequence of near-wall turbulence activities recorded on video tape. The interval of each of the two consecutive scenes was two fiftieths of a second, and the field of view was approximately 300 wall units in the spanwise (vertical) direction. The flow was from right to left and the smoke wire, which was placed at $y^+=4$, can be seen on the

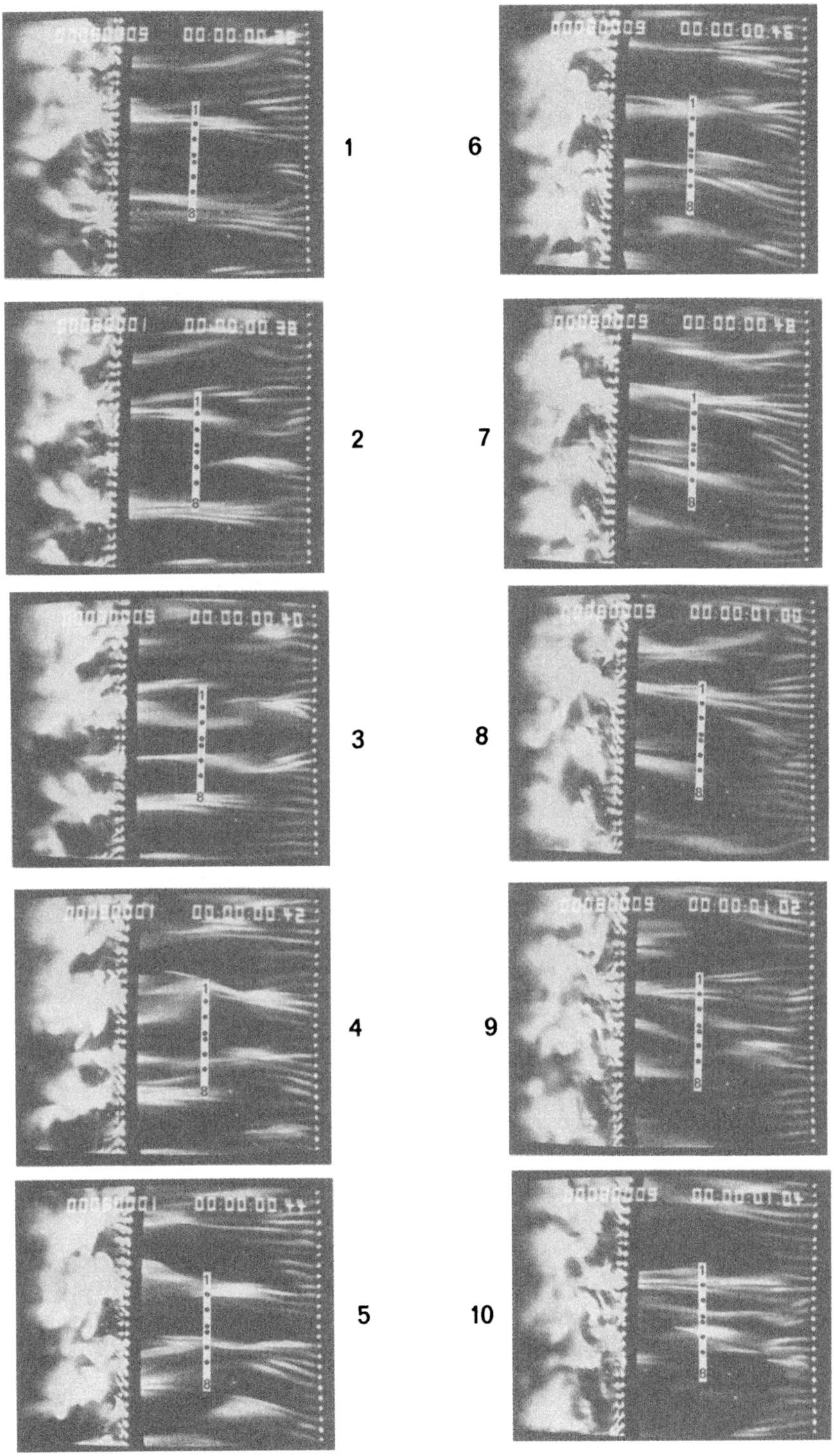

Fig. 1 Typical sequence of near-wall turbulence activities

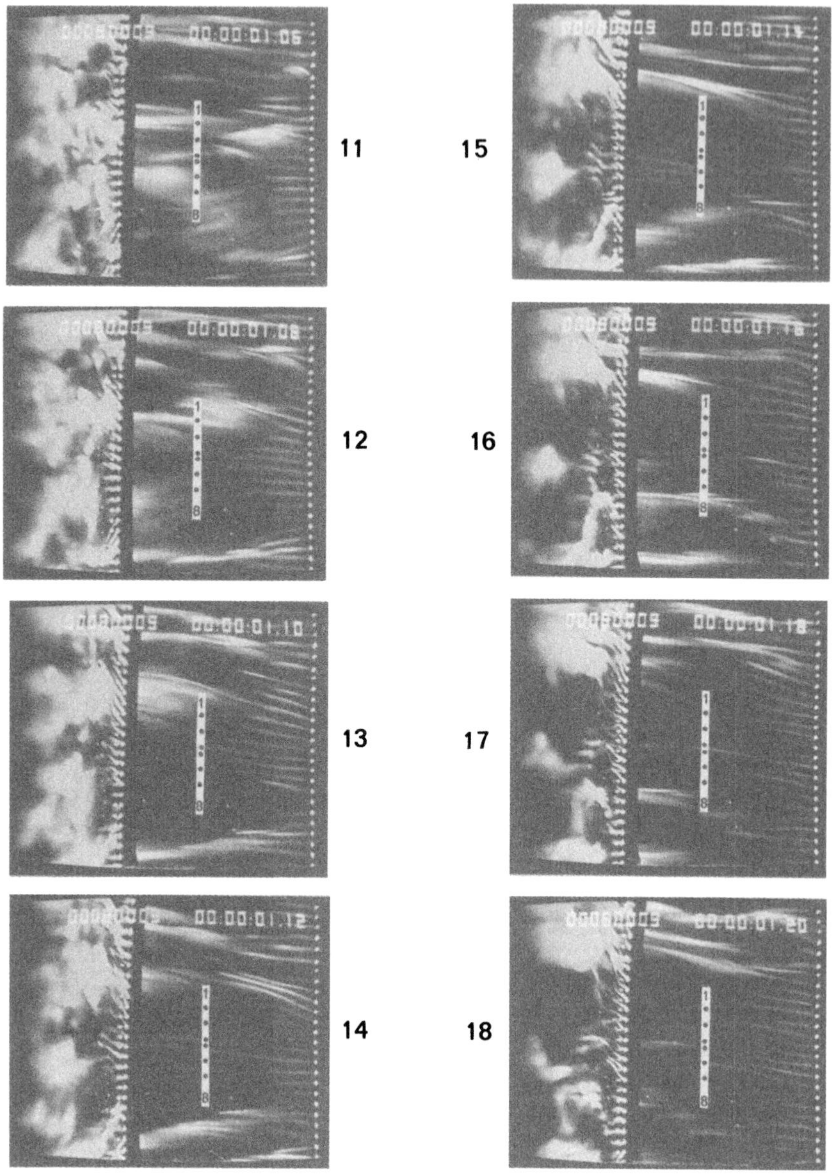

Fig. 1 (continued)

right-hand edge of each picture. The end view was superimposed over the planview in the left half of each picture, both of which were synchronised and physically aligned with each other throughout the video recording. Positions of sensors were indicated by their numbers shown in the centre of each picture, and the time code was located near the upper right hand corner.

It can be seen in the sequence of pictures that smoke filaments generated from the wire were tangled up into thick smoke tubes by vortices in the

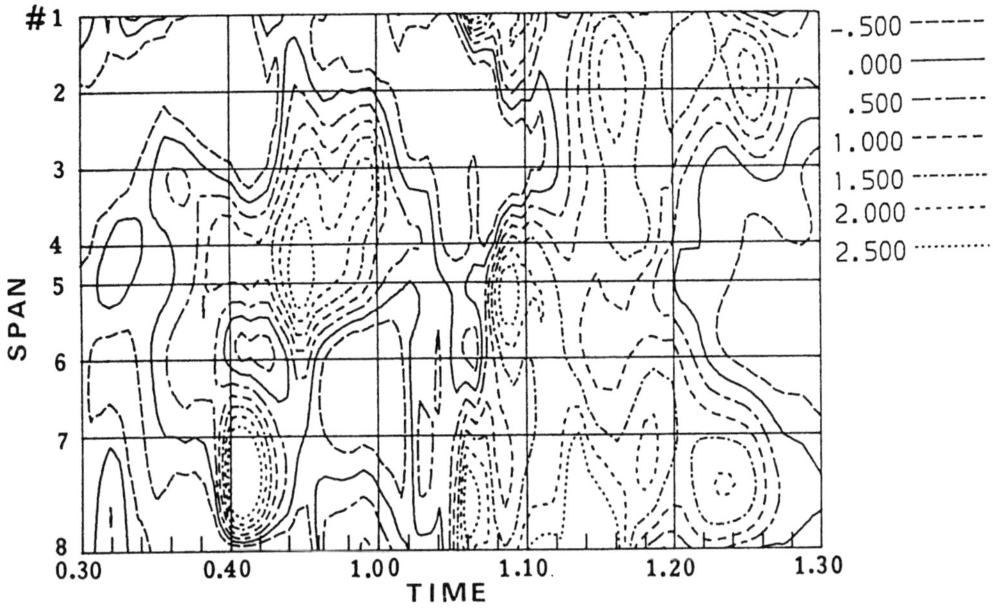

Fig. 2 Iso-skin friction map corresponding to Fig. 1 normalised with the standard deviation

flow. They were seen brightly illuminated as they were carried away from the wall into the plane of laser light which was placed higher than the wire at $y^{+}=8$.

The normalised signal of fluctuating skin friction is given in Fig. 2 and the running standard deviation used for the burst detection is plotted in Fig. 3. Since the "near-wall bursts" are associated with a large and rapid increase in skin-friction fluctuation, they can be identified with those having dense iso-skin friction contours in Fig. 2. Similarly, they can be found on the peaks in Fig. 3.

Although there is some limitation of resolution in video reproduction in Fig. 1, one can see a correlation between longitudinal smoke tubes and the "near-wall bursts" identified in Figs. 2 and 3. The correlation is particularly clear between the time 1:00 and 1:10 second (picture numbers 8 to 13) where three bursts were observed sequentially yet nearly simultaneously (see Figs. 2 and 3), firstly at the #7 and #8 sensors and then between the #4 and #6 sensors and finally at the #1 and #2 sensors.

One can also observe a large-scale fluid displacement away from the wall following the series of "near-wall bursts". The displacement seemed to have started right after the "near-wall burst" at the #7 and #8 sensors at time 1:06 second (picture number 11) and it spread forward to the next position of the "near-wall burst" in the spanwise direction (upward).

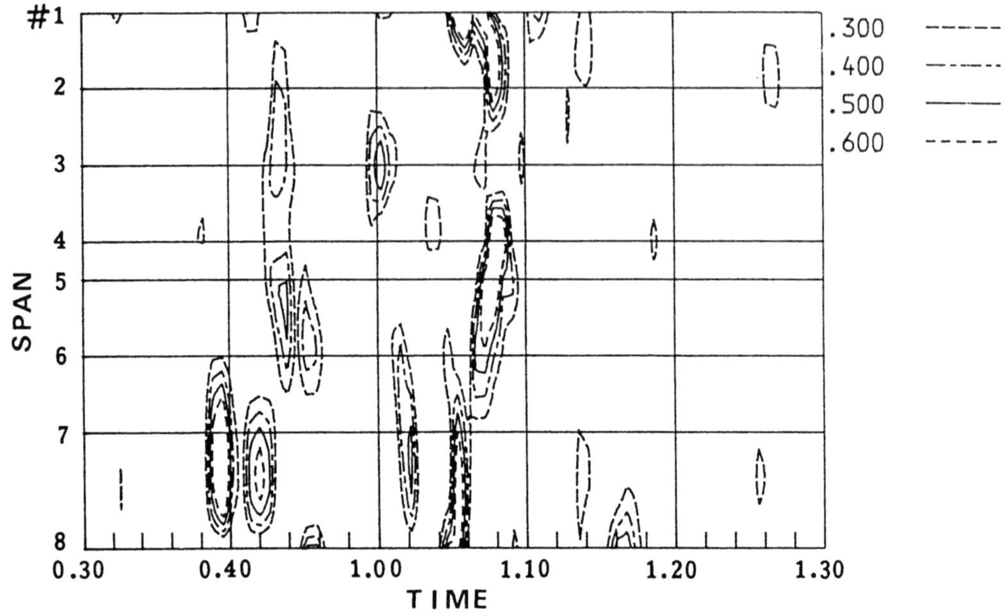

Fig. 3 Contour of the running standard deviation used for the burst detection

After this large-scale displacement of fluid, a period (t^{+}=40 to 80) of positive skin-friction fluctuation was observed which was followed by a period of negative skin-friction fluctuation.

Throughout the sequence of near-wall turbulence activities, rotation of smoke tubes was noticed particularly in the end views. This may suggest that the smoke tubes are manifestations of longitudinal vortices.

5. CONCEPTUAL MODEL

Figure 4 is a sequence of near-wall activities showing only the plan view. Since the end view was not superimposed in this sequence, the field of view was nearly doubled compared to Fig. 1. From this figure, a clear sequence of activities can be seen which seemed to have repeated in a staggered fashion. At the beginning of the sequence (picture number 7) there seemed to be pairs of longitudinal smoke tubes. The pairs deformed into a loop-shape in the following scene (picture number 10), whose upstream ends then became another pair of smoke tubes joining with their neighbouring counterparts. It is also observed that the sequences were taking place nearly in phase with the neighbouring sequences, showing a surprising spanwise regularity of near-wall activities.

A conceptual model for the burst sequence was proposed by Choi [1] based on an extensive amount of flow visualisation studies and is shown in Fig. 5. The model is similar to the one proposed by Wallace [4] whose dynamics was basically due to vortex stretchings. However, Choi's model is different from

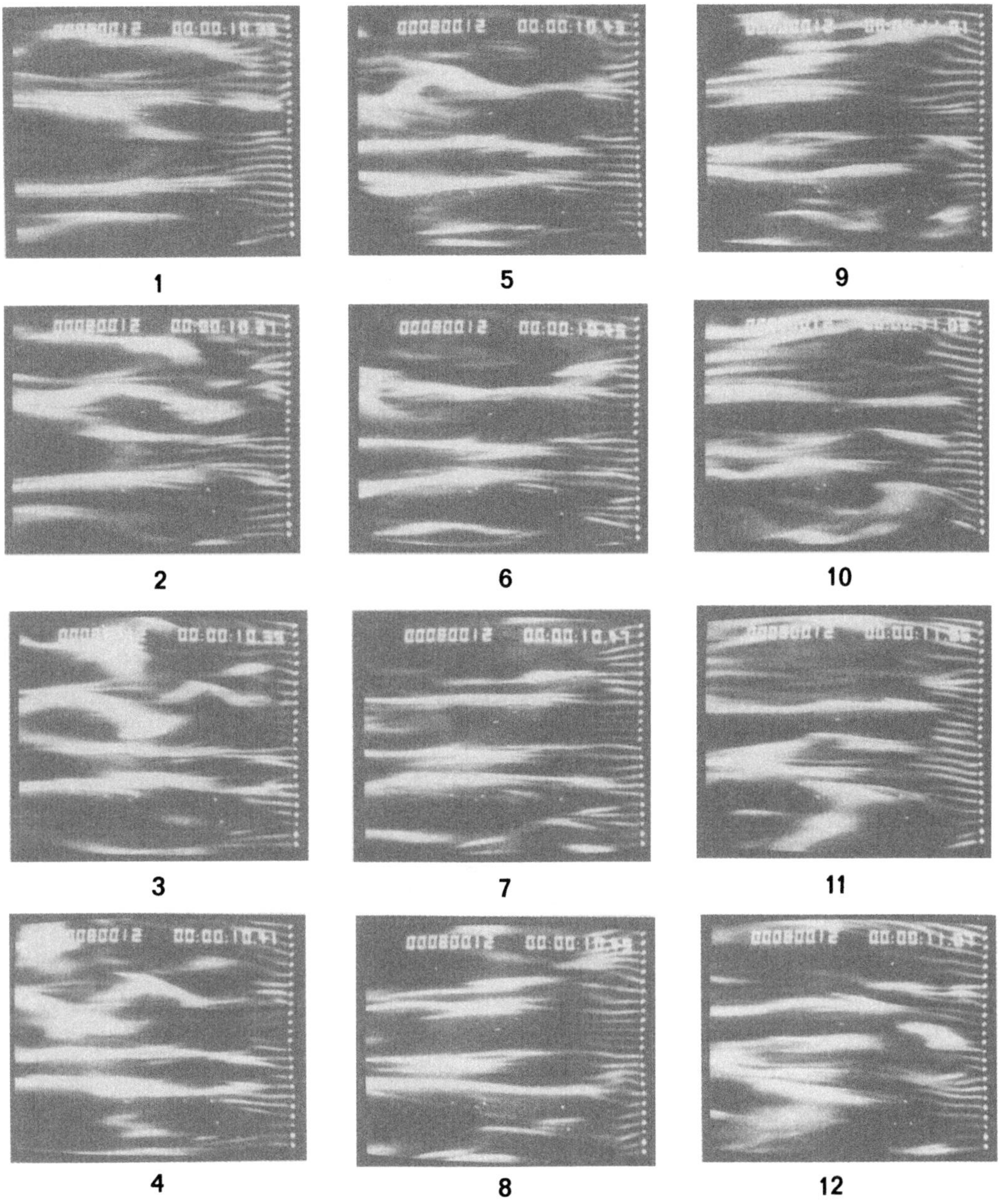

Fig. 4 Sequence of near-wall turbulence activities on a flat plate

Wallace's in the dynamics of near-wall activities, especially in the description of the "near-wall bursts". In Choi's model the "near-wall bursts" are observed between pairs of counter-rotating vortices, where the higher-momentum fluid is induced downward between the pairs giving rise to instantaneously large skin friction. The most recently acquired experimental data by Choi [7], who measured the conditionally sampled streamwise and

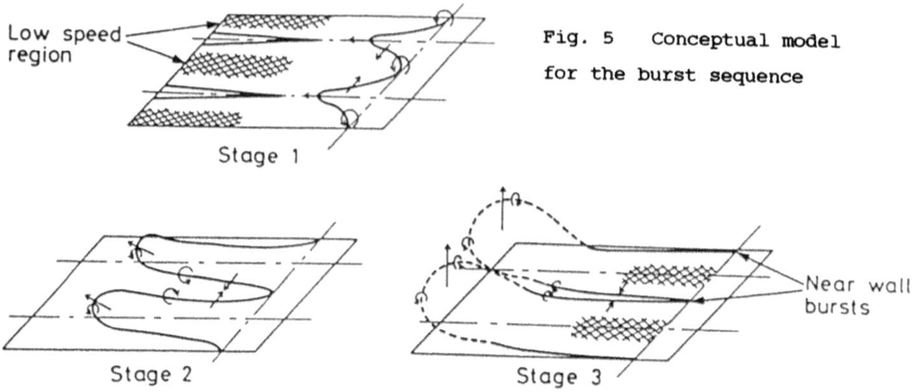

Fig. 5 Conceptual model for the burst sequence

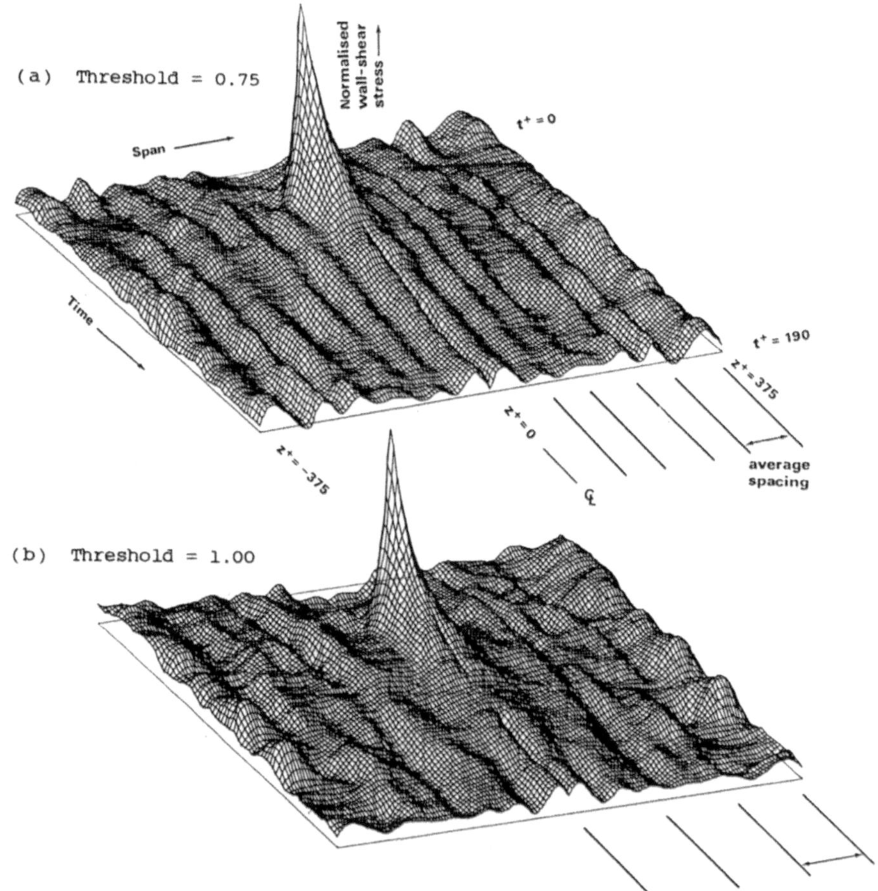

Fig. 6 Isometric view of conditionally sampled "near-wall bursts" with three different threshold values

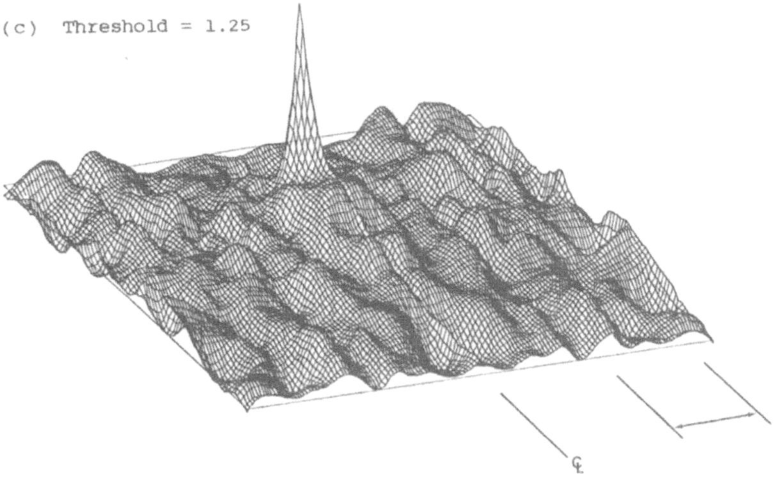

vertical velocity components during the "near-wall bursts", seemed to support Choi's model in his description of the near-wall activities.

6. CONDITIONAL SPACE CORRELATION

Isometric views of conditionally sampled "near-wall bursts" are shown in Figs. 6(a)-(c) for three different threshold values. The skin-friction signal from each sensor mounted flush with the wall surface in the spanwise direction was conditionally sampled and ensemble averaged when the trigger sensor at $Z^+=0$ detected the bursts. Hence the longitudinal ridges seen in Figs. 6(a)-(c) can be interpreted as the positive correlations between the "near-wall bursts" at the centreline ($Z^+=0$) and the spanwise locations where the ridges are observed. Here, the threshold value was given with respect to the standard deviation of the skin-friction signal from the trigger sensor.

It can be seen in the figure that the average spacing between the ridges was narrower at smaller threshold value; it was approximately 70 wall units at the threshold value of 0.75. With an increase of the threshold value to 1.0 and 1.25 the spacing became wider to 115 and 160 wall units, respectively. These results indicated that the spanwise spacing of the "near-wall bursts" was an increasing function of the threshold value. They were consistent with the observation of the flow visualisation that "strong" bursts were formed as a result of amalgamation between two or more "weak" bursts and that their physical size was proportional to their intensity.

This work has been carried out with the support of the Department of Trade and Industry, U.K.

1. Choi, K.-S. (1985) Near-wall turbulence structure on a riblet wall. BMT Report, BMT, Feltham, Middlesex, U.K.

2. Kline, S.J. et al. (1967) The structure of turbulent boundary layers. J. Fluid Mech. 30, 741.

3. Kim, H.T. et al. (1971) The production of turbulence near a smooth wall in a turbulent boundary layer. J. Fluid Mech. 50, 133.

4. Wallace, J.M. (1982) On the structure of bounded turbulent shear flow : a personal view. In Development in Theoretical and Applied Mechanics, vol. XI, ed. by T.J. Chung and G.R. Karr, Univ. of Alabama, Huntsville.

5. Smith, C.R. (1984) A synthesized model of the near-wall behaviour in turbulent boundary layers. In Proceedings of Eighth Symposium on Turbulence, ed. by G.K. Patterson and J.L. Zakin, Univ. of Missouri, Rolla.

6. Blackwelder, R.F. and Kaplan, R.E. (1976) On the wall structure of the turbulent boundary layer. J. Fluid Mech. 76, 89.

7. Choi, K.-S. (1986) Drag reduction by manipulation of near-wall turbulence structure. Presented at the European Meeting on Turbulent Drag Reduction in Lausanne, Switzerland, 1-3 September 1986.

On the Evolution of Shear-Layer Structures in Near-Wall Turbulence

A.V. Johansson[1], *P.H. Alfredsson*[1], *and H. Eckelmann*[2]

[1]Department of Mechanics, The Royal Institute of Technology, S-10044 Stockholm, Sweden
[2]Max-Planck Institut für Strömungsforschung, D-3400 Göttingen, Fed. Rep. of Germany

Lift-up of low-speed fluid from the viscous sublayer creates shear-layer structures in the buffer region, that have been shown to be dynamically important for the generation of turbulence in wall-bounded flows. The present experiments, carried out in the Göttingen oil channel, focus on the evolution of shear-layers, by means of two-probe measurements in the buffer region of turbulent channel flow. Various separations in the streamwise and normal directions were used to obtain data on the inclination angles and progation velocity of the shear-layer, as well as the coherence of the flow structure.

1. Introduction

The formation and evolution of shear-layer like flow structures in the buffer region of wall-bounded turbulent shear flows are intimately coupled to the process of turbulence production. These structures are formed when low-speed fluid from the viscous sublayer is ejected outwards from the wall. Flow visualization studies have shown that there is a tendency of the viscous sublayer in all wall bounded shear flows to evolve a streaky pattern of narrow low and high speed regions. The ejection of fluid from the viscous sublayer is often initiated as a slow outward migration of a low-speed streak, or a part thereof, followed by a rapid outward motion. The resulting flow structure will, hence, have a relatively small spanwise extent, and can be characterized as an internal shear-layer which essentially is confined to the near-wall region. In the long-time evolution of the shear-layer the velocity component normal to the wall (v) will be quite small as compared with the deviation (u) from the local mean velocity (U). It will be sharpened due to the mean shear and, in its final stage, it may break up into small scale turbulence. This process is responsible for a substantial part of the turbulence production.

To obtain quantitative data on the flow field associated with the shear-layer BLACKWELDER & KAPLAN [1] used conditional sampling of signals from a rake of ten hot-wires covering the near-wall region up to $y^+=100$ (superscript + is used for normalization with wall variables in the usual manner, e.g. $y^+=yu_\tau/\nu$, where y is the direction normal to the wall, u_τ is the friction velocity and ν is the kinematic viscosity). They used a trigger, or detection, probe at $y^+=15$ and were able to show that the coherence of the detected structure extends out to about

y^+=100 and, as expected, that the structure is fairly narrow. The width was found to be roughly 50 ℓ_* ($\ell_* = \nu/u_\tau$).

Despite the inherent importance of the spatial structure of these shear layers, only few researchers have attempted to quantify their properties. From the data of BLACKWELDER & KAPLAN [1] one may infer that the shear layer is inclined at about 22° in the buffer region (using a Taylor's hypothesis with an assumed propagation velocity of $13u_\tau$) and that this angle decreases further out from the wall. The same general behaviour is found in the conditionally sampled data of NAKAGAWA & NEZU [2]. From long-time correlation measurements, however, the inclination angle in the buffer region has been found to be about 10° (see, e.g. KREPLIN & ECKELMANN [3]. The smaller angle obtained with the latter technique should be expected, since in that case all types of turbulent structures contribute, including, e.g., large scale sweep motions with small inclination angles.

In the present work the evolution of shear-layer structures in the near-wall region was studied, including their coherence, inclination angle and propagation velocity. For this purpose a detector probe at y^+=15 and a second probe displaced in the x and y directions were used. An enhanced conditional sampling scheme, in which phase-jitter associated with the conditional averaging process was removed, was applied to the signals from these probes. The experimental procedure and conditional sampling scheme are described in section 2, and the results are given in section 3. The conclusions are summarized in section 4.

2. Experimental procedure

For the main part of the experiments a fixed detector probe at y^+=15 was used together with a traversable probe at various distances from the wall (y^+=20, 30, 50, 100) and at various streamwise positions (Fig. 1a). Also a (fixed) wall shear stress sensor was used in conjunction with a traversable trigger probe at y^+=15 for various streamwise separations between the two probes (Fig. 1b).

The experiments were carried out in a fully developed turbulent channel flow in the Göttingen oil channel at a Reynolds number of 7500 based on centreline velocity and channel width. For this case the viscous lengthscale ($\ell_* = \nu/u_\tau$) is as large as 0.55 mm, which ensures minimally disturbed measurements even when the two hot-film probes are only 5ℓ_* apart (in the y-direction). This facility is, hence, well suited for the study of shear-layer characteristics. However, the low velocity of this flow necessitates a long sampling time at each position. It was here chosen as 40 minutes or about 2200 b/U_{CL}. For details of the facility and the basic flow the reader is referred to ECKELMANN [4].

The Variable-Interval Time-Averaging (VITA) technique (BLACKWELDER & KAPLAN, [1]) was used to detect the presence of a shear-layer at y^+=15. The correspondence

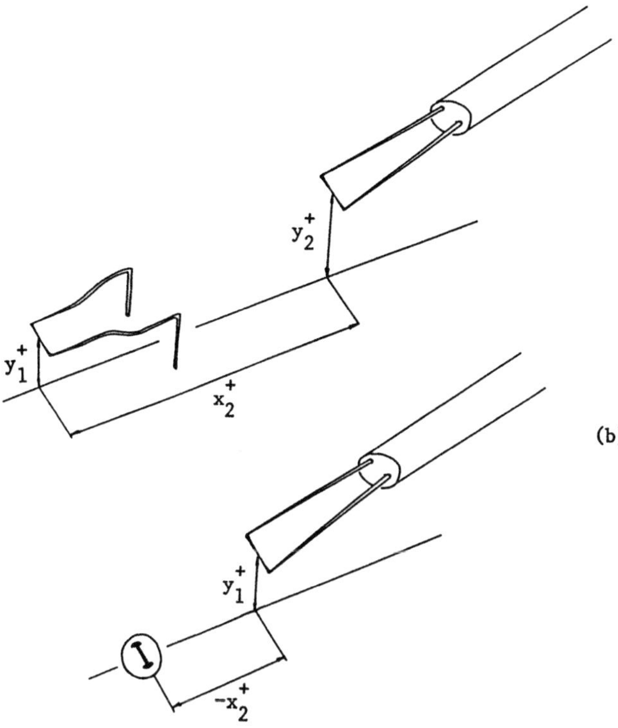

Figure 1 Probe configuration. a) fixed detector probe at $y_1^+=15$ and traversable probe at y_2^+ (sensor lengths: $2\ell_*$). b) traversable detector probe at $y_1^+=15$ and wall shear stress probe

between VITA events and shear-layers was substantiated by e.g. JOHANSSON & ALFREDSSON [5] and JOHANSSON, HER & HARITONIDIS [6]. In this detection scheme the short-time variance of the u-signal over an averaging time T is calculated, and an event is considered to occur when this exceeds a chosen threshold level (k) times the long-time variance (u_{rms}^2). Futhermore, u is here demanded to be increasing at the detection times. In this averaging process for the signal from the non-detector probe there is a smearing effect caused by phase jitter, i.e. the variation in arrival time of the shear-layer at this position. To remove this effect a second iteration in the conditional averaging was performed. The average of the first iteration was then only used to determine the mean arrival time of the events, whereafter the individual arrival times were determined by finding the maximum short-time variance within a chosen time-window around the mean arrival time. The length of the window was here taken as 15 viscous time units (t_*). The individual events were then given appropriate time-shifts before construction of the final conditional average. A similar technique was tried by BLACKWELDER & KAPLAN [1], whereas JOHANSSON et al. [6] determined the time-shifts from the cross-correlation between individual events and the ensemble average. The removal of phase jitter is here essential, particularly in order to get a correct picture of the amplitude variation in the x and y directions of the associated velocity disturbance, i.e. the coherence of the event.

3. Results

Enhanced conditional averages of the wall shear stress and the streamwise velocity at various near-wall positions are shown in Fig. 2, which gives an overall picture of the development of the shear-layer structure in the x-y plane. The signatures from $y^+=20$ are not shown here, but are similar to those at $y^+=30$, albeit somewhat larger in amplitude. At both these positions the identity of the structure is well

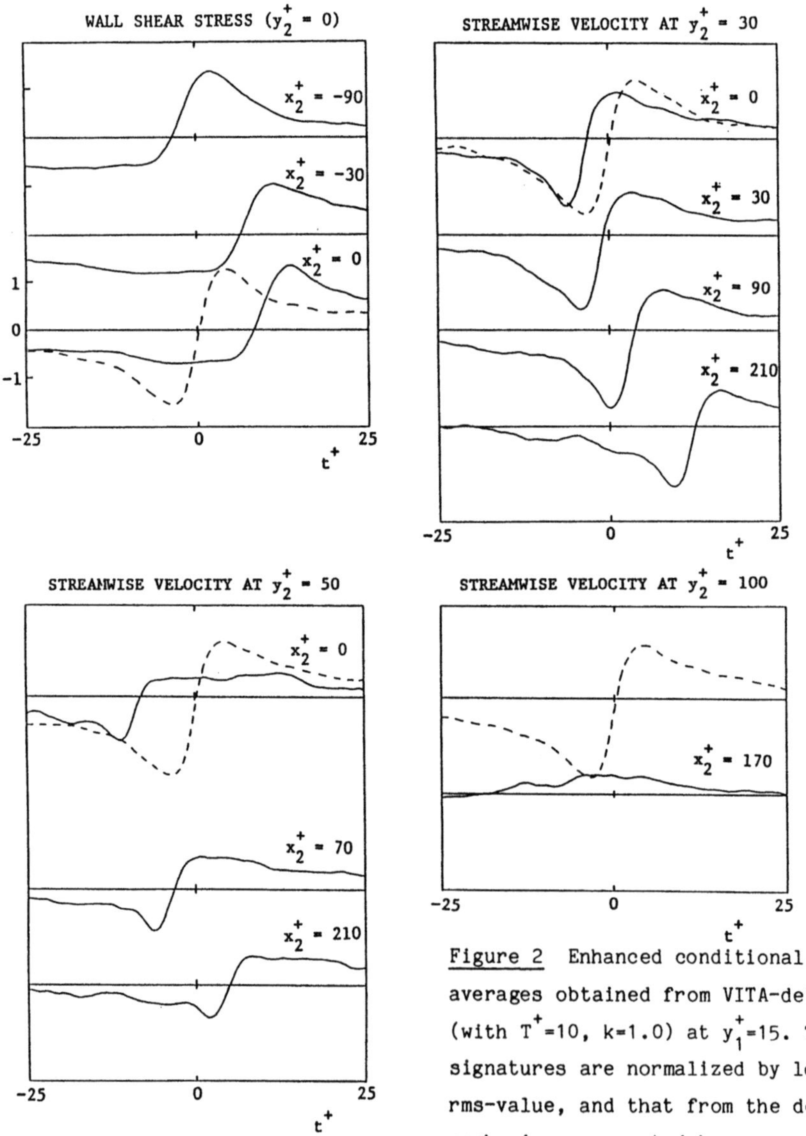

Figure 2 Enhanced conditional averages obtained from VITA-detection (with $T^+=10$, $k=1.0$) at $y_1^+=15$. The signatures are normalized by local rms-value, and that from the detector probe is represented by ---

preserved over a travelling distance of more than 500 ℓ_*. Indeed, the amplitude of the signature at y^+=30 only varies 22 % over the 210 ℓ_* long travelling distance shown in Fig. 2.

The peak-to-peak amplitude of the signatures decreases rapidly with increasing distance from the wall for the second probe (Fig. 3), which clearly shows that the structure essentially is confined within the buffer region. At y^+=100 the signature has practically lost its shear-layer character, and only a fairly long period of excess streamwise velocity is seen. The enhancement method may then introduce some artificiality, wherefore this signature is shown without enhancement in Fig. 2.

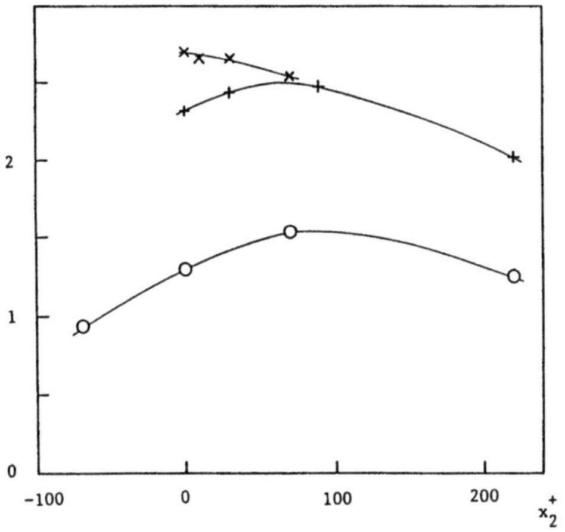

Figure 3 Peak-to-peak amplitude (normalized with local rms-value) of the signatures at x: y_2^+=20, +: y_2^+=30, o: y_2^+=50, as function of x_2^+

The fact that the shear-layer signature is seen earlier above the detector probe indicates a forward leaning structure. Its propagation velocity and inclination angle can easily be obtained from the time-difference between the zero-crossing of the actual signature and that from the detector probe. Such time-differences at y^+=30 are shown in Fig. 4 as function of the downstream distance. The inverse of the slope defines the propagation velocity (c_{prop}=$\Delta x^+/\Delta t^+$). For y^+=0 and y^+=20 it was found to be 13.0, at y^+=30 it was slightly higher (13.6) and at y^+=50 it was about 15. The latter is close to the local mean velocity at that position.

It is interesting to note that the propagation velocity is practically constant up to y^+=30. The higher value of c_{prop} further out means that the structure becomes stretched by the mean shear. This is also reflected in variation of the shear-layer angle with distance from the wall. It is determined from $\Delta y^+/\Delta x^+$ with Δx^+ taken as the value when the signature overlaps with the reference dito, or equivalently, the Δx^+ value for which Δt^+=0 (see Fig. 4). Closest to the wall the angle was found to be 7° (when determined from y^+=0 and 15), whereas in the buffer

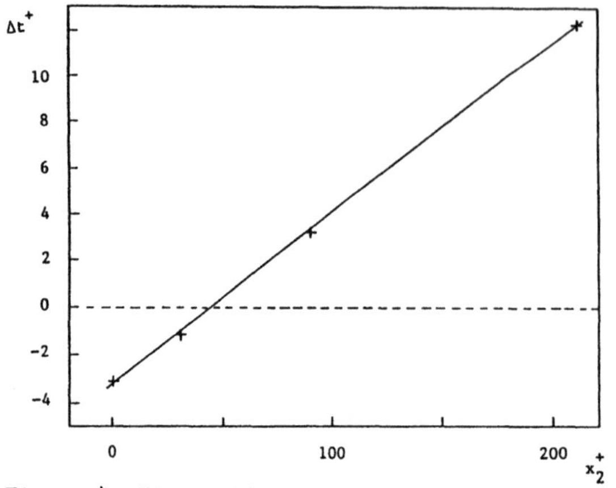

Figure 4 Time-shift of zero-crossing from the signatures at $y_2^+=30$ as function of downstream distance

region between $y^+=15$ to 30 it was 20°. At $y^+=50$ the angle again is smaller (14°) in consistency with the higher propagation velocity found at this position. The stretching of the shear-layer also means that the structure cannot extend far beyond the buffer region.

KREPLIN & ECKELMANN [3] determined inclination angles of the front as defined from long-time correlation measurements. They found an angle of 4.7^o in the viscous sublayer, which is in reasonable accordance with the present finding of 7° as a mean value between the wall and $y^+=15$. However, the angle obtained from long-time correlation measurements in the buffer region [3] is about a factor of two smaller than the angle found here for the shear-layer "front". This indicates that the shear-layers probably are the most highly inclined structures of any dynamical significance in the near-wall region.

The streamwise extent of the structure may be estimated as the propagation velocity times the distance in time between the minimum and the maximum of the velocity signature. The latter is closely coupled to the averaging time used in the detection criterion (see JOHANSSON & ALFREDSSON [5]). However, since the value of T chosen here ($10t_*$) is close to the value for which the maximum number of detections are found, the estimated streamwise extent will at least be a representative value. From Fig. 2 we get 100 ℓ_* to be compared with the length over which it retains its identity, which is at least 500 ℓ_*.

To facilitate the understanding of the above presented results the data of Fig. 2 were replotted into lines of constant disturbance velocity (normalized with the local rms-level) in the x-y plane using the measured propagation velocities to transform the time to an x-coordinate (Fig. 5). An interpretation near at hand is that a part of a low-speed streak has been lifted from the wall, and that high-speed fluid is moved toward the wall on the upstream side of the shear-layer. LANDAHL [7] showed that a patch of small scale Reynolds stress tends to produce a

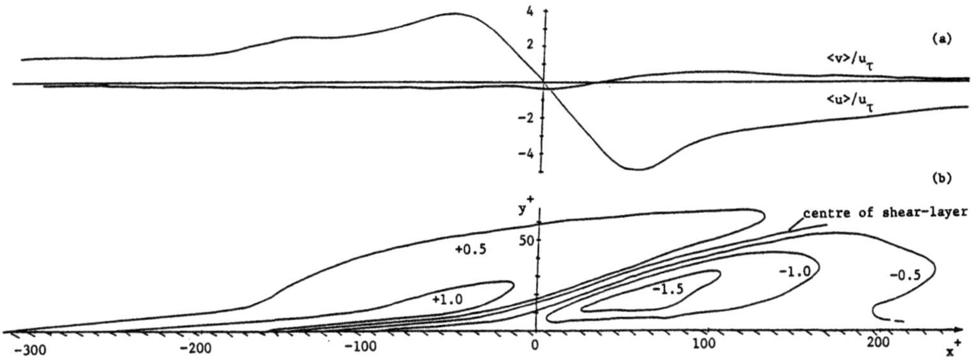

Figure 5 Shear-layer flow structure. a) Velocity signatures in absolute scale from $y^+ = 15$ (as function of x^+). b) Iso-velocity curves, normalized with local rms-value, in the x^+-y^+-plane

large scale disturbance with a sense of rotation opposite to that of the mean vorticity, as indeed is the case for the large scale structure depicted in Fig. 5. We can also observe the maximum in inclination angle in the region $y^+=15-30$.

Also included in Fig. 5 are the streamwise and normal velocity signatures (as function of x) from $y^+\approx 15$. They are shown in an absolute scale (i.e. normalized by u_τ) and we can note that the v-component is very small as compared with the u-component. This illustrates the fact that the displacement of fluid elements in the vertical direction generates a streamwise velocity disturbance that remains even for times when the v-component has died out. The dominance of horizontal inertia for long times has also been shown theoretically by LANDAHL [8] in an analysis of the evolution of a flat coherent structure in the near wall region.

The u, v and uv signatures (normalized by the respective rms-values) from $y^+=15$ are shown in Fig. 6. One may note that the v-component changes sign earlier than u, and that the largest contribution to the Reynolds stress (and turbulence production) occurs during the ejection phase. There is a waviness in the v-signature during the period of negative v. It is of moderate amplitude, but was observed also at other near-wall positions and in similar data from $y^+=15$ measured in the water channel at the Royal Institute of Technology (Stockholm). It may be the evidence of a small dispersive wave-part in the otherwise advection dominated evolution of the shear-layer structure. The waviness was most clearly seen for the sharpest shear-layers, i.e. those detected with small values of the averaging time T. This is also in accordance with what one should expect from theoretical considerations (Landahl, private communication).

4. Conclusions

The conclusions may be summarized as follows

1. Shear-layers are formed through lift-up of low-speed streaks, which forms a coherent flow structure that is essentially confined within the near-wall region (below $y^+\approx 100$).

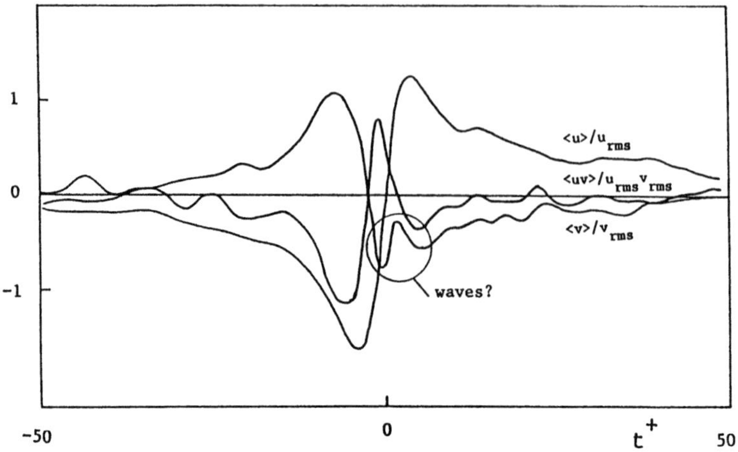

Figure 6 Conditional averages of u, v and uv, normalized with rms-values, at $y^+=15$. The mean value of uv has been subtracted from <uv>

ii. The streamwise extent of the shear-layer is of the order of 100 ℓ_* and it retains its identity over a travelling distance of at least 500 ℓ_*

iii. The propagation velocity is almost constant ($\approx 13u_\tau$) up to $y^+=30$. Further out it increases and the flow structure becomes highly stretched.

iv. The inclination angle of the shear-layer structure is shallow close to the wall and attains a maximum of 20^0 in the buffer region ($y^+\approx 20$).

Acknowledgement

The present study was carried out with financial support from the Max-Planck Institut für Strömungsforschung and the Swedish State Board for Technical Development, which we gratefully acknowledge. We also want to thank Mårten Landahl for illuminating discussions and Catherine Lagercrantz for word processing.

REFERENCES

1. R.F. Blackwelder & R.E. Kaplan: J. Fluid Mech. 76, 89 (1976).
2. H. Nakagawa & I. Nezu: J. Fluid Mech. 104, 1 (1981).
3. H.-P. Kreplin & H. Eckelmann: J. Fluid Mech. 95, 305 (1979).
4. H. Eckelmann: J. Fluid Mech. 65, 439 (1974).
5. A.V. Johansson & P.H. Alfredsson: J. Fluid Mech. 122, 295 (1982).
6. A.V. Johansson, J. Her & J.H. Haritonidis: Accepted for publication in J. Fluid Mech (1986).
7. M.T. Landahl: Ph. Fluids 20, S55 (1977).
8. M.T. Landahl: In Proc. IUTAM Symposium on Turbulence and Chaotic Phenomena in Fluids, Kyoto 1983, ed. by T. Tatsumi, p. 467 (North-Holland 1984).

Coherent Structures in Turbulent Pipe Flow

A. Dinkelacker and M. Sieber

Max-Planck-Institut für Strömungsforschung,
D-3400 Göttingen, Fed. Rep. of Germany

Wall pressure fluctuations and velocity fluctuations were measured simultaneously in fully-developed turbulent pipe flow. Signal evaluation was done using the method of conditional signal averaging. The measurements show that characteristic velocity signatures in the flow are correlated to high or low pressure events at the pipe wall. The convection of these velocity signatures can be followed over comparatively long distances along the pipe (at least 13 pipe diameters) and even in the jet leaving the pipe these signatures can be detected.

1. Introduction

Several investigations on turbulent wall pressure fluctuations have been performed at the Max-Planck-Institut für Strömungsforschung. The aim of these investigations was better to understand the flow processes by which the wall pressure fluctuations are generated. The experiments reported here are a continuation of earlier work by LANGEHEINEKEN and DINKELACKER /1/, /2/ in turbulent pipe flow. While in the earlier work velocity fluctuations in the neighbourhood of a wall pressure transducer were investigated, in the work reported here the field of measurements has been extended to a larger region in the pipe, and into the flow field of the jet leaving the pipe. Furthermore the methods of measurement and data evaluation have been improved. In the doctoral thesis of the second author, submitted to the University of Göttingen, the new part of the investigations is described in detail /3/.

2. Experiment

Figure 1 shows the flow facility, the main part of which is a long straight pipe with a pipe diameter D = 5 cm and a pipe length L = 9.8 m, so that L/D = 196. At the pipe inlet early transition is ensured by means of a strip of sandpaper. The experimental arrangement was so conceived as to get low disturbance levels from noise and vibration. Fluctuations of the wall pressure and of the streamwise velocity were investigated in the experiments. The wall pressure fluctuations were measured with a "pinhole microphone" consisting of a hole in the pipe wall of 0.5 mm diameter and a 1/8" Brüel and Kjaer microphone. The pinhole was 110 mm upstream of the pipe outlet. The velocity fluctuations were measured with a hot wire probe with a sensitive wire length of 0.5 mm. Evaluation of the measurements was mainly done using the method of "conditional signal averaging". The analogue data were fed directly into a computer, where comparatively large data sets were stored for further evaluation.

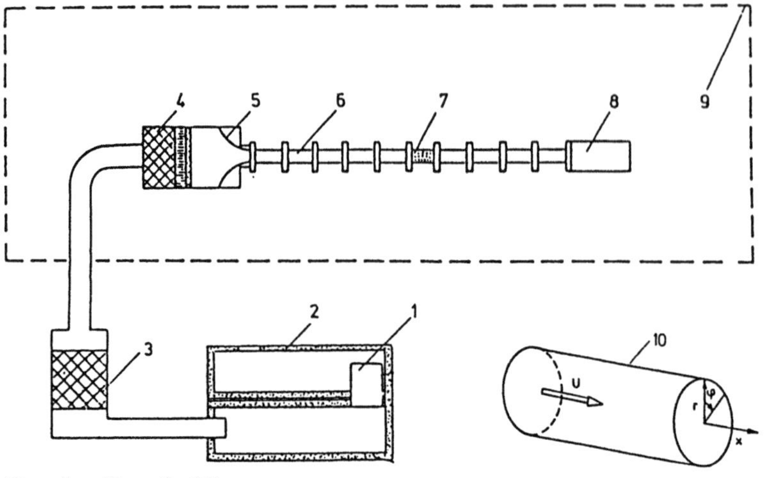

Fig. 1. Flow facility
(1 low noise fan, 2, 3, 4 sound insulation, 5 flow contraction, 6 pipe, 7 sound damping pipe section, 8 measuring section with pinhole microphone and hot wire, 9 elastically suspended frame, 10 sketch of the coordinate system used for the measurements)

The measurements in the pipe were performed mainly at three different air velocities. Higher velocities give a better signal to noise ratio. Lower velocities give a better spatial resolution in terms of wall units. The following table gives a survey of the flow data in the pipe:

centerline velocity	U_{cl}	2.9 m/s	8.5 m/s	11.0 m/s
Reynolds number	$Re = \frac{U_{cl} \cdot D}{\nu}$	9600	28500	36000
shear velocity	u_τ	0.147 m/s	0.39 m/s	0.48 m/s
length scale	$l_w = \frac{\nu}{u_\tau}$	0.105 mm	0.040 mm	0.032 mm
characteristic time	$t_w = \frac{\nu}{u_\tau^2}$	0.710 ms	0.100 ms	0.067 ms

Cylindrical coordinates r, φ, x were used (Fig. 1). The origin is positioned so, that the pinhole microphone has the coordinates $r = R$; $\varphi = 0$; $x = 0$. The position of the hot wire probe is denoted by (y_h, φ_h, x_h). The distance from the pipe wall $y_h = R - r_h$ is used instead of the coordinate r_h.

3. Results

3.1 The basic observation

In /1/ and /2/ it was shown that events of instantaneous high pressure at the pipe wall are accompanied by the occurrence of characteristic velocity patterns in the turbulent flow. Such a pattern consists of a low speed streak followed by a high speed streak. In the region between the two streaks a steep rise in the streamwise

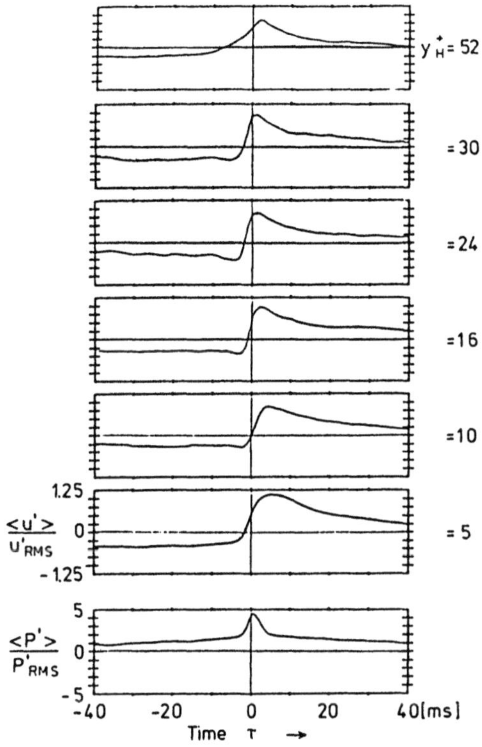

Fig. 2. Conditionally averaged streamwise velocity signatures at different distances from the pipe wall; (U_{cl} = 2.9 m/s; condition: $p \geq 4\, p_{rms}$: position of hot wire: y_h varying, $\varphi_h = 0$, $x_h = 0$; here $y_h^+ = 52$ corresponds to y_h = 5 mm or $y_h/R = 0.2$; 1 ms corresponds here to $\tau^+ = 1.4$; figure taken from /1/)

velocity occurs. This rise is related to the pressure peak at the wall. The wall pressure peak can thus be interpreted as a kind of "stagnation pressure" generated inside the flow by the impact of the high speed streak on the low speed streak. Fig. 2 gives an example of measured curves from /1/. The lowest curve is the conditionally averaged wall pressure signature <p> (normalized with p_{rms}) with the pressure peak at time $\tau = 0$. The condition used for the averaging procedure was $p \geq 4 p_{rms}$. The pressure peak has a duration of about 5 ms corresponding here to $\tau^+ = 7$ (the deviation from the zero line outside the peak is background noise). The other six curves in Fig. 2 show conditionally averaged streamwise velocity signatures <u> (normalized with the local u_{rms}) measured at different distances from the wall ($5 \leq y^+ \leq 52$ corresponding here to $0.02 \leq y/R \leq 0.2$ where R is the pipe radius). These curves clearly show the low and high speed parts of the flow pattern and the velocity rise between both. The rise-time in the <u>-signatures measured close to the wall is about 5 ms (similar to the duration of the pressure peak), further away from the wall the rise-time in the <u>-signatures increases.

While in /1/ mainly the flow in the neighbourhood of the pressure transducer was investigated, the region of flow measurements has been extented considerably in the experiments reported in /3/. Fig. 3 shows such a series of measurements. The hot wire is now positioned further away from the wall. It should be noted that these measurements were performed with a higher flow velocity (here U_{cl} = 11.0 m/s) in

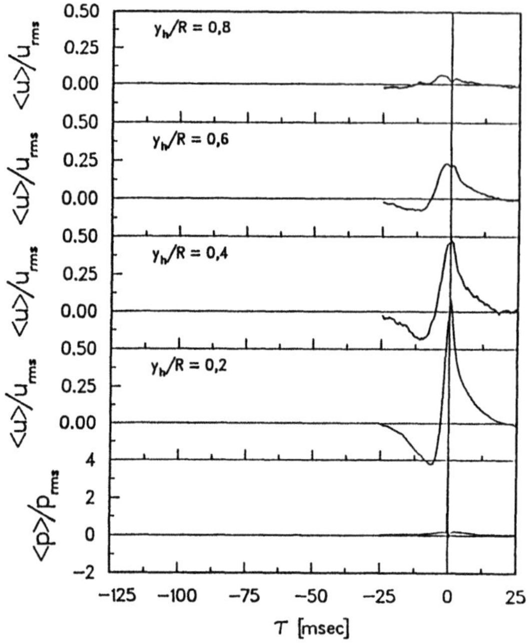

Fig. 3. Same as Fig. 2, but the measurements were performed with a higher flow velocity and at larger distances from the pipe wall (U_{cl} = 11.0 m/s; condition: $p \geq 4\,p_{rms}$; y_h varying; $\varphi_h = 0$, $x_h = 0$; 1 ms corresponds here to $\tau^+ = 15$)

order to get a better signal to noise ratio. At this velocity a time of 1 ms corresponds to $\tau^+ = 15$. The condition used for the conditional averaging is again $p \geq 4p_{rms}$; the lowest curve in Fig. 3 shows the <p>-signature. The pressure peak used for signal detection is now so slim that it can hardly be seen in Fig. 3. The four upper curves again show <u>-signatures, now measured at larger distances from the wall ($0.2 \leq y_h \leq 0.8$ corresponding here to $156 \leq y^+ \leq 625$). As one can see the <u>-signatures follow the same pattern as described in the context of Fig. 2. The amplitudes of the signatures decrease with increasing wall distance. At a distance of $y_h/R = 0.8$ the <u>-signature has nearly disappeared. The duration of the observed <u>-signatures is about 40 ms. With a convection velocity of about 8 m/s this duration corresponds to a length of 320 mm, which is about 5 pipe diameters or $\Delta x^+ = 10\ 000$. In /1/ a length of the streaks of 4 pipe diameters or $\Delta x^+ = 2000$ was reported. This indicates that the length of the pattern is more a function of the pipe diameter D than of the length scale l_w.

Figure 4 shows <u>-signatures measured with the hot wire situated at an angle $\varphi_h = 180°$. The condition for sampling is again $p \geq 4p_{rms}$. As one can see the pattern of the <u>-signature measured at $y_h/R = 1$ (center of the pipe) and that measured at $y_h/R = 0.5$ are completely different to the <u>-signatures shown in Fig. 3. Regions of high and low velocity are interchanged. The signature measured close to the wall ($y_h/R = 0.2$) again shows a change in the pattern. It is more like the signatures plotted in Fig. 3. This could indicate that the observed <u>-signatures are related to ring-like structures in the pipe flow. Details of this, however, have not yet been investigated.

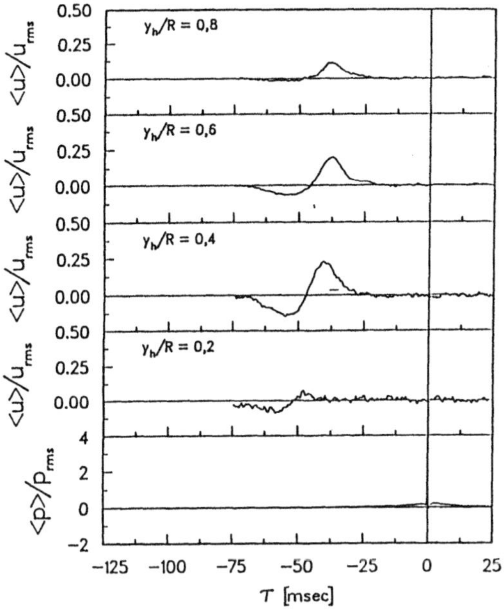

Fig. 4. Same as Fig. 3, but the measurements were now performed in the opposite half-space in the pipe (U_{cl} = 11.0 m/s; condition: $p \geq 4p_{rms}$; y_h varying; φ_h = 180°, x_h = 0)

3.2 Convection of the patterns

In /1/ it was shown already that convection of the patterns can be observed over distances of at least 2 pipe diameters. Figures 5 and 6 show now that this convection of the characteristic <u>-signatures can be observed over a distance of at least 13 pipe diameters. As this distance is limited by the experimental possibilities and not by the disappearance of the <u>-signatures it can be concluded that the flow

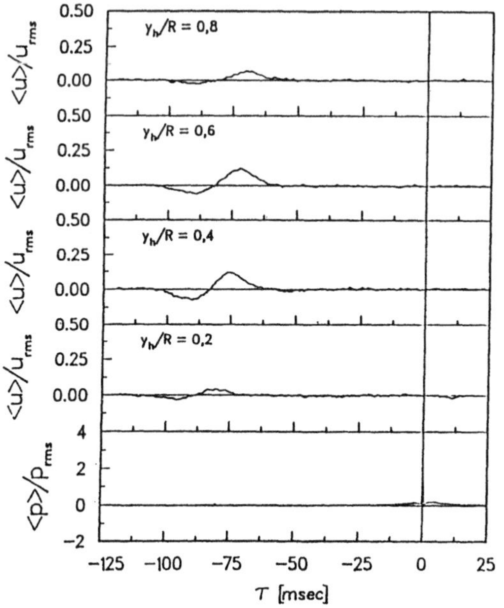

Fig. 5. Same as Fig. 3, but the position of the hot wire is now 350 mm upstream of the wall pressure transducer (i.e. $x_h/R \doteq -14$; all other data same as in Fig. 3)

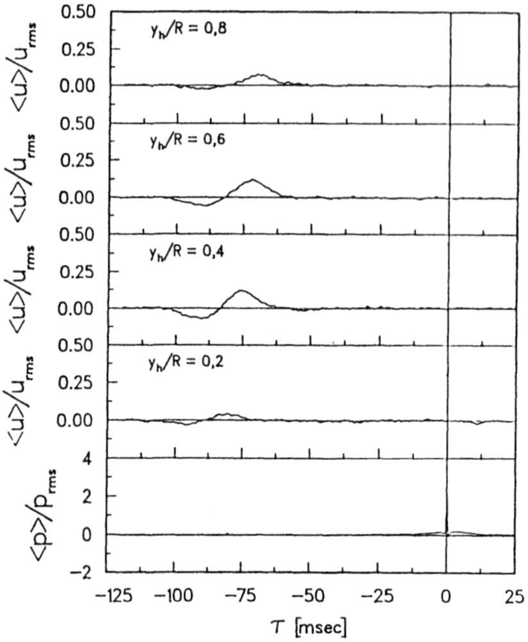

Fig. 6. Same as Fig. 3, but the position of the hot wire is now 650 mm upstream of the wall pressure transducer (i.e. $x_h/R = -26$; all other data same as in Fig. 3)

structures related to the observed <u>-signatures persist for more than 13 pipe diameters (13 pipe diameters correspond here to $\Delta x^+ = 20\ 000$). Besides this astonishingly long lifetime of the <u>-signatures three more things are noteworthy:

(a) One can calculate convection velocities for the <u>-signatures by comparing Figs. 3, 5 and 6. They are different for different wall distances. In /3/ it is shown that they lie about 15 % below the local mean velocity.

(b) In Figs. 5 and 6 the <u>-signatures at $y_h/R = 0.8$ are well developed while in Fig. 3 the corresponding <u>-signature has disappeared. This indicates - as was already shown in /1/ for smaller values of x and y - that the pattern moves towards the wall.

(c) The duration of the <u>-signatures does not change very much along the path from $x_h/R = -26$ (Fig. 6) to $x_h/R = 0$ (Fig. 3).

Figure 7 shows measurements of <u>-signatures in the jet leaving the pipe. The hot wire is now positioned at $x_h = 130$ mm ($x_h/R = 5.2$). As mentioned earlier the microphone is positioned 110 mm from the pipe exit. Hence the hot wire position $x_h = 130$ mm is in the jet 20 mm downstream of the pipe exit. As can be seen from Fig. 7 the characteristic <u>-signatures corresponding to events of high wall pressure in the pipe can also be found in the jet leaving the pipe. The fact that the amplitudes of the <u>-signatures at $x_h = 130$ mm are still fairly large indicates that the <u>-signatures can certainly be followed further downstream.

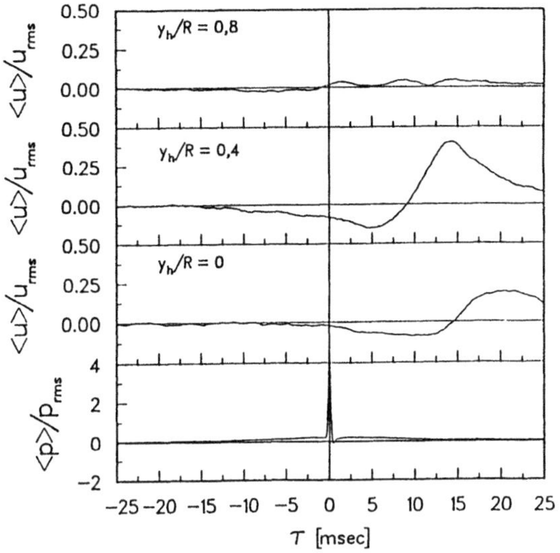

Fig. 7. Conditionally averaged streamwise velocity flucuations measured in the jet leaving the pipe (U_{cl} = 11.0 m/s; condition: $p \geq 4p_{rms}$; y_h varying $\varphi_h = 0°$; $x_h/R = 5.2$ corresponding to a distance of 20 mm from the pipe exit)

3.3 Changes in the azimuthal direction

Figure 8 finally shows some <u>-signatures measured with the hot wire in positions at different angles φ_h and at a distance y_h = 1 mm from the wall corresponding here to $y_h^+ = 25$. (For technical reasons the curves shown in Fig. 8 are measured with a smaller flow velocity, a smaller number of samples and a smaller trigger level. These changes, however, are not significant for the items discussed here.) From Fig. 8 one sees that the amplitudes of the <u>-signatures decrease rapidly with

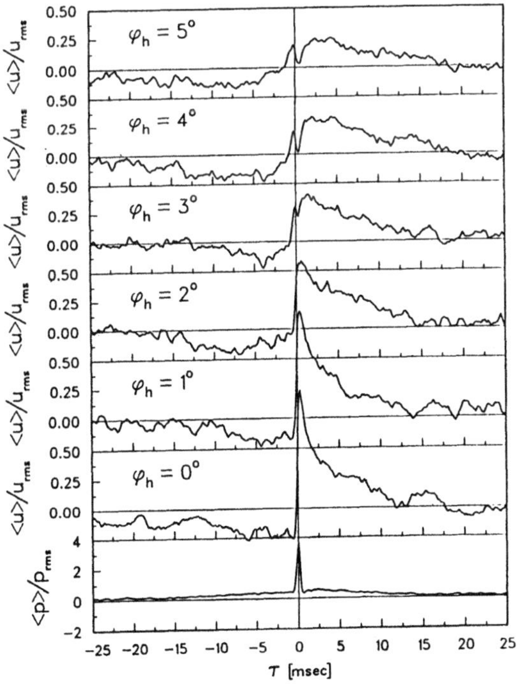

Fig. 8. Conditionally averaged streamwise velocity fluctuations measured at different azimuthal positions of the hot wire (U_{cl} = 8.5 m/s; condition: $p \geq 3p_{rms}$; $y_h/R = 0.04$; φ_h varying, $x/R = 0$; here 1 ms corresponds to $\tau^+ = 10$)

increasing angle φ_h, which indicates that the flow structures which generate the <u>-signatures have a rather small extension in spanwise direction. Furthermore from Fig. 8 one can see that a systematic change occurs in the <u>-signatures. For $0° \leq \varphi_h \leq 2°$ the <u>-signatures have a positive slope at the time $\tau = 0$. For $\varphi_h > 2°$ the slope at the time $\tau = 0$ becomes negative. (The spanwise distance z measured along the wall which corresponds to $\Delta \varphi = 1°$ is $\Delta z = 0.44$ mm or here $\Delta z^+ = 11$.) It is interesting that LANGEHEINEKEN /1/ observed a similar dependence of the <u>-signatures on the angle φ. Langeheineken, however, had a flow velocity of $U_{cl} = 2.9$ m/s, and found the change at an angle $\varphi_h = 9°$. If one recalculates these angles in terms of the spanwise distance at the pipe wall z^+ the change occurs in both cases at about $z^+ = 35$. It is noteworthy that such <u>-signatures as found in Fig. 8 for angles $\varphi_h > 2°$ are measured at angles $\varphi_h = 0°$ if one uses not "high" pressure events but "low" pressure events as sampling condition /1/, /3/.

4. Conclusions

The measurements show that in developed turbulent pipe flow large coherent structures exist which move downstream over distances of at least 13 pipe diameters. The longitudinal extension of these structures is comparatively large - of the order of 5 pipe diameters - and seems to scale more with the pipe diameter than with wall units. In spanwise direction systematic changes of the structures are observed over spanwise distances of a small fraction of the pipe diameter. These changes seem to scale with wall units. As was shown already in /1/ and /2/ the conditional averaged <u>-signature found here for the condition "high" wall pressure is the same as the conditional averaged <u>-signature found by BLACKWELDER and KAPLAN /4/ for "bursts" in the turbulent boundary layer of a flat plate. It can be expected that the large scale structures investigated here will have implications for a better understanding of the bursting phenomena in turbulence.

References

1. Th. Langheineken: Zusammenhänge zwischen Wanddruck- und Geschwindigkeitsschwankungen in turbulenter Rohrströmung. Mitt. MPI für Strömungsforschung, Göttingen Nr. 70 (1981)
2. A. Dinkelacker, Th. Langeheineken: Relations between wall pressure fluctuations and velocity fluctuations in turbulent flow. In: Structure of Complex Turbulent Shear Flow (IUTAM Symposium Marseille 1982) ed. by R. Dumas and I. Fulachier (Springer, Berlin, Heidelberg, New York 1983) p. 1
3. M. Sieber: Großräumige Strukturen in der turbulenten Rohrströmung. Doctoral thesis submitted to the University of Göttingen 1986
4. R.F. Blackwelder, R.E. Kaplan: On the wall structure of the turbulent boundary layer. J. Fluid Mech. (1976), vol. 76, part 1, pp. 89-112

Behaviour of Coherent Structures in a Turbulent Boundary Layer with Wall Suction

L. Fulachier[1], *T. Benabid*[1], *F. Anselmet*[1], *R.A. Antonia*[2], *and L.V. Krishnamoorthy*[2]

[1]Institut de Mécanique Statistique de la Turbulence, Unité Mixte Université C.N.R.S. No. 380033, 12 Av. du Général Leclerc, F-13003 Marseille, France
[2]Department of Mechanical Engineering, University of Newcastle, New South Wales 2308, Australia

1. Introduction

The primary motivation of the present work is to gain some insight into the behaviour of coherent structures in a turbulent boundary layer over a smooth surface. One way to increase our understanding of the dynamics of these structures is to consider their response to a sudden change in boundary condition. In the present case, a self-preserving boundary layer which develops on a smooth impervious surface is subjected to the influence of a suddenly applied wall suction. Also of interest is the role that the structures play with respect to their contribution to the transfer of momentum and heat.

Previous experiments /1-5/ on a turbulent boundary layer with wall suction, have shown that suction decreases the Reynolds normal and shear stresses, the temperature variance and the turbulent heat fluxes. The above measurements did not extend to a region very close to the wall. In a turbulent heated pipe flow, ELENA /6-7/ found, from measurements at one point, that suction reduced the inflow number, i.e. the arrival of fluid from regions further away from the wall than the measurement point. This finding is in agreement with that of RAMNEFORS and NYDEN /8/ who used the quadrant analysis method to show that the mean period of sweeps was increased by suction. Further, the pipe flow measurements /6/ inside the viscous sublayer suggested that large amplitude temperature fluctuations were more affected by suction than large amplitude longitudinal velocity fluctuations. Since temperature fluctuations depend on all three components of the velocity vector /3,9/ and the magnitude of the velocity fluctuation normal to the wall is one order smaller than that of the other two components, it was conjectured that suction has a more significant effect on the spanwise velocity component than on the longitudinal velocity component. Suction would therefore decrease the instability of the sublayer by diminishing spanwise perturbations, thereby resulting in a lamellar-type flow /6,10/.

We present in this paper the first series of results related to the near-wall structure and the behaviour of coherent structures in the fully turbulent flow region. The experiments were carried out at the I.M.S.T., using both a wind tunnel and a water tunnel. Hot and cold wire measurements were done in the wind tunnel while flow visualisations and laser velocimetry measurements were done in the water tunnel.

2. Experimental Arrangement, Conditions and Measuring Techniques

A schematic representation of the wind tunnel and water tunnel boundary layers is shown in Fig. 1. The effect of suction can be quantified by either the parameter $A = -\rho_w V_w/\rho_1 U_1$ or the parameter $A^+ = -V_w/U_\tau$ where ρ is the density, U_τ is the friction velocity and the subscripts w and 1 correspond to wall and free stream conditions respectively.

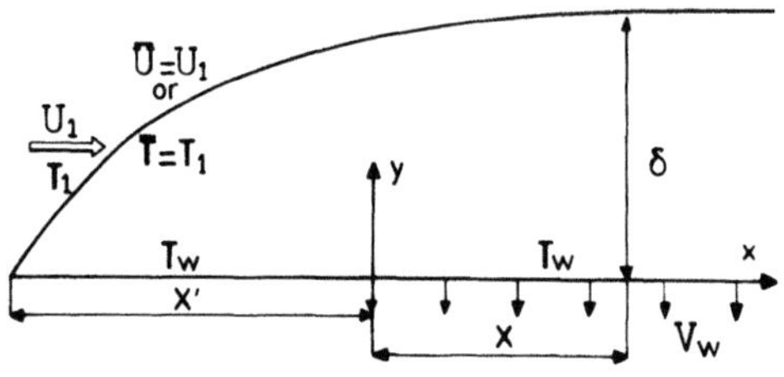

Fig. 1
Schematic experimental configuration (air or water)

2.1 Water Tunnel

A porous plate (25 cm high, 16 cm wide) was installed in one of the vertical walls of the 20 x 20 cm working section of the I.M.S.T. water tunnel /11/. The plate leading edge was located at a distance X' = 64.2 cm (case H1) from the boundary layer trip (three-dimensional roughness elements). The latter permitted the establishment of a fully developed turbulent boundary layer, as inferred from both visualisations and mean velocity measurements. Dye could be injected through small holes in the wall upstream of the porous plate. It could also be injected via a narrow slit (longitudinal extent ≅ 0.2 mm, width = 18 cm), located 3 cm upstream of the porous plate. A few visualisations were made downstream of the porous plate when the leading edge of the latter was approximately 4 cm downstream of the roughness trip (case H2).

The free stream velocity U_1 was nominally 16 cm/s and the pressure gradient was zero. The flow was isothermal with $T_w = T_1$. In the case H1, measurements were made at several distances X from the leading edge but especially at X = 10 cm where the boundary layer thickness δ is 50 mm and the friction velocity U_{τ_0} = 0.78 cm/s without suction (the corresponding skin friction coefficient c_{f_0} is 47.7×10^{-4}). The Reynolds number R, based on the momentum thickness, was equal to 620. Three suction rates were used for the velocity measurements : A = 0.0014 , 0.0028 , 0.0048 (A^+= 0.026, 0.047 , 0.072). A fourth suction rate, A = 0.0007 (A^+ = 0.014), was used in flow visualisations. Profiles of $\overline{U}$ were obtained with a 15 mW He-Ne laser fitted with a Bragg cell.

In flow visualisations, both plan and side views were used. For side views, the boundary layer was generally tagged with green dye (fluorescein) injected immediately upstream of the roughness trip; a single dye filament injected in the contraction enabled, depending on the precise location at which it was introduced, the visualisation of either the irrotational flow outside the boundary layer or the flow within the boundary layer. A white dye (rhodorsil or silicone oil emulsion), injected at a point 1 cm upstream of the porous plate, enabled visualisation of the flow behaviour close to the wall. For the majority of plan views, rhodorsil was injected through the wall slit. We estimated that, for the slit injection, dye was introduced at an effective y^+ less than 0.1 ($y^+ = yU_\tau/\nu$, where ν is the kinematic viscosity). Synchronised flash photographs were taken in the x-y plane (main shear) and the x-z plane (spanwise) using illumination by relatively thin sheets of light. A few simultaneous views of both planes were made with two synchronised cameras. Films were also made with a Super 8 mm camera, with views taken in either the x-y or x-z planes.

2.2 Wind Tunnel

The porous surface (1.8 m long, 0.56 m wide) which has been installed in the wind tunnel S2 at the I.M.S.T. has already been described in detail /2,3/. Upstream of this plate is an impervious surface of length X' equal to 3.05 m. The complete surface was heated such that $T_w - T_1 = 10[K]$ (in previous experiments /3/ a value of $20[K]$ was used). As in previous experiments, most of the measurements were made at a station located a distance X = 0.64 m from the beginning of the porous surface, with a slightly negative streamwise pressure gradient. The nominal free stream velocity U_1 is 12.2 m/s with $U_{\tau_0} = 0.48$ m/s , $c_{f_0} = 32 \times 10^{-4}$ and R = 4900. Most of the results were obtained for $A = 0.003$ ($A^+ = 0.055$) but a few results are also given for $A = 0.0015$ ($A^+ = 0.032$).

The mean velocity was measured with total and static head tubes. Measurements of the longitudinal u and spanwise w velocity fluctuations were made with a V-probe operated with constant temperature circuits, for values of y^+ as small as 1. The temperature fluctuation θ was obtained with a 0.63 µm cold wire operated with a constant current (0.1 mA) circuit. A rake of as many as eight cold wires (with a nominal separation of 0.9 mm between adjacent wires) was used for the study of temperature fronts.

3. Results and Discussion

3.1 One Point Measurements

With suction, mean velocity profiles measured in the water tunnel for $y^+ \gtrsim 3$ were found to be in good agreement with those measured in the wind tunnel. In the sublayer, the results followed the distribution $\overline{U}^+ = (1/A^+)[1-\exp(-A^+y^+)]$, with $\overline{U}^+ = \overline{U}/U_\tau$ and, throughout the boundary layer, $\overline{U}^+$ decreased as A^+ increased. The friction coefficient, estimated by the wall slope method, increased with A^+. This increase is due to the wall-deviated momentum, caused by suction, the shear stress distribution in the inner region assuming the form

$$\tau - \rho_w V_w \overline{U} = \tau_w \qquad (1)$$

where τ is the total shear stress. In the fully turbulent region, (1) can be written, after integration and with the assumption that the Karman constant κ is unaffected by suction /2,4,12/, as

$$-(2/A^+)(1-A^+ \overline{U}^+)^{1/2} = \kappa^{-1} \log y^+ + C_0 \qquad (2)$$

With respect to the total heat flux q, the relation analogous to (1) is

$$q - \rho_w V_w C_p (T_w - \overline{T}) = q_w \qquad (3)$$

If it is assumed that the turbulent Prandtl number Pr_T is constant in the fully turbulent region, the pseudo mean temperature

$$(2/A^+)\left[1 - (1-A^+ \overline{T}^+)^{1/2 Pr_T}\right] = \kappa^{-1} \log y^+ + F_q \qquad (4)$$

should be independent of suction /2,3,4/. Note that $\overline{T}^+ = (T_w - \overline{T})/T_\tau$ is the friction temperature, and F_0 is independent of suction.

Figure 2 shows that (2) is satisfied for both water tunnel and wind tunnel data, albeit over a narrow range of y^+ in the water tunnel because of the small Reynolds number. We also verified that (4) is closely followed by the wind tunnel data (Fig. 2).

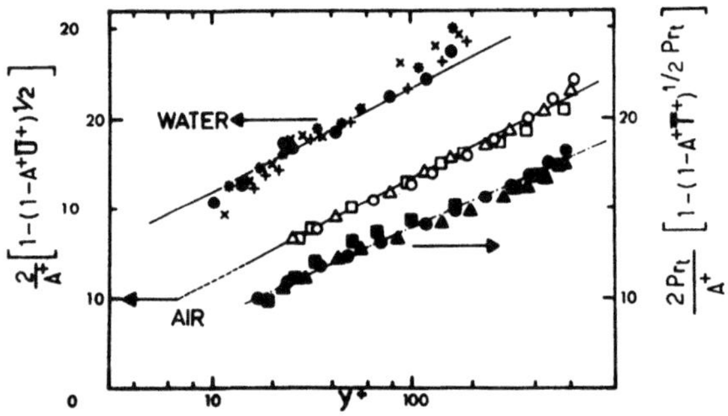

Fig. 2
Pseudo mean velocity [Eq.(2)] and pseudo mean temperature [Eq.(4)] distributions. Water: ●, $A^+=0$; ×, $A^+=0.026$; +, $A^+=0.047$ *, $A^+=0.072$. Air : o,●, $A^+= 0$; Δ,▲, $A^+= 0.032$; □,■, $A^+= 0.055$

Figure 3 shows distributions of the rms longitudinal (u') and spanwise (w') velocity fluctuations, normalised by the friction velocity U_τ, in the near-wall region. There is a significant decrease, when suction is applied, in both quantities, the decrease being accentuated in Fig.3 due to the increase in U_τ (with suction). For A = 0.003, the decrease, in the viscous sublayer, of u' and w' is about 24% and 19% respectively. It should also be noted that the correlation coefficient between u and w was less than 0.07, supporting the two-dimensionality of the mean flow. Further, the probability density functions (pdf) of u and w are practically unchanged by suction. Suction would therefore decrease, in almost the same way, longitudinal and spanwise perturbations. This result contradicts the conjecture first advanced by ELENA /6/ and later taken up in /10/.

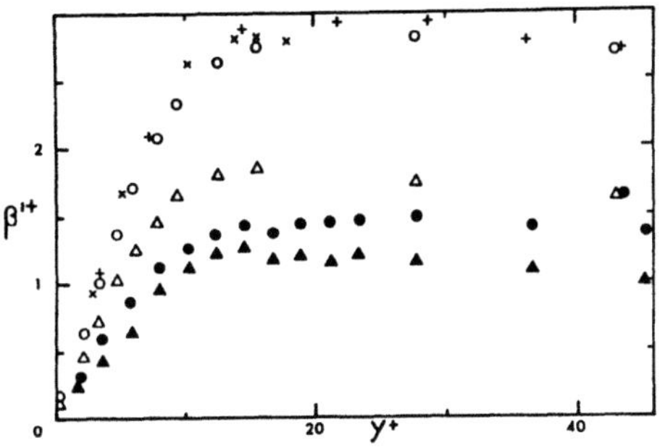

Fig. 3
Effect of suction on the rms longitudinal and spanwise velocity fluctuations in the near-wall region.
$A^+ = 0$:o,+ /13/, × /14/ , u' ;
Δ , ⊕ /13/ , w'.
$A^+ = 0.055$: ●,u' ; ▲ ,w' .

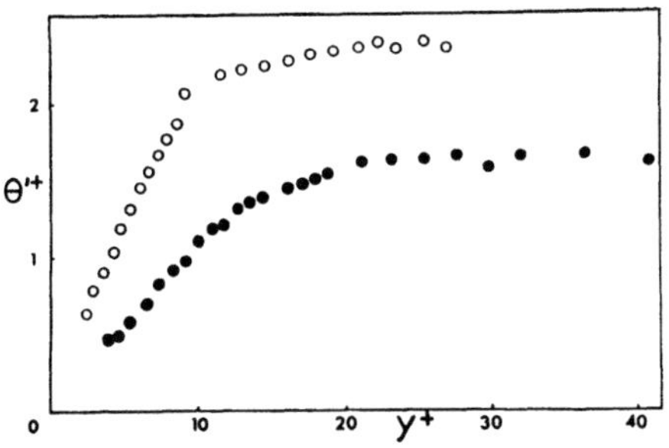

Fig. 4
Effect of suction on the rms temperature fluctuation in the near-wall region
$A^+ = 0$: o ;
$A^+ = 0.055$: ●

Temperature fluctuations are reduced, as a result of suction, even in the sublayer as is shown in Fig. 4. Further, experiments indicate that the pdf of θ is slightly reduced for large negative fluctuations, implying that the average frequency of sweeps is decreased with suction. Support for this observation is provided by the pdf of u which indicates a decrease for large positive velocity fluctuations.

3.2 Visualisations

In the absence of suction, the films that were made point to the existence of structures (in the form of vortex loops) that first appear very near the wall and are often stretched out into the external part of the boundary layer, their tips rotating

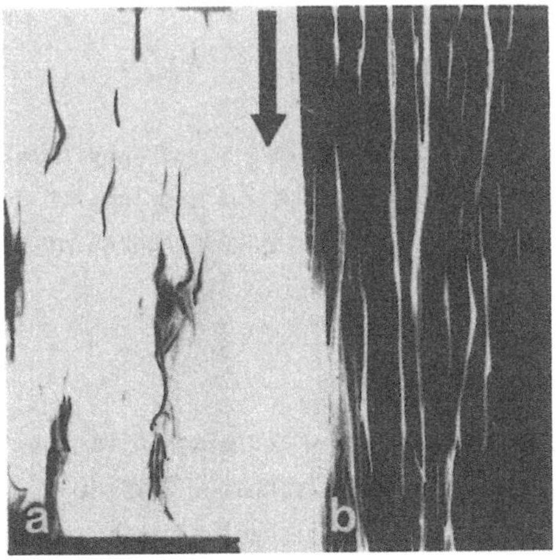

Fig. 5
Influence of suction on low-speed streaks (dye injection through wall slit upstream of the porous plate).
(a) A = 0 ;
(b) $A^+ = 0.0007$

with the external large scale bulges. This suggests that a link exists between the wall and the external layer. The flow visualisation also highlights (Fig. 5) the stabilisation by suction of the low-speed streaks near the wall. There is also a significant reduction in the frequency of ejection of wall dye by the suction. Both the stabilisation of the streaks and the reduction in the ejection frequency increase as suction is increased. Plan views of the wall dye emphasise the greater streamwise coherence and the smaller spanwise oscillations of the streaks with suction. However as suction increases, some dye is sucked through the wall and, even when dye is visible, one cannot be certain of the effective distance from the wall corresponding to the visualisation. To overcome these difficulties, the removable wall was inverted so that the boundary layer first developed on the porous plate and the visualisations were carried out on the impervious surface downstream of the porous section. Figure 6 shows simultaneous plan and elevation views of a single dye streakline with and without suction. The spanwise stabilisation is evident (plan views) and the dye ejections are, on average, decreased with suction although violent ejections do occur occasionally. This would explain the decrease, due to suction, of the production of turbulent energy /1/.

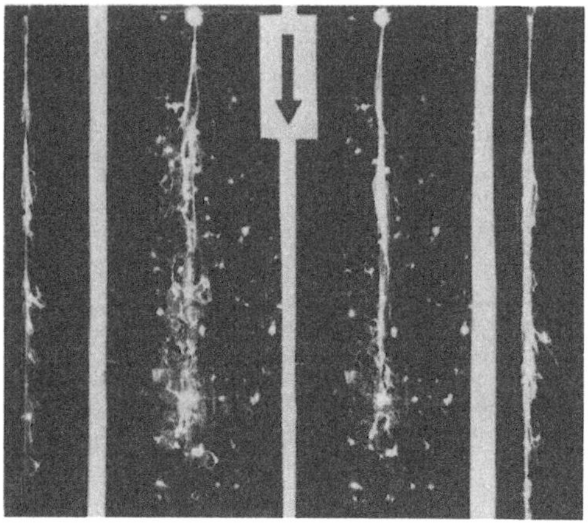
(a) (A) (b) (a) (B) (b)

Fig. 6
Flow stabilisation by wall suction (dye injection through hole downstream of the porous plate). Note that, for this figure, the porous plate is upstream of the impervious surface.
(A) $A^+ = 0$: (a) elevation ; (b) plan view.
(B) $A^+ = 0.0048$: (a) plan view ; (b) elevation

With or without suction, films and photographs indicate that dye moves away from the surface at a relatively shallow angle before reaching average inclinations of about 45°. This latter inclination was that reported by HEAD and BANDYOPAKHYAY /15/ for vortex loops or hairpin vortices.

3.3 Average Location of Temperature Fronts in the x-y Plane

Simultaneous temperature signals obtained with a rake of cold wires aligned in the y direction are shown in Fig. 7. The wire closest to the wall (signal θ_1) is at $y^+ = 16$. There are relatively sudden hot-cold temperature changes which can be iden-

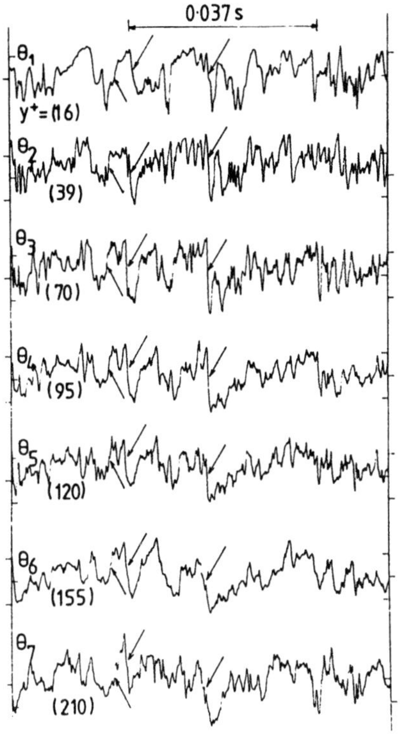

Fig. 7
Simultaneous temperature signals from the y-aligned rake ($A^+ = 0$)
↓ : hot-cold fronts ;
↑ : cold-hot fronts .

tified on all signals. These changes have been associated /16/ with internal temperature fronts and can be interpreted as the boundary between ejections and sweeps. There are also cold-hot temperature fronts (Fig. 7) which demarcate the boundary between sweeps and ejections /17/. These latter fronts are not as coherent, in space, as the hot-cold fronts and have not been considered in this paper. A future detailed study of these fronts is envisaged.

To delineate the average location of a hot-cold front in the x-y plane, it is necessary to convert the time interval between the arrival of the front at a particular y^+ (corresponding to θ_n) and that at a reference location (say $y^+ = 16$, corresponding to θ_1) to a longitudinal separation X. This conversion can be implemented with the use of a convection velocity $\overline{U}_c$, determined from a rake of cold wires aligned in the x direction. The shape and magnitude of $\overline{U}_c$ were found to be significantly affected by suction in the near-wall region. The average inclination $\alpha = dy^+/dx^+$ of these fronts is displayed in Fig. 8 as a function of y^+, with $y^+ = 16$ as the reference location, for $A^+ = 0$ (the data for $A^+ = 0.055$ have not yet been reduced). The angle α varies linearly in the region $20 \lesssim y^+ \lesssim 45$ ($14° \lesssim \alpha \lesssim 38°$). For comparison, KREPLIN and ECKELMANN /18/ obtained an inclination of about 10° over the same y^+ range. For $y^+ \gtrsim 80$, we find that α is essentially constant, equal to about 47°, in agreement with our visualisations. We also note that CHEN and BLACKWELDER /14/ obtained an inclination of about 45° for their hot-cold fronts, HEAD and BAN-

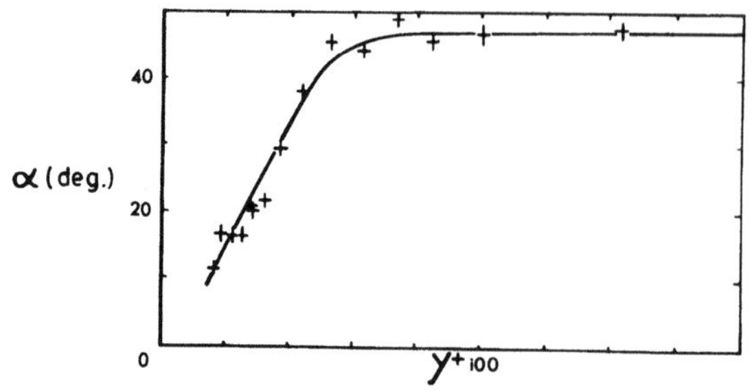

Fig. 8
Average inclination of hot-cold temperature fronts ($A^+ = 0$)

DYOPAKHYAY /15/ reported a value between 40° and 50° for the inclination of vortex loops while MOIN and KIM /19/ predict a value of 45° on the basis of large eddy simulations.

4. Conclusions

Measurements and visualisations indicate that suction increases the stability of the viscous sublayer by decreasing the amplitude of both longitudinal and spanwise oscillations. We feel confident that a detailed study of the temperature fronts will quantify the influence of suction on coherent structures. Specifically, we aim to delineate the topology of these structures and estimate their contribution to the average momentum and heat transfers.

5. Acknowledgements

We are grateful to M. Astier for his technical assistance and A. Morand for the photography. We also thank P. Collini and F. Djenidi for their contributions to the experiments.

6. References

1. A. Favre, R. Dumas, E. Verollet, M. Coantic : *J. de Mécanique*, 5 , 3 (1966)
2. E. Verollet : Thèse de Doctorat ès Sciences, Université de Provence, I.M.S.T. (1972) and Rap. CEA-R-4872 (1977)
3. L. Fulachier : Thèse de Doctorat ès Sciences, Université de Provence, I.M.S.T. (1972)
4. E. Verollet, L. Fulachier, I. Dekeyser : *Int. J. Heat Mass Transfer*, 20, 107 (1977)
5. L. Fulachier, E. Verollet, I. Dekeyser : *Int. J. Heat Mass Transfer*, 20, 731 (1977)
6. M. Elena : Thèse de Doctorat ès Sciences, Université d'Aix-Marseille II, I.M.S.T. (1975)
7. M. Elena : *Phys. Fluids*, 27, 861 (1984)

8. M.O. Ramnefors, O.B. Nyden : In *Proc. 9th Symp. on Turbulence*, University of Missouri-Rolla, 4-1 (1984)
9. L. Fulachier, R. Dumas : *J. Fluid Mech.*, 77, 257 (1976)
10. L. Fulachier, M. Elena, E. Verollet, R. Dumas : In *Structure of Turbulence in Heat and Mass Transfer*, ed. by Z.P. Zarić, Vol. 14 (Hemisphere, New York, 1982) p. 193
11. R. Dumas, C. Domptail, P. Bonmarin : Rap. ATP-CNRS, Instabilités dans les Fluides et les Plasmas, I.M.S.T. (1977)
12. T.N. Stevenson : Rep. Aero. No. 166, College of Autonautics, Cranfield (1963)
13. P.S. Klebanoff : N.A.C.A. Rep. 1247 (1954)
14. H.P. Kreplin, H. Eckelmann : *Phys. Fluids*, 22, 1233 (1979)
15. M.R. Head, P. Bandyopadhyay : *J. Fluid Mech.*, 107, 297 (1981)
16. C-H. P. Chen, R.F. Blackwelder : *J. Fluid Mech.*, 89, 1 (1978)
17. C.S. Subramanian, S. Rajagopalan, R.A. Antonia, A.J. Chambers : *J. Fluid Mech.*, 123, 335 (1982)
18. H-P. Kreplin, H. Eckelmann : *J. Fluid Mech.*, 95, 305 (1979)
19. P. Moin, J. Kim : *J. Fluid Mech.*, 155, 441 (1985)

Identification of Sweeps with the Help of the Instantaneous Velocity Gradient dU/dy

M. Randolph [1], *H. Eckelmann* [1], *and S.G. Nychas* [2]

[1]Max-Planck-Institut für Strömungsforschung, D-3400 Göttingen, Fed. Rep. of Germany
[2]Department of Chemical Engineering, Aristotle University of Thessaloniki, GR-54006 Thessaloniki, Greece

In a fully developed turbulent channel flow, measurements of the instantaneous spatial gradient $\frac{dU}{dy}$ were carried out. The instantaneous gradient can take on positive and negative values, which are, in the buffer layer $(11 \leq y^+ \leq 35)$, of the order of magnitude of the mean wall gradient $\left.\frac{d\overline{U}}{dy}\right|_W$. Inspired by the visual studies of CORINO & BRODKEY and NYCHAS, HERSHEY & BRODKEY which connected large positive spatial as well as large positive temporal gradients to a sweep, correlations of the temporal gradient $\frac{dU}{dt}$ and the spatial gradient $\frac{dU}{dy}$ were calculated. It was found that these gradients were strongly correlated. Through conditional correlation with appropriately chosen spatial gradients, criterion which allow sweep detection are determined.

1. Introduction

Events in a turbulent flow which displace the surrounding media due to their higher velocity, so called *sweeps*, were first described by CORINO & BRODKEY [1] in their visual studies and further expanded on by NYCHAS, HERSHEY & BRODKEY [2]. A definition of sweeps in an Eulerian coordinate system was later given by WALLACE, ECKELMANN & BRODKEY [3]. According to this, sweeps are events having a positive u- (u in meanflow direction) and negative v-fluctuation (v normal to the wall). This definition does not describe the essential characteristic of sweeps, namely that sweeps are fronts of accelerated fluid. As already known from the above visual studies, these fronts are associated with large positive temporal velocity gradients $\frac{dU}{dt}$ and with large positive spatial velocity gradients $\frac{dU}{dy}$ (y normal to the wall). The aim of this paper is to exploit this characteristic of the sweep front and establish, with the help of direct measurement of the spatial and temporal gradients, a criteria with which sweeps in turbulent bounded flows can be identified.

2. Facility

The measurements were carried out in the Göttingen oil channel, described in ECKELMANN [4]. Only the relevant parameters for the present measurements will be repeated here. At an constant temperature of $25.3°C$ the kinematic viscosity of the oil measured $\nu = 5.74 \cdot 10^{-2}\ cm^2/s$. The Reynolds number based on a centre line velocity of $U_{\mathbb{C}} = 19.6\ cm/s$ and a channel width of $2b = 22\ cm$ was $Re \approx 7500$. For a measured wall gradient $\left.\frac{d\overline{U}}{dy}\right|_W$ of $18.5\ s^{-1}$ the friction velocity at this Re-number amounted to $u_\tau = \sqrt{\nu \cdot \left.\frac{d\overline{U}}{dy}\right|_W} = 1.03\ cm/s$.

The gradient measurements were carried out with a TSI–1244 20W two sensor gradient probe. The hot film probe was run by two DISA 55M01 constant temperature anemometers at an overheat ratio of 2% corresponding to a temperature difference between the sensors and the fluid of about $8°C$. Data acquisition and evaluation was done with a digital computer.

The gradient probe was introduced into the flow such that the two sensors were parallel to the wall and perpendicular to the mean flow direction. The sensors measured the instantaneous velocities $U_1(y - \frac{\Delta y}{2})$ and $U_2(y + \frac{\Delta y}{2})$. The velocity U_1 is assigned to the sensor closest to the wall. The gradient is given by:

$$\frac{dU}{dy} = \frac{U_2 - U_1}{\Delta y} \quad .$$

Since the velocity difference decreases with increasing wall distance, reaching an almost zero value at the channel centre line at $y^+ = yu_\tau/\nu = 200$, high demands on the accuracy of the measurements must be placed. For the error in the gradient to be less than 10%, the velocity measurements must have an error of less than 1%. On the one hand the sensor spacing must be chosen large enough to allow measurements with a sufficient accuracy, on the other hand the sensor spacing must be small enough to allow the interpretation of the velocity differences as a gradient. For the present measurements the sensor spacing chosen was $\Delta y = 2.9\ mm$, corresponding to $\Delta y^+ \approx 5$. Presently, it is being investigated as to what extent measurements with a smaller sensor spacing are possible.

The mean gradient measured with probes such as those used here is systematically lower than the true value calculated from the velocity profile. This phenomenon was investigated by BÖTTCHER & ECKELMANN [5]. They determined the systematic deviation of the measured from the calculated gradient for various gradient probes in a plane laminar couette flow. The result was a calibration curve with which measured gradients can be corrected. This calibration, although determined in a laminar flow, was also used for the measurements in the turbulent flow discussed here. So, the instantaneous values of the gradient $\frac{dU}{dy}$ were corrected before being averaged. Measurements at wall distances between $y^+ = 11.6$ and $y^+ = 200$ (℄) were carried out. Because of the relatively large sensor spacing of $\Delta y^+ \approx 5$, measurements for y^+ less than 11.6 were not undertaken.

3. Results

The lower halves of Fig. 1a–d show velocity signals at four different wall distances. A strong similarity between the time signals from the two sensors of the gradient probe can be recognized. The difference between them becomes less with an increasing wall distance. The upper halves of Fig. 1a–d display the spatial gradient $\frac{dU}{dy}$ and the two temporal gradients $\frac{dU_1}{dt}$ and $\frac{dU_2}{dt}$. The latter were calculated from $U_1(t)$ and $U_2(t)$. It can be seen that large positive spatial gradients are accompanied by large positive temporal gradients. With increasing wall distance, the instantaneous values of the spatial gradient decrease while the instantaneuos values of the temporal gradients hardly change. Still, a clear relationship between the spatial and temporal gradients can be observed. Contrasting to this, it is not possible to infer a large spatial gradient

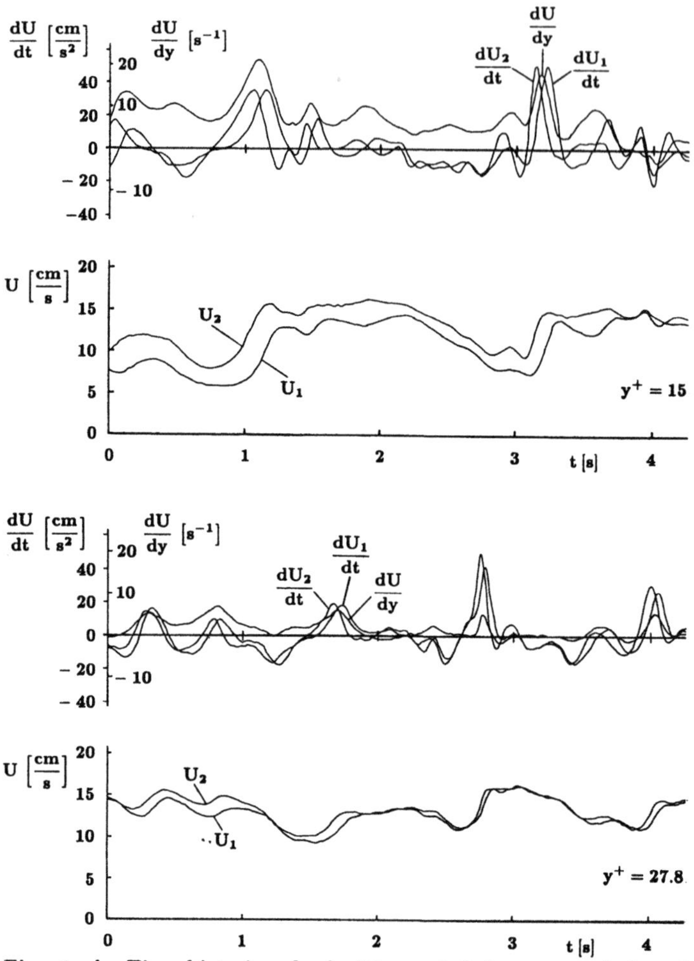

Fig. 1a–b. Time histories of velocities and their space and time derivatives at wall distances of $y^+ = 15$ and $y^+ = 27.8$

from a large positive temporal gradient. In fact, $\frac{dU}{dy}$ can even be zero or negative, as can be seen, for example, on Fig. 1b for $y^+ = 89$ and at $t \approx 1$ s. Compared to the actual data acquisition time of two hours, Fig. 1a–d only show short excerpts of 4 s corresponding to a dimensionless time of $t^+ = tu_\tau^2/\nu = 74$.

Probability density distributions of the spatial gradients $\frac{dU}{dt}$ are shown in Fig. 2. Distributions for measurements at nine wall distances, including those of Fig. 1a–d, are plotted. The probability density distributions for wall distances from $y^+ = 148$ to $y^+ = 200$ (℄) are almost symmetrical and are only slightly different from the one at $y^+ = 89$. With decreasing wall distance, the fluctuation range increases and the distribution shifts towards a postive gradient.

Figure 3 shows the mean gradient $\frac{d\overline{U}}{dy}$ (solid line) obtained by differentiation of the mean velocity profile, as well as the measured mean gradient $\overline{\frac{dU}{dy}}$ (solid dots). The measured mean

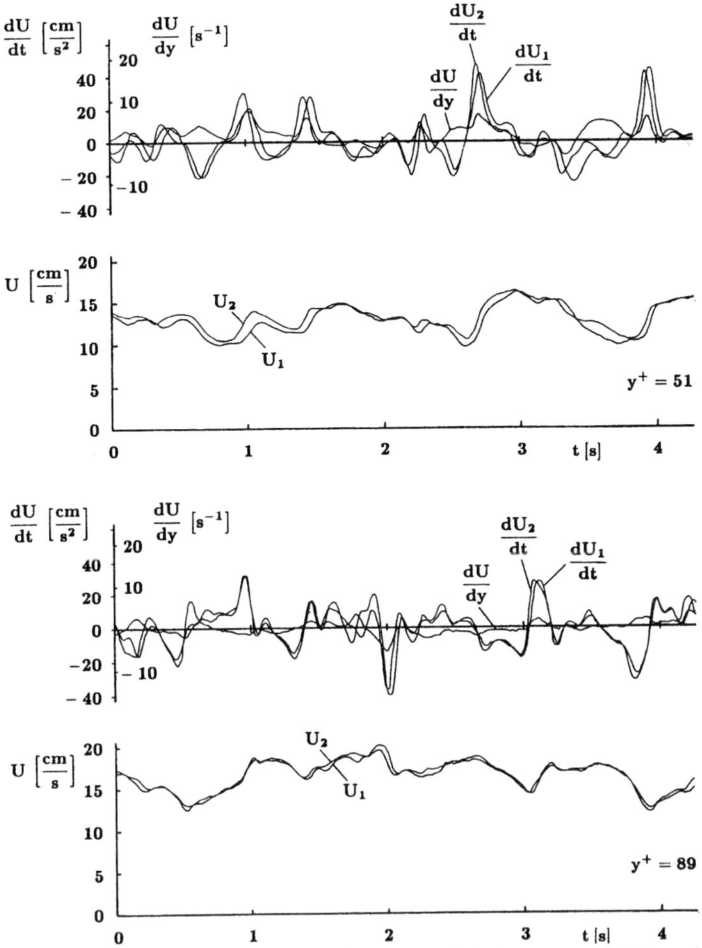

Fig. 1c–d. Time histories of velocities and their space and time derivatives at wall distances of $y^+ = 51$ and $y^+ = 89$

gradients follow the calculated ones fairly well. The mean velocity profile was determined independently from the gradient measurements with a single sensor probe.

Figure 3 also shows the fluctuation range of the gradient $\frac{dU}{dy}$ as a function of the wall distance. The two outer curves, which are nearely symmetrical to the mean, trace the rarely reached extreme values of the gradient. In the buffer layer ($11.6 \leq y^+ \leq 35$) these extreme values are -0.7 and 1.7 times the mean wall gradient $\left.\frac{d\overline{U}}{dy}\right|_W$. In the logarithmic region ($y^+ > 35$) they become smaller, but still have values of $\pm 0.3 \left.\frac{d\overline{U}}{dy}\right|_W$ at the channel centre line. The two, also symmetrical, inner curves shown in Fig. 3 mark the bounds inside of which the the gradient varies for 99% of the time. For comparison, in Fig. 3 the results of VUKOSLAVĈEVIĆ & WALLACE [6] are plotted.

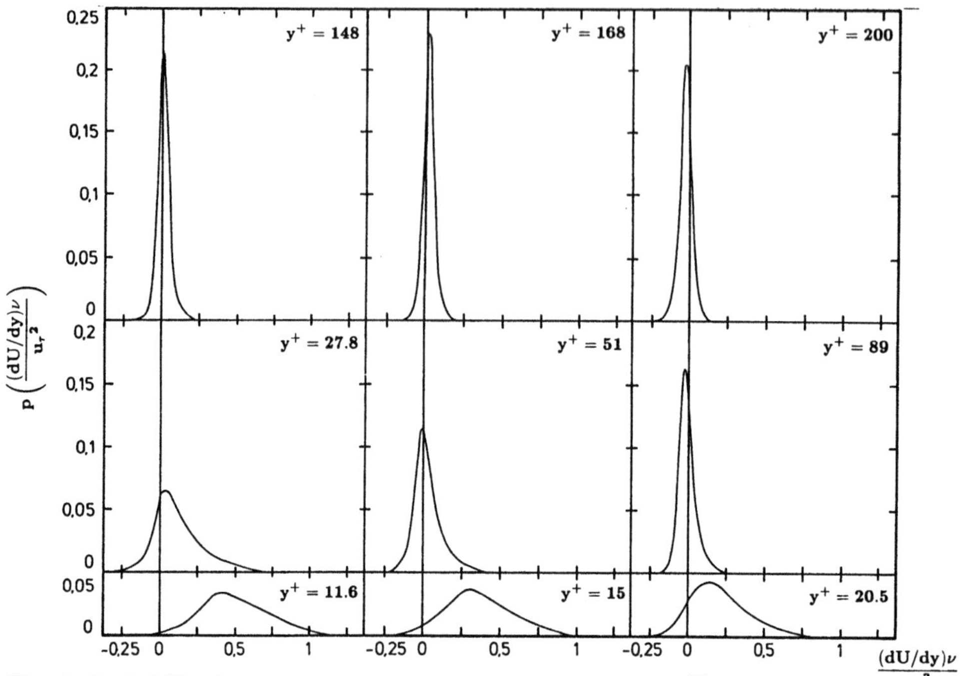

Fig. 2. Probability density distributions of the velocity gradient, $\frac{dU}{dy}$

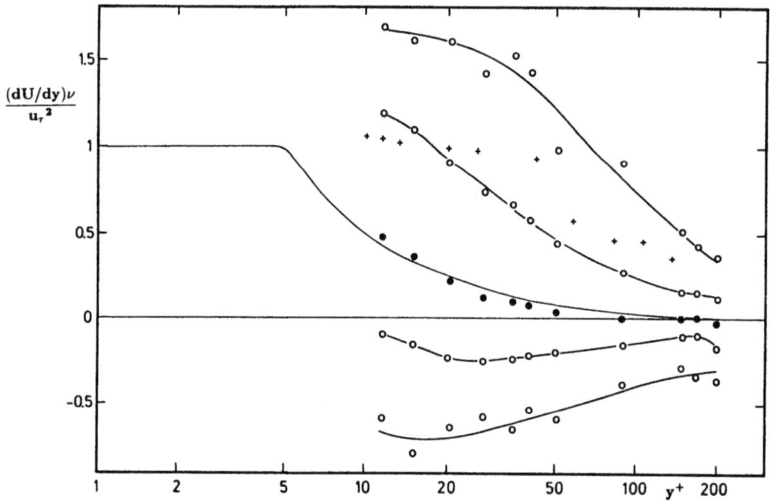

Fig. 3. (∘, outer curve), maximum and minimum gradients, $\frac{dU}{dy}$; (∘, inner curve), bounds containing 99% of $\frac{dU}{dy}$ occurrences; (•), mean gradients, $\frac{d\overline{U}}{dy}$, measured with two sensor gradient probe; (——), mean gradients calculated from slope of mean velocity profile;
(+), VUKOSLAVĈEVIĆ & WALLACE [7]

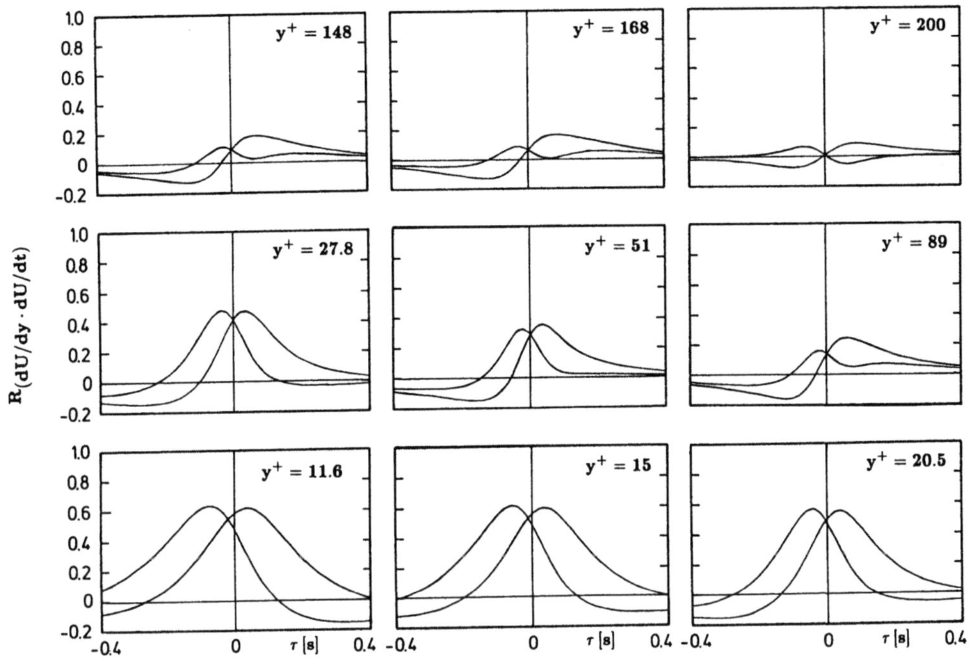

Fig. 4. Correlation of the spatial gradient, $\frac{dU}{dy}$, with the temporal gradients, $\frac{dU_1}{dt}$ (right curve) and $\frac{dU_2}{dt}$ (left curve)

To quantitavely grasp the relationship between the spatial and temporal gradient, the correlation functions of $\frac{dU}{dy}$ and $\frac{dU_1}{dt}$ as well as $\frac{dU}{dy}$ und $\frac{dU_2}{dt}$ were calculated. The results for nine different wall distances are displayed in Fig. 4. The correlation coefficient decreases mildly in the buffer layer but strongly in the logarithmic region as the wall distance becomes larger. Whereas the maximum correlation in the buffer layer is 0.6, it only reaches 0.1 in the channel centre line. It is noteworthy that the correlation functions at wall distances of $y^+ \leq 27.8$ are nearly symmetrical and have maxima of equal size, while at wall distances $51 \leq y^+ \leq 200$ the correlation functions are assymmetrical and, exept at $y^+ = 200$ (℄), have maxima of different size.

In agreement with the visual studies of CORINO & BRODKEY [1] and NYCHAS et. al. [2], results obtained here by anemometric means show that the spatial and temporal gradients are strongly correlated. It is therefore obvious to use the simultaneous occurance of these gradients as a criteria for the recognition of sweeps in an Eulerian coordinate system.

The spatial and temporal gradients are correlated even better if restricted to positive gradients $(\frac{dU}{dy})_c$, whose magnitude is chosen from the seven different ranges given in the ordinate of Fig. 5. To demonstrate this, the conditioned correlation function

$$R_{(\frac{dU}{dy}\cdot\frac{dU}{dt})_C} = \frac{\overline{\left(\frac{dU}{dy}\right)_C \cdot \left(\frac{dU}{dt}\right) \cdot F}}{\left(\frac{dU}{dy}\right)_{rms} \cdot \left(\frac{dU}{dt}\right)_{rms}} \quad ,$$

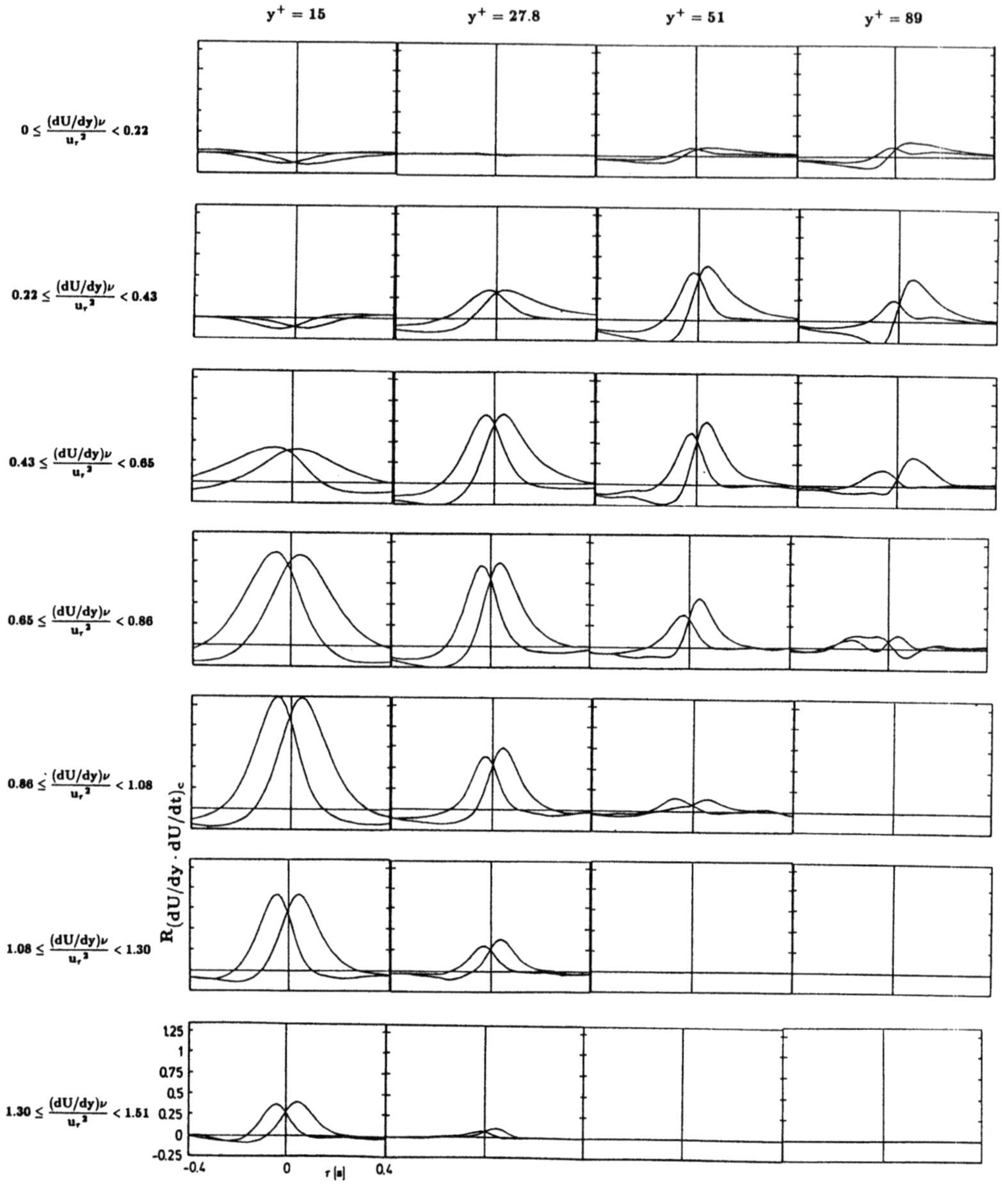

Fig. 5. Conditioned correlations of the spatial gradient, $\frac{dU}{dy}$, with the temporal gradients, $\frac{dU_1}{dt}$ (right curve) and $\frac{dU_2}{dt}$ (left curve). The correlation function is multiplied by the fraction of time the condition is fullfilled

F beeing the fraction of time the condition is fullfilled, was calculated for four different wall distances (Fig. 5). It can be seen that at $y^+ = 15$ the temporal gradients correlate most strongly with the spatial gradients which are of similar size to the mean wall gradient. With increasing wall distance, the value of the spatial gradient with which best correlation is achieved, decreases, but is still in the order of magnitude of the mean wall gradient.

The ranges of the spatial gradient, shown in Fig. 5, which correspond to the highest correlation for each wall distance, are suitable for recognizing sweeps. At $y^+ = 15$, for example, the instantaneous spatial gradient $\frac{dU}{dy}$ would be chosen from the interval $0.86 \leq \frac{(dU/dy)\nu}{u_\tau^2} \leq 1.08$.

Sweeps first introduced by CORINO & BRODKEY [1] in Lagrangian coordinates, which feature fronts of high and low speed fluid, can be characterized best in Eulerian coordinates by spatial and temporal gradients. Therefore, the detection of sweeps by the spatial gradient is physically better justified than detection by the sign of the u- and v-fluctuation as used in the quadrant split method of WALLACE et al. [3], moreso because of the difficulty in defining the fluctuating u-component (long time or short time average).

4. Conclusion

- In the buffer layer instantaneous values of the gradient $\frac{dU}{dy}$ can reach -0.7 to 1.7 times that of the mean wall gradient.
- Correlation calculations confirm the visual studies of CORINO & BRODKEY [1] and NYCHAS et al. [2] showing that large spatial gradients always occur together with a large temporal gradient.
- The value of the spatial gradient at which sweeps can be detected best is given (Fig. 5).

References

1. E.R. Corino & R.S. Brodkey: J. Fluid Mech. 37, p. 1 (1969)
2. S.G. Nychas, H.C. Hershey & R.S. Brodkey: J. Fluid Mech. 61, p. 513 (1973)
3. J.M. Wallace, H. Eckelmann & R.S. Brodkey: J. Fluid Mech. 54, p. 39 (1972)
4. H. Eckelmann: J. Fluid Mech. 65, p.439 (1974)
5. J.M. Wallace, R.S. Brodkey & H. Eckelmann: J. Fluid Mech. 37, p. 673 (1977)
6. J. Böttcher & H. Eckelmann: Exp. Fluids 3, p. 87 (1985)
7. P. Vukoslavĉević & J.M. Wallace: Rev. Sci. Instrum. 52, p. 85 (1981)

Turbulence Structure in a Cylinder Wake

M. Hayakawa[1] *and A.K.M.F. Hussain*[2]

[1]Department of Mechanical Engineering II, Hokkaido University, Sapporo 060, Japan

[2]Department of Mechanical Engineering, University of Houston-University Park, Houston, TX 77004, USA

Large-scale coherent structures in the fully turbulent wake of a circular cylinder in air are successfully educed up to the downstream distance of 40d (d is the diameter of the cylinder) at a Reynolds number of 1.3×10^4. We have employed a newly developed conditional sampling technique, which is based on the recognition of spatially correlated vorticity, i.e. coherent vorticity from instantaneous vorticity maps. Topological features of the coherent structure and incoherent turbulence characteristics are documented; these suggest an important role of vortex stretching in the mechanisms of production and entrainment. Significant three-dimensionality of coherent structures is observed even in the near wake region through a more objective measurement than has been done heretofore.

1. Introduction

A large number of investigations in the past two decades have shown that large-scale organized structures play a central role in the dynamics of turbulent shear flows /1-4/. This fact was found largely by flow visualization techniques which have intrinsic limitations in revealing the details of turbulence mechanisms. Conventional (time average) correlation measurements, that have been widely used in attempts to quantify the spatial characteristics of large-scale motions, lose the instantaneous information and dilute the structure details. The most extensively used technique so far in quantitative measurement of coherent structures is the phase-locked measurement via some controlled excitations or appropriate triggering techniques /3/. However, such kind of measurements are useful either in the region where the marked periodicity of flow is retained or in the region close to the trigger location.

The present work employed an alternative conditional sampling technique which is able to discriminate the centers of large-scale structures. The basic idea is that vorticity should be the most relevant flow variable for identifying coherent structures. On this basis, we used a rake of X-wires and calculated the instantaneous vorticity normal to the plane of the rake and the flow direction, from which the coherent structures were recognized as phase-correlated vorticity concentrations.

Measurements were made in the turbulent plane wake of a circular cylinder placed normal to the uniform flow. Compared to other shear flows, the most prominent feature of the cylinder wake is the initial formation of fairly periodic, alternating vortices. Utilizing this nature, a number of periodic sampling measurements /5-7/ have been made in the near wake region mostly up to the downstream distance of x/d = 8, beyond which the periodicity of the flow is progressively reduced, producing significant smearing. The present experiment covered the range of x/d = 10 to 40. In principle, however, our detection scheme employed could be applied to any shear flow, even in the self-preserving region, as long as coherent structures occur in the flow.

2. Experimental Arrangement

Experiments were carried out in a low-speed wind tunnel with a test section of 92 cm x 48 cm cross section and 8 m length. A 2.7 cm diameter circular cylinder was installed on the centerplane of the test section. Eight X-wire probes were used for the velocity measurement, and the voltage signals from a 16-channel anemometer set were stored on digital tapes for later analysis. In order to take into account the three-dimensional nature of wake structures, four different configurations of probe arrangement were employed as shown in Fig. 1: that is, a single rake S1 or T1 and two parallel rakes S2 or T2. The coordinates and main notations to be used are also shown in the figure.

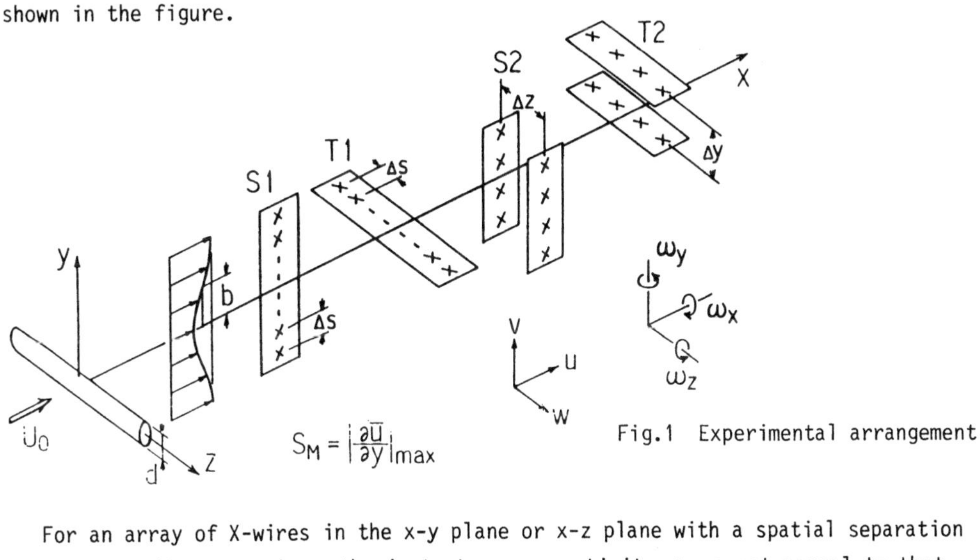

Fig.1 Experimental arrangement

For an array of X-wires in the x-y plane or x-z plane with a spatial separation Δs between adjacent probes, the instantaneous vorticity component normal to that plane can be computed from velocity signals by adopting the Taylor's hypothesis;

the spanwise component $$\omega_z = \frac{\partial v}{\partial x} - \frac{\partial u}{\partial y} \simeq -\frac{1}{U_c}\frac{\Delta v}{\Delta t} - \frac{\Delta u}{\Delta s}$$

the transverse component $$\omega_y = \frac{\partial u}{\partial z} - \frac{\partial w}{\partial x} \simeq \frac{\Delta u}{\Delta s} + \frac{1}{U_c}\frac{\Delta w}{\Delta t}$$

Here, Δt is the time step between successive data points and U_c is the streamwise structure celerity. In the present analysis, the mean velocity at the average location of the passing structure centers was chosen as U_c. The identification of structures was made from appropriately smoothed vorticity signals which were obtained by a short time averaging technique. Data were taken at x/d = 10, 20, 30 and 40 under a constant free-stream velocity of U_0 = 7 m/s, the Reynolds number being 1.3×10^4.

3. Eduction Procedure

Temporal traces of smoothed velocity components and spanwise vorticity (measured by the single rake S1) revealed instantaneous behaviours of velocity field and vorticity concentrations. An example of equi-vorticity maps taken at x/d = 20 is shown in Fig.2. The time lapse is taken from right to left, and contour levels are normalized by the local maximum mean shear S_M; clockwise and counter-clockwise vorticity levels are denoted by broken and solid lines, respectively. One can recognize alternating vortices across the centerplane. As noticed even at a relatively small distance from the cylinder, considerable dispersion in structure size, strength, shape and displacement has appeared as is particularly evident in the left hand side portion of the figure. Such an irregularity was observed to increase progressively downstream. Thus, we first selected the transverse detection center which was determined from histograms of occurrence of structure centers as a function of y at each x-station (Fig.3). The peak location of histogram (connected by solid line in Fig.3) gives the optimal detection center y_c which is located much closer to the wake centerplane than the half-width of the wake b (dotted line).

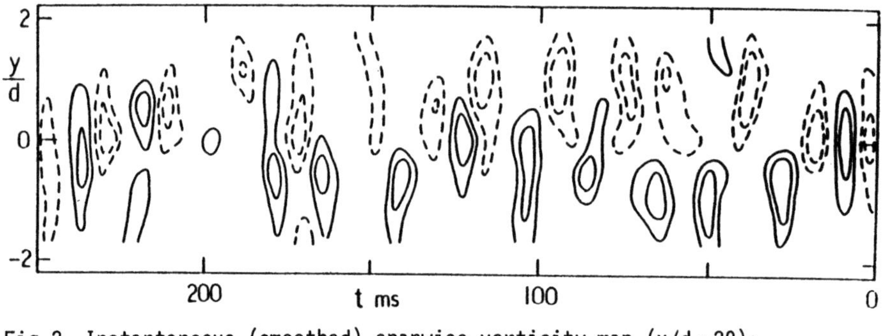

Fig.2 Instantaneous (smoothed) spanwise vorticity map (x/d = 20); contour levels: $\omega_z = \pm 4, \pm 2$

The eduction scheme has employed two principal conditions: one for the structure strength and the other for the structure size. The strength of the structure was discriminated by applying a threshold to the smoothed vorticity signals. Then, the location of each structure center was assigned the coordinate (y_c, t_c). The selection of the structure size was made by requiring that the smoothed vorticity has a spatial extent (l_x, l_y) around each center in x and y. The criterion is such that

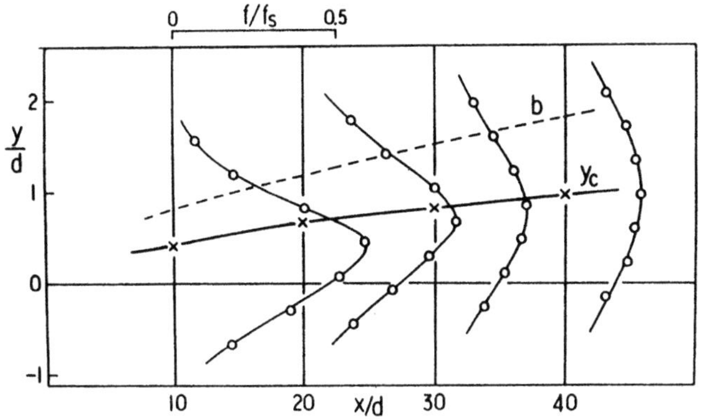

Fig.3
Histogram of passage of structure centers

the vorticities at $(y_c \pm l_y, t_c)$ and $(y_c, t_c \pm l_x/U_c)$ have values higher than another threshold but lower than the center value.

With the structure center as a phase reference, accepted structures satisfying the above criteria were aligned and ensemble-averaged. Before calculating the final result, a further improvement of eduction was made by taking the cross-correlation between individual structure vorticity and the initial ensemble-averaged vorticity. Each realization was re-aligned by a time shift equal to that of the peak correlation from zero time. The final ensemble average involves computations of coherent and incoherent components, $\langle f \rangle$ and f_r, of any quantity f; that is,

$$f(\mathbf{x}, t) = \langle f(\mathbf{x}, t) \rangle + f_r(\mathbf{x}, t) \; ; \quad \langle \; \rangle \text{ denotes the ensemble average.}$$

Note that each structure property was calculated from aligned but unsmoothed (raw) signals. For further details of the eduction procedure, see Ref./8/.

4. Mechanism of Large-scale Structure

Various measures characterizing the coherent structure and the associated incoherent turbulence have been successfully educed for x/d = 10 to 40. It was found that the number and the size of educed large-scale structures decrease with increasing x; the peak vorticity value (i.e. the strength of structure) decreases also, but remains in constant proportion to the local maximum mean strain rate. Here, for space conservation, we show only typical ensemble-averaged quantities obtained at x/d = 20 (Fig.4). These are: (a)coherent velocity perturbation vector, based on $u_c = \langle u \rangle - U_c$ and $v_c = \langle v \rangle$, (b)incoherent turbulence intensity, $\langle q_r^2 \rangle = \langle u_r^2 \rangle + \langle v_r^2 \rangle$, (c)incoherent Reynolds stress, $\langle -u_r v_r \rangle$, and (d)coherent strain rate, $\langle S \rangle$. Topologically, the flow pattern is characterized by two critical points: the center and the saddle. In each figure, the center, the saddle and the boundary of the structure are denoted by a + mark, a x mark and a dotted line, respectively. The structure boundary is suitably defined with coherent vorticity contour level of $\langle \omega_z \rangle / S_M = 1.0$.

The velocity vector pattern relative to the frame advected with a velocity U_c (Fig.4a) clearly shows well organized vortical motion around the center and also

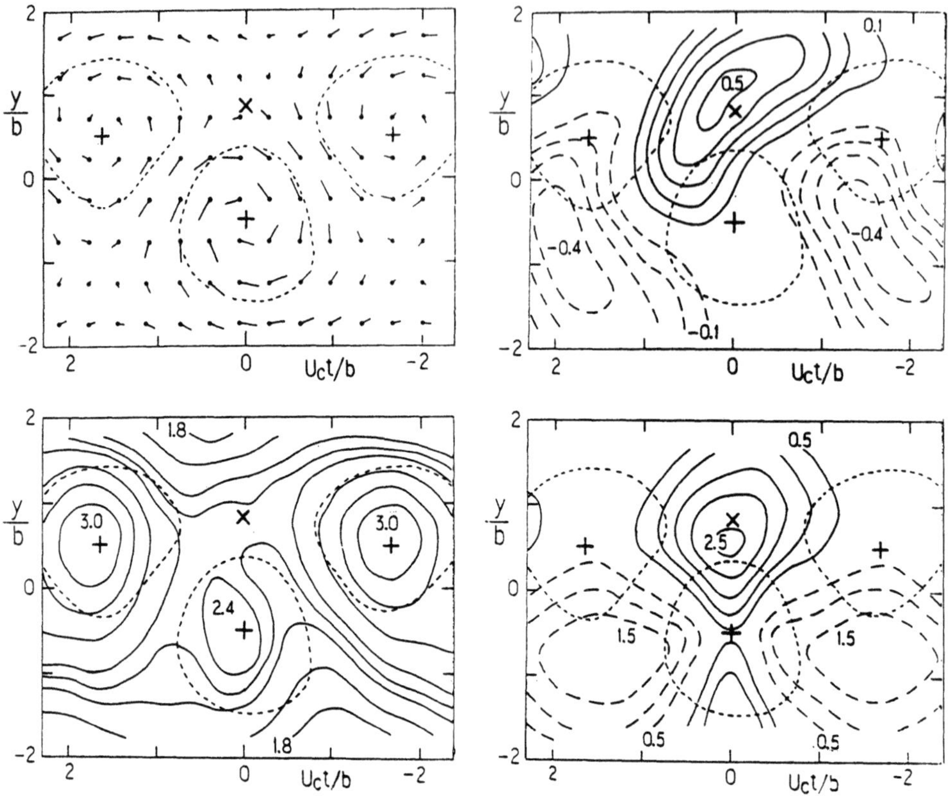

Fig.4 Coherent structure properties; (a)velocity vector pattern, (b)incoherent turbulence intensity $10^2 \times \langle q_r^2 \rangle / U_0^2$, $\langle q_r^2 \rangle = \langle u_r^2 \rangle + \langle v_r^2 \rangle$, (c)incoherent Reynolds stress $10^2 \times \langle -u_r v_r \rangle / U_0^2$, (d)coherent strain rate $\langle S \rangle / S_M$

specifies saddle points (denoted by x). Since non-turbulent fluid outside the structure is engulfed into the wake region mostly at the front (downstream side) of the structure, the turbulence intensity at the front is much lower than at the back, and lower contour levels at the front extend toward the wake centerplane (Fig.4b). This implys that significant amount of fluid can be entrained from the opposite side of the wake across the centerplane and also suggests that the braid is not a contineous sheet but perforated. Contours of the incoherent Reynolds stress (Fig.4c) and the coherent strain rate (Fig.4d) have their maxima located near the saddle point. The intense strain field in the saddle region appears to be the consequence of vortex stretching induced by successive like-signed vortices. Thus, the coherent production (not shown here) has a similar contour pattern to the incoherent Reynolds stress contour; that is, significant production of smaller-scale turbulence due to large-scale structures occrs again in the saddle region.

These results support the ideas of Cantwell & Coles /6/ and Hussain /9/ that the basic mechanism of turbulence production is the vortex stretching along the braid connecting successive spanwise rolls. This implies further that the braid consists of longitudinal coherent substructures, i.e. ribs /9/. From these results, the mechanism of large-scale structure and the associated flow physics of a fully turbulent wake emerges; these can explain most of our experimental data satisfactorily. The strong spanwise vorticity induces the large-scale motion of external fluid along the converging separatrix. The engulfed fluid is advected to the saddle region where most production of smaller-scale turbulence takes place due to the direct consequence of vortex stretching along the diverging separatrix. The turbulence produced is continuously advected from this region into the structure and also away from central portion of the wake. The most effective turbulence mixing is very likely to occur at the back of structure due to the three-dimensional vorticity interaction between ribs with primarily streamwise vorticity and large-scale structure with spanwise vorticity. Further discussions on the mechanism of the wake structure are included in Ref./8/.

5. Three-dimensional Aspect

The increasing dispersion of individual vortex size, strength, etc. as well as the presence of ribs, suggests that some appreciable three-dimensionality should have occurred even in the short distance from the cylinder.

An example of temporal traces of the transverse vorticity ω_y measured by a single rake T1 at x/d = 20 is shown in Fig.5. The y-position is close to the wake half-width location, and the contour levels are again normalized by S_M. Indeed, there are many vorticity concentrations with the peak values comparable with those of spanwise vorticity concentrations. These concentrations sometimes form paired pattern with opposite-signed vorticity, suggesting the presence of counter-rotating vortices. However, the large-scale ω_y-concentrations do not occur periodically in the spanwise and/or streamwise directions. This irregular behaviour is quite different from the visualization picture of Cimbala /10/ who observed aperiodic wavy smoke patterns developing downstream from the initially two-dimensional smoke pattern in the x-z plane in the low Reynolds number wake flow.

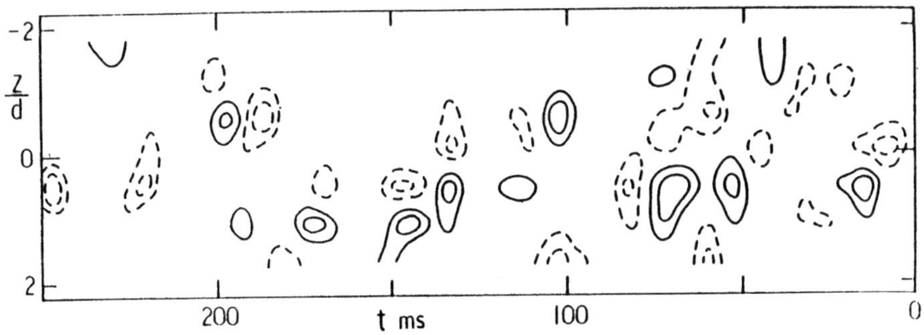

Fig.5 Instantaneous (smoothed) transverse vorticity map (x/d = 20, y/b = 0.94); contour levels: ω_y = ±3, ±1.5

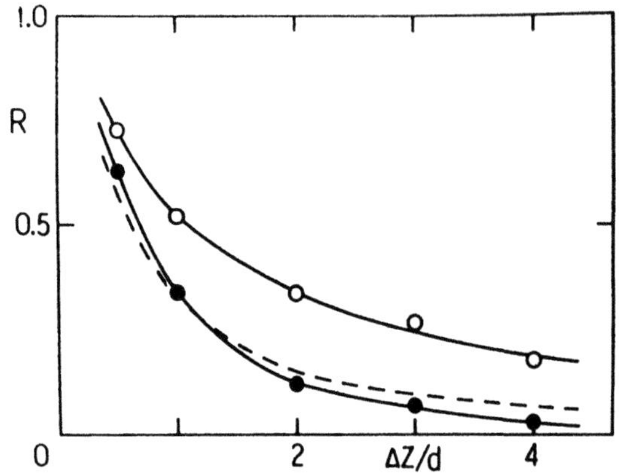

Fig.6 Spanwise vorticity correlation;
○ : x/d = 20, y/b = 0.47
● : x/d = 40, y/b = 0.45
---: x/d = 20, y/b = 0.94

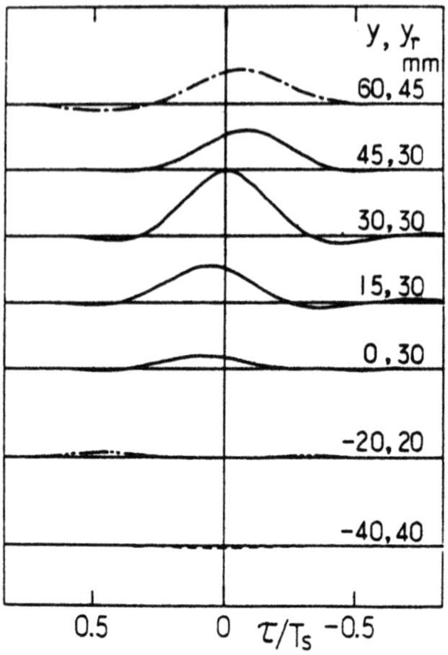

Fig.7 Ensemble averages of ω_y (x/d = 40); y_r denotes the location of reference rake

Due to the three-dimensionality seen in Fig.5, the spanwise correlation of spanwise (smoothed) vorticity signals, measured by two rakes S2, decreases considerably with increasing spanwise separation Δz even at x/d = 20 (Fig.6).

The coherence of transverse vorticity concentrations across the wake region was examined by the parallel rakes T2; that is, we detected the ω_y-concentrations at one x-z plane (the reference position) and took the ensemble average of ω_y at other x-z planes. The result obtained at x/d = 40 is shown in Fig.7 which includes data for four different reference positions denoted as y_r. The phase-shift of ω_y-peak in each trace with respect to the detection center ($\tau = 0$) indicates that the transverse structure is inclined to the downstream side as one moves away from the wake centerplane. The most probable oblique angle was found to be roughly 30 degrees from the y-axis. Further, structures extending over both sides of the wake centerplane, like the so-called Grant's model /11/, have not been observed with significant probability in the moderately far wake region covered by the present experiment.

6. Closing Remarks

An alternative conditional sampling measurement based on the recognition of large-sacle vorticity concentrations revealed the detailed turbulence mechanism in the plane wake of a circular cylinder up to the downstream distance of 40d that being far beyond the region covered by earlier quantitative measurements. The result strongly suggests the important role of vortex stretching in the saddle region in the mechanisms of turbulence production and entrainment process.

With several combinations of hot-wire arrays, it has been demonstrated that significant three-dimensionality has appeared in the very near field of the norminally plane wake. However, a further look into the dynamical role and significance of the three-dimensional nature remains to be explored via future work which would require the simultaneous measurement of spanwise and streamwise vorticities.

This work was supported by ONR Grant N00014-85-K-0126 monitored by Dr. M.M. Reischmann.

References

1. A. Roshko: AIAA J. 10, 1349 (1976)
2. B.J. Cantwell: Ann. Rev. Fluid Mech. 13, 457 (1981)
3. A.K.M.F. Hussain: Phys. Fluids 26, 2816 (1983)
4. D. Coles: The Dryden Research Lecture, AIAA Paper 85-0506 (1985)
5. M.E. Davies: J. Fluid Mech. 75, 209 (1976)
6. B.J. Cantwell & D. Coles: J. Fluid Mech. 136, 321 (1983)
7. M. Kiya & M. Matsumura: Bulletin of JSME 28, 2617 (1985)
8. A.K.M.F. Hussain & M. Hayakawa: to appear in J. Fluid Mech. (1986)
9. A.K.M.F. Hussain: In Turbulence and Chaotic Phenomena in Fluids, ed. by T. Tatsumi, IUTAM (North-holland 1984) p.453
10. B. Cimbala: In Turbulent Shear Flows 5, 4.1 (1985)
11. H.L. Grant: J. Fluid Mech. 4, 149 (1958)

Coherent Structures and Their Relation to Instability Processes in a Round Free Jet

S. Drobniak and J.W. Elsner

Technical University of Czestochowa, al. Zawadzkiego 25,
PL-42-200 Czestochowa, Poland

1. Introduction

The coherent structures in free axi-symmetric jet appear in the region where the laminar-turbulent transition occurs. The loss of stability of this kind of flow, leading eventually to the formation of organized vorticity, may be examined in two ways. The first one has been proposed by MICHALKE [1]. He treated the laminar-turbulent transition of a round jet as a loss of stability of the thin shear layer in the close vicinity to the jet outlet (Fig.1). As the thickness δ of the shear layer is much smaller than the nozzle radius R so the whole problem may be considered as a two-dimensional one. The solution of the Rayleigh equation - either temporal or spatial - gives in consequence the amplification factor "a" and the convection velocity U_k of the perturbation.

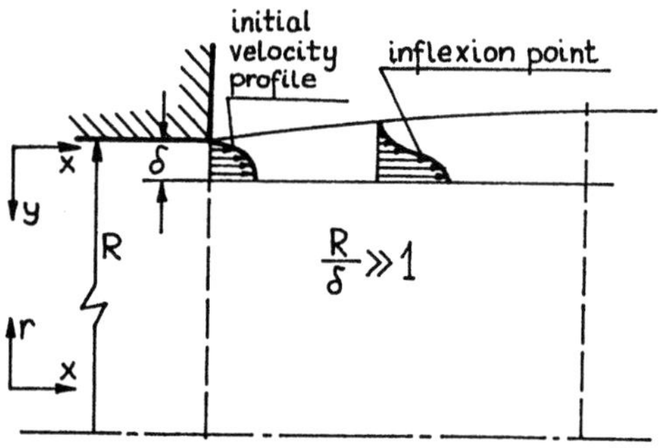

Fig.1. The geometry of the jet shear layer close to the outlet.

The second possible treatment of this problem has been proposed by BATCHELOR and GILL [2] who analysed the laminar-turbulent transition of the round, free jet as a loss of stability of the cylindrical discontinuity surface (Fig.2) surrounding the potential core. Authors anticipated the possibility of other than axisymmetric perturbation which may be derived from their equations. The final solution gives also the growth rate "a" and the convection velocity U_k of the perturbation. It must be noticed here that both these quantities may easily be measured thus enabling us to verify experimentally the theories mentioned above.

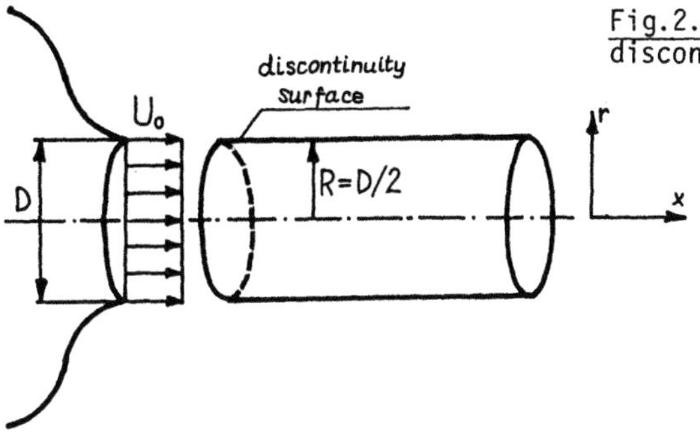

Fig.2. The geometry of the cylindrical discontinuity surface.

2. Experimental Results

The existence of coherent structures in a round free jet has been proven beyond any doubt but the conditional sampling procedures do not usually reveal them. It appears that they remain hidden under the background noise resulting from the interaction of different forms of organized vorticity existing in the initial region of the jet. One needs therefore some means of extracting the particular forms of coherent structures from behind the large jitters of the signal by introducing a certain degree of order into the flow field. Since the time of early papers by GINEWSKI et al. [3] the acoustic field is regarded to be the best amplifier of the organized vorticity. The method of acoustic excitation has been used for the large variety of flows [4], [5] and the present research also follows this way.

The experiment was carried out in the wind tunnel (Fig.3) with the circular outlet. The whole test rig was placed inside the anechoic chamber to avoid the external disturbances. The loudspeaker producing the pure tone acoustic wave was placed outside the wind tunnel allowing therefore for independent variation of Strouhal and Reynolds numbers. The data from the DISA 55M hot-wire bridge were fed via the linearizer to the Hewlett-Packard 9826 B digital signal analyzer. The

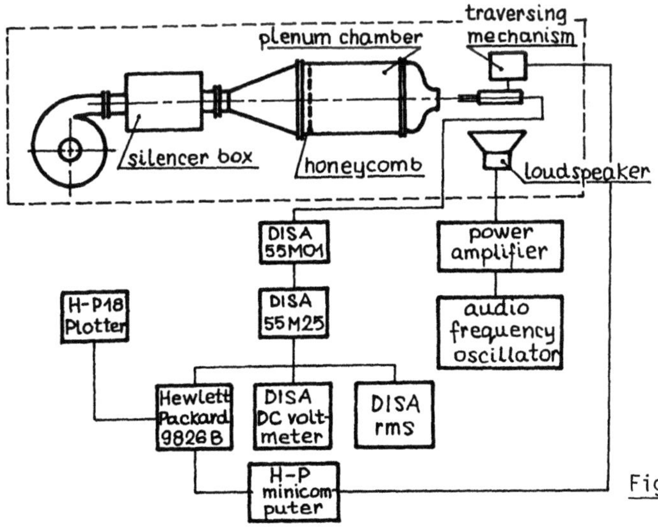

Fig.3. The experimental setup.

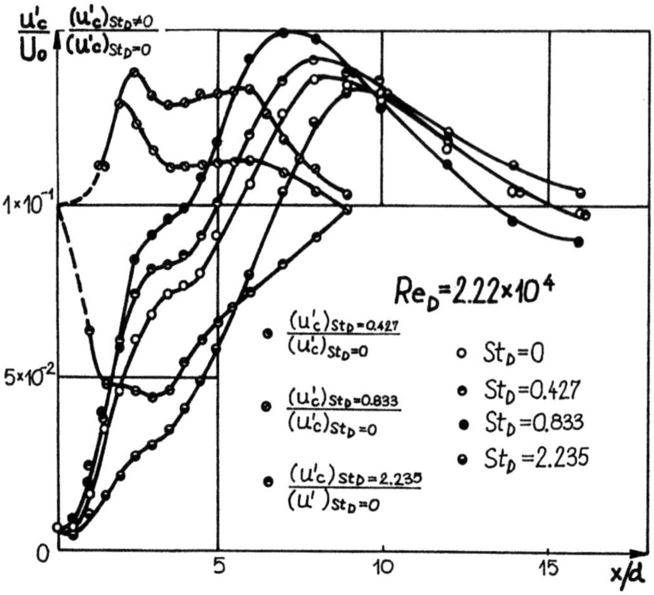

Fig.4. The evolution of velocity fluctuations along the jet axis.

sampling frequency was up to 1.2 MHz depending on the bandwidth of the signal analysed. The data could be stored on the digital casette memory for further processing. The two different jet nozzle diameters D = 0.04 [m] and D = 0.0254 [m] have been used in order to widen the Reynolds number range considered.

The influence of external forcing has been illustrated in Fig.4 presenting a downstream evolution of longitudinal velocity fluctuations measured at the jet axis. The curves shown here correspond to the non-excited case as well as to the Strouhal numbers for which the maximum effect of forcing is obtained. The quantity u_c', according to the conventional time averaging procedure applied, contains the joint information on the periodic as well as random fluid motion. The data collected in Fig.4 suggest that the external excitation of the round, free jet may either increase or damp the intensity of longitudinal velocity fluctuations in the near flow region (say $x/D < 7$). It may be interesting to note that the characteristic "plateau" existing in the area x = 3 - 4 D of the nonexcited jet is still visible in the case of low frequency excitations ($St_D < 1$). It has been shown [5] that the location tion of this "plateau" accords with the position where the maximum amplitude of organized motion is observed. As can be seen in Fig.4 the curves corresponding to low frequency excitations are simply shifted up in comparison with the natural jet. It means therefore that this kind of forcing really amplifies the existing coherent structures without introducing any new phenomena into the flowfield. On the other hand, the high frequency forcing ($St_D > 1$) damps the longitudinal velocity fluctuations and smoothes out the "plateau". This may suggest a possible elimination of some forms of organized motion. The more detailed explanation of this phenomenon will be given later.

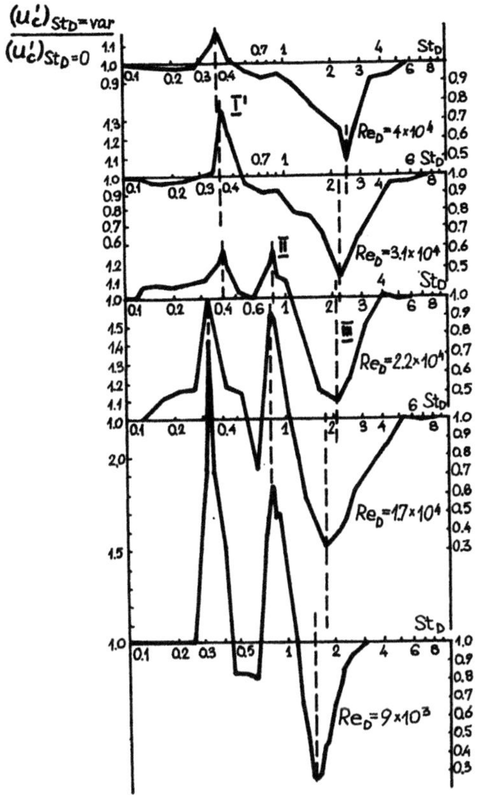

Fig.5. The jet response to the external excitation.

The summary of the jet response to acoustic excitation is shown in Fig.5 for several Reynolds numbers Re_D confined within the range $9 \cdot 10^4 - 5 \cdot 10^5$. It presents the amplification of longitudinal velocity fluctuations measured in the point placed at the jet axis in the distance x = 2.5 D from the nozzle. The amplification of turbulence level observed for low frequency forcing (denoted I' in Fig.5) is the more visible the lower is the Reynolds number. As the Re_D value is decreased the second peak corresponding to St_D = 0.8 appears (symbol II in Fig.5) and the response of velocity fluctuations to acoustic excitation also follows the previous rule (the higher degree of amplification the lower is the Re_D value). The high frequency forcing applied in the range St_D = 1.6 - 2.5 (symbol III in Fig.5) results in the well marked damping of turbulence intensity, the more distinct the lower is the value of Re_D applied.

Following the previous considerations concerning the mechanism of acoustic excitation one may conclude that reaction of the fluctuating velocity field shown in Fig.5 is a consequence of amplification of particular forms of organized motion existing in the initial region of the jet. The most important problem now is to explain the origin of those peaks of amplification or damping. The first indications could be obtained from scaling laws of particular characteristic frequencies. As it can be seen from Fig.6 the Strouhal numbers of both low frequency peaks (i.e. I and

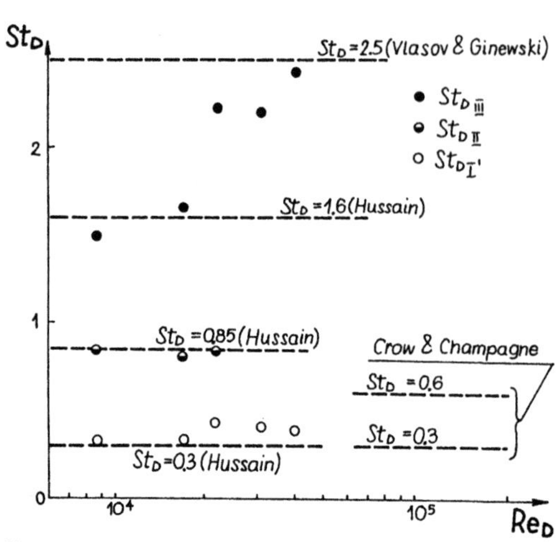

Fig.6. The behaviour of characteristic St_D numbers.

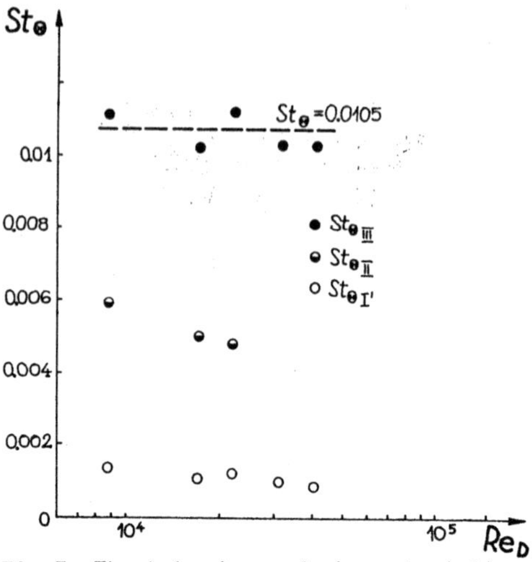

Fig.7. The behaviour of characteristic St_θ numbers.

II) become independent of the Reynolds number if the jet exit diameter D is applied as a linear dimension. As this quantity (i.e. diameter D) is a characteristic linear scale in the solution of BATCHELOR and GILL one may conclude that the forms of organized motion marked as I and II have something in common with the instability of cylindrical discontinuity surface (jet column mode - this term has been proposed by CROW and CHAMPAGNE [6]).

The high frequency forcing (form III) scales on the momentum thickness θ of the initial shear layer. As it can be seen from Fig.7 when θ is applied as a characteristic linear dimension then the resulting Strouhal number St_θ becomes independent of the Reynolds number Re_D. It seems therefore that the III-mode of organized motion originates from the instability of the initial layer described by MICHALKE (θ is the characteristic linear scale in his solution).

The physical picture of all the forms of organized motion present in this type of flow has been obtained from the smoke-wire visualization performed at the lowest Reynolds number $Re_D = 9 \cdot 10^3$. The frozen flow-patterns of three different forms of organized motion selected by proper frequency forcing are presented in Fig.8. As can be seen they are created by rolling up of the axi-symmetric shear layer and travel downstream with the frequency determined by the external forcing. The coherent structures of the II and III-type emerging as axi-symmetric vortex rings are somewhat similar even if their spacing is quite different. Coherent structures of the III type revealed by the high frequency forcing are formed as vortex rings closely spaced inside the mixing zone of the jet (see Fig.8c). After a very short distance one can observe the pairing process leading eventually to the formation of larger vortex rings with twice the previous spacing. A similar picture can be observed in the case of the II-mode of organized motion (Fig.8b) where the pairing process can also be noticed but the vortices maintain their coherence at the much

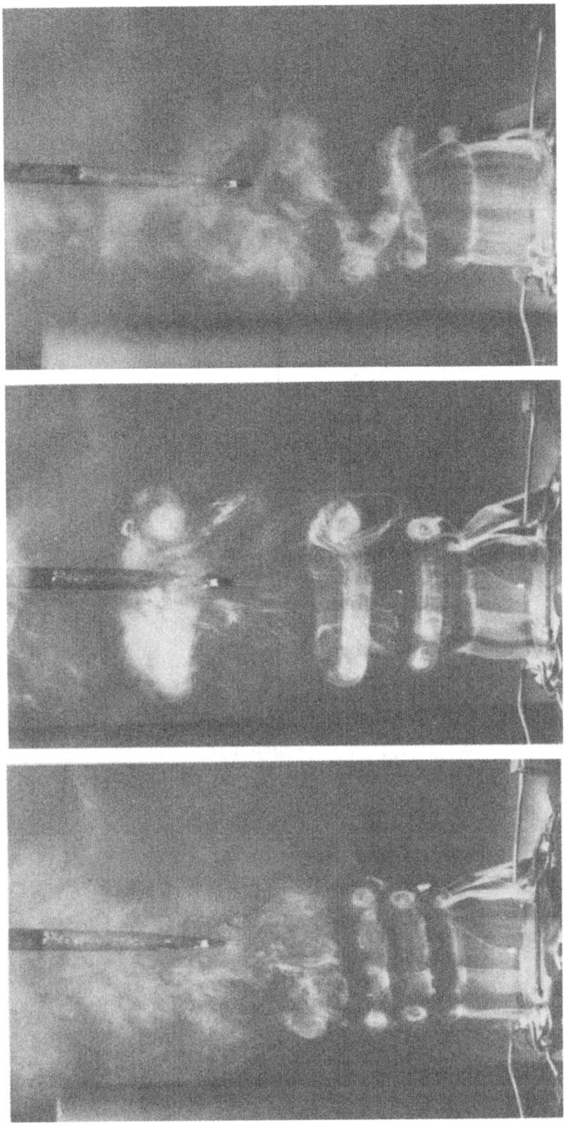

I

II

III

Fig.8. The smoke-wire pictures of particular coherent structures in a forced jet (flow is from the right to the left).

larger distance than it could be seen for the III-mode. The most unexpected however is the picture of the I-mode which appears as a large helix resulting from the non-symmetric rolling up of the cylindrical shear layer (see Fig.8a).

The knowledge of the physical picture of coherent structures obtained from the visualization experiment allows us to understand the different response of fluctuating velocity field to the low and high frequency forcing. The low frequency excitation causes the amplification of the large scale structures and the substantial enlargement of the mixing zone. In consequence, it leads to the diminution of the potential core and to the growth of velocity fluctuations observed at the jet axis.

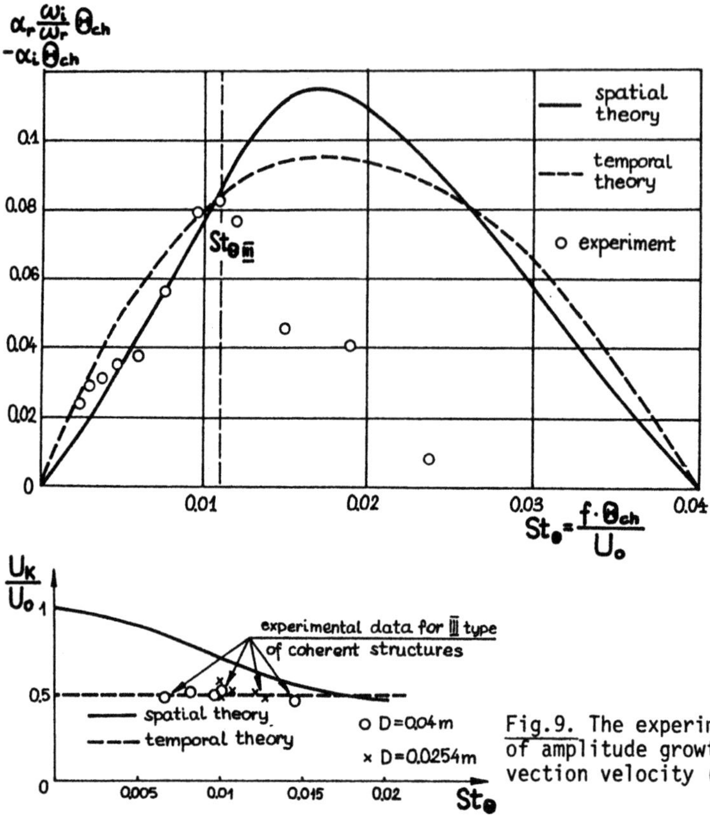

Fig.9. The experimental verification of amplitude growth rate (a) and convection velocity (b) with the theory [1].

On the other hand, the high frequency forcing amplifies the "tightly packed" vortex rings of the III-mode (having much smaller cross-sections comparable with shear layer thickness) leading finally to the elongation of potential core and to the decrease of velocity fluctuations at the jet axis.

The more convincing proof of the validity of the hypothesis stated above has been obtained from the comparison of theoretical predictions cited above with the results of the present experiment. Figure 9a presents the spatial as well as temporal amplification factors of the perturbation amplitude calculated by MICHALKE versus the Strouhal number St_θ. In both cases the maximum growth rate should be observed for $St_\theta = 0.017$ while in fact the maximum amplification of the III-mode takes place for $St_\theta = 0.011$. On the other hand, the convection velocity of these coherent structures corresponds well with the temporal solution of Michalke as can be seen in Fig.9b.

In the case of low frequency structures (the I and II-modes) the magnitude of the amplification factor predicted by BATCHELOR and GILL is much higher than observed in the present experiment (see Fig.10). This is probably due to the rather crude assumptions applied in their theory (i.e. infinitely thin shear layer). On the other hand, the convection velocities of both the I and II-mode agree quite

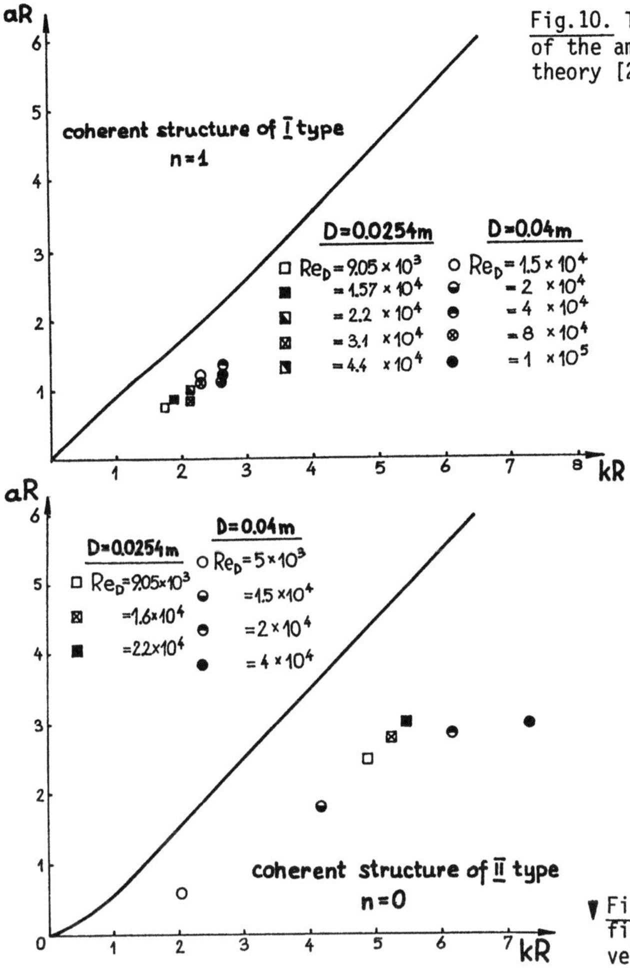

Fig.10. The experimental verification of the amplitude growth rate with the theory [2].

▼ Fig.11. The experimental verification of the convection velocity with the theory [2].

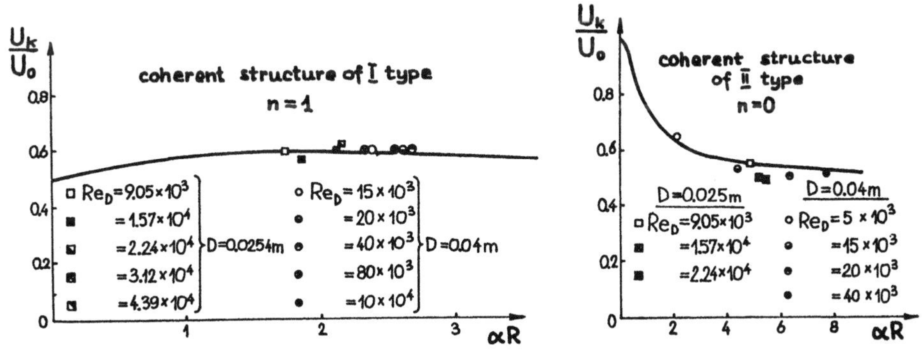

well with the theoretical predictions as shown in Fig.11. It is worth noticing that in the range of $Re_D = 10^4 - 5 \cdot 10^5$ the convection velocity of the helical mode is substantially larger ($U_k/U_o = 0.6$) than that of both the symmetric ones ($U_k/U_o = 0.5$).

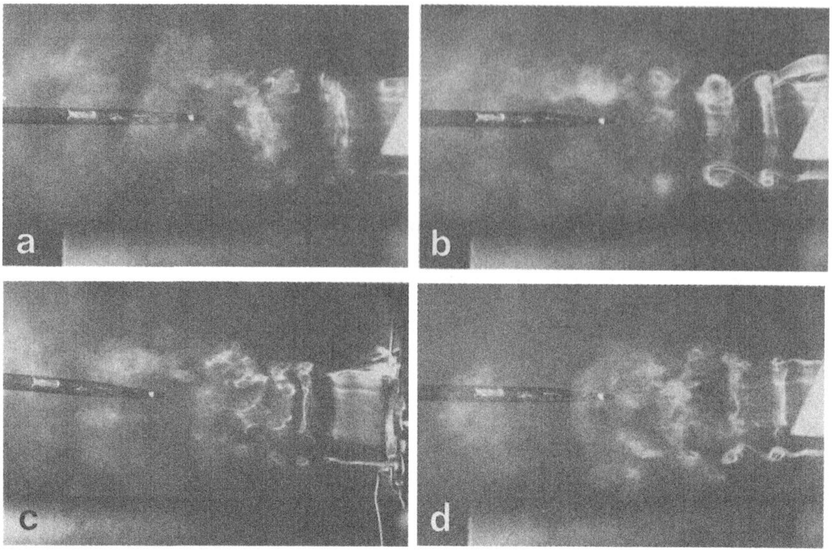

Fig.12. The instantaneous smoke-wire pictures of coherent structures in a natural jet (the flow is from the right to the left).

The three observed forms of coherent structures do also exist in the natural jet as can be seen in Fig. 12. It presents the instantaneous pictures of "frozen" flow patterns selected from a large number of photos taken at random without any forcing. One may easily recognize that the consecutive pictures 12a, 12b, 12c show the I, II and III-mode of coherent structures, respectively. An interesting phenomenon may be noticed in Fig.12d presenting in the middle of the picture the helical mode neighbouring with the II-type symmetric coherent structure emerging from the jet nozzle (the right side of the picture). In consequence, the evolution of mean and fluctuating velocity fields of the round, free jet is a result of momentary balance between the particular coherent structures. It does not mean however that the three forms of organized motion described above are equally likely to be met in this type of flow. The frequency analysis of velocity fluctuations performed along the jet axis (Fig. 13) revealed two peaks of power spectral density (P.S.D) at the frequencies characteristic for the I-mode. It means therefore that the helical coherent structures have the largest share in the energy of velocity fluctuations and one may expect that the helical mode dominates the natural jet. One of the most important parameters of the organized motion, namely the wavelength λ determined as $\lambda = U_k/fch$ (fch - characteristic frequency of the structure), has been presented in Fig.14 versus the Reynolds number Re_D. As can be seen the wavelengths λ_{II} and λ_{III} of the II and III mode decrease with the growth of the Reynolds number while λ_I is practically constant in the entire range of Re_D analysed. Even more surprising is the dependence of the λ_I value upon the jet nozzle diameter while this tendency could not be noticed for the symmetric modes. It seems that this phenomenon may be caused by two different shapes of the nozzles used in the present experiment. The

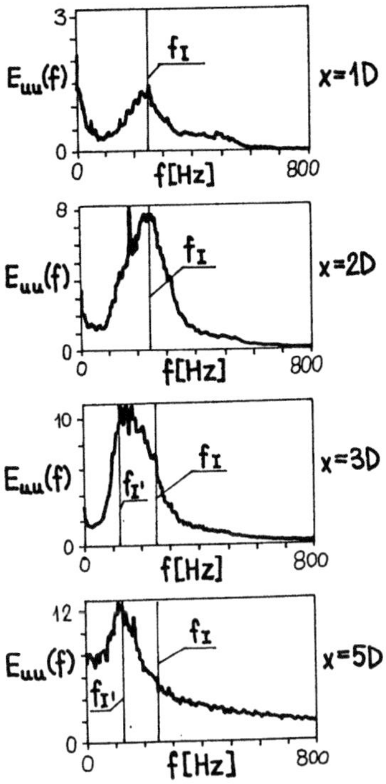

Fig.13. The downstream evolution of P.S.D. of longitudinal velocity fluctuations in a natural jet.

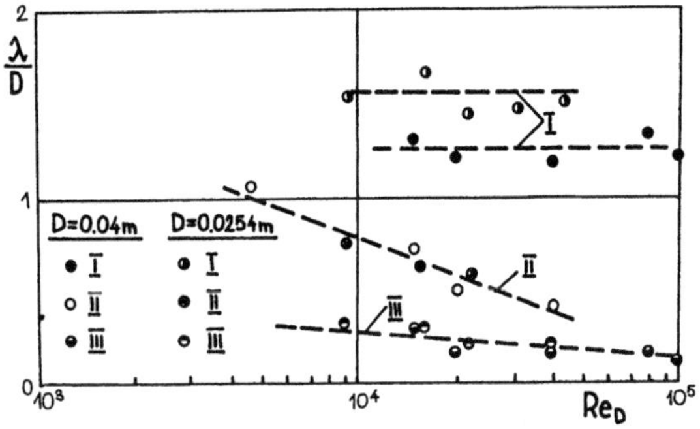

Fig.14. The wavelengths of particular forms of coherent structures.

nozzle of the diameter D = 0.04 [m] was profiled according to Witoszyński formula [7] while the design of the smaller one (D = 0.0254 [m]) was based on the procedure recommended by MOREL [8]. This dependence of the parameters of organized motion on the geometry of the nozzle is probably due to the influence exerted by the wall curvature upon the frequency of the Tollmien-Schlichting wave existing in the exit boundary layer. A similar observation has been reported by Drubka and Nagib [9] who have tested the different polynomial nozzles but the whole problem is far from clear and certainly needs further research.

As can be seen from Fig.14 the wavelength λ_{II} depends most evidently upon the actual value of Re_D that may answer the question why the II-mode exists only in the limited range of the Reynolds number. The results of the present experiment as well as the literature data [5] suggest that the highest Re_D value for which the II-mode may still be observed is $5 \cdot 10^4$. As can be seen in Fig.14 for $Re_D = 5 \cdot 10^4$ the wavelength λ_{II} reaches twice the value of λ_{III}. It may suggest that for higher values of Reynolds number one gets the subharmonic amplification of the III-mode with the simultaneous elimination of the III-mode.

3. Conclusions

The results obtained so far suggest that the two different types of instability lead eventually to the formation of three separate modes of coherent structures -

either symmetric or helical - which may exist in the same initial region of the round, free jet. It means therefore that due to the specific geometry of the flow considered the physical picture of the laminar-turbulent transition is much more complex than in other flows.

Bibliography

1. Michalke, A.: On Spatially Growing Disturbances in an Inviscid Shear Layer, J.F.M., 1965, 23; also: On the Inviscid Instability of the Hyperbolic-Tangent Velocity Profile, J.F.M., 1964, 19.
2. Batchelor, G.K., Gill, A.E.: Analysis of the Stability of Axisymmetric Jets, J.F.M., 1962, 14
3. Ginewski, A.S.: Aeroakusticeskije vzaimodejstvia, Masinostrojenie, Moscow 1978
4. Freymouth, P.: On Transition in a Separated Laminar Boundary Layer, J.F.M. 1, 1966, 25
5. Hussain, A., Zaman, K.: Vortex Pairing in a Circular Jet Under Controlled Excitation, J.F.M., 1980, 101, p.1 and 2
6. Crow, S. Champagne, F.: Orderly Structures in Jet Turbulence, J.F.M., 1971, 48
7. Gukasowa et al.: Aerodinamiceskoje soversestvovanije ..., Moscow, 1960
8. Morel T.: Comprehensive Design of Axi-Symmetric Wind Tunnel Contractions, Trans. ASME, 1975
9. Drubka, R., Nagib, H.: Instabilities in Near Field of Turbulent Jets and Their Dependence on Initial Conditions and Reynolds Number, II T Fluids & Heat Transf. Rep. R81-2, Chicago, 1981

Experimental Study of an Incompressible, Plane Mixing Layer by Temporal and Spectral Analysis

J. Delville, Z. Chahine, and J.P. Bonnet

Centre d'Etudes Aérodynamiques et Thermiques, Laboratoire d'Etudes Aérodynamiques, UA C.N.R.S. 191, Université de Poitiers, 43, rue de l'Aérodrome, F-86000 Poitiers, France

1. Introduction

For the general purpose of turbulent structure determination in any fluid flow, a great deal of work has recently been devoted to conditionnal measurements ; this approach has been largely discussed in many papers; in any situations it requires some decision criteria which are not universal. On other hand, extracting informations from statistical values is a challenging idea developped some years ago by GRANT [1], TOWNSEND [2] and LUMLEY [3]. The approach of Lumley, the so-called orthogonal decomposition, is more direct than the a-posteriori verification of basis structure description of Grant and Townsend. The aim of this paper is not to discriminate between one or other approaches, but to give detailed space correlation measurements either in the time or in the frequency domains. By using a powerfull experimental measurement system and data reduction procedure, a complete set of correlations may be obtained in the classical statistical sense allowing to make use of the orthogonal decomposition approach as adopted recently by GLAUSER and GEORGE [4]. The choosen configuration is here the asymptotic region of a fully turbulent plane mixing layer in incompressible flow. The data collected are not yet used for the orthogonal decomposition, but still discussed in terms of general characteristics of the mixing layer.

2. Apparatus and Data Reduction

2.1. Experimental Setup

Experiments are performed in an open wind tunnel of square section 30x30 cm^2 with a test section of 1.2 m long. The plane mixing layer is created by a flat plate splitting two air streams at respectively U_A = 42.8 m/s and U_B = 25.2 m/s ; these two velocities are obtained by putting different head-losses devices in the settling chamber. The flat plate is bevelled with 3% slope from 50 mm upstream of the trailing edge where the boundary layers are fully turbulent with a thickness of about 7 mm. The temperature of the air is stabilized nearly the ambiant with a precision of 0.5°C. The coordinates are X,Y and Z respectively in the streamwise direction (from the trailing edge), perpendicular to the plane of the splitter plate (Y=0 at the center line) and parallel to the trailing edge (the flow is two-dimensional following Z).

Measurements are performed by hot-wire anemometry using TSI Constant Temperature Anemometers (T.S.I. 1750) operating 2.5 μm diameter wires ; the length of the wires is 1.25 mm. Miniature X wires probes with typical 45 degrees of yaw are used. Special calibration of the sensitivity coefficients is performed with a home-built calibration tunnel giving a good accuracy in the coefficient measurements. Two independent probe holders can be displaced by stepping motors in 2 directions (X and Y) for the "stationary" (reference) probe and for 3 directions (X, Y and Z) for the "moving" probe. The combination of the displacement is driven by a micro-computer. The coordinates of the stationary probe are denoted X_0, Y_0, Z_0, where X_0 is fixed to 600 mm from the trailing edge and Z_0 is zero (i.e. on the axis of the test section) ; the moving probe is then located at X_0+x', Y_0+y' and Z_0+z'. The external turbulence level is 0.3%.

2.2. Data Collection and Numerical Treatment

The use of an array-processor linked to a micro-computer allows to perform the measurements and numerical treatments in Real Time. Signals from the four anemometers are collected by the micro-computer at an A/D conversion rate of 50 kHz per channel to form sets of 8 k words per channel. The collected data are send to the array-processor via a serial data link, with a transfer rate of 200 k words per second. With this method, one collected set of samples can numerically be treated by the array-processor during the next set is collected by the micro--computer. The numerical procedure uses Fast Fourier Transform techniques. In a first step the hot wires law for velocity and yaw is taken into account, then cross-correlations, cross-spectrum and spectra for two of the four velocity components are accumulated. Preliminary checks of the numerical process have been made by comparing digital results with correlations obtained using analog apparatus in the case of single hot wires.

3. One Point Measurements

The classical shear layer thickness evolution is given in Fig. 1. The linear behaviour of $\delta(X-X_V)$, where X_V is the virtual origin of the layer, can be observed. All measurements presented here are performed at a streamwise location X_0 = 600 mm or X_0-X_V = 680 mm. At this location the mean velocity profiles satisfy the well known similarity profiles ; the turbulence quantities are plotted in Fig. 2, normalized by the local velocity scale $\Delta U = U_A-U_B$. All the one-point measurements agrees with other workers results within the discrepancies mainly attributable to the external turbulence level as pointed out by RODI [5].

The mean characteristics of the mixing layer are the following :

velocity ratio $r = 0.59$,

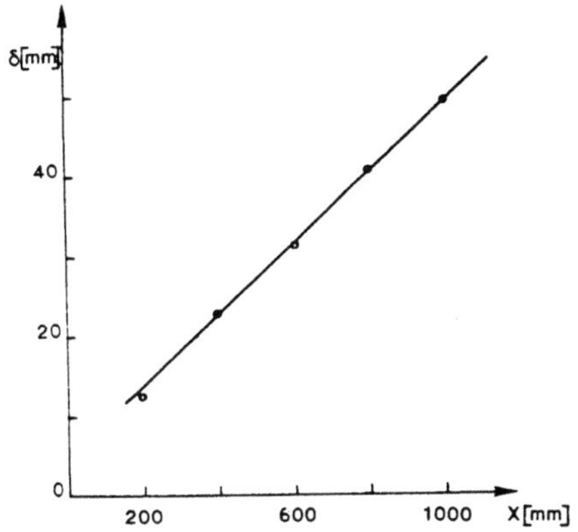

Fig. 1
Streamwise evolution of the layer thickness

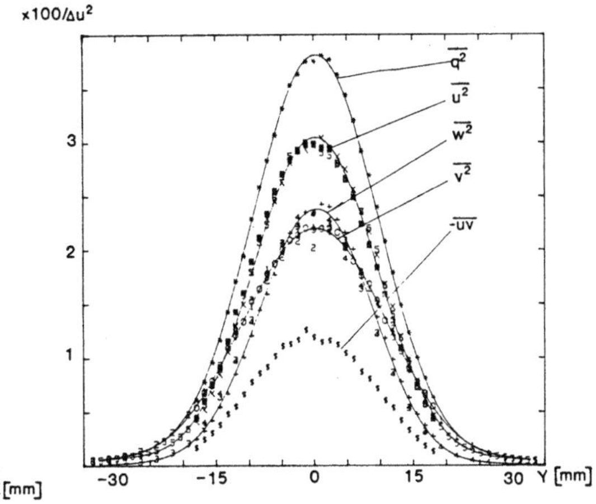

Fig. 2
Lateral distribution of turbulence quantities

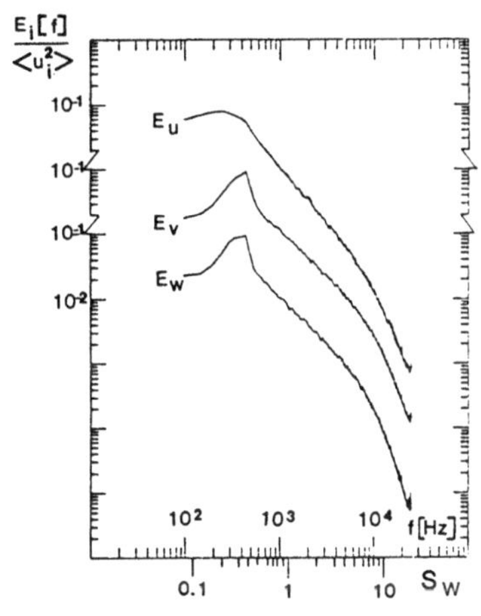

Fig. 3
Power spectral density of fluctuating velocity components

spread parameter $\sigma = 43.1$,

and at location X=600 mm :

vorticity thickness $\delta_W = 28$ mm,

momentum thickness $\theta = 5.7$ mm.

The power spectrum densities of the three velocity components u,v and w are plotted in Fig. 3 for a transverse location of $Y/\delta = 0.3$. No characteristic frequency can be detected on the u spectra (E_u) ; however for a transverse location corresponding to the irrotational part of the flow, a maximum of energy of E_u is obtained at about 300 Hz. The local Strouhal number based on the vorticity thickness, defined by $S_W = f\delta_W/ U_0$, where $U_0 = \frac{1}{2}(U_A+U_B)$, is equal to 0.22 ; this value is close to those given by BROWAND and HO [6] ($S_W \approx 0.2$). The power spectrum density of the v and w component clearly show characteristic frequency ; the maximum of energy is obtained in the neighborhood of 400 Hz. The ratio between the characteristic frequency of u and v component is nearly 0.75 ; this ratio can be compared to the result of JONES [7] which obtains a value of 0.67.

4. Two-points measurements : time domain

Correlation coefficients R_{ij} $(Y_0,x',y',z';\tau)$ are defined by

$$R_{ij}(Y_0,x',y',z';\tau) = \frac{\langle u_i(X_0,Y_0,Z_0,t) \, . \, u_j(X_0+x',Y_0+y',Z_0+z';t+\tau) \rangle}{\langle u_i^2(X_0,Y_0,Z_0;t) \rangle^{1/2} . \langle u_j^2(X_0+x',Y_0+y',Z_0+z';t) \rangle^{1/2}}$$

X_0 and Z_0 being respectively fixed at 600 mm and = 0. Previous measurements (BOURIOT et al. [8]) having shown the local validity of the Taylor hypothesis, only y, z and τ separations are imposed. The measurements presented here are plotted as iso-correlation coefficient contours which give an overall view of the results. In the τ-Y (or X-Y) plane, we plot in Fig. 4a to 4d the contours of

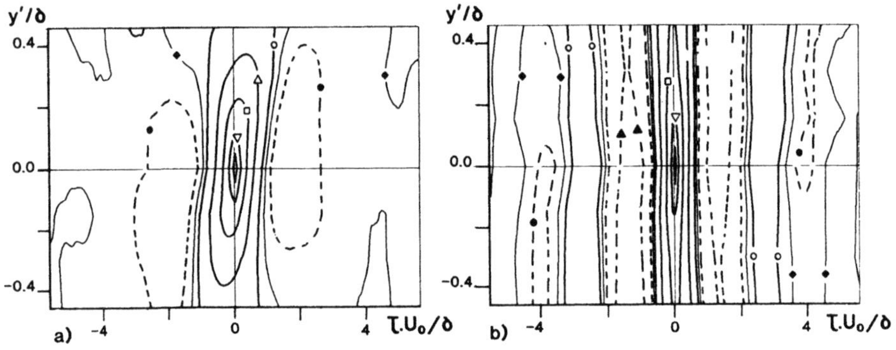

	■	▲	●	◆	○	△	□	▽
Rij	-0.40	-0.20	-0.05	0.00	0.05	0.20	0.40	0.60

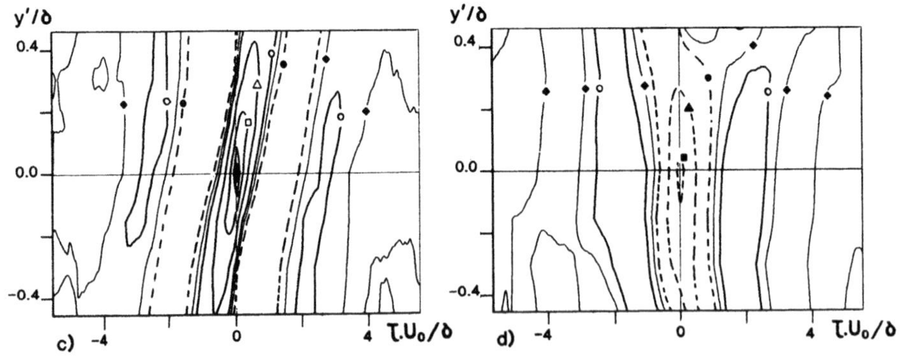

Fig. 4 Iso-correlation coefficients contour plots in τ-Y (X-Y) plane. stationary probe on the centerline (Y_0 = 0). a) R_{11} ; b) R_{22} ; c) R_{33} ; d) R_{12}

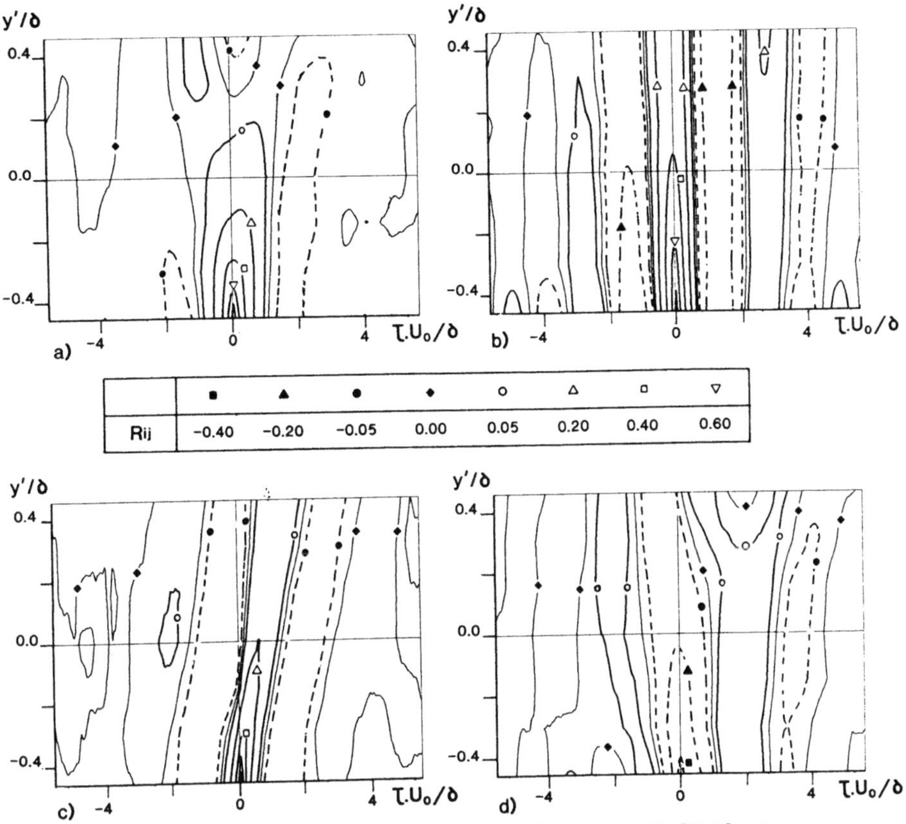

Fig. 5 Iso-correlation coefficients contour plots in τ-Y (X-Y) plane. stationary probe at Y_0 =-0.45δ (low-speed part). a) R_{11} ; b) R_{22} ; c) R_{33} ; d) R_{12}

respectively R_{11}, R_{22}, R_{33} and R_{12} when the stationary probe is located on the centerline (Y_0 = 0), when in Fig. 5a to 5d the same quantities are obtained at Y_0 = -0.45δ (i.e. in the low-speed part of the mixing layer). Notice that Fig. 4 and 5 present a dilatation of Y axis compared to τ (or X) one. All the curves look like ellipsoïds with different aspect ratios and inclinations with respect to the Y axis. For the u components (R_{11}), Fig. 4a and 5a show negative correlations when the time delay increases ; for y separations, the correlation coefficient decreases rapidly, reaching negative values when the two probes are symetrically placed with respect to the centerline (Fig. 5a) ; there is a general inclination of the ellipsoïds corresponding to the classical features of this flow. For v components (R_{22}), the results are quite different. First, the inclination is no more sensitive ; the ellipsoïds are nearly aligned with the Y axis. Second, there is a very important periodicity for τ (or X) separations, giving high positive and negative correlation levels. This time periodicity corresponds to the characteristic frequency detected on the v spectra (Fig. 3). For large Y separations, either on the centerline or off this position, the correlation coefficient remains very high, even though the correlation itself can be low as in Fig. 5b. For w component, the contour plots (Fig. 4c and 5c) show an intermediate situation between u and v component (due to the link imposed by continuity). An inclination of the same order than for u component can be observed, but the time periodicity is close to that of v component. For all the three components, a good antisymmetry following y separations is observed when the stationary probe is on the centerline (Fig. 5). For u-v correlations (R_{12}, Fig. 4d and 5d), the situation is more complex, particularly at Y_0 = -0.45δ (Fig. 5d)). Moreover, it seems that the overall feature of R_{22} is close to that of R_{12} : very small inclination and short time periodicity.

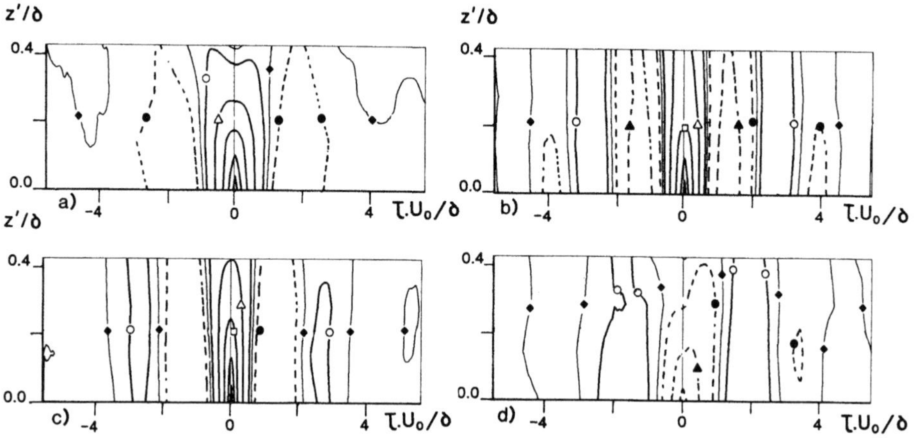

Fig. 6 Iso-correlation coefficients contour plots in τ-Z (X-Z) plane. probes located on the centerline. a) R_{11} ; b) R_{22} ; c) R_{33} ; d) R_{12}

In the τ-Z (or X-Z) plane, Fig. 6 shows the general behaviour already observed in X-Y plane for the periodicity in time. For u component, (Fig. 6a), here also the correlation coefficient R_{11} falls rapidly for large z separation, but remains high for v or w components (Fig. 6b and 6c) ; for the u-v correlation R_{12} (Fig. 6d) the values decrease rapidly when z separation increases.

The observations of space-time correlation leads to conclude that the behaviour of transverse velocity components (particularly v) is quite different than that of streamwise velocity. The integral scales deduced in Y and Z direction is larger for R_{22} than for R_{11} or R_{12} ; in X (from τ) directions, there is a marked periodicity for all correlations involving almost one transverse velocity component.

5. Two-points measurements : spectral domain

The coherence function for the three components are given in fig.7a, 7b and 7c respectively for u (Γ_{11}), v (Γ_{22}) and w (Γ_{33}) components ; all these results are obtained when the stationary probe is on the centerline. The repartitions are

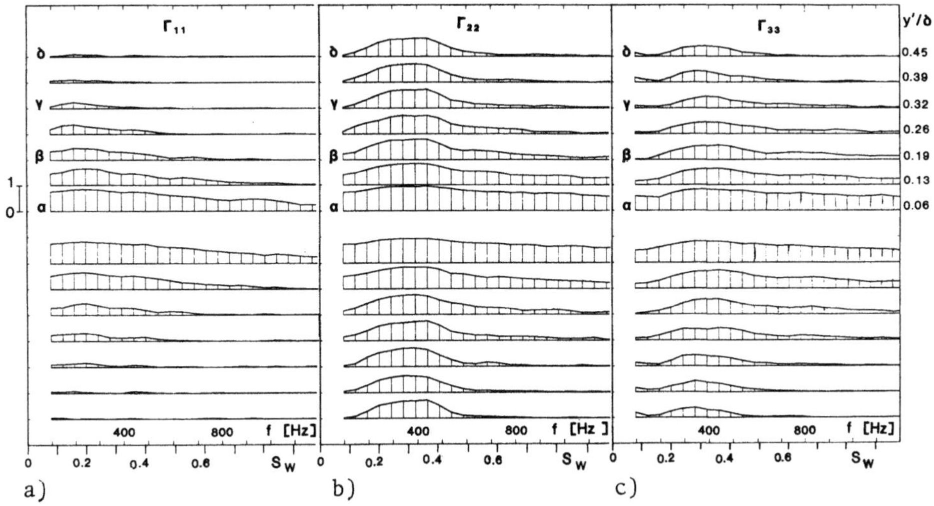

Fig. 7 Coherence function. stationary probe on the centerline (Y_0=0).
a) Γ_{11} ; b) Γ_{22} ; c) Γ_{33}

given either in terms of frequency or in local Strouhal based on vorticity thickness δ_w(S_w= f.d_w/ U_0). The phases of the cospectra are given in Fig. 8 only for 4 separations in Y direction given in Fig. 7.

The evolution of Γ_{11} is characterized by a rapid decrease of the coherence when the Y separation increases ; for y' $\geqslant$ 0.4δ, the coherence is nearly non-significative. At the contrary, for v or w components (Γ_{22} and Γ_{33}, Fig .7b and c), the coherence is largely significative for the largest y separation ; the frequency domain of coherence lies then between 200 and 500 Hz. The characteristic frequency of 400 Hz observed in v or w spectra are no longer evident on the coherence function. The observation of the phase angles shows a rapid evolution for u component (Fig. 8a) ; this evolution of ϕ_{11} shows an increasing slope with respect to frequency when the y separation increases. For v components (ϕ_{22}, Fig. 8b), the phase angle decreases for frequencies below 400 Hz but the values are never less than -35 degrees ; at higher frequencies (lower wavelengths) the v signals are nearly in phase whatever the y separation be. The behaviour of ϕ_{33} is more intricate (Fig. 8c). Here also, the typical frequency of 400 Hz seems to play an important role ; ϕ_{33} is minimum for this frequency when the space (y) separation increases. The interpretation of these different behaviours is not clear. Figure 9 shows the phase angles ϕ^*_{ii} obtained for the frequency corresponding at the maximum of coherence function ; one can observe a slow increase of ϕ^*_{11} (Fig. 9a) when y increases up to the separation y' ≈ 0.3δ ; for higher separations, the coherence is nearly irrelevant, so do the phase angle. When the coherence is significative, the phase ϕ^*_{11} remains low (typically less than 15 degrees). For v component, ϕ^*_{22} (Fig. 9b) is very low (of order of -5 degrees) with negative values. The behaviour of ϕ^*_{33} (Fig.9c) is quite different ;

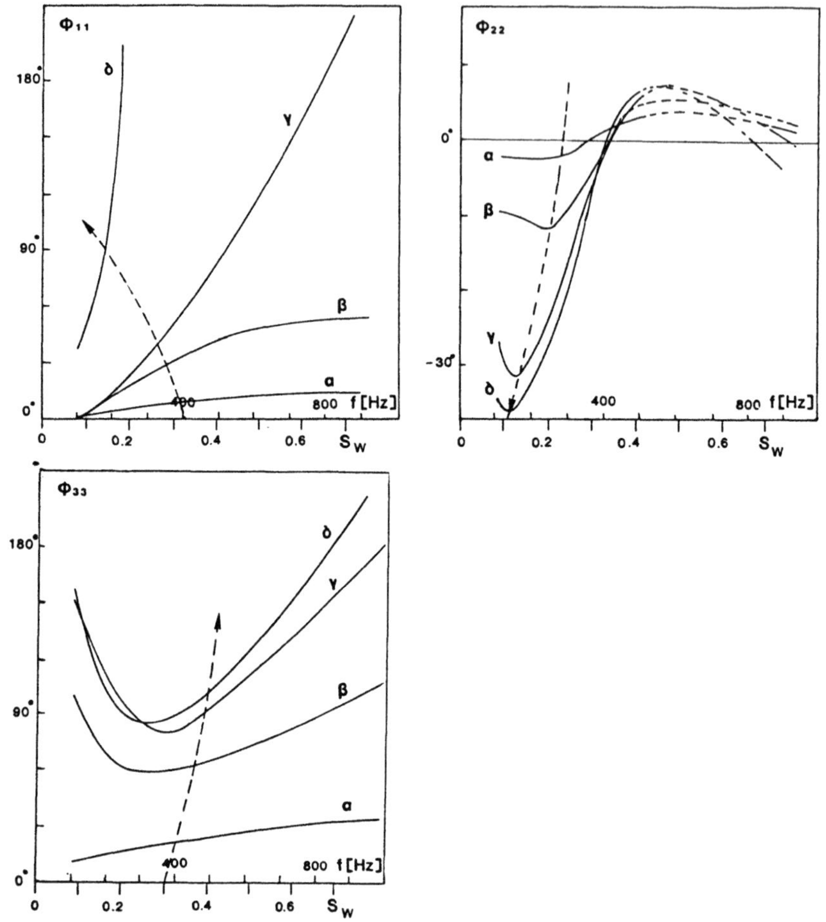

Fig. 8 **Phase angles. stationary probe on the centerline ($Y_0 = 0$). a) $\Phi 11$; b) $\Phi 22$; c) $\Phi 33$ (curves α, β, γ, δ, refers to the Y separations given in Fig. 7)**

one can observe that when the y separation increases, the phase of maximum coherence (always significative) seems to reach an asymptotic value of 90 degrees.

Then, the transverse velocity components w are very coherent and mainly in phase opposition when the stationary probe is on the center line and the second probe placed towards the external part of the layer, when v components are also highly correlated and, in mean, practically in-phase and the u component very weekly coherent.

6. Conclusion

The measurement performed in the asymptotic part of the mixing layer with velocity ratio of 0.6 shows that space correlations measured in the classical statistical way are very intricate. In space-time domain, v and w correlations

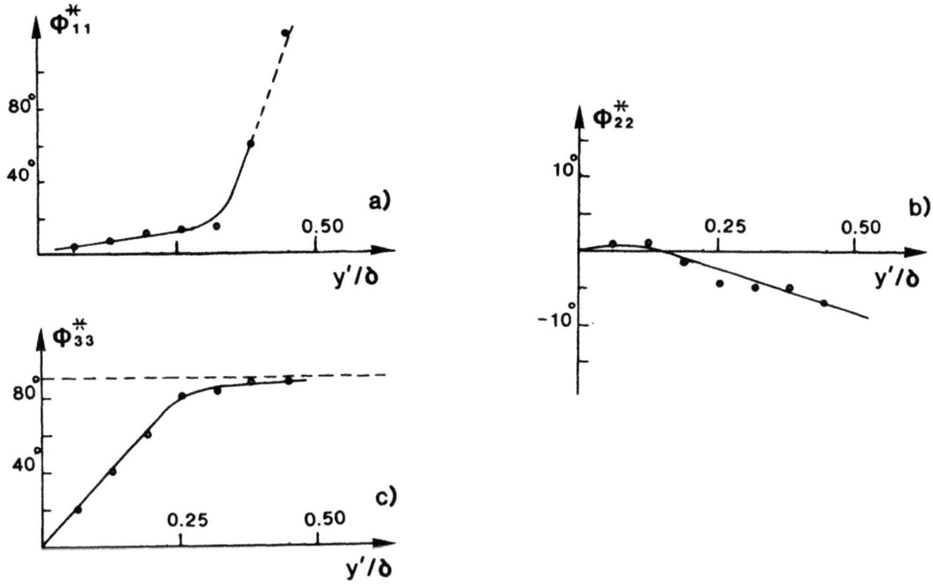

Fig. 9 Phase angles for the frequencies of maximum coherence. stationary probe located on the centerline . a) ϕ^*_{11} ; b) ϕ^*_{22} ; c) ϕ^*_{33}

behave in a quite different manner than streamwise component does. The integral scales are quite different ; v components are strongly correlated on larger distances, with well structured features. The general picture of the iso-correlation patterns are so different following the concerned velocity component that direct interpretation of these results should be hazardous. One can however conclude that only a very high degree of organization of the turbulent structures, as already observed by several authors, can explain theses statistical results. For any detailed interpretation of these measurement, the orthogonal decomposition proposed by Lumley should be the right way ; for this purpose, it appears that the spatial extend of the separations are not sufficient in the present study, particularly in z direction where no periodicity can be assumed as in the work in axisymmetric mixing layers of GLAUSER and GEORGE [4]. In the spectral domain, the results obtained complete the overal description of the flow, but the data are more complicated to interpret ; one can deduce that v components evolves practically in phase in a large part of the layer when w component are in phase opposition. As in time domain, the different velocity components seem to behave in quite different ways, confirming a strong three-dimensional structuration of the turbulent field.

Aknowledgements

This work was financially supported by the Direction de Recherches Etudes et Techniques under Grant number 84/057-1. M. Henry Garem is gratefully aknowledged for his usefull participation.

References

1. M.L. Grant : J. Fluid Mech. 4, 149-190 (1958)
2. A.A. Townsend : The Structure of Turbulent Shear Flows (Cambridge University Press 1956)
3. J. Lumley : Coherent Structure in Turbulence in "Transition and Turbulence" R.E. Meyer Edit. Academic Press (1981)
4. M.N. Glauser, W.K. George : European Turbulence Conference, Lyon, July 86, P. 165-166
5. W. Rodi : a Review of Experimental Data of Uniform Density Free Turbulent Boundary Layers in "Studies in Convection" B.E. Launder Edit. Academic Press (1975)
6. F.K. Browand, C.M. Ho : Journal de Méc. Théorique et Appliquée, Special Issue, 99-120 (1983)
7. I.S.F. Jones : J. Fluid Mech. 74, part 2 , 269-295 (1975)
8. M. Bouriot, J. Delville, H. Garem : Rapport Final, Contrat D.R.E.T. 82/169-1 (1983)

Part VII

Experimental Techniques: Hot-Wire Anemometry, Vorticity Meters, Electrochemical Methods, Image Analysis

A Critical Analysis of Transverse Vorticity Measurements in a Large Plane Shear Layer

J.F. Foss, S.K. Ali, and R.C. Haw

Department of Mechanical Engineering, Michigan State University, East Lansing, MI 48824, USA

1. Introduction

A four-wire array of (hot-wire) probes, configured to measure the transverse vorticity (ω_z) in a turbulent shear flow ($\bar{u} = \bar{u}(x,y)$) has been described in a series of publications: FOSS [1,2,3]; FOSS, et.al. [4] provides the most recent description of the probe, the computing algorithms, and representative results. Considering the conceptual and the computational importance of vorticity, the relative paucity of methods for its direct measurement is considered to represent the difficulty of the task and not the inherent interest in the results of such measurements. Companion papers in the present volume include a presentation of vorticity measurements in MHD flows: Tsinober, et.al., and some quite interesting results of vorticity/velocity measurements in a turbulent boundary layer by Balint, et.al. The latter continues the efforts by Wallace and co-workers to utilize a 9 wire probe of exceptionally small dimensions. A general review of vorticity measurements is provided by WALLACE [5].

The present communication considers four basic attributes of the four-wire vorticity probe:

- i) the influence of a transverse velocity component (w) on the pitch angle measurement (γ_{xy}) provided by the X-array,
- ii) the influence of an instantaneous spatial gradient in the pitch angle ($\partial\gamma/\partial z$) on the computation of ω_z,
- iii) the influence of the uncertainties in the magnitude of ($\partial u_n/\partial s$) in the computation of ω_z where "n" is the normal and "s" is the streamwise quasi-instantaneous coordinate pair as described in Section 2, and
- iv) the influence of the spatial dimension of the micro-circulation domain (ca $1 mm^2$) and the evaluation of ω_z.

The objective of the present communication is to assess the role of these four attributes in the measurement of the transverse vorticity and to establish the bases for an improved technique.

Detailed considerations of the four attributes are provided in Section 3 following a description of the probe and computing algorithm in Section 2. Characteristic results, that demonstrate the technique, are presented in FOSS, et.al. [4].

2. Basic Operating Principle of the Four-Wire Transverse Vorticity Probe

Consider a flow field whose time-mean properties permit the definition of a streamwise (x) and a lateral (y) direction. The non-zero time average velocity components are then $\bar{u}(x,y)$, $\bar{v}(x,y)$. Instantaneously, streamwise (s) and normal (n), coordinates can be defined in the xy plane and the transverse vorticity component (ω_z) at a point (x,y) can be approximated by evaluating the circulation around a small domain: $\delta x \delta y \approx \delta s \delta n = \delta A$, in this plane. The measurement strategy adopts this approach and utilizes the nomenclature

$$\langle \omega_z \rangle = \left\langle \frac{\partial u_n}{\partial s} \right\rangle - \left\langle \frac{\partial u_s}{\partial n} \right\rangle = \delta \Gamma / \delta A \tag{1}$$

where $\langle\ \rangle$ represents a spatial average over the area δA.
The four-wire array is shown in Fig. 1 (from FOSS [1]) and a schematic representation for the area (δA) of the micro-circulation domain is shown in Fig. 2 (from FOSS et al. [4]).

The measurement principle is summarized as:

i) the X-array is used to evaluate $\gamma (\equiv \tan^{-1} v/u)$ at $z + \delta z$.

ii) it is assumed that $\gamma(z) \simeq \gamma(z + \delta z)$ at each instant in time.

iii) a convective segment (Δs) is defined from the four hot-wire measurements at t_j and t_{j-1} as

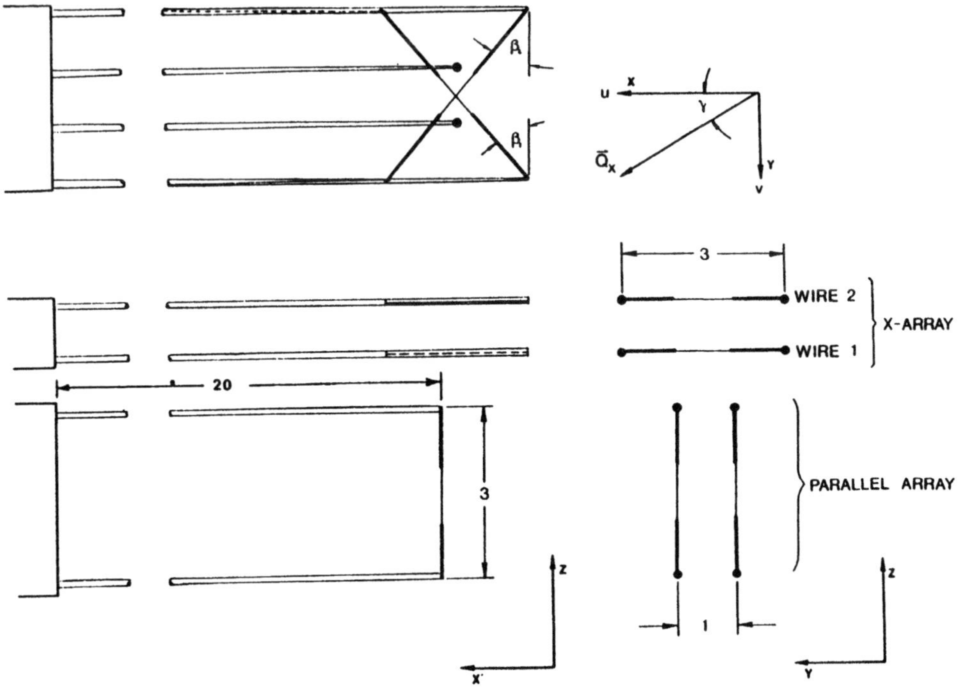

Fig. 1 The transverse vorticity probe (Foss[2])

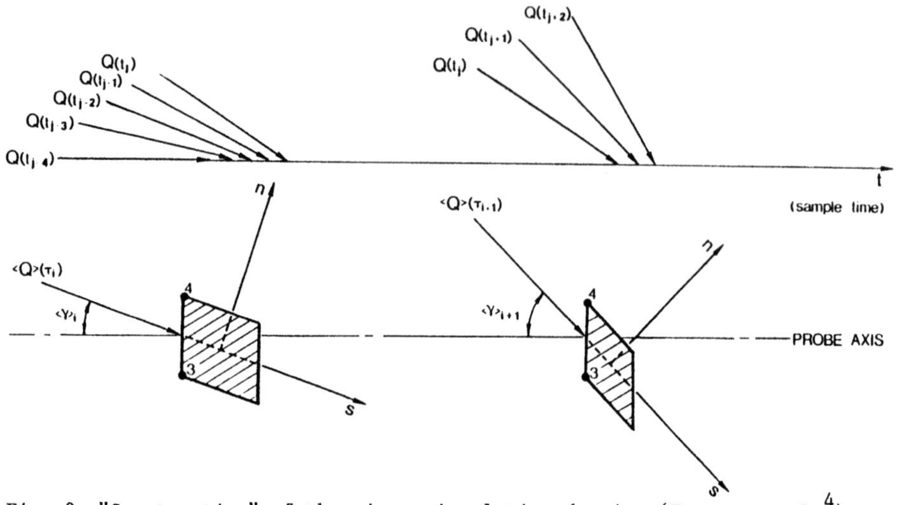

Fig. 2 "Construction" of the micro-circulation domains (Foss, et. al.[4])

$$\Delta s = u_c \ \delta t_j \ \cos\left[\gamma'(t_j) - \langle \gamma(\tau_k) \rangle\right] \tag{2}$$

$$u_c = 0.5 \left\{\frac{Q_3(t_{j-1}) + Q_4(t_{j-1})}{2} + \frac{Q_3(t_j) + Q_4(t_j)}{2}\right\} \tag{3}$$

$$\gamma'(t_j) = \frac{\gamma(t_j) + \gamma(t_{j-1})}{2} \tag{4}$$

where Q is the xy plane velocity magnitude and $\langle\gamma(\tau)\rangle$ is the cumulative average value of γ since the last microcirculation domain was established.

iv) the accumulated length: $\delta s = \Sigma \ \Delta s$, is checked (for each time step) against the projected width δn:

$$\delta n = \Delta \cos \langle\gamma(\tau)\rangle \tag{5}$$

where Δ is the width between the parallel wires. (Note, $\Delta \simeq$ 1mm). The algorithm selects the final time value: $t_j = \tau_k$, such that

$$\delta n \simeq \delta s \tag{6}$$

Hence, the micro-circulation domain size can be characterized as

$$0.5 \ \Delta^2 \leq \delta s \delta n \leq \Delta^2 \approx 1\text{mm}^2 \tag{7}$$

where $\gamma_{max} \simeq \pi/4$ and hence $\cos^2 \gamma \gtrsim 0.5$.

The circulation ($\delta\Gamma$) around the micro-domain can be retrospectively computed once $\langle\gamma\rangle$ is known. The parallel array wires (3,4) provide the δs segments of the contour integral as

$$\pm\int_s^{s+\delta s} Q_i \cos[\gamma - \langle(\gamma\tau_k)\rangle] \ ds \simeq$$

$$\pm\sum_j \frac{Q_i(t_j) + Q_i(t_{j-1})}{2} \cos\left\{0.5\left[\gamma(t_j) + \gamma(t_{j-1})\right] - \langle \gamma(\tau_k) \rangle\right\} \Delta s(t_j) \tag{8}$$

where the positive value is associated with "i=3" and the negative value with "i=4". The ends of the micro-circulation domain provide the values $u_n(s + \delta s)$ and $u_n(s)$ to complete the circulation integral, specifically

$$\int_n^{n+\delta n} u_n(s+\delta s)\, dn - \int_n^{n+\delta n} u(s)\, dn \simeq u_n(s+\delta s)\, \delta n - u_n(s)\delta n. \qquad (9)$$

Note that the sum of equations (8) and (9) define the value of $\langle \omega_z \rangle$ given in (1).

3. Four Attributes of the Measurement Procedure--A Critical Evaluation

3.1 Effects of the transverse velocity (w) on the X-array response

3.1.1 Problem statement

The two voltages of the X-array are used to evaluate the velocity magnitude (Q) and pitch angle (γ) in the x-y plane for an instant of time. However, the true velocity is a three-dimensional quantity and two polar angles would be required to specify its direction. Alternatively, one can consider the present (Q,γ) values to be supplemented by a transverse velocity: w. This latter component, whose presence increases the convective heat loss from both wires, causes the calculated two dimensional Q to be too large and the angle to be too small. The latter effect arises because the wire with the smaller normal component (the one whose $|\beta - \gamma|$ value is the larger) will be most strongly biased by the presence of w and its excessive E value will dominate the γ calculation.

3.1.2 Experimental Investigation

Two X-arrays were mounted on the vorticity probe support shaft in an opposed configuration. Namely, one x-array was used to compute γ_{xy} and the second was used to compute γ_{xz}. The four hot-wire voltages were then iteratively corrected for their respective transverse velocity effects: E_1, E_2 were corrected for w and E_3, E_4 were corrected for v, until convergence. The error term:

$$\Delta\gamma = \gamma_{x,y}(\text{final}) - \gamma_{x,y}(\text{initial}) \qquad (10)$$

was then computed and a histogram of $\Delta\gamma$ values was developed.

3.1.3 Results and Conclusions

Figure 3 presents a histogram of approximately 1500 independent samples of $\Delta\gamma$ for the physical location: x=200 cm = 320 $\theta(0)$ and y = 12 cm = 19.2 $\theta(0)$ where $(y - y_{\frac{1}{2}})/(x - x_0) = 0.036$, and $\bar{u}/U_o = 0.495$. As evident from the data which summarize the histogram: mean value = 0.34 and standard deviation = 2.7 degrees, the effect of the transverse velocity is not strongly significant for this measurement location. Although general conclusions cannot be drawn from one experiment (i.e., this test would have to be repeated for each measurement location for each

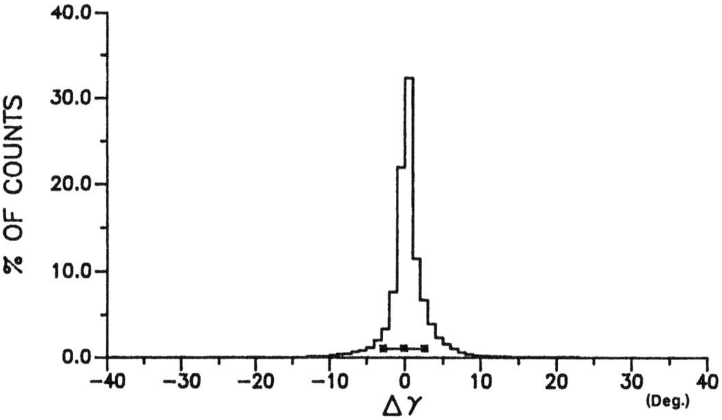

Fig. 3 Influence of transverse velocity w on the inferred γ value $\{\Delta\gamma \equiv \Gamma_{corr} - \gamma_{meas}\}$

subject flow field) it is inferred that this contribution to the ω_z uncertainty is less important than the other attributes considered below.

3.2 Spatial Gradient of the Pitch Angle: $\partial\gamma/\partial z$.

3.2.1 Problem Statement

An inherent feature of the present probe geometry is the transverse displacement between the X and parallel arrays; the physical dimension is typically of order 3.5 mm for the center-to-center distance. It is then assumed, for the ω_z evaluation, that $\partial\gamma/\partial z \simeq 0$ at the instant of the voltage samples. This error is clearly flow field dependent (as is the effect of w in 3.1) and the free shear layer is again recognized to be a relatively benign test flow field.

3.2.2 Experimental Program and Results

Two parallel X-arrays were used to investigate this effect; their geometry can be readily visualized by replacing wires 3 and 4 (Fig. 1) with an xy plane X array. This configuration allows the value of $\delta\gamma$:

$$\delta\gamma = \gamma(z + \Delta z) - \gamma(z) \tag{11}$$

to be evaluated for each sample time. A representative histogram of such values is presented in Figure 4; these data are from the location identified in 3.1.3 and the mean value is -0.33 and the standard deviation is 8.97 degrees.

3.2.3 Conclusions

The mean value of the $\delta\gamma$ population is satisfyingly small; the standard deviation gives definite cause for concern. As suggested by this latter value, there is a significant probability of experiencing a relatively large error in the $\gamma(z)$ value from using the simultaneously recorded $\gamma(z + \Delta z)$ value.

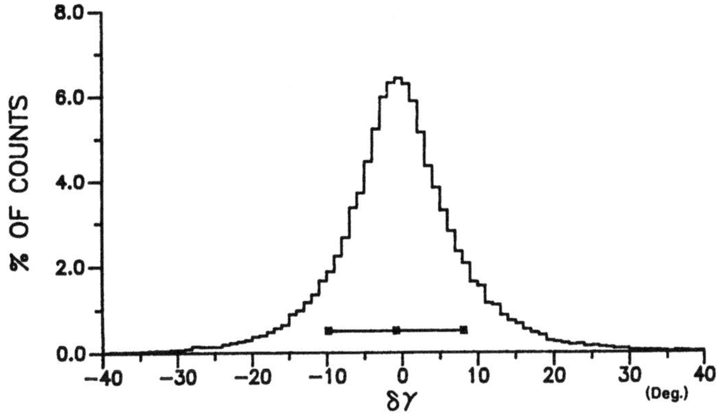

Fig. 4 Influence of spatial gradient ($\partial\gamma/\partial z$) on the inferred γ value $\{\delta\gamma \equiv \gamma(z+\Delta z) - \gamma(z)\}$

It is concluded that a modified probe geometry that would cause the slant wires to "straddle" the parallel is in order. Such a probe is currently under development at MSU.

3.3 Influence of $\partial u_n/\partial s$ on the ω_z calculation

3.3.1 Problem Statement

Several factors combine to render the partial derivative: $\partial u_n/\partial s$, as a significant source of uncertainty in the $\langle\omega_z\rangle$ calculation. Namely,

i) this quantity is dependent upon the $\partial\gamma/\partial z \approx 0$ assumption (Section 3.2),

ii) the use of the quasi-instantaneous Taylor hypothesis can be criticized: LANG and DIMOTAKIS [6] and LANG [7], albeit the criticism can also be criticized; see Appendix A.

iii) single values: $u_n(s + \delta s)$ and $u_n(s)$, must be used in equation (9) unlike the stepwise (Δs) contributions that are possible for the equivalent of the ($\partial u_s/\partial n$) term; see equa (8).

Although the accuracy of the $\partial u_n/\partial s$ term could not be assessed, it was possible to assess its influence on the computed ω_z value; specifically, the two segments of the contour integral were expressed as partial derivatives: $\partial u_n/\partial s$ and $\partial u_s/\partial n$, and a joint histogram of these values, was used to characterize the relative importance of the less certain derivative.

3.3.2 Results and Conclusions

The four-wire data from the physical location of Section 3.1.3 was utilized to create the joint histogram presented in Fig. 5. It is apparent that the two partial derivatives are essentially uncorrelated and that the level surfaces form approximate circles. The $\partial u_n/\partial s$ values are thus observed to make a significant contribution to the $\langle\omega_z\rangle$ values.

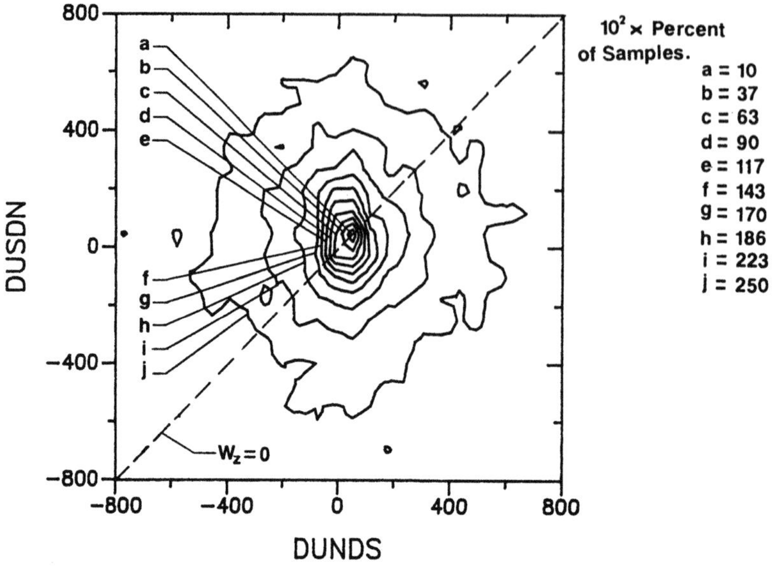

Fig. 5 Joint histogram of the spatial derivatives that form ω_z $\{= \partial u_n/\partial s - \partial u_s/\partial n\}$

The line which represents $\omega_z = 0$ is shown as a dashed segment on the figure. Note that the viscous shear is a maximum along this line (for a given radius from the origin of the joint histogram) and that the $\omega_z = 0$ values at large radii are interpreted as the zero crossings in the highly fluctuating time series: $\langle\omega_z(\tau)\rangle$, as shown in FOSS et.al. [4].

It is inferred that large ω_z values are relatively little effected by $\partial u_n/\partial s$ uncertainties; however, the desired zero-identification for the "intermittency decision" may be strongly influenced by the uncertainties in this derivative. The small ω_z values require additional study for the probe to gain its desired operating status.

3.4 Spatial Dimension of the Micro-circulation Domain

3.4.1 Problem Statement

The vorticity (like the velocity) is a point function and all attempts to measure it with discrete velocity measurements will inherently lead to a low pass filter effect wherein some smoothing of the true $\omega_z(t)$ distribution is unavoidable. Clearly, this effect is flow field dependent and the present observations are offered as representative data.

The spatial low pass filtering effect can be assessed by defining measures of the small scale velocity fluctuations and by comparing these with the observed vorticity values. Since wire 3 can be used to provide a time-series measure of the dissipative scales as

$$\frac{\partial u}{\partial x} \sim \frac{\partial u_s}{\partial s} = -\frac{1}{\langle u_{conv}\rangle}\,\frac{Q_3(t+\delta t) - Q_3(t-\delta t)}{2\delta t}\,, \qquad (12)$$

and since these values can be compared with the vorticity values, it is possible to provide a direct evaluation of this low-pass filtering effect.

A histogram of the $\partial u_s/\partial s$ data gave evidence that the discretized voltage samples: 1LSB = 2.5×10^{-3} volts and $\delta t = 20 \times 10^{-6}$ sec., resulted in a false discretization of the $(\partial u_s/\partial s)$ values. That is, no entries were recorded between the large number of zero values (for $\partial u_s/\partial s$) and values of approximately 100. Since the larger $\partial u_s/\partial s$ values were not strongly influenced by these problems, and since the central issue is to compare the largest entries of the two populations, the ratio (R) of: i) the percentage of $\partial u_s/\partial s$ magnitudes to ii) the percentage of ω_z magnitudes that were larger than 100 and 300 sec^{-1}, were evaluated. That is, R is defined as:

$$R \equiv \frac{\text{percentage of samples: } |[\partial u_s/\partial s]| > K \times 10^2}{\text{percentage of sample: } |\omega_z| > K \times 10^2} \tag{13}$$

where K = 1,3 for the present study.

3.4.2 Results and Discussion

The measure, provided by R of (10), shows values which suggest that the statistical data are not strongly influenced by the 1 mm dimension for the micro-domain. Namely, a nominal R value of 0.7 was obtained for the location defined in Section 3.1 and for K = 1 and 3.

It is concluded that some measure such as R should be included with every processed data set. It is not feasible to alter the micro-domain size in an experiment, but its influence on the ω_z result should be assessed at representative locations.

4. Conclusions

The following conclusions are based upon the observed results for the flow in the present large (single stream) shear layer. It is to be noted that this is a relatively benign flow field and that more hostile environments may further limit the usefulness of the present probe.

a.) The transverse velocity (w) effect on the X-array does not appear to represent a serious error source for the present measurement technique.

b.) The spatial separation between the X and parallel arrays provides a significant uncertainty in the γ value at the parallel array. A modified probe, wherein the slant wires of the X-array straddle the wires of the parallel array is recommended (and is under development at the time of this writing).

c.) Large ω_z values are not too vulnerable to $(\partial u_n/\partial s)$ uncertainties; ω_z values near zero (such as those required for intermittency discrimination) are susceptible to such uncertainties and more sophisticated data analysis techniques are warranted for such evaluations.

d.) The effect of the micro-circulation domain size has been considered using a stochastic ratio: R (see equa (13)). Additional measures of this effect may be appropriate and such measures should always accompany reported $\langle\omega_z\rangle$ results.

5. Acknowledgements

The support of the Alexander von Humboldt Foundation (for the senior author) during a portion of this investigation and the grant which made it possible: NAG-3-671, from the NASA Lewis Research Center, Dr. P. Durbin, grant monitor, are gratefully acknowledged.

6. References

1. Foss, J.F., "Accuracy and uncertainty of transverse vorticity measurements in a free shear layer," EB4 Bull. Am. Phys. Soc. Ser. II, 21:10 (November 1976).
2. Foss, J.F., "Transverse vorticity measurements" (an invited paper) Dynamic Flow Conference '78, The Disa Corp. Skovlunde Denmark, 1979.
3. Foss, J.F., "Advanced Techniques for Transverse Vorticity Measurements," 1981, 7th Biennial Symp. on Turb., Rola, MO, Sept.
4. Foss, J.F., Klewicki, C.L., Disimile, P.J., "Transverse Vorticity Measurements Using An Array of Four Hot-Wire Probes," NASA C.R. 178098, Langley Res. Center, May 1986.
5. Wallace, J.M., "Methods of Measuring Vorticity in Turbulent Flows," Exp. in Fluids, Vol. 4, pp. 61-71, 1986.
6. Lang, D.B. and Dimotakis, P., "Measuring Vorticity Using the Laser Doppler Velocimeter," Bull. Am. Phys. Soc., Vol. 27, p. 1166, 1984.
7. Lang, D.B., "Laser Doppler Velocity and Vorticity Measurements in Turbulent Shear Layers," Ph.D. Thesis, California Institute of Technology, 1985.

7. Appendix A

Comments on the Lang [7] Evaluation of $\partial v/\partial x$ and the Four-Spot LDA Transverse Vorticity Probe Technique.

The four spot vorticity measurement technique that has been developed at C.I.T. (Refs 6,7) offers two distinct advantges: i) the four measurements are made in one plane and ii) the LDA technique provides a direct evaluation of the streamwise (u) and transverse (v) velocity components. The technique is also inherently limited since simultaneous samples from the four spots are statistical rarities and a smoothing procedure is (must be) applied to the individual time series. The smoothing (i.e., the low pass filtering operation) is fully described by Lang [7] including a second "smoothing" of the ω_z pdf. Clearly, in the limit, the low pass operations will recover the $\partial\bar{u}/\partial y$ value and hence the filtered data, at a minimum, faithfully represent some attributes of the ensemble of ω_z values. A comparison of the histograms of [4] and [7] suggests that the smoothing (filtering) operations may remove some of the largest members of the ω_z ensemble. It is inferred that comparative studies, using the two probes in similar flow fields, may provide insight into the attributes of both methods.

Reference 7 includes an evaluation of various forms of the Taylor Hypothesis to recover $\partial v/\partial x$ from the v(t) record. The smoothed data are used to provide a true value for $\partial v/\partial x$ (within the limits established by the smoothing procedure) and this value is compared to $\{[-1/u_c]\,[\partial v/\partial t]\}$ for each step in the time series; different selections for u_c were evaluated. Since only $v(x+\delta x)$ and $v(x-\delta x)$ are available from the LDV measurements, Lang utilized a "convected frame" approximation to define v(x) in terms of the separately measured (and smoothed) values: $v(x\pm\delta x)$. The time transport: $\tau=h/2\bar{u}$, was used for the "space-time" transformation. It is inferred that the evaluation of the Taylor hypothesis will be influenced by this transformation and that a definitive evaluation of the hypothesis is yet to be accomplished. It also should be noted that the present (MSU) technique utilizes the local streamwise: "s" vs the laboratory coordinate: "x", direction; this Taylor Hypothesis cannot be tested in general; only those instants for which $s \simeq x$ are subject to the indirect evaluation scheme of [7].

A Study of the Vortical Structure of the Turbulent Boundary Layer

J.-L. Balint[1], *P. Vukoslavĉević*[2], *and J.M. Wallace*[1]

[1]University of Maryland, Mechanical Engineering Department, College Park, MD 20742, USA

[2]University of Titograd, YU-81000 Titograd, Yugoslavia

1. Introduction

One of the defining characteristics of turbulence is its rotational nature, but the determination of vorticity, the measurement of rotation, has been an elusive goal for experimental researchers. The severe requirements on spatial resolution and accuracy in estimating velocity differences make this type of measurement extremely difficult in typical turbulent flows. Innovations in hot-wire anemometry techniques, the appearance of new electronic circuitry and versatile mini-computers, the use of laser-doppler velocimetry and various optical methods have resulted in recent partially successful attempts to measure the vorticity field characteristics [1].

Knowledge of vorticity dynamics is of great importance in the understanding of turbulent shear flows. In his review of trends in experimental turbulence research, Laufer [2] stated that, "vorticity measurements suggest themselves as the most promising method for quantitative study of the ordered (turbulent) motion." Willmarth [3], in a survey of turbulent boundary layer research pointed out that, "turbulence structure cannot easily be understood without the concept of vorticity." Many experimental investigations have shown that the vortical structures observed in turbulent shear flows are strongly related to the Reynolds shear stress production. They are responsible for most of the energy transfer between scales of motion, and they play an important role in the transport mechanisms of the fluid properties. Several physical models for the dynamics of such vortical structures have been suggested, but a complete and comprehensive theory of the subject is still needed. The experimental measurements of vorticity can provide an adequate kinematic description of the phenomena and also the possibility of critical confirmation of dynamical theories. In this paper we describe a 9-sensor hot-wire probe which is capable of simultaneously providing estimates of all three components of vorticity.

2. Construction and Testing of the Probe

The design and construction of the probe have been initiated at the University of Maryland Turbulence Research Laboratory by Wassman and Wallace [4] and greatly improved over the past few years by Vukoslavĉević, Wallace and Balint [5,6,7].

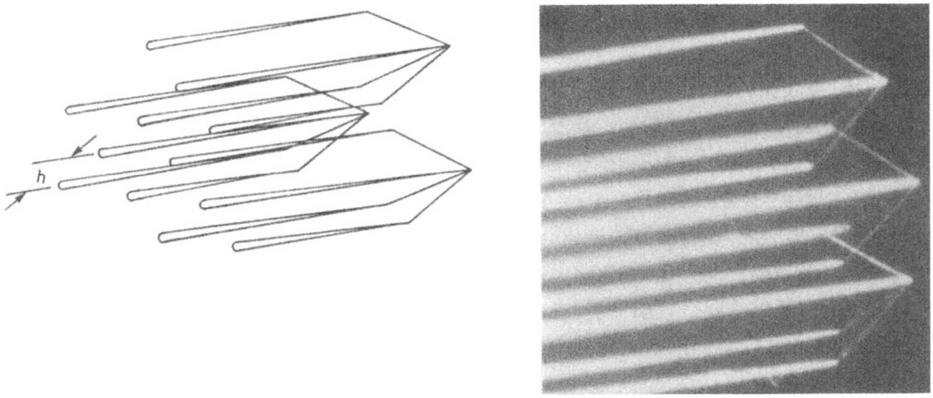

Figure 1. Sketch and photograph of the 9-sensor probe

The probe arrangement, sketched and photographed in Fig. 1, consists of three arrays of 2.5 μm diameter hot-wire sensors. The distance between the supporting prongs is about 0.6 mm and the distance between arrays is about 1.3-1.5 mm, thus having an effective sensing diameter of about 2.5 mm. With this dimension, the probe has a spatial resolution of 9.5 viscous lengths or 9.7 Kolmogorov scales at $y^+=14$ in the boundary layer study described herein. The probe construction was a tedious task because of its size, complex design and fragility. The main problem was to fit twelve supporting prongs in the smallest feasible space and still avoid problems such as electrical shorts, feedback instabilities in the anemometer electronics due to common ground connections, and thermal interference between the wires. To overcome these difficulties, small ceramic tubes have been used to insulate the prongs, the common ground resistances have been increased by nickel plating the common prongs, and the overheat ratio of the anemometer circuits has been lowered to a value of 1.2.

Equation (1) is the nonlinear algebraic form taken by the three nonlinear expressions for the squares of the effective cooling velocities acting at the centers of each of the wires of each array in terms of the three velocity components at the centroid of the arrays, expanded in a Taylor series to first order. The array number is indicated by the subscript i and the wire number is indicated by the subscript j.

$$U^2_{e_{ij}} = [K_{ij1}(U_0 + C_{ij1}\frac{\partial U}{\partial z} + C_{ij2}\frac{\partial U}{\partial y}) + K_{ij2}(v_0 + C_{ij3}\frac{\partial v}{\partial z} + C_{ij4}\frac{\partial v}{\partial y})]^2 + [K_{ij3}(w_0 + C_{ij5}\frac{\partial w}{\partial z} + C_{ij6}\frac{\partial w}{\partial y})]^2 \qquad (1)$$

The K_{ijk}'s are calibration coefficients and the C_{ijk}'s are constant coefficients for a given array geometry which are positive or negative fractions of h (see Fig. 1). U_0, v_0 and w_0 are the unknown velocity components at the centroid of the probe, $\partial U/\partial y$, $\partial U/\partial z$, $\partial v/\partial y$, $\partial v/\partial z$, $\partial w/\partial y$ and $\partial w/\partial z$ are the velocity gradients in the plane transverse to the axis of the probe, and h is the average of the projected wire lengths on this plane. The effective cooling velocities are related to the measured bridge voltages through calibration relations of the King's Law form thus leaving nine unknowns to be found by the solution of these nine equations. This is accomplished by an iterative procedure which is quite convergent. One of the three streamwise gradients can be obtained directly from the continuity equation, and the other two are estimated using Taylor's hypothesis, $\partial/\partial x = -1/\bar{U}\ \partial/\partial t$, in the flow region where it is found to be a good approximation. The adequacy of this approximation is assessed directly with the probe by comparing the value of $\partial U/\partial x$ from continuity with the value found from Taylor's hypothesis.

It was necessary to develop efficient methods to numerically solve such equations, and also to examine the problems of non-uniqueness of solutions and non-convergence of the Newton-Raphson-type solution algorithm. More details about the probe characteristics and the solution method as well as an analysis of the measurement accuracy are to be published elsewhere. Suffice it to say here that the spurious values of vorticity, found in the irrotational calibration and test flow, which are due to measurement error, do not exceed 20% of the rms streamwise vorticity measured close to the wall of the boundary layer at the same mean speed. The error analysis showed that a relative error of about 1% in the velocity measurement can induce such spurious values of vorticity. However, the vorticity statistics have been obtained on fluctuating quantities; they are not affected by mean spurious values.

3. Experimental Results

The measurements have been carried out in a zero pressure gradient turbulent boundary layer. All the components of the instantaneous velocity and vorticity vectors have been determined for positions normal to the wall - in the buffer layer, the logarithmic region and the velocity defect region of the boundary layer. Each point has been sampled at the Nyquist digitizing frequency of 500 Hz during 3 mn.

3.1 Velocity Statistics

The statistics of the average of all three components of velocity have been compiled and compared to previously reported boundary layer data as well as channel flow data in the wall region as an indicator of the probe's performance. The boundary layer studied grows on the lower wall of the 8.5 m long wind tunnel test section and was tripped by a 5 mm diameter wire to initiate and stabilize it. The principal characteristics of the boundary layer investigated are listed in the following table:

U_∞(m/s)	δ(m)	θ(m)	Re_θ	δ^*	H	Re_x	u_τ/U_∞
2.39	0.1397	0.0136	2100	0.0175	1.28	$1.3\mathrm{x}10^6$	0.042

where δ is the boundary layer thickness, θ is the momentum thickness, Re_θ is the Reynolds number based on θ, δ^* is the displacement thickness, $H=\delta^*/\theta$ is the shape factor, and Re_x is the Reynolds number based on the streamwise measurement location x.

The measured mean velocities exhibit the established logarithmic and "wake" regions very well. The y/δ distributions of the turbulence intensities of all three velocity components are shown in Fig. 2. They compare quite well with the boundary layer data of Klebanoff [8] and Gupta and Kaplan [9] ($Re_\delta \simeq 80{,}000$ & 17,700 respectively). These RMS velocity fluctuations have also been normalized with the friction velocity U_τ and plotted against y^+ for the wall region as is also shown in Fig. 2. Here, in addition to comparison with [8] and [9], com-

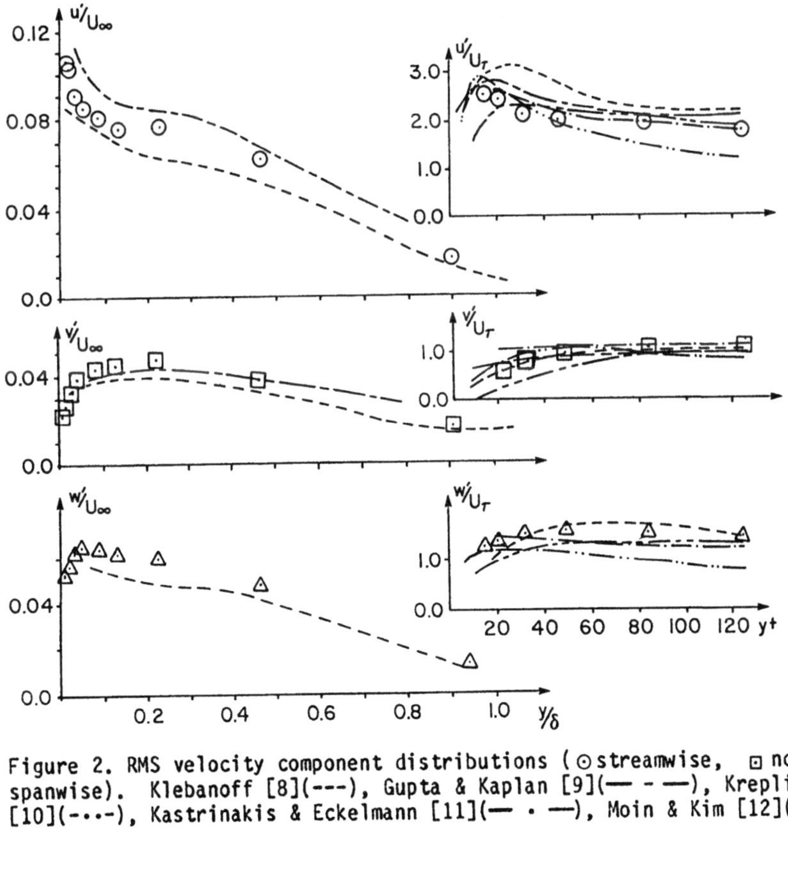

Figure 2. RMS velocity component distributions (⊙ streamwise, ⊡ normal, △ spanwise). Klebanoff [8](---), Gupta & Kaplan [9](— - —), Kreplin & Eckelmann [10](-•••-), Kastrinakis & Eckelmann [11](— • —), Moin & Kim [12](— -- —)

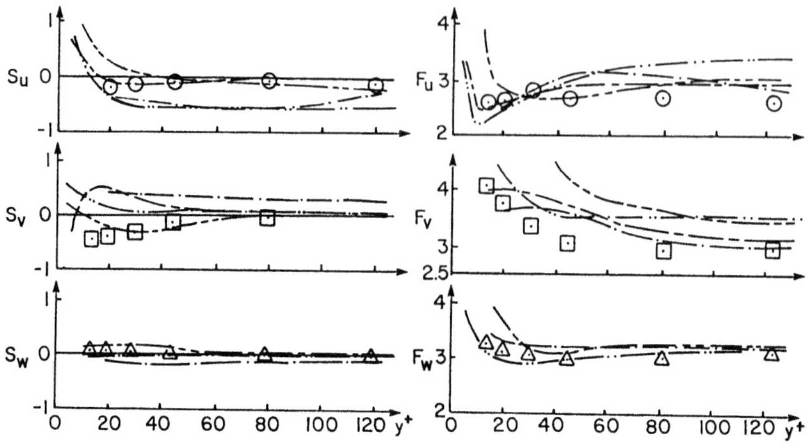

Figure 3. Velocity component skewness and flatness factor distributions (symbols and lines as in Fig. 2)

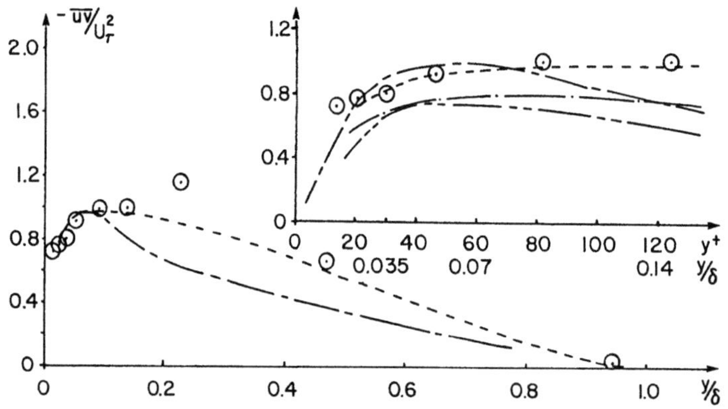

Figure 4. Reynolds stress distribution (lines as in Fig. 2)

parison with the channel flow experimental data of Kreplin and Eckelmann [10] and Kastrinakis and Eckelmann [11] as well as with the large-eddy simulation channel flow data of Moin and Kim [12] is excellent in the region near the wall (Re_b based on the channel half-width, is equal to 3,850; 12,000; 13,800 respectively). Comparison with the rms streamwise vorticity distribution of [11] and the rms distributions of all three vorticity components of [12] will be made in Fig. 5.

The skewness and flatness factors of all three velocity components are shown in Fig. 3. They are compared to the boundary layer data of [9] and the channel flow data of [10],[11] and [12]. The streamwise skewness factors follow the same trends as the comparison data. The skewness values of the normal component differ somewhat from the experiments in [9], [10] and [11]; they have negative values in the region $15 < y^+ < 80$, which also appears in the large-eddy simulation data [12]. The spanwise skewness factors should be zero because the flow symmetry, and the measured values are consistent with this expected result. The flatness fac-

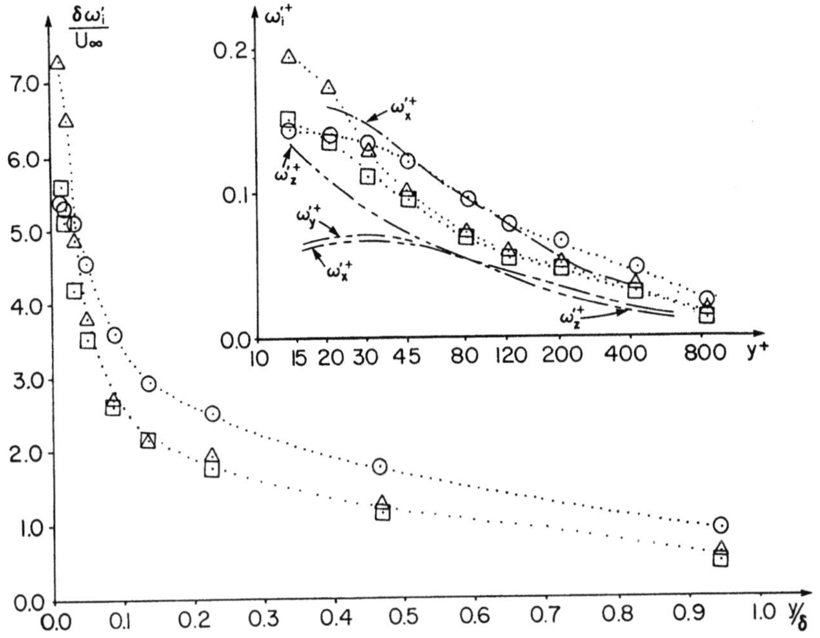

Figure 5. RMS vorticity component distributions (symbols and lines as in Fig. 2, (••••) connect the measured data)

tors of all three components show slightly lower values than the comparison data over most of the range.

The velocity correlations $-\overline{uv}$ normalized by the friction velocity squared U_τ^2 are presented and compared in Fig. 4. Again here, the present results show the same trends and magnitudes, both when compared to the boundary layer data of [8] and [9] for the full range of y/δ and in the wall region when compared to all the channel flow data [10],[11] and [12] as well.

We conclude that the overall comparison with the velocity measurements of other authors show satisfactory agreement and indicate the ability of the probe to adequately measure the velocity field.

3.2 Vorticity Statistics

The RMS values of all three vorticity fluctuation components, ω_i, are shown in Fig. 5 normalized with the outer variables δ and U_∞ and with the wall variables ν and U_τ. Compared in the latter plot are the streamwise component measurements of Kastrinakis and Eckelmann [11] and the large-eddy simulation calculations of all three components by Moin and Kim [12]. Our data agree reasonably well with those of [11], but only show the same trends as [12], being over twice the magnitude throughout the range.

From our data, one can conclude that most of the vorticity fluctuation activity is concentrated near the wall. Beyond a vertical location y/δ of about 0.1, the RMS distribution of all three components decrease monotonically; the normal and spanwise values are virtually equal whereas the streamwise component is slightly higher. In [12] the vorticity fluctuations appear to approach isotropy in the region away from the wall. Close to the wall ($y^+ \approx 14$) the measurements show that the most intensive fluctuations are in the spanwise direction, the streamwise and normal RMS fluctuations being almost equal. But in the region $15 < y^+ < 30$, a dramatic decrease in spanwise fluctuations occurs. The streamwise fluctuations virtually maintain their magnitude in the same range so that at $y^+ \approx 30$ their intensity is as large as the spanwise fluctuation intensity, the normal fluctuations being the lowest. This appears to indicate a transfer of spanwise vorticity into streamwise vorticity in the region $15 < y^+ < 30$, this observation supports the hair-pin vortex model of the wall layer given by Wallace [13] drawing on the composite results of many investigations. Close to the wall, the preferred orientation of the vortex lines is in the spanwise direction; with increasing distance from the wall, the activity in the streamwise direction due to stretching, lifting and reorientation of the vortex lines becomes predominant.

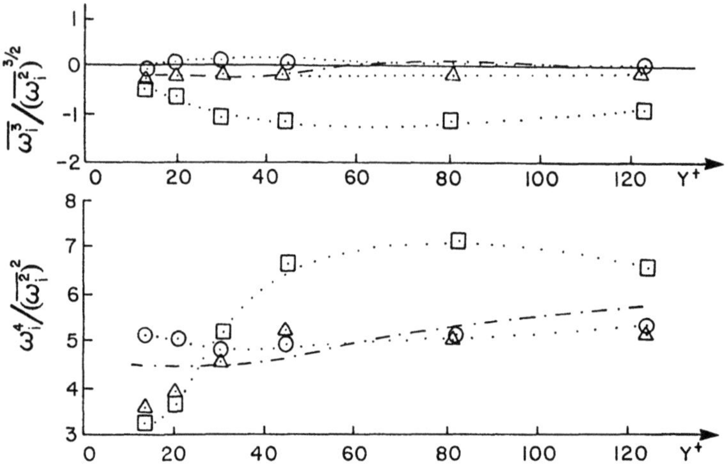

Figure 6. Vorticity component skewness and flatness factor distributions (symbols and lines as in Fig. 2)

The skewness factors of vorticity fluctuations in all three directions are presented in Fig. 6. Because of flow symmetry, the streamwise and normal vorticity fluctuations have an equal probability to be positive or negative. Their skewness values should be zero as the measurements demonstrate quite well. The persistent negative value of the spanwise vorticity skewness factor throughout the extent of the boundary layer shows that, although the most probable value of the fluctuation ω_z is positive, the largest fluctuations are negative values of ω_z, i.e. the instantaneous component Ω_z is more negative than the already negative value of the mean value Ω_z. This probably reflects spanwise vortex stretching activity.

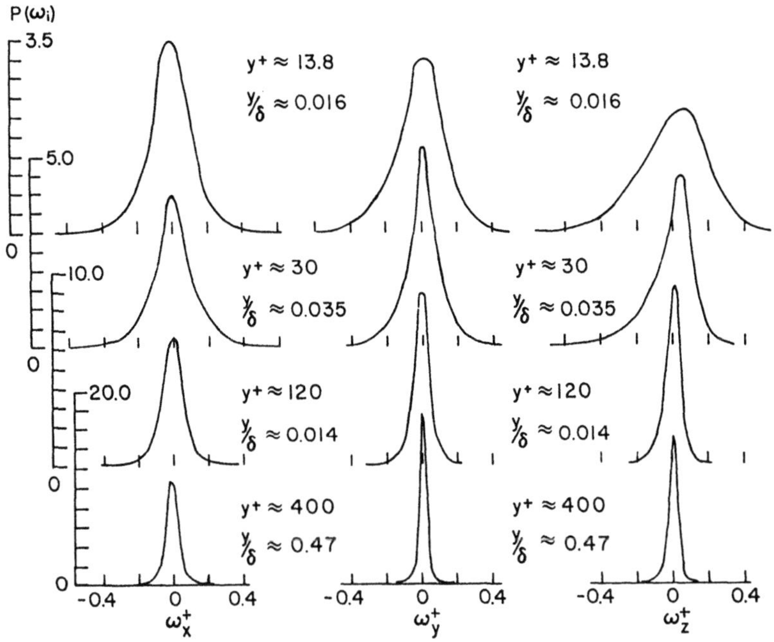

Figure 7. Vorticity component probability density functions

Also shown in the figure are the vorticity component flatness factors. It can be seen that the extreme values of the fluctuating vorticity component ω_z are relatively larger for locations away from the wall, increasing from about 3.2 at $y^+ \approx 14$ to about 7.15 at $y^+ \approx 80$. Both flatness and skewness factors of the streamwise vorticity fluctuations are compared to the channel flow data of [11], and the agreement in the wall region is quite good.

Figure 7 shows the probability density distributions of the three components of dimensionless vorticity, $\omega_i^+ = \omega_i \nu / U_\tau^2$, at four boundary layer positions. The shapes are as expected from the RMS, skewness and flatness results. The spanwise distribution is assymetric, and all the component distributions narrow with distance from the wall.

4. Conclusions

The data presented herein is the first set of experimental measurements of the statistics of all three components of the turbulent vorticity field. The measurements were carried out in a moderate Reynolds number turbulent boundary layer. They reveal some structural features of this flow. Planned further analysis of this data using conditional analysis of the terms of the vorticity transport equation is expected to reveal more details about the vorticity dynamics of the flow.

5. Acknowledgement

This research was supported by NSF Grant MEA 82-14078.

6. References

1. J.M. Wallace: Exp. in Fluids 4, 61 (1986)
2. J. Laufer: Ann. Rev. Fluid Mech. 7, 307 (1975)
3. W.W. Willmarth: In Advances in Applied Mechanics, ed by C.S. Yih, Vol.15, (Academic Press, New York, 1975) p. 159
4. W.W. Wassmann, J.M. Wallace: Bull. Am. Phys. Soc. 24, 1142 (1979)
5. P. Vukoslavcevic, J.M. Wallace: Rev. of Sci. Instr. 52, 869 (1981)
6. J.M. Wallace: Bull. of the Am. Phys. Soc. 24, 1401 (1983)
7. J.-L. Balint: Thèse Docteur D'Etat ès Sciences, Univ. of Lyon (1986)
8. P.S. Klebanoff: NACA Tech. Notes N. 3178 (1954)
9. A.K. Gupta, R. Kaplan: Phys. Fluids 15, 981 (1972)
10. H.P. Kreplin, H. Eckelmann: J. Fluid Mech. 95, 305 (1979)
11. E.G. Kastrinakis, H. Eckelmann: J. Fluid Mech. 137, 165 (1983)
12. P. Moin, J. Kim: J. Fluid Mech. 118, 341 (1982)
13. J.M. Wallace: Lect. Notes in Phys. 235, 253 (1985)

Preliminary Vorticity Measurements by a Method Based on a MHD Principle

A. Tsinober, E. Kit, and M. Teitel

Department of Fluid Mechanics, Faculty of Engineering, Tel-Aviv University, Ramat Aviv, Tel-Aviv 69978, Israel

The idea of the method comes from MHD and is based on the equation div $\mathbf{j} = 0$ and the Ohm's law $\mathbf{j} = \sigma(\nabla\phi + \mathbf{v} \times \mathbf{B})$, where $\mathbf{j}$ - is the electric current, - the potential of the electrical field, $\mathbf{v}$ - velocity, $\mathbf{B}$ - magnetic field intensity. The consequence is $\nabla^2\phi$ = **Brotv** (the term **vrotB** is several orders smaller). This enables one to measure the component of the vorticity ω_B parallel to the magnetic field, by measuring the $\nabla^2\phi$ which in turn could be measured using a 7 electrodes probe realizing a central difference approximation of the Laplacian. Thus, the method is an absolute one and for this reason no calibration procedure is necessary.

To the best of our knowledge this idea was suggested by Grossman et. al. (1958). The first qualitative attempt to implement this method was made by Baker (1971) using a five-electrode probe in a laminar flow.

Facility, Instrumentation, Data Acquisition and Processing

The experiments were performed with salted water (salinity ~ 1%) in a constant head facility with the test section 5cm x 5cm in cross section and 1.2m in length. The use of salted water (low electrical conductivity) ensured that the dynamic interaction of the flow and the magnetic field is entirely negligible since for salted water flow even in very strong magnetic fields the electromagnetic force is several orders of magnitude smaller than the viscous and inertia forces.

The scheme of experimental arrangement of the test section is shown in Fig. 1. We used a grid composed of circular rods of diameter 5mm spaced a distance 10mm between their centers. The turbulence level in the oncoming flow before the grid was reduced below 1% of the mean velocity by a system of small scale screens. Most of the experiments were performed with the mean velocity u = 67cm/sec. A small number of experiments were made for velocities between 30cm/sec and 115cm/sec. The imposed DC magnetic field B = 0.5 tesla.

The induced electrical field was measured by multielectrode probes consisting of two, four and seven electrodes. Specially processed silver-silver chloride electrodes were used. This made it possible to reduce their electrochemical offset potential to the noise level of the electronics used in the experiments (0.3 μV to the input in the frequency range up to 200Hz).

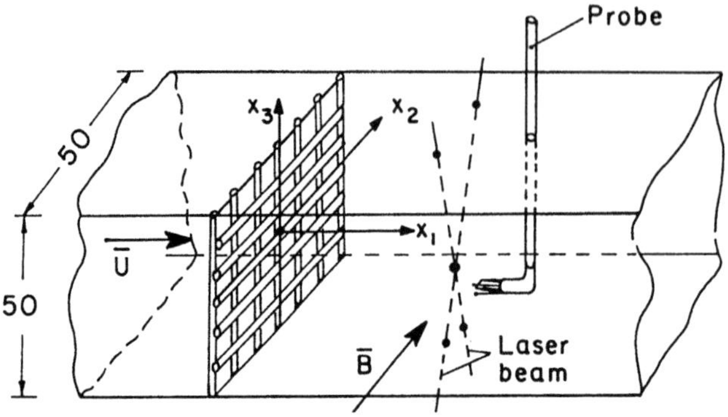

Fig. 1 Scheme of experimental arrangement of the facility test section.

The seven electrode probe was used to produce a central-difference approximation of the Laplacian $\nabla^2\phi$ and in this way allowing to measure the vorticity component parallel to the imposed magnetic field. To minimise the mechanical contamination of the signal the approximating grid was turned 22.5° in the x_1x_3 plane (Fig. 2a). This is possible due to the invariance of Laplace operation to rotation. The distance between the tips of the electrodes was 1mm (the electrode diameter being 0.1mm) for the vorticity probe (seven electrodes, Fig. 2b). All other probes had the distance of 2mm between the tips of the electrodes.

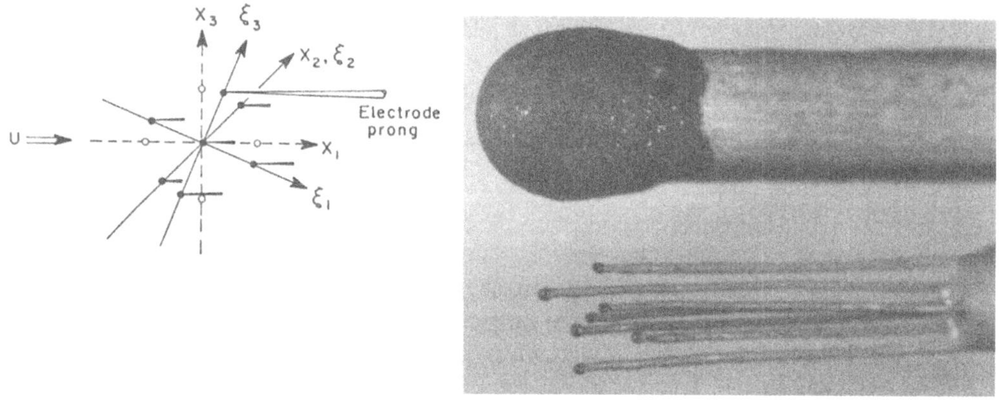

Fig. 2 a) Scheme of the tip of a 7-electrode vorticity probe. The tips of the electrodes in the plane X_1OX_3 are turned 22.5° relative to the OX_1 axis
b) Tip of a probe shown together with the head of a match.

The data (the total up to eight channels) was acquired by a PDP 11/23+ minicomputer and recorded on a magnetic tape. The sampling rate was from 250Hz to 1kHz per channel. The minimal buffer length was 512 points per channel to facilitate the data processing by the fast Fourier transform in obtaining the spectra and correlation functions. The record of almost every experimental point consisted of 50 buffers, so that all the data (intensities, correlations and spectra) were obtained as the result of ensemble averaging over these 50 buffers.

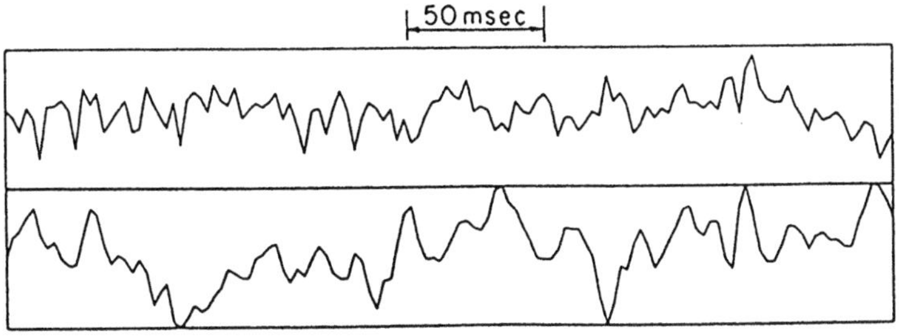

Fig. 3 Comparing the oscillograms of simultaneous signals ω_2 and. u_1 illustrating larger content of high frequencies in the ω_2 signal.

Preliminary measurements of vorticity and related quantities

Oscillograms of simultaneous signals of ω_2 - vorticity component parallel to the magnetic field and the u_1 - velocity component are shown in Fig. 3. It is clearly seen that the ω_2 signal contains much more high frequencies. The same is seen from an example of one dimensional spectra (Fig. 4).

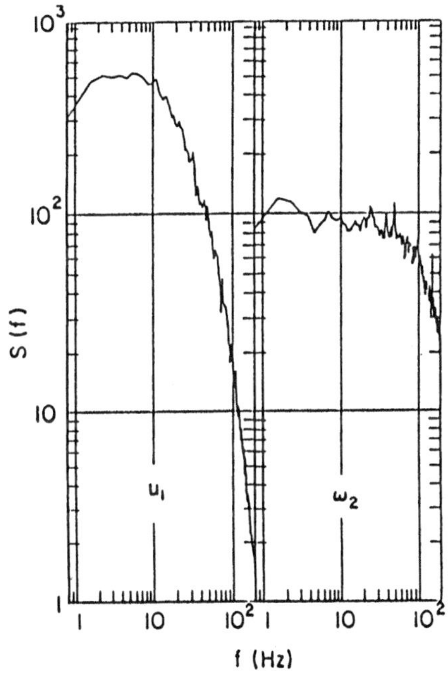

Fig. 4 One dimensional spectra of velocity u_1 and vorticity ω_2.

As a test of the vorticity probe performance the isotropy relations between second derivatives of ϕ in the fame 0 ξ_1 ξ_2 ξ_3 (see Fig. 2a) were checked from the measurements. The results are shown in Table 1 and are quite reasonable bearing in mind that we made a rather crude approximation of the second derivatives. Presented in this Table are also results for the derivatives $\partial^2\phi/\partial_1\partial x_i$ $(i = 1,2,3)$ in which the $\partial/\partial x_1$ was obtained using the Taylor hypothesis and $\partial/\partial x_i$ $(i = 1,3)$ from the relations $\partial\phi/\partial x_1 = (\partial\phi/\partial\xi_1)\cos 22.5° - (\partial\phi/\partial\xi_3)\sin 22.5°$; $\partial\phi/\partial x_3 =$

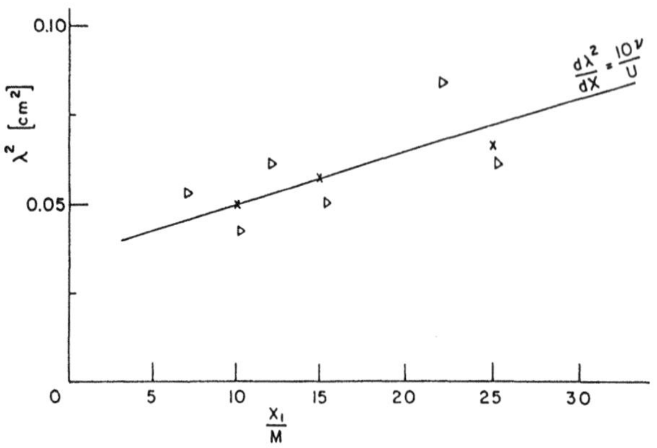

Fig. 5 Variation of dissipation length scale during decay. +- hot film, x- vorticity probe.

$(\partial\phi/\partial\xi_1)\sin 22.5° + (\partial\phi/\partial\xi_3)\cos 22.5°$.Z The results for the dissipation length scale from the relations $\lambda^2 = \overline{u_1^2}/\overline{\partial u/\partial x_1)^2}$ (+) , $\lambda^2 = 5\overline{u_1^2}/\overline{\omega_2^2}$ (x) are shown in Fig. 5 together with a line of theoretical slope $d\lambda^2/dx = 10\nu/U$. Though the scatter is considerably large the tendency seems to be the right one.

Concluding Remarks

The most pressing problem is that the resolution scale of the potential difference probes (~ 1mm) was not small enough in comparison to the grid mesh (5mm). This is especially important when vorticity and related quantities (like $\partial^2\phi/\partial x_i\partial x_j$) are attempted to be measured. For this reason the presented results of vorticity measurements should be considered as preliminary only. However, these measurements seem to show right trends qualitatively as, for example, could be seen from the spectra shapes in high frequency region and the dissipation length scale behaviour. To obtain more reliable results it is necessary to increase the scale separation. (A large scale facility is currently under construction). It is

Table 1. Verifying the isotropy relations for the second derivatives of the potential of the electrical field

$\frac{X}{M}$	$\overline{(\frac{\partial^2\phi}{\partial\xi_1^2})^2}$ / $\overline{(\frac{\partial^2\phi}{\partial\xi_3^2})^2}$	$\overline{(\frac{\partial^2\phi}{\partial\xi_2^2})^2}$ / $\overline{(\frac{\partial^2\phi}{\partial\xi_3^2})^2}$	$\overline{\frac{\partial^2\phi}{\partial\xi_1^2}\frac{\partial^2\phi}{\partial\xi_3^2}}$ / $\overline{(\frac{\partial^2\phi}{\partial\xi_3^2})^2}$	$\overline{\frac{\partial^2\phi}{\partial\xi_1^2}\frac{\partial^2\phi}{\partial\xi_2^2}}$ / $\overline{(\frac{\partial^2\phi}{\partial\xi_3^2})^2}$	$\overline{\frac{\partial^2\phi}{\partial\xi_2^2}\frac{\partial^2\phi}{\partial\xi_3^2}}$ / $\overline{(\frac{\partial^2\phi}{\partial\xi_3^2})^2}$	$\overline{(\frac{\partial^2\phi}{\partial x_1\partial x_2})^2}$ / $\overline{(\frac{\partial^2\phi}{\partial x_1^2})^2}$	$\overline{(\frac{\partial^2\phi}{\partial x_1\partial x_3})^2}$ / $\overline{(\frac{\partial^2\phi}{\partial x_1^2})^2}$
10	0.80	0.360	0.332	0.188	0.223	0.376	0.44
15	0.82	0.351	0.327	0.192	0.214	0.361	0.43
25	0.92	0.324	0.362	0.224	0.222	0.322	0.40
Theoretical Value	1	1/3	1/3	2/9	2/9	2/9	1/3

noteworthy that the main advantage of the presented method of vorticity measurements is that it does not require calibration and it is applicable to any flow (not only isotropic).

References

Baker, R.C. 1971 On the electromagnetic vortex probe. J. Phys. E4, 99-101.

Baker, R.C. 1983 A review of recent developments in electromagnetic flow measurement. In: Progress in Astronautics and Aeronautics, Vol. 84, 225-259.

Grossman, L.M.; Li, H. & Einstein, H. 1957 Turbulence in civil engineering. Investigation of liquid shear flow by electromagnetic induction. Proc. Amer. Soc. Civ. Engrs., J. Hydr. Div. 83, No. 1394, 1-15.

Skin-Friction Measurements in Two- and Three-Dimensional Highly Turbulent Flows with Separation

P. Dengel, H.H. Fernholz, and M. Hess

Hermann-Föttinger-Institut für Thermo- und Fluiddynamik, Technische Universität Berlin, D-1000 Berlin 12, Germany

1. Summary

This paper presents comparative skin-friction measurements obtained by a Preston tube, a surface fence, a single film and a McCROSKEY hot-film probe, and a wall pulsed wire in two- and three-dimensional highly turbulent flows (near wall turbulence level > 40%) with separation and instantaneous reverse flow. Fluctuating values of skin friction were obtained by the wall pulse wire and the hot films in addition to the mean values and the direction of the wall stream line. Differences in the results of the measuring techniques are explained and the failure of devices under these severe conditions is displayed.

2. Introduction

There are many different techniques used to measure skin friction in wall bounded flows (cf. WINTER [1]). However, when the flow is highly turbulent, the time mean value of the skin friction is low or instantaneous reverse flow may occur, the number of valid measuring techniques is greatly reduced. This is due on the one hand to the design of the probes and the physical laws they are based on and on the other to specific features of the flow such as instantaneous reverse flow or the structure of the turbulence in the near wall region.

A particularly versatile instrument for measuring skin friction is the pulsed wall probe (GINDER & BRADBURY [2], EATON & JOHNSON [3], CASTRO & DIANAT [4]). It can measure the mean and fluctuating values of the shear stress τ_w, and also the sense and direction. By reliably determining zero mean-stress it can indicate separation and reattachment lines. Its operation is independent of the turbulence level and structure of the flow being investigated. Within its operating range ($|u_\tau| < 0{,}8$ m/s for the devices tested so far) we have used the wall pulsed wire to check data obtained with the other sensors listed above in three different incompressible air test flows. (Fig. 1):

(1) An axisymmetric turbulent boundary layer with a severe adverse pressure gradient on the verge of separation (DENGEL et al. [5]).

(2) A highly turbulent and disturbed flow over a normal bluff plate with a long splitter plate on the plane of symmetry behind it. The flow separates at the sharp

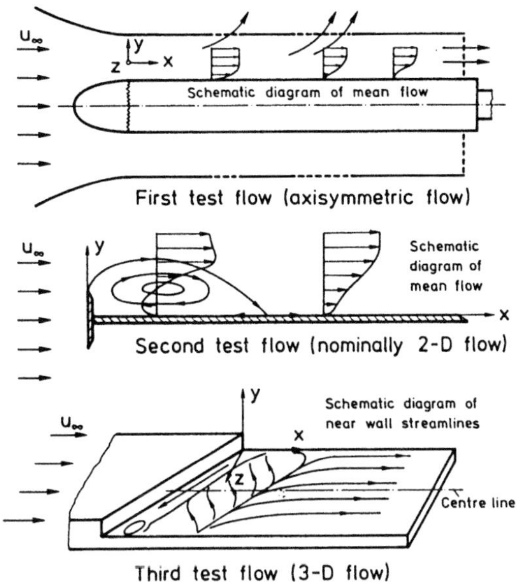

Fig. 1: Schematic diagrams of the test flows.

bevelled edge of the bluff plate, forms a free shear layer above the reverse-flow region, which is bounded on its other side by the splitter plate, and reatteaches to the splitter plate. It then adjusts slowly to the wall boundary conditions downstream (RUDERICH & FERNHOLZ [6]).

(3) A three-dimensional turbulent shear flow downstream of a swept backward facing step (with sweep angle α 34.5° and a ratio of width to step height of 13) generated in the same test section as the second test flow. The oncoming boundary layer upstream of the step was turbulent.

Figure 2 presents the oil-flow pattern in the region of the step (flow from left to right) showing a markedly three-dimensional flow field downstream of the step.

Fig. 2: Oil-flow picture on the floor (Re_h = 2.5 · 10^4).

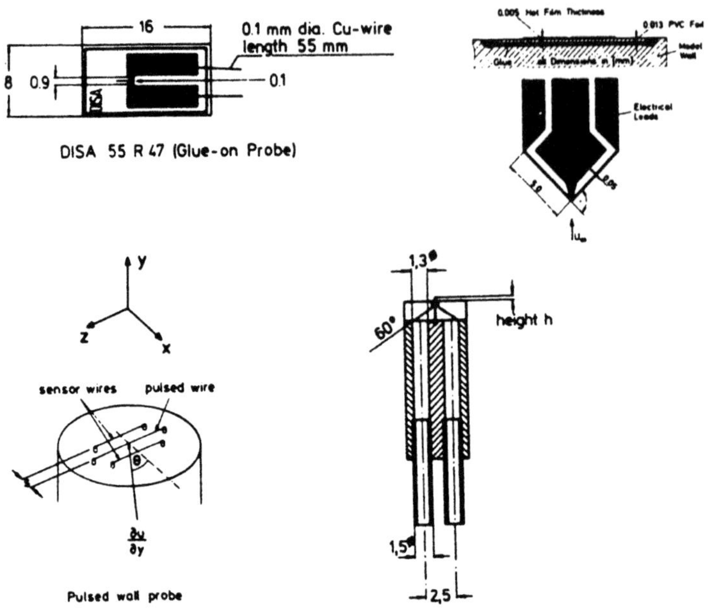

Fig. 3: Skin friction gauges: Single hot film, McCROSKEY hot film, wall pulsed wire, surface fence.

3. Measuring Devices and Calibration Techniques

Four of the five skin friction gauges are shown in Fig. 3. In the axisymmetric flow the probes could be moved continuously along a slot parallel to a generator of the cylinder. Otherwise they were mounted on or in plugs which could be inserted flush with the plane wall (test flows 2 and 3). The Preston tubes (not shown) had outer dia. between 0.9 and 2 mm and were mounted on an electrically driven traverse gear which was introduced into the test section through a slot in the ceiling. Patel's calibration curve was used.

The wall pulsed wire, a modified form of that of Ruderich & Fernholz [6], is shown in Fig. 4. It consists essentially of three fine parallel tungsten wires about 50 μm above the probe surface plate carrying the wire support. CASTRO [7] noted the influence of the wire length on the intensity of the wall shear stress. Therefore the active length of the sensor wires was varied between 0.2 and 2 mm. The calibration measurements were performed in a two-dimensional zero pressure gradient turbulent boundary layer (Fig. 5) and gave a maximum value for the ratio $(\overline{\tau_w'^2})^{1/2}/\,\bar{\tau}_w$ of 0.37 agreeing well with data of ALFREDSON et al. [8]. It was concluded that an active wire length of 0.5 mm was small enough to avoid spatial-averaging effects which would result in apparently low values of the intensity $(\overline{\tau_w'^2})^{1/2}$.

For the electronic equipment and the calibration procedure the reader is referred to [6].

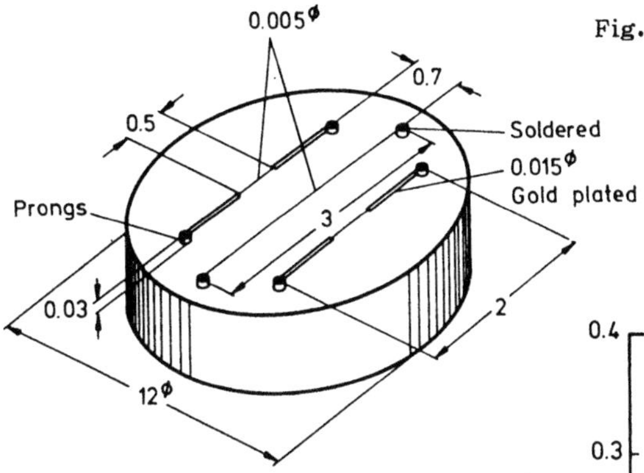

Fig. 4: Wall pulsed wire.

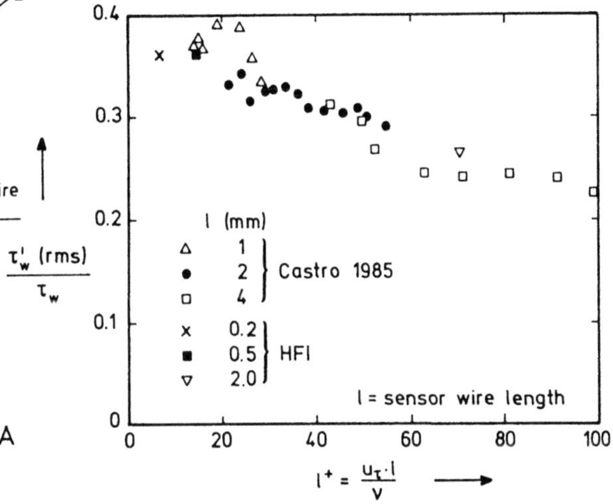

Fig. 5: Influence of sensor-wire length on τ'_ω.

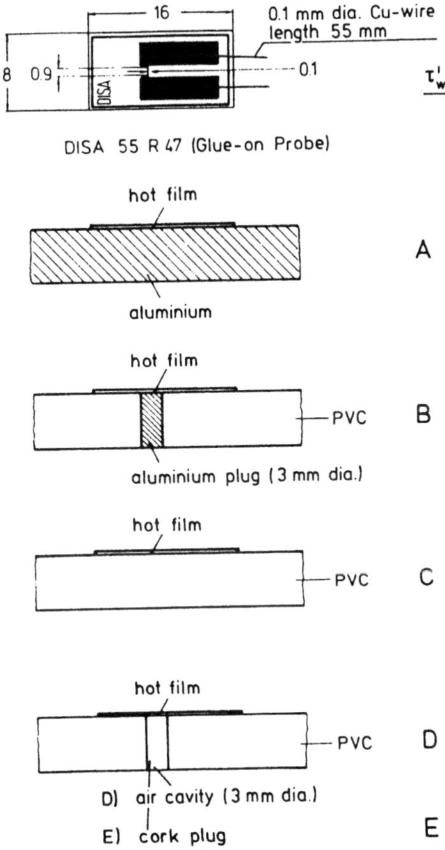

Fig. 6: Test configurations for single hot film.

The hot-film measurements were performed by using single hot films (DISA 55 R 47) and a McCROSKEY V-probe [9]. The former were glued on the different wall materials or covered an air-filled cavity to vary the amount of heat transferred to the substrate (Fig. 6). After calibrating the single hot films against a Preston tube in the axisymmetric turbulent boundary layer, the ratio of the intensity of τ'_ω divided by $\bar{\tau}_\omega$ was compared with the equivalent value - 0.37 - obtained by the

Table 1

Wall material	Aluminum	Aluminum	PVC	Air cavity	Cork
	(A)	plug (B)	(C)	(D)	(E)
$\tau'_{w\,r.m.s}/\bar{\tau}_w$	0.30	0.18	0.13	0.11	0.11

wall pulsed wire [14]. Depending on the substrate the hot-film values vary between 0.11 and 0.30 (cf. Table 1).

The trend of these data is confirmed - though not explained - by the results of sine-wave tests which provide information about the frequency response of the hot films (Fig 7). Here the curve with a peak at higher frequency corresponds to a higher value of $\tau'_{w r.m.s}/\bar{\tau}_w$ in table 1. Alfredson et al. [8] have apparently made an attempt to explain the low response of the hot film as dependent on the flow medium (air) since measurements in water were shown to be largely unaffected by this problem.

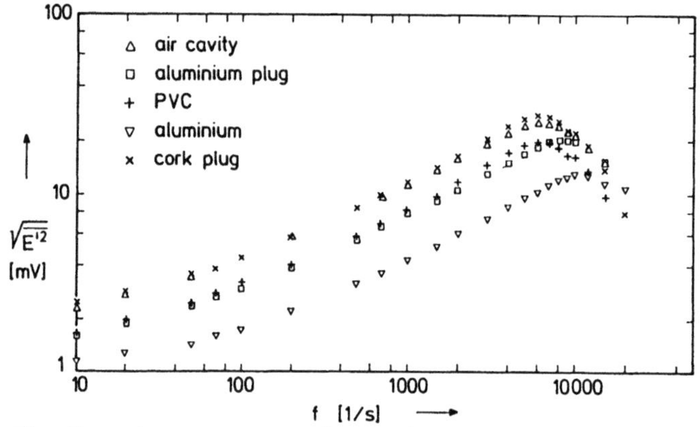

Fig. 7: Sine-wave tests for various single hot-film configurations.

The surface fence has been used in our laboratory for more than 20 years to measure the magnitude and direction of skin friction (e.g. VAGT & FERNHOLZ [10]) and its behaviour would appear to be well known. In order to determine the skin friction, the pressure difference upstream and downstream of the fence has to be measured. This is difficult to obtain, however, if the mean value of τ_w approaches zero, as at separation or reattachment, and if the fluctuations τ'_w are large. Large fluctuations destroy the worth of the static calibration curve for the surface fence, especially if the pressure transducer used is not symmetrical in the arrangement of its pressure chambers (in our case a STATHAM Typ PM 97 TC with a range of approximately ± 35 mm water). An incorrect calibration could probably be avoided if the transducer is calibrated dynamically or if a symmetric design is used. The whole effect does not occur if a micromanometer (FROEBEL & VAGT [11]) is used instead of the transducer. At these small pressure differences the

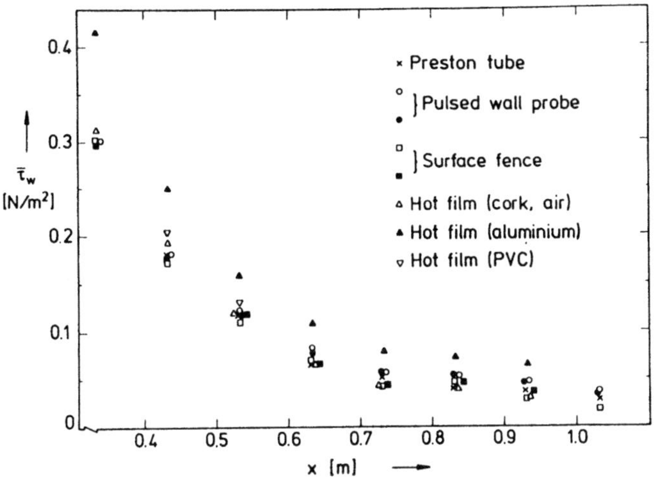

Fig. 8: Comparative measurements of mean skin friction in an APG boundary layer.

elasticity of the tubing between pressure sensor and measuring device becomes important. A preliminary test led us to use rigid tubes (hard polyethylene), and all surface fence measurements were made with these and the micromanometer.

4. Discussion of the Measurements

The measurements of $\bar{\tau}_w$ shown in Fig. 8 were taken in the axisymmetric APG turbulent boundary layer. With the exception of the hot film on the aluminum substrate all other techniques give results which agree well over the whole range of pressure gradients down to separation. Error estimates for Preston-tube measurements in APG were developed by BROWN & JOUBERT [12] and FREI & THOMANN [13] both based on measurements by means of floating element balances. These estimates give maximum values of 1 and 1.5%, respectively and thus confirm the reliability of the majority of the measuring techniques for mean values of the skin friction in APG boundary layers without instantaneous reverse flow.

The second test flow contains separation, a reverse-flow region, reattachment and a readjusting wall-bounded shear flow with very high levels of skin friction intensity $(\overline{\tau_w'^2})^{1/2}$. Figure 9 presents the surface-fence and the wall pulsed-wire data plotted against x/x_R (x_R beeing the length of the reverse-flow region) which agree very well with each other. $c_f' = 2(\overline{\tau_w'^2})^{1/2}/\rho_\infty u_\infty^2$ was obtained with two wall pulsed-wire probes, ∞ indicating the undisturbed upstream condition. The first probe had an active length of 2 mm [6] and the second one of 0.5 mm. Both probes give identical results over most of the test flow region, indicating a larger turbulence structure than in a boundary layer. In the flow region where the wall bounded shear flow adapts more and more to that of a turbulent boundary layer, the flow structure becomes smaller and the data from the 2 mm probe are too low.

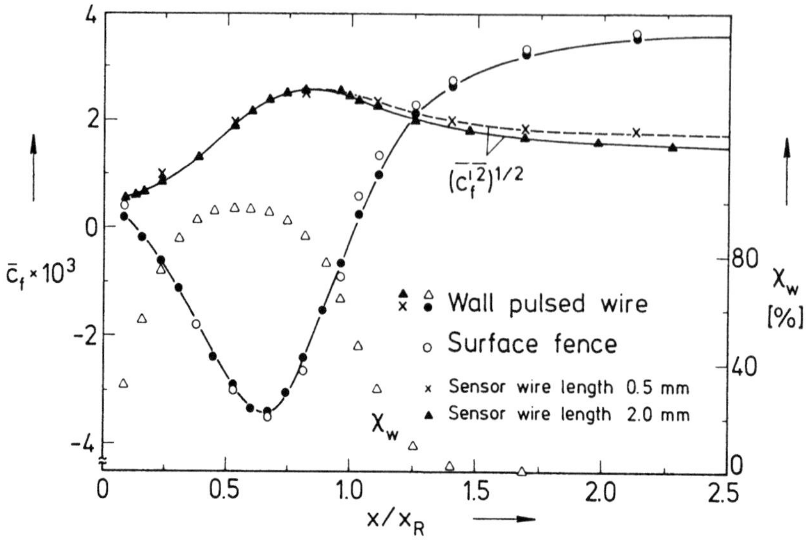

Fig. 9: Comparative measurements of $\bar{c}_f$, c'_f and χ_w in test flow 2.

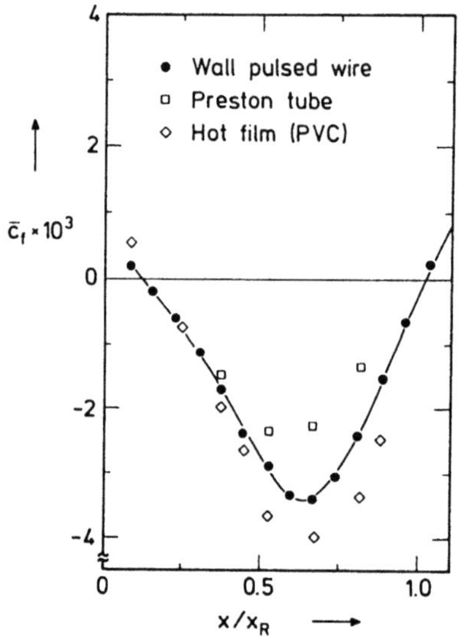

Fig. 10: Measurements of $\bar{c}_f$ in test flow 2.

χ_w is the reverse-flow parameter which is defined as the ratio of the wall pulsed-wire signals with a negative sign to the sum of all signals. This parameter is helpful in the explanation of the discrepancies between the hot-film and Preston tube data on the one hand and surfacefence and wall pulsed-wire measurements on the other. A comparison of these data is given in Fig 10.

The hot-film measurements are too high over the larger part of the flow region investigated. This is due to rectification effects caused by the instantaneous reverse flow indicated by the distribution of the reverse-flow parameter χ_w.

The Preston tube data disagree most in a flow region where it was found [6] that the logarithmic low of the wall was no longer valid.

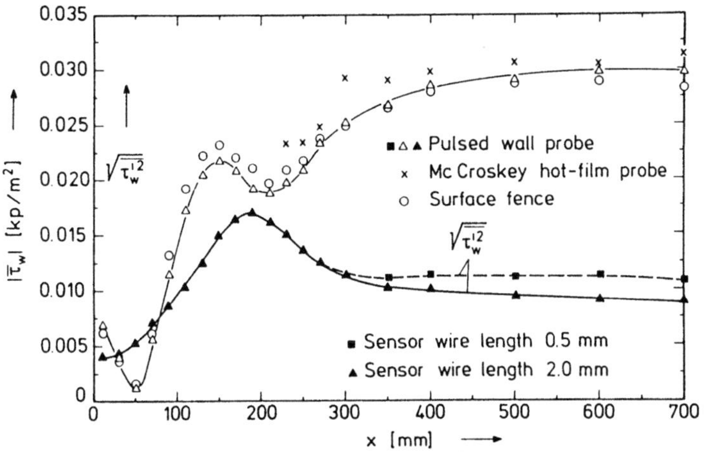

Fig. 11: Comparative measurements of $\overline{\tau}_w$ and τ_w' in test flow 3.

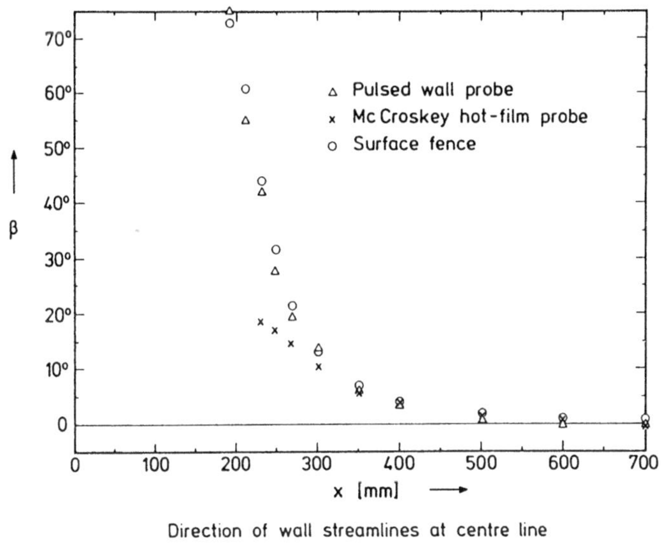

Fig. 12: Comparative measurements of the angle of the wall streamlines in test flow 3.

This, together with the effect of instantaneous reverse flow, should suffice to prevent anyone performing skin-friction measurements in such complex flows with hot films or Preston tubes.

The magnitude and direction of $\overline{\tau}_w$ were also measured by the surface fence, the wall pulsed wire and the McCROSKEY hot-film probe in the three-dimensional flow downstream of a swept backward facing step. These measurements are compared in Fig. 11 and 12. As in the flow presented in Fig. 9 agreement between the data of the wall pulsed wire and the surface fence is good. The differences in the τ_w'-curves are caused by the different wire length of the sensors (cf. explanation for Fig. 9). For a comparison with other skin friction gauges (e.g. [10]) Fig. 13

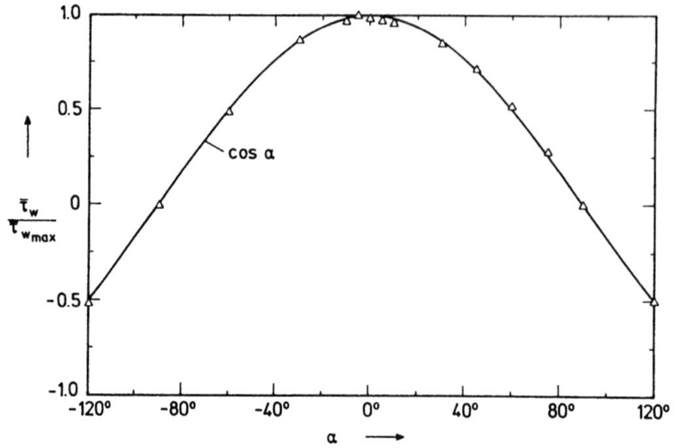

Fig. 13: Yaw response of wall pulsed wire at a near wall turbulence level of 90%.

shows the yaw characteristic of the wall pulsed wire taken at a position in the flow where $\tau'_{w r.m.s}/\bar{\tau}_w$ was 90%. The McCROSKEY hot-film probe which has been disigned specially to measure both magnitude and direction gives magnitudes which are too high and flow angles which are too low if the crossflow angle increases beyond about 15°.

5. Conclusions

The wall-pulsed wire is an excellent sensor for measuring the mean value, fluctuation intensity, sense and direction of skin friction. It will cope with highly turbulent flows including both instantaneous and mean reverse flow, separation and reattachment. If special care is taken with the pressure measurements (and excepting the fluctuation intensity) the same holds for the surface fence. The two types of hot-film sensor can be used for mean flow measurements well away from "complex flow regions". If the flow is strongly three-dimensional the accuracy of the McCROSKEY probe is sharply reduced.

Acknowledgements

The authors thank the Deutsche Forschungsgemeinschaft for the financial support of this investigation.

6. References

1. K.G. Winter: A outline of the techniques available for the measurements of skin friction in turbulent boundary layers. Prog. Aerospace Sci. 18, 1-57, 1977.
2. R.B. Ginder, L.J.S. Bradbury: Preliminary investigation of a pulsed-gauge technique for skin friction measurements in highly turbulent flow. ARC Paper 34 448, 1973.

3. J.K. Eaton, J.P. Johnston: Turbulent flow reattachment. An experimental study of the flow and structure behind a backward-facing step. Rep. MD-39 Stanford Univ., 1980.

4. I.P. Castro, M. Dianat: Surface flow patterns on rectangular bodies in thick boundary layers. J. Wind Engng. Indust. Aerodyn. II, 107-119, 1983.

5. P. Dengel, H.H. Fernholz, J.-D. Vagt: Turbulent and mean flow measurement in an incompressible axisymmetric boundary layer with incipient separation. In Turbulent Shear Flows 3, ed. Bradbury et al. Springer Verlag, 1982.

6. R. Ruderich, H.H. Fernholz: An experimental investigation of a turbulent shear flow with separation, reverse flow, and reattachment. J. Fluid Mech. 163, 283-322, 1986.

7. I.P. Castro, M. Dianat: The measurement of skin friction with a pulsed-wire gauge. Final report on University of Surrey/ RAE agreement 2064/069 XR Aero.

8. P.H. Alfredson, A.V. Johansson, J.H. Haritonidis, H. Eckelmann: On the fluctuating wall shear stress and the velocity field in the viscous sublayer. Submitted to J. Fluid Mech., 1986.

9. The McCroskey probe was kindly provided by Dr. H. U. Meier of DFVLR Göttingen.

10. J.-D. Vagt, H.H. Fernholz: Use of surface fences to measure wall shear stress in three-dimensional boundary layers. Aero. Quart. 24, 87-91, 1973.

11. E. Froebel, J.-D. Vagt: Ein automatisch abgleichendes Flüssigkeitsmanometer mit digitaler Anzeige. DLR-FB 74-10, 1974.

12. K.C. Brown, P.N. Joubert: The measurements of skin friction in turbulent boundary layers with adverse pressure gradients. J. Fluid Mech. 35, 737-757, 1969.

13. D. Frei, H. Thomann: Direct measurements of skin friction in a turbulent boundary layer with strong adverse pressure gradient. J. Fluid. Mech. 101, 79-95, 1980.

14. X. Gu: Diplomarbeit am Hermann-Föttinger-Institut für Thermo- und Fluiddynamik, Technische Universität Berlin.

Numerical Experiments on Static Non-Linear Effects in a Single Hot-Wire Measuring Technique

M. Hoffmeister

Institut für Mechanik der Akademie der Wissenschaften der DDR, DDR-9010 Karl-Max-Stadt, PSF 408, GDR

1. Introduction and Aims

The evaluation of hot-wire measurements is usually carried out on the basis of the static calibration properties of the measuring system applied. Even the simplest case, i.e. the determination of the velocity vector in an isothermal, homogeneous fluid flow leads to a non-linear relation between the input quantities at the measuring element - the three components of the velocity vector - and the output voltage of the anemometer. Although, in principle, this non-linearity can be taken into account by the use of a three-wire probe, numerous evaluation techniques employ a linearisation of the static calibration relationship. Generally, this is also done in the known methods of the rotated single wire probe yielding the complete Reynolds stress tensor at measuring points in stationary turbulent shear flows. Unfortunately, there is only little information on the errors due to the linear approximation of the calibration function. This holds especially for the uncertainty of the single measured components of the Reynolds stress tensor and their dependence on the conditions at the measuring point. Since the problem cannot be solved with the aid of calibration experiments in sufficiently defined turbulent shear flows - they are not available - in the following a numerical simulation is carried out: On the basis of an assumed probability density function (pdf) for the velocity vector at the measuring point and with the aid of a known (or assumed) calibration function, those output quantities of an interesting measuring system are calculated which are needed as starting values for the evaluation method to be examined with respect to linearisation errors. The comparison of the results ensuing from the evaluation method with the corresponding original values given by the pdf, yields the required information. Parameter dependencies can be studied by the alteration of the pdf and the calibration function.

The numerical procedure described above was worked out and programmed for Gaussian pdfs by BIRKNER /1/. The mean velocity vector and the Reynolds stress tensor at the measuring point can be chosen. The calibration function of the measuring system - e.g. consisting of a Constant Temperature Anemometer (CTA) in connection with a slanted hot-wire probe - was assumed to have the form of King's law with the cosine law describing the direction dependency of the heat transfer at the sensing element. The angle between the hot-wire and the probe axis can also be chosen. The numerical simulation model yields the DC- and RMS-voltages of the measuring system as function of the rotation angle of the probe. In this paper these quantities serve as "measuring values" for the examination of a special linear or non-linear single hot-wire evaluation technique developed by HOFFMEISTER /2/,/3/. This method is capable of determining the mean velocity vector and the complete Reynolds stress tensor with the aid of only one hot-wire probe and only one position of the probe axis in the flow field. Comparisons with the initial values of the simulation model elucidate the behaviour of the evaluation techniques and, particularly, the systematic errors appearing with growing turbulence intensities.

2. The Numerical Simulation Model

Assuming the pdf $p(\vec{u})$ for the velocity vector $\vec{u}$ at the measuring point the mean output voltage $\bar{E}$ of a measuring system follows from

$$\bar{E} = \iiint_{R^3} E(\vec{u})\, p(\vec{u})\, d\vec{u} \,, \tag{1}$$

and the RMS-value $(\overline{E'^2})^{1/2}$ from

$$\begin{aligned} \overline{E'^2} &= \iiint_{R^3} [E(\vec{u}) - \bar{E}\,]^2 p(\vec{u})\, d\vec{u} \\ &= \iiint_{R^3} E^2(\vec{u})\, p(\vec{u})\, d\vec{u} - \bar{E}^2 \,. \end{aligned} \tag{2}$$

The static calibration function $E(\vec{u})$ shall be given by

$$E = (A_W + B_W |\vec{u}_N|^{1/2})^{1/2} \,, \tag{3}$$

where the index $(\ )_N$ denotes the velocity component normal to the hot-wire, and A_W, B_W are calibration constants.

The numerical calculation of the integrals (1),(2) over the three-dimensional space R^3 requires comparatively little effort and yet is accurate, if the pdf is prescribed in the form of a Gaussian normal distribution

$$p(\vec{u}) = \frac{1}{\sqrt{(2\pi)^3 |D|}} \exp\left\{ -\frac{1}{2} (\vec{u} - \bar{\vec{u}})^T D^{-1} (\vec{u} - \bar{\vec{u}}) \right\} , \qquad (4)$$

and the Gaussian quadrature formulas are used. This was done by BIRKNER /1/. The covariance matrix $D = (\overline{u_i' u_j'})$ and the mean velocity vector $\bar{\vec{u}} = (\bar{u}_i)$ in (4) can be chosen. Figure 3 shows the definition of the space fixed Cartesian x_i-co-ordinate system (i = 1,2,3) with the velocity components $\bar{u}_i$. The hot-wire position in the x_1,x_2-plane is characterized by the origin of the rotation angle α_m, i.e. by $\alpha_m = 0$.

E x a m p l e : CTA with a 45[°] slanted hot-wire probe;

$$A_w = 25.673\ [V^2] \ ; \quad B_w = 8.733\ [V^2/(m/s)^{1/2}] \ ;$$

$$\frac{D}{A^2 u_d^2} = \begin{pmatrix} 4.806\cdot10^{-3} & -0.342\cdot10^{-3} & 0.753\cdot10^{-3} \\ -0.342\cdot10^{-3} & 5.772\cdot10^{-3} & 0.936\cdot10^{-3} \\ 0.753\cdot10^{-3} & 0.936\cdot10^{-3} & 1.188\cdot10^{-3} \end{pmatrix} ;$$

$$\frac{\bar{\vec{u}}}{u_d} = \begin{pmatrix} 0.0322 \\ -0.2761 \\ 0.6813 \end{pmatrix} ; \quad \begin{matrix} |\bar{\vec{u}}| = 6.100\ [m/s] \ ; \\ \hat{\alpha}_x = 112.061\ [°] \ ; \\ \hat{\beta}_x = 2.508\ [°] \ ; \end{matrix}$$

(reference velocity: $u_d = 8.29$ [m/s]).

Figures 1 and 2 indicate $\bar{E}$ and $(\overline{E'^2})^{1/2}$ versus α_m from (1) and (2) with A^2 as parameter. Defining the turbulence intensity by

$$Tu = \left\{ \frac{\overline{u_1'^2} + \overline{u_2'^2} + \overline{u_3'^2}}{3 |\bar{\vec{u}}|^2} \right\}^{1/2}$$

the example yields $Tu = 0.0853\ A$, i.e. the turbulence intensity varies from 8.53[%] to 85.3[%] . It is remarkable, that the amplitudes of the E-plots decrease monotonously with growing Tu while the amplitudes of the $(\overline{E'^2})^{1/2}$-curves initially increase and then decrease.

3. Examination of a Single Hot-Wire Measuring Technique

The basic relations of the evaluation method proposed by HOFFMEISTER /2/,/3/ are

$$(\hat{a}_i + \hat{b}_{ij} \frac{\Delta\hat{w}_j}{\hat{c}}) \frac{\Delta\hat{w}_i}{\hat{c}} + \hat{b}_{ij} \frac{\overline{w_i' w_j'}}{\hat{c}^2} = \bar{F} - \hat{F} , \qquad (5)$$

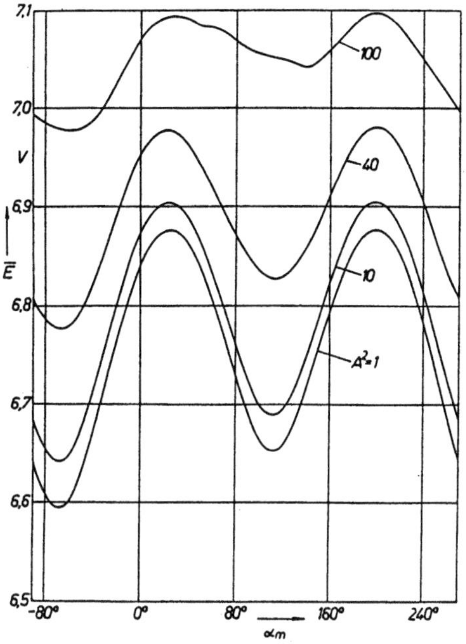

Fig. 1 Mean output voltages as function of the rotation angle and the turbulence intensity

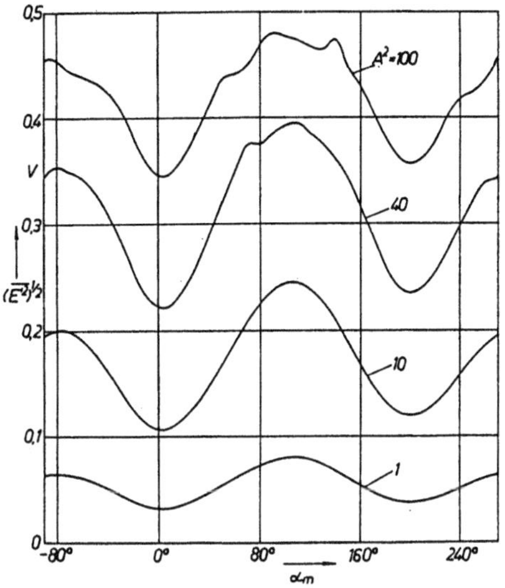

Fig. 2 RMS output voltages as function of the rotation angle and the turbulemce intensity

$$\left(\hat{a}_i\hat{a}_j + 4\hat{a}_i\hat{b}_{jk}\frac{\Delta\hat{w}_k}{\hat{c}} - \hat{b}_{ij}\hat{b}_{kl}\frac{\overline{w'_k w'_l}}{\hat{c}^2} \right) \frac{\overline{w'_i w'_j}}{\hat{c}^2} + 2\hat{a}_i\hat{b}_{jk}\frac{\overline{w'_i w'_j w'_k}}{\hat{c}^3}$$

$$+ \left(2\hat{a}_i\hat{c}_{jkl} + \hat{b}_{ij}\hat{b}_{kl} \right) \frac{\overline{w'_i w'_j w'_k w'_l}}{\hat{c}^4} = \overline{F'^2} \ . \qquad (6)$$

Here, w_i denotes the velocity components in a Cartesian co-ordinate system which is space fixed and oriented at the mean velocity vector. At the beginning of an evaluation calculation only a first approximation of the mean velocity vector is available. This estimate shall have the components $\hat{w}_i$, so that the required mean velocity components $\overline{w}_i$ follow from $\overline{w}_i = \hat{w}_i + \Delta\hat{w}_i$. The correction terms $\Delta\hat{w}_i$, like the Reynolds stresses $\overline{w'_i w'_j}$ and the velocity moments of the third and fourth order appear as unknowns in (5),(6).

The linear, quadratic, and cubic sensitivity coefficients - $\hat{a}_i$, $\hat{b}_{ij}$, and $\hat{c}_{ijk}$ - follow from finite differences formulas with a characteristic mesh size Δw. The dimensionless CTA output voltage is defined by $F = 2E\hat{E}/(\hat{c}^{1/2}\breve{B})$, where $\hat{E}$ denotes the output voltage following from the calibration function for the (estimated) mean velocity vector having the magnitude $\hat{c}$. $\breve{B}$ is a constant related to B_w. For details of the numerical treatment of (5),(6) the reader is referred to the original papers. The term $2\hat{a}_i\hat{c}_{jkl}$ accounting for certain cubic effects of the calibration function is added in (6) compared with HOFFMEISTER /3/.

In the following, three cases are described, namely, the linear approach ($\hat{b}_{ij} \equiv 0$, $\hat{c}_{ijk} \equiv 0$), the quadratic approach ($\hat{c}_{ijk} \equiv 0$), and the quadratic approach with cubic terms (the complete equations). In the cases of a non-linear evaluation the interesting unknowns $\Delta\hat{w}_i$ and $\overline{w_i w_j}$ are calculable only in the form of certain linear combinations with those velocity moments of higher order which - within the accuracy - cannot be separated from the equation system. These linear combinations make it possible to estimate the influence of the higher order moments involved on the interesting unknowns $\Delta\hat{w}_i$ and $\overline{w_i' w_j'}$. In the following example this effect is neglected.

E x a m p l e : "Measuring values" $\bar{E}$ and $(\overline{E'^2})^{1/2}$ from Figs. 1 and 2 at α_m = -60, -50,..., 280, 290 [°] ;
$\tilde{B} = B_w(\sin 45)^{1/2} = 7.344$ $[V^2/(m/s)^{1/2}]$;
rows of the matrices (5),(6) at
$\hat{\alpha}_y$ = -140, 130,..., 130, 140 [°] with
$\hat{\alpha}_y = \alpha_m - \alpha_{mo}$;
(α_{mo} denotes the approximate hot-wire position in the plane formed by the mean velocity vector and the probe stem);
the uncertainty of the simulated measurements and calibrations is taken from real conditions, see HOFFMEISTER /4/;
the characteristic mesh size for the determination of the sensitivity coefficients follows from
$\Delta w/\hat{c} \approx Tu$.

The three versions of the single hot-wire evaluation technique (linear approach, quadratic approach, and quadratic approach with cubic terms) yield the results indicated in Figs. 3 and 4 as functions of A^2. The comparison with the original values of the model flow shows that the magnitude of the mean velocity vector ensuing from the linear approach deviates from the true value by approximately 6 [%] already at $A^2 = 10$ ($Tu \approx 27$ [%]). This error increases and reaches approximately 21 [%] at $A^2 = 40$ ($Tu \approx 54$ [%]). Obviously, much better results are obtained with the aid of the quadratic approach: the error of the magnitude remains smaller than 2 [%] at $A^2 = 40$, and also

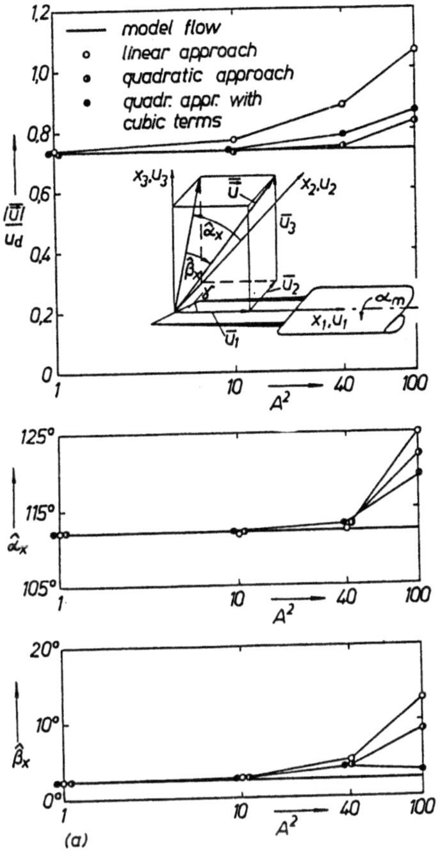

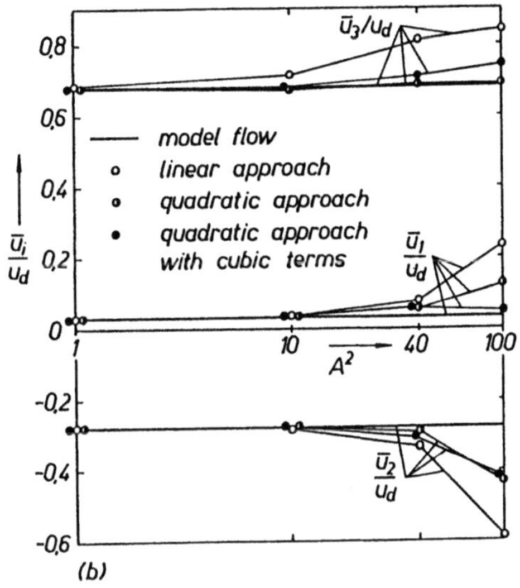

Fig. 3 Mean velocity vector comparison: Model flow - linear and non-linear evaluation techniques, (a) magnitude and angles, (b) Cartesian components

the direction of the mean velocity vector can be determined with an accuracy of about 2[°] at this turbulence level. The addition of the cubic terms to the quadratic equations does not yield clear advantages. This is also reflected in Fig. 3b . Still higher turbulence intensities ($A^2 = 100 \rightarrow Tu \approx 85$[%]) lead to rapidly growing systematic errors. Figure 4 illustrates that the components of the Reynolds stress tensor respond in different ways to the increasing turbulence intensity. The linear approach generates an error of about 40[%] in $\overline{u_2'^2}$, an error of about 70[%] in $\overline{u_1'u_2'}$, and an error of about 100[%] in $\overline{u_2'u_3'}$ at $A^2 = 10$, while the other components are closer to the true values. With the aid of the non-linear approaches these deviations can be substantially reduced. At $A^2 = 40$ strong systematic errors of all approaches appear in $\overline{u_1'^2}$, and the linear approach even leads to a change of the $\overline{u_1'u_2'}$-sign. Without further discussion it may be concluded, that - at least under the given conditions - a reasonable application of the linear approach is restricted to turbulence intensities below 10...15[%], while the quadratic approach can be used

Fig. 4 Reynolds stress tensor comparison: Model flow - linear and non-linear evaluation techniques, (a) normal stresses, (b) shear stresses

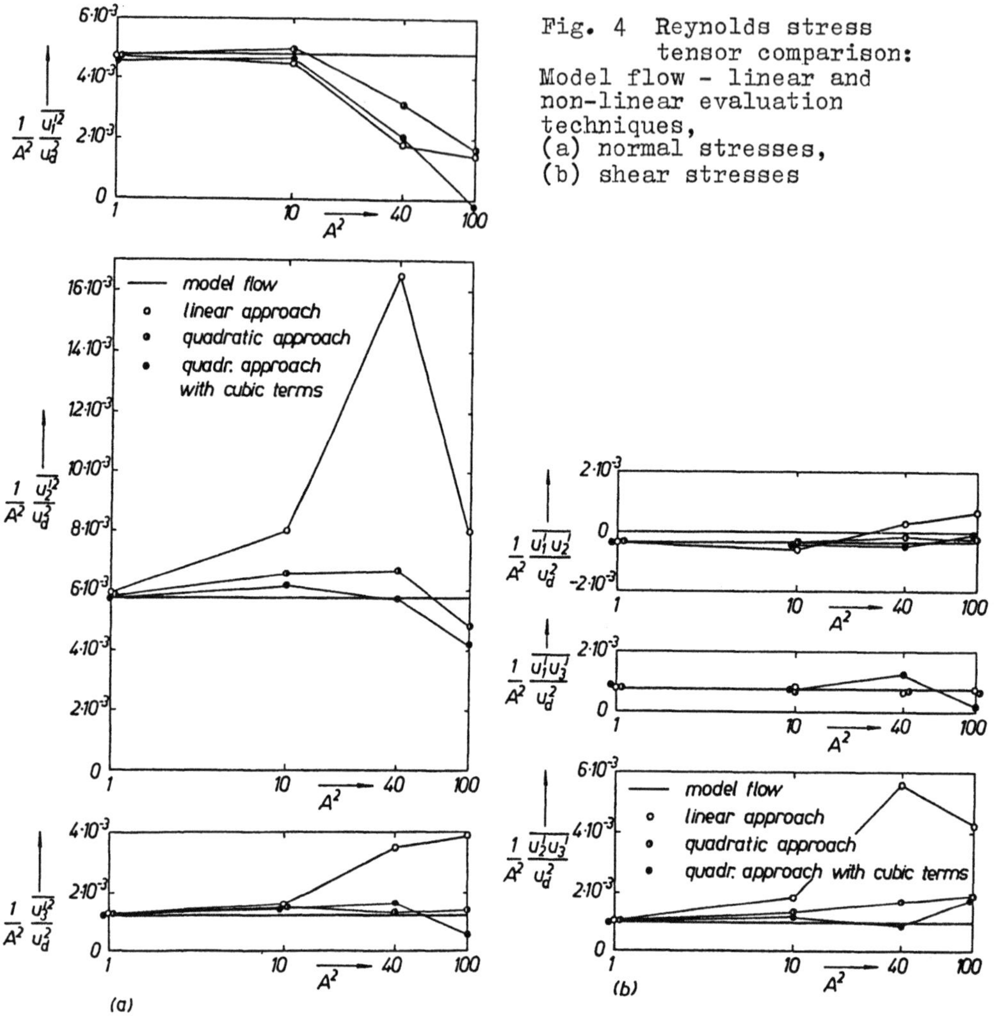

up to 25...30[%], and with respect to the determination of the mean velocity vector up to 50...55[%].

4. Summarizing Remarks

The numerical simulation of the input/output relationship of hot-wire measuring systems proves to be a useful tool for studying systematic errors in certain evaluation methods due to approximations of the real static calibration properties. A special solution depending on the assumption of a Gaussian pdf of the velocity vector at the measuring point was developed and programmed for single hot-wire measuring techniques exploiting the method of the rotated probe. The calibration function can be chosen at will. The main advantage consists in providing - at least within the numerical accuracy - exact and complete input/output data without any other additional error sources (e.g. from the dynamical response, from the spatial averaging of the hot-wire, etc.). In this way, the single components of the mean velocity vector and the Reynolds stress tensor ensuing from the measuring and evaluation technique to be examined, can be compared with true values. The numerical experiment carried out in this paper shows the influence of a growing turbulence intensity on the results of a special linear or non-linear single hot-wire measuring technique. Further developments should take into account generalized forms of the pdf in order to provide the possibility of examining the effect of skewness and flatness factors.

R e f e r e n c e s

1. D. Birkner: Diplomarbeit, Humboldt-Universität Berlin, Sektion Mathematik (1979, unpublished)
2. M. Hoffmeister: Maschinenbautechnik B19, 8 (1970) or - shortened - in DISA Information No. 13 (1972)
3. M. Hoffmeister: Maschinenbautechnik B23, 12 (1974)
4. M. Hoffmeister: Maschinenbautechnik B24, 1 (1975)

Hot Wire in Wall Proximity

G. Janke

Hermann-Föttinger-Institut für Thermo- und Fluiddynamik,
Technische Universität Berlin, D-1000 Berlin 12, Germany

1. Summary

Experiments were carried out on the behaviour of hot wires near conducting walls. They include tests in laminar and tubulent flows, "wind-off" response and rotation tests. Special attention was given to the calibration procedure at low velocities and, finally, a new method to determine the wall distance was developed.

2. Introduction

Hot wires which are calibrated in the free stream will give erroneous readings in measurements near conducting walls. Since - to our knowledge - there are no solid materials with lower values of conductivity than that of air, this will always be the case in the near wall region of air boundary layers.

Quite a number of studies on this problem have been performed with very diverse results.

Probably best known is that of WILLS [1] who, for laminar channel flow, obtained the relationship

$$Nu \left(\frac{T_w}{T_a}\right)^{-0.17} = A + B \cdot K_w \left(2\frac{y}{d}\right) + B \cdot Re_d^{0.45} \qquad (1)$$

based on the free-stream calibration of COLLIS & WILLIAMS [2]. The function K_w $(2^y/_d)$ was introduced by WILLS to account for the wall effect. Although (1) does not hold for $Re_d = 0$, it implies that heat is transfered from the wire to the wall by conduction only and can thus be considered a modified version of the "wind-off" correction of VAN DER HEGGE ZYJNEN [3].

For turbulent flow WILLS found that (1) was not applicable but suggested that half the value of the laminar correction should be used.

A completely different approach was first taken by OKA & KOSTIC [4] and later confirmed by HEBBAR [5]. They found that velocities measured in the viscous sublayer of turbulent boundary layers for different friction velocities fall on one curve when plotted in wall coordinates, i.e.

$$u_m^+ = f(y^+) \quad \text{with} \quad u_m^+ = \frac{u_m}{u_\tau}, \quad y^+ = \frac{y \cdot u_\tau}{\nu} \quad \text{and} \quad u_\tau = \sqrt{\frac{\tau_w}{\rho}} \qquad (2)$$

u_m : measured velocity, y : wall distance,
τ_w : wall shear stress

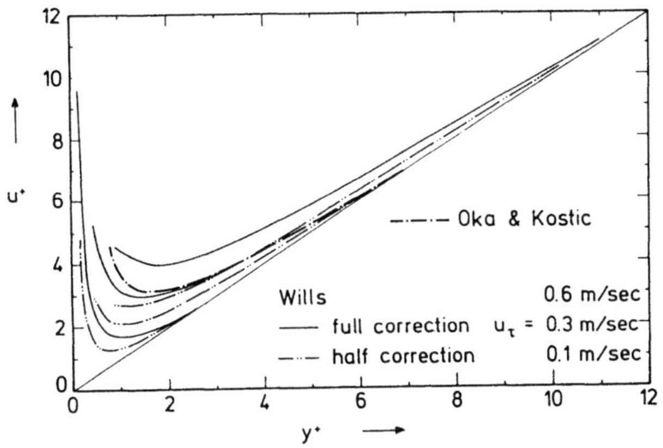

Fig. 1: WILLS' correction in wall coordinates

The fundamental incompatibility of (1) and (2) is shown in Fig. 1 where WILLS' formula is replotted in wall coordinates using his data. However, it may be noticed that for certain values of u_τ, agreement is fair, which might explain why WILLS' correction works in some cases and not in others.

The main concern of the present study is to clarify these differences, which makes it necessary to repeat WILLS' experiment in light of the findings of OKA & KOSTIC [4].

Although the laminar channel flow used by WILLS allows to infer low velocities from large pressure differences and to observe the hot wire through a microscope with high resolution, it was not used as a testflow in the present investigation. The reason for this is that the results of DIANAT & CASTRO [6] indicate severe interference effects in a similar channel flow.

Experimental set up

The test flows were a laminar and a turbulent boundary layer generated on a flat plate with an elliptical nose. The plate formed one wall of the test section and was placed downstream of the contraction with a gap to peel off the nozzle boundary layer (Fig. 2). The measuring port was situated 100 mm downstream of the leading edge. Here the laminar boundary layer had a characteristic maximum fluctuation level of 1.1% at $y^+ = 10$. Skin friction was determined using that part of the linear region of the laminar boundary layer where measurements were not yet affected by wall effects (Fig. 3). This determination can be justified by the following consideration. If in the range $0.3 < u/u_\delta < 0.5$ the measuring error in the velocity is estimated conservatively to be 4%, the maximum divergence from linearity of a BLASIUS profile to be 3% and the error in measuring the wall distance is negligible, then the uncertainity in u_τ is 4%. This latter value is comparable to errors made in skin friction measurements of turbulent boundary layers. The turbulent boundary layer (Fig. 4) was generated by a V-shaped tripping band just downstream of

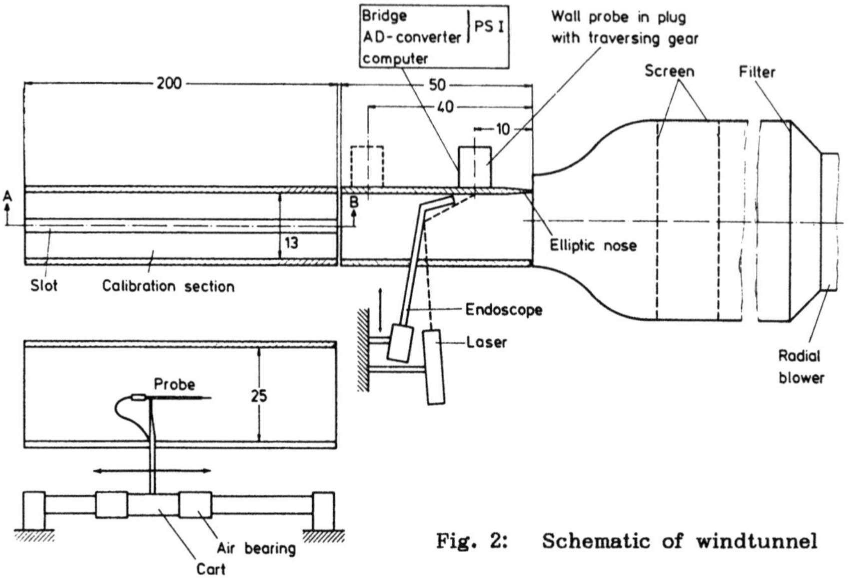

Fig. 2: Schematic of windtunnel

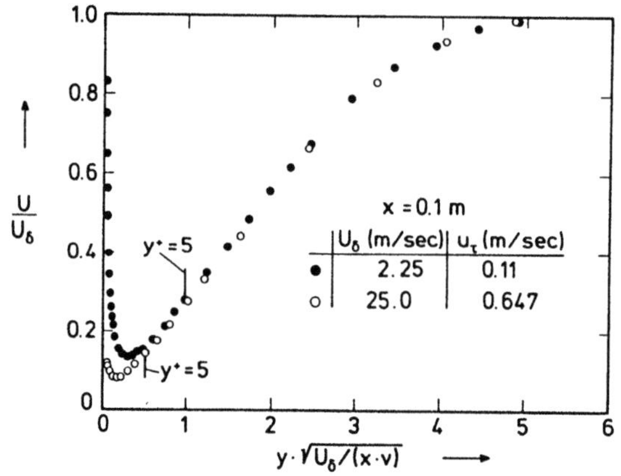

Fig. 3: Laminar boundary layer

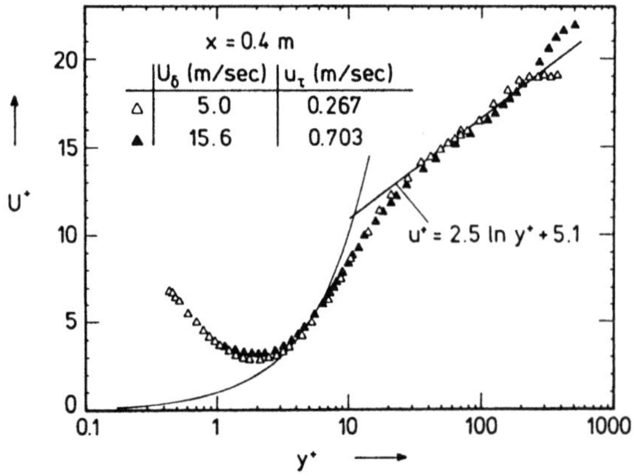

Fig. 4: Turbulent boundary layer

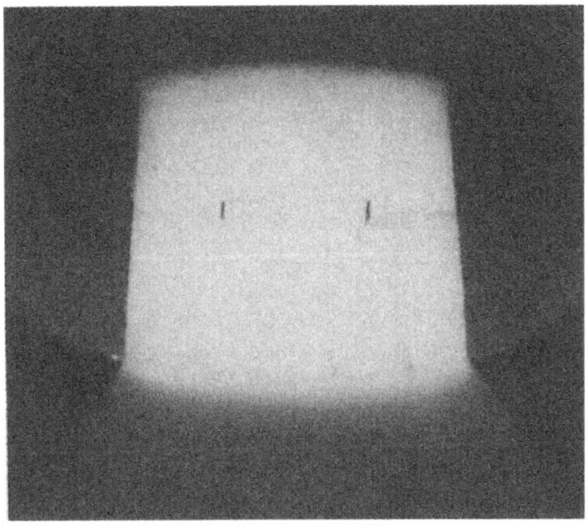 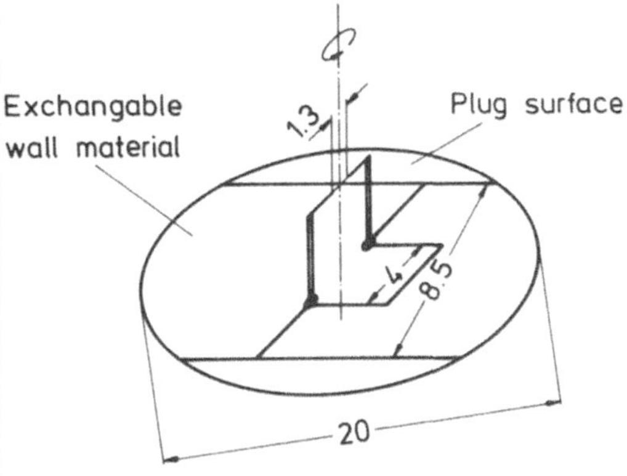

Fig. 5: Wall probe
wire: platinum coated tungsten ⌀ 5 μm, gold plating ⌀ 20 μm
prongs: tapered, at tip ⌀ 40 μm
overheat: 70%
wall: Macor (tradename), ceramics, $\lambda = 0.04$ cal/(cm sec grad)

the leading edge and skin friction was measured by PRESTON tubes using PATEL'S calibration [7].

The hot-wire probe is shown in Fig. 5. The wall consisted of a small ceramic plate with high thermal and low electrical conductivity in order to protect the hot wire in the CT mode from being destroyed when touching the wall. Preliminary tests showed that with this wall material the hot-wire response was hardly distinguishable from that on a copper wall, although the conductivity of the two wall materials differed by more than an order of magnitude. Although it was possible to remove the hot-wire probe from the wall plug and calibrate it in the freestream, it proved to be more convenient to calibrate the "built in" hot-wire at a wall distance of 4 mm by means of a calibrated second probe.

Since the results of these experiments depend to an exceptionally high degree on the accuracy of the measurements, much time had to be spent on minimizing errors from two main sources, namely the free-stream calibration of the hot wire at low velocities, and the determination of the wall distance.

The principle of the distance-measuring method and a possible implementation are displayed in Fig. 6. If a laser beam hits the wire, the reflection from the wire itself and the mirrorimage reflection produce an interference pattern which can be observed through a lense system. The fringe spacing is then proportional to the inverse of the wall distance. The advantages of this method are:

- it is applicable when the wind is on, thus taking into account the deflection of the wire and the traverse gear due to the aerodynamic load,
- it is insensitive towards vibrations of laser and optics
- a high resolution is obtained with inexpensive optical elements, e.g. a wall distance of 30 μm produces a fringe spacing of 0.5 mm at a distance of 50 mm from the wire. The error is conservatively estimated to be ± 2.5 μm.

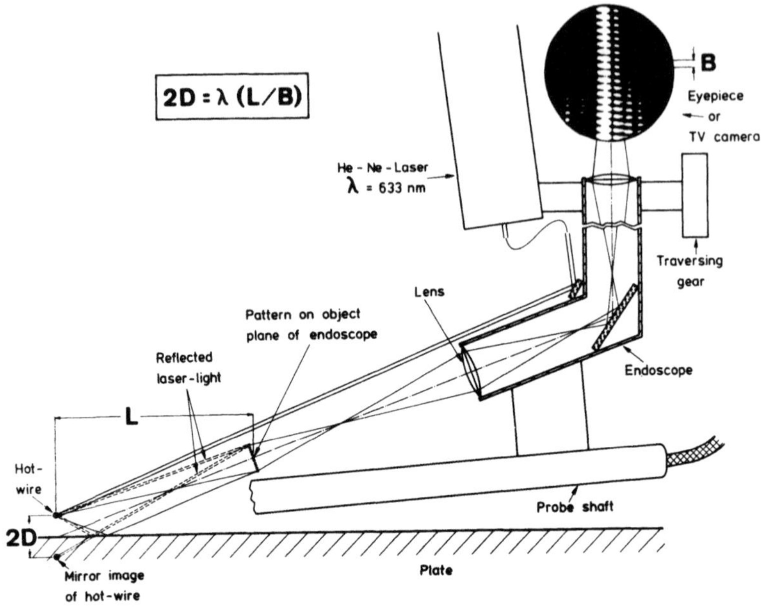

Fig. 6: Determination of wall distance

- the determination of the fringe spacing can be automated.

The main disadvantage of this method is that the receiving optics has to be aligned with the main reflection direction at some mechanical expense.

Various methods to calibrate hot wires at low speeds are known (cf. COMTE-BELLOT [8]). The problems encountered in the present experiments are described briefly.

(1) Pitot tubes as a calibration device could not be used, mainly because of the insufficient sensitivity of the micromanometer (FV-System, least count 0.005 mm of water) and the drift of the pressure transducer (Statham Δp_{max} = 7 mm of water).

(2) The "vortex-shedding" method ($50 < Re_D < 180$, cf. BERGER [9] gave fair agreement over most of the range (Fig. 7) but showed a systematic divergence for $u \to 0$, probably due to the decreasing aspect ratio of the cylinders with increasing diameter (e.g. GERICH [10]).

(3) Use of the "wind-off" response (Fig. 10). This method will be discussed below.

(4) The flying hot-wire calibration was used as a reference method. The test section for the calibration is shown in Fig. 4. It consisted of a rectangular duct with a belt-driven sled mounted on air bearings to eliminate vibrations. The voltage output of the anemometer was calibrated against that of a tachometer, the linearity of which was checked by two photo-diodes and an electronic clock.

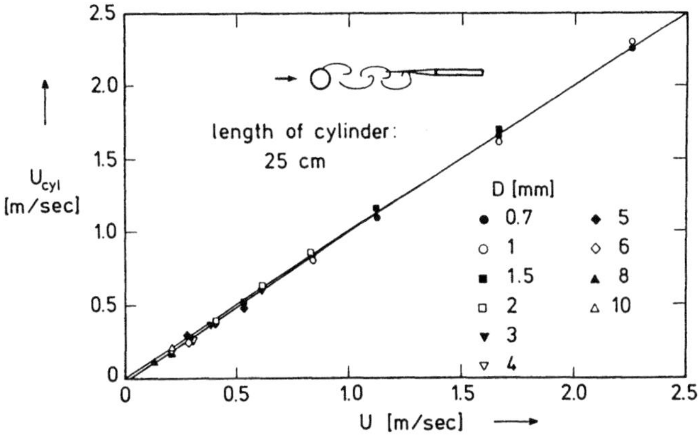

Fig. 7: Vortex shedding calibration

Results and Discussion

The mean velocities in the viscous sublayer and in the laminar boundary layer are plotted in wall coordinates in Fig. 8. It is observed that

(1) a relation $u_m^+ = f(y^+)$ holds over the range $0.11\ ^m/_{sec} < u_\tau < 0.7\ ^m/_{sec}$.

(2) For mean values the hot-wire response does not depend on the fluctuation level of the boundary layer.

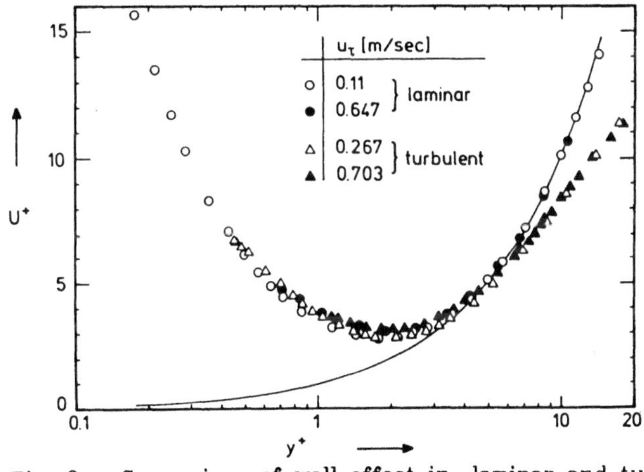

Fig. 8: Comparison of wall effect in laminar and tubulent boundary layers

The first observation contradicts the general trend of WILLS' results. It confirms the findings of OKA & KOSTIC [4] and extends them to laminar boundary layers. It is further in agreement with numerical results of BATHIA et al. [11] and of WACKER [12] who solved the Navier-Stokes equations and the energy equation for constant flow properties for a cylinder in a laminar shear flow along a wall.

The second observation can be explained on the basis of the model of BATHIA et al. [11], extending a consideration of ACHARYA & ESCUDIER [13]. For a linear velocity profile near the wall the following relationship holds for the skin friction

$$\tau_w = \rho \cdot \nu \cdot {}^{u}/_{y}$$

and thus

$$u_m^+ = \frac{u_m \cdot y}{\nu} \cdot \frac{1}{\sqrt{\frac{u \cdot y}{\nu}}} \quad \text{and} \quad y^+ = \sqrt{\frac{u \cdot y}{\nu}} \; .$$

Inserting this into eqn. (2) yields

$$\frac{u_m \cdot y}{\nu} = \sqrt{\frac{u \cdot y}{\nu}} \cdot f\left(\sqrt{\frac{u \cdot y}{\nu}}\right) = g\left(\frac{u \cdot y}{\nu}\right) \; . \tag{3}$$

For a fixed wall distance this is an explicit relationship between the true velocity u and the measured velocity u_m. Now, if it is assumed that 1.) the velocity profile in the viscous sublayer is instantaneously linear and that 2.) eqn. (3) holds also for every instant, then noticable differences in the hot-wire response between laminar and turbulent flow are only possible if $g({}^{u \cdot y}/_{\nu})$ is strongly nonlinear, because then higher order moments of u(t) would affect the mean value of $u_m(t)$.

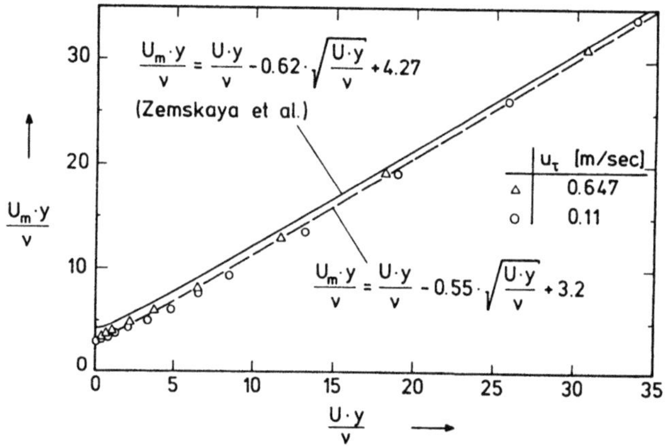

Fig. 9: Direct correlation between true and measured velocity. - - - least square fit of measurements.

Figure 9 reveals that $g({}^{u \cdot y}/_{\nu})$ is an almost linear function. Thus, if the assumptions 1.) and 2.) of BHATIAS' model are correct, then - in accordance with observation (2) - the hot-wire response in its mean value should not depend on the turbulence level.

Another interesting feature of Fig. 9 is that $g({}^{u \cdot y}/_{\nu})$ assumes a finite value for u = 0. This confirms the observation of ZEMSKAYA et al. [14] that $u_m^+ \sim {}^{1}/_{y^+}$ as $y^+ \rightarrow 0$. Furthermore, if equ. (2) is to hold even for u = 0 and $u_m \cdot y/\nu$ = const for u = 0 than it follows that

$$u_m \sim {}^{1}/_{y} \quad \text{for} \quad u = 0 \tag{4}$$

This inversely linear relationship is confirmed by the measurements shown in Fig. 10. It may be used for the calibration of hot wires at low velocities ("wind-off" response) if the wall distance can be measured very accurately.

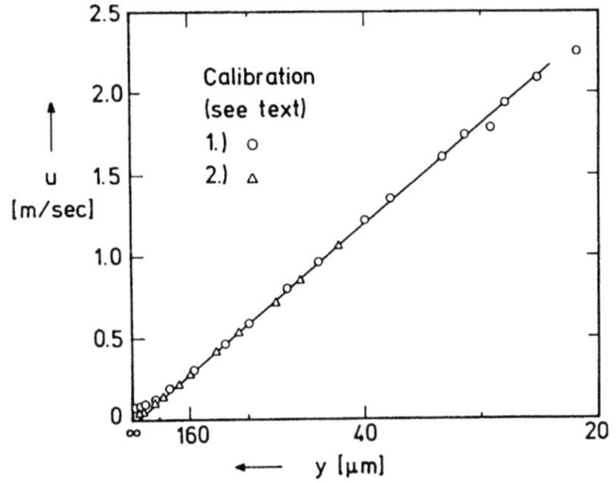

Fig. 10: "Wind-off" response of hot wire approaching a conducting wall

One may note, however, that there is a deviation from the above relationship (curve 1.) at velocities below approximately 0.3 m/s. This is due to an inaccuracy of the linearization, because the exponent $n = 0.45$ in the calibration law

$$E^2 = C + D \cdot U^n \tag{5}$$

was used over too wide a velocity range. If the calibration curve is approximated by a polynomial fit and over a smaller range, curve 2.) is obtained which now provides the correct trend.

It can be readily shown that the inversely linear behaviour is not the result of curve fitting but has a physical background:
COLE & ROSHKO [15] have shown theoretically that the following equation holds for "long" hot wires if $Re_d \ll 1$.

$$Nu = 2/[\ln (8/\gamma \; Pr \; Re_d)] \quad , \; \gamma = 0.577 \ldots \tag{6}$$

This is compared with the theoretical result for pure conduction obtained by WACKER [12] who found for a cylinder near a perfectly conducting wall

$$Nu = 2 \cdot \frac{2 \frac{y}{d}}{\sqrt{\left(2\frac{y}{d}\right)^2 - 1} \cdot \ln \left[2 \frac{y}{d} + \sqrt{\left(2\frac{y}{d}\right)^2 - 1}\right]} \tag{7}$$

which for $\frac{y}{d} > 3$ reduces to

$$Nu \approx 2/ \ln (4 \; y/d) \; . \tag{8}$$

Substituting u in Re_d of (6) by u_m and equating (6) and (8) results in eqn. (4). For future theoretical investigations it may be interesting to underline that the asymptotic solution (8) is a good approximation of the full solution (7), already for very small wall distances.

Since the hot-wire probe protruded through the wall, prong-interference effects were negligible and it was possible to perform the rotation test for the yaw response of the probe in the vicinity of the wall. This experiment revealed the following result (Fig. 11). If the measured velocity u_m is scaled on the effective skin

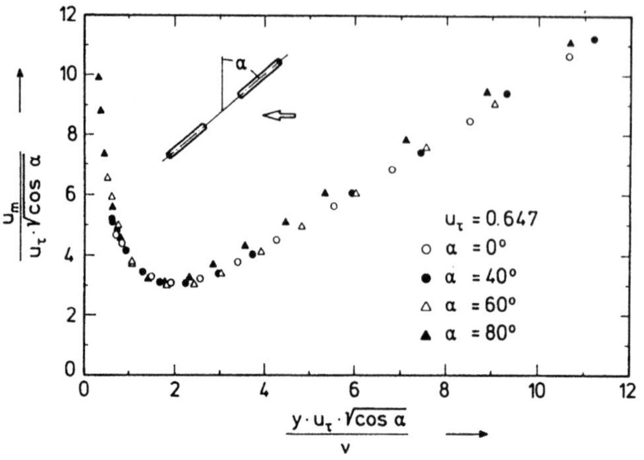

Fig. 11: Rotation test in wall vicinity

friction velocity $u_\tau \cdot \sqrt{\cos\alpha}$, in laminar flow a single curve is obtained for different yaw angles α. The deviation for $\alpha = 80°$ is also observed in measurements taken in the free stream. Such close agreement with the cosine law was unexpected since this relationship was assumed to hold only for wires with large ratios of active wire length to wall distance. Whereas the maximum value in the present experiment was only 55.

Finally, it is necessary to discuss the problem of establishing a "universal correction" for mean velocities in the near-wall region. The excess velocity is plotted and compared in Fig. 12 with the results of other authors, who state - except for KRISHNAMOORTHY et al. [16] - that a function of the type given in eqn. (2) holds if only u_τ is varied.

The data show a wide scatter. This is due to the broad range of parameters used in the various experiments, such as probe configuration, wall material and overheat ratio. It is unlikely that all these parameters can be taken into account to generate a universal field of correction curves with sufficient accuracy. It is

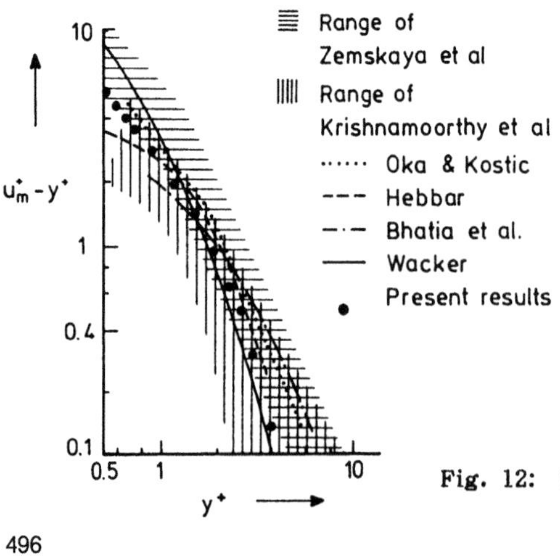

Fig. 12: Excess velocity $\Delta u^+ = u_m^+ - y^+$

therefore recommended to "calibrate" individual probes for specific wall material and overheat ratio in a flow with known wall shear stress in order to obtain a specific "calibration" curve (eqn.(2)) which then can serve for a consistency check in unknown flows. Via this calibration the skin friction could be determined directly from the relationship given in Fig. 9 if u_m and y are measured, since u is proportional to τ_w. This procedure yields, however, values of τ_w with an accuracy of only ± 8%, even if u_m and y can be measured very accuratly.

4. Conclusion

The main results of the present work can be summarized as follows:

1. There is no difference in the mean value of the hot-wire response near conducting walls in laminar and turbulent flows. For a specific wall material, overheat ratio and probe geometry, the relationship $u_m^+ = f(y^+)$ holds to a good degreeof accuracy.
2. The yaw response follows the above relationship.
3. For "wind-off" the hot-wire shows a relationship $u_m \sim {}^1/_y$.
4. The wall distance can be determined with a resolution of at least ± 2.5 μm by an interferometric method.

Acknowledgements

The author thanks the Deutsche Forschungsgesellschaft for financial support of this investigation.

References

1. J. A. B. Wills: The correction of hot wire readings for proximity to a solid boundary, J. Fluid Mech. 12, 388-396, 1962.
2. D. C. Collis & M. J. Williams: Two dimensional convection from heated wires at low Reynolds numbers, J. Fluid Mech. 6, 357-84, 1959.
3. B. G. van der Hegge Zijnen: Measurement of the velocity distribution in the boundary layer along a plane surface, Thesis, Delft, 1924.
4. S. Oka, Z. Kostic: Influence of wall proximity on hot wire velocity measurements, DISA Information 13, 29-33, 1972.
5. K. S. Hebbar: Wall poximity corrections for hot wire readings in turbulent flows. DISA Information 25, 15-16, 1980.
6. M. Dianat, I. P. Castro: The measurements of skin friction with a pulsed-wire gauge, Final report on University of Surrey/ RAE a agreement 2064/ 069 XR Aero, 1983.
7. V. C. Patel: Calibration of the Preston tube and limitations on its use in pressure gradients, J. Fluid. Mech. 23, 185-208, 1965.
8. G. Comte-Bellot: Hot-wire and hot-film anemometers. Measurement of unsteady fluid dynamic phenomena, Hemisphere Publ. Corp. 123-162, 1977.

9. E. Berger: Bestimmung der hydrodynamischen Größen einer Kármánschen Wirbelstraße aus Hitzdrahtmessungen bei kleinen Reynoldschen Zahlen, ZFW 12, Heft 2, 41-59, 1964.

10. D. Gerich: Über die Veränderung der Kármánschen Wirbelstraße durch Endscheiben an einem Kreiszylinder. Dissertation, Göttingen, 1985.

11. J. C. Bathia, F. Durst, J. Jovanovic: Correction of hot wire anemomenter measurements near walls, J. Fluid Mech. 122, 411-431, 1982.

12. D. Wąacker: Die numerische Lösung der Navier-Stokesschen Gleichungen für einen beheizten Zylinder in Wandnähe mit Berücksichtigung des Wärmeübergangs, Dissertation, TU Berlin, 1985.

13. M. Acharya, M. P. Escudier: A direct formulation of correlations for wall-proximity corrections of hot-wire measurements. Experiments in Fluids 2, 135-136, 1984.

14. A. S. Zemskaya,V. N. Levitsky, V. N. Repik, U. Ye, P. Yu Sosedko: Effects of the poximity of the wall on hot wire readings in laminar and turbulent boundary layers, Fluid Mech. Soviet Res. 8, 133-141, 1979.

15. J. Cole & A. Roshko: Heat transfer from wires at Reynolds numbers in the Oseen range, Proceedings of Heat Transfer and Fluid Mechanics Institute, Univ. of Calif., 13-23, 1954.

16. L. V. Krishnamoorthy, D. H. Wood, R. A. Antonia, A. J. Chambers: Effect of wire diameter and overheat ratio near a conducting wall, Experiments in Fluids 3, 121-127, 1985.

Wall Velocity Gradient Fluctuations in Turbulent Flow (Newtonian Fluid and Dilute Polymer Solutions)

A. Ambari, C. Deslouis, S. Robin, and B. Tribollet*

LP 15 du C.N.R.S., "Physique des Liquides et Electrochimie", Tour 22, Associé à l'Université Pierre et Marie Curie, F-75252 Paris Cedex 05, France

Electrochemical methods are able to provide local and instantaneous values of the mass transfer rate at a wall-fluid interface. In the framework of a linear theory, assuming the time-dependent fluctuations as ergodic and the system as stationary from a statistical standpoint, the power spectrum density (PSD) of the mass flux $W_{\tilde{J}}$ and that of the velocity gradient $W_{\tilde{\alpha}}$ are linked through the following relationship [1,2] :

$$W_{\tilde{J}} = ||H(jf)||^2 \cdot W_{\tilde{\alpha}} \tag{1}$$

The transfer function H(jf) represents formally the linear response of the time-dependent mass transfer rate $\tilde{J}$ to a sine-wave modulation of low level of the wall velocity gradient $\tilde{\alpha}$. Recently [2], H(jf) was numerically calculated in the case of a two-dimensional velocity field near the wall and the frequency analysis of the experimental transfer function fulfilled very satisfactorily the theoretical predictions.

Two applications of the electrochemical method will be presented here : a study of wall turbulence in a rectangular channel and a study of the polymer-flow interactions close to a wall in elongation and turbulent flows.

1. Principle of the Electrochemical Method

The mass fluxes are measured by means of a redox reaction evolving at an electrode embedded at the wall. If a "high" enough overvoltage is applied at the electrode, one species can be consumed at its interface with the fluid, fast enough so that a perfect mass sink is obtained at the wall ($C_W = 0$).

$$O_x + ne^- \rightleftarrows Red$$

The limiting diffusion current I measured in such conditions is proportional to the mass transfer rate J ($J = D\frac{\partial C}{\partial y}$) as follows : I = nFJ. D is the molecular diffusion coefficient and F is the Faraday number.

* Adresse actuelle : Ecole Hassania des Travaux Publics, B.P. 8108, Casablanca 02 - Maroc

In this work, two electrochemical reactions were used on platinum electrodes :

$I_3^- + 2e^- \rightleftarrows 3I^-$ (KI N/10)

for wall turbulence studies in a rectangular channel and

$Fe(CN_6)^{3-} + e^- \rightleftarrows Fe(CN_6)^{4-}$ (KCl N)

for the study of polymer-flow interactions.

2. Mass Transfer Analysis

For high Schmidt numbers the concentration boundary-layer near the wall is so thin that the velocity field can be represented by the first terms in a Taylor series expansion. For a fully developed flow in the x-direction the three velocity components are then given as :

$$\bar{U} = \bar{\alpha}y + \alpha(x,z,t)y$$
$$v = \beta(x,z,t)y^2$$
$$w = \gamma(x,z,t)y$$

The mass balance equation relates the time dependent concentration field $C(x,y,z,t)$ to these velocity components :

$$\frac{\partial C}{\partial t} + \vec{V}.\nabla C = D\Delta C \qquad (2)$$

By an appropriate choice of the microelectrode configuration some symplifing hypothesis can be proposed. The concentration derivatives (first and second) with respect to the transversal direction z tends towards a zero value. If the width of the microelectrode is thin enough, the concentration profile cannot be developed far from the wall and the effect of the turbulent diffusivity may be ignored. Thus, the equation (2) depends only on the velocity gradient α :

$$\frac{\partial C^+}{\partial t^+} + U^+ \frac{\partial C^+}{\partial x^+} = \frac{1}{S_c} \left(\frac{\partial^2 C^+}{\partial y^{2+}} + \frac{\partial^2 C^+}{\partial x^{2+}}\right) \qquad (3)$$

All quantities have been made dimensionless using the friction velocity v_0, the kinematic viscosity ν and the bulk concentration C_∞.

Wall Turbulence in a Rectangular Channel

The experiments were conducted in a two dimensional rectangular smooth channel which is 2.85m long, 12cm wide, and either 1.2cm or 1.7cm high. The electrochemical section is located at 2m from the upstream end to ensure fully developed flow conditions.

By measurements of limiting diffusion currents on the microelectrode, we obtain as variation for the coefficient of skin friction :

$$C_f = 0.065.Re^{-1/4} \quad \text{for } h = 1.2\text{cm} \tag{4}$$

and $$C_f = 0.051.Re^{-1/4} \quad \text{for } h = 1.7\text{cm} \tag{5}$$

in the Reynolds number range : $10^4 < Re < 2.5.10^4$. Re is defined by : $Re=V.h/\nu$; where V is the flow rate velocity, ν the kinematic viscosity. The equations (4) and (5) are in good agreement with Dean's measurements [3] ($0.073.Re^{-1/4}$) for aspect ratio greater than 7 and with JOHNSTON'S [4] previous suggestion ($0.0706.Re^{-1/4}$). In the second relation where the aspect ratio is 7, the significant difference of the friction factor variation might be explained by the assumption that a two dimensional flow is no more available.

One dimensional longitudinal wall velocity gradient spectrum functions $W_{\tilde{\alpha}}(f)$ have been measured in various turbulent flows with the means of $W_{\tilde{J}}(f)$. Examples of a PSD of limiting diffusion current fluctuations and the corresponding one to the PSD of wall velocity gradient fluctuations for Re = 12800 are showed respectively in figures 1A and 1B. In figure 1B, we find a "-1.67" slope for the line portion in the high frequency range.

For all measured velocity gradient spectra, one obtains slopes included in the following range : -1.66 ± 0.03

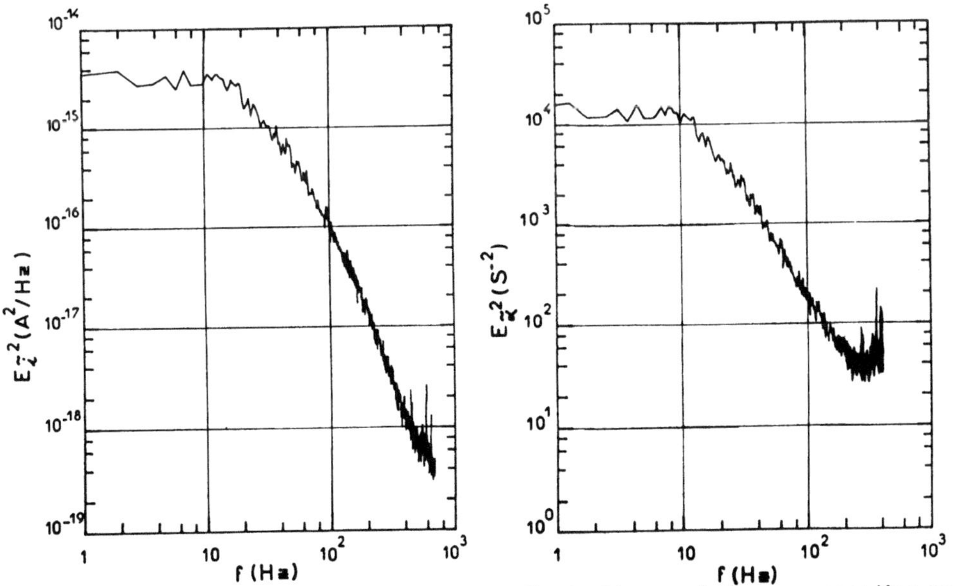

Fig. 1 : PSD of limiting diffusion current fluctuations and the corresponding one to the PSD of wall velocity gradient fluctuations (Re = 12800)

This value is in very good agreement with the second hypothesis of the Kolmogorov's theory for isotropic turbulence, providing the existence of an inertial subrange near the wall, where the spectra have a "-5/3" slope.

In these experimental velocity gradient spectra, we observed a substantial part of the inertial subrange but the dissipation range at higher frequencies where the spectrum falls off much more rapidly with an appreciable effect of the viscosity, was not obtained.

We may use Taylor's hypothesis, assuming that the disturbances are transported with the flow rate velocity without distortion. So, the wave number k is related to the frequency by :

$$k = \frac{2\pi f}{V} \tag{6}$$

Using the equation (6), we can obtain the gradient velocity spectra in function of k and they are showed in all the Reynolds number range for the depths h = 1.2cm in figure 2A and h = 1.7cm in figure 2B. Here in these two precedent figures, we see that the cut-off wave number k_c is independent of the Reynolds number and depends on the depth of the channel. This wave number is related to the size of the largest disturbances. So, we can read in these curves :

$k_c = 0.655 \pm 0.050 cm^{-1}$ for h = 1.2cm

$k_c = 0.515 \pm 0.035 cm^{-1}$ for h = 1.7cm

Roughly, if we assume the disturbances as circular eddies transported with the flow rate velocity V, the diameter of the largest disturbances is equal to half the depth h of the duct. The wave number associated is about :

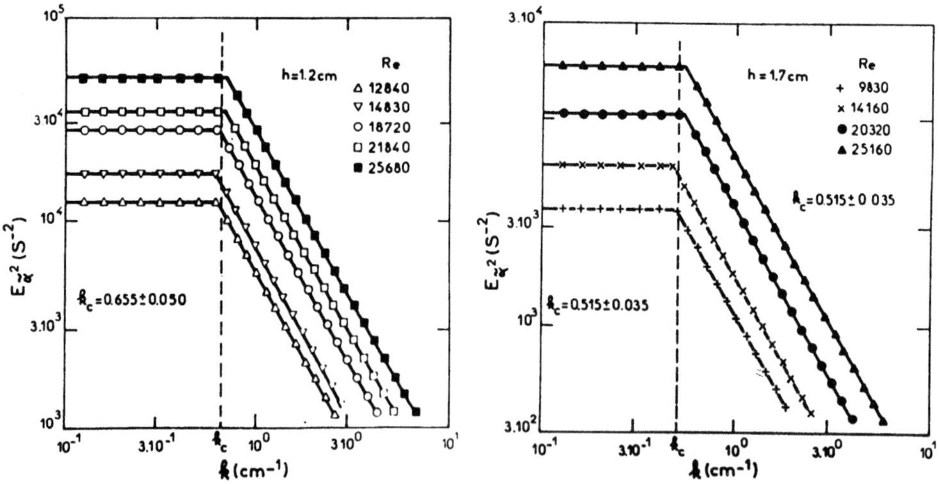

Fig. 2 : Gradient velocity spectra for two differents depths h

$$k_1 = \frac{2}{\pi . h} \qquad (7)$$

This wave number varies only with h. We have :

$k_1 = 0.531 cm^{-1}$ for $h = 1.2 cm$

$k_1 = 0.374 cm^{-1}$ for $h = 1.7 cm$

The divergence between k_c and k_1 is less than 25%, which confirms that the first attenuation on the PSD of the diffusion current fluctuations (see figure 1) is actually ascribed to hydrodynamics and is not due to a filtering by the concentration boundary layer.

Mass transfer in dilute polymer solution

All data in this section have been obtained with a microelectrode embedded on a rotating disk.

1. Laminar Flow Field

The laminar flow field created by an infinite disk rotating in an incompressible fluid with constant physical properties was first studied by VON KARMAN [5]. The velocity components can be written in the following form by use of quantities characteristic of the boundary layer : $V_r = r\Omega F(\xi)$, $V_\theta = r\Omega G(\xi)$ and $V_y = \sqrt{\nu\Omega}\, H(\xi)$ with $\xi = y\sqrt{\Omega/\nu}$. r, θ and y are the radial, circumferential and axial coordinates in a cylindrical system (Ω is the angular velocity). The rate of strain tensor may be derived and thus the elongational gradient S may be calculated and is equal to $2F\Omega$. This is the most salient feature for the polymer study : all the fluid particles entering the boundary layer bear the same elongational gradient at a given distance from the wall.

From [5], the maximum value of $F(\xi)$ is of the order of 0.18 and therefore $S_{max} \sim 0.36\Omega$. According to theoretical predictions by DE GENNES [6] and HINCH [7] for an elongational gradient S high enough (exceeding a critical value S^*) a linear flexible chain polymer, originally under a coil shape, could be sharply stretched into a fully extended shape. This coil $\rightleftharpoons$ stretch transition is expected when the criterion $S^*.\tau_Z > 1$ is fulfilled [6] where τ_Z is the Zimm relaxation time of the polymer coil in an infinitely dilute solution.

According to earlier works, namely by GREGORY, STEWART and WALKER [8], three domains can be characterized in a newtonian fluid, i.e. the laminar, transition and turbulent flows.

From the time average data for the mass transfer rate on the microelectrode reported in figure 3, only two domains appear : the first one below a Reynolds number

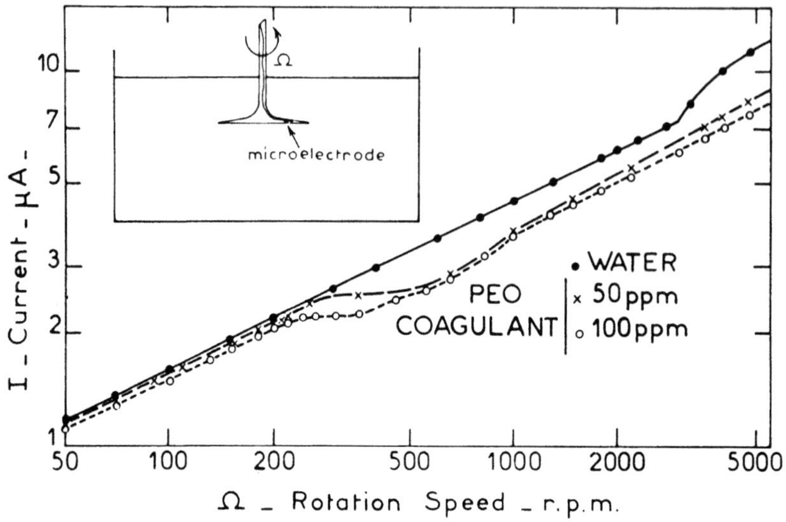

Fig. 3 : Time average current versus the rotation speed

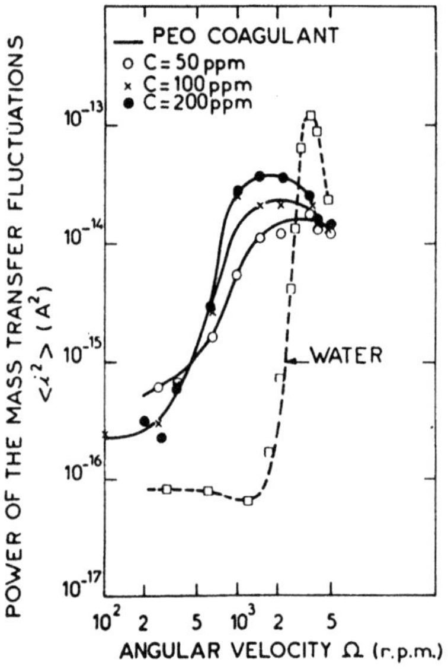

Fig. 4 : rms values of the mass transfer fluctuations for water and PEO solutions vs the angular velocity

Re ($Re = \Omega r^2/\nu$) of 3.2 10^5 (Ω = 3800r.p.m.) corresponds to a mass transfer equation : $I \propto \Omega^{1/2}$ previously established in laminar flow [9].

The second domain above this Re value corresponds to a fully developed turbulent flow. However fluctuations occur much below the critical Reynolds number, and have been ascribed to the development of Stewart vortices [8]. In fact, when conside-ring the turbulent mass transfer intensity of these fluctuations (figure 4) in water a second critical Re number of 1.4 10^5 (Ω = 1700r.p.m.) marking the onset of these vortices can be measured. It may be noticed in the same figure, that the maximum turbulent intensity corresponds to Re close to the critical Reynolds number (3.2 10^5).

When small amounts of a polyethylene oxide "PEO coagulant" ($\overline{M}_W \sim 5.10^6$) are added to the solution, a deviation with respect to the solvent curve is observed, starting from small Ω values ($\sim$300r.p.m). At the same time, and when increasing Ω, instabilities take place, in a velocity range far below the onset of the Stewart vortices in water. The critical Ω^* -and therefore S^*- values were determined by considering the velocity for which the derivative of the fluctuations power with Ω is maximum (point of inflection on the curves (o), (x), and (•) in figure (4)). The obtained values for S^{*-1} lie between 3 and 4.10^{-2}s and are of the same order of magnitude as the Zimm relaxation time, $\tau_Z \sim 10^{-2}$s.

2. Power Spectrum Densities

a) Laminar Flow

In what follows, the PSD of the gradient velocity fluctuations only will be presented, after the procedure defined previously of taking out the effect of the mass transfer attenuation. The data are given in figure 5, the PSD corresponding to 100ppm shows a lower cut-off frequency than the 50ppm solution.

b) Transition Flow

In figure 6 are plotted the spectra for water and for a 50ppm PEO solution at = 2200rpm, i.e. in transition flow for water. The spectrum in water shows a rather complex shape with a marked shoulder centered around 400Hz which contains a significant part of the energy of the overall fluctuations. This frequency band is related to the Stewart vortices.

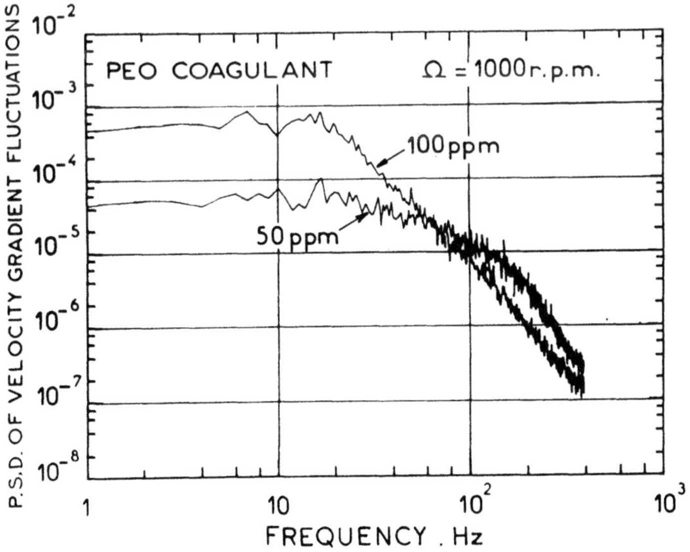

Fig. 5 : PSD of the gradient velocity fluctuations in laminar flow

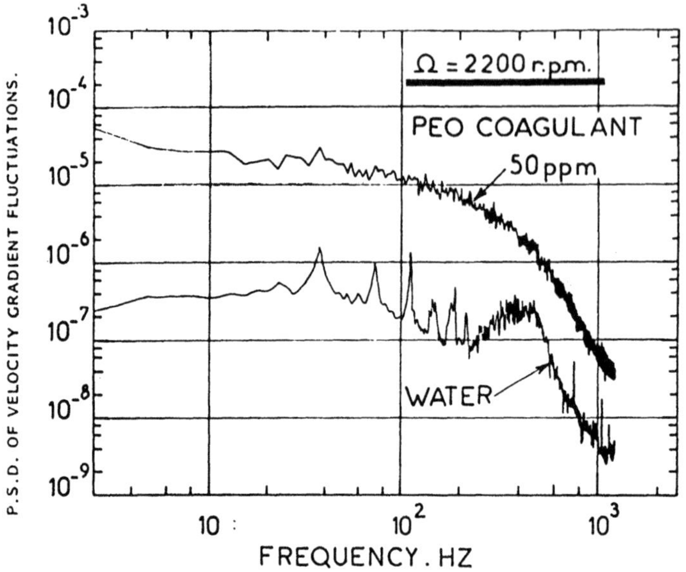

Fig. 6 : PSD of the gradient velocity fluctuations in transition flow

In PEO solution, the spectrum appears more smoothed, the peaks assigned to the fundamental (40Hz) and harmonies of the angular velocity as well as that of the Stewart vorticies having vanished. On the same time, the energy level is considerably increased at all frequencies.

c) Turbulent and Drag Reduction Regimes

The spectra for water and PEO solution in figure 7 exhibit similar features. However, a cross-over of the curves happens at about 1kHz where begins the inertial subrange, the energy contained in the large structures being higher for the PEO solution and on a contrary the smaller structures appearing more damped in the PEO solution.

In the inertial subrange, a dependence of the form $W_{\tilde{\alpha}} \propto f^{-2}$ for water is found, at variance with the usual "five-third" Kolmogorov law. In the PEO solution, the slope of the HF range portion of the curve equals about -4 as for the laminar and transition flows.

The power of the mass transfer fluctuations is shown in figure 7 for water and different PEO solutions (50,100,200 and 1200ppm) as functions of the angular velocity. The sharp increase of the power of the fluctuations corresponds to the point where the time average mass transfer rate decreases as regards to that in the solvent alone ; the maximum value for the polymer is approached whereas no Stewart vorticies are detected from the solvent curve examination.

The most striking feature concerning the sequence of the different spectra in the PEO solutions is that they display a quasi identical shape in the three hydrodynamic regimes, with the same -4 slope at high frequencies, their cut-off frequency

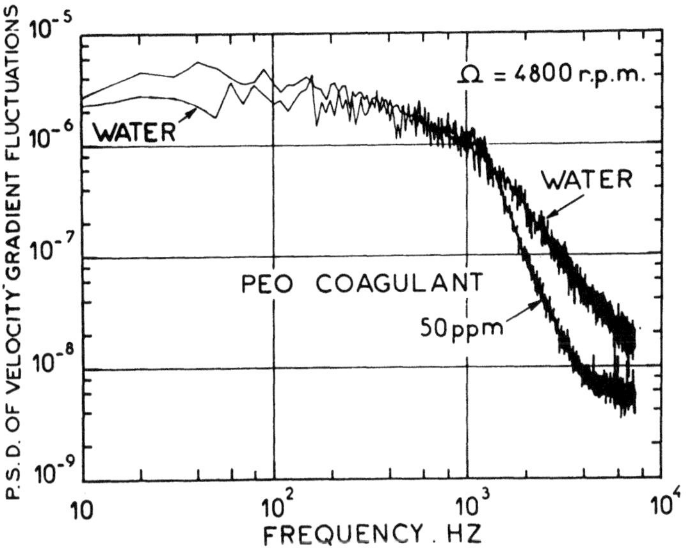

Fig. 7 : PSD of the gradient velocity fluctuations in turbulent flow

depending linearly on the angular velocity from 1000 to 4800r.p.m. This observation tells in favour of a shift of the onset of the transition to turbulence and also a severe damping of the inertial subrange.

References

1. C. Deslouis, B. Tribollet and L. Viet : Ann. N.Y. Acad. Sci. Vol.404 p.471 (1983)
2. A. Ambari, C. Deslouis and B. Tribollet : I.J.H.M.T. Vol.29 n°1 p.35 (1986)
3. R.B. Dean : I.C. Aer. Report 74-11, Imp.Coll. of Sci. and Tech., England (1974)
4. J.P. Johnston : Fluid Dyn. Panel Spec. Meeting on Turb. Shear. Flows, London (1971)
5. Von Karman : Zeit Angew. Math. und Mech. Vol.1 p.233 (1921)
6. P.G. de Gennes : Scaling concepts in polymer physics, Part A, Cornell Univ. Press (1979)
7. E.J. Hinch : In Polymère et Lubrification, Coll. Int. CNRS, Brest p.241 (1974).
8. N. Gregory, J.T. Stuart and W.S. Walker : Phil. Trans. of the Roy. Soc., London, A248, p.155 (1955).
9. L. Mollet, P. Dumargue, M. Daguenet and D. Bodiot : Electrochim. Acta, Vol.12, p.841 (1974).

Image Analysis of the Spinning Cylinder near Wake

J. Massons, J. Escoda, Jna. Gavaldà, and F. Diaz

Dpt. Fisica, Fac. Ciències Quimiques, Universitat de Barcelona,
E-43005 Tarragona, Spain

An experimental determination of statistical properties of the interface separating turbulent and non-turbulent fluid in a stationary and spinning cylinder near wake has been made by means of flow visualization and image processing techniques. The results, which comprise intermittency factor, auto and cross-correlations, auto and cross-spectra and phase spectra, indicate that the rotation of the cylinder provides substantial changes in the structure of the flow, and the interface position may be regarded as a stationary gaussian random variable.

1. Introduction

In recent years image processing and flow visualization techniques have been of considerable interest for the analysis of different situations in fluid mechanics, (SHOKR et al., [1], HERNAN et al., [2]). These experimental techniques can provide interesting statistical information about the boundary between turbulent and non-turbulent flow, which essentially characterizes the turbulent shear flows and justifies the entrainment mechanism in such flows.

The results given by LARUE et al. [3], suggest that the interface position of the fully developped turbulent cylinder wake is a stationary gaussian random variable. The present paper provides experimental information about the interface of a stationary cylinder near wake and also shows the changes introduced in the interface when the wake is generated behind a rotational cylinder. Cine pictures were taken of the wake, in a flow visualization water channel. These pictures were analyzed by means of image processing techniques in order to establish the most significant features of the interface. The properties of the turbulent / non-turbulent interface were investigated using statistical functions, i.e., intermittency factor, auto and cross-correlation functions, auto and cross-spectra and phase spectra.

2. Experimental Details

The flows analyzed in this paper were generated by the interaction between a stationary or rotating cylinder and a potential flow. These were produced in a flow visualization channel described by MASSONS et al. [4]. A definition sketch of the flow configuration is given in Fig. 1.

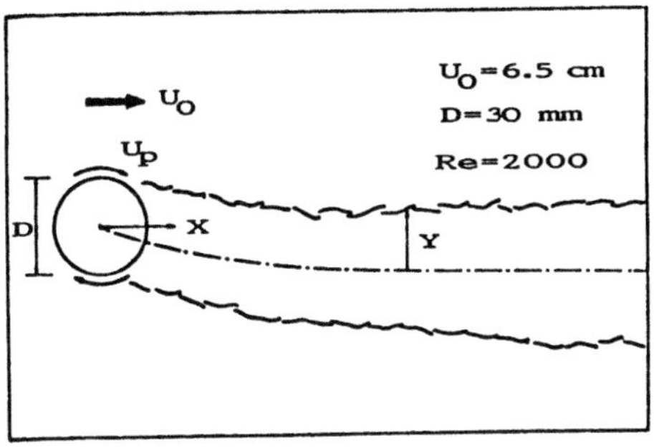

Fig. 1 Definition sketch of the flow

Flow visualization experiments were carried out introducing dye into the flow and cine pictures were taken for two experimental situations, $U_p/U_0=0$ (stationary cylinder case) and $U_p/U_0=2$ (rotating cylinder case). Figure 2 shows two frames corresponding to the two different wakes studied in the present work. The film rate was 24 frames per second and 600 frames were used for analyzing each configuration. The pictures were digitalized in a PERKIN-ELMER microdensitometer, and analyzed using the specific software of the IBAS/IPS image equipment. The generated digital data were finally processed by the IBM 3083 computer at the Universitat de Barcelona.

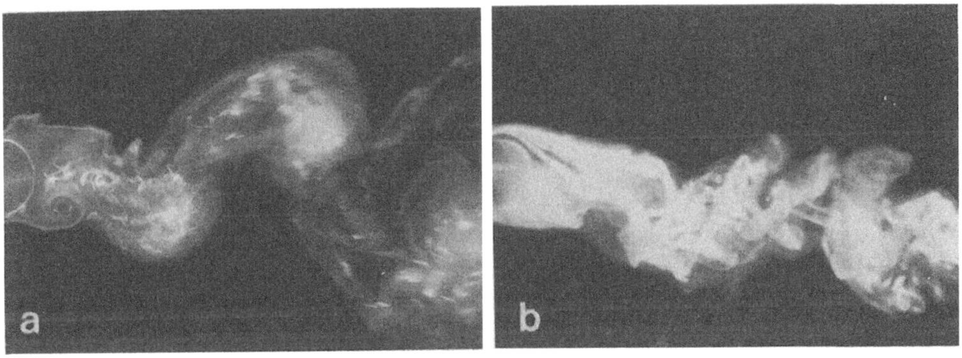

Fig. 2 Two frames of the flow visualization film a) $U_p/U_0=0$ b) $U_p/U_0=2$

3. Analytical Considerations

The time evolution of the position of the interface in the upper boundary of the wake, Y(tj), is studied in this paper. Statistical properties are examined by means of the intermittency factor, the auto and cross-correlation functions, the auto and cross-spectra and the phasé spectra.

The intermittency factor, γ, points to the probability of occurrence of turbulent flow at a specified location. In the present study, this definition is equivalent to the probability of the lateral interface coordinate being equal to or exceeding a specific value, Y_0,

$$\gamma = \text{Prob} (Y_0 \leq Y) \qquad (1)$$

The auto and cross-correlation functions and the auto and cross-spectra for the interface coordinates are defined as

$$\rho_{nn'}(k\Delta t)=\frac{1}{N-k}\sum_{j=1}^{N-k}(y_n(t_j)y_{n'}(t_{j+k})) \qquad \rho_{n'n}(-k\Delta t)=\rho_{nn'}(k\Delta t) \qquad \begin{matrix}\text{if } n=n' \text{ auto-correlation}\\ \text{if } n\neq n' \text{ cross-correlation}\end{matrix} \qquad (2)$$

$$E_{nn'}\left(\frac{k}{N\Delta t}\right)=\left|C_{nn'}\left(\frac{k}{N\Delta t}\right)\right|=\left|\Delta t\sum_{m=1-N}^{N-1}\rho_{nn'}(m\Delta t)\exp(-2\pi i\, m k/N)\right| \qquad \begin{matrix}\text{if } n=n' \text{ auto-spectrum}\\ \text{if } n\neq n' \text{ cross-spectrum}\end{matrix} \qquad (3)$$

where N is the number of frames digitalized, n and n' are the values of X/D corresponding to the correlated stations, $\Delta t=1/24$ s, $k=0,1,2,\dots,K$, with $K<<N$, $y=(Y-\bar{Y})/y'$ and $\bar{Y}$ and y' are the mean value and the root-mean-square of the interface position.

Finally, the phase spectrum is given by

$$\phi_{nn'}=\text{arc tg}\left[-\text{Imag}(C_{nn'})/\text{Real}(C_{nn'})\right] \qquad (4)$$

4. Results and Discussion

Figure 3 shows the lateral distribution of the intermittency factor for stationary and spinning cylinder cases with respect to the standarized gaussian variable $(Y-\bar{Y})/y'$, at different longitudinal positions. The dashed curve represents the distribution function

$$\gamma(z) = \left(1/\sqrt{2\pi}\right)\int_z^{\infty}\exp(-s^2/2)\,ds \qquad (5)$$

From these results it is seen that the basic processes governing the motion of the stationary cylinder near wake are not gaussian, because of the coherent structures present in this flow. One can also observe a displacement between the point $(Y-\bar{Y})/y'=0$ and the corresponding $Y(\gamma=1/2)$. On the other hand, in the spinning cylinder situation

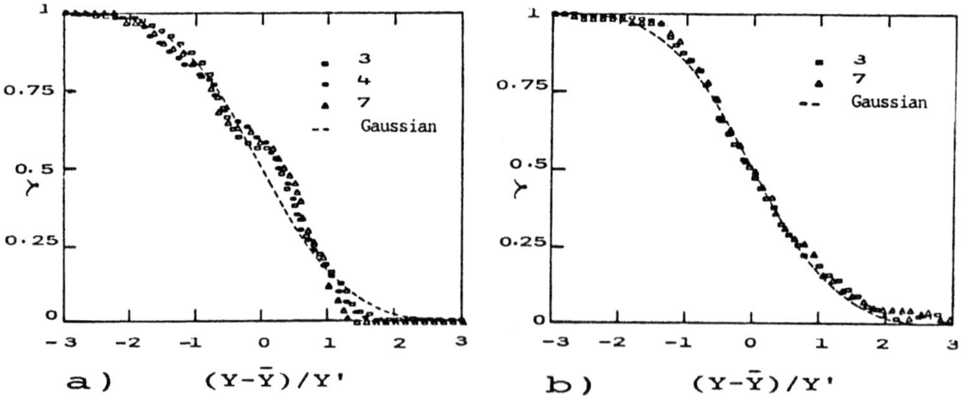

Fig. 3 Normalized intermittency factor lateral distributions

a) $U_p/U_0 = 0$ b) $U_p/U_0 = 2$

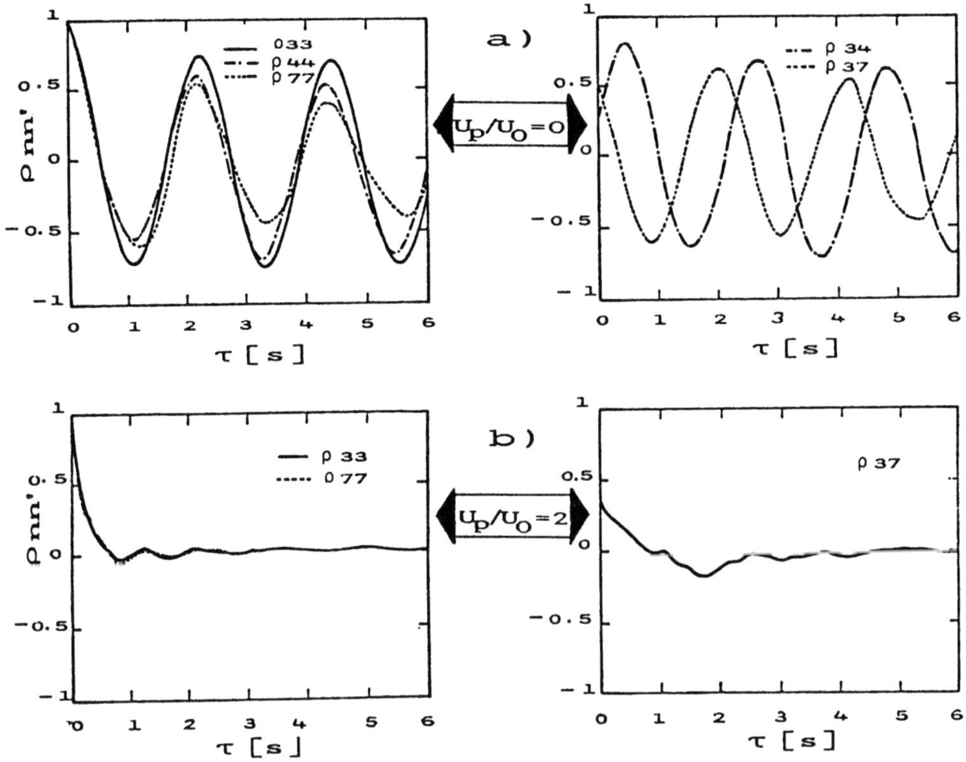

Fig. 4 Auto and cross-correlation functions a) $U_p/U_0=0$ b) $U_p/U_0=2$

the process becomes partially gaussian as no significant structures are shed from the cylinder (DIAZ et al.,[5]).

The auto and cross-correlation functions for different longitudinal positions are illustrated in Fig. 4. In the stationary case the results contain regular oscillations indicating that a well-defined Karmán vortex street exits in the near wake region. The slow but significant decay of the amplitude of these oscillations shows that the vortex shedding process is randomly modulated in frequency and phase, as KEFFER et al. [6] have demonstrated. Figure 4 shows a decay of the Kármán activity and a increase in the random modulation of the shedding process with rotational speed, which must be linked to the thickening of the layer of rotating fluid near the cylinder with rotation, consequently the separation process occurs away from the surface of the cylinder and under different lateral velocity gradients in the regions above and below the cylinder. We note that these findings are consistent with those published by DIAZ et al. [5].

A frequential characterization ot the flows is carried out translating the auto and cross-correlation results to the frequency domain. Figure 5 presents the auto and cross-spectra. These results confirm those mentioned above. A well-defined peak, which is related to the shedding frequency of the Kármán vortex (Strouhal frequency), exits for stationary cylinder wake. The effects of the cylinder rotation on the

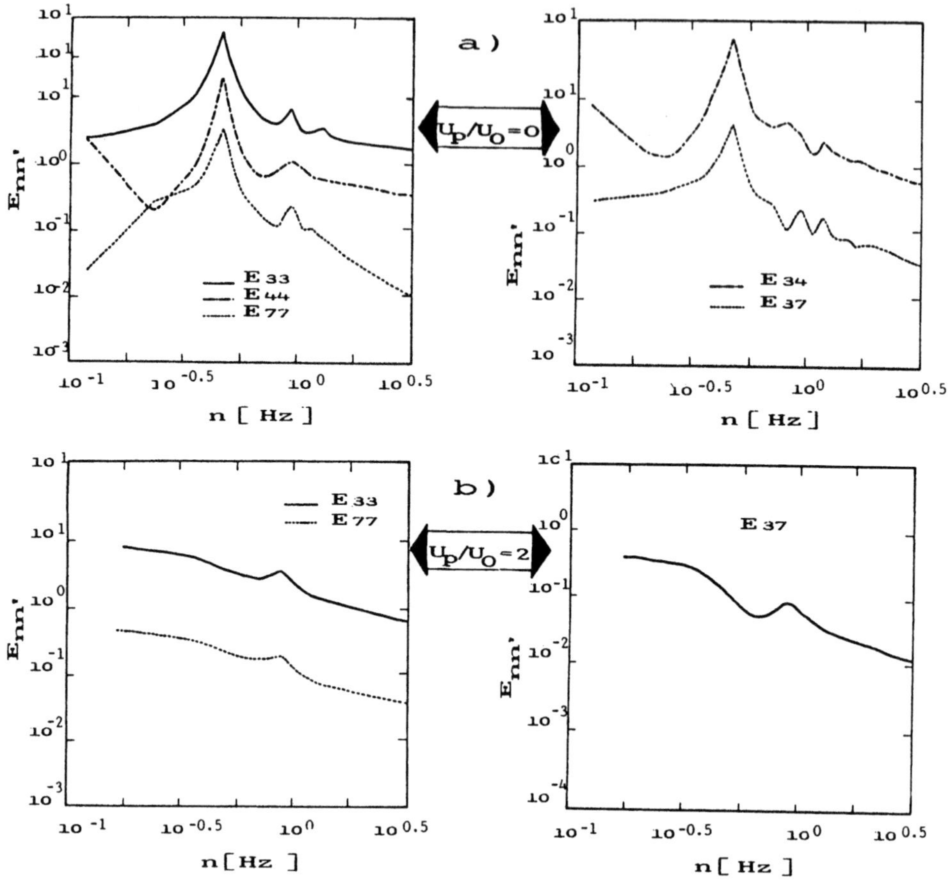

Fig. 5 Auto and cross-spectra functions a) $U_p/U_0=0$ b) $U_p/U_0=2$

spectral distribution of the rotating cylinder wake are illustrated in Fig. 5b. As can be seen, the rotation of the cylinder intensifies the inherent random modulation in the shedding process. A redistribution and homogenization of the structures is produced and, thus, the spectral peak becomes substantially lower than that observed in the stationary case. The associated frequencies are 0.48 Hz and 0.87 Hz, which yields a Strouhal number of about 0.22 and 0.40 for the stationary and spinning cylinder cases respectively, in agreement with those obtained by DIAZ et al. [5] using hot wire anemometry.

Finally, Figure 6 shows the phase spectra for the two experimental situations. The phase values are therefore consistent with those imposed by the free-stream velocity and the displacement between positions. It can also be observed that no-coherent contributions to the signal exist for large frequencies, showing a random oscillation of the associated phase values.

5. Conclusions

The present work shows that flow visualization in combination with image processing techniques is a suitable experimental method for research in turbulence.

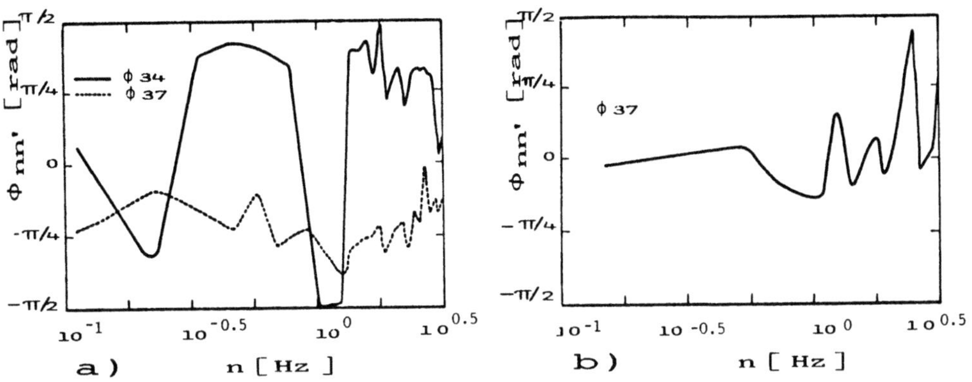

Fig. 6 Phase spectra functions a) $U_p/U_0=0$ b) $U_p/U_0=2$

The experimental results obtained in this study establish that shedding of Karman vortices in a cylinder-generated wake is modified by rotation of the cylinder. A non-gaussian behaviour governs the motion in the stationary cylinder near wake, whereas statistical analysis of the interface separating turbulent and non-turbulent fluid demostrates that the interface position of the near wake may be regarded as a stationary gaussian random variable, for the rotational cylinder situation , ($U_p/U_0=2$).

Acknowledgements

The authors wish to thank Dr. Ll. Sole for his scientific contributions in the image processing technique. The technical facilities provided by TAQSA and ALCUDIA S.A. are acknowledged as well.

References

1. M. Shockr, J.F. Keffer : Structure of Complex Turbulent Shear Flows, ed. by R. Dumas and L. Fulachier, IUTAM Symp., Marseille, (Springer, Berlin, Heidelberg 1982) p. 165
2. M. Hernan, J. Jiménez : J. Fluid Mech. 119,323 (1982)
3. J.C. LaRue, P.A. Libby: Phys. Fluids 19, 1864 (1976)
4. J. Massons, F. Díaz, Jna. Gavaldà, X. Ruiz: Proceedings of the XX Reunion Bienal de la RSEF, Sitges, Barcelona, (1985) p. 317
5. F. Díaz, J. Gavalda, J.C. Kawall, J.F. Keffer, F. Giralt: Phys. Fluids 26, 3454 (1983)
6. J.F. Keffer and J.G. Kawall: Proceedings of the AIAA-13Th Fluid & Plasme Dynamics Conference, Snowmass, Colorado (1980) P. 15.18

Part VIII

Engineering Applications of Turbulence and the Effect of External Disturbances

Applications Industrielles de la Turbulence, ses Progrès Récents, son Avenir

P. Perrier

Chef du Département Aérodynamique, Avions Marcel Dassault, 78 Quai Carnot, F-92210 St. Cloud, France

0. INDUSTRIE DE POINTE ET RECHERCHE FONDAMENTALE

Les remarques que nous allons formuler sur les progrès dans les applications industrielles des recherches récentes dans le domaine complexe de la modélisation des écoulements turbulents sont valables essentiellement pour l'industrie de pointe disposant et d'équipes de haut niveau, de moyens de calcul et de tests expérimentaux assurant un transfert efficace des recherches de base. Cependant nous pensons que ces progrès préfigurent un usage plus général de ces méthodes en milieu industriel, grâce à la diminution probable d'un à deux ordres de grandeur du coût des calculs dans la décennie à venir ; de plus, les méthodes de calcul a priori des écoulements turbulents et leur efficacité, ne cesseront pas de justifier, par leurs progrès, une meilleure rentabilité de la prévision initiale des écoulements turbulents complexes. Ainsi les voies ouvertes actuellement par l'industrie de pointe et spécialement l'industrie aéronautique, ainsi que les problèmes posés, peuvent utilement orienter et préparer l'avenir et servir ainsi de banc d'essai aux progrès de la recherche fondamentale.

1. LES ECOULEMENTS TURBULENTS A SIMULER ET L'ETAT SOMMAIRE DE LEUR SIMULATION

L'industrie aéronautique est caractérisée par un poids extrêmement élevé de l'aérodynamique dans les performances et même la simple viabilité de ses programmes d'avions, d'engins et l'intérêt renouvelé pour les problèmes hypersoniques, ouvre les problèmes de mécanique des fluides à résoudre vers les hautes températures et les effets de raréfaction et de dissociation, c'est-à-dire au-delà des équations de Navier-Stokes vers les équations de Boltzmann et vers leur couplage avec les équations de la chimie.

Ainsi le spectre des applications, tant en écoulement externe, ce qui était le domaine habituel des avionneurs, qu'en écoulement interne, s'est élargi à l'ensemble des domaines couverts dans l'industrie mécanique et thermique. Trois grands axes de progrès se dégagent actuellement et ont chacun renouvelé l'intérêt pour les recherches fondamentales en turbulence.

Dans le domaine des avions civils, la recherche d'une économie maximale a conduit à reprendre complètement la viabilité des gains de traînée possible par traitement des couches limites. D'abord, s'est posé le problème d'une prévision valable de la transition à la turbulence et de la longueur, et donc du processus physique de celle-ci. Des calculs se mettent en place pour la prévision systématique des instabilités des couches limites laminaires et sur le processus de croissance des perturbations initiales ou sur l'amortissement des grosses structures turbulentes. Ces études supposent un calcul très précis des couches limites tridimensionnelles sur l'ensemble d'un avion, tant en laminaire qu'en turbulent, et une compréhension renouvelée du phénomène de croissance des perturbations, de l'intermittence et de l'interaction entre le frottement de paroi et l'écoulement turbulent extérieur à la frontière très inhomogène de la sous-couche laminaire. Les progrès sur la prévision des instabilités pour un champ de vitesse donnée ont été rapides mais les progrès sur la connaissance des processus de formation, voire le traitement des structures turbulentes des couches limites, sont plus lents et loin de déboucher sur des prévisions théoriques mais se cantonnent actuellement dans l'accumulation de résultats expérimentaux, souvent encore loin de la qualité nécessaire à un progrès de la compréhension fine des processus.

Un autre domaine où la pression de la nécessité conduit à accroître les progrès de la modélisation est le traitement des zones décollées. Si les progrès des calculs des couches limites tridimensionnelles permettent une prévision de plus en plus fine des lignes de décollement, voire même du rotationnel émis le long de ces lignes de décollement, le traitement du début de transition à une turbulence établie, spécialement en présence de forts rotationnels, est encore médiocrement modélisable car insuffisamment compris. C'est tout le problème du retour à l'isotropie d'une turbulence fortement hors d'équilibre qui est posé. De plus, on sait mieux maintenant que le champ de pression dépend fortement des dissipations dans cette zone initiale ; les progrès conditionnent donc la simple prévision de la traînée ou de la portance des avions à grande incidence, voire des voitures ou de tous véhicules fonctionnant normalement avec des zones décollées importantes, même si la connaissance du détail de l'écoulement instationnaire ne semble pas nécessaire. Pour les écoulements internes, une même exigence de compréhension conditionne les progrès sur la connaissance des écoulements dans les conduits d'entrée d'air et leur fonction de transfert vis-à-vis de la part des fluctuations turbulentes de fréquence significative pour les zones de compresseur en aval.

Enfin, la modélisation des écoulements complètement hors d'équilibre à la fois chimique et thermodynamique, qui caractérise les exigences de prévision des flux thermiques sur les avions hypersoniques comme Hermès au cours de sa rentrée (fig. 1), va dans un avenir proche combler une part de l'intervalle qui séparait les aérodynamiciens des fluides parfaits des thermiciens ou spécialistes de combustion. L'introduction des modes entropiques de la turbulence et des instabilités de nature thermique met à son maximum l'exigence de compréhension des cas plus simples de turbulence incompressible.

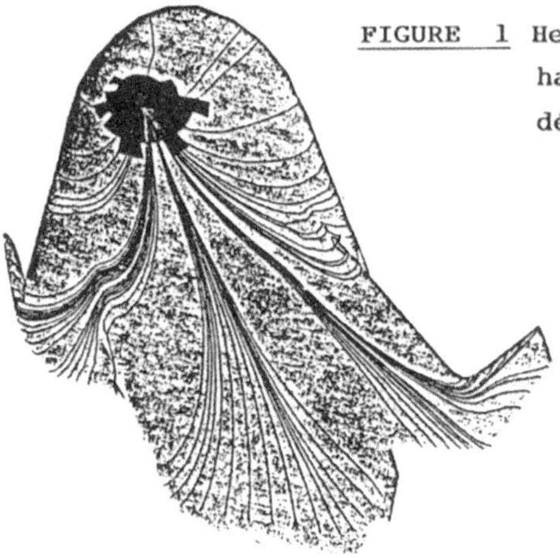

FIGURE 1 Hermès : couches limites turbulentes à haute enthalpie avec dissociation et déséquilibre thermodynamique

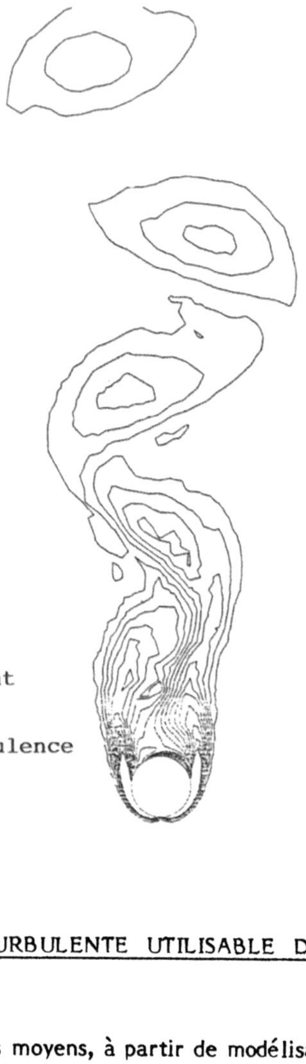

FIGURE 2 Simulation directe d'un écoulement turbulent derrière un cylindre par homogénéisation sur une turbulence tridimensionnelle en équilibre

2. PROGRES RECENTS DE LA MODELISATION TURBULENTE UTILISABLE DANS L'INDUSTRIE

La phase préalable de calcul des écoulements turbulents moyens, à partir de modélisation locale simple, est maintenant bien gérée. Elle consiste essentiellement en la maîtrise des trois premiers niveaux du calcul des écoulements moyens : calcul des solutions des équations de Navier-Stokes avec une viscosité turbulente équivalente locale, calcul d'équations complémentaires exprimant la convection et la diffusion des quantités turbulentes k, ε , ou équations exprimant le comportement des composantes du tenseur de Reynolds et leur modélisation dans l'espace physique (fig. 4). Si l'emploi de ces méthodes est peu systématique, cela tient plus souvent au coût des calculs ou à la complexité des cas pratiques d'utilisation, c'est-à-dire à la complexité insuffisante des codes de calcul petit à petit levée par l'emploi de plus en plus courant des méthodes d'éléments finis.

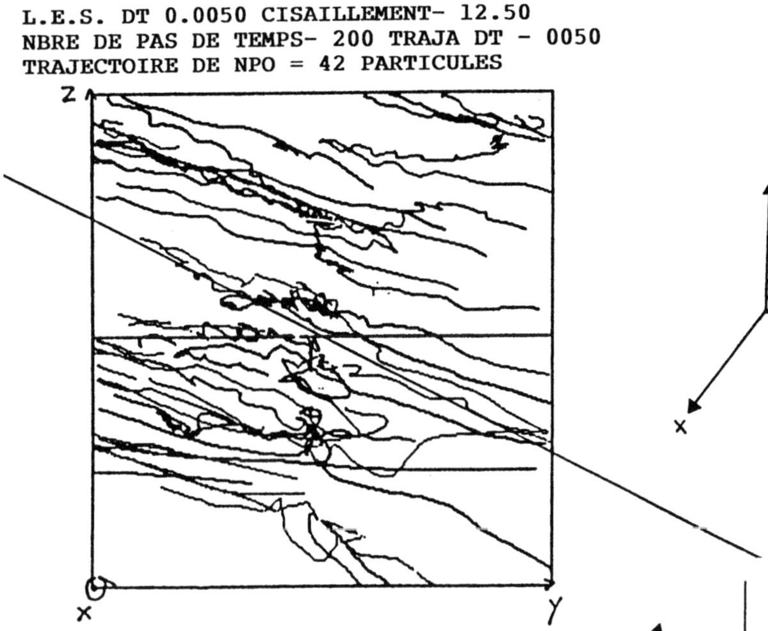

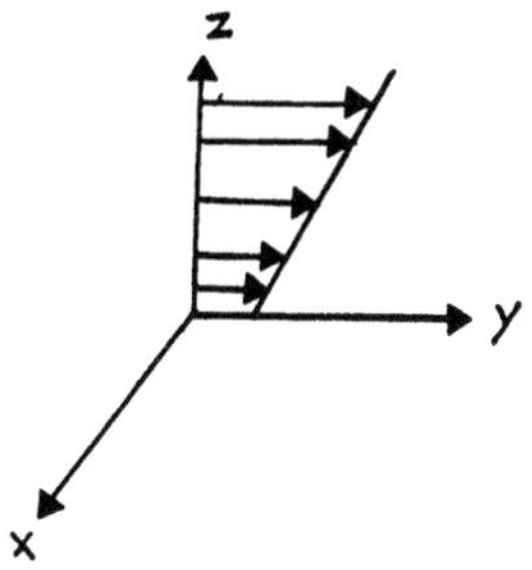

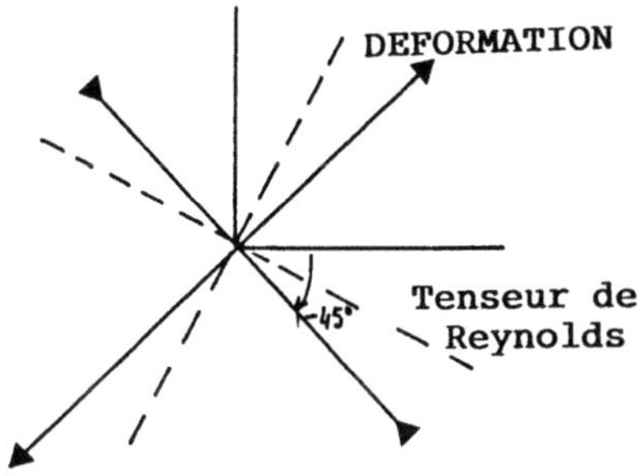

FIGURE 3

Simulation directe de turbulence trajets de particules en écoulement cisaillé

Mais la principale limitation à l'emploi unique de telles méthodes vient de leur limitation à des cas de turbulence bien établie, c'est-à-dire où il ne reste plus de trace notable des conditions initiales par nature inhomogènes.

Une telle hypothèse suppose une longueur d'établissement au moins égale à 3 à 5 fois la taille des plus grosses structures générées, ce qui exclut la plupart des zones d'intérêt majeur proche des décollements, des culots, voire d'une simple interaction onde de choc-couche limite, ce qui restreint la modélisation à des écoulements de couche limite, sillage, jets dans leurs phases établies ou aux écoulements confinés bien après la génération des plus grosses structures à condition de diminuer son intérêt au comportement macroscopique, voire moyen de l'écoulement.

Il importait donc de faire progresser la modélisation vers une prise en compte d'une turbulence moins en équilibre (réf. 1) et deux voies complémentaires se sont ouvertes effectivement au cours des vingt dernières années.

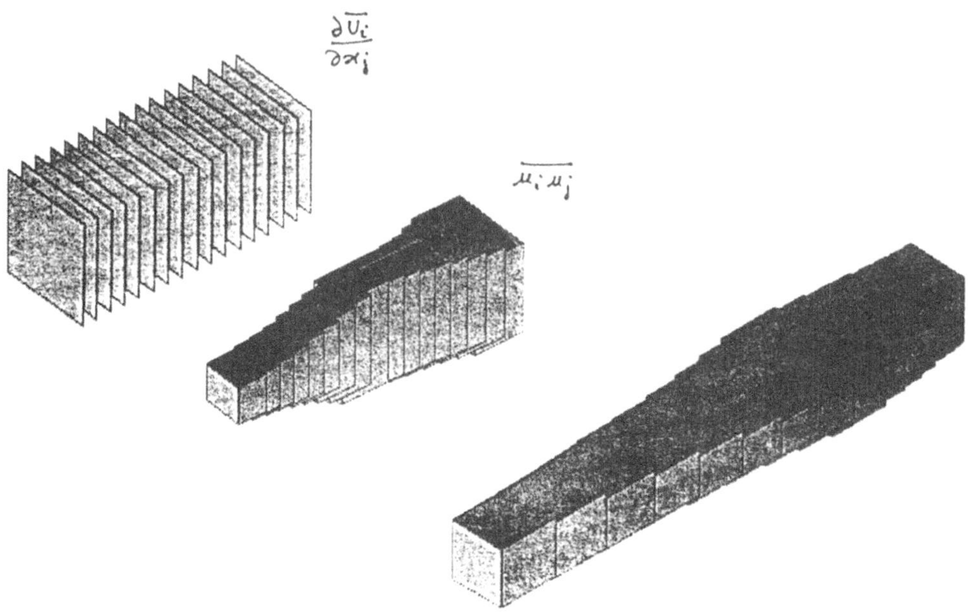

Figure 4 **Suivi des tenseurs de déformation et de Reynolds sur une ligne de courant dans un conduit variable.**

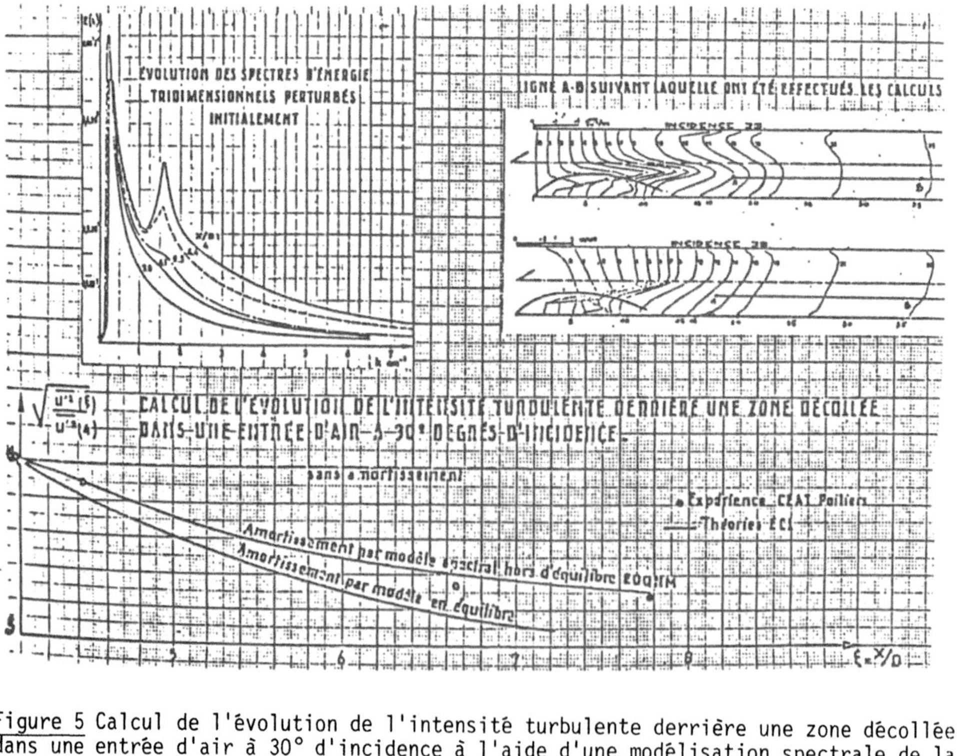

Figure 5 Calcul de l'évolution de l'intensité turbulente derrière une zone décollée dans une entrée d'air à 30° d'incidence à l'aide d'une modélisation spectrale de la turbulence.

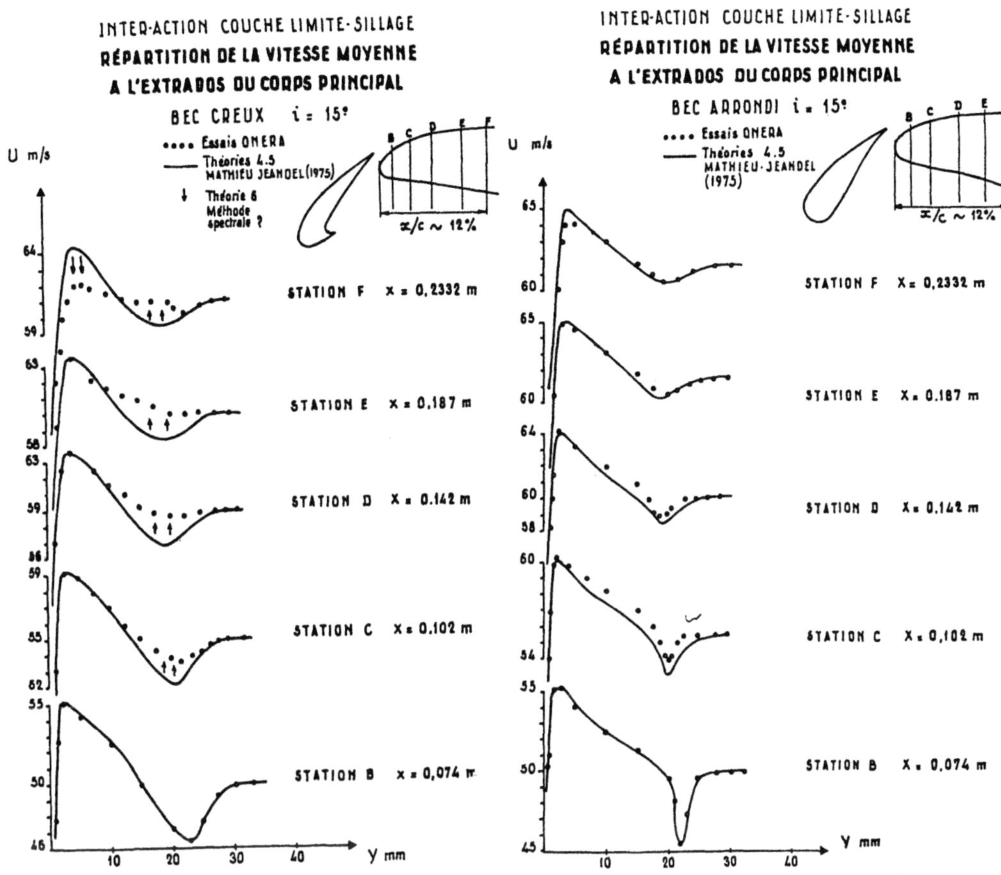

Figure 6 Interaction couche limite sillage sur des hypersustentateurs d'avion.

Ou bien l'on gardait le formalisme des équations de Navier-Stokes portant sur l'écoulement moyenné et l'on chargeait les hypothèses de fermeture de descendre à une plus grande finesse de modélisation, ou bien on coupait franchement le champ en champ macroscopique, dit des grosses structures et champ microscopique.

Vu par l'utilisateur, la première approche revient à traiter le fluide turbulent comme un matériau aux propriétés particulières face aux déformations et avec une mémoire exprimant le temps de retour à l'isotropie ou à l'équilibre spectral de l'énergie turbulente. Une modélisation simple de ce type a permis de prendre en compte, avec succès, des réarrangements d'une turbulence dans une autre (mélange sillages-couches limites, par exemple) d'utilisation industrielle courante par exemple dans le problème des hypersustentateurs d'avions (fig. 6) (réf. 6).

Par contre, une étude plus détaillée des non équilibres spectraux a mis en évidence la richesse d'une telle approche, les capacités d'une formulation à l'aide de l'équation de Craya à prendre en compte raisonnablement des turbulences déjà en fort déséquilibre spectral avec des fermetures statistiques assez efficaces comme l'EDQNM. Si l'on suit l'écoulement moyen, on peut ainsi évaluer le retour à l'isotropie d'une turbulence assez marquée initialement par les conditions aux limites et les succès remportés dans le suivi de l'amortissement spectral des perturbations, induites par le décollement à l'entrée dans une conduite d'entrée d'air d'avion à grande incidence, indiquent bien la puissance d'une telle approche (fig. 5) (réf. 2). Il importe essentiellement à une telle modélisation de prendre maintenant en compte les perturbations créées non par un simple déséquilibre spectral mais par la présence majoritaire de rotationnels émis au décollement, de façon à correctement prédire le comportement de retour à un sillage turbulent établi d'une turbulence fortement marquée par des rotationnels colinéaires à la vitesse donnant une structure en cigare.

Le passage à l'isotropie d'une turbulence généralement initialement fortement marquée de bidimensionnel par les obstacles et de rotationnel en "cigare" plutôt qu'en "crêpe" est le principal axe d'amélioration de telles méthodes, mais on connaît les limitations industrielles que risque d'imposer la génération de champs initiaux complets cohérents pour initier de tels calculs complexes mais assez riches pour prendre en compte une grande part des échecs actuels des méthodes simples moyennes à fermeture dans le plan physique.

Cependant l'amélioration des efficacités des codes de simulation quasi-directe et les réductions des coûts des calculs en ordinateur laissent espérer que le voisinage immédiat des obstacles pourrait être pris en compte par ce type de calcul. Deux voies d'améliorations de ceux-ci se présentent, soit que l'on modélise les petites structures et la coupure avec des termes turbulents issus de modélisations plus efficaces qu'une simple dissipation (fig. 3) (réf. 3), soit que l'on suive le formalisme mathématique de l'homogénéisation des petites structures dont l'intérêt principal réside dans l'absence d'hypothèse supplémentaire de fermeture, (fig. 2) (réf. 4-5). Ces deux voies complémentaires devront être complétées par des générateurs de condition aux limites cohérentes.

3. REMARQUES POUR LES RECHERCHES A VENIR

Ainsi l'approche industrielle rejoint peu à peu les recherches scientifiques apparemment les plus détachées de l'application immédiate dans un souci de modélisation de haut niveau suffisamment fiable, ce qui n'a pas été le cas dans les fermetures turbulentes pratiques initiales. Pour cela, les chercheurs, comme les industriels, doivent s'appuyer sur des expériences orientées vers la mesure très précise de turbulences notablement hors d'équilibre, soit spectralement, soit anisotropes comme les expériences récentes sur l'effet de l'anisotropie créé par les parois et les rotationnels d'ensemble ou localisés.

Une grande zone exagérément inconnue subsiste également dans les effets de frontière turbulente et de contamination, ainsi que dans l'effet des couplages pression-vitesse et des instabilités de l'écoulement moyen.

Que des études, à la fois expérimentales et numériques, s'imposent pour faire progresser notre connaissance est évident pour beaucoup. A tous d'y participer, maintenant qu'un nombre de plus en plus élevé d'industriels majeurs prend conscience de l'importance de la modélisation turbulente en aérodynamique, aérothermique, combustion, etc... Un lien possible de participation serait une structure d'échange et de travail sur les plus gros outils numériques disponibles et de comparaison avec des expériences de base de plus en plus fines, structure souhaitée par l'industrie de pointe comme la recherche universitaire avancée, structure pouvant devenir un pôle d'excellence et un lien d'échange privilégié entre industries et universités européennes.

REFERENCES

1 - MATHIEU J., JEANDEL D.
Turbulence et approche spectrale
Ecole d'été d'analyse numérique EDF-CEA-INRIA 1982.

2 - COURTY J.C., CAMBON C., HAMADICHE M.
"La turbulence dans les manches à air" - 21ème Colloque d'Aérodynamique Appliquée - Ecully 7, 8, 9 Novembre 1984.

3 - Rapport de contrat DRET/AMD-BA
Méthodes spectrales de prévision de la turbulence utilisables pour des calculs à finalité aéronautique n° 83.34 349 lots 1 et 2, 11/04/84.

4 - P. PERRIER, O. PIRONNEAU
Couplage des grosses et petites structures turbulentes par l'homogénéisation.
CRAS, 13 Février 1978.

5 - C. BEGUE, T. CHACON, O. PIRONNEAU, D.W. Mc LAUGHLIN, G.C. PAPANICOLAOU
Computing methods in applied sciences and engineering, VI, Versailles, 12 - 16 Décembre 1983.

6 - COURTY J.C.
Problème de confluence sillage couche limite à l'aide d'un modèle élaboré de turbulence, 13ème Colloque d'Aérodynamique Appliquée,
Ecully 8, 9, 10 Novembre 1976.

Manipulated Turbulence Structure in Flight

A. Bertelrud

Department of Mechanics, Royal Institute of Technology, Stockholm, Sweden and FFA, The Aeronautical Research Institute of Sweden, S-16111 Bromma, Sweden

During the past years a considerable effort has been directed towards reducing drag through manipulation of turbulent boundary layers, and several techniques are presently under investigation in laboratory conditions.

One method receiving considerable attention is the use of thin transverse ribbons (or Large Eddy BreakUp, LEBU, devices) beleived to modify the large-eddy structure: CORKE et. al [1], BUSHNELL [2], BERTELRUD et. al [3] and COUSTOLS, COUSTEIX [4]. Experiments have shown that a net drag reduction may be obtained for certain ranges of geometry and Reynolds numbers, but so far no full-scale test has been performed to prove the performance on a real ship, aircraft etc.

In a cooperative effort by the FFA, Bromma, Sweden, and FMV-PROV, Malmslätt, Sweden, an investigation has been performed on an aircraft in flight, modifying the turbulence structure using transverse manipulators. Time-averaged as well as turbulence information was gathered, indicating a drastic reduction in vertical convective velocity, well in-line with laboratory tests at lower Reynolds numbers, under idealized conditions.

1. Experiment

The experiment had several objectives; with emphasis on the feasability of the technique: "Do the devices break, and do they pose any operational problems?" The investigation was not intended to explore whether net drag reduction is possible or not. Compared with laboratory tests, the flight tests presented extrapolations in several flow conditions:

- Mach number
- Reynolds number
- Pressure gradient
- Three-dimensionality.

Figure 1, from BERTELRUD et. al [3], shows the development of momentum thickness for flat plate flow using thin transverse ribbons under low-speed laboratory conditions. The result is a few percent net drag reduction, interpreted directly from the momentum thickness under the assumption of 2D flow. The figure illustrates the typical feature of most experiments, i.e. a large device drag which it takes a long distance to compensate. Nevertheless, the result is encouraging, as proper shapes of

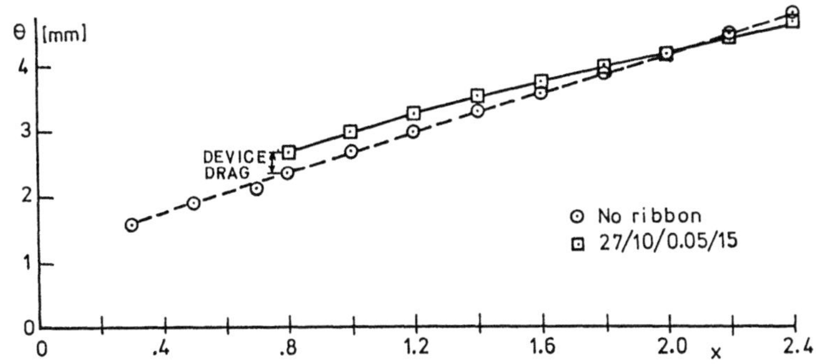

Fig. 1 Momentum thickness development with and without LEBU device. Small net drag reduction is indicated. Device notation is Spacing/Chord/Thickness/Wall distance = 27mm/10mm/0.05mm/15mm. From BERTELRUD et.al [3].

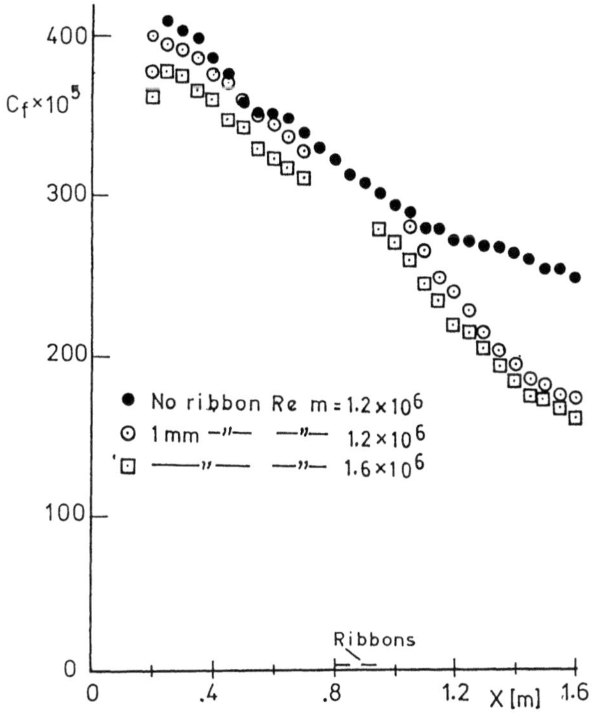

Fig. 2 Local skin friction versus streamwise position. From BERTELRUD et al. [3].

ribbons may have considerably lower drag. Figure 2 illustrates the development of local skin friction in a strong adverse pressure gradient with and without transverse manipulators. The reduction is similar to the results from the flat plate tests, and therefore it was considered reasonable that meaningful results could be obtained even on a swept wing in flight, where the pressure gradients may be large, but the three-dimensionality in general is small.

The flight tests were performed with a SAAB 32A Lansen swept wing attack aircraft already available as a flying laboratory for aerodynamic tests. Characteristics of the flight envelope and the flow parameters are:

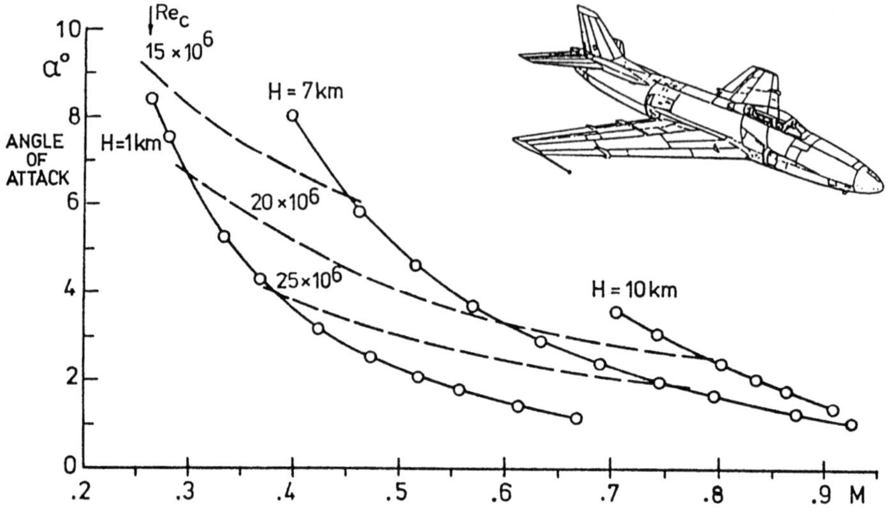

Fig. 3 Aircraft and flight envelope.

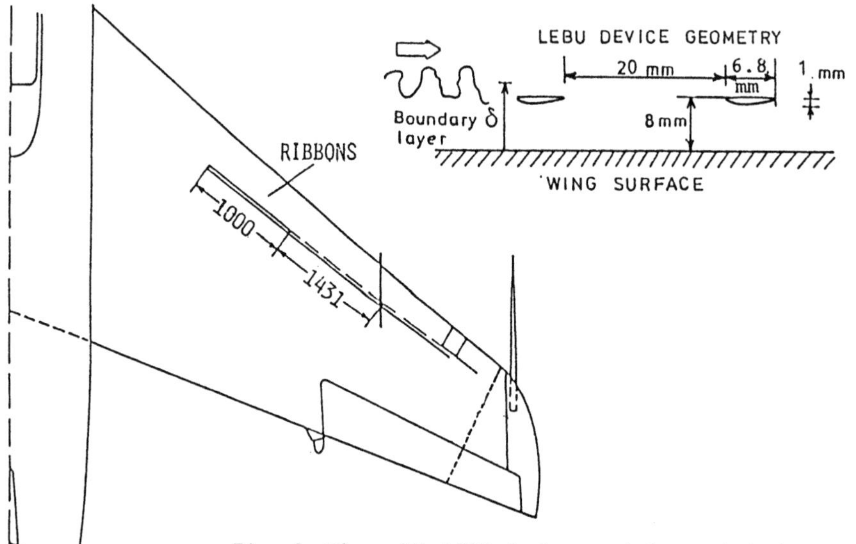

Fig. 4 Wing with LEBU device position and device geometry.

M = 0.25 – 0.95

H = 1 – 11 km

Re_c = 8 – 30 millions

Re_{LEBU} = 55 – 130 000

Figure 3 shows the aircraft and the flight envelope.

The LEBU's were mounted on the wing, as indicated in Fig. 4. The device position was chosen to have an almost constant boundary layer thickness throughout the flight regime. Due to aircraft safety the length of the ribbons was restricted to 1 meter. Several different ribbons were used, initially rectangular with different relative

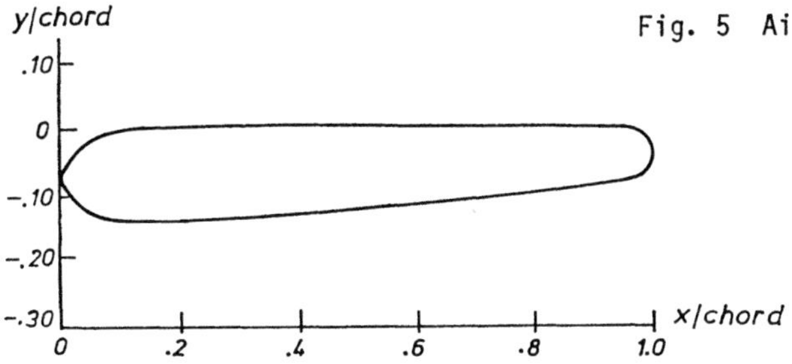

Fig. 5 Airfoil shape – 32 LEBU.

thickness, later profiled. In the present paper results from two series of tests with profiled, pre-stressed LEBU's are presented. Figure 5 shows the shape of the profiles; they were manufactured from rectangular bands being rounded in the leading edge and cambered on one side at the trailing edge. Two LEBU spacings were used, 20 and 80 mm which for the main part of the flight envelope corresponded to 2 and 8 boundary layer thicknesses.

2. Results

As mentioned, one of the main objectives of the investigation was to ascertain whether or not the main characteristics of the flow remained unaltered. Figure 6 shows that this was indeed the case, illustrating the pressure coefficient at one location on the wing chord as function of Mach number for three different altitudes. There is no change in pressure coefficient, neither at close-to-stall conditions pertinent to low Mach numbers, or at the shock-present conditions for high Mach numbers. However, the local skin friction was reduced appreciably in the entire Mach

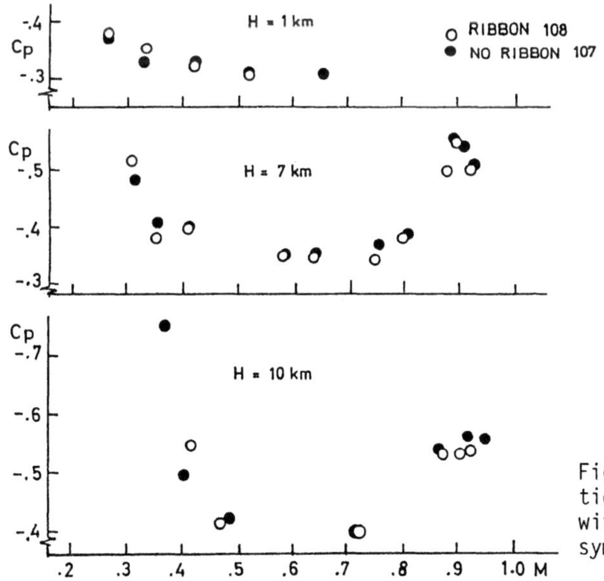

Fig. 6 Static pressure C_p as function of Mach number. Open symbols with LEBU device present, filled symbols are baseline.

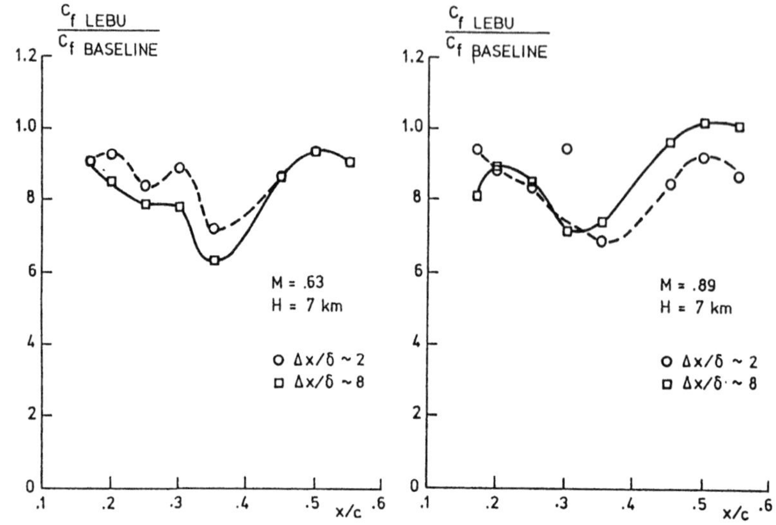

Fig. 7 Reduction of local skin friction.

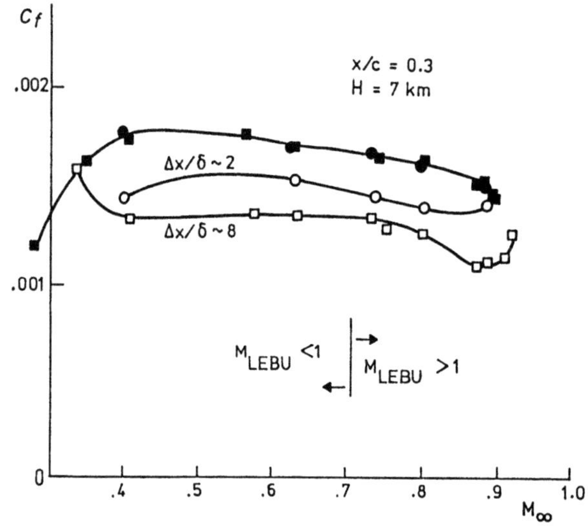

Fig. 8 Effect of Mach number on C_p at one location.

number regime, as Fig. 7 shows. Here it may be observed that the case with $\Delta x/\delta=8$ separation has an appreciably larger reduction; this is the case for supersonic as well as subsonic flow at the location of the LEBU device.

Exploring the local skin friction development downstream, it turns out that the larger ribbon separation exhibits a larger reduction in local skin friction – in this case actually as large as 35%. For the illustrated flight condition, the return to baseline value occurs as fast for the larger separation as for the smaller one. However, in Fig. 7b it can be seen that for this supersonic case, the skin friction reduction extends further downstream, possibly a result of thinner boundary layer at the LEBU device. The effect of ribbon separation on the reduction in local skin frictin may be seen clearly in Fig. 8, where C_f for 30% chord has been plotted against flight Mach number. For $M_\infty \simeq 0.79$ the local flow at the LEBU becomes sonic, and this may be an explanation for the "bucket" in C_f-reduction for the $\Delta x/\delta \simeq 8$ case. The filled symbols are baseline values for the two different series of tests, and may be seen as an indication of the repeatability of the tests.

Proper reduction to velocity of hot-wire data at compressible conditions is difficult and at transonic conditions almost impossible. However, the convective velocity may be estimated using the correlation between wires on a hot wire rake. Figure 9 shows that this vertical velocity is drastically reduced for the LEBU case, a clear indication that the LEBU far upstream has a lasting effect on the turbulence structure.

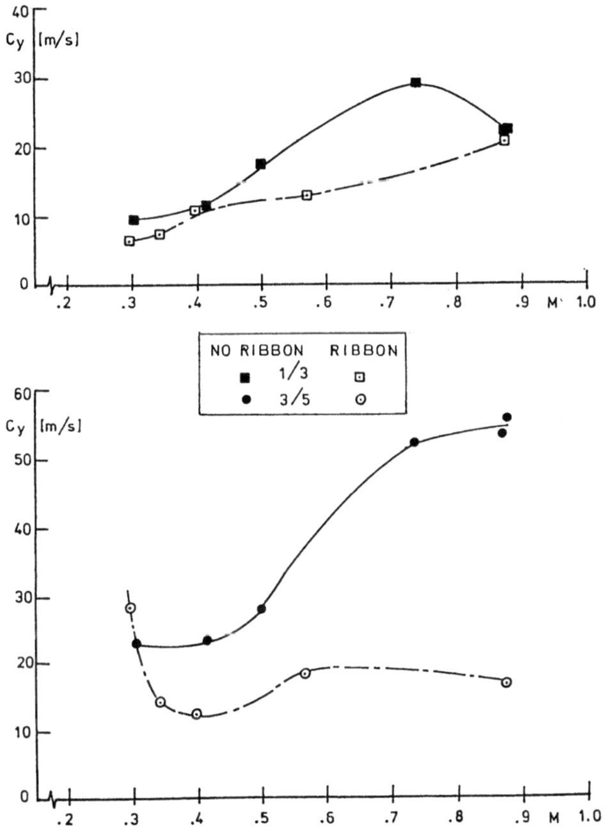

Fig. 9 Average phase velocities C_y between wires for cases with and without LEBU device. $\xi = 0.8$, $\Delta x/\delta \simeq 2$.

3. Device Optimization

The object of device optimization work is to maximize net drag reduction. This is obtained through a combination of minimizing device drag while the drag reduction capability of the device downstream is retained. A variety of possibilities exist:

- Use of single or tandem devices; $\Delta x/\delta = 7 - 8$
- Airfoil shape for minimum drag
- Non-dimensional wall distance; $h/\delta = 0.6 - 0.9$
- Non-dimensional chord; $c/\delta > 1$

- Angle of attack α or lift coefficient C_L
- Viscous effects characterized by Re_{LEBU}

The single most important parameter is the device drag. This may be split into:

* Pressure drag
* Skin friction drag
* Induced drag

The pressure drag may be calculated with an inviscid code, subsonic or transonic, will be zero, but the resulting non-zero estimated leading edge thrust will be modified by:

- Viscous effects

  These may be estimated through an empirical expression, CARLSON [5]:

  $C_t = f(M, Re)$

- Quasi-stationary effects

  This effect, called the Katzmayer effect, may be estimated using

  $\alpha = \alpha_0 + \Delta\alpha_0 * \sin(\Omega \cdot t)$, see KATZMAYER [6].

The following section shows how this has been performed for the profiled shape used in the flight tests. Figure 10 shows the computed inviscid pressure distribution at 0 and 1 degrees angle of attack. Only 35 panels have been used in the calculations, but although the levels may be a matter of discussion, the general trends are considered correct. The cambering close to the trailing edge is seen to yield an appreciable asymmetry even in the leading edge region. The computed inviscid leading edge thrust is an integration of the pressure forces around the profile in the direction of the chord, and is shown in Fig. 11. The Katzmayer effect has been computed for an oscillating outer stream with $\Delta\alpha_0$ of 1 and 2 degrees amplitude respectively, and the two curves in the figure indicate a large effects. However, these estimated inviscid effects are not realistic. A proper analysis would involve computation of the boundary layer effects as well as local separations on the pressure field, but in the present case use is made of simpler empirical formulae

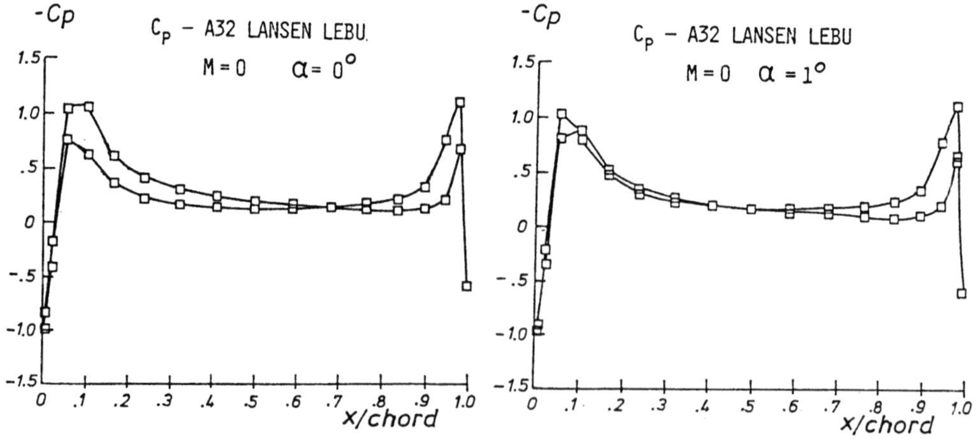

Fig. 10 C_p – A32 Lansen LEBU.

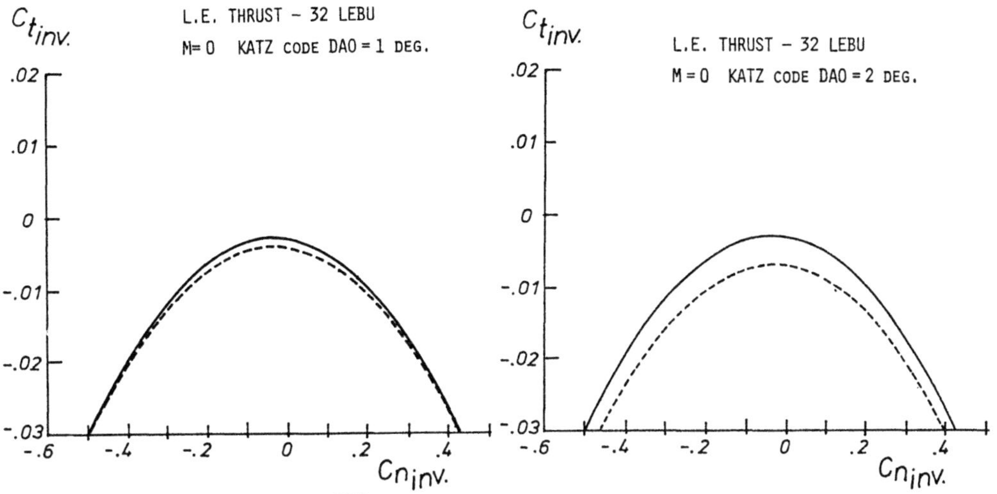

Fig. 11 L.E. Thrust – 32 LEBU.

CARLSON [5]. As Figure 12 indicates a drag bucket exists for the pressure drag on a standard NACA airfoil. As long as the lift coefficients are sufficiently small in magnitude, the theoretical leading edge thrust is obtainable, whereas use of larger lift coefficients will yield appreciable drag increases.

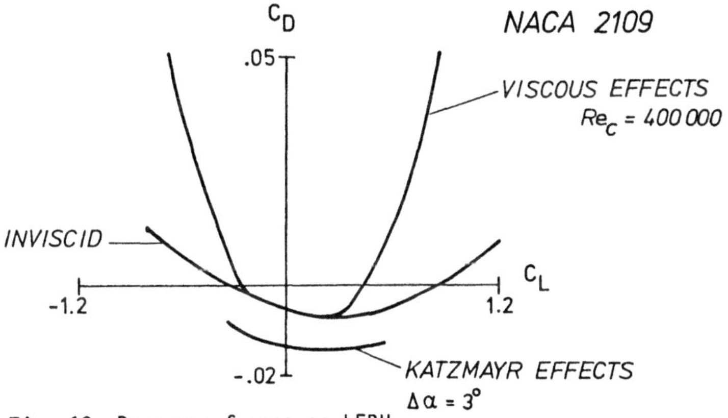

Fig. 12 Pressure forces on LEBU.

4. Conclusions

Several conclusions have emerged from the present investigation:

- LEBU's (manipulators) are feasible, i.e. can be mounted and maintained on subsonic aircraft in normal operation.
- Local reduction of skin friction is confirmed for transonic, pressure gradient cases at flight Reynolds numbers.
- Use of $\Delta x/\delta$ = 8 yields larger skin friction reduction than $\Delta x/\delta$=2.
- The turbulence structure is altered even as far downstream as regions where the time-averaged skin friction reduction is no longer visible.

5. References

1. T.C. Corke, Y.G. Guezennec, H.M. Nagib: "Modification in drag of turbulent boundary layers resulting from manipulation of large scale structures". In Viscous Flow Drag Reduction, Progress in Astronautics and Aeronautics, Vol. 72, (1980) pp. 128-143.
2 D.M. Bushnell: "Turbulent Drag Reduction for External Flows". AIAA Paper 83-0227 (1983).
3 A. Bertelrud, T.V. Truong, and F. Avellan: "Drag Reduction in Turbulent Boundary Layers Using Ribbons". AIAA Paper 82-1370 (1982).
4 E. Coustols, and J. Cousteix: "Réduction du Frottement Turbulent: Modérateurs de Turbulence". 22ème Colloque d'Aérodynamique Appliquée Lille, 13-15 Nov (1985).
5 H.W. Carlson, and R.J. Mack: "Studies of Leading-Edge Thrust Phenomena". AIAA 18th Aerospace Sciences Meeting, Pasadena, California, Jan 14-16, (1979).
6 R. Katzmayer: "Effect of Periodic Changes of Angle of Attack on Behavior of Airfoils". NACA PM-147, Oct 1922.

On the Manner in Which Outer Layer Disturbances Affect Turbulent Boundary Layer Skin Friction

A.M. Savill

Engineering Department, University of Cambridge, Cambridge CB2 1PZ, UK

Summary

This paper presents results, from both experimental studies and turbulence modelling, of the manipulated flow behind flat plate devices introduced into the outer region of turbulent boundary layers. An analysis of mean velocity profiles, in conjunction with drag balance and other measurements for u_τ, reveals only small changes in the log law and little influence of plate thickness on skin friction, except in the immediate vicinity of the manipulator where variations in the wall static pressure are recorded. However the drag reducing efficiency of the prefered tandem plate arrangement is found to depend simultaneously on plate length and gap. These observations are discussed together with recent turbulence measurements and previous flow visualisation. Implications emerge regarding the relative importance of the various proposed drag reduction mechanisms, and, in the light of these, alternative 'triplet' and 'quadruplet' configurations are investigated. It is concluded that larger nett drag reductions may be attainable, while the skin friction distribution downstream of a single plate is shown to be predicted quite accurately by an Algebraic Stress Model.

1. Introduction

There is now considerable interest in the possibility of controlling the eddy structure of turbulent shear flows. In the case of boundary layers attention has focused on two particular techniques aimed specifically at drag reduction: surface modifications in the form of longitudinal ribbing or riblets, and devices mounted within the flow. The riblets have been shown to consistently reduce surface drag by up to 8% under a variety of conditions [1] and are close to practical commercial application. The situation with the outer layer manipulators is not so clear partly because they have an inherent drag penalty associated with them and therefore must produce sufficient skin friction reduction over a large enough downstream extent to recoup this before nett drag reductions can be achieved. However they remain a serious alternative approach because they may offer the possibility of larger nett reductions [2] and can apparently be combined successfully with riblets; the drag reductions being additive [1].

Such studies are also of more general value for the information they provide on the manner in which the boundary layer responds to disturbances; thereby revealing further details about its structure and potential for control. An intensive research effort by a very large number of research groups has indicated that flat plate (and more recently [3] thin aerofoil section) manipulators do not act solely as Large Eddy Break-Up devices (LEBUs), but influence the flow structure in several ways [4-15]. The proposed drag reduction mechanisms have been divided into two types [8,9],

immediate or 'plate' effects which act at the device site:
(a) Suppression of large scale eddies by 'vortex unwinding' due either to approaching transverse vortices generating cancelling starting vortices at the device [4] or as a result of the unsteady circulation associated with a fluctuating angle of attack [5]
(b) Suppression of large scale motions due to the restriction imposed on the vertical velocity component by the rigid surface of the device [5] (alternatively described as direct severance of hairpin vortices leading to disruption of the bulge scale collective motion of these structures [6]), extending to longitudinal scales larger than the boundary layer thickness [7];
(c) 'Downwashing' of the plate wake due to the lift it generates which directs the momentum deficit towards the wall [8], and leads to a more efficient redistribution of turbulence energy [9];
(d) Possible non-linear acoustic effects which may modify shedding frequencies [10].
And persistent or 'wake' effects:
(e) Introduction of new, energetic small scales [5], which promote an enhanced energy cascade from existing larger scales [7];
(f) An interaction between wake vortices of opposite sign to the mean flow vorticity (these perhaps being enhanced by the plate/large scale interaction [7]) and the near wall (hairpin) structure downstream which leads to a reduction in momentum transfer outside the sublayer [6]. Alternatively an interaction between the vorticity shed into the wake and the oncoming vorticity which reduces normal velocity fluctuations near the wall [11];
(g) Transfer of energy from the turbulence to the mean flow beneath and in the lower half of the wake partly balanced by transfer back in the upper part [5];
(h) A 'blocking effect' of the wake by which it shields the flow beneath it from incursions of external irrotational fluid and the influence of turbulence fluctuations in the outer region [7].

Emphasis has been placed either primarily on the suppression of the large scales [3,8,12] (by what might be termed the 'LEBU action') or the influence of the wake [6,13,14,15], the persistence of the latter being regarded either as an effect or a cause respectively. The best evidence for a strong wake influence has so far been provided by a close comparison of flow visualisation with drag balance data [6,14]. For single (tandem) plates mounted at a range of heights within the layer these revealed that the position where the maximum local skin friction reduction occurred was in all cases virtually coincident with the point at which the (second) wake penetrated to the viscous sublayer, and the magnitude of this reduction seemed to be closely related with the strength of the wake vortices at this stage. The shape of the Cf distributions were thus determined by the height of the device together with the flow velocity. The largest overall skin friction reduction appeared to be with the plate(s) mounted at a height $h=0.75\delta$ at the $Re_\Theta=2000$ of the present experiments and this has subsequently been confirmed by an analysis of integrated Cf data [14]. At higher Reynolds numbers the indication was that the devices should be moved closer to the wall to maintain the same effect and others have since reported $h=0.68\delta$ at $Re_{\Theta}=3500$ [2,16] and $h=0.5\delta$ at $Re_{\Theta}=5000$ [2]. The flow visualisation also suggested that the apparent optimum tandem plate spacing s (leading edge to leading edge) of 5 to 10 was connected with the distance over which the initial vortex street remained essentially fully coherent, since the degree of amplification of the second wake seemed to depend on the influence of the first plate wake at the origin of the boundary layers on the second, but in that case the plate gap g (from trailing edge to leading edge) should be a more relevant parameter. However all these observations were made using relatively thick (0.4 to 1mm) plates with blunt trailing edges which would tend to over-emphasise the importance of the wake, and it has therefore been suggested that interpretations based on these are not relevant to the understanding of the very thin (0.1mm) plates and thin aerofoil sections required to achieve nett drag reductions. The new data presented here was obtained using the same wind tunnel facility ana drag balance as the earlier thick plate studies.

2. Experimental Results With Discussion

2.1. Mean Velocity Profiles

Mean velocity data obtained from total head traverses behind a single plate are shown plotted in logarithmic form in Fig. 1, using values of u_τ obtained with the drag balance. These have been compared with other similar data from pitot traverses made in conjunction with alternative drag balance and laser oil film interferometry (Lofi) measurements for tandem devices [17]. The straight line corresponds to the normal flat plate law-of-the-wall with $A=5.62$ ($\kappa=0.41$), $B=5.24$. Near the device the profiles are strongly influenced by the imposition of the wake defect and are similar to those deduced from Lofi data in the flow behind a thicker cylinder mounted in a boundary layer [18] although the implied changes in κ and B are smaller; both increasing in contradiction to earlier findings based on momentum balance measurements [5,19] and wall similarity techniques [20]. However, beyond 15δ from the device only small

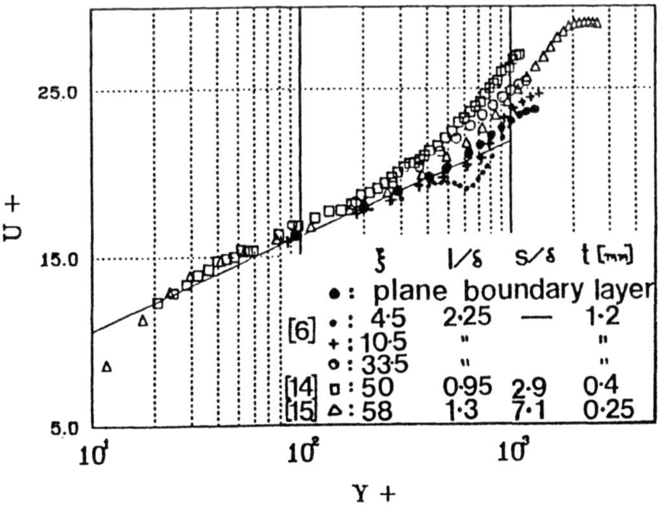

Fig. 1 Logarithmic profiles behind single and tandem manipulator plates

changes are evident in agreement with other findings [19,21]. The manipulated boundary layer exhibits a logarithmic region with the same slope as the undisturbed layer, but 5-10% larger intercept B which could be interpreted as indicating a thicker sublayer. There is also an increase in the strength of the wake component as reported by others [15,20]. According to Banyopadhyay [22] this appears to be a general feature of drag reducing flows while the parallel upward shift of the logarithmic region is similar to that seen riblet surfaces and in weak polymer solutions and presents the appearance of 'negative roughness'.

2.2. Skin friction measurements

The results presented above suggest that the previous use of wall similarity techniques to evaluate skin friction in such flows is acceptable beyond the initial region (a conclusion supported by a comparison of Preston tube and drag balance measurements [14]). Preston tube and Clauser Plot data are therefore compared with more recent drag balance data for cylinders, single and tandem plates in Figs. 2 and 3. For the same device drag the three different geometries offer progressively larger total skin friction reduction showing that this cannot be due just to the passive effect of the momentum deficit introduced into the flow; active mechanisms must be

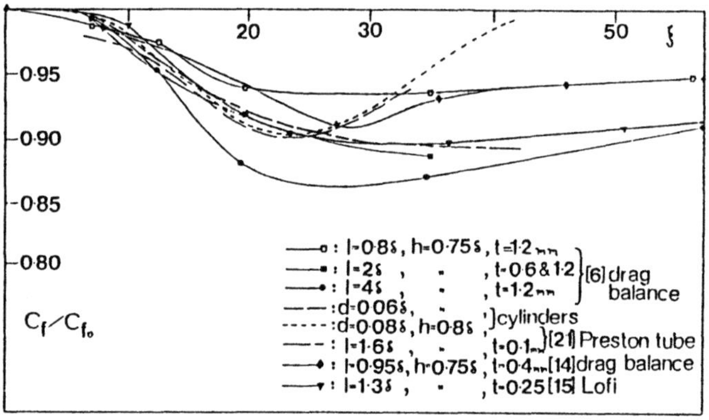

Fig.2 Cf distributions behind cylinders and single plates of various thicknesses

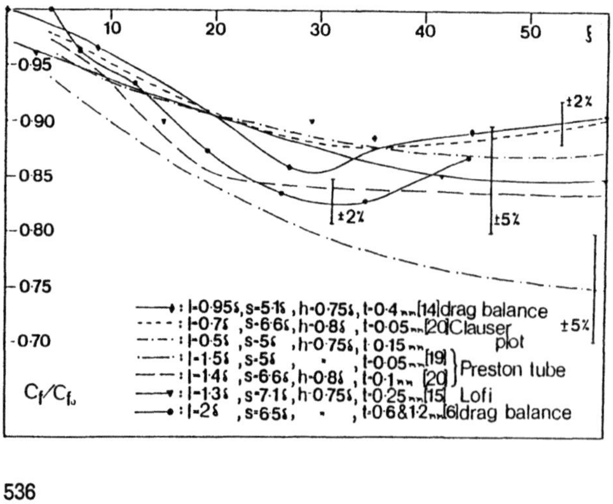

Fig.3 Cf distributions behind tandem plates of various thicknesses

involved. The results also confirm an earlier observation [6] that the plate efficiency for single elements improves considerably when the plate length exceeds δ, and indicate the same holds true for tandem devices, a finding supported by results obtained elsewhere [2,3,14,20]. Flow visualisation has suggested [7] that plates which are greater than half the approximately 2δ streamwise length of a typical boundary layer bulge are more effective at disrupting the associated large scale motion. An optimum length of $l>1.3\delta$ was indicated by the earlier studies and Nguyen (private communication) has recently found that a value of $1.3-1.5\delta$ produces the maximum overall reduction for both single and tandem plates. Equally significantly, within the experimental accuracy quoted by individual researchers, the skin friction distributions presented in Figs. 2 & 3 appear to be little affected by plate thickness variations from 1.2mm down to 0.1mm, except in the vicinity of the device itself. This suggests that the earlier conclusions based on experiments with relatively thick plates are relevant to nett drag reduction manipulators because, although it is possible that the relative significance of individual mechanisms may be different in the two cases, it seems extremely unlikely that a completely different set of mechanisms or even a major change in the balance of contributions of the proposed set could produce the same Cf response in both instances.

The drag balance data do show that with thicker devices the initial skin friction reduction is delayed beyond the second element of a tandem arrangement rather than originating at the first plate, but it appears that this is due to perturbations of the wall static pressure beneath these higher flow-resistance devices. Cp variations

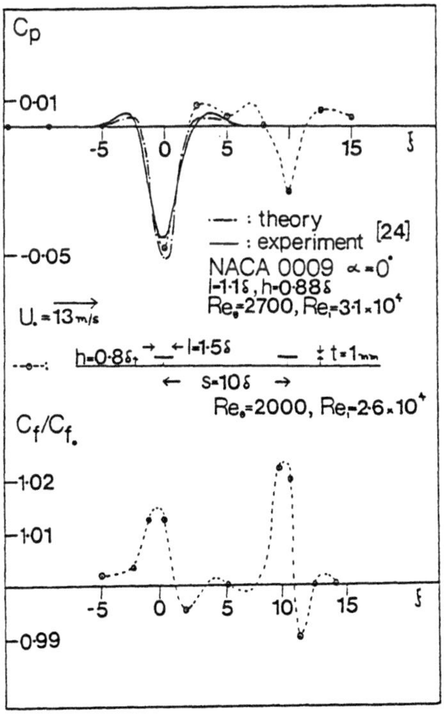

Fig.4 Cp and Cf variations beneath thick manipulator plates in tandem

have been detected from pressure tappings and the effect on skin friction revealed by a series of measurements using a sub-layer scale razor blade which was calibrated against the drag balance (The use of such a technique to measure skin friction has been discussed by Winter [23] who provides an otherwise unpublished calibration due to Gaudet which closely fitted the present data and appears to be largely independent of the detailed razor blade geometry). Figure 4 presents Cp and associated Cf distributions beneath a tandem configuration. The former is in good agreement with experimental and theoretical findings for a single element [24], confirming that a gap of approximately 5 chords is sufficient to isolate the individual pressure fields of the two devices. However the Cf distribution is complex and results in the postponement of the downstream Cf reduction (similar experimental results and turbulence model predictions have recently been obtained at CERT/ONERA [25]).

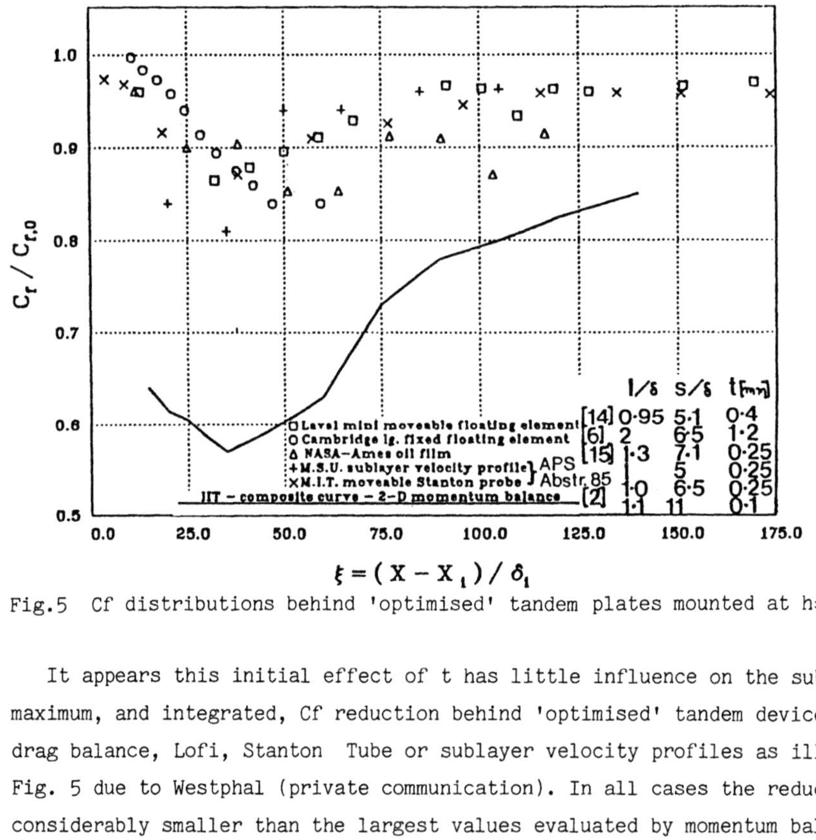

Fig.5 Cf distributions behind 'optimised' tandem plates mounted at h=0.75δ

It appears this initial effect of t has little influence on the subsequent maximum, and integrated, Cf reduction behind 'optimised' tandem devices determined by drag balance, Lofi, Stanton Tube or sublayer velocity profiles as illustrated by Fig. 5 due to Westphal (private communication). In all cases the reductions are considerably smaller than the largest values evaluated by momentum balance [2]. The direct measurements are in line with other momentum balance measurements [3] which indicate a nett reduction of 5%. By contrast the most beneficial results point to nett gains in excess of 25%.

It seems there are several probable explanations for this difference, apart from the possibility of spanwise non-uniformities affecting the results, and these have been investigated. First the larger reductions were obtained at a higher chord

Reynolds number using very carefully set-up, machined and tensioned thin plates that may have avoided detrimental vibration and transient flow separation which would degrade the 'plate' effects. Although there was no visual evidence of vibration with the 1mm plates used in the current work they should be more susceptible to such effects. Secondly the best results were achieved by setting the plates at a small positive angle of attack of less than three degrees (known to be beneficial while negative angles are detrimental [3]) to enhance downwashing. Tests showed that it was not possible to similarly improve the performance of thicker plates. Both positive and negative angles of attack proved detrimental probably due to the immediate onset of leading edge separation. Thirdly a larger plate separation of 11 or 14δ was employed with the devices mounted closer to the wall. Other results [3] have indicated that the general level of nett reductions may be improved to 8% by increasing s from 5 to 10δ. It has been argued [2] that the limited variation in nett drag observed over the wider range of separations from 2 to 15δ is due to an effective balance between reductions in device drag through shielding of the second element in the wake of the first at small s and improvements in Cf reducing efficiency as s is increased. However direct device drag measurements [6,26] indicate a saving only for 1-2δ spacings at low chord Reynolds numbers, and none at higher speeds which is consistent with observations of a deterioration in both skin friction and nett drag benefits [3]. Furthermore Veuve and Truong [16] have reported that the value of Cf measured at a fixed position downstream of the device varies in an oscillatory manner as s is increased. For plates with $l>\delta$ there is an indication of a similar response in the nett drag results also. In Fig.6 these two sets of data have been replotted in terms of g, which it has previously been argued should be related to the length over which the initial vortex street appears fully coherent (or in the case of very thin plates the distance over which the spectral peak and u' excess remain identifiable [27]). As part of the present investigations a study has been made of the effect of g on the maximum Cf reduction and the results have been included in Fig.6 together with the trends indicated by some recent integrated Cf measurements of Veuve and Truong [28]. Since the four data sets refer to different quantities the vertical scale is arbitrary and has been adjusted so that their mean values coincide to allow a comparison of the relative changes. Tentatively all the data follow trend of the dotted line which indicates perhaps an optimum gap of $7\frac{3}{4}\delta$. It is interesting to note that Veuve (private communication) has independently arrived at a figure of 7.7δ on the basis of momentum balance surveys. The absolute variation is small, probably within the experimental error, but the fact that these four different types of measurement all show the same effect is convincing and seems to provide evidence for a weak beating of the two wakes which has previously only been suspected by Falco (private communication).

The flow visualisation [6] showed that sufficiently small gaps resulted in the generation of a single greatly amplified wake and might therefore have advantages, but the present drag measurements indicated this was only true for plates with $l<\delta$

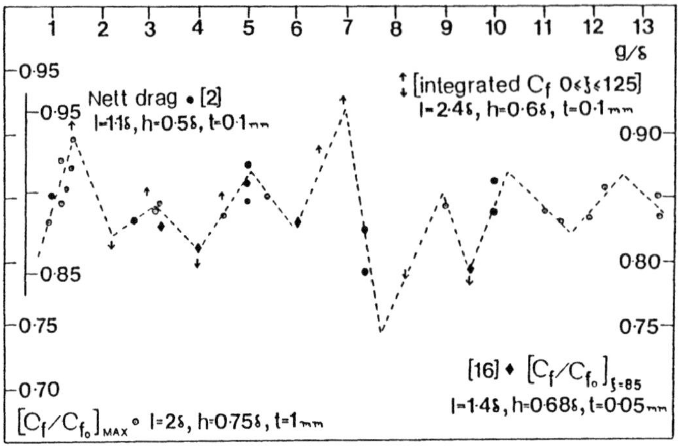

Fig.6 Effect of gap on tandem performance

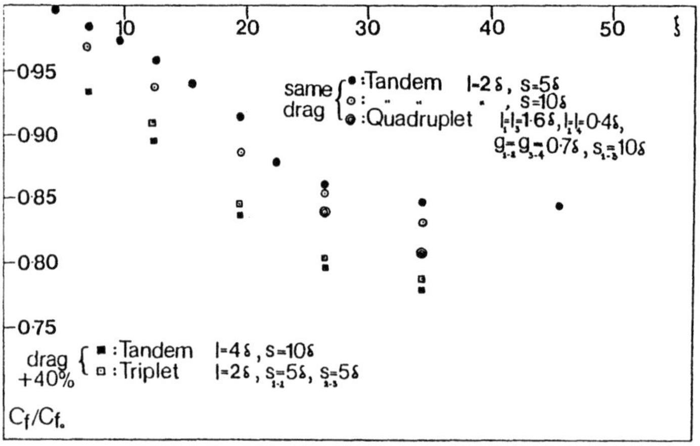

Fig.7 Alternative device configurations

when the maximum reduction was achieved with $g<2\delta$. Similar results have been found for integrated Cf reduction [14,16] and a nett reduction reported with these shorter plates only for $g=0.85\delta$ [19]. It may be that the main influence of g in this case is to increase the effective length of the device in order to improve the plate effects. With the longer tandems the influence of two wakes seems to outweigh that of a single amplified one. These ideas are supported to an extent by Rapid-Distortion Theory (RDT) calculations [29] of the reduction in v' fluctuation behind single and tandem devices. The RDT results indicate that the gain from employing two closely spaced plates is roughly eqivalent to that achieved with a single plate of the same total length. However when the two plates are far enough apart (approximately 10δ) for the second plate to be considered as interacting separately with the approaching boundary layer flow and the shed vorticity from the first plate then the effect on v' is squared. (The RDT analysis also suggests that the effectiveness of the devices diminishes as they are moved closer to the wall while the device drag data shows their drag coefficient then increases, so it is not clear that reducing h has any advantages).

The effect of increasing only the spacing of a tandem configuration is further illustrated by the data presented in Fig.7. A single unconfirmed experiment by Anders (private communication) indicated considerable further improvement might be possible using tandem plates several δ in length. However it was found that increasing the plate lengths from 2 to 4δ produced insufficient improvement to compensate for the associated increase in device drag. Probably the apparent optimum length of 1.3δ results from a compromise between device drag penalty and performance. It has also been suggested [9] that another reason why the tandem proves better than a single element of the same total length is because the momentum deficit is introduced in small increments and that a triplet arrangement might therefore prove better still. The results obtained for a parametrically simple triplet device are compared with the tandem data in Fig.7 and indicate that for the same drag penalty the performance is actually marginally worse with this more complex configuration. Optimisation has not been attempted, but Nguyen (private communication) has recently arrived at a similar conclusion. Of course if complexity was not a consideration the above results might suggest that it would be an advantage to replace each of the elements of a tandem by two smaller closely spaced plates in order to generate two amplified wakes. Such a 'quadruplet' system was tested and for nominally the same drag penalty as the original tandem an average improvement in Cf reduction of 4% was recorded.

2.3. Turbulence Data and Modelling

Streamwise turbulence intensity profiles measured with a cross-wire probe are presented for several stations behind a single element in Fig.8. These have also been normalised by the wall shear stress as determined with the drag balance and are compared with data recorded closer behind a similar device in a weak adverse pressure gradient [19]. The latter shows the second peak associated with production in the

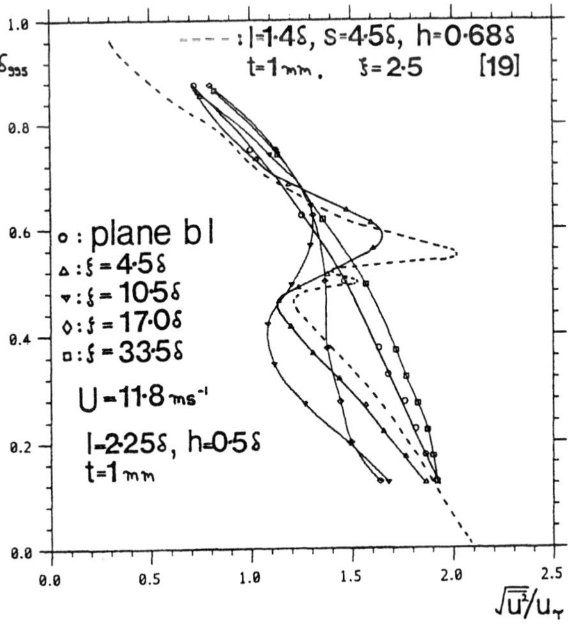

Fig.8 Streamwise turbulence intensity profiles behind single elements

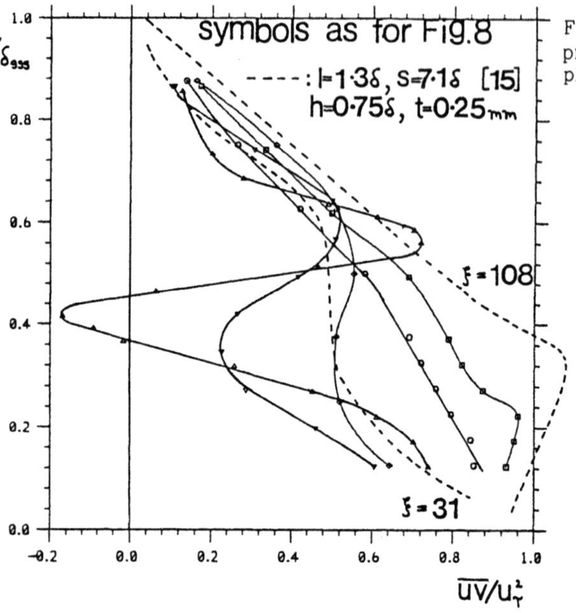

Fig.9 Reynolds shear stress profiles behind single and tandem plate devices

lower half of the wake which has disappeared by the first station in the present case. Otherwise the results are in good agreement and the maximum intensity is close to that expected [30] for the probe sensing length l+=30 used in the present measurements. There is no evidence of the overshoot near the wall reported by Veuve and Truong [16] which they attributed to a movement of the location of maximum turbulence production out into the flow. The corresponding Reynolds shear stress data is presented in Fig.9. Negative values are recorded close to the device and the correlation is immediately suppressed over a much greater vertical extent than the streamwise intensity; an effect noted by several other researchers [8,15,25,28,31]. The comparison with tandem data near the maximum reduction location and far downstream is excellent. As has been pointed out by Westphal [15] the overshoot in the outer part of the flow is consistent with the existence of an increased wake component and leads to a very flat or even reverse sign gradient of $\overline{uv}$ near the wall. Superposition of the shear stress profiles onto those for the mean velocity reveals only very small, local 'negative production' regions and it seems very unlikely these could drive any redistribution of energy.

Regarding other mechanisms the mean and turbulence profiles closest to the device indicate a mild downwashing of the wake towards the wall of approximately 1° in agreement with other estimates [8,9]. Several u' spectra have been recorded near the wall and in the vicinity of the irrotational interface. They show that the the damping of the large scales and enhancement of the energy cascade close to the device, which were revealed by similar data recorded within its wake [7], results in a suppression over a very wide range of scales at the position of maximum skin friction reduction.

Correlation measurements [7,27,31] show that the usual inclined u' contours in the xy plane are severed by the plate/wake at least up to maximum Cf reduction location (vertical integral scale reduced by 50%), although the flow above and below the wake retains its correlation. This is consistent with the view that the plate cuts through hairpin vortices and that its wake continues to separate the two regions of boundary layer structure (Thomas [32] has pointed out that this interpretation is in turn consistent with the continued existence of the log law). The conjecture that the wake acts as a passive barrier preventing interactions between these two regions is supported by the RDT calculations [29] and Navier-Stokes simulations [33], and the indication is that this effect perists as long as the wake remains identifiable within the flow which visualisation suggests can be greater than 70δ. Recovery of the integral scales is certainly very slow [27,31]. The spectra near the interface confirm flow visualisation observations that the bulge scale motions are replaced by smaller scale wake eddies. It seems these continue to prevent incursions of external fluid and to inhibit weak wallward sweep type motions, thus interrupting the turbulence regeneration cycle in the manner described by Falco [13], long after the maximum reduction has been achieved.

Aerofoil section devices will necessarily replace thin tensioned plates in practical applications of such manipulators. However existing aerofoil data is strongly influenced by low Reynolds number effects [3]. This provides one justification for applying turbulence models to such flows. Previous attempts at this have been restricted to simple mixing length [34] or two-equation [4] modelling of the influence of assumed length scale changes and/or the momentum deficits. While these have proved instructive they have tended to over- or under-estimate the observed drag reductions. Figure 10 presents preliminary skin friction predictions obtained with an Algebraic Stress Model (ASM) employing a parabolic forward marching scheme which has previously been used successfully [35] to model wake/boundary layer interactions where larger wake producing bodies were mounted at h>δ above the wall. With precisely the same set of model constants the predictions were found to be in remarkably good agreement with drag balance results for the single element when the experimental mean velocity and turbulence intensity data at the first station 2.25δ downstream of the trailing edge were used as input. (It could be argued that these

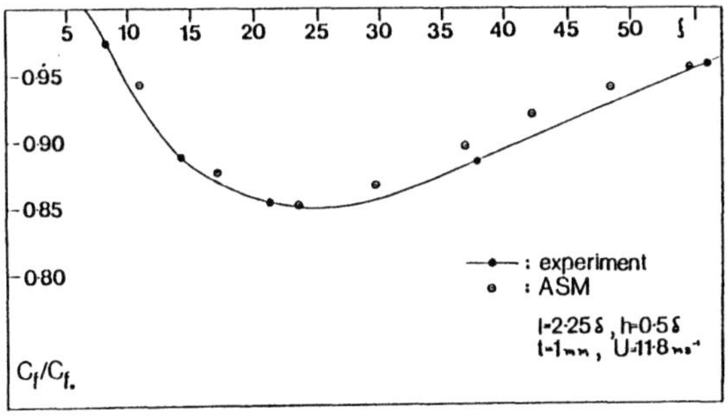

Fig.10 Comparison of ASM predictions with experimental Cf results for a single manipulator plate

supplied much of the necessary information on the immediate effects of the plate and initial influence of its wake). Discrepancies occured in the initial region identified by large deviations from the log law, because the model made use of wall functions, and in the relaxation region beyond the maximum reduction position where no allowance was made for any blocking influence of the wake.

3. Concluding Remarks

It seems that the skin friction reduction extending downstream of manipulator plates results from a sequence of effects. The initial reduction must be due solely to the direct influence of the imposed boundary condition on the vertical velocity component of large scale motions. However it appears that the maximum Cf reduction is determined both by the immediate effects of the plate(s) on the approaching flow and the interaction between the modified (hairpin) eddy structure and the wake vortices introduced into the flow behind the manipiulator. It also appears that the persistant action of the wake in partly blocking the normal interplay between the inner and outer layers helps to ensure that the subsequent relaxation is delayed beyond the typical 40δ recovery length. (There is some evidence to suggest that complete recovery may not be attained [36]). The optimum parameters for tandem flat plate devices at relatively low chord and momentum thickness Reynolds numbers are apparently l=1.3δ, g=7.7δ, h=0.75-0.5δ (Re_Θ =1000-5000), t=0.1mm, and it seems these can be related to the optimisation of specific mechanisms. With such configurations set up under carefully controlled laboratory conditions it seems nett drag reductions of 10-20% are attainable and further gains may be possible with alternative configurations designed to maximise the influence of particular mechanisms. However device complexity may be a limitation and it is questionable whether the same parameters will apply to the application of aerofoil devices at high speed, although they may provide a better guide than low Re aerofoil studies. In disturbed boundary layers the relative contributions of the different mechanisms and hence optimum configurations may be changed. Thus in flows containing stronger transverse or longitudinal large scale eddies, transitional flows, and boundary layers developing on concave surfaces one would speculate that the 'LEBU action' of the device might be particularly important, whereas in the presence of free-stream turbulence the blocking effect of the wake may be more beneficial.

The emphasis of this paper has been on how the devices modify the outer flow. Only limited information is currently available on the manner in which the near wall structure responds to these changes despite the fact that it is this that ultimatley determines the drag reduction. (The insensitivity of the wall shear stress to plate thickness, and hence wake type, and perhaps the accuracy of the ASM predictions, suggests the inner layer responds in a rather uniform manner to a range of diffferent outer layer disturbances). A parallel study [37] has suggested that the reason manipulators can be constructively combined with riblets is because their drag reducing mechanisms compliment one another. It appears that the manipulators indirectly induce a greater regularity in the intensity of turbulence generation events and that the riblets, which act directly on the near wall flow, reinforce this effect while imposing an additional spanwise spatial ordering. It is possible the riblets also generate a stable sheet of longitudinal eddies which acts as a second shield for low speed regions at the bottom of the grooves.

Acknowledgement

The author wishes to thank Dr. John Mumford for his assistance in the recording of experimental data, Professor B.E. Launder for making available the ASM code, all those who have provided information as yet unpublished, and Rolls-Royce plc for supporting this work under University Research Project No. 239.

References

1. M.J. Walsh, A.M. Lindemann: AIAA-84-0347 (1984)
2. M.W. Plesniak, H.M. Nagib: AIAA-85-0518 (1985)
3. J.B. Anders, R.D. Watson: AIAA-85-0520 (1985)
4. J.N. Heffner, L.M. Weinstein, D.M. Bushnell: Prog. Astro. & Aero. 72, 110 (1979)
5. T. Corke, H.M. Nagib, Y.G. Guezennec: NASA CR 165861 (1982)
6. J.C. Mumford, A.M. Savill: in Laminar Turbulent Boundary Layers, ed. by E.M.Uram & H.E.Weber, ASME FED Vol. 11 (1984) pp.41-51
7. A.M.Savill: EUROMECH 181 Abstracts (1984) (see also FFA TN 1984-60)
8. Y.G. Guezennec, H.M. Nagib: AIAA-85-0519 (1985)
9. R. Narasimha, K.R. Sreenivasan: AIAA-85-0517 (1985)
10. D.M. Bushnell: AIAA-84-1527 (1984)
11. A.P. Dowling: J. Fluid Mech. 160, 447 (1985).
12. D.M. Bushnell: AGARD Report No.723 5 (1985)
13. R.E. Falco: AIAA-83-0377 (1983)
14. J. Lemay, D. Provencal, R. Gordeau, V.D. Nguyen, J. Dickinson: AIAA-85-0521 (1985)
15. R.V. Westphal: AIAA-86-0283 (1986)
16. M. Veuve, T.V. Truong: EUROMECH 181 Abstracts (1984)
17. V.D. Nguyen, A.M. Savill, R.V. Westphal: AIAA TN (to be published)
18. U. Muller: Poster presentation at this Conference (1986)
19. A. Bertelrud, T.V. Truong, F. Avellan: AIAA-82-1370 (1982)
20. S. Takagi: Proc. 15th. Turbulence Symposium, Tokyo (1983)
21. S. Takagi: Proc. 2nd. Asian Congr. Fluid Mech., Beijing (1983).
22. P.R. Bandyopadhyay: Trans. ASME JFE 108, 127 (1986)
23. K.G. Winter: Prog. Aerospace Sci. 18, 1 (1977)
24. P.R. Bandyopadhyay, R.D. Watson: Submitted to Experiments in Fluids
25. E. Coustols, J. Cousteix: To be presented at the European Drag Reduction Meeting, Lausanne (1986)
26. S.P. Govindaraju, F.W. Chambers: AIAA-86-0284 (1986)
27. R.F. Blackwelder, S.-I. Chang: AIAA-86-0287 (1986)
28. M. Veuve, T.V.Truong: To be presented at the European Drag Reduction Meeting, Lausanne (1986)
29. P. Balakumar, S.E. Widnall: Physics of Fluids 29 1779 (1986)
30. A.V. Johansson, P.H. Alfredsson: J.F.M. 137, 409 (1983)
31. J.Lemay, J. Delville, H. Garem, J.-P. Bonnet: To be presented at the European Drag Reduction Meeting, Lausanne (1986)
32. A.S.W. Thomas: AGARD Report No.723 1 (1985)
33. R.B. Kinney, M.E. Taslim, S.C. Mung: University of Arizona Dept. Aeosp. & Mech. Eng. Rep. No. NAG-1-141 (1984)
34. B. van den Berg: AGARD CP No.365 (1984)
35. D. Agoropoulos: Ph.D. thesis, University of Cambridge (1986)
36. I. Tani, T. Motohashi: Proc. 16th. Turbulence Symposium, Tokyo (1984)
37. A.M. Savill: Proc. 4th International Symposium on Flow Visualisation, Paris (Hemisphere Publishing Corporation, Washington 1986)

Response of Turbulence to Large Amplitude Oscillations in Channel Flow

S. Tardu, G. Binder, and R. Blackwelder

Institut de Mécanique de Grenoble, B.P. 68,
F-38402 Saint-Martin-d'Hères Cedex, France

1. Introduction

Allusions and conjectures about the influence of imposed oscillations on the bursting mechanism in wall flows are not uncommon in the litterature on unsteady turbulent boundary layers or channel flows. Since the bursting accounts for most of the turbulence production some workers in the field have somewhat hastily concluded that the mean flow characteristics of the velocity and of the turbulence must be affected. However, most experimental evidence demonstrates that these characteristics are remarkably insensitive to imposed oscillations. Conversely should it then be concluded that the same is true for the bursting ? Considering the complexity of the phenomenon the answer to this question is certainly not obvious and probably not unique but depending upon the frequency and amplitude of the imposed oscillations.

In a more fundamental context the question raised is the response of bursting to periodic accelerations and decelarations of the flow. Physical insight into this problem requires, of course, more detailed information than simple time-mean values such as the phase averaged characteristics. This question is related but not identical to the response of bursting to spacial acceleration or deceleration. There may, in particular, be important differences because the longitudinal length scales may not be of the same order.

Experimental results published on this subject are not only very limited but also contradictory. MIZUSHINA et al (1, 2) have measured an increase in the mean bursting frequency $\overline{f}_b$ when the frequency of the imposed oscillations f exceeded a critical value and they conclude that there is a resonance between the bursting and these oscillations. But their method of burst detection based on a weak peak in the auto-correlation function is open to criticism so that that their results must be considered with caution. RICHTER and RONNEBERGER (3,4) have used the high-pass filtered u'-signal to detect the bursts. They find that the rate of occurence is modulated and lags the outer velocity oscillation by the same amount of time as the high turbulent frequencies. The well established VITA detection technique (5) was used by KOBASHI and HAYAKAWA (6) and more recently by COUSTEIX and HOUDEVILLE (7,8). The former find that $\overline{f}_b$ is the same as in the corresponding steady flow while the latter unfortunatly do not mention this comparison. However they report that $\overline{f}_b$ is approximately the same for two different values of f. Both groups find that the phase averaged bursting frequency $\langle f_b \rangle$ is modulated in phase with the turbulent intensity $\langle u'^2 \rangle$ when a constant threshold $k\overline{u'^2}$ is used in the detection scheme. But when COUSTEIX and HOUDEVILLE use a modulated threshold $k\langle u'^2 \rangle$ which seems physically more meaningful, $\langle f_b \rangle$ is 180° out of phase with $\langle u'^2 \rangle$ for both oscillation frequencies. These authors then rightly caution against any conclusions which could be drawn from these results.

Clearly the response of bursting to imposed oscillations is at present a confused issue and even a partial clarification calls for new experimental data. In this paper such measurements taken in a 2-D water channel (2h = 10 cm) are presented. This facility has the capability of producing nearly established turbulent flow with imposed oscillations over a wide range of amplitude and frequency (9, 10).

2. Experimental Conditions

The standard notations are used : $\overline{q}$, $\tilde{q}$, q' are respectively the time-mean, periodic oscillation and turbulent fluctuation of quantity q : $q = \overline{q} + \tilde{q} + q'$; designates the phase average, e.g. $\langle q \rangle$, $\langle q'q' \rangle$, $\langle f_b \rangle$; $A_{\langle q \rangle}$ and $\phi_{\langle q \rangle}$ designate the amplitude and phase-shift of the first Fourier mode of q and similarly $A_{\langle q'q' \rangle}$ and $\phi_{\langle q'q' \rangle}$; q^+ is the variable normalized with wall scales - u_τ , ν / u_τ, u_τ^2/ν, velocity, length and frequency-based on the time-mean wall shear-stress $\overline{\tau}$.

In previous work (8-11) it was shown that the similarity parameter for the oscillating inner layer is the Stokes length ℓ_s^+ ($\ell_s = \sqrt{2\nu/\omega}$, $\omega = \pi f$) or equivalently the frequency $\omega^+ = 2/\ell_s^{+2}$ and that the effects of the oscillations are most clearly felt at intermediate frequencies, say $10 \lessapprox \ell_s^+ < 20$. At higher frequencies the oscillating shear $\partial\tilde{u}/\partial y$ is confined in the viscous wall layer and at lower frequencies the flow tends toward the quasi-steady regime. For instance the modulation of the turbulent fluctuation of the wall shear-stress $A_{\langle \tau'\tau' \rangle}$ has a frequency cut-off when $\ell_s^+ \lessapprox 10$, i.e. $\omega^+ \gtrsim .02$, which shows that the turbulence can then no longer respond to the imposed oscillations (11). On the other hand it was found that the phase lag of the turbulence with respect to the oscillation $\phi_{\langle \tau'\tau' \rangle} - \phi_{\langle \tau \rangle}$ tends rapidly toward zero when $\ell_s^+ \gtrsim 20$.

Since the aim at the start of this work is to establish whether the bursting responds to imposed oscillations the first experimental conditions selected were those for which an effect was most likely to be observed. On the basis of the preceeding considerations the value $\ell_s^+ = 16$ was selected. On the other hand the oscillation amplitude should for the same reason be as large as possible. However, since the response of the wall hot film gages used here is very complicated when the flow reverses, the amplitude had to be selected below the critical value for which such reversal ceases. This requires for $\ell_s^+ = 16$ (see Fig. 5 of Ref. 11) :

$A_{\tilde{u}_\mathfrak{C}} / \overline{U}_\mathfrak{C} < 50\ \%$

A first set of data was taken and analysed for these conditions and $\overline{U}_{\mathfrak{c}}$ = 18,5 cm/s, $u_{\tilde{\mathfrak{c}}}$ = 1 cm/s, T = 8 s, $A_{\langle U_{\mathfrak{c}}\rangle}/\overline{U}_{\mathfrak{c}}$ = 0.39. Under these conditions the amplitudes of $\langle\tau\rangle$ and $\langle\tau'\tau'\rangle$ were respectively 49 % and 80 % and there was thus a risk of instantaneous flow reversals due to turbulence when $\langle U_{\mathfrak{c}}\rangle$ was minimum. Moreover, at $\langle U_{\mathfrak{c}}\rangle$-minimum ($\sim$11 cm/s) the Reynolds number was rather low and the corresponding steady flow may not have been far enough from the transitional regime to be considered fully turbulent. Hence, a decrease in bursting frequency about $\langle U_{\mathfrak{c}}\rangle$-min could have been caused by the natural trend of the flow to return to the laminar regime at low Reynolds numbers. The result would thus not have been conclusive about the modulation of bursting in fully turbulent flow.

Because of these objections the experiment was respeated at a higher mean Reynolds number and at a smaller amplitude. All the results given below are from this second experiment characterized by the following parameters :

$\overline{U}_{\mathfrak{c}}$ = 30.33 cm/s ; Re_h = 15200 ; T = 3 s ; $A_{\langle U_{\mathfrak{c}}\rangle}/U_{\mathfrak{c}}$ = 0.29

$\overline{\tau}$ = 2.37 cm^2/s^2 ; $A_{\langle\tau\rangle}/\overline{\tau}$ = 0.37 ; $\phi_{\langle\tau\rangle} - \phi_{\langle U_{\mathfrak{c}}\rangle}$ = 17°

$\sqrt{\overline{\tau'\tau'}}/\overline{\tau}$ = 0.30 ; $A_{\langle\tau'\tau'\rangle}/\overline{\tau'\tau'}$ = 0.59 ; $\phi_{\langle\tau'\tau'\rangle} - \phi_{\langle\tau\rangle}$ = -74°

The Reynolds number corresponding to the minimum velocity is about 10^4. It was further checked that at this velocity in steady flow $\sqrt{\overline{\tau'\tau'}}/\overline{\tau}$ = 0.35 as expected in fully turbulent flow. The evolutions of $\langle\tau\rangle$ and $\langle\tau'\tau'\rangle$ during the cycle are shown on Fig. 1 et 2. The nearly sinusoïdal variations of $\langle\tau\rangle$, the large amplitude and low level during nearly half the cycle of $\langle\tau'\tau'\rangle$ should be underlined. The time lag of $\langle\tau'\tau'\rangle$ with respect to $\langle\tau\rangle$ computed from the above phase lag is Δt^+ = 150 which is consistent with previous measurements (11).

3. Burst Detection

The quantity used here to study the bursting is the instantaneous wall shear-stress sensed with a hot-film wall gage. The dimensions of the sensing element is Δx^+ = 3 and Δz^+ = 15. The VITA scheme (5) is applied to this signal to detect the bursts.

Let us recall that this consists in computing the short time variance of a turbulent signal q' :

$$\hat{\sigma}_{q'}(t) = \frac{1.}{\hat{T}} \int_{t-\frac{1}{2}\hat{T}}^{t+\frac{1}{2}\hat{T}} q'^2 dt' - \left[\frac{1.}{\hat{T}} \int_{t-\frac{1}{2}\hat{T}}^{t+\frac{1}{2}\hat{T}} q' dt' \right]^2$$

and in selecting the events producing a value $\hat{\sigma}_{q'}(t)$ which exceeds a certain threshols, i.e. $\hat{\sigma}_{q'}(t') \geq k\overline{q'q'}$. The detection may be further improved with the additional requirement that the slope of τ' is positive. In most previous

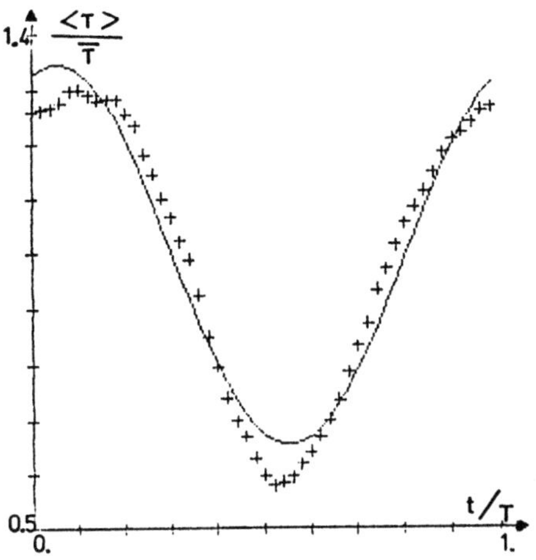

Fig. 1. Phase averaged wall shear-stress. The full line is the fundamental Fourier mode

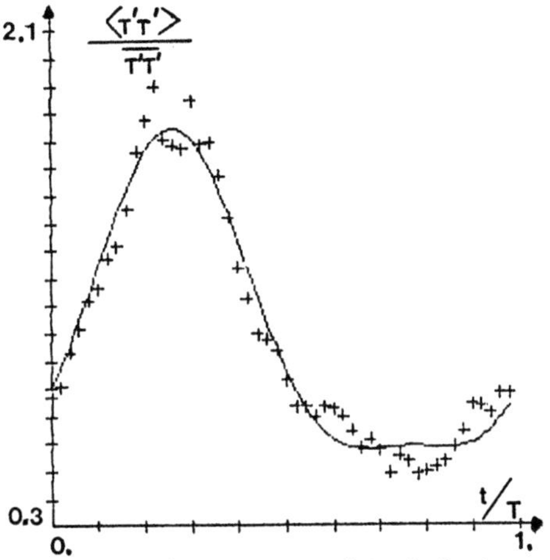

Fig. 2. Phase averaged turbulent fluctuation of wall shear-stress. The full line is the least square function computed with the first phase Fourier modes

investigations the u'-signal was used to this effect. They have demonstrated that the best detection is obtained when the hot wire is placed at about $y^+ = 15$ and that VITA with positive slope singles out the sweeps. However, it has been shown in particular by COUSTEIX and HOUDEVILLE (7, 8) and CHAMBERS et al (12) that VITA applied to the wall shear-stress signal τ' detects the bursts just as well as with u'.

The selection of events with VITA depends upon two adjustable parameters, the integration time $\hat{T}$ and the threshold level k. In order to improve the objectivity of the present results the values chosen were the same as those of CHAMBERS et al (12), i.e. $\hat{T}^+ - 10$, $k - 0.3$.

For easy reference the flow and detection parameters of KOBASHI-HAYAKAWA, COUSTEIX-HOUDEVILLE and of the present investigation are gathered in Table 1.

In order to avoid any possible spuriousinterference of the imposed oscillation frequency with those of the true turbulence the real $\tau'(t)$ signal was computed and recorded. This required the following sequence of operations :

TABLE 1 : Flow and bursting detection parameters

	ℓ_s^+	$A_{\langle u_{\mathfrak{C}}\rangle}/\bar{U}_{\mathfrak{C}}$	Detection signal	$\hat{T}^+$	k
KOBASHI-KAYAKAWA	18	.12	u' at $y^+ = 18$	10	1.2.
COUSTEIX-HOUDEVILLE	14	.08	u' at $y^+ = 15$	25	1.2.
	22	.12			
Present	16	.29	τ'	10	0.3

- digitizing and recording of the hot-film set output, sampling frequency $f_s^+ = 2{,}5$ (600 Hz), total record length 20 mn = 400 T = 2.8 10^5 ν/u_τ^2
- simultaneous recording of synchronization pulse phase locked with the device generating the forced oscillations which marks the beginning of each cycle. Any phase drift is thus avoided in the processed signals.
- conversion to $\tau(t)$, by use of the calibration curve determined just prior to the data recording and checked just afterwards - Drift is linearly distributed over the record length
- computation of $\langle\tau\rangle$
- computation and storage of $\tau'(t) = \tau(t) - \langle\tau\rangle$

With $\tau'(t)$ in storage various treatments could be applied to the same record, in particular VITA, conditional averages, zero-crossing, filtering, etc.

4. Results and Discussion

Sample records of $\tau(t)$, $\tau'(t)$ and $\hat{\sigma}_{\tau'}(t)$ are shown on Fig. 3. It is seen that the pronounced peaks in $\hat{\sigma}_{\tau'}$ corresponds as expected from previous work to sharp increases in τ'.

4.1. Bursting frequency

The bursting frequency f_b is the first quantity of interest. It was determined in 50 different intervals equally distributed over the cycle with the constant threshold $k\overline{\tau'\tau'}$(Fig. 4) and with the phase averaged threshold $k\langle\tau'\tau'\rangle$(Fig. 5). The same value k = 0.3 has been used in both cases. The main features of these figures are summarized in Table 2.

TABLE 2 Bursting characteristics

Threshold	$\overline{f_b}$ (Hz)	$\overline{f_b^+}$	$A_{\langle f_b\rangle}/\overline{f_b}$	$\phi_{\langle f_b\rangle} - \phi_{\langle\tau'\tau'\rangle}$
$k\ \overline{\tau'\tau'}$	0.88	.0037	1.19	1°
$k\langle\tau'\tau'\rangle$	0.78	.0033	.56	10°

Furthermore the bursting frequency measured in the steady flow corresponding to the same $\overline{U_\ell}$ is $(\overline{f_b^+})_{steady} = 0.0034$ - a value which agrees well with the measurements of CHAMBERS et al (12) and of others who have used u' for detection. The total number of events detected is equal to 926.

These results clearly show that in the present conditions $\langle f_b\rangle$ is strongly modulated not only when a constant threshold criteria is used but also when the threshold follows the variations of the turbulent intensity $\langle\tau'\tau'\rangle$ during the cycle. As may be expected the modulation of $\langle f_b\rangle$ is smaller in the second case

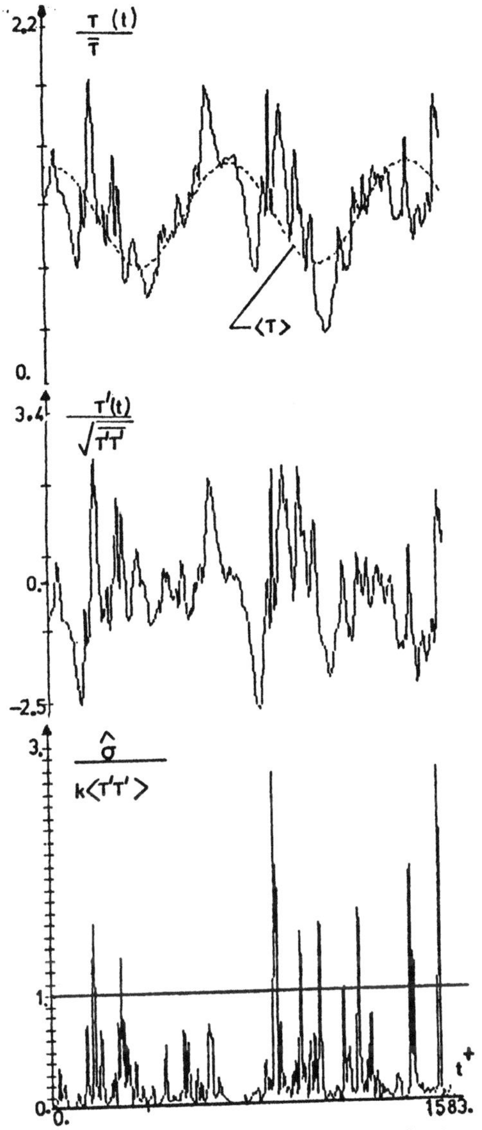

Fig. 3. Wall shear-stress, turbulent fluctuation and VITA variance signal

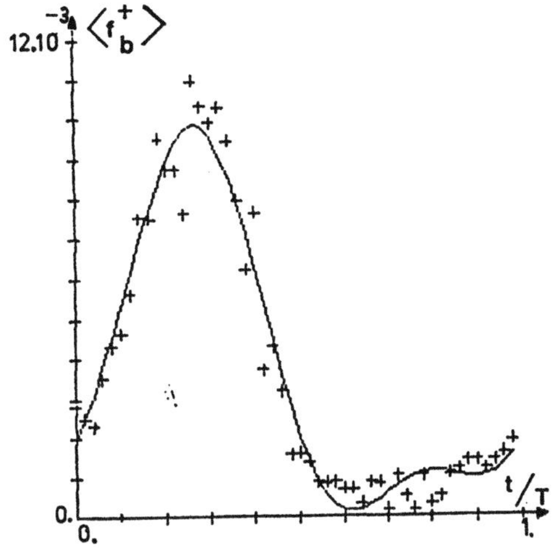

Fig. 4. Phase averaged bursting frequency. Threshold : $k\,\overline{\tau'\tau'}$
Full line as in Fig. 2.

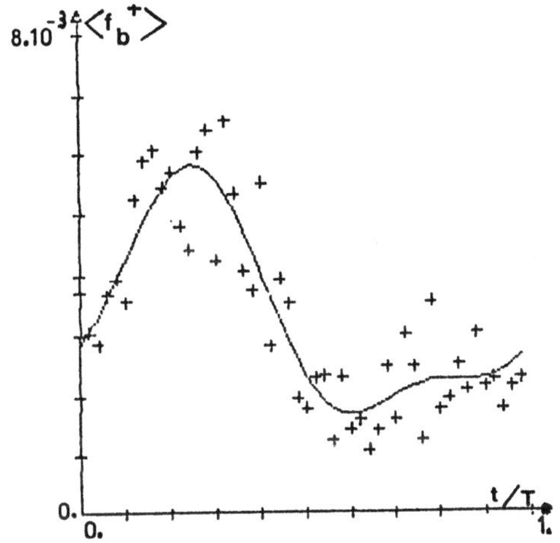

Fig. 5. Phase averaged bursting frequency. Threshold : $k\langle\tau'\tau'\rangle$
Full line as in Fig. 2.

than in the first one but it still reaches 56 %. This value should be compared to the 60 % amplitude of the $\overline{\tau'\tau'}$ - modulation. The near equality of these two figures may, however, be a shear coincidence, (Note that since the amplitude referred to here is that of the first Fourier mode, the 119 % amplitude modulation with the constant threshold does not mean that $\langle f_b\rangle$ has negative values !). This large modulation with variable threshold is physically quite significant since it reveals an increase in activy more important than the simple increase in turbulent intensity. This result is thus in contradiction with the finding of COUSTEIX and

HOUDEVILLE mentioned above. A possible reason for this discrepancy may be sought in the VITA parameters used by these authors, namely $\hat{T}^+ = 25$ and $k = 1.2$, which are both large compared to those commonly used, and which yieled a mean bursting frequency about four times smaller than those reported elsewhere. It seems, thus, that with these parameters they made a strict selection among the bursts based on a slower rise time and larger amplitude which could well account for the differences in behaviour of $\langle f_b \rangle$.

Moreover, it is observed that the mean bursting frequency whether the threshold is constant or not is hardly different from the value in steady flow in agreement with KABASHI-KAYAKAWA and COUSTEIX-HOUDEVILLE. Since the flow and the turbulent intensity are strongly modulated this constancy of $\overline{f}_b$ despite the large modulation of $\langle f_b \rangle$ during the cycle is quite remarkable. It reveals a surprizing balance between the cyclic increase and decrease in bursting activity.

Finally Fig. 4 and 5 and Table 2 show that $\langle f_b \rangle$ is practically in phase with the turbulent intensity $\langle \tau'\tau' \rangle$ at the wall (while $\langle \tau'\tau' \rangle$ lags $\langle \tau \rangle$ by 74°) -

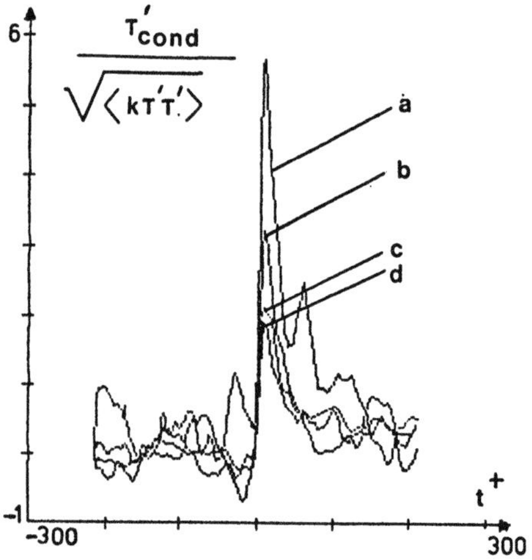

Fig. 6. Conditional averages of events a) t/T = 0.24 , b) 0.52 , c) 0.08 , d) 0.88

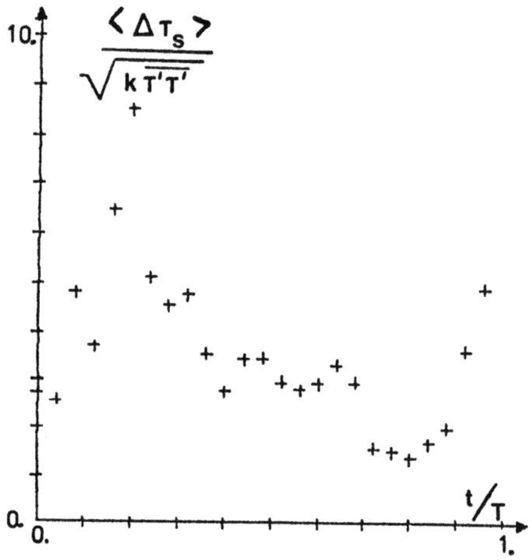

Fig. 7. Phase average of sweep magnitude scaled with time mean threshold : $\overline{\Delta\tau_s} / \sqrt{k\overline{\tau'\tau'}} = 3.65$

4.2. Conditionally Averaged Sweep Pattern

Conditional averages $\langle \tau'_{cond} \rangle$ of the sweep pattern at different times of the cycle are shown on Fig. 6. The shapes are very similar to those of CHAMBERS and al (A2) with a very steep rise and an exponential like decay.

From these pattern the sweep time t_s and magnitude $\Delta\tau_s$ occuring during the cycle and obtained. In order to reduce the statistical scatter the cycle is devided into 25 intervals. The sweep time is defined as by CHAMBERS et al as the time interval during which $\tau'_{cond} \geq (1-e^{-1})(\tau'_{cond})_{max}$ while the magnitude is defined as the difference between the maximum and the minimum preceeding the sharp rise which is detected.

4.3. Magnitude of Sweeps

The magnitude of the sweeps is a random quantity whose distribution is not related in a simple way to the selection threshold. One would, however, expect that this magnitude is larger on the average at the time in the cycle where the frequency of occurence is larger. This is approximatively the case as shown by Fig. 6 where $\langle \Delta \tau_s \rangle$ is normalized with the constant threshold $\sqrt{k \overline{\tau' \tau'}}$. The peak of $\langle \Delta \tau_s \rangle$ is however, much narrower than the one of $\langle fb \rangle$. If $\langle \Delta \tau_s \rangle$ is scaled by $\sqrt{k \langle \tau' \tau' \rangle}$ (Fig. 7) the trend of the variations are obscured by the statistical scatter but the peak is still distinguishable indicating thus that the sweeps are also relatively stronger when their frequency is higher.

The magnitudes found here are larger by a factor of about 3 than those reported by CHAMBERS et al. The turbulent intensities measured by these authors are also much lower than the one measured here (our results being in agreement with the value commonly accepted). These consistently lower amplitudes could be due to the relatively large size of their sensing element.

4.4. Sweep Time

The sweep time $\langle t_s^+ \rangle$ versus phase is presented on Fig. 8. It may seen that this time is roughly constant within the statistical scatter with an average value $\overline{t_s^+} =$ 20 which compares well with the results of CHAMBERS et al. The question arises whether the time scale $\nu / \langle \tau \rangle$ would be more appropriate than the mean viscous time scale $\nu / \overline{\tau}$. Since $\langle \tau \rangle$ varies here by roughly a factor of two from the minimum to the maximum, $\langle t_s \rangle \langle \tau \rangle / \nu$ varies in the same proportion. One may in this respect further ask whether the integration time in VITA should not be modulated with $\langle \tau \rangle$ and be taken equal to $\hat{T} = 10 \nu / \langle \tau \rangle$.

Finally the sweep-time ratio $\langle f_b \rangle \langle t_s \rangle$ which is the percentage of time occupied by the sweeps is presented on Fig. 9. The average value is quite comparable again to that of CHAMBERS et al. The variations of $\langle f_b \rangle \langle t_s \rangle$ reflect mainly those of f_b since $\langle t_s \rangle$ is nearly constant.

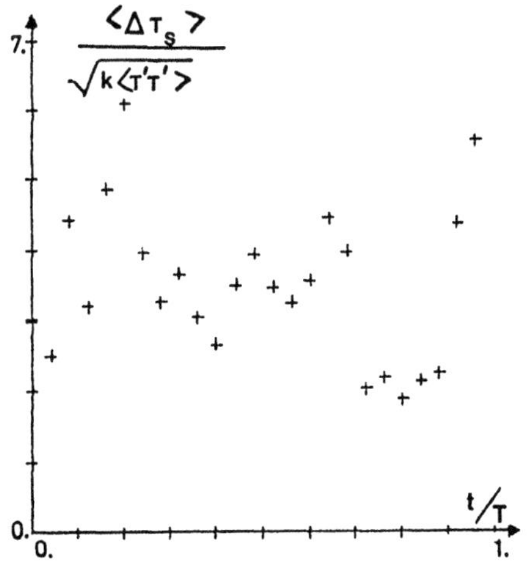

Fig. 8. Phase average of sweep magnitude : $\overline{\Delta \tau_s} / \sqrt{k \langle \tau' \tau' \rangle} = 3.6$ scaled with phase averaged threshold

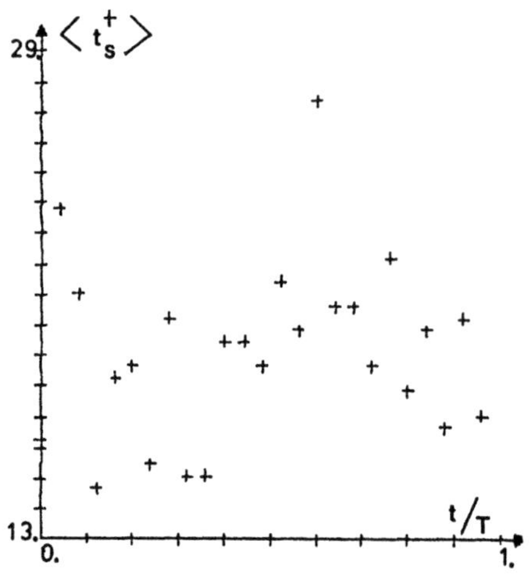

Fig. 9. Phase averaged sweep time $t_s^+ = 20$

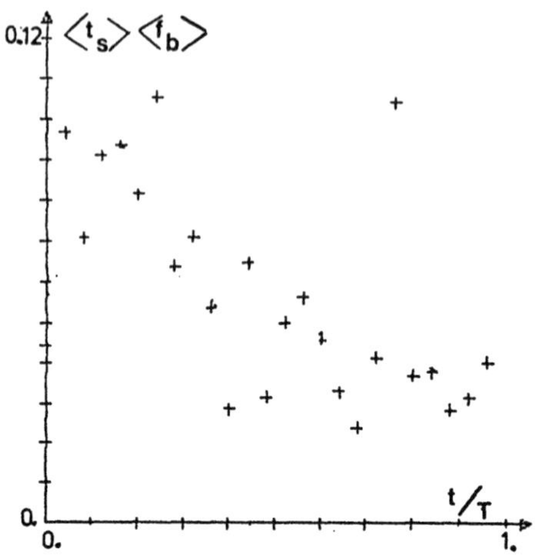

Fig. 10 Phase averaged sweep time ratio $\langle t_s \rangle \langle f_b \rangle = 0.06$

5. Conclusion

It is shown in the present investigation that oscillations, imposed upon a turbulent channel fow at a frequency such that $\ell_s^+ = 16$ and a center-line amplitude of 29 %, modulate the bursting process. In particular, the bursting frequency varies with an amplitude of 56 % during the cycle and in phase with the turbulent intensity even if the detection threshold is modulated with this intensity. The time-mean bursting frequency is, however, the same as in the corresponding steady flow.

The structure within the sweep itself is not strongly affected by the oscillations. If the magnitude of the conditionnally averaged signal is somewhat larger when the turbulence is maximum, the sweep-time, on contrary, remains constant during the cycle.

Much more work remains, of course, to be done before such questions as the influence of frequency and amplitude of the imposed oscillations on the response of the bursting process, the time scaling, the response of the shear-stress, the effect of flow reversal, will receive the beginning of an answer.

REFERENCES

1. T. MIZUSHINA, T. MARUYAMA, Y. SHIOZARI : J. Chem. Engr Japan 6, 6 (1973)
2. T. MIZUSHINA, T. MARUYAMA, H. HIRASAWA : J. Chem. Engr Japan, 8, 3 (1975)
3. K. RICHTER, D. RONNEBERGER : In Unsteady Turbulent Shear Flow (IUTAM Symp. Toulouse), ed. R. MICHEL et al (Springer, Berlin-Heidelberg, 1981) p.77.
4. D. RONNEBERGER : 5-th Symp. Turbulent Shear Flows, Cornell Un. (1985), p.18.19.
5. R. BLACKWELDER, R.F. KAPLAN : J. Fluid Mech. 76 (1976)

6. Y. KOBASHI, M. HAYAKAWA : In Unsteady Turbulent Shear Flow (IUTAM Symp. Toulouse) ed. R. MICHEL et al (Springer, Berlin-Heidelberg, 1981) p. 67
7. J. COUSTEIX, R. HOUDEVILLE : ONERA-CERT report OA 53/2259 (Toulouse, France) 1983
8. J. COUSTEIX, R. HOUDEVILLE : 5th Symp. Turbulent Shear Flows, Cornell Un. (1985), p. 18.7.
9. G. BINDER, J.L. KUENY : In Unsteady Turbulent Shear Flow (IUTAM Symp. Toulouse) ed. R. MICHEL et al. (Springer, Berlin-Heidelberg, 1981)
10. G. BINDER, S. TARDU, R. BLACWELDER, J.L. KUENY : Proc. Unsteady Aerodynamics AGARD-CP 386 (Göttingen 1985) p. 10.1
11. G. BINDER, S. TARDU, R. BLACKWELDER : 5th Symp. Turbulent Shear Flows, Cornell Un. (1985), p. 16.1
12. F.W. CHAMBERS, H.D. MURPHY, D.M. Mc ELIGOT : J. Fluid Mech 123 (1983)

Turbulence in Oscillatory Boundary Layers

B.M. Sumer, B. L. Jensen, and J. Fredsøe

Technical University of Denmark, Institute of Hydrodynamics and Hydraulic Engineering, DK-2800 Lyngby, Denmark

This paper reports the results of an experimental investigation on the turbulent oscillatory boundary layer flows which occur over both the smooth and rough walls, the rough wall case having a roughness parameter, the amplitude-to-roughness-height ratio of 1800. The boundary layer properties are found to be strongly dependent on the phase as well as on the wall category.

INTRODUCTION

Wave-induced oscillatory boundary layers are of great practical importance as to their applications in various fields of coastal and offshore engineering; for example in sediment transport and in pipeline-related problems. Although much progress has been made in the understanding of steady turbulent boundary layer flows in recent years, little is known about the unsteady turbulent boundary layer flows despite its practical importance. Therefore, investigation of turbulent oscillatory boundary layers, which is a sub-class of the unsteady boundary layer flows, is well justified.

The present study is complementary to Hino et al.'s [1] work in the sense that it covers both smooth- and rough-wall boundary layer flows. Whereas, in Hino et al. only the smooth wall case has been investigated. As far as the rough-wall turbulent boundary layer flows are concerned, there are two major contributions in the literature, namely Jonsson and Carlsen [2] and Kemp and Simons [3]. However, both studies suffer from the fact that the ratio of the flow amplitude to the rougness height, a/k, is too small. Indeed, in the former work, the value of the parameter a/k is 124 in one test and 28 in the other, and in the latter it is in the range of 5 - 8. Furthermore, both studies investigate two-dimensional roughness form. However, in real-life flow situations under storm conditions in the sea, the bed is usually rough and covered by three-dimensional roughness elements (sand), Further the bed is plane at large values of the dimensionless Shields' parameter whereby the parameter a/k is in the range of 10^3 -10^4 because the roughness now only is due to the sand elements and not to the bedforms.

The rough wall reported in the present paper has a more realistic value of a/k, namely 1800. In the experiments, the amplitude and the period of the motion are kept constant, irrespective of the type of the wall studied to be able to make comparison on the same basis.

Earlier results of this study have been presented at the 3rd International Symposium on Applications of Laser Anemometry to Fluid Mechanics held in Lisbon, Portugal (Sumer, Jensen and Fredsøe [4]).

EXPERIMENTAL FACILITY

Experiments were carried out in a U-shape oscillating water tunnel shown in Fig. 1.

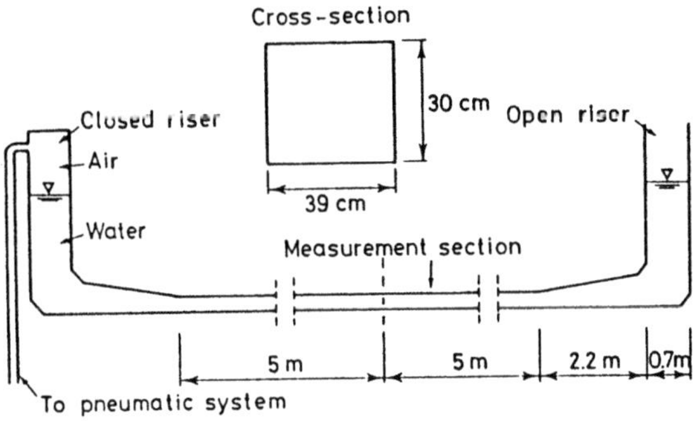

Fig.1 Oscillating water tunnel.

Table 1

	WALL	
	Hydr. smooth	rough
k	-	1,5 mm
kU_{fm}/ν	-	184
a	2,71 m	2,71 m
U_m	2,1 m/s	2,1 m/s
ω	0,774 s^{-1}	0,774 s^{-1}
U_f	9,8 cm/s	13,6 cm/s
T	15° C	15° C
aU_m/ν	5×10^6	5×10^6
a/k	-	1800

The flow conditions are summarized in Table 1 where k = the roughness height from the base bottom, U_{fm} = the maximum wall shear velocity, a = the amplitude of the oscillatory motion, U_m = the maximum free stream velocity, ω = the angular frequency (see also Fig. 2), T = the temperature of the water. The bottom wall of the tunnel was painted with a plastic paint for the smooth-wall experiments. For the

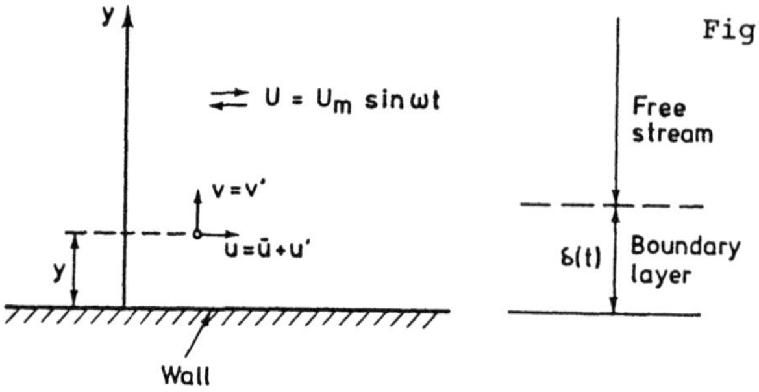

Fig. 2 Definition sketch.

rough wall experiments, the bottom was covered with sand of fairly uniform size; the sand was glued one layer deep to the bottom in a closely packed manner.

A Laser Doppler Anemometer (LDA) was used to measure the streamwise and transverse velocity components, u and v, respectively (Fig. 2). The velocity recordings were made for 50 cycles of the oscillating flow at a sampling interval of 14 milliseconds. Mean values of the quantities are calculated through ensemble averaging according to

$$\bar{\phi}(y,\omega t) = \frac{1}{N} \sum_{j=1}^{N} \phi(y,\omega[t+(j-1)T]) \qquad (1)$$

where N = the number of cycles and T = the period.

More detailed information about the experimental facility can be found in Sumer, Jensen and Fredsøe [4].

RESULTS AND DISCUSSION

Fig. 3 shows sample records, the top record representing the free-stream velocity and the bottom ones the two components of the velocity in the boundary layer.

Fig. 4 gives the mean velocity profiles as the oscillatory motion progresses in time while Fig. 6 presents the corresponding shear-stress profiles. The shear stress is calculated from the velocity profiles through the momentum integral equation

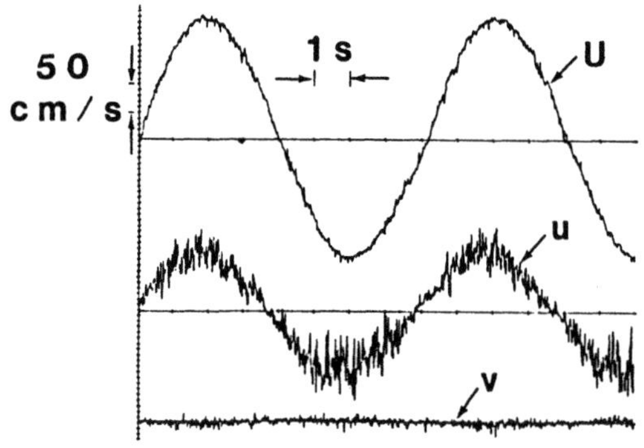

Fig. 3 Sample velocity records. Smooth wall. y = 0.5 mm.

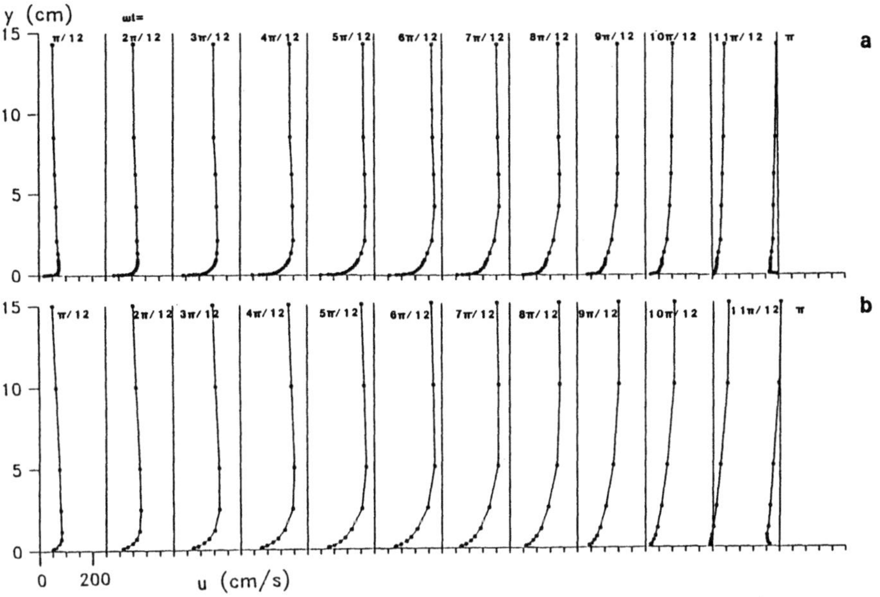

Fig. 4 Mean velocity distributions. a) smooth wall. b) rough wall.

$$\frac{\tau}{\rho} = \int_y^d \frac{\partial}{\partial t} (U - \bar{u})\, dy \tag{2}$$

Here d is any distance from the wall at which the condition $\tau = 0$ is satisfied.

Fig. 5 presents the semi-log plots of the velocity distributions where the straight lines represent the logarithmic law:

$$\frac{\bar{u}}{U_f} = \frac{1}{\kappa} \ln \frac{yU_f}{\nu} + B \qquad ; \quad \text{Smooth wall} \tag{3}$$

$$\frac{\bar{u}}{U_f} = \frac{1}{\kappa} \ln \frac{30y}{k_s} \qquad ; \quad \text{Rough wall} \tag{4}$$

Here U_f = the wall shear velocity, κ = von Kármán constant (= 0.4), ν = the kinematic viscosity, k_s = the Nikuradse's equivalent sand roughness and B = the universal constant corresponding to the hydraulically smooth-wall boundary category. Now, as to the data presented in this figure, the following points appear to be noteworthy.

(1) The experiments show that the logarithmic layer does not exist until the phase reaches a value of about $\omega t = 1.5\ \pi/12$. This is simply because the boundary layer is not thick enough to house the logarithmic layer in it until this moment has been reached.

(2) The experiments also indicate that the constant stress layer displaces markedly towards the outer flow regions for phases larger than approximately $\omega t = 8.5\ \pi/12$, as seen in Fig. 6. Therefore the lo-

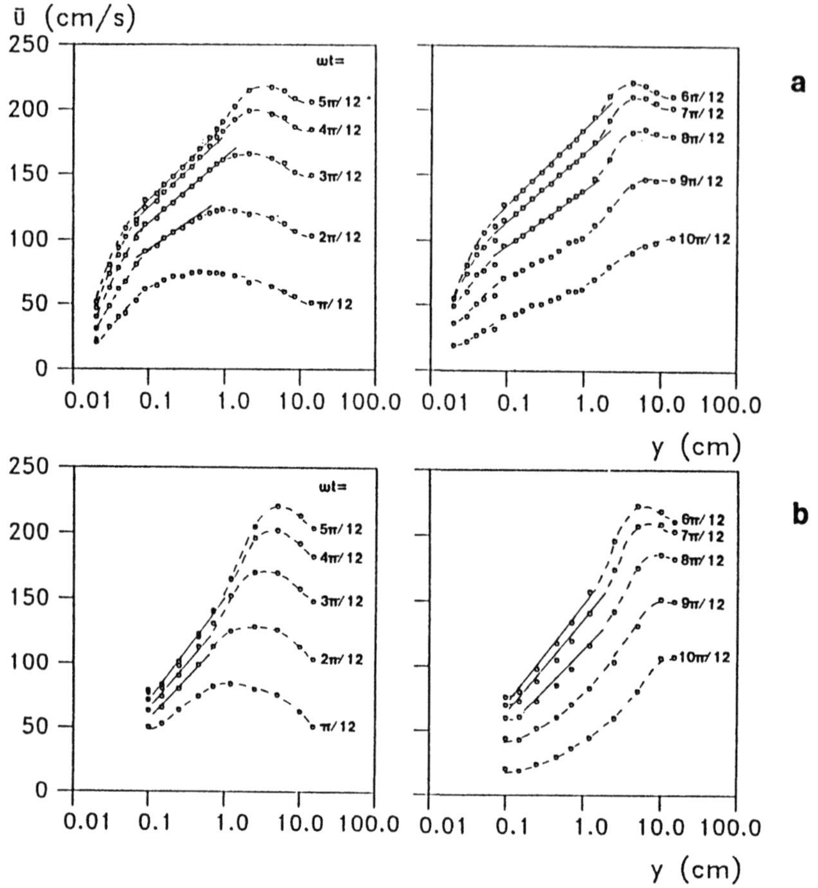

Fig. 5 Mean velocity distributions in semi-log plot. a) smooth wall. b) rough wall. Straight line: Logarithmic law (Eqs. 3 and 4)

garithmic layer should also be displaced to these higher elevations for $\omega t > 8.5\ \pi/12$. The velocity distribution may still be represented by the logarithmic law. However U_f in such a representation should no longer be the wall shear velocity, but a velocity defined by $(\tau/\rho)^{1/2}$ in which τ is the constant shear stress occurring at these elevations rather than at the wall.

From the considerations in (1) and (2), it is clear that the logarithmic law as given in Eqs. (3) and (4) is valid in the phase interval $1.5\ \pi/12 < \omega t < 8.5\ \pi/12$.

By fitting straight lines to the logarithmic-layer portion of the velocity data in Fig. 5, the wall shear velocity U_f, the Nikuradse's equivalent sand roughness k_s and the constant B have been calculated. The shear velocity is plotted in Fig. 8 and it appears to be in good agreement with the shear velocity determined from the momentum integral equation.

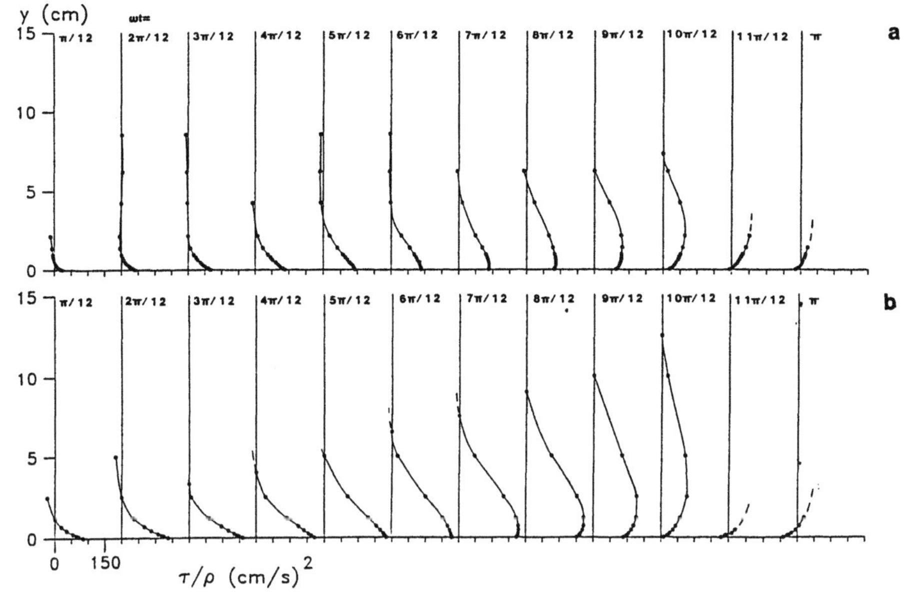

Fig. 6 Shear stress distributions. a) smooth wall. b) rough wall

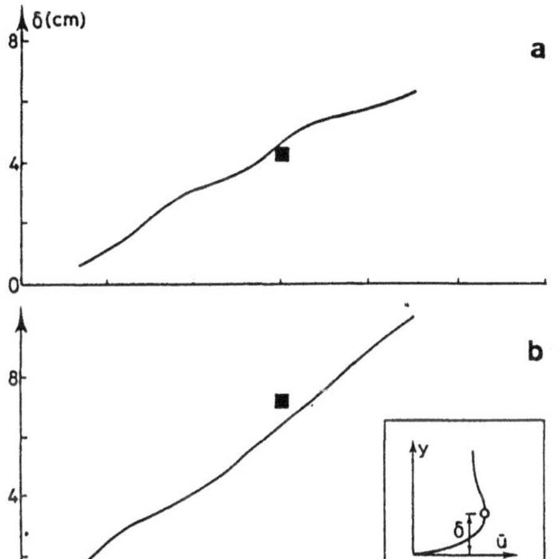

Fig. 7 Boundary layer thickness.
a) smooth wall. b) rough wall
■ Fredsøe's (1984) theory.

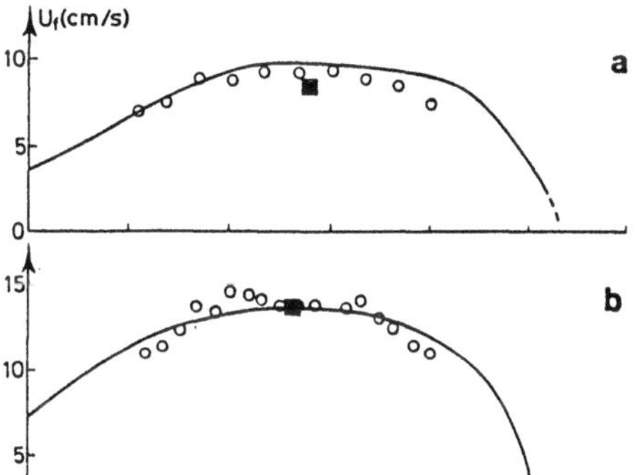

Fig. 8 Wall shear velocity.
a) smooth wall. b) rough wall.
Solid curve: from momentumintegral equation (Eq. 5). o: from semi-log plot of velocity distributions (Eqs. 3 and 4).
■ Fredsøe's (1984) theory

On the other hand, the k_s value is obtained to be 4 mm, which approximately reveals the relation $k_s = 2.5$ k known from the steady flow boundary-layer research.

The value obtained for the constant B is approximately 3.5, which is a relatively low value with respect to that reported in literature for steady boundary layer flows, namely 5. This can be attributed to the fact that the wall in the tests did not act as a mirror-shine smooth surface, but rather as a hydraulically smooth one due to very high shear velocities experienced. Indeed, for the value of 3.5 of the constant B, Nikuradse's data (Monin and Yaglom, [5], p. 289) indicate that the wall should have a roughness value like $k_s U_f/\nu = 10$, or alternatively $k_s = 0.1$ mm, considering $U_f = 10$ cm/s. The latter can represent a reasonable value for the Nikuradse roughness of the surface used in the experiments.

Fig. 7 illustrates how the boundary layer thickness grows as the flow progresses. This figure contains also the values predicted through Fredsøe's [6] analytical study, which compares favourably well with the present test results.

It is not surprising that the boundary layer thickness in the rough wall case is considerably larger than in the smooth wall. This is simply because the momentum transfer in the rough wall case is enhanced by the effect of the boundary roughness.

Fig. 8 presents the variation of the wall shear stress with respect to ωt. The solid curve in the figure represents the shear velocity determined from the momentum integral equation

$$U_f^2 = \frac{\tau_0}{\rho} = \int_0^d \frac{\partial}{\partial t} (U - \bar{u})\, dy \qquad (5)$$

using the velocity distributions of the present experiments.

This figure also includes the shear velocity values determined from the semi-log plots of the velocity distributions, which has been discussed already. The figure contains also the maximum U_f values calculated from Fredsøe's [6] analytical study. The latter values are plotted on the diagram, taking into consideration the particular phases that the theory predicts for the maximum U_f values, namely $\phi = 7^o$ for the case when the wall is smooth and $\phi = 11.5^o$ when the wall is rough. Here, ϕ = the phase lead of the shear velocity over the free-stream velocity. Both the maximum value and the phase are in good agreement with the present test results.

Although the free-stream conditions are maintained exactly the same in both tests, the shear velocity for the rough-wall case appears to be substantially larger than the shear velocity in the case of smooth wall. This is an obvious result and is merely caused by the presence of the roughness elements on the boundary.

It is apparent from Fig. 8 that the shear-velocity variation is certainly not sinusoidal, but rather it is flat. This can be attributed to the following fact. The experiments show that the variation of the streamwise velocity $\bar{u}$ with respect to time becomes flat near the wall. Now, since U_f can be expressed as $U_f \propto \bar{u}$ very near the wall, then U_f variation should also be expected to be flat.

As to why the variation of $\bar{u}$ with respect to time becomes flat near the wall, this can be attributed to the combined action of the following effects: (1) the growth of the boundary layer combined with increased outer flow velocity (for $\omega t < \pi/2$) and (2) the enhanced momentum transfer in the boundary layer as the flow velocity takes high enough values during the course of the oscillatory motion.

Fig. 9 gives the turbulence data plotted as a function of the distance from the wall, while Fig. 10 gives the detailed structure of the turbulence data very close to the bottom in the case of smooth wall. In Fig. 9, the dashed line represents the upper edge of the boundary layer.

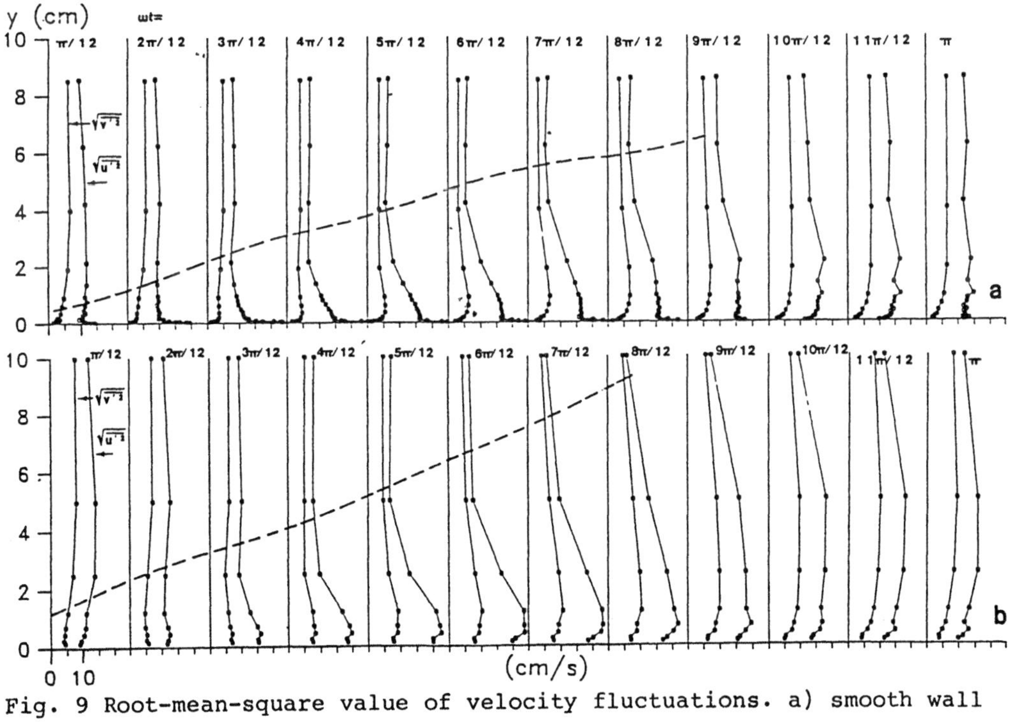

Fig. 9 Root-mean-square value of velocity fluctuations. a) smooth wall b) rough wall. Dashed line: Edge of the boundary layer.

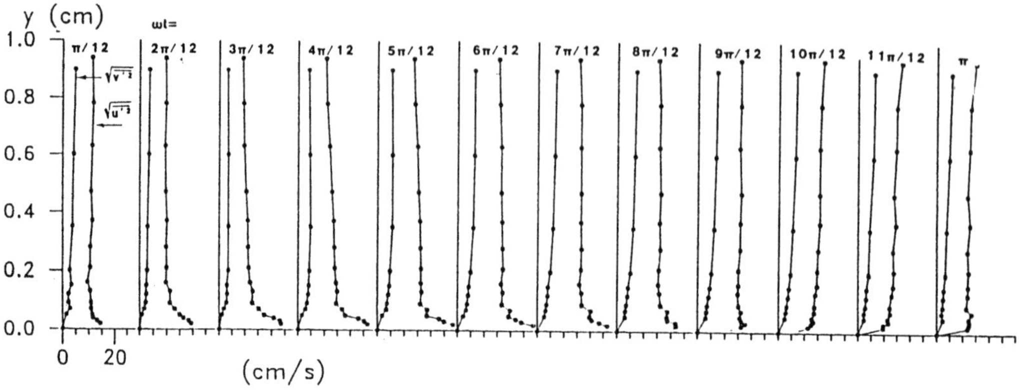

Fig. 10 Close-up picture of the root mean-square value of the velocity fluctuations in smooth wall.

Figs. 9 and 10 illustrate how the turbulence evolves as the flow progresses in time. First, let us focus on what happens in the boundary layer. The figures indicate the following: (1) The turbulence builds up near the wall, as the boundary layer develops; (2) This turbulence constantly diffuses away from the wall across the boundary layer; (3) The near-wall built-up of turbulence almost comes to an end by the time of approximately $\omega t = 10\ \pi/12$; and (4) the turbulence becomes almost uniformly distributed across the depth as the time of flow reversal is reached.

Now, let us consider the free-stream region. The flow in the free-stream region, just after the instant of flow reversal $\omega t = 0$, will see a uniformly distributed turbulence, which is put in the region in the previous half cycle of the oscillatory motion. The data in Fig. 9 indicate that this turbulence in the free-stream region undergoes decaying as the flow progresses.

Hino et al.'s [1] smooth-wall experiments also confirm the description given above.

The mixing length l defined by

$$\frac{\tau}{\rho} = \left(\nu + l^2 \left|\frac{\partial \bar{u}}{\partial y}\right|\right) \frac{\partial \bar{u}}{\partial y} \qquad (6)$$

has been determined by using the experimental profiles $\tau(y)$ and $\bar{u}(y)$ shown in Figs. 4 and 6, respectively. Fig. 11 shows the time evolution of the mixing length as the flow progresses. In Fig. 6, also the familiar relation from the steady boundary layer research,

$$l = \kappa y \qquad (7)$$

is plotted as a reference line where κ = von Kármán constant. The hor-

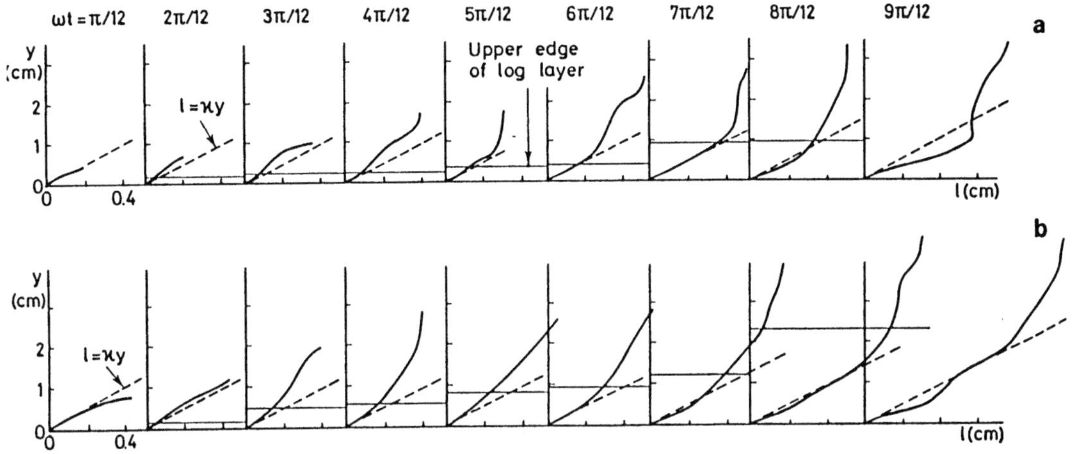

Fig. 11 Mixing length. a) Smooth wall. b) Rough wall

izontal lines in the figure represent the upper edge of the region in which the logarithmic law as given by Eqs. 3 and 4 is applicable.

Fig. 11 indicates that the experimentally determined profiles are in good agreement with the relation $l = \kappa y$ in the region where the logarithmic law is applicable.

This figure also indicates that there is an increase in the mixing length for phase values larger than about $\omega t = 8\pi/12$. This can be attributed to the fact that the boundary layer at this stage of the oscillating motion experiences an adverse pressure gradient, thus the momentum transfer in the transverse direction is enhanced, leading to an obvious increase in the mixing length.

Fig. 12 presents two spectral densities of the streamwise component of the velocity u', one corresponding to the accelaration stage of the oscillating motion where $\omega t = 3\pi/12$ and the other corresponding to the deceleration stage with $\omega t = 9\pi/12$. Fig. 12a shows the smooth wall data and Fig. 12b the rough wall data. The spectral density here is the normalized spectral density defined by

$$\int_0^\infty S(f)\, df = 1 \qquad (8)$$

For the spectral density analysis, the velocity recordings were made at a sampling interval of 1 millisecond for 80 cycles of the oscillating flow. The spectral density was obtained through the Fast Fourier Transform (FFT) of the velocity signal at various values of ωt. The length of the velocity signal for which the FFT analysis was applied was maintained at 200 milliseconds. The spectral densities were smoothed through ensemble averaging according to Eq. (1). The spectral density diagrams presented in Fig. 12 are the smoothed ones.

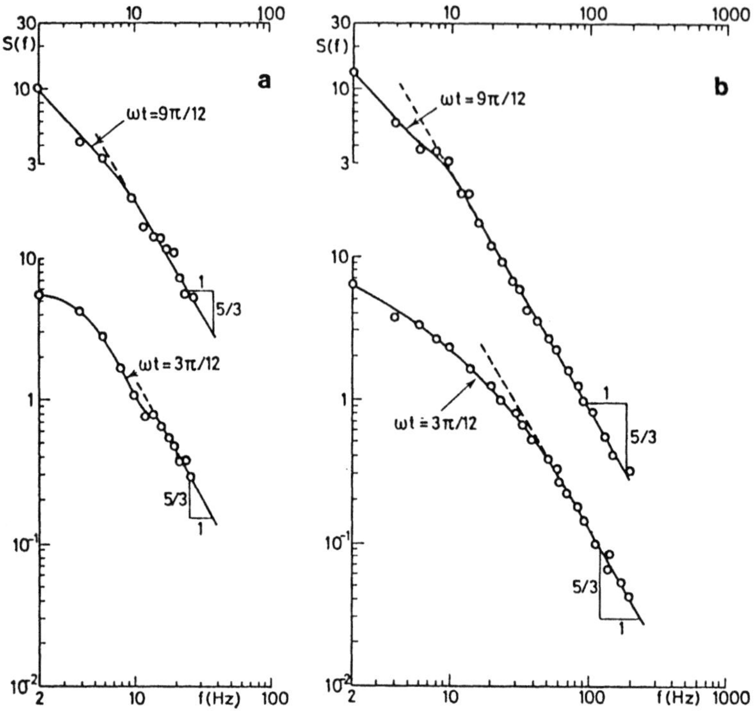

Fig. 12 Spectral density. a) Smooth wall b) Rough wall. y = 2.5 mm.

Note that the high frequency part of the spectral density curve where the so-called laser ambiguity effect is dominant has been cut off to avoid confusion.

From Fig. 12, the following three points appear to be noteworthy.

(1) The inertial subrange (in which $S(f) \sim f^{-5/3}$) exists for both the acceleration and deceleration stages.

(2) The upper limit of the energy containing range experiences a shift towards the lower frequencies, as the flow proceeds from the acceleration stage to the deceleration stage. This means that the scales of "energy containing" eddies become larger and larger as the flow develops through its phase space. It should be noted that this is consistent with the picture given in Fig. 11.

(3) It is known that the value of the spectral density at zero frequency is proportional to the macro time scale of turbulence. Now, an inspection of Fig. 12 indicates that the macro time scale of turbulence is substantially larger at the deceleration stage than at the acceleration stage. This again appears to be consistent with the information presented in Fig. 11.

CONCLUSIONS

An oscillatory turbulent boundary layer exhibits highly interesting flow regimes throughout its phase.

It was found that the familiar logarithmic law is applicable to a wide range of the phase for each half cycle of the motion.

The present study indicates that the turbulence in the boundary layer evolves steadily throughout one half cycle of the motion. It was found that the boundary layer turbulence gets spread across the depth by the time that the flow reverses. It was observed that this turbulence, which actually constitutes the free-stream turbulence in the next half cycle, undergoes a steady decay as the flow progresses in this next half cycle.

The experiments indicated that the scale of turbulence steadily increases as the flow develops through its phase space.

Although the outer flow parameters (the amplitude and the period of the oscillating motion) are maintained exactly the same, marked differences were observed in the boundary layer properties of the smooth and rough wall cases studied.

ACKNOWLEDGEMENT

This study is partially supported by the FTU project "Turbulence around offshore structures" of the Danish Scientific Council (STVF). The LDA used in this study is a donation by Dansk Investeringsfond (DIFKO).

REFERENCES

1. Hino, M., Kashiwayanagi, M., Nakayama, A. and Hara, T. Experiments on the turbulence statistics and the structure of a reciprocating oscillatory flow. J. Fluid Mech. Vol. 131, p. 363.

2. Jonsson, I.G. and Carlsen, N.A. 1976 Experimental and theoretical investigations in an oscillatory rough turbulent boundary layer. J. Hyd. Res. Vol. 14, No. 1, p. 45.

3. Kemp, P.H. and Simons, R.R. The interaction between waves and a turbulent current: Waves propagating with the current. J. Fluid Mech. Vol. 116, p. 227.

4. Sumer, B.M., Jensen B.L., and Fredsøe, J. 1986. Experimental investigation of turbulent ocillatory boundary layer flows. 3rd International Symposium on Applications of Laser Anemometry to Fluid Mechanics, 7-9 July 1986, Lisbon, Portugal.

5. Monin, A.S. and Yaglom, A.M. 1973 Statistical fluid mechanics: Mechanics of turbulence. The MIT Press.

6. Fredsøe, J. 1984 Turbulent boundary layer in wave-current motion. ASCE, J. Hyd. Engrg. Vol.110, No. 8, p. 1103.

The Structure of Turbulence Measured in a Relaxing Boundary Layer

U.R. Müller

Aerodynamisches Institut, Rheinisch-Westfälische Technische Hochschule Aachen, D-5100 Aachen, Fed. Rep. of Germany

Introduction

The experiments /1, 2/ studied the interaction of a turbulent boundary layer and a cylinder wake, and documented the relaxation towards flat-plate conditions by measured mean velocities and some components of the Reynolds stress tensor. The more recent investigation /3/ continued that work. By measuring all components of the mean velocity vector and those of the second and third order tensors of the fluctuations, results of sufficient detail were provided for analysing the balances of the Reynolds stresses $\overline{u_i u_j}$ according to the transport equations. The nonmeasurable terms were derived as closing entries to the balances. In /4/ the data have already been used for testing a computational method employing the three-equation Reynolds stress model of /5/.

In addition to /3/, the present paper provides further experimental results. In particular, those triple velocity correlations transporting kinetic turbulence energy and shear stress in transverse or longitudinal direction as well as the corresponding convection velocities were analysed. Then a number of structural functions like the stress-energy ratio - $\overline{uv}/k$ or the ratios of the intensities $\overline{u^2} : \overline{v^2} : \overline{w^2}$, which are frequently included in Reynolds stress models, were derived from the turbulence measurements.

Experimental arrangement and hot-wire techniques

In the experimental investigation, a two-dimensional incompressible flat-plate turbulent boundary layer with a momentum-thickness Reynolds number of 3800 and free-stream Reynolds number of $1.8 \cdot 10^6$/m was disturbed by a circular cylinder lying parallel to the surface and normal to the free-stream direction, Fig. 1. The axis of the cylinder, which had a diameter of D = 10 mm, was positioned at a wall distance of y_c = 13 mm, which was equal to half the thickness of the undisturbed shear layer. The pressure distribution measured at the wall showed negligible gradients beyond x/D = 5 indicating a momentum balance of inertia and turbulent shear forces.

Mean and fluctuating velocities were measured by hot-wire anemometry at 7 stations within the range $10 \leq x/D \leq 70$. A single hot-wire probe, Fig. 2, was used for near-wall measurements. Throughout the investigation, sensors were used with an active length of 1 mm, a diameter of 5 μm and gold-plated ends with a diameter of 20 μm. For obtaining all components of the mean velocity vector and of the tensors of the double and triple velocity

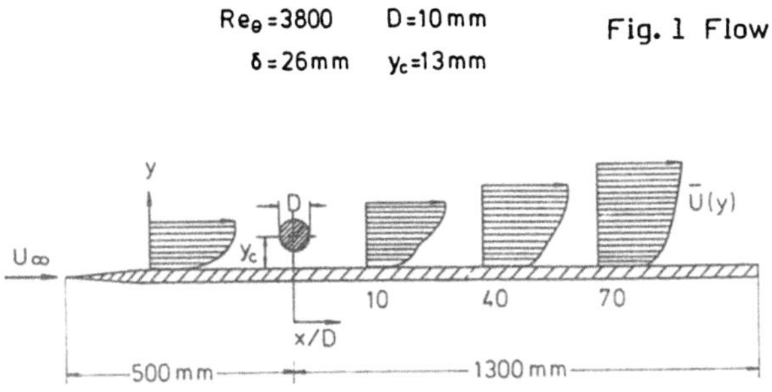

Fig. 1 Flow field investigated (not to scale)

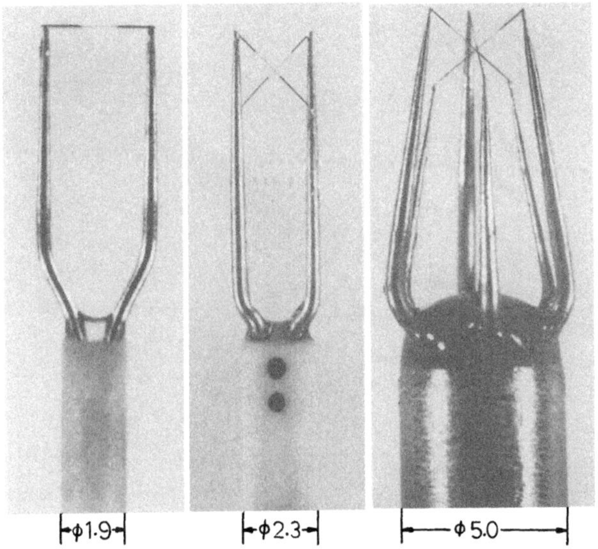

Fig. 2 Single, X and triple sensor hot wire probes

correlations, four traverses with an X-wire probe were necessary /6/, while a single traverse was sufficient when applying our newly developed triple hot-wire probe, Fig. 2. Comparison of single, X- and triple wire measurements yielded good agreement of the results for $x/D \geq 20$, /7/. The hot-wire signals were sampled, stored and processed with a PDP 11/34 computer.

Results and discussion

Flow development

The development of measured profiles of the streamwise mean velocity, the Reynolds shear stress $-\overline{uv}$, and the kinetic turbulence energy $\overline{k} = \overline{u_i u_i}/2$ (summation over repeated indices) is displayed in Figs. 3 - 5. The velocities are normalized by the free-stream velocity U_∞. The mean velocity was accelerated in the inner and decelerated in the outer layer, with the boundary layer thickness δ increasing from 45 to 62 mm. For $x/D \geq 20$ and $y = 29$ mm, the velocity was found to be constant.

According to the boundary layer equations, flow acceleration requires positive gradients of the Reynolds shear stress $\partial(-\overline{uv}) / \partial y > 0$. The measured shear stress profiles, Fig. 4, showed the corresponding behaviour and resembled those of adverse-pressure-gradient flows. The wall

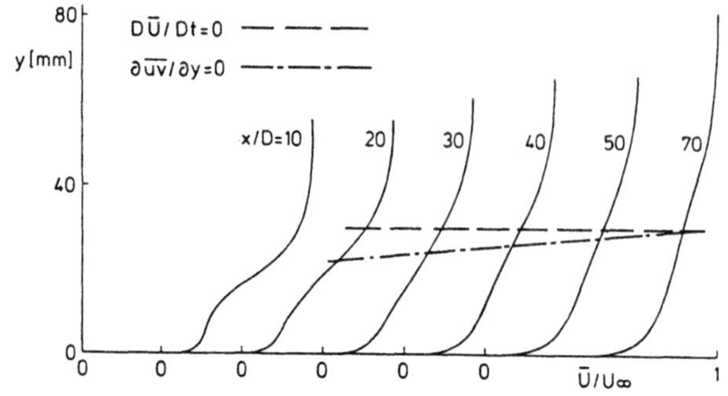

Fig. 3 Measured streamwise mean velocity

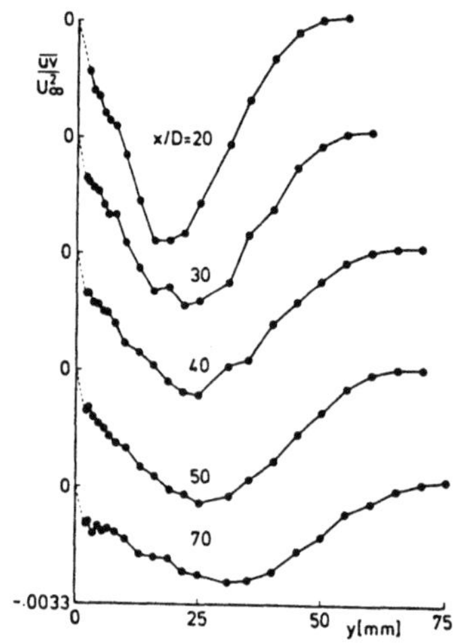

Fig. 4 Reynolds shear stress

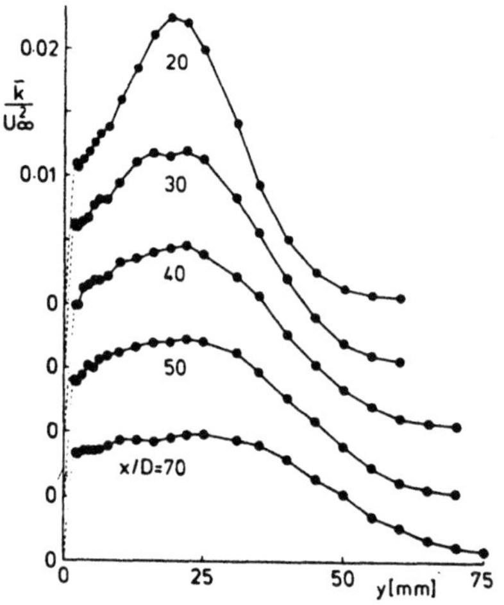

Fig. 5 Kinetic turbulence energy

distance of the maxima of $-\overline{uv}$ increased from approximately 20 mm at x/D = 20 to 30 mm at x/D = 70. For generating the vanishing gradient of the mean velocity mentioned above, the action of turbulent shear stress had been enhanced by normal stress diffusion $-\partial\overline{u^2}/\partial x$, /3/.

The development of the profiles of the kinetic energy, Fig. 5, showed a rapid decrease of the peak values measured in the central part of the boundary layer; the inner-layer gradient $\partial\bar{k}/\partial y$ had nearly dropped to zero at x/D = 70.

Transverse turbulent diffusion of kinetic energy is accomplished by the triple velocity correlations $\overline{u_i^2 v}$, Fig. 6. With $\overline{u^2v} : \overline{v^3} : \overline{w^2v}$ being approcimately 2 : 2 : 1, the correlation $\overline{kv} = \overline{u_i u_i v}/2$, see /3/, was about 5/4 $\overline{v^3}$. While the outer-layer positive-valued peaks of the correlations $\overline{u_i^2 v}$ decreased by more than fifty percent within $20 \leq x/D \leq 70$, the negative values in the near-wall region dropped to zero rapidly for $x/D < 40$. A corresponding results had already been found in /8/ and /9/ for relaxing shear flows beyond reattachment.

Turbulent diffusion of kinetic energy in longitudinal direction is due to the triple correlations $\overline{u_i^2 u}$, Fig. 7, which were found to have the order of magnitude of $-\overline{u_i^2 v}$. Corresponding to the latter correlations, the value of $\overline{v^2u}$ measured closest to the wall was about zero in contrast to $\overline{u^3}$ and $\overline{w^2u}$ having positive values near the wall throughout the flow field. The measured profiles indicate that $\overline{u^3}$ was approximately equal to the sum of $\overline{v^2u}$ and $\overline{w^2u}$. Therefore the correlation $\overline{ku} = \overline{u_i u_i u}/2$, see /3/, was close to $\overline{u^3}$.

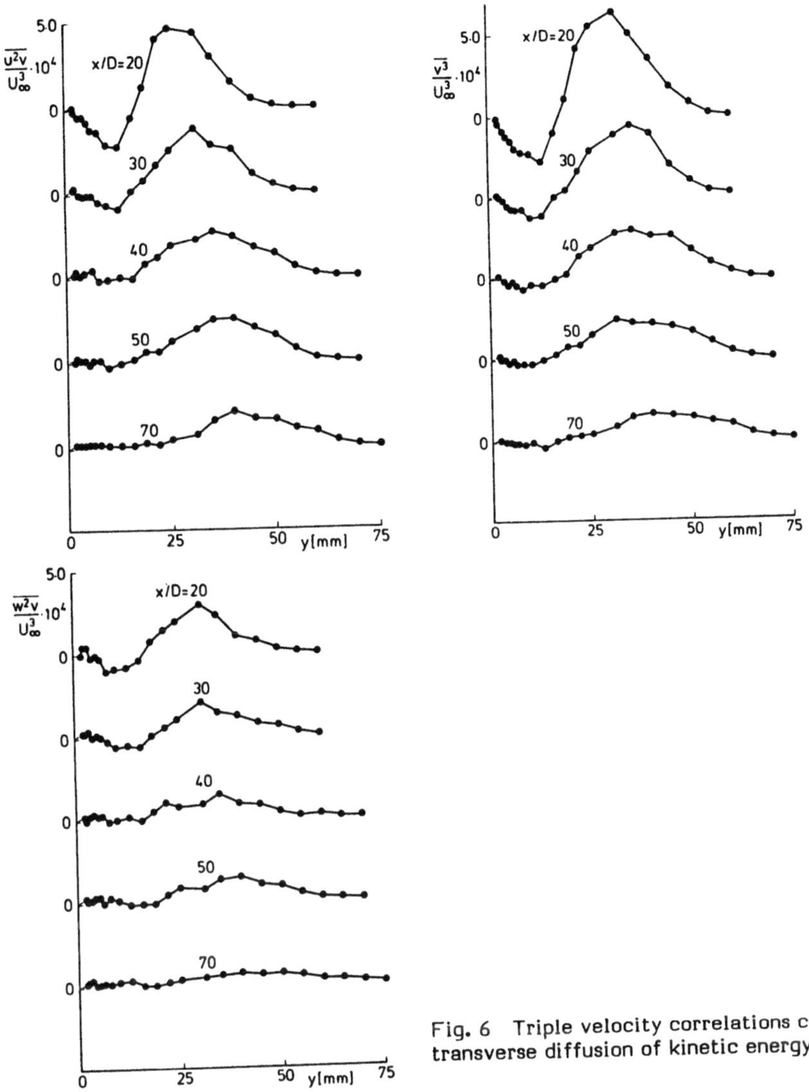

Fig. 6 Triple velocity correlations contributing to transverse diffusion of kinetic energy

The measurements of the triple products $\overline{u^3}$ and $\overline{v^3}$ yielded positive values of $\overline{u^3}$ in the inner layer and negative ones in the outer and opposite signs for $\overline{v^3}$. Such a behaviour indicated that the shape of the joint probability density function (pdf) of the u- and v-fluctuations changed in y-direction. From the time-dependent measurements the skewness $S_u = \overline{u^3}/(\overline{u^2})^{3/2}$ and flatness factors $F_u = \overline{u^4}/(\overline{u^2})^2$ of the u-fluctuation as well as the corresponding ones for the v-component were derived, Figs. 8 and 9. The joint pdf, Fig. 10, was obtained for a mesh size of $1x1(m/s)^2$, equivalent to a grid size of $u_i/U_\infty = 1/28$. For x/D = 20 and y = 2 mm, the contours were slightly tilted in the u-v-plane, for larger wall distances the plane of symmetry was approximately given by v/u = -2/3. For y = 8 mm, steep gradients of the pdf distributions in the quadrant v>0, u<0 and small ones in the quadrant u<o, v>0 yielded skewness factors $S_v<0$ and $S_u>0$. For y $\rightarrow$ 25 mm, the peak value of the pdf moved towards the quadrant u>0,

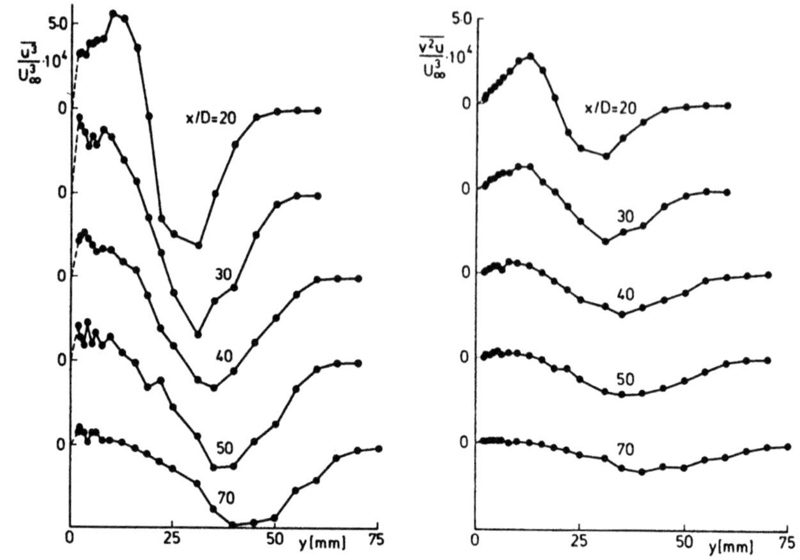

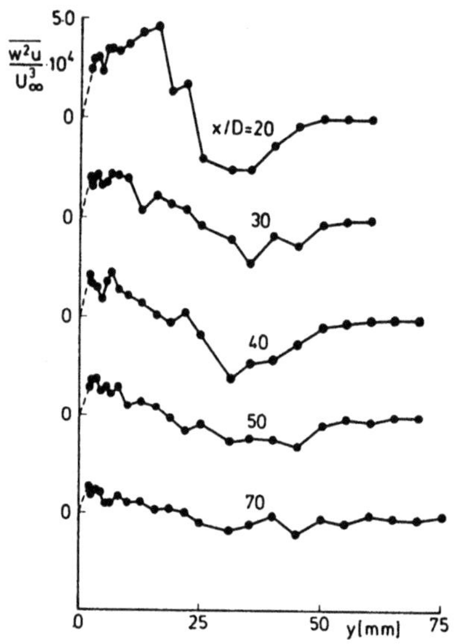

Fig. 7 Triple velocity correlations contributing to longitudinal diffusion of kinetic energy

$v<0$ resulting in $S_u < 0$ and $S_v > 0$. For x/D = 70, the shape of the pdfs was obtained close to Gaussian with $S_u = S_v \approx 0$ in the inner part of the shear layer. In the outer part, the behaviour was similar to that measured for x/D = 20. The flatness factors of both velocities, Fig. 11, were about four close to the wall at x/D = 20, while for x/D≥40 they were close the Gaussian value of three. In the outer layer the results increased to values of about six.

Structural functions

The measured Reynolds stresses indicated structural changes during the relaxation process. Close to the wall the ratio $\overline{w^2}/\overline{u^2}$, Fig. 12a, was about 1.1 at x/D = 20 and dropped to 0.9 for x/D→70. For larger wall distances the profiles approximately yielded $\overline{w^2}/\overline{u^2} = 0.8$ for the

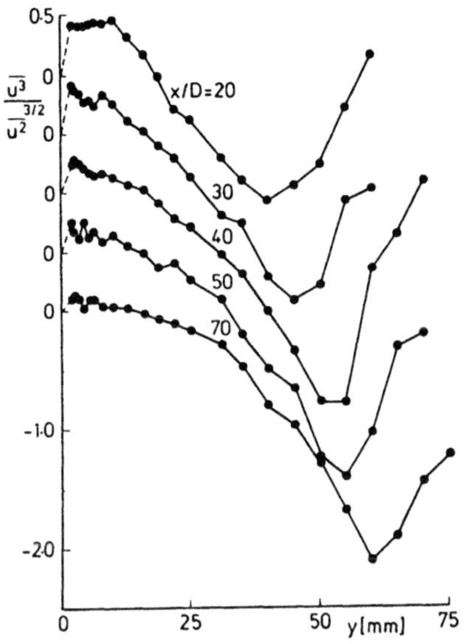

Fig. 8 Skewness factors of u-fluctuation

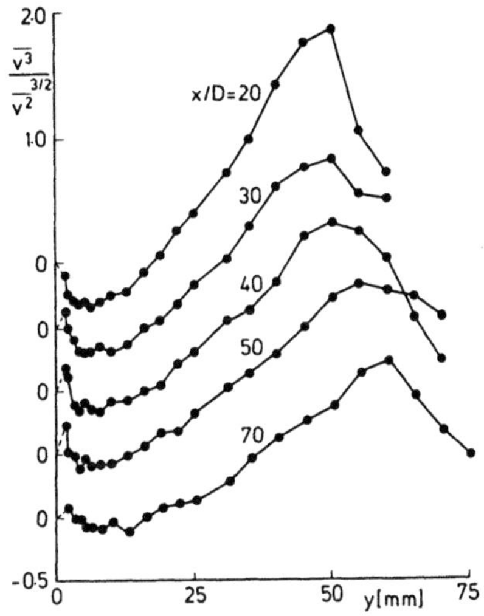

Fig. 9 Skewness factors of v-fluctuation

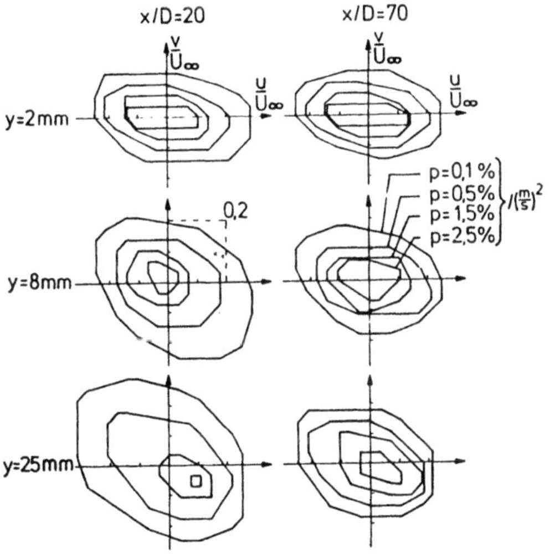

Fig. 10 Iso-contours of joint probability density functions of u- and v-fluctuations

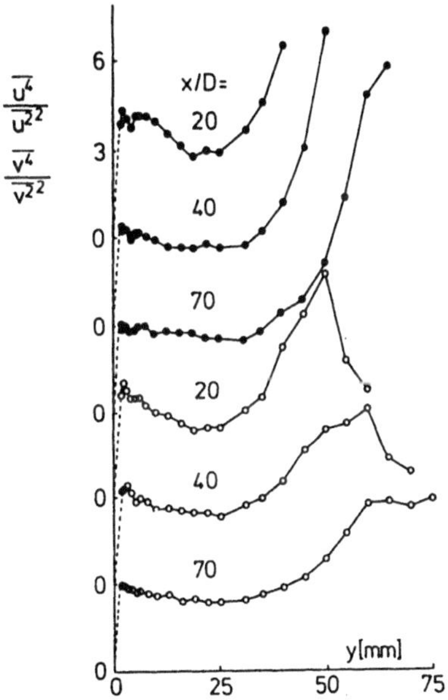

Fig. 11 Flatness factors of u- and v-fluctuations

central part at x/D = 20 and 0.7 for x/D = 70. At the boundary-layer edge the accuracy of the results is expected to decrease. The structural function $\overline{v^2}/\overline{u^2}$, Fig. 12b, rose from 0.25 close to the wall to constant values in the central layer, which fell rapidly from 0.8 at x/D = 20 to 0.6 at x/D = 40 and then levelled off to 0.5 at x/D = 70. From the results the contribution of the individual normal stresses to the kinetic energy $\bar{k}$ can be analysed. Near the wall the ratio $\overline{u^2} : \overline{w^2} : \overline{v^2}$ developed from 10 : 11 : 3 at x/D = 20 to 10 : 9 : 3 at x/D = 70, in the central

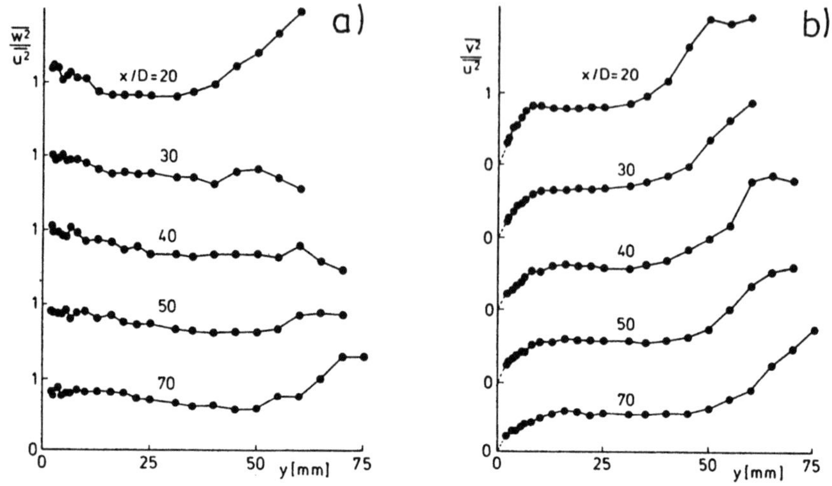

Fig. 12 Ratios of spanwise and transverse intensities to longitudinal one

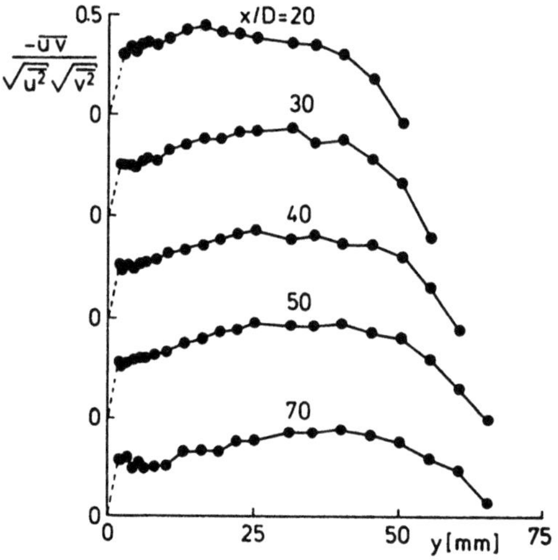

Fig. 13 Correlation coefficient

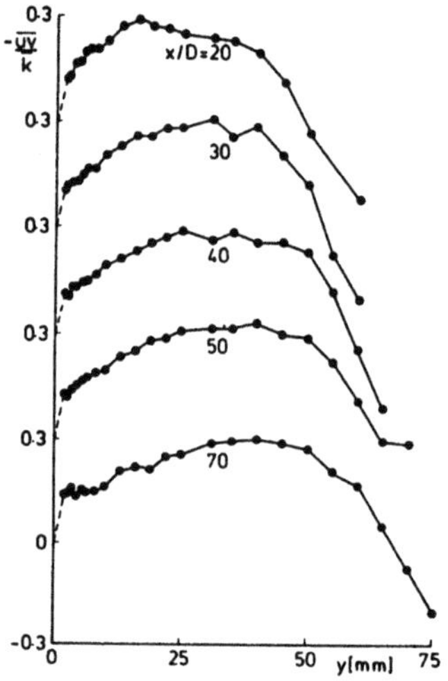

Fig. 14 Shear stress-energy ratio

layer the contribution to $\bar{k}$ steadily altered from 10 : 8 : 8 to 10 : 7 : 5, which was close to the flat-plate values of 4 : 3 : 2.

The profiles of the correlation coefficient $-\overline{uv}/\sqrt{\overline{u^2}\,\overline{v^2}}$, Fig. 13, was found to be self-similar throughout the flow field. The near-wall values of 0.25 rose to 0.45 in the central part of the shear layer, which value is typical for undisturbed boundary layers. Structural similarity was also found for the shear stress-energy ratio $-\overline{uv}/\bar{k}$, Fig. 14. For $x/D \geq 30$, the profiles yielded values of 0.1 close to the wall and flat maxima of 0.3 in the central layer.

Turbulent diffusion of kinetic energy by the correlation $\overline{kv}$, whose individual contributions were discussed above, may be described by the transport velocity $v_k = \overline{kv}/\bar{k}$, /8/. The profiles of v_k, Fig. 15, closely followed those of $\overline{kv}$. Negative near-wall velocities indicated a

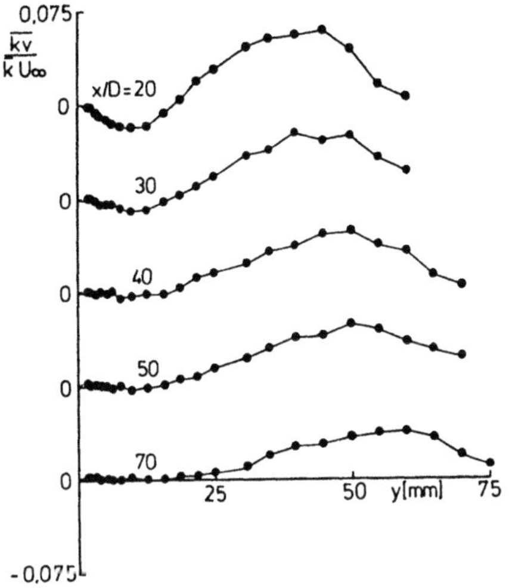

Fig. 15 Transverse convection velocity for kinetic energy

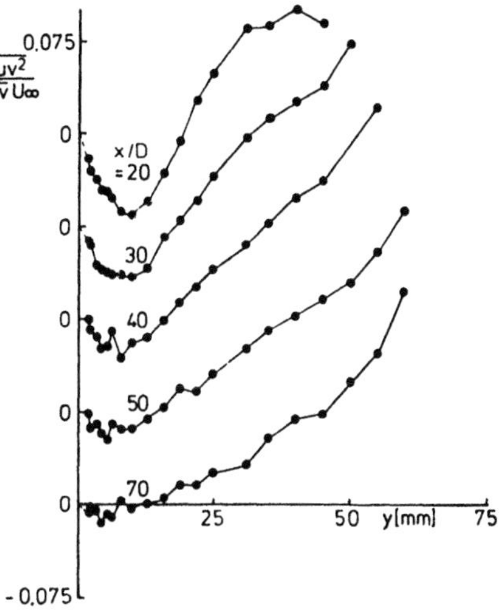

Fig. 16 Transverse convection velocity for shear stress

transport of energy towards the surface, in the outer layer transport was in the opposite direction. At x/D = 20 and 30, v_k was zero exactly at the wall distances of $\overline{k}_{max}$, thus pointing towards the adequacy of a gradient-type model for the diffusion $\overline{kv} = v_t\, \partial \overline{k}/\partial y$; v_t is the eddy viscosity. Corresponding to the results of /8/ and /9/, the near-wall data of the triple correlation $\overline{kv}$ decreased to zero faster in the relaxation process than any second-order correlation or its gradient did. Therefore a gradient-diffusion assumption is not appropriate for simulating the experimental results.

The transport velocity for the Reynolds shear stress $-\overline{uv}$ being $v_\tau = \overline{uv^2}/\overline{uv}$, Fig. 16, behaved quite different from that for $\overline{k}$. Even at the stations farthest downstream, the near-wall results were negative indicating a transport of shear stress to the wall. For x/D = 20, the wall distance of $v_\tau = 0$ was identical to the location of $-\overline{uv}_{max}$, but for x/D = 70 the coordinate $y/\delta(v_\tau = 0)$ was about 1/4 as opposed to $y/\delta(-\overline{uv}_{max}) \cong 1/2$. The results rule out any simple model for describing the transverse turbulent convection of $-\overline{uv}$.

Concluding remarks

The relaxation of a two-dimensional, incompressible shear flow, generated by the interaction of a turbulent boundary layer (Re_θ = 3800) and the wake of a circular cylinder (D = 10 mm), was analysed from extensive hot-wire measurements. In addition to /3/, the development of several triple velocity correlations and structural functions are reported, which have an impact on Reynolds-stress modeling /5,4/. The maxima of the profiles of the triple products $\overline{u_i^2 v}$ and $\overline{u_i^2 u}$, diffusing kinetic turbulence energy in transverse and longitudinal direction, decreased by 50 percent within $20 \leq x/D \leq 70$, while the inner layer values dropped to zero for x/D < 40. The triple correlations $\overline{u^3}$ and $\overline{v^3}$ changing signs across the boundary layer reflected the corresponding development of the skewed joint probability density functions of the u- and v-fluctuations. The profiles of the transport velocities for kinetic energy,

$v_k = \overline{kv}/\overline{k}$, and the corresponding ones for the turbulent shear stress indicated that the measured turbulent convection generally cannot be described by a simple gradient-diffusion closure model. In the central part of the boundary layer, the ratios of the velocity intensities $\overline{u^2} : \overline{w^2} : \overline{v^2}$ developed from 10 : 8 : 8 at x/D = 20 to 10 : 7 : 5 at x/D = 70, while the correlation coefficient $-\overline{uv}/\sqrt{\overline{u^2}\,\overline{v^2}} \approx 0.45$ as well as the stress-energy ratio $-\overline{uv}/\overline{k} \approx 0.3$ behaved self-similar.

References

1. E. Marumo, K. Suzuki, T. Sato: A turbulent boundary layer disturbed by a cylinder. J. Fluid Mech. 87, p. 1, 121 (1978).

2. E. P. Tsiolakis, E. Krause, U. R. Müller: Turbulent boundary layer-wake interaction. Proc. 4th Intern. Symposium on Turbulent Shear Flows, Techn. University Karlsruhe (1983).

3. U. R. Müller: Measurements of mean flow and Reynolds stress dynamics in a relaxing boundary layer. To be submitted to J. Fluid Mech. (1986).

4. U. R. Müller: Berechnung einer relaxierenden Grenzschicht mit einem Reynolds Stress Model. ZAMM, Band 67, Heft 4/5, T 884 (1986).

5. K. Hanjalić, B. E. Launder: Contribution towards a Reynolds-stress closure for low-Reynolds-number turbulence. J. Fluid Mech. 74, p.4, 593 (1976).

6. U. R. Müller: On the accuracy of turbulence measurements with inclined hot wires. J. Fluid Mech. 119, 155 (1982).

7. U. R. Müller: Comparison of turbulence measurements with single, X and triple hot-wire probes. Submitted to Exp. in Fluids (1985).

8. B. Chandrsuda, P. Bradshaw: Turbulence structure of a reattaching mixing layer. J. Fluid Mech., 110, 171 (1981).

9. D. M. Driver, H. L. Seegmiller: Features of a reattaching turbulent shear layer in a diverging channel flow. AIAA J. 23, 163 (1985).

List of Participants

ALCARAZ E. (F)/ Lab. Meca. Fluides & Acoustique, E.C.L., 69131 ECULLY CEDEX.

ANSELMET F. (F)/ I.M.S.T., 12 avenue General Leclerc, 13003 MARSEILLE.

ANTONIA R. (AUSTR.)/ Dept. of Mechanical Engineering, Univ. of NEWCASTLE, 2308 NEW SOUTH WALES.

AUPOIX B. (F)/ ONERA/CERT, 2 av. Ed. Belin, 31055 TOULOUSE CEDEX.

AZAD R.S. (CAN.)/ Univ. of Manitoba, R3T 2N2, WINNIPEG, MANITOBA.

BALINT J.L. (USA)/ Univ. Maryland, College Park, 20742 MARYLAND,

BATAILLE J. (F)/ Lab. Meca. Fluides & Acoustique, E.C.L., 69131 ECULLY CEDEX.

BATCHELOR G.K. (GB)/ Dept. Appl. Math. & Th. Physics, Univ. of Cambridge, Silver street, CB3 9EW CAMBRIDGE.

BECHERT D. (RFA)/ D.F.V.L.R. Muller Breslau strass 8, D 1000 BERLIN 12.

BELORGEY M. (F)/ Univ. du Havre, Place R. Schuman, 76610 LE HAVRE.

BERGE P. (F)/ C.E.A. SACLAY, S.P.S.R.M, 91191 GIF/YVETTE CEDEX.

BERTELRUD A. (S)/ A. Res Inst., FFA, Box 11021, S-161-11, BROMMA.

BERTOGLIO J.P. (F)/ Lab. Meca. Fluides & Acoustique E.C.L., 69131 ECULLY CEDEX.

BETCHOV R. (CH)/ 53 avenue du Lignon, 1219 GENEVE.

BINDER G. (F)/ Inst. de Méca. de Grenoble, BP 68, 38402 ST MARTIN D'HERES CEDEX.

BIRCH N.T. (GB)/ Rolls-Royce ltd, P.O. Box 31 moor lane, DERBY DE2 8BJ.

BONNET J.P (F)/ C.E.A.T., 43 rue de l'Aérodrome, 86000 POITIERS.

BOOY R. (NL)/ Lab. Fluid Mech., Stevinweg 4, 2628 CN DELFT.

BORLETEAU J.P. (B)/ Inst. Von Karman, 72, Chaussée de Waterloo, 1640 RHODE ST GENESE.

BRASSEUR J.G. (USA)/ Univ. Clemson, Dept. Mech. Eng., SC 29634-0921 CLEMSON.

BRINDLEY J. (GB)/ Univ. of LEEDS, Dept. Applied Math. Studies, LEEDS LS2 9JT.

BROWN E. (GB)/ Off. Naval Research, 223 old Marylebone Road, NW1 5TH LONDON.

BRUN G. (F)/ METRAFLU, 64 chemin des mouilles, 69130 ECULLY.

BRUNET Y. (F)/ I.N.R.A. Bioclimat., Grignon, 78850 THIVERNAL.

BUHLER K. (RFA)/ Univ. Karlsruhe, Kaiserstrasse 12, D 7500 KARLSRUHE 1.

BULL M. (RFA)/ Max-Planck-Inst., Busenstrasse 10, D 3400 GÖTTINGEN.

CAMBON C. (F)/ Lab. Meca. Fluides & Acoustique, E.C.L., 69131 ECULLY CEDEX.

CAPERAN P. (F)/ Institut de Mécanique, Domaine universitaire B.P. 68, 38402 SAINT MARTIN D'HERES.

CHACON-REBOLLO T. (E)/ Fac. Math., Dept. Ecuacions Func., SEVILLA.

CHAMPAGNE F. (USA)/ Univ. Arizona, Aerosp. & Mech. Eng. Dept., TUCSON AZ 85721.

CHEVRAY R. (USA)/ Depart. Mech. Eng., Univ. Columbia, N.Y. 10027 NEW YORK.

CHOI K.S (GB)/ British Mar. Techno. Ltd, 67 Stanton av., TEDDINGTON, MIDDLESEX.

CHOLLET J.P. (F)/ Instit. Mécanique Grenoble, BP 68, 38402 ST MARTIN D'HERES CEDEX.

CHRISTAKIS P. (G)/ Inst. Turbulence, Univ.de PATRAS, Ell. Str. 3 PATRAS.

COANTIC M. (F)/ I.M.S.T., 12 av. Gal. Leclerc, 13003 MARSEILLE.

COGNET G. (F)/ L.E.M.T.A., ENSEM BP 850, 54011 NANCY CEDEX.

COMTE-BELLOT G. (F)/ Lab. Meca. Fluides & Acoustique, E.C.L., 69131 ECULLY CEDEX.

COURTY J.C. (F)/ Avions Marcel Dassault, 78, quai Carnot, 92214 ST CLOUD CEDEX.

DARBY M.S. (GB)/ Depart. of Applied Math. Studies, University of Ledds, LEEDS LS2 9JT.

DELVILLE J (F)/ C.E.A.T., 43 rue de l'Aérodrome, 86000 POITIERS.

DENGEL P. (RFA)/ Technische Univer., Muller Breslau strasse 8 D 1000 BERLIN 12.

DINKELACKER A. (RFA)/ Max-Planck-Institut, Bunsenstr. 10, D 3400 GOTTINGEN.

DRAZIN P.G. (GB)/ School Math., Univ. Walk, BS8 1TW BRISTOL.

DROBNIAK S. (POL)/ Inst. Maszyn Cieplnych, Al. Zawadzkiego 25, 42-200 CZESTOCHOWA.

DRUMMOND I.T. (GB)/ D.A.M.T.P., Silver street, CB3 9EW CAMBRIDGE.

DUBOIS (F)/ D.R.E.T., 26, Bd Victor, 75996 PARIS ARMEES.

DUBOIS M. (F)/ S.P.S.R.M., Orme des merisiers, CEA, 91191 GIF SUR YVETTE CEDEX.

DUSSAUGE (F)/ I.M.S.T., 1é Avenue du Gal Leclerc, 13003 MARSEILLE.

ECKELMANN H. (RFA)/ Max-Planck-Inst., Bunsenstrasse 10, D 3400 GOTTINGEN.

ELSNER J.W. (POL)/ Inst. Maszyn Cieplnych, Al Zawadzkiego 25, 42-200 CZESTOCHOWA.

ETLING D. (RFA)/ Univ. Hannover, Herrenauser Street 2, 3000 HANNOVER 21.

FANNELOP T.K. (CH)/ Inst. fur Aerodynamik, Eid. Technische Hjochschule, Ethz, 8092 ZURICH.

FARGE M. (F)/ Ecole Norm. Sup. L.M.D., 24 rue Lhomond, 75231 PARIS CEDEX 5.

FERNHOLZ H.H. (RFA)/ Hermann Fottinger Instit., Strasse des 17 Juni 135, D 100 BERLIN.

FIEDLER H.E. (RFA)/ Hermann Fottinger Inst., Muller-Breslau str. 8, 1000 BERLIN 12.

FLEURY F. (F)/ Univ. du Havre, B.P. 67, 76130 MONT SAINT AIGNAN.

FOSS J. (USA)/ Michigan State Uni., East Lansing, 48824 Michigan.

FRIEDRICH R. (RFA)/ Techni. Univer MUNCHEN, Arcisstr 21, 8000 MUNCHEN 2.

FRISH U. (F)/ Observatoire de Nice, B.P. 139 06003 NICE CEDEX.

FULACHIER L. (F)/ I.M.S.T., 12 av. Général Leclerc, 13003 MARSEILLE.

GAGNE Y. (F)/ I.M.G., BP 68 38402 St MARTIN D'HERES CEDEX.

GAY B. (F)/ Lab. Méca. Fluides & Acoustique, E.C.L., 69131 ECULLY CEDEX.

GERMANO M. (I)/ Politech. Dip. Ing. Aeron. Spazi., Corso Duca degli Abruzzi 24, 10129 TORINO.

GERZ T. (RFA)/ Inst. Phys. Atmos., D.F.V.L.R., Oberpfaffenhofen, D-8031 WESSLING/OSSI.

GLAUSER M.N. (USA)/ SUNY Univ. Buffalo, turbu., 339 Engineering East, Buffalo, 14260 NEW-YORK.

GRIGNAN P. (F)/ G.E.R.D.S.M Chemin de la gardiole, le Brusc, 83140 SIX FOURS.

HILL J.C. (USA)/ Dept. of Chemicall Engng., 231 SWEENEY HALL AMES, AMES IOWA 50011.

HILL R.J. (USA)/ Nat. Oceanic & Atm. Adm. Wave Propag. Lab., R/E/WPI US Dept Commerce 325 BROADWAY, BOULDER CO 80303.

HOFFMEISTER M. (DDR)/ Inst. für Mech., IT, Rudower Chaussee 5, 1199 BERLIN.

HORGAN R.R. (GB)/ Dept. Appl. Math. & Th. Physics, Univ. of Cambridge, Silver Street, CB3 9EW CAMBRIDGE.

HUNT J.C. (GB) Dept. Appl. Math. & Th. Physics, Univ. of Cambridge, Silver Street, CB3 9EW CAMBRIDGE.

HUQ P. (GB)/ Univ. of Cambridge, Trumpington Str, CAMBRIDGE CB2 IPZ.

HUSSAIN A.K.M.F. (USA)/ Dept. of Mechanical Engng, 4800 CALHOUN ROAD, HOUSTON TEXAS 77004.

JACQUIN L (F)/ O.N.E.R.A., 29 av. Division Leclerc, 92320 CHATILLON.

JANKE G. (RFA)/ Hermann Fottinger Inst., Müller Breslau Strasse 8, D 1000 BERLIN 12.

JEANDEL D. (F)/ Lab. Méca. Fluides & Acoustique, E.C.L., 69131 ECULLY CEDEX.

JOHANSSON A.V. (S)/ Dept. of Mechanics, Royal Inst. Technology, S-100 44 STOCKHOLM.

JUVE D. (F)/ Lab. Meca. Fluides & Acoustique, E.C.L., 69131 ECULLY CEDEX.

KATZ W. (ISRAEL)/ Dept. Fluid Dynamics, TEL AVIV UNIV., RAMAT AVIV 699720.

KHOUDLI M. (F)/ L.N.H., E.D.F., 6, quai Watier, 78401 CHATOU CEDEX.

KRAUSE E. (RFA) Aero. Institut der RWTH Aachen, Wullnerstr. zwischen 5 und 7, 5100 AACHEN.

LARCHEVEQUE M. (F)/ Labo. Météo. Dynamique, 24 rue Lhomond, 75231 PARIS CEDEX 05.

LE BAS (F)/ Univ. du Havre Place R. Schuman, 76610 LE HAVRE.

LEUCHTER O. (F)/ O.N.E.R.A., 29 av. Division Leclerc, 92320 CHATILLON/S/BAGNEUX.

LIANDRAT J. (F)/ S.T.P.A.-EG2 4 av. de la porte d'Issy, 75996 PARIS ARMEES.

LUMLEY J.L. (USA). Cornell Univ., 238 Upson Hall, ITHACA, 14853 NEW YORK.

MALIK N.A. (GB)/ D.A.M.T.P., Univ. Cambridge, Silver street, CB3 9EW CAMBRIDGE.

MARION Y. (F)/ Lab. Méca. Fluides & Acoustique, E.C.L., 69131 ECULLY CEDEX.

MASSAGUER J.M. (E)/ E.T.S.E. Camins, Jorge Girona Salgado 31, 08034 BARCELONA.

MASSONS J. (E)/ Fac. Depart. Fisica, Univ. Barcelona, Pça. Imperial Tarraco 1, 43005 TARRAGONA, CATALUNYA.

MATHIEU J. (F)/ Lab. Meca. Fluides & Acoustique, E.C.L., 69131 ECULLY CEDEX.

MENEGUZZI M. (F) Serv. Astrophysique, Centre Etudes Nucléaires de Saclay, 91191 GIF SUR YVETTE.

MENTER F. (RFA)/ D.F.V.L.R., BUNSENSTR. 10, 3400 GOTTINGEN.

MERCADER (E)/ E.T.S.E., Camins, BARCELONE.

MERZKIRCH W. (RFA)/ Univ. ESSEN, Postfach 103764, D-4300 ESSEN.

MICHARD M. (F)/ Lab. Meca. Fluides & Acoustique, E.C.L., 69131 ECULLY CEDEX.

MOFFATT H.K. (GB)/ Dept. Appl. Math. & Th. Physics, Univ. of Cambridge, Silver Street, CB3 9EW CAMBRIDGE.

MOREL R. (F)/ Lab. Meca. Fluides & Acoustique, E.C.L., 69131 ECULLY CEDEX.

MULLER U.R. (RFA)/ Aerod. Inst., Wullnerstrasse 5-7, D-51 AACHEN.

NAGLE P. (F)/ CERDAN, D.C.A.N. BP77, 83800 TOULON NAVAL.

NIEUWSTADT F. (NL)/ Delft Univ. Techno., Rotherdam Weg 145, 2628 DELFT.

NOCERA L. (I)/ I.F.A.M.-C.N.R., Via del giardino, I - 56100 PISA.

NYCHAS S.G. (G)/ Univ. Thessaloniki, Box n° 453, 54006 THESSALONIKI.

O'BRIEN E. (USA)/ State Univ. New York, Dep. of Mechn. Engnrg, STONY BROOK NY 11794.

OLSSON (S)/ Aeron. Research Institute, Box 11021 S-16111, STOCKHOLM.

OOMS G. (NL)/ Koninklijke/Shell, P.O. Box 3003, 1003 AA AMSTERDAM.

ORTEGON-GALLEGO F. (F)/ I.N.R.I.A., Domaine de voluceau Rocquencourt, 78153 LE CHESNAY CEDEX.

PANCHEV S. (BULG)/ Fac. Physics, Univ. of Sofia, 5A Ivaniv. Bld, 1126 SOFIA.

PAUL S. (F)/ C.E.A., Centre de Vaugiours, 77181 COURTRY.

PERRIER P. (F)/ Avions Marcel Dassault, 78 Quai Carnot, 92210 St CLOUD.

PICART A. (F)/ Lab. de Thermodynamique, UA 230 CNRS, BP 67, 76130 MONT SAINT AIGNAN.

PIRONNEAU O. (F)/ I.N.R.I.A. Domaine de Voluceau Rocquencourt, 78153 LE CHESNAY CEDEX.

POLITANO H. (F)/ OBS. DE NICE, BP 139, 06003 NICE CEDEX.

PY B. (F)/ I.U.T. Mécan. Phys., Rue de Kergoat, 29283 BREST CEDEX.

RANDOLPH M. (RFA)/ Max-Planck-Inst., Bunsenstr. 10, D-3400 GOTTINGEN.

RATNANATHER J.T. (GB)/ NAG Oxford Univ. Computing Lab., 8-11 Keble RD, OXI 3QD OXFORD.

REY C. (F)/ E.N.S.M., 1 rue de la Noé, 44072 NANTES CEDEX.

RICHTER K. (RFA)/ Technische Univ. Munchen, Arcis street 21, 8000 MUNCHEN 2.

ROBIN S. (F)/ Univ. Paris 7. Phys. Liqui.Electro., 4 place Jussieu, tour 22, 75258 PARIS CEDEX 05.

RODET J.C. (F)/ Laboratoire de la Marine, GERDSM Chemin de la gardiole, LE BRUSC 83140 SIX FOURS.

ROSANT J.M. (F)/ E.N.S.M., 1 rue de la Noé, 44072 NANTES CEDEX.

RYHMING I.L. (CH)/ Ecole Polytechnique Fédérale, Instit. Machines hydr. et Mécanique, 1015 LAUSANNE.

SAVILL M. (GB)/ Univ. Cambridge, Trumpington Street, CB2 1PZ CAMBRIDGE.

SAVULESCU S.N. (AUT)/ Gartergasse, 5 A-2320 SCHWECHAT, KLEDERING.

SCHMITZ D. (RFA)/ M.B.B., Postfach 801160, 8000 MUNCHEN 80.

SCHUMANN U. (RFA)/ D.F.V.L.R., Instit. Physik Atmosphäre, D-8031 OBERPFAFFENHOFEN.

SEIFERT (ISRAEL)/ Dept fluid dynamics, TEL AVIV UNIV., RAMAT AVIV 6997.

SHEPHERD T.G. (GB)/ Univ. Cambridge, Silver street, CB3 9EW CAMBRIDGE.

SIEBER M. (RFA)/ Max-Planck-Institut, Busenstr. 10, D 3400 GOTTINGEN.

SMITH F.T. (GB)/ Univ. Coll. Math. Dep., Gower St, WCIE 6BT LONDRES.

SOUFIANI (F)/ Lab. Energétique Moléculaire, Gr. voie des vignes, 92290 CHATENAY-MALABRY.

STRETCH D. (GB)/ Engng. Dept., Univ. of Cambridge, Trumpington ST, CB2 1PZ CAMBRIDGE.

STUART J.T. (GB)/ Imperial Col. Science Techno., 180 Queen's Gate, SW7 2BZ LONDON.

SULEM C. (F)/ CNRS/CMA Ecole Norm. Sup., 45 rue d'Ulm, 75250 PARIS 5ème.

SULEM P.L. (F)/ C.N.R.S., Observatoire de Nice B.P. 139, 06003 NICE CEDEX.

SUMER M. (DK)/ Inst. Hydrodyn. Hydro. Eng., DU-2800 LYNGBY.

SUNYACH M. (F)/ Lab. Méca. Fluides & Acoustique, E.C.L., 69131 ECULLY CEDEX;

TEISSEDRE C. (F)/ O.N.E.R.A, Division de l'Aerodyn., 29 av. de la Division Leclerc, 92322 CHATILLON CEDEX.

TEZDUYAR T. (USA)/ Dept. Mechnics Eng., Univ. Houston Texas, 77004 HOUSTON.

THUAL O. (F)/ C.N.R.S./GAME, 42 av. Gustave Coriolis, 31057 TOULOUSE CEDEX.

TICHY J. (USA)/ Dt. of Mech. Engng, 2022 Rensselaer Polytech. Inst., TROY NY 12181.

TOWNSEND A.A. (GB)/ Univ. Cambridge, Cavendish Labo., Emmanuel College, CB2 3AP CAMBRIDGE.

TRIBOLLET B. (F)/ LP15 CNRS. Phy. Liqui. Electro., 4 place Jussieu, tour 22, 75252 PARIS CEDEX 05.

TRUONG T.V. CH/ Ecole Polytechnique Fédérale, Instit. machines hydr. et mécanique, 1015 LAUSANNE.

TSINOBER A. (ISRAEL)/ Fac. Eng., Tel Aviv Univ., RAMAT AVIV 69978.

TSIRIKOGLOU (G)/ Univ. de PATRAS, PATRAS.

UITTENBOGAARD R.E. (NL)/ Delft Hydrau. labo., P.O. Box 177, 2600 MH DELFT.

VALIHO L. (E)/ Lab. Fluid Mech., E.T.S.I.I University of ZARAGOZA, ZARAGOZA.

VAN DEN BERG B. (NL)/ N.L.R., P.O. BOX 153, 8300 AD EMMELOORD.

VAN DER GELD C. (NL)/ Univ. Techno., Dept Aerospace Eng., Kluyverweg 1, 2629 HS DELFT.

VASANTA RAM V.I. (RFA)/ Ruhr Univ. Bochum, Postfach 102148, D-4630 BOCHUM 1.

WEISENBORN T. (NL)/ Koninklijke/Shell-lab. Amsterdam, Badhuisweg 3, 1031 CM AMSTERDAM.

Index of Contributors

Turbulent Shear Flows 5

Selected Papers from the Fifth International Symposium on Turbulent Shear Flows, Cornell University, Ithaca, USA, August 7–9, 1985

Editors: F. Durst, B. E. Launder, J. L. Lumley, F. W. Schmidt, J. H. Whitelaw

1986. 260 figures. Approx. 380 pages.
Hard cover DM 228,–. ISBN 3-540-16885-0

Contents: Homogeneous and Simple Flows. - Free Shear Flows. - Wall Flows. - Reacting Flows. - Index of Contributors.

This volume contains twenty-five papers presented at the Fifth Turbulent Shear Flows Symposium. The contributions have been considerably extended and updated to be of long-lasting value to the scientific community. They give a representative statement of current research in homogeneous and simple flows, free flows, wall flows and reacting flows. Each section begins with an introduction by a distinguished worker in the field. These surveys should be a great help not only to experts but also to newcomers to turbulent shear flow research. The reader will find coverage of the following topics: modelling of pressure terms of the scalar fluxes, inhomogeneous turbulence and applications to boundary layer flows, interactions of homogeneous turbulent scales, structural considerations, weak shear layer interactions, mixing layers and jet measurements. Experimental and numerical results on wall boundary layer flows and contribution flows are presented. The book should become a standard reference for physicists and engineers working in this exciting field of research.

Springer-Verlag
Berlin Heidelberg
New York London
Paris Tokyo

Springer